Handbook of Parallel Constraint Reasoning

Youssef Hamadi · Lakhdar Sais
Editors

Handbook of Parallel Constraint Reasoning

Editors
Youssef Hamadi
Laboratoire d'informatique (LIX)
de l'École polytechnique
Palaiseau
France

Lakhdar Sais
CRIL, CNRS UMR 8188
Université d'Artois
Lens
France

ISBN 978-3-030-09694-6 ISBN 978-3-319-63516-3 (eBook)
https://doi.org/10.1007/978-3-319-63516-3

Softcover re-print of the Hardcover 1st edition 2018

Printed on acid-free paper

This Springer imprint is published by the registered company Springer International Publishing AG part of Springer Nature
The registered company address is: Gewerbestrasse 11, 6330 Cham, Switzerland

Foreword

Many computational challenges in artificial intelligence, machine learning, and data science can be formulated as optimization problems subject to a set of constraints. Depending on the details of the underlying optimization task and the given constraints, a range of different constraint optimization and modeling formalisms have been introduced. Over the last two decades, the development of ever more efficient constraint solvers has led to a shift from a largely academic endeavor to an area with significant real-world impact. Successful applications are in machine learning, hardware and software verification and synthesis, AI planning and scheduling, and combinatorial optimization. In the early 1990s, typical constraint solvers could handle problems involving a few hundred variables and constraints. Current solvers can handle instances with hundreds of thousands of variables and millions of constraints, making real-world applications feasible. Novel algorithmic techniques combined with a series of increasingly sophisticated implementations are key drivers behind these advances. Note that given the formal worst-case intractability of constraint optimization, this level of progress was completely unanticipated. However, we now understand that most constraint problems of interest have significant internal structure that can be automatically uncovered and exploited effectively by modern solvers. The effectiveness of current solvers continues to open up new areas of applications in artificial intelligence and computer science, in general. In fact, anyone encountering a computational task that is NP-complete or even lies beyond NP should consider whether modern constraint solvers can be of use in their domain.

The natural next step in the development of ever more powerful constraint optimization methods is the use of parallelism through multi-core processors or distributed cloud computing platforms. The *Handbook of Parallel Constraint Reasoning* provides an exhilarating and comprehensive collection of chapters dealing with all aspects of parallel constraint reasoning and optimization. The handbook is remarkable in its coverage and shows the richness of topics in constraint reasoning, while also highlighting the many connections between subareas. Topics include, among others, Boolean Satisfiability (SAT) solving, Maximum Satisfiability (MaxSAT) solvers, Satisfiability Modulo Theory (SMT) solvers, Automated Theorem Proving (ATP), Quantified Boolean Formulas (QBF) solvers, Answer Set Programming (ASP), Logic

Programming (LP), Integer Linear Programming (ILP) methods, Model Checking for Linear Temporal Logic (LTL), and algorithm configuration. The handbook also includes several chapters on real-world applications. The wealth and selection of material make this a truly seminal handbook. The chapters are written in a highly comprehensive manner, which means that they can also be read independently. Moreover, in addition to a detailed discussion of parallelization techniques, each contribution also includes a concise introduction to the basic techniques underlying each area (such as SAT solving, SMT solving, etc.). The handbook therefore also provides a valuable resource for anyone who wants to get an understanding of the opportunities modern constraint reasoning and optimization provide, while, for researchers in the area, the book covers the state-of-the-art in parallel techniques and the remaining challenges. Parallelization provides the framework for the next level of constraint solving. This handbook leads the way.

Ithaca, July 2017 *Bart Selman*

Preface

Constraint reasoning, the ability to draw inferences or decisions from available knowledge using a wide range of techniques that spans logic, computer science, operations research, and artificial intelligence, benefits from several powerful formalisms and algorithms that cover large classes of combinatorial problems at the heart of science, engineering, and business. Modern approaches for tackling these intractable problems are usually declarative as they divide the task into two major interdependent steps: modeling and solving. Depending on the complexity and nature of the problem to be handled, choosing the most appropriate formalism that combines expressiveness and solving efficiency is an important issue across several disciplines.

We can divide these formalisms into four categories.

- Those having their roots in propositional logic, Satisfiability (SAT), the prime NP-complete problem, and its extensions, which cover optimization (MaxSAT), quantified Boolean formulas (QBF), and several first-order theories (SMT).
- Those built on higher-order logics, Automated Theorem Proving (ATP), and Answer Set Programming (ASP), which is based on the stable model semantics of Logic Programming.
- Those related to operations research and artificial intelligence, Mixed-Integer Linear Programming (MILP), Constraint Programming (CP), and Stochastic Local Search (SLS).
- Those that are part of Computer Science as specific algorithms and data structures for important classes of problems: A* for optimal path finding and graph traversal problems, Model Checking for Linear-time Temporal Logic (MC/LTL), Binary Decision Diagrams (BDD) for checking given specifications over complex systems, and Model-Based Diagnosis (MBD) to explain complex systems behaviors.

The large search spaces explored through these paradigms have raised early parallelizing endeavors. However, the scarcity of computational resources made these attempts marginal and did not really impact practitioners. In the last ten years, the situation has dramatically changed. First, the end of Moore's law, the widespread use of graphics-card processing, and the ubiquity of cheap cloud-computing resources have raised new scenarios where any practitioner can quickly harness a large amount of cost effective computational resources to tackle its problems. Second, the previous paradigms and methods have matured to the point of reaching industrial strength and are now essential to the development of many services and products (including software and hardware verification, very-large-scale logistics problems, complex networks design, etc.). Third, the transition to a new information and digital era has resulted in the emergence of new data-intensive technologies leading to even more complex and hard combinatorial problems out of the reach of sequential computing.

These three factors have reignited and rejuvenated research on parallel constraint reasoning, and the aim of the first *Handbook of Parallel Constraint Reasoning* is to capture the full breadth and depth of these efforts. It presents work demonstrating the use of multiple resources from single-machine multi-core and GPU-based computations to very-large-scale distributed execution platforms of up to 80,000 processing units. The intended audience of the handbook consists of researchers, graduate students, and practitioners who wish to learn about the state of the art in parallel constraint reasoning. Each of the seventeen chapters is intended to be a self-contained survey of one of the previous paradigms, and is written by leading researchers in the area.

The book is divided into two main parts. The first one presents the theories and algorithms associated with the previous formalisms and problems. It is subdivided into several sub-parts which group together related formalisms and methods. In the first sub-part, SAT and its MaxSAT, QBF, and SMT extensions are presented. Since parallelizing techniques are typically generalized from SAT to its extensions, we encourage readers to read these chapters successively. The second sub-part considers first-order logic and logic programming with chapters devoted to parallel automated theorem proving, and parallel answer set programming. The third sub-part describes mathematical-optimization, and artificial-intelligence-based formalisms, with chapters on parallel mixed-integer linear programming, constraint programming, and stochastic local search. Finally, the last sub-part collects methods for parallel breadth-first search, including A* exploration, model checking for linear temporal logic specifications, parallel binary decision diagrams data structures, and model-based diagnosis.

The second part presents tools and applications. It starts with a chapter devoted to the automatic composition of parallel portfolios algorithms through machine learning techniques. The ideas and the underlying concepts are sufficiently general to be extended and exploited in all the representation and solving models presented in the first part. Afterwards, two applications are presented. The first one shows how parallel satisfiability can speed up the verification of complex embedded systems by orders of magnitude. The second one shows how parallel stochastic local search can provide higher-quality long-reach optical network designs.

The parallel algorithms presented here are built around a common set of mechanisms and ideas, including divide-and-conquer with work stealing, portfolios of competing techniques, knowledge sharing of new lemmas and objective-function bounds, heuristic information exchanges, and efficient synchronizing for determinization. We thus hope that the importance of these mechanisms as presented in this handbook will inspire researchers from parallel and distributed computing and act as a means for cross-fertilization.

The start and completion of this project would not have been possible without much support and encouragement. This handbook is an effort of people from the different sub-communities of constraint reasoning, and we take this opportunity to express our gratitude to the fifty-four co-authors who contributed to this book.

We are indebted to Ronan Nugent at Springer-Verlag for his great patience and renewed interest in materializing all these efforts into the book you are holding now.

Paris, Lens,
March 2017

Youssef Hamadi
Lakhdar Saïs

Contents

Part I Theory and Algorithms

1 Parallel Satisfiability ... 3
Tomáš Balyo and Carsten Sinz
1.1 Introduction ... 3
1.2 Preliminaries ... 4
1.2.1 Satisfiability (SAT) ... 4
1.2.2 Local Search Algorithms for SAT ... 5
1.2.3 The DPLL Algorithm ... 5
1.2.4 Resolution Refutation ... 7
1.2.5 The CDCL Algorithm ... 7
1.2.6 Parallel Computing Architectures ... 9
1.2.7 Measuring Speedups ... 9
1.3 Divide-and-Conquer Approaches ... 10
1.3.1 Problem Decomposition and Load Balancing ... 10
1.3.2 Implementations of Search-Space-Splitting Solvers ... 13
1.3.3 Search Space Splitting in CDCL ... 14
1.3.4 Cube and Conquer ... 15
1.4 Parallel Portfolios – Diversify and Conquer! ... 15
1.4.1 Virtual Best Solver ... 15
1.4.2 Pure Portfolio Solvers and Diversification ... 16
1.4.3 Clause-Sharing Portfolios ... 16
1.4.4 Impact of Diversification and Clause Sharing ... 17
1.4.5 Examples of Parallel Portfolio Solvers ... 19
1.5 Parallel Local Search ... 23
1.5.1 Multiple Flips ... 23
1.5.2 Portfolios ... 24
1.6 Future Challenges ... 24
References ... 25

2 Cube-and-Conquer for Satisfiability ... 31
Marijn J.H. Heule, Oliver Kullmann, and Armin Biere
2.1 Introduction ... 31
2.2 Preliminaries ... 33
2.3 Combining CDCL and Lookahead ... 33
2.4 Creating Cubes: The Basic Method ... 36
2.5 Creating Cubes: a General Methodology ... 38
2.5.1 General Framework ... 38
2.5.2 Cutoff Heuristic ... 39
2.5.3 Heuristics for Splitting ... 41
2.6 Solving Cubes ... 42
2.6.1 Sequential Solving ... 42
2.6.2 Solving Cubes in Parallel ... 44
2.7 Interleaving the Cube and Conquer Phases ... 45
2.7.1 Ineffective Lookahead Heuristics ... 46
2.7.2 Concurrent Cube-and-Conquer ... 46
2.7.3 Cubes on Demand ... 48
2.8 Proofs of Unsatisfiability ... 50
2.9 Experimental Evaluation ... 52
2.9.1 Application Benchmarks ... 52
2.9.2 The Boolean Pythagorean Triples Problem ... 54
2.10 Conclusions ... 56
References ... 57

3 Parallel Maximum Satisfiability ... 61
Inês Lynce, Vasco Manquinho, and Ruben Martins
3.1 Introduction ... 61
3.2 Maximum Satisfiability ... 65
3.2.1 Sequential MaxSAT Algorithms ... 66
3.2.1.1 Linear Search Algorithms ... 67
3.2.1.2 Unsatisfiability-Based Algorithms ... 69
3.2.1.3 Other Algorithmic Solutions and Implementation Issues ... 71
3.3 Parallel MaxSAT ... 71
3.3.1 Portfolio Approaches ... 72
3.3.1.1 Parallel Unsatisfiability-Based Algorithms ... 73
3.3.1.2 Parallel Linear Search Algorithms ... 74
3.3.1.3 Implementation Issues ... 75
3.3.2 Search Space Splitting ... 75
3.3.2.1 Interval Splitting ... 75
3.3.2.2 Guiding Paths ... 79
3.3.2.3 Other Splitting Schemes and Implementation Issues ... 80
3.3.3 Clause Sharing ... 81
3.3.3.1 Conditions for Safe Clause Sharing ... 81

3.3.3.2 Clause-Sharing Heuristics ... 82
3.3.3.3 Comparison Between Clause-Sharing Heuristics ... 84
3.4 Deterministic Parallel MaxSAT ... 85
3.4.1 Motivation ... 86
3.4.2 Deterministic Solver ... 87
3.4.2.1 Standard Synchronization ... 89
3.4.2.2 Period Synchronization ... 90
3.4.2.3 Dynamic Synchronization ... 91
3.4.3 Comparison Between Non-deterministic and Deterministic Solvers ... 92
3.5 Research Directions ... 92
3.5.1 Scalability ... 92
3.5.2 Clause Sharing ... 93
References ... 94

4 Parallel Solving of Quantified Boolean Formulas ... 101
Florian Lonsing and Martina Seidl
4.1 Introduction ... 101
4.2 Background ... 105
4.3 Sequential Search-Based QBF Solving ... 108
4.4 Parallel QBF Solving at a Glance ... 111
4.5 Parallel QBF-Solving Approaches ... 115
4.6 Challenges and Potential of Parallel QBF Solving ... 127
4.7 Conclusion ... 131
References ... 132

5 Parallel Satisfiability Modulo Theories ... 141
Antti E.J. Hyvärinen and Christoph M. Wintersteiger
5.1 Introduction ... 141
5.2 General Preliminaries ... 142
5.2.1 Theories ... 142
5.2.2 The Underlying Conflict-Driven, Clause-Learning SAT Solver ... 143
5.2.3 Theory Combination ... 144
5.2.4 Interpolants ... 145
5.2.5 SMT Solvers ... 145
5.3 Portfolios of SMT Solvers ... 146
5.3.1 Parallel SMT Based on Algorithm Portfolios ... 148
5.3.2 Lemma Sharing in Portfolios ... 148
5.3.3 Centralized Lemma Databases ... 149
5.3.4 Experiments on the Algorithmic Framework ... 150
5.3.5 Lemma Sharing and Partitioning ... 150
5.4 Search-Space Partitioning in SMT ... 151
5.4.1 Plain Partitioning ... 152
5.5 Decomposition ... 157
5.5.1 Experimental Evidence ... 158

5.5.2 Variations and Extensions . . . 159
5.6 Combinations of Parallelization Algorithms . . . 161
5.6.1 The Parallelization Tree . . . 161
5.6.2 Iterative Partitioning with Partition Trees . . . 163
5.6.3 Safe and Repeated Partitioning . . . 165
5.6.4 Constructing Partitions . . . 168
5.7 Further topics . . . 171
References . . . 172

6 Parallel Theorem Proving . . . 179
Maria Paola Bonacina
6.1 Introduction . . . 179
6.2 Theorem Proving Strategies . . . 181
6.2.1 Subgoal-Reduction Strategies . . . 182
6.2.2 Ordering-Based Strategies . . . 184
6.2.2.1 Expansion-Oriented Strategies . . . 186
6.2.2.2 Contraction-Based Strategies . . . 187
6.2.3 Instance-Based Strategies . . . 189
6.3 Parallelization of Theorem Proving . . . 190
6.3.1 Parallelism at the Term or Literal Level . . . 191
6.3.1.1 Parallelism at the Literal Level for Subgoal-Reduction Strategies . . . 191
6.3.1.2 Parallelism at the Term Level for Ordering-Based Strategies . . . 191
6.3.2 Parallelism at the Clause Level . . . 193
6.3.2.1 Parallelism at the Clause Level for Subgoal-Reduction and Instance-Based Strategies . . . 193
6.3.2.2 Parallelism at the Clause Level for Ordering-Based Strategies . . . 194
6.3.3 The Rise of Parallel Search . . . 196
6.3.4 Multi-search . . . 198
6.3.4.1 Multi-search for Subgoal-Reduction Strategies . . . 199
6.3.4.2 Multi-search for Ordering-Based Strategies . . . 200
6.3.5 Distributed Search . . . 202
6.3.5.1 Distributed Search for Ordering-Based Strategies . . . 202
6.3.5.2 The Basic Clause-Diffusion Mechanisms . . . 203
6.3.5.3 The Subdivision of Clauses in Clause-Diffusion . . . 204
6.3.5.4 The Subdivision of Inferences in Clause-Diffusion . . . 205
6.3.5.5 Distributed Global Contraction, Distributed Fairness, and Distributed Proof Reconstruction . . . 206
6.3.5.6 The Clause-Diffusion Provers . . . 207
6.4 Discussion . . . 211

6.4.1 Parallel Theorem Proving and Parallel Satisfiability 211
6.4.2 Parallelism and First-Order Model-Based Reasoning 215
References 217

7 Parallel Answer Set Programming 237
Agostino Dovier, Andrea Formisano, and Enrico Pontelli
7.1 Introduction 237
7.2 Background 240
7.2.1 Definite Logic Programming 240
7.2.2 Normal Logic Programs and Answer Set Programming .. 242
7.2.3 Datalog 244
7.2.4 Alternative ASP Computation Models 244
7.2.4.1 Program Completion 244
7.2.4.2 Conflict-Driven Search 245
7.2.4.3 ASP Computation 246
7.3 Parallelizing the Grounding Phase 247
7.3.1 Introduction 247
7.3.2 Naive Parallel Grounding 248
7.3.3 Multi-level Parallel Grounding 248
7.4 Parallelizing the Inference Phase I: Parallel Datalog 252
7.5 Parallelizing the Inference Phase II: Parallel ASP 253
7.5.1 Parallelizing the Search Process 253
7.5.1.1 General Idea and Seminal Work 253
7.5.1.2 Techniques for Task Sharing 255
7.5.1.3 Scheduling and Load Balancing 258
7.5.1.4 Parallelizing Lookahead 261
7.5.2 GPU-Based Parallelism 264
7.5.2.1 GPU-Based Datalog Solving 265
7.5.2.2 GPU-Based Conflict-Driven ASP Solving 266
7.5.3 Moving Towards Large-Scale Architectures 270
7.5.3.1 The Map-Reduce Programming Model 270
7.5.3.2 Datalog and Map-Reduce 271
7.5.3.3 Towards ASP: Well-Founded Semantics and Map-Reduce 272
7.5.3.4 Other Relevant Applications of Map-Reduce 274
7.5.4 Portfolio Approaches for ASP 274
7.6 Discussion and Conclusions 276
References 277

8 Parallel Solvers for Mixed Integer Linear Optimization 283
Ted Ralphs, Yuji Shinano, Timo Berthold, and Thorsten Koch
8.1 Introduction 284
8.2 Sequential Algorithms 286
8.2.1 Basic Components 286
8.2.2 Advanced Procedures 288
8.3 Parallel Algorithms 290

8.3.1 Scalability and Performance 291
8.3.1.1 Scalability 292
8.3.1.2 Performance 294
8.3.2 Properties 295
8.3.2.1 Abstraction and Integration 295
8.3.2.2 Granularity 297
8.3.2.3 Adaptivity 298
8.3.2.4 Knowledge Sharing 299
8.3.2.5 Load Balancing 299
8.3.2.6 Synchronization and Coordination 303
8.3.2.7 Determinism 304
8.3.3 Implementation 305
8.3.3.1 Platform 305
8.3.3.2 Frameworks and Solvers 308
8.3.3.3 Coordination Mechanisms 308
8.4 Software 316
8.4.1 Solvers 317
8.4.2 Frameworks 321
8.5 Performance Measurement 324
8.5.1 Performance Variability 325
8.5.2 Comparisons 326
8.5.3 Instance Selection 327
8.5.4 Alternative Performance Measures 328
8.5.5 Summary Measures 328
8.6 Concluding Remarks 329
References 329

9 Parallel Constraint Programming 337
Jean-Charles Régin and Arnaud Malapert
9.1 Introduction 337
9.1.1 Filtering + Propagation 339
9.1.2 Search 339
9.1.2.1 Search Methods in Solvers 341
9.1.3 Parallelism and Constraint Programming 342
9.1.3.1 Parallel Propagators and Propagation 342
9.1.3.2 Search Space Splitting 342
9.1.3.3 Portfolio Algorithms 344
9.1.3.4 Distributed CSPs 345
9.1.3.5 Problem Decomposition 345
9.2 Background 346
9.2.1 Parallelism 346
9.2.1.1 Parallelization Measures and Amdahl's Law 346
9.2.2 Embarrassingly Parallel Computation 347
9.2.3 Internal and External Parallelization 348
9.2.4 Constraint Programming 349

9.3 Parallel Search Tree 350
9.3.1 Static Partitioning 350
9.3.2 Dynamic Partitioning 351
9.3.2.1 Local Subtree Solving 352
9.3.2.2 Subtree Definition 353
9.4 Problem Decomposition 356
9.4.1 Principles 356
9.4.1.1 Sub-problems Generation: a Top-Down Method 358
9.4.1.2 Sub-problems Generation: a Bottom-Up Method 361
9.4.1.3 Implementation 362
9.4.1.4 Size of the Partition 363
9.4.2 Determinism 364
9.5 Comparison Between the Work-Stealing Approach and EPS 364
9.6 Experiments 365
9.6.1 Benchmark Instances 365
9.6.1.1 Implementation Details 366
9.6.1.2 Execution Environments 366
9.6.2 Multi-core 367
9.6.3 Data Center 368
9.6.4 Cloud Computing 369
9.6.5 Comparison with Portfolios 370
9.7 Conclusion 371
References 372

10 Parallel Local Search 381
Philippe Codognet, Danny Munera, Daniel Diaz, and Salvador Abreu
10.1 Introduction 381
10.2 Local Search Metaheuristics 383
10.3 Sources of Parallelism 386
10.3.1 Single-Walk and Multiple-Walk Methods 386
10.3.2 Parallel Speedups and Runtime Distributions 387
10.4 Single-Walk Approaches 389
10.5 Independent Multiple-Walk Approaches 390
10.5.1 Early Independent Multiple-Walk Methods 390
10.5.2 Recent Experiments and Performance Results 392
10.6 Cooperative Multiple-Walk Approaches 394
10.6.1 Metaheuristic Parallelization Approaches 395
10.6.2 Agent-Based Approaches 398
10.6.3 Framework Approaches 401
10.7 Parallelism at Work 403
10.7.1 Stable Matching Problem 403
10.7.2 The Quadratic Assignment Problem 405
10.8 Conclusion 408
References 409

11 Parallel A* for State-Space Search 419
Alex Fukunaga, Adi Botea, Yuu Jinnai, and Akihiro Kishimoto
11.1 Introduction 419
11.2 Preliminaries: Review of A* 421
11.2.1 The A* Algorithm 423
11.3 Parallel Best-First Search Algorithms 424
11.3.1 Parallel Overheads 425
11.3.2 Centralized Parallel A* 425
11.3.3 Decentralized Parallel A* 428
11.3.3.1 Termination Detection in Decentralized Parallel Search 429
11.4 Hash-Based Decentralized A* 430
11.4.1 Hash Distributed A* 431
11.5 Decentralized Search Using Structure-Based Search Space Partitioning) 432
11.6 Hash Functions for Hash-Based Decentralized Work Distribution 433
11.6.1 Multiplicative Hashing 433
11.6.2 Zobrist Hashing 434
11.6.3 Operator-Based Zobrist Hashing 434
11.6.4 Abstraction 435
11.6.5 Abstract Zobrist Hashing 436
11.6.6 Hyperplane Work Distribution 437
11.6.7 Empirical Comparison of Hash Functions 439
11.6.8 Domain-Independent, Automatic Generation of Hash Functions 441
11.6.9 Hash-Based Work Distribution in Model Checking 441
11.7 Parallel Portfolios Using A* 442
11.8 Parallel, Limited-Memory A* (Parallel IDA*, TDS, PRA*) 443
11.8.1 Transposition Table-Driven Scheduling (TDS) 444
11.8.2 Work Stealing for IDA* 444
11.8.3 Parallel Window Search 445
11.8.4 Parallel Retracting A* (PRA*) 446
11.9 Parallel A* in Cloud Environments with Practically Unlimited Available Resources 446
11.9.1 Iterative Allocation Strategy 447
11.10 Parallel A* and IDA* on Graphics Processing Units 448
11.11 Other Approaches 449
References 450

12 Parallel Model Checking Algorithms for Linear-Time Temporal Logic 457
Jiri Barnat, Vincent Bloemen, Alexandre Duret-Lutz, Alfons Laarman, Laure Petrucci, Jaco van de Pol, and Etienne Renault
12.1 Introduction 458
12.2 Preliminaries: LTL Model Checking and Automata 461

12.2.1 Automata-Theoretic Model Checking 461
12.2.2 Sequences and ω-Words 461
12.2.3 Linear-Time Temporal Logic 462
12.2.4 Kripke Structures 463
12.2.5 Büchi Automata 463
12.2.6 The Emptiness-Check Problem 466
12.2.7 Implicit Models and Automata 469
12.2.8 Simpler Subclasses 471
12.3 Basic Sequential LTL Model Checking Algorithms 473
12.3.1 On-the-Fly Algorithms 473
12.3.2 Depth-First Search 474
12.3.3 Nested-DFS 476
12.3.4 Algorithms Based on SCC Decomposition 478
12.4 Multi-core, DFS-Based Solutions 481
12.4.1 Terminal and Weak Acceptance 481
12.4.2 CNDFS 484
12.4.3 Multi-core/DFS-Based SCC Decomposition 486
12.5 Distributed, BFS-Based Solutions 492
12.5.1 One-Way-Catch-Them-Young 492
12.5.2 MAP 494
12.5.3 Combining OWCTY and MAP 497
12.6 Conclusion 498
References 499

13 Multi-core Decision Diagrams 509
Tom van Dijk and Jaco van de Pol
13.1 Introduction 509
13.2 Preliminaries 510
13.2.1 Boolean Logic and Notation 511
13.2.2 Binary Decision Diagrams 511
13.2.3 Multi-terminal Binary Decision Diagrams 513
13.2.4 Algorithms on Decision Diagrams 514
13.2.5 Parallelism 516
13.2.6 Historical Perspective 517
13.3 Parallel Decision Diagrams 519
13.3.1 Work-Stealing 519
13.3.2 Parallel Operations with Work-Stealing 522
13.3.3 Conclusion 525
13.4 Concurrent Data Structures 525
13.4.1 Representation of Nodes 525
13.4.2 Unique Table 526
13.4.3 Computed Table 531
13.5 Garbage Collection 533
13.6 Empirical Results 535
13.6.1 Symbolic Model Checking 536

13.6.2 Symbolic On-the-Fly Reachability ... 536
13.6.3 Symbolic Bisimulation Minimisation ... 537
13.6.4 Probabilistic Model Checking ... 540
13.7 Conclusions ... 540
References ... 541

14 Parallel Model-Based Diagnosis ... 547
Kostyantyn Shchekotykhin, Dietmar Jannach, and Thomas Schmitz
14.1 Introduction ... 547
14.1.1 Background ... 547
14.1.2 Outline of the Chapter ... 548
14.2 Reiter's Diagnosis Framework ... 549
14.2.1 Example: A Diagnosis Problem Instance ... 549
14.2.2 Diagnoses and Conflicts ... 551
14.2.3 The Hitting Set Tree Algorithm ... 553
14.2.4 Example: Hitting Set Tree Construction ... 554
14.2.5 Complexity Considerations ... 556
14.3 Alternative Approaches to Compute Diagnoses ... 556
14.4 Parallelization of Tree Search Algorithms ... 559
14.4.1 General Parallelization Strategies ... 559
14.4.2 Applying Domain-Independent Parallelized Search Techniques ... 561
14.5 Parallelized Hitting Set Tree Construction Schemes ... 562
14.5.1 Computing Multiple Hitting Set Tree Nodes in Parallel ... 562
14.5.1.1 Level-Wise Parallelization ... 563
14.5.1.2 Full Parallelization ... 564
14.5.2 Computing Nodes and Conflicts in Parallel ... 565
14.5.2.1 Background: QUICKXPLAIN and MERGEXPLAIN ... 566
14.5.2.2 Strategies for Combining Node and Conflict Computation ... 567
14.6 Effectiveness of Computing Multiple Nodes in Parallel ... 569
14.6.1 General Considerations ... 569
14.6.2 Results for Standard Electronic Circuit Benchmark Problems ... 571
14.6.3 Systematic Variation of Problem Characteristics ... 572
14.6.3.1 Method ... 572
14.6.3.2 Results ... 574
14.7 Alternative Model-Based Diagnosis Parallelization Approaches ... 574
14.7.1 Tree-Based Approaches To Find One or Few Diagnoses ... 574
14.7.2 Distributed Hitting Set Algorithms with Known Conflicts ... 576
14.8 Summary ... 577
References ... 578

Part II Tools and Applications

15 Selection and Configuration of Parallel Portfolios 583
Marius Lindauer, Holger Hoos, Frank Hutter, and Kevin Leyton-Brown
15.1 Introduction 584
15.2 Per-Instance Selection of Parallel Portfolios 586
15.2.1 Problem Statement 586
15.2.2 Parallelization of Sequential Algorithm Selectors 588
15.2.2.1 Performance-Based Nearest Neighbor (PNN) . 589
15.2.2.2 Distance-Based Nearest Neighbor (DNN) 589
15.2.2.3 Clustering 590
15.2.2.4 Regression 590
15.2.2.5 Pairwise Voting 591
15.2.3 Parallel Presolving Schedules 591
15.2.4 Empirical Study on Satisfiability Benchmarks 591
15.2.5 Other Parallel Portfolio Selection Approaches 594
15.3 Automatic Construction of Parallel Portfolios from Parameterized Solvers 595
15.3.1 Problem Statement 596
15.3.2 Automatic Construction of Parallel Portfolios (ACPP) ... 598
15.3.2.1 Multiplying Configuration Space: GLOBAL ... 599
15.3.2.2 Iterative Approach: PARHYDRA 599
15.3.2.3 Comparing GLOBAL and PARHYDRA 601
15.3.2.4 Empirical Study on SAT 2012 Challenge 602
15.3.2.5 ACPP with Multiple Solvers 602
15.3.3 Automatic Construction of Parallel Portfolios from Parallel Parameterized Solvers 603
15.3.3.1 Configuration of Clause Sharing 603
15.3.3.2 Portfolio Construction Using Parallel Solvers 604
15.3.3.3 Empirical Study on 2012 SAT Challenge 606
15.4 Conclusions and Future Work 607
References 609

16 An Application of Parallel Satisfiability Solving to the Verification of Complex Embedded Systems 617
Orlando Ferrante, Alberto Ferrari, Christos Sofronis, Leonardo Mangeruca, and Luca Benvenuti
16.1 Introduction 617
16.2 FormalSpecs Verifier Verification Framework 618
16.3 Integration of the ManySAT Solver 619
16.4 Cruise Control Use Case 620
16.5 Simulink Model and Specification 622
16.5.1 Continuous-Time Non-linear Model 622
16.5.1.1 ECU Subsystem 622
16.5.1.2 Engine Subsystem 623

16.5.1.3 Vehicle Dynamics Subsystem 624
16.5.2 Discrete-Time Discrete-Value Model 625
16.5.2.1 Verification Subsystems 627
16.6 Experimental Results 628
16.6.1 Cruise Control Model 628
16.6.1.1 Bounded Model Checking Verification with Incremental Bounds 629
16.6.1.2 Bounded Model Checking Verification with Fixed Bound Value 629
16.6.2 Additional Experiments 630
16.7 Conclusions 631
References 631

17 Parallel Constraint-Based Local Search: An Application to Designing Resilient Long-Reach Passive Optical Networks 633
Alejandro Arbelaez, Deepak Mehta, Barry O'Sullivan, and Luis Quesada
17.1 Introduction 634
17.2 Formal Specification and Complexity 635
17.3 A Mathematical Model for ERDCMST 637
17.4 Iterated Constraint-Based Local Search 638
17.4.1 Move Operators 639
17.4.2 Operations and Complexities 640
17.5 Sequential Algorithm 644
17.6 Parallel Algorithm 646
17.6.1 Multi-Walk and Single-Walk 646
17.6.2 Parallel Moves for ERDCMST 647
17.7 Application: Long-Reach Passive Optical Networks 650
17.8 Empirical Evaluation 652
17.8.1 ERDCMST Results: Sequential LS 653
17.8.2 ERDCMST Results: Parallel LS 656
17.9 Conclusions and Future Work 662
References 662

List of Algorithms 669

Index 671

List of Contributors

Adi Botea
IBM Research, Dublin, Ireland, e-mail: adibotea@ie.ibm.com

Agostino Dovier
University of Udine, Dept of Mathematics, Computer Science, and Physics, Italy, e-mail: agostino.dovier@uniud.it

Akihiro Kishimoto
IBM Research, Dublin, Ireland, e-mail: akihirok@ie.ibm.com

Alberto Ferrari,
Advanced Laboratory on Embedded Systems - United Technologies Research Center, East Hartford, CT, USA, e-mail: name.surname@utrc.utc.com

Alejandro Arbelaez
Insight Centre for Data Analytics, University College Cork, Ireland, e-mail: alejandro.arbelaez@insight-centre.org

Alex Fukunaga
The University of Tokyo, Japan, e-mail: fukunaga@idea.c.u-tokyo.ac.jp

Alexandre Duret-Lutz
LRDE, Epita, Paris, France, e-mail: adl@lrde.epita.fr

Alfons Laarman
TU Wien, Vienna, Austria, e-mail: alfons@laarman.com

Andrea Formisano
University of Perugia, Dip. di Matematica e Informatica, Italy, e-mail: andrea.formisano@unipg.it

Antti E. J. Hyvärinen
Università della Svizzera italiana, Lugano, Switzerland, e-mail: antti.hyvaerinen@usi.ch

Armin Biere
Johannes Kepler University, Linz, Austria, e-mail: biere@jku.at

Arnaud Malapert
Université Côte d'Azur, CNRS, I3S, France, e-mail: arnaud.malapert@unice.fr

Barry O'Sullivan
Insight Centre for Data Analytics, University College Cork, Ireland, e-mail: barry.osullivan@insight-centre.org

Carsten Sinz
Karlsruhe Institute of Technology (KIT), Karlsruhe, Germany, e-mail: carsten.sinz@kit.edu

Christoph M. Wintersteiger
Microsoft Research, e-mail: cwinter@microsoft.com

Christos Sofronis,
Advanced Laboratory on Embedded Systems - United Technologies Research Center, East Hartford, CT, USA, e-mail: name.surname@utrc.utc.com

Daniel Diaz
University Paris 1/CRI, France, e-mail: daniel.diaz@univ-paris1.fr

Danny Munera
University of Antioquia, Medellin, Colombia, e-mail: danny.munera@udea.edu.co

Deepak Mehta
Insight Centre for Data Analytics, University College Cork, Ireland, e-mail: deepak.mehta@insight-centre.org

Dietmar Jannach
Department of Computer Science, TU Dortmund, Germany, e-mail: dietmar.jannach@tu-dortmund.de

Enrico Pontelli
New Mexico State University, Dept. of Computer Science, USA, e-mail: epontell@cs.nmsu.edu

Etienne Renault
LRDE, Epita, Paris, France e-mail: renault@lrde.epita.fr

Florian Lonsing
Institute of Information Systems, TU Wien, Austria, e-mail: florian.lonsing@tuwien.ac.at

Frank Hutter
University of Freiburg, Germany, e-mail: fh@cs.uni-freiburg.de

Holger Hoos
University of British Columbia, Canada & Leiden University, Netherlands, e-mail: hoos@cs.ubc.ca

Inês Lynce
INESC-ID / Instituto Superior Técnico, Universidade de Lisboa, Portugal, e-mail: ines.lynce@tecnico.ulisboa.pt

Jaco van de Pol
Formal Methods and Tools, University of Twente, Enschede, The Netherlands, e-mail: j.c.vandepol@utwente.nl

Jean-Charles Régin
Université Côte d'Azur, CNRS, I3S, France, e-mail: jcregin@gmail.com

Jiri Barnat
Masaryk University, Brno, Czech Republic, e-mail: xbarnat@fi.muni.cz

Kevin Leyton-Brown
University of British Columbia, Canada, e-mail: kevinlb@cs.ubc.ca

Kostyantyn Shchekotykhin
Institute for Applied Informatics, Alpen-Adria-Universität Klagenfurt, Austria, e-mail: konstantin.schekotihin@aau.at

Laure Petrucci
LIPN, CNRS, Paris, France, e-mail: Laure.Petrucci@lipn.univ-paris13.fr

Leonardo Mangeruca
Advanced Laboratory on Embedded Systems - United Technologies Research Center, East Hartford, CT, USA, e-mail: name.surname@utrc.utc.com

Luca Benvenuti
"Sapienza" University of Rome, Italy, e-mail: luca.benvenuti@uniroma1.it

Luis Quesada
Insight Centre for Data Analytics, University College Cork, Ireland, e-mail: luis.quesada@insight-centre.org

Maria Paola Bonacina
Dipartimento di Informatica, Università degli Studi di Verona, Italy, e-mail: mariapaola.bonacina@univr.it

Marijn J.H. Heule
The University of Texas at Austin, USA, e-mail: marijn@cs.utexas.edu

Marius Lindauer
University of Freiburg, Germany, e-mail: lindauer@cs.uni-freiburg.de

Martina Seidl
Institute for Formal Models and Verification, JKU Linz, Austria, e-mail: martina.seidl@jku.at

Oliver Kullmann
Swansea University, UK, e-mail: O.Kullmann@swansea.ac.uk

Orlando Ferrante,
Advanced Laboratory on Embedded Systems - United Technologies Research Center, East Hartford, CT, USA, e-mail: name.surname@utrc.utc.com

Philippe Codognet
University Pierre & Marie Curie/LIP6, France, e-mail: philippe.codognet@upmc.fr

Ruben Martins
University of Texas at Austin, USA, e-mail: rmartins@cs.utexas.edu

Salvador Abreu
University of Évora/LISP/CRI, Portugal, e-mail: spa@di.uevora.pt

Ted Ralphs
Lehigh University, Bethlehem, PA, USA, e-mail: ted@lehigh.edu

Thomas Schmitz
Department of Computer Science, TU Dortmund, Germany, e-mail: thomas.schmitz@tu-dortmund.de

Thorsten Koch
Zuse Institute, Berlin, Germany, e-mail: koch@zib.de

Timo Berthold
Fair Isaac Germany GmbH, Berlin, Germany, e-mail: timoberthold@fico.com

Tom van Dijk
Institute for Formal Methods and Verification, Johannes Kepler University, Linz, Austria, e-mail: tom.vandijk@jku.at

Tomáš Balyo
Karlsruhe Institute of Technology (KIT), Karlsruhe, Germany, e-mail: biotomas@gmail.com

Vasco Manquinho
INESC-ID / Instituto Superior Técnico, Universidade de Lisboa, Portugal, e-mail: vmm@sat.inesc-id.pt

Vincent Bloemen
University of Twente, Enschede, The Netherlands, e-mail: v.bloemen@utwente.nl

Yuji Shinano
Zuse Institute, Berlin, Germany, e-mail: shinano@zib.de

Yuu Jinnai
The University of Tokyo, Japan, e-mail: ddyuudd@gmail.com

Part I
Theory and Algorithms

Chapter 1
Parallel Satisfiability

Tomáš Balyo and Carsten Sinz

Logic is the beginning of wisdom, not the end –
Leonard Nimoy

Abstract The propositional satisfiability problem (SAT) is one of the fundamental problems in theoretical computer science, but it also has many practical applications. Parallel algorithms for the SAT problem have been proposed and implemented since the 1990s. This chapter provides an overview of current approaches and their evolution over recent decades towards efficiently solving hard combinatorial problems on multi-core computers and clusters.

1.1 Introduction

SAT is one the most important problems in computer science. It was the first problem proven to be NP-hard [16]. Despite its complexity there are very efficient SAT solvers which make it possible to design successful algorithms for hard problems by translating them to SAT.

Parallelizing algorithms for combinatorial decision problems, such as SAT, is not an easy task, as the search space is highly irregular and different search heuristics can have a tremendous effect on the observed run-time. Theoretical results vary from super-linear speedups for random problems [55] on the positive side to profound proof-theoretic limitations [34] on the negative.

Nevertheless, important parallelization techniques for SAT have been developed, including divide-and-conquer approaches, portfolio solvers, and parallel local search solvers.

As the increase in compute power of a single processor core has been stagnating over recent years, it has become even more important to invent and engineer parallel algorithms that can make optimal use of current and future computer architectures.

Karlsruhe Institute of Technology (KIT)
Karlsruhe, Germany

Y. Hamadi und L. Sais (eds.), *Handbook of Parallel Constraint Reasoning*,
https://doi.org/10.1007/978-3-319-63516-3_1

This chapter is organized as follows: after an introduction to basic notions and algorithms for the SAT problem, parallel computing architectures and the problem of measuring speedups are discussed. Then the current main lines for parallel SAT algorithms are presented, namely divide-and-conquer (also known as search space partitioning), portfolios (diversify-and-conquer), and local search solvers. The chapter closes with a look at future challenges.

1.2 Preliminaries

In this section we give the basic definitions and properties of the satisfiability problem, which can also be found in any SAT-related textbook (for example the Handbook of Satisfiability [7]).

1.2.1 Satisfiability (SAT)

We start with the definition of a formula, which is the input of the SAT problem.

Definition 1 (CNF Formula). A *Boolean variable* is a variable with two possible values, *True* and *False*. A *literal* of a Boolean variable x is either x or $\neg x$, i.e., *positive* or *negative literal* of x. A *clause* is a disjunction (OR) of literals. A *conjunctive normal form (*CNF*) formula* is a conjunction (AND) of clauses. We can also regard a clause as a set of literals and a CNF formula as a set of clauses, since the ordering is not important in either case.

In the remainder of the chapter we will just use the term formula instead of CNF formula. Next we define what is a satisfying assignment.

Definition 2 (Satisfying Assignment). A *truth assignment* ϕ of a formula F assigns a *truth value* to its variables. The assignment ϕ satisfies

- a positive literal if it assigns the value True to its variable,
- a negative literal if it assigns the value False to its variable,
- a clause if it satisfies at least one of its literals,
- a CNF formula if it satisfies each one of its clauses.

If ϕ satisfies a formula F, then ϕ is called a *satisfying assignment* for F.

A clause with no literals is called an *empty clause*. Such a clause cannot be satisfied by any truth assignment. The definition of satisfiability follows.

Definition 3 (Satisfiability). A formula F is said to be *satisfiable* if there is a truth assignment ϕ that satisfies F, i.e., ϕ is a satisfying assignment of F. Otherwise, the formula ϕ is *unsatisfiable*.

The *problem of satisfiability (SAT)* is to determine whether a given formula F is satisfiable or unsatisfiable.

A *SAT solver* is a procedure that solves the SAT problem. For satisfiable formulas we also expect a SAT solver to produce a satisfying assignment. An example of a satisfiable formula with its satisfying assignment follows.

Example 1. $F = (x_1 \vee x_2 \vee \neg x_4) \wedge (x_3 \vee \neg x_1) \wedge (\neg x_1 \vee \neg x_2)$ is a CNF formula with three clauses: $\{(x_1 \vee x_2 \vee \neg x_4), (x_3 \vee \neg x_1), (\neg x_1 \vee \neg x_2)\}$ and six literals $\{x_1, \neg x_1, x_2, \neg x_2, x_3, \neg x_4\}$ on four variables $\{x_1, x_2, x_3, x_4\}$. F is satisfiable with $\phi = \{x_1 \rightarrow False, x_2 \rightarrow True, x_3 \rightarrow True, x_4 \rightarrow True\}$ being a satisfying truth assignment of F.

1.2.2 Local Search Algorithms for SAT

The simplest approach to SAT solving is local search. A generic local search algorithm starts with a truth assignment (usually random, i.e., each variable has a random truth value assigned) and then iteratively selects a variable whose value is flipped (changed to False if it was True and vice versa) until a satisfying assignment is reached. The pseudo-code of this generic local search is presented as Algorithm 1.1.

Obviously, this algorithm only works for satisfiable formulas; for unsatisfiable instances it does not terminate. The performance of the algorithm depends on the initial truth assignment and the way of selecting variables for flipping. In the best-case scenario the initial assignment is already satisfying and we are finished. Also, if the variable selection were ideal, we could reach a satisfying assignment from any initial assignment in at most n steps (where n is the number of variables). In practice we need to use heuristics for both these steps and the main loop (the variable flipping) is executed only a limited number of times after which the algorithm gives up.

The initial truth assignment is often chosen randomly and for the variable selection a heuristic minimizing the number of unsatisfied clauses is used. Two historically important examples of local search algorithms are GSAT [49] and WalkSat [48]. GSAT select a variable that reduces the number of unsatisfied clauses most when flipped. WalkSat first randomly selects a clause that is not satisfied under the current assignment and flips one of its literals based on the number of clauses that become satisfied and unsatisfied after the flip. WalkSat additionally performs a random selection of the literal to flip in a certain percentage of the flips to emulate random walk, hence the name WalkSat.

Local search algorithms are usually the best choice for randomly generated satisfiable formulas and some combinatorial problems encoded to SAT.

1.2.3 The DPLL Algorithm

Most of the current state-of-the-art SAT solvers are based on the CDCL (conflict-driven clause learning) algorithm [41], which is in turn based on the Davis Putnam

Algorithm 1.1: A Generic Local Search Algorithm

```
1 Function LS(Formula f)
2     φ ← generate truth assignment
3     while F not satisfied by φ do
4         v ← pick a variable
5         φ[v] ← ¬φ[v]
6     return true
```

Logemann Loveland (DPLL) algorithm [17]. Before we give a description of CDCL in the following section we review DPLL here. The DPLL algorithm is basically a depth-first search of partial truth assignments (truth assignments where some variables remain unassigned) with three additional enhancements. The explanation of these enhancements follows.

- *Early termination.* If all literals are False in some clause, we can backtrack since it is obvious that the current partial truth assignment cannot be extended into a satisfying assignment. If all clauses are satisfied we can stop the search. The remaining unassigned Boolean variables can be assigned arbitrarily.
- *Pure literal elimination.* Given a partial truth assignment ϕ a pure literal is a literal the negation of which does not appear in any of the clauses not satisfied by ϕ. The variable corresponding to a pure literal can be assigned to make each clause where it appears true. This might lead to the appearance of new pure literals.
- *Unit propagation.* A clause is called unit if all but one of its literals are false under ϕ and the remaining literal is unassigned. The unassigned literal of a unit clause must be assigned to be true. This can make other clauses unit and thus force new assignments. The cascade of such assignments is called unit propagation.

In the DPLL procedure the enhancements are used after each decision assignment of the depth-first search. First we check the termination condition. If the formula is neither satisfied nor unsatisfied by the current partial assignment, we continue by unit propagation. Finally we apply the pure literal elimination. Unit propagation is called before pure literal elimination because it can cause the appearance of new pure literals. The other way around, pure literal elimination will never produce a new unit clause, since it does not make any literals false. Pseudo-code of DPLL is presented as Algorithm 1.2.

We can see that DPLL is a sound and complete algorithm (always terminates and answers correctly) from the fact that DPLL is a systematic depth-first search of partial truth assignments. The enhancements only filter out some branches that represent non-satisfying assignments.

The time complexity of this procedure is exponential in the number of variables. That corresponds to the number of vertices of a binary search tree with depth n, where n is the number of variables. However, in practice, thanks to unit propagation

Algorithm 1.2: The DPLL Algorithm

```
1 Function DPLL(Formula F, Assignment φ)
2     doUnitPropagation(F,φ)
3     if all literals false in some clause then
4         return false
5     doPureLiteralElimination(F,φ)
6     if all clauses satisfied then
7         return true
8     x ← choose an unassigned variable
9     return DPLL(F, φ[x] = True) or DPLL(F, φ[x] = False)
```

and early termination, the DPLL procedure never goes as deep as n in the search tree. The maximal depth reached during search is often a fraction of n. This makes DPLL run much faster on instances with n variables than one would expect from the formula 2^n.

1.2.4 Resolution Refutation

Resolution is a rule of inference which produces a new clause from clauses containing complementary literals. Two clauses C and D are said to contain complementary literals if there is a Boolean variable x such that $x \in C$ and $\neg x \in D$. The produced clause (containing all the literals from C and D except for x and $\neg x$) is called the *resolvent* of C and D (notation: $R(C,D)$).

A formula F containing C and D is satisfiable if and only if $F \wedge R(C,D)$ is satisfiable. This implies that if the empty clause can be resolved from the clauses of a formula then this formula is unsatisfiable (since the empty clause cannot be satisfied). The *Resolution Refutation* algorithm keeps adding resolvents to its input formula until either the empty clause is added (which means the input formula is unsatisfiable) or no more new resolvents can be added (in which case the input formula is satisfiable). Note that resolvents added in one step can be used as input clauses for resolutions in later steps.

Although the resolution refutation algorithm is sound and complete it is not very efficient in practice since it has exponential memory complexity (in general there are exponentially many possible resolvents for a formula).

1.2.5 The CDCL Algorithm

The conflict-driven clause learning (CDCL) algorithm is the state-of-the-art algorithm for solving SAT problems. It was first implemented in the SAT solver Grasp [41].

Algorithm 1.3: The CDCL Algorithm

```
1  Function CDCL(Formula f)
2      decLev ← 0
3      φ ← ∅
4      if doUnitPropagation(f,φ) = CONFLICT then
5          return false
6      while not all variables assigned do
7          decVar ← pick decision variable
8          decVal ← pick a truth value
9          decLev ← decLev + 1
10         φ[decVar] = decVal with decision level decLev
11         if doUnitPropagation(f,φ) = CONFLICT then
12             (learnedClause, backLev) ← analyze conflict
13             if backLev ≥ 0 then
14                 decLev ← backLev
15                 f ← f ∧ learnClause
16                 φ ← unassign variables with decision level ≥ backLev
17             else
18                 return false
19     return true
```

In this subsection we describe only the basic concepts behind CDCL. For a more detailed comprehensive description please refer to [7].

The CDCL algorithm combines ideas of DPLL search and resolution refutation. The pseudo-code of CDCL is presented in Algorithm 1.3. Similarly to DPLL the algorithm performs depth-first search of partial truth assignments and uses improvements such as unit propagation and early termination. Additionally, CDCL performs a procedure called *conflict analysis* each time a conflict state is reached, i.e., every literal becomes false in some clause under the current partial assignment.

The conflict analysis determines which decisions and which clauses (via unit propagation) are responsible for the conflict. The clauses responsible for the conflict are called *reason clauses*. By resolving the reason clauses of a conflict we get new clauses that can be added to the formula. Clauses added this way are called *learned clauses*.

In the CDCL algorithm each truth value assignment to a variable has an attribute called its *decision level*. The assignments implied by the initial unit propagation have decision level zero, the assignments coming from the first branching decision and the unit propagation that follows it have decision level one, and so on. In DPLL the decision level represents the depth of the recursive call during which the variable was assigned. The decision level increases by one after every branching decision and is decreased by one after a conflict is encountered and we backtrack to the previous decision.

In CDCL the decision level can decrease by more than one during backtracking. This is called *non-chronological backtracking* or *backjumping*. The decision level to which the algorithm "backjumps" is calculated during conflict analysis.

1.2.6 Parallel Computing Architectures

In this subsection we review the basic notions related to parallel computing, such as parallel architectures, memory models, and definitions of speedup and parallel efficiency.

Based on the access to the main memory used in a parallel system we can distinguish two kinds of parallel architectures.

- *Shared Memory Architectures*. The main memory is shared between all processing elements in a single address space. It is used on single computers with multiple (multi-core) processors. The advantages of this approach are that all processes have very fast access to the shared data and less total memory is used, which allows the solution of larger problems. The disadvantage is that race conditions must be addressed (usually with locks), which may lead to parallel slowdown or even deadlocks, and this makes implementations error prone.
- *Distributed Memory Architectures*. Each processing element has its own address space and communication is usually done by message passing. This approach can be used on single computers or on grids/clusters of computers. The speed of communication is lower than in the case of shared-memory architectures but the design and implementation of such a system is usually simpler.

A parallel system can also use a combination of these architectures. For example the parallel solver HordeSat[6], which was designed to run on clusters of multi-core computers, uses shared-memory communication inside the nodes and distributed-memory communication between the nodes.

1.2.7 Measuring Speedups

The speedup of a parallel solver P compared to a sequential solver S for a given benchmark is the ratio of run times that the solvers need to solve that benchmark, i.e., $s = t_P/t_S$, where t_P and t_S are the runtimes of the parallel and sequential solver respectively.

In parallel processing, one usually wants good scalability in the sense that the speedup over the best sequential algorithm goes up near linearly with the number of processors. Measuring scalability in a reliable and meaningful way is difficult for SAT solving since running times are highly nondeterministic. Hence, we need careful experiments on a large benchmark set chosen in an unbiased way.

By averaging the speedups for each benchmark instance we can compute the *average speedup*. The average speedup is not a very robust measure since it is highly dependent on a few very large speedups that might be just due to luck. For this reason we often get very large average speedup values that are not representative for the entire benchmark set. Calculating the median of the speedups gives us the *median speedup*. The value of the median speedup is often very small if the benchmark set contains a large number of easy benchmarks where parallelization does not bring any benefit, and therefore it is not an ideal measure either. A better measure is the *total speedup* which is the sum of runtimes for the parallel solver divided by the sum of runtimes for the sequential solver on the benchmark set.

Nevertheless, all these measures can treat a massively parallel solver (a solver designed for hundreds or thousands of processors) unfairly when most instances are actually too easy to justify investing in a lot of hardware. Indeed, in parallel computing, it is usual to analyze the performance on many processors using *weak scaling* where one increases the amount of work involved in the considered instances proportionally to the number of processors. Therefore the set of benchmarks considered for calculating the average, median, and total speedups is usually restricted to those instances where the sequential solver needs at least $c \times p$ seconds where p is the number of processors used by the parallel solver and c is constant.

1.3 Divide-and-Conquer Approaches

Historically, the first parallelization approaches for the SAT problem were based on splitting the search space. Here, different tasks search for a satisfying assignment in disjunct portions of the search space. Different ways to split the search space have been proposed [10, 11, 12, 14, 15, 20, 30, 32, 33, 38, 46, 50, 58, 59]. Splitting the search space should preferably yield portions of potentially equal size to balance the search evenly among different tasks. Predicting the size of the search space for a DPLL or CDCL search is extremely hard and no satisfactory solutions exist up to now, even though some promising attempts have been made [36, 37].

Thus, the search space is typically not split up statically (at the start of the algorithm), but dynamically, as soon as one processor involved in the search becomes idle.

1.3.1 Problem Decomposition and Load Balancing

Problem decomposition plays a central role within the design process of parallel algorithms, since it influences all other design phases. In this stage, the whole problem is divided into appropriate subproblems (called *tasks*) which can be executed in parallel by the available processors. Problem decomposition must achieve two (typically conflicting) goals:

- Minimize idle times of available processors.
- Minimize overhead due to communication and excess computation.

Basically, problem decomposition can be carried out statically (i.e. tasks are defined at compile time) or in a dynamic manner, where tasks are generated (on demand) at run-time. In the latter case, tasks are explicit objects within the parallel program which can be dynamically assigned to processors for execution.

Due to the sophisticated heuristics employed by contemporary DPLL-based SAT solvers it is virtually impossible to predict the time needed to solve a specific SAT instance. Accordingly, the run-time of an individual task cannot be predicted and the run-times of different tasks may vary considerably. For SAT instances exhibiting such a highly irregular problem structure a static approach to decomposition can result in significant processor idling. Thus, for realizing a robust parallel SAT-solving method, dynamic problem decomposition becomes mandatory.

For problems based on heuristic search, typically an *exploratory* approach to problem decomposition is employed where tasks represent untried branches of the search tree. Specifically, this technique enables a running solving task to efficiently split off a part of its own search space, generating a new task. The parallel computation terminates when all generated tasks have been completed or when a task reports a solution. In the latter case, the remaining tasks can be canceled.

Technically, exploratory decomposition can be accomplished by a transformation of the assignment stack of a running solving process. It narrows the search space of the solving process by fixing the decision and the corresponding implications of the first level. The released search space is defined by the top-level assignments and the flipped decision. In this way, tasks can be represented by a set of assignments. The splitting procedure is depicted in Figure 1.1. This technique was first described by Chrabakh and Wolski [15]. It represents a refinement of the guiding-path approach developed by Zhang et al. [59].

In order to enable dynamic problem decomposition, a sequential solver must be adapted to support the discussed transformation of the assignment stack and it must also be capable of initializing the level 0 of the assignment stack according to the set of assignments delivered by a task.

The parallel computation starts with a single task that is responsible for the whole search space (i.e. the task is defined by an empty set of assignments). When idle processors are detected (either initially or upon completion of a task), search space splitting is performed to induce additional parallelism. This procedure is steered by the load-balancing process, which we discuss next.

Generally, dynamic problem decomposition requires explicit load balancing, i.e., tasks have to be assigned to processors at run-time. Especially for problems with high irregularity, the *task pool* model should be employed. It decouples problem decomposition and load balancing by using an explicit data structure holding tasks resulting from dynamic decomposition operations.

The task pool model can be organized in either a centralized or a distributed fashion. In a centralized approach, a master processor maintains a global task pool from which processors can request new tasks when they become idle. The master also

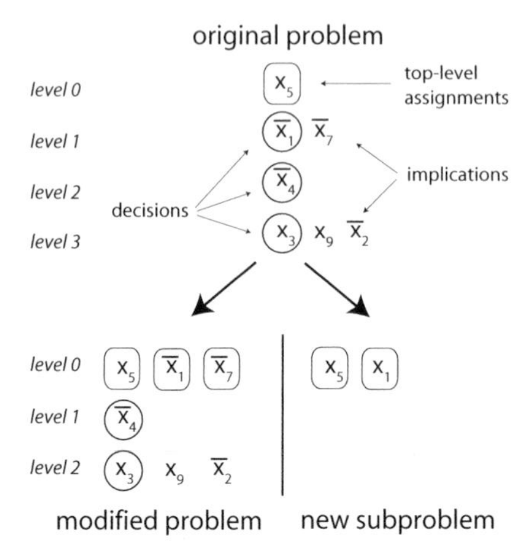

Fig. 1.1: Problem Decomposition

keeps track of the activity of each processor. Thus it can select an active processor to perform problem decomposition when the size of the pool falls below a given threshold. This ensures that the task pool is sufficiently filled to serve task requests in a timely fashion.

Fully distributed load balancing requires that every processor maintains its own task pool. In this setting, problem decomposition and load balancing must be accomplished autonomously by the processors. If a processor runs out of tasks it chooses another processor (e.g., by a round-robin or a randomized scheme) from which to request new tasks. When a predefined amount of time has elapsed without a reply, a request to a different processor is issued. On the other hand, active processors perform problem decomposition when the size of the task pool falls below a certain threshold. In order to prevent parallelism being generated in an uncontrolled way, the number of splitting operations a processor performs must be limited, e.g., by a minimum time interval between two consecutive split operations. In the distributed task model, choosing appropriate threshold and timing values is a subtle task, which can in practice only be managed by extensive experimentation. Due to the lack of a central controller, detecting the end of a parallel computation (i.e., all generated tasks have been executed) requires explicit protocols, e.g., Dijkstra's token-ring-based termination detection algorithm [18].

In general, a centralized approach can establish a more accurate view of the state of the processors and is more easy to implement, particularly on shared-memory architectures. However, with an increasing number of processors, centralized components soon become a sequential bottleneck of a parallel computation, which can seriously limit the overall efficiency. Thus, at least for distributed-memory architectures with a large number of processors a distributed design should be preferred.

The decomposition procedure we have discussed in this section represents the approach taken by most of the existing parallel SAT solvers. However, in the light of the latest generation of sequential SAT-solving methods, exploratory decomposition can become a source of work anomalies. The search spaces of the generated subproblems are mutually disjoint, but their union isn't necessarily identical to the search space covered by the sequential algorithms (e.g., due to the failure-driven assertion technique). Consequently, the total amount of work carried out may differ significantly between the sequential and the parallel algorithm. On the one hand, this can result in poor speedups (due to excess computation) and on the other hand super-linear speedups are possible.

1.3.2 Implementations of Search-Space-Splitting Solvers

Table 1.1 shows implementations of search-space-splitting parallel DPLL SAT solvers. (Abd El Klalek *et al.* [19] also provide an overview and classification of many parallel SAT solvers.) For each solver the target infrastructure is indicated as well as whether the implementation provides fault detection and clause exchange.

Solver / Author(s)	Year	Ref.	Infra-structure	Fault Det.?	Clause Exch.?	Comment
Böhm & Speckenmeyer	1994	[14]	Transputer	no	no	First parallel SAT implementation
PSATO	1994	[58]	Cluster	yes	no	Introduced notion of "Guiding Path"
PSolver	1998	[38]	Grid	yes	no	Master / slave approach allowing integration of different sequential solvers
PaSAT	2001	[50]	SMP	no	yes	First solver with clause exchange
//Satz	2001	[32]	Cluster	no	no	Defined the notion of "ping-ping phenomenon"
GridSAT / GradSAT	2003	[15]	Grid	no	yes	Refined search space splitting
ySAT	2005	[20]	SMP	no	yes	Focus on cache performance
ZetaSAT	2005	[12]	Grid	yes	no	Runs on heterogeneous grids
NorduGrid	2006	[30]	Grid	yes	no	Splitting based on the "scattering rule"
PaMiraXT	2009	[47]	SMP & Cluster	no	yes	Master/client model

Table 1.1: Some early DPLL-based search-space-splitting SAT Solvers

Problem decomposition via dynamic search space splitting results in highly irregular run-times. This is shown in Figure 1.2. For a thousand runs on each instance, the run-time distribution is depicted; the two instances on top are satisfiable, the ones on the bottom unsatisfiable.

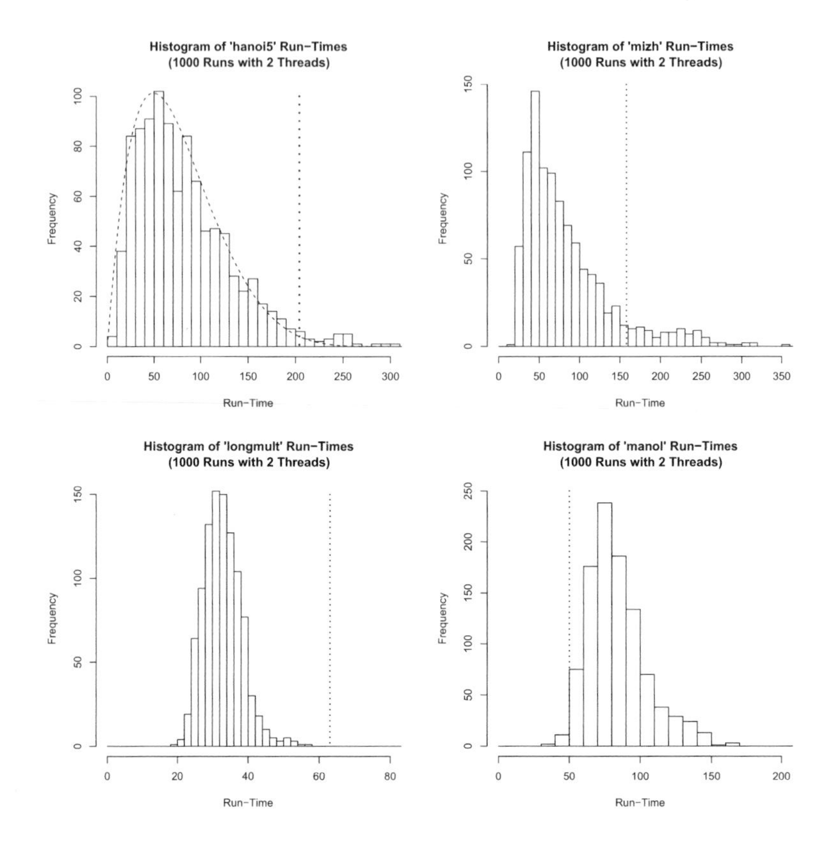

Fig. 1.2: Run-time distributions for 1,000 runs of the parallel solver PaSAT [50] on four selected instances (towers of Hanoi; cryptanalysis; hardware multiplier; pipelined microprocessor). The two instances on the top (Hanoi and mizh) are satisfiable, the ones on the bottom are unsatisfiable. On the x-axis, the run-time of PaSAT on two cores is shown. The y-axis indicates the number of times the run-times were in the given interval. The vertical line depicts the sequential run-time of the solver (one core). In the top left figure, the dashed curve indicates a beta distribution with $\alpha = 2.0$ and $\beta = 6.5$

1.3.3 Search Space Splitting in CDCL

As CDCL traverses the search space in a less structured way than DPLL, approaches based on the guiding path and dynamic decomposition cannot be directly applied for search-space-splitting CDCL solvers.

Adaptations have been implemented in PCASSO [40], Treengeling [9], and Ampharos [3]. The employed techniques are similar to the Cube and Conquer approach described below.

1.3.4 Cube and Conquer

The basic idea of the Cube and Conquer [28] approach is to use look-ahead techniques to split the problem into a large number (thousands) of subproblems, which can then be solved in parallel. The Cube and Conquer approach is discussed in detail in Chapter 2.

1.4 Parallel Portfolios – Diversify and Conquer!

In this section we discuss the simplest and yet (currently) most powerful approach to parallelizing SAT – parallel portfolios. We start by explaining the concept of the virtual best solver that served as the inspiration for the portfolio approach. Then we discuss clause sharing, which is an important component of any portfolio SAT solver. We conclude the section by reviewing some of the existing portfolio SAT solvers.

1.4.1 Virtual Best Solver

In a SAT competition a collection of SAT solvers submitted by researchers from all over the world is run on a pre-selected set of benchmark problems with some time limit (usually 1 hour or 5,000 seconds per instance). The results of a solver are defined as the set of run times for each problem solved by that solver. The solver solving the highest number of problems (within the time limit) is the winner of the competition; ties are broken by comparing the average run times.

When a SAT competition is organized and the results are published it is common to include the results for the *virtual best solver* (VBS) along with the results of the actual solvers participating in the competition. The results of the VBS are calculated as follows. For each benchmark that was solved by at least one of the participating solvers we take the best run time from the run times of the solvers on that benchmark. This implies that no solver has better run time than the VBS on any of the benchmarks or solves a benchmark not solved by the VBS.

Is it possible to have a real solver that is as good as the VBS? Such a solver would need to have the ability to instantly select the best SAT solver for any benchmark. This seems to be rather difficult; however, if we have a parallel architecture and only care about wall-time, there is a simple solution. We run all the available solvers in parallel on the given problem and as soon as one of the solvers finds a solution we terminate all the remaining solvers. This parallel solver would clearly achieve the same results as the VBS. A solver like this is called a *parallel portfolio solver*.

1.4.2 Pure Portfolio Solvers and Diversification

In the 2011 SAT Competition the PPfolio [45] solver demonstrated that it is possible to win several tracks of the competition by just taking the best solvers from the previous competition and trivially combining them using a shell script into a portfolio. The author of PPfolio argues that such a simple portfolio solver can serve as an approximation of the virtual best solver. But he also "shamelessly claims" [31] that “it's probably the laziest and most stupid solver ever written” which “does not even parse the CNF” and “knows nothing about the clauses”. This most basic kind of portfolio solver is called a *pure portfolio* and the results obtained by this portfolio are only due to the base solvers selected.

A pure portfolio solver winning the competition can be very demotivating for the developers of the included solvers since someone else is winning with their solver.[1] To avoid this situation the following SAT competitions restricted or completely prohibited the participation of such portfolios.

A portfolio can be also created by using just one SAT solver, which is run several times in parallel with different configuration settings. The motivation behind this approach is that the performance of SAT solvers is heavily influenced by a high number of different settings and parameters of the search such as the heuristic used to select a decision literal in DPLL/CDCL, different restart policies or clause learning and deleting schemes in CDCL. Numerous parameter configurations are possible but none of them dominates all the other configurations on each problem instance.

The process of selecting good configurations for a portfolio solver is called *diversification*. Similarly to stock market portfolios a SAT solver portfolio should be diversified to achieve variety and increase the robustness of the solver. In a well diversified parallel portfolio solver each core solver explores a different region of the search space and therefore the overlap, i.e., redundant work, is minimized. The usual parameters that are diversified are related to decision heuristics (for example community branching [53] and block branching [52]), restart heuristics [51], and clause deletion strategies [22]. These configurations are often selected by hand but methods for automatic configuration of SAT solvers for portfolios are also studied [57].

1.4.3 Clause-Sharing Portfolios

If a portfolio is based on CDCL (conflict-driven clause learning) solvers then learned-clause exchange can be implemented. This grants a considerable boost to the solvers performance. Together with diversification it is an important mechanism to reduce duplicate work, i.e., parallel searches working on the same part of the search space.

[1] It should be noted that non-portfolio solvers are often derived from existing solvers too. However, they typically make reference to the original solver, e.g., by having a name derived from the original solver's name. Moreover, some competitions include a “Hack Track”, in which small modifications to an existing solver can be submitted.

A clause learned from a conflict by one CDCL instance distributed to all the other CDCL instances will prevent them from doing the same work again in the future.

The problems related to clause sharing are to decide how many and which clauses should be exchanged. Exchanging all the learned clauses is unfeasible especially in the case of large-scale parallelism due to communication overhead. Also having too many clauses slows down a CDCL solver. A simple solution is to distribute all the clauses that satisfy some conditions. The conditions are usually related to the length of the clauses (number of literals in them) and/or their glue value [4] (the number of different decision levels associated with the literals of the clauses). A technique to dynamically adjust the size of shared clauses has been proposed in [25].

An interesting technique called "lazy clause exchange" was introduced in a recent paper [5] and used in the parallel version of the SAT solver Glucose [4]. In this policy a solver does not share a clause immediately after it is learned, but only after it proves its worth by being useful locally. Being useful locally means that the clause appears in conflicts as a reason clause at least a given number of times. This restriction does not apply to short clauses (at most two literals) and clauses with a low glue value. The policy also contains a strategy for importing clauses from other solver instances. The incoming clauses are put in "probation" before a potential entry into the clause database. This limits the negative impact of importing too many clauses. The probation phase is implemented by watching only one literal in these clauses, which means that they are not used for unit propagation and are only detected when they become unsatisfied. At that point they leave probation and are promoted to the regular learned-clause status. The experimental data in [5] show that only 10% of the imported clauses leave probation on average, which demonstrates how well-founded this strategy is.

Similarly to sequential SAT solvers, clause-sharing portfolio solvers can produce proofs of unsatisfiability and therefore be validated [27].

Clause sharing can be implemented in a lockless fashion as demonstrated by the SAT solver SArTagnan [25, 35].

Clause sharing in a parallel environment introduces non-determinism to the solver, which might not be desirable for practitioners who expect run time reproducibility. This issue has been addressed in [23] where a fully deterministic parallel portfolio solver was designed.

1.4.4 Impact of Diversification and Clause Sharing

Diversification and clause sharing are both essential components of a successful CDCL portfolio SAT solver. But to better understand them let us take a look at the impact of these techniques in isolation for satisfiable and unsatisfiable random 3-SAT instances.

By looking at the cactus plots in Figure 1.3 we can observe that clause sharing is essential for unsatisfiable instances while not very beneficial and even slightly

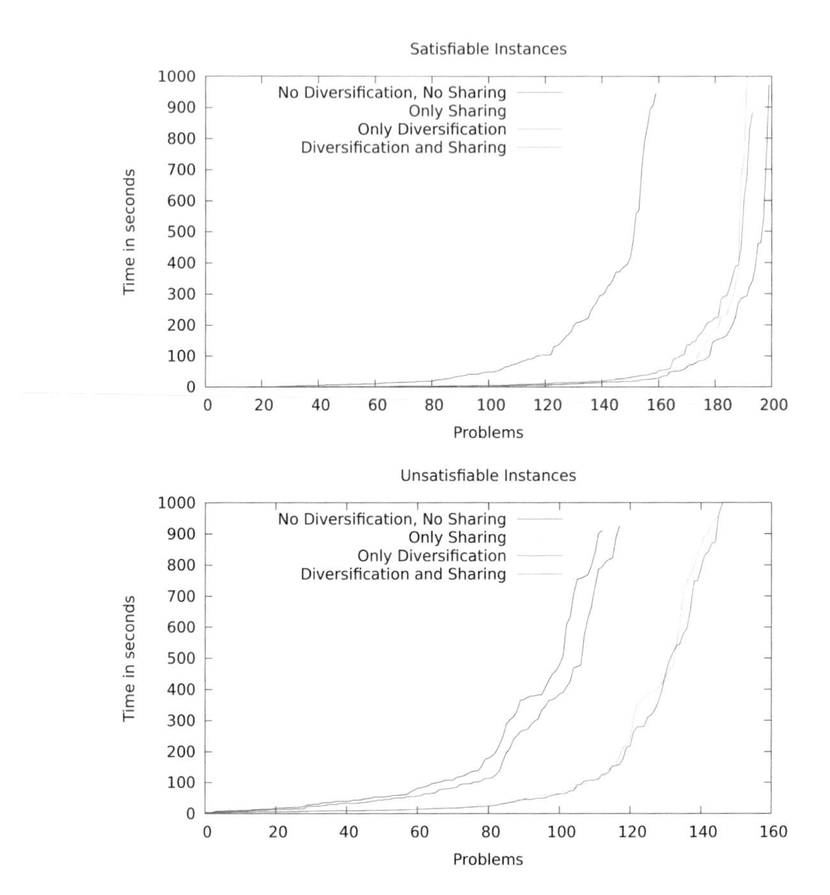

Fig. 1.3: The influence of diversification and clause sharing on the performance of HordeSat[6] on random 3-SAT problems. Plot is taken from [6]

detrimental for satisfiable problems. On the other hand, diversification has only a small benefit for unsatisfiable instances but high impact for satisfiable ones.

This is actually in accordance with what one would expect. To solve a satisfiable formula we only need to find a satisfying assignment anywhere in the search space. To do this efficiently we only need to diversify the search (which we also do when only allowing clause sharing). On the other hand, for unsatisfiable problems we must actually construct a resolution proof. The different solvers in the portfolio construct different segments of the proof and to get the complete proof we join these segments via clause sharing. Without clause sharing each solver must construct the complete proof alone.

1.4.5 Examples of Parallel Portfolio Solvers

ManySat

ManySat [24] was the first successful clause-sharing parallel portfolio SAT solver. It was developed in 2008 and won first place in the Parallel Track of the 2008 SAT Race and 2009 SAT Competition. Most previous parallel SAT solvers were designed using the divide-and-conquer paradigm but since ManySat the parallel tracks of all SAT Competitions and SAT Races have been dominated by portfolio solvers.

ManySat was implemented on top of the well-known sequential SAT solver MiniSat [54] and the basic idea is that each parallel process should exploit a particular parameter set such that their combination represents a set of orthogonal yet complementary strategies.

In the original version the authors defined four different strategies. Each of the four strategies featured a different restart scheme and different decision literal polarity heuristic. All four strategies used the VSIDS [54] decision variable selection heuristic with a different percentage of random choices. Additionally half of the strategies employed extended clause learning, which allows for bigger backjumps.

ManySat was designed for parallel systems with the shared-memory architecture – basically for multi-core/multi-CPU computers. The clause sharing was organized via lockless queues containing the clauses a particular solver wants to share. Unit clauses were imported only at restarts while longer clauses were imported immediately on the fly. Overall, all the clauses with eight or fewer literals were shared. The value eight was determined based on experimental evaluation using SAT Competition benchmarks.

The performance of ManySAT was evaluated (in the 2008 SAT Race) on four-core computers where it achieved a super-linear average speedup of 6.02 [24]. ManySAT was the most successful parallel solver (considering SAT Competitions/Races results) until 2010 when Plingeling [9] took over.

Plingeling

Plingeling is the parallel version of the CDCL SAT solver Lingeling [8]. Both solvers first appeared in the 2010 SAT Race, where Lingeling placed second in the Main Track and Plingeling won the Parallel Track. In most of the following SAT Competitions and SAT Races[2] Plingeling won and Lingeling placed among the top three solvers. Plingeling constantly evolved during these years and in the remainder of this subsection we will describe the changes since the 2010 version up to the 2016 version.

2010. Similarly to ManySAT, Plingeling is a portfolio solver implemented on top of Lingeling using Pthreads. A boss thread reads the input formula and generates

[2] At least up until the 2016 SAT Competition, which was the latest competition at the time of writing this text.

separate solver instances for worker threads. The diversification between the worker threads is achieved by setting different random seeds, preprocessing effort, and decision heuristics. The workers share clauses, but only unit clauses (clauses with one literal) are exchanged. This is done via the boss thread at regular intervals. The boss thread maintains a global unit table that is lazily synchronized among the workers.

2011. Additionally to unit clauses, literal equivalences are shared in the 2011 version of Plingeling. An equivalence between literals l_i and l_j can be viewed as a pair of binary clauses $(l_i \vee \neg l_j)$ and $(\neg l_i \vee l_j)$, therefore Plingeling now shares unit and some binary clauses. However, this is not how Plingeling implements this feature. The sharing of equivalences is implemented via a global union-find data structure of equivalences maintained by the boss thread.

2012. There is not much change regarding the features of the portfolio, however, there are changes in implementation. Now the boss thread alone does the input parsing and preprocessing and only then the worker thread solver instances are created. This reduces the memory consumption and makes the solver more robust for large instances, since worker thread addition can be stopped if the amount of available memory is running low.

2013. The sharing of longer clauses is added to Plingeling. Clauses with up to 40 literals are exchanged if their glue value[3] is at most 8. The sharing is implemented via a global clause stack maintained by the boss thread. The worker threads read the clauses from the global stack in the oldest first order, therefore it acts like a queue.

2014. Local Search is added to Plingeling. First the formula is examined in order to figure out whether it resembles a uniform random instance. This is done by looking at the average number of literal occurrences and its standard deviation. If the formula looks random several worker threads (or even all but one) run local search instead of CDCL (Lingeling). This was done to make Plingeling competitive also on random satisfiable problems that are still best solved by local search algorithms.

2015. Diversification is improved. Automatic parameter configuration techniques [29] are applied to find optimal parameter settings for various families of benchmarks from previous competitions. Each one of the worker threads uses one of these configurations.

2016. The parallel front-end is identical to the previous version, i.e., no changes besides use of the newest version of Lingeling by the worker threads.

In Figure 1.4 we plot the number of problems solved per time limit for each version of Plingeling. Bear in mind that each different Plingeling version uses a different Lingeling version at its core. We can see that the 2015 and 2016 versions perform best and are very similar, which is not at all surprising based on their description. The third best is the 2013 version followed by a large gap and then 2014 and the remaining versions. The natural question here is what went wrong with the 2014 version that it fell behind so much compared to the 2013 version. We believe it was caused by local search being used on instances that were not uniform random. Based on experimental logs we know that version 2014 solved significantly fewer

[3] The number of distinct decision levels associated with the literals of the learned clause.

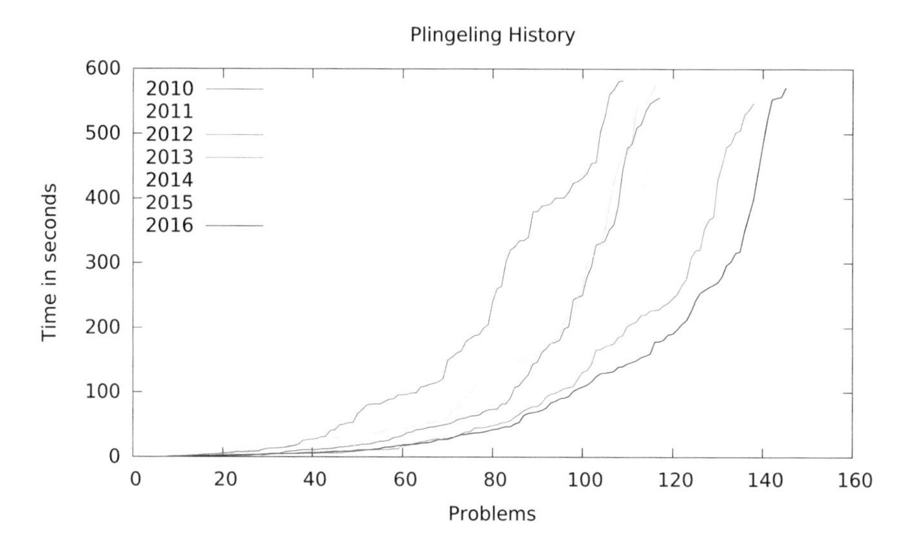

Fig. 1.4: The performance of Plingeling versions 2010 - 2016 on the benchmark of problems of the 2016 SAT Competition

unsatisfiable problems than 2013, 2015, and 2016 while solving a similar number of satisfiable problems.

HordeSat

HordeSat is a portfolio SAT solver designed for massively parallel architectures, i.e., computer clusters with hundreds of multi-core computers. An overview of the high-level design decisions made when designing HordeSat follows.

Modular Design. Rather than committing to any particular SAT solver HordeSat uses an interface that is universal and can be efficiently implemented by current state-of-the-art SAT solvers. This results in a more general implementation and the possibility to easily add new SAT solvers to the portfolio.

Decentralization. All the nodes in the parallel system are equivalent. There is no boss or central node that manages the search or the communication. Decentralized design allows more scalability and also simplifies the algorithm.

Overlapping Search and Communication. The search and the clause exchange procedures run in different (hardware) threads in parallel. The system is implemented in such a way that the search procedure never waits for any shared resources, at the expense of losing some of the shared clauses.

Hierarchical Parallelization. HordeSat is designed to run on clusters of computers (nodes) with multiple processor cores, i.e., we have two levels of parallelization. The first level uses the shared-memory model to communicate between solvers running on the same node and the second level relies on message passing between the nodes of a cluster.

HordeSat defines a C++ interface that is used to access the instances of the core solvers. This interface has the following methods.

- `addClause(vector<int> clause)` add clauses of the input formula
- `solve()` start solving, returns SAT/UNSAT/UNKNOWN
- `setInterrupt()` tell the solver to stop the search
- `unsetInterrupt()` allow the solver to continue solving
- `setPhase(int var, bool val)` suggest a truth value for a variable. This is just a recommendation and can be ignored by the solver.
- `diversify(int rank, int size)` tell the core solver to diversify its settings. The specifics of diversification are left to the solver. The provided parameters can be used by the solver to determine how many solvers are working on this problem (`size`) and which one of those is this solver (`rank`). A trivial implementation of this method could be to set the pseudo-random number generator seed of the core solver to `rank`.
- `addLearnedClause(vector<int> clause)` add a learned clause to the core solver. The solver can decide whether and how long this clause is useful for it.
- `setLearnedClauseCallback(LCCallback* lcc)` the solver calls the callback function when it learns a clause to share it.

The interface is designed to closely match current CDCL SAT solvers, but any kind of SAT solver can be used. For example a local search SAT solver could implement the interface by ignoring the calls to the clause-sharing-related methods.

Since HordeSat can only access its core solvers via the interface defined above, the only tools for diversification are setting phases using the `setPhase` method and calling the solver-specific `diversify` method.

The `setPhase` method allows the partitioning of the search space in a semi-explicit fashion. An explicit search space splitting into disjoint subspaces is usually done by imposing phase restrictions instead of just recommending them. The explicit approach is used in parallel solvers based on the divide-and-conquer methodology described in Section 1.3.

In HordeSat each variable in each core solver gets a random phase recommendation with a probability of $(\#solvers)^{-1}$, where $\#solvers$ is the total number of core solvers in the portfolio. This is done in conjunction with the `diversify` method whose behavior is defined by the core solvers.

The clause sharing in HordeSat happens periodically in rounds. Each round a fixed sized (1,500 integers in the implementation) message containing the literals of the shared clauses is exchanged by all the processes in an all-to-all fashion. Each process prepares the message by collecting the learned clauses from its core solvers. The clauses are filtered to remove duplicates. The fixed-sized message buffer is filled up with the clauses; shorter clauses are preferred. Clauses that do not fit are discarded.

The detection of duplicate clauses is implemented by using Bloom filters [13]. A Bloom filter is a space-efficient probabilistic set data structure that allows false-positive matches, which in this case means that some clauses might be considered to be duplicates even if they are not.

Although important learned clauses might get lost, we believe that this relaxed approach is still beneficial since it allows a simpler and more efficient implementation of clause sharing.

1.5 Parallel Local Search

There are two kinds of approaches to parallelizing local search. One is doing multiple flips in parallel and the other is the portfolio approach described above for CDCL algorithms. A special kind of local search called Survey Propagation has also been parallelized using GPU computation [39].

1.5.1 Multiple Flips

The parallel version of the local search solver GSAT [49] called PGSAT [44] first divides the set of variables into k groups (typically k is the number of processors) and then in each iteration each processor flips one of the variables from its group until a solution is found. The variable is selected using the GSAT heuristic.

Experiments with PGSAT have shown that speedup is achieved only up to a specific value of k. After this optimal value of k (denoted by k^*) the performance drops. An interesting observation is that the value of k^* depends on the instance we want to solve and appears to be correlated to the average connectivity of the variable-clause graph of the instance [44].

The solver PGWSAT [43] is a combination of PGSAT [44] and WalkSat [48]. In each iteration PGWSAT either acts like WalkSat (flipping a literal from one of the unsatisfied clauses) or PGSAT. The behavior is chosen randomly with the WalkSat strategy being used on between 50% and 70% of the steps. PGWSAT is shown to outperform PGSAT on random 3-SAT instances.

A parallel version of the solver genSat (generalized GSAT) [56] flips all the variables that have the best GSAT score (number of unsatisfied clauses after the flip) in each iteration. This is in contrast to other GSAT-style algorithms where only one of the variables with the best score is flipped (ties are broken randomly). Parallel genSat is experimentally shown to require fewer flips to solve a problem than the original GSAT algorithm.

1.5.2 Portfolios

The basic idea of portfolios (running several different solvers in parallel) can be applied to local solvers the same way as it is used for CDCL. The first local search solver to do this was gNovelty+ (v. 2)[42] in the 2009 SAT Competition, where it achieved first place in the parallel random category. The solver did not do any kind of sharing so it was pure portfolio.

It is not clear what kind of information should be shared in a portfolio of local search solvers. The sharing of learned clauses cannot be adopted from CDCL portfolios since local search solvers cannot produce them.

Several strategies of information sharing in local search portfolios were suggested in [2]. The solvers exchange their best assignment (satisfying the highest number of clauses) and based on these assignments a new starting assignment is constructed. This construction is different for each cooperation strategy. The strategies range from certain voting mechanisms to various probabilistic constructions. The most successful strategy, called *Prob-NormalizedW*, constructs the assignment using a probabilistic method that ensures that better variable values (w.r.t. satisfied clauses) have a higher chance of being adopted.

In a follow-up work [1] it was shown that this approach does not scale well for massively parallel systems and the proposed solution was to split the solvers into smaller groups (e.g., 16 solvers) that cooperate internally but do not exchange information between the different groups.

1.6 Future Challenges

In 2013 Hamadi and Wintersteiger published a paper listing seven challenges in parallel SAT solving [26]. The first challenge is to design a way to automatically estimate the number of parallel processes that should be used to solve a formula. The second challenge is about finding new ways to decompose the input formulas or the search space of a SAT-solving algorithm that outperform current techniques. The third challenge is about parallelizing preprocessing techniques used in modern SAT solvers. The next challenge is related to clause sharing and it asks for better techniques for estimating the local quality of learned clauses coming from other solvers. The technique called "lazy clause exchange" [5] used in the parallel version of the SAT solver Glucose [4] is a step towards solving this challenge. Challenges five and six ask for new encodings that would be specifically designed for parallel SAT solvers. Finally, the seventh challenge is to design a completely new parallel SAT algorithm from scratch, i.e., not based on existing algorithms, that performs on a par with or better than the current state of the art. We extend this list by adding the following three new challenges:

Massively Parallel Sat Solving. Design SAT solvers scalable in highly parallel environments, i.e., computer clusters with thousands or even millions of processors. Such solvers could potentially be used to solve large hard problem instances coming from computational biology and chemistry and even resolve open problems from fields such as combinatorics and number theory.

Utilizing Graphics Processing Units (GPUs) for SAT. Modern graphics cards are highly parallel computing units with hundreds of cores. General-purpose computing on GPUs is useful to accelerate various algorithms (notably for problems involving matrices) but has not yet been successfully used for SAT solving. Although there have been attempts to adapt existing SAT algorithms for GPU we are yet to see a GPU-based solver outperform standard CPU solvers. It seems that completely new algorithms need to be developed for the GPU.

Parallel Incremental SAT Solving. Many applications of SAT are based on solving a sequence of very similar SAT instances that often only differ in a few clauses. Although these instances can be solved independently, it can be very inefficient compared to an incremental SAT solver, which can reuse knowledge acquired while solving the previous instances (for example some of the learned clauses). Several SAT solvers support incremental SAT solving, however none of them is parallel. Since incremental solvers are very useful for improving performance in practical applications it is very important to develop highly scalable parallel incremental SAT solvers.

Acknowledgements

We would like to thank Wolfgang Blochinger for allowing us to use material from an unpublished draft in Section 1.3.1.

References

[1] Arbelaez, A., Codognet, P.: Massively parallel local search for SAT. In: 2012 IEEE 24th International Conference on Tools with Artificial Intelligence. vol. 1, pp. 57–64. IEEE (2012)

[2] Arbelaez, A., Hamadi, Y.: Improving parallel local search for SAT. In: International Conference on Learning and Intelligent Optimization. pp. 46–60. Springer (2011)

[3] Audemard, G., Lagniez, J.M., Szczepanski, N., Tabary, S.: An adaptive parallel SAT solver. In: International Conference on Principles and Practice of Constraint Programming. pp. 30–48. Springer (2016)

[4] Audemard, G., Simon, L.: Predicting learnt clauses quality in modern SAT solvers. In: International Joint Conference on Artificial Intelligence (IJCAI). vol. 9, pp. 399–404 (2009)

[5] Audemard, G., Simon, L.: Lazy clause exchange policy for parallel SAT solvers. In: Theory and Applications of Satisfiability Testing (SAT), pp. 197–205. Springer (2014)

[6] Balyo, T., Sanders, P., Sinz, C.: Hordesat: A massively parallel portfolio SAT solver. In: Heule, M., Weaver, S. (eds.) Theory and Applications of Satisfiability Testing (SAT), Lecture Notes in Computer Science, vol. 9340, pp. 156–172. Springer International Publishing (2015)
[7] Biere, A., Heule, M., van Maaren, H., Walsh, T.: Handbook of Satisfiability: Volume 185 Frontiers in Artificial Intelligence and Applications. IOS Press, Amsterdam, The Netherlands (2009)
[8] Biere, A.: Lingeling, plingeling, picosat and precosat at SAT race 2010. In: Technical Report 10/1, FMV Reports Series, Institute for Formal Models and Verification, Johannes Kepler University (2010)
[9] Biere, A.: Lingeling, plingeling and treengeling entering the SAT competition 2013. In: Proceedings of SAT Competition 2013, University of Helsinki. pp. 51–52 (2013)
[10] Blochinger, W., Sinz, C., Küchlin, W.: Distributed parallel SAT checking with dynamic learning using DOTS. In: Gonzales, T. (ed.) Proc. of the IASTED Intl. Conference Parallel and Distributed Computing and Systems (PDCS 2001). pp. 396–401. ACTA Press, Anaheim, CA (Aug 2001)
[11] Blochinger, W., Sinz, C., Küchlin, W.: Parallel propositional satisfiability checking with distributed dynamic learning. Parallel Computing 29(7), 969–994 (2003)
[12] Blochinger, W., Westje, W., Küchlin, W., Wedeniwski, S.: ZetaSAT – boolean SATisfiability solving on desktop grids. In: IEEE International Symposium on Cluster Computing and the Grid. pp. 1079–1086 (2005)
[13] Bloom, B.H.: Space/time trade-offs in hash coding with allowable errors. Communications of the ACM 13(7), 422–426 (1970)
[14] Boehm, M., Speckenmeyer, E.: A fast parallel SAT-solver – efficient workload balancing. Annals of Mathematics and Artificial Intelligence 17(3-4), 381–400 (1996)
[15] Chrabakh, W., Wolski, R.: GridSAT: A Chaff-based distributed SAT solver for the grid. In: Proc. of Supercomputing 03. Phoenix, Arizona, USA (2003)
[16] Cook, S.A.: The complexity of theorem-proving procedures. In: ACM Symposium on Theory of Computing. pp. 151–158. ACM, New York, NY, USA (1971)
[17] Davis, M., Logemann, G., Loveland, D.: A machine program for theorem-proving. Communications of the ACM 5(7), 394–397 (1962)
[18] Dijkstra, E.W., W.H.J.Feijen, van Gasteren, A.: Derivation of a termination detection algorithm for distributed computations. Inf. Proc. Letters 16, 217–219 (1983)
[19] El Khalek, Y.A., Safar, M., El-Kharashi, M.W.: On the parallelization of sat solvers. In: Computer Engineering & Systems (ICCES), 2015 Tenth International Conference on. pp. 119–128. IEEE (2015)
[20] Feldman, Y., Dershowitz, N., Hanna, Z.: Parallel multithreaded satisfiability solver: Design and implementation. Electr. Notes Theor. Comput. Sci. 128(3), 75–90 (2005)

[21] Gu, J.: The multi-sat algorithm. Discrete Applied Mathematics 96-97, 111–126 (1999)
[22] Guo, L., Jabbour, S., Lonlac, J., Saïs, L.: Diversification by clauses deletion strategies in portfolio parallel SAT solving. In: Tools with Artificial Intelligence (ICTAI), 2014 IEEE 26th International Conference on. pp. 701–708. IEEE (2014)
[23] Hamadi, Y., Jabbour, S., Piette, C., Sais, L.: Deterministic parallel DPLL. Journal on Satisfiability, Boolean Modeling and Computation 7, 127–132 (2011)
[24] Hamadi, Y., Jabbour, S., Sais, L.: Manysat: a parallel SAT solver. In: Satisfiability, Boolean Modeling and Computation. vol. 6, pp. 245–262 (2008)
[25] Hamadi, Y., Jabbour, S., Sais, L.: Control-based clause sharing in parallel sat solving. In: Twenty-First International Joint Conference on Artificial Intelligence (2009)
[26] Hamadi, Y., Wintersteiger, C.: Seven challenges in parallel SAT solving. AI Magazine 34(2), 99 (2013)
[27] Heule, M., Manthey, N., Philipp, T.: Validating unsatisfiability results of clause sharing parallel SAT solvers. In: POS@ SAT. pp. 12–25 (2014)
[28] Heule, M.J., Kullmann, O., Wieringa, S., Biere, A.: Cube and conquer: Guiding cdcl SAT solvers by lookaheads. In: Haifa Verification Conference. pp. 50–65. Springer (2011)
[29] Hutter, F., Hoos, H.H., Leyton-Brown, K., Stuetzle, T.: ParamILS: an automatic algorithm configuration framework. Journal of Artificial Intelligence Research 36, 267–306 (October 2009)
[30] Hyvärinen, A.E., Junttila, T., Niemelä, I.: A distribution method for solving SAT in grids. In: International Conference on Theory and Applications of Satisfiability Testing (SAT'06). pp. 430–435 (2006)
[31] Järvisalo, M., Le Berre, D., Roussel, O.: The SAT 2011 Competition – Results of Phase 1 – slides. `http://www.cril.univ-artois.fr/SAT11/phase1.pdf` (2011), accessed: 2015-12-18
[32] Jurkowiak, B., Li, C., Utard, G.: Parallelizing Satz using dynamic workload balancing. In: Kautz, H., Selman, B. (eds.) LICS 2001 Workshop on Theory and Applications of Satisfiability Testing (SAT 2001). Electronic Notes in Discrete Mathematics, vol. 9. Elsevier Science Publishers, Boston, MA (Jun 2001)
[33] Jurkowiak, B., Li, C.M., Utard, G.: A parallelization scheme based on work stealing for a class of SAT solvers. Journal of Automated Reasoning 34(1), 73–101 (2005)
[34] Katsirelos, G., Sabharwal, A., Samulowitz, H., Simon, L.: Resolution and parallelizability: Barriers to the efficient parallelization of SAT solvers. In: Proceedings of the Twenty-Seventh AAAI Conference on Artificial Intelligence, July 14-18, 2013, Bellevue, Washington, USA. (2013)
[35] Kaufmann, M., Kottler, S.: Sartagnan parallel portfolio SAT solver with lockless physical clause sharing. In: Pragmatics of SAT. Citeseer (2011)
[36] Kilby, P., Slaney, J., Thiébaux, S., Walsh, T.: Estimating search tree size. In: Proceedings of the 21st National Conference on Artificial Intelligence - Volume 2. pp. 1014–1019. AAAI'06, AAAI Press (2006)

[37] Knuth, D.E.: Estimating the efficiency of backtrack programs. Mathematics of Computation 29(129), 121–136 (Jan 1975)
[38] Kokotov, L.: Distributed SAT solver framework (1998)
[39] Manolios, P., Zhang, Y.: Implementing survey propagation on graphics processing units. In: International Conference on Theory and Applications of Satisfiability Testing. pp. 311–324. Springer (2006)
[40] Manthey, N.: Towards Next Generation Sequential and Parallel SAT Solvers. Ph.D. thesis, Technischen Universität Dresden, Fakultät Informatik (Jan 2014)
[41] Marques-Silva, J.P., Sakallah, K.A.: GRASP: A search algorithm for propositional satisfiability. IEEE Transactions on Computers 48(5), 506–521 (1999)
[42] Pham, D.N., Gretton, C.: gnovelty+ (v. 2). In: Proceedings of SAT Competition 2009, Artois University. pp. 9–10 (2009)
[43] Roli, A., Blesa, M., Blum, C.: Random walk and parallelism in local search. In: Proceedings of MIC'2005 – Meta–heuristics International Conference. Vienna, Austria (2005)
[44] Roli, A.: Criticality and parallelism in structured SAT instances. In: International Conference on Principles and Practice of Constraint Programming. pp. 714–719. Springer (2002)
[45] Roussel, O.: Description of ppfolio 2012. Proc. SAT Challenge p. 46 (2012)
[46] Schubert, T., Lewis, M., Becker, B.: PaMira - a parallel SAT solver with knowledge sharing. In: 6th International Workshop on Microprocessor Test and Verification (2005)
[47] Schubert, T., Lewis, M.D.T., Becker, B.: Pamiraxt: Parallel SAT solving with threads and message passing. JSAT 6(4), 203–222 (2009)
[48] Selman, B., Kautz, H.A., Cohen, B.: Noise strategies for improving local search. In: AAAI. vol. 94, pp. 337–343 (1994)
[49] Selman, B., Levesque, H.J., Mitchell, D.G., et al.: A new method for solving hard satisfiability problems. In: AAAI. vol. 92, pp. 440–446 (1992)
[50] Sinz, C., Blochinger, W., Küchlin, W.: PaSAT - parallel SAT-checking with lemma exchange: Implementation and applications. In: Kautz, H., Selman, B. (eds.) LICS 2001 Workshop on Theory and Applications of Satisfiability Testing (SAT 2001). Electronic Notes in Discrete Mathematics, vol. 9. Elsevier Science Publishers, Boston, MA (Jun 2001)
[51] Sonobe, T., Inaba, M.: Counter implication restart for parallel SAT solvers. In: Learning and Intelligent Optimization, pp. 485–490. Springer (2012)
[52] Sonobe, T., Inaba, M.: Portfolio with block branching for parallel SAT solvers. In: International Conference on Learning and Intelligent Optimization. pp. 247–252. Springer (2013)
[53] Sonobe, T., Kondoh, S., Inaba, M.: Community branching for parallel portfolio SAT solvers. In: International Conference on Theory and Applications of Satisfiability Testing. pp. 188–196. Springer (2014)
[54] Sorensson, N., Een, N.: Minisat v1.13 a SAT solver with conflict-clause minimization. Tech. rep., Chalmers University of Technology, Sweden (2005)

[55] Speckenmeyer, E., Monien, B., Vornberger, O.: Superlinear speedup for parallel backtracking, pp. 985–993. Springer Berlin Heidelberg, Berlin, Heidelberg (1988)
[56] Strohmaier, A.: Multi-flip networks: parallelizing gensat. In: Annual Conference on Artificial Intelligence. pp. 349–360. Springer (1997)
[57] Xu, L., Hoos, H., Leyton-Brown, K.: Hydra: Automatically configuring algorithms for portfolio-based selection. AAAI Conference on Artificial Intelligence (2010)
[58] Zhang, H., Bonacina, M.P.: Cumulating search in a distributed computing environment: A case study in parallel satisfiability. In: Proc. of the First Int. Symp. on Parallel Symbolic Computation. pp. 422–431. Linz, Austria (1994)
[59] Zhang, H., Bonacina, M.P., Hsiang, J.: PSATO: A distributed propositional prover and its application to quasigroup problems. Journal of Symbolic Computation 21, 543–560 (1996)

Chapter 2
Cube-and-Conquer for Satisfiability

Marijn J.H. Heule[1], Oliver Kullmann[2], and Armin Biere[3]

Abstract Satisfiability (SAT) is considered to be one of the most important core technologies in formal verification and related areas. Even though there is steady progress in improving practical SAT solving, there are limits on the scalability of SAT solvers. In this chapter, we present the *cube-and-conquer* paradigm which addresses this issue and targets reducing solving time on hard instances. This two-phase approach partitions a problem into many thousands (or millions) of cubes using lookahead techniques. Afterwards, a conflict-driven solver tackles the problem, using the cubes to guide the search. On several hard competition benchmarks, our hybrid approach outperforms both lookahead and conflict-driven solvers. Moreover, because cube-and-conquer is natural to parallelize, it is a competitive alternative for solving SAT problems in parallel. We demonstrate the strength of cube-and-conquer on the Boolean Pythagorean Triples problem, a recently solved challenge from Ramsey Theory. Cube-and-conquer achieves linear-time speedups on this problem even when using thousands of cores. Moreover, we show how to compute a proof for such a hard problem when solving it using cube-and-conquer.

2.1 Introduction

Satisfiability (SAT) solvers have become very powerful tools to tackle problems ranging from industrial formal verification [4] to hard combinatorial challenges [37]. The most successful tools are known as *conflict-driven clause learning* (CDCL) solvers [31]. These solvers have data structures optimized for huge instances and focus reasoning on learning new clauses from emerging conflicts. Although there exist several approaches to parallelize CDCL solvers [12], it appears hard to significantly improve performance on most industrial problems.

[1] The University of Texas at Austin, United States · [2] Swansea University, United Kingdom · [3] Johannes Kepler University, Linz, Austria

Y. Hamadi und L. Sais (eds.), *Handbook of Parallel Constraint Reasoning*,
https://doi.org/10.1007/978-3-319-63516-3_2

On the other hand, *lookahead* solvers [18] focus on small hard problems that require sophisticated heuristics to solve them efficiently. These solvers can be parallelized naturally and effectively. Yet, even with many cores at hand, they cannot compete with single-core CDCL solvers on industrial problems.

While developing a method for computing van der Waerden numbers, Kullmann observed that CDCL and lookahead solvers can be interleaved in such a way that the combination outperforms both pure methods. In short, lookahead is used to assign a certain fraction of the variables, and afterwards CDCL tackles the reduced problem. For optimal performance the lookahead solver partitions the original problem into thousands (sometimes millions) of cubes. The CDCL solver iteratively assumes each cube to be true and solves the simplified instance. This was systematically developed in [21]. The most recent successful application of this method, called *cube-and-conquer*, in this area (Ramsey theory), is solving the Boolean Pythagorean Triples problem [20], a long-outstanding mathematical problem.

In order to apply this method on a large spectrum of problems, we present a mechanism that determines dynamically when to cut off a branch in the search tree of a lookahead solver to send it to a CDCL solver. Using this mechanism, several hard industrial problems can be solved more efficiently using the combination of solvers than with a stand-alone SAT solver. Additionally, the combined solving method can be parallelized naturally as well. Therefore, using a parallel implementation of our method, we are able to solve some hard instances faster than alternative methods.

Our approach is based on the following intuition. Obviously the reduced formulas, after applying some decisions, become easier to solve. Furthermore, at least empirically, CDCL solvers are effective at solving instances which are rather easy for their size (but possibly impossible to solve by lookahead solvers), utilizing *local* heuristics including those based on variable activities. On the other hand, lookahead solvers are considered to be better at picking good decisions at the top-level, by using more global heuristics. There has to be a transition between hard and easy subproblems. So we try to switch from lookahead to CDCL solving when the subproblem seems to become easy (for the CDCL solver).

The outline of this chapter is as follows. After some preliminaries in Section 2.2, an overview of the cube-and-conquer method is provided in Section 2.3 as well as a description of both solver types. Section 2.4, discussing the above application to Ramsey theory, offers a motivating study of the method. Then a general methodology is developed. The details of the first phase, the "cube"-phase (partitioning the problem) are discussed in Section 2.5, and the details of the second phase, the "conquer"-phase (solving the subproblems) in Section 2.6. Two approaches to interleave the cube and conquer phases are presented in Section 2.7. A framework to produce proofs of unsatisfiability using cube-and-conquer is presented in Section 2.8. Experimental results are presented in Section 2.9 and some conclusions are drawn in Section 2.10.

2.2 Preliminaries

For a Boolean variable x, there are two *literals*, the positive literal, denoted by x, and the negative literal, denoted by $\neg x$. A *clause* is a disjunction of literals, and a *CNF formula* is a conjunction of clauses. A clause can be seen as a finite set of literals, and a CNF formula as a finite set of clauses. A special clause is the empty clause $\perp$, containing no literal. A *unit clause* contains exactly one literal. A truth assignment for a CNF formula F is a function φ that maps variables in F to $\{\mathbf{t}, \mathbf{f}\}$; in general φ is partial, and might even assign no variable, while a total assignment assigns all variables in F. If $\varphi(x) = v$, then $\varphi(\neg x) = \neg v$, where $\neg\mathbf{t} = \mathbf{f}$ and $\neg\mathbf{f} = \mathbf{t}$. A clause C is satisfied by φ if $\varphi(l) = \mathbf{t}$ for some $l \in C$. So the empty clause is never satisfied by any assignment. An assignment φ satisfies F if it satisfies every clause in F. A *cube* is a conjunction of literals and a *DNF formula* a disjunction of cubes. A cube can be seen as a finite set of literals and a DNF formula as a finite set of cubes. If $c = (l_1 \wedge \cdots \wedge l_k)$ is a cube, then $\neg c = (\neg l_1 \vee \cdots \vee \neg l_k)$ is a clause. A truth assignment φ can be seen as the cube of literals l for which $\varphi(l) = \mathbf{t}$. A cube c is satisfied by φ if $\varphi(l) = \mathbf{t}$ for all $l \in c$. An assignment φ satisfies DNF formula D if it satisfies some cube in D. A DNF formula D is called a *tautology* if every total assignment φ satisfies D. For a CNF formula F, *Boolean constraint propagation* (BCP) (or *unit propagation*) propagates all unit clauses, i.e., repeats the following until fix-point: if there is a unit clause $(l) \in F$, remove from F all clauses that contain the literal l, and remove the literal $\neg l$ from all remaining clauses in F. The resulting formula is referred to as BCP(F). If $\perp \in$ BCP(F), we say that BCP derives a conflict.

2.3 Combining CDCL and Lookahead

The main complete SAT solver types are *conflict-driven clause learning* (CDCL) solvers [31] and *lookahead* solvers [18]. In short, CDCL solvers are optimized for large industrial problems and consequently use inexpensive decision heuristics. In contrast, lookahead solvers focus on small hard problems on which it pays off to compute sophisticated decision heuristics. This section describes the main features of these solvers, and how we want to combine the two types.

Overview

The central approach in this chapter deals with a lookahead solver that partitions a formula into many subformulas, which in turn are solved by a CDCL solver. The sophisticated decision heuristics of lookahead solvers are used to compute important decision variables. These decisions are provided to the CDCL solver to guide the search process.

Figure 2.1 illustrates this approach by an example. The left shows a binary search tree produced by a lookahead solver. Internal nodes contain a decision variable. On

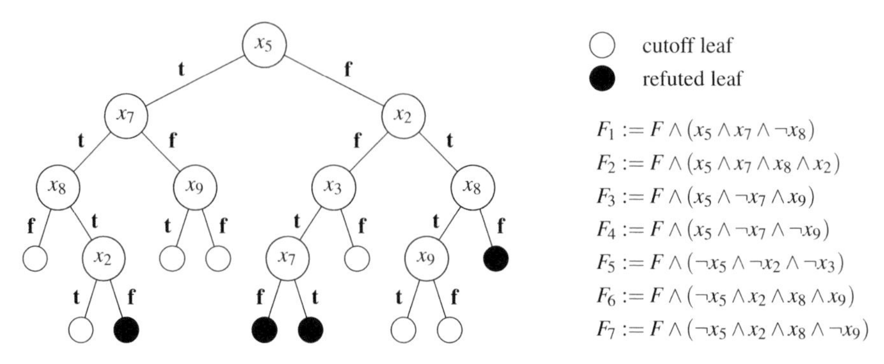

Fig. 2.1: A partition of a CNF formula F into seven subformulas F_i. The binary search tree on the left is constructed by a lookahead solver. It shows in the internal nodes the decision variable, and on the edges the truth value of a branch. Black leaves represent refuted leaves, while white leaves are cutoff leaves. The decisions of cutoff leaves yield a cube of assumptions that together with F forms a subformula F_i

the edges the truth value is shown to which a decision variable is set to reach a child node. There are two possible leaf nodes. Either the lookahead solver refutes the branch because a conflict emerges, or the *cutoff heuristic* suggests that this branch should be solved by a CDCL solver. This heuristic (discussed in detail in Section 2.5) is crucial for the effectiveness of the approach.

The cutoff branches can be described as the cubes of the decisions on the path to the leaf. A CDCL solver can solve the branch by either adding the decisions as unit clauses, or by adding them as *assumptions* (see the Incremental SAT solving paragraph below). In case one of the branches is satisfiable, the original formula is satisfiable (and hence remaining branches can be neglected). If all cutoff branches are unsatisfiable, the original formula is unsatisfiable.

The use of lookahead heuristics to partition a formula has been proposed by Hyvärinen *et al.* [22], who proposed to partition formulas into dozens of subformulas which are distributed on a grid to be solved in parallel. The starting point of this chapter is the discovery, discussed in Section 2.4, that some hard combinatorial problems can be efficiently solved by partitioning them into many thousands of subformulas (millions, or even billions for harder problems). Inspired by these results we focus on the latter approach. We also use more sophisticated lookahead techniques as employed in state-of-the-art lookahead solvers.

Lookahead solvers

Since CDCL is currently the dominant approach in practical SAT solving, we assume the reader already knows how CDCL solvers work, and otherwise refer to [31] for more details.

Lookahead solvers combine the David-Putnam-Logemann-Loveland (DPLL) algorithm [7] with *lookaheads*; for a general discussion see [18, 26], while we describe here an exemplary scheme. Given a CNF formula F, a lookahead on literal x works as follows. First, x is assigned to **t**, followed by BCP. Second, in case there was no conflict, the difference between F and the reduced formula F' is measured. The quality of lookahead techniques depends heavily on the used measurement. A frequently used method weighs the clauses in $F' \setminus F$ (the ones that are reduced but not satisfied). Third, all simplifications are reversed to get back to F. If a conflict was detected during the lookahead, then x is forced to **f** and is called a *failed literal*. The measurements are used to determine the decision variable in each node of the search tree. In general a variable x is chosen for which the lookahead both on x and $\neg x$ results in a large reduction of the formula. We remark that this scheme combines reduction (elimination of failed literals) and lookahead (estimating the quality of a branch by considering its development in the future), while in general these processes can be different.

State-of-the-art lookahead solvers are KCNFS [8], MARCH [32], OKSOLVER [25], and SATZ [29]. These solvers show strong performances on hard random k-SAT formulas, but they cannot compete with CDCL solvers on large industrial instances. Apart from random instances, lookahead techniques are also useful for combinatorial problems; these problems have fsome form of structure to be exploited, and yield relatively small but typically very hard SAT problems.

While measuring the reduction of the formula F, most lookahead solvers also perform *local learning*. In contrast to the learning in CDCL solvers, local learning computes clauses (mostly unary and binary) that can be added to the formula for further reduction, but that have to be removed again during backtracking to the parent node in the search tree. An example of local learning is hyper binary resolution [2]. Current state-of-the-art lookahead solvers do not implement conflict clause learning as in CDCL solvers, and most do not not even implement backjumping (except for the OKSOLVER). For an overview of local learning we refer to [18].

Incremental SAT solving

A frequently used feature of CDCL solvers is *incremental SAT solving* [11]. The solver provides an interface to (i) add clauses to the formula and (ii) solve the formula under a cube of assumptions (decisions at level 0). Both techniques are very useful for tools that integrate SAT solvers. The input of an incremental solver can be seen as a sequence consisting of both clauses and cubes, where each cube defines a *job*, which is the conjunction of that cube and all clauses preceding it in the sequence. In the context of cube-and-conquer we solve one formula under a set of cubes, thus all clauses precede all cubes in the solver input. A useful feature of incremental SAT solvers is that if a formula has no solutions under a given cube c, then the solver returns a subset $c' \subseteq c$ that was required to prove unsatisfiability. The clause $\neg c'$ can then be added to the formula to improve performance on other cubes.

As an example of the above, let us return to Figure 2.1. Now, consider a CDCL solver solving F_2, which is F assuming cube $(x_5 \wedge x_7 \wedge x_8 \wedge x_2)$. If however only $(x_8 \wedge x_2)$ is required to prove unsatisfiability, then we can add $(\neg x_8 \vee \neg x_2)$ to the formula. This binary clause conflicts with F_6 and F_7, so by adding it, these cubes are immediately refuted.

2.4 Creating Cubes: The Basic Method

In this section we describe cube-and-conquer in its simplest form, as it came out of investigations into van der Waerden-like numbers ([1, 28]). The principal aim is to solve extremely hard instances, which would take many years on a single machine. Thus a natural splitting of the problem into subproblems is applied, and since lookahead solvers are competitive on these instances, it is natural to use lookahead for this task. The great surprise now is that on these (easy) subproblems, conflict-driven solvers are very fast, and via this collaboration a *total speedup* (regarding the *total running time*) of at least a factor of two (compared always to the best single solver available) is achieved. So even on a single machine the problems are solved at least twice as fast, and additionally the splitting is ideal for parallelization (via clusters for example; no communication is needed between the processes). This was the birth of "cube-and-conquer". The lookahead solver is the OKSOLVER, which participated successfully at the SAT 2002 competition and aims at being as "theoretically clean" as possible; see [25, 26] for further information, and see the OKLIBRARY [27] for the renovated source code. It uses complete elimination of failed literals, and autarky reduction for the partial assignment at hand [24]. The distance along a branch is, as discussed above, a weighted sum of the number of new clauses, while the heuristic is the product of these values for the two branches (to be maximized); again (as for the reduction), all variables are (always) considered.

Computing the cubes is rather simple: cubes are partial assignments, corresponding to initial parts of the paths from the root to leaves in the splitting (branching) tree, and the task is to "cut off" these paths at the right place. Two methods are implemented, interpreting a depth parameter $D \geq 0$: the branches are either cut off when exactly D decisions have been made (method A), or when the total number of assigned variables (decisions, unit propagations, failed literals, autarkies) is at least D (method B).

The interface to the sub-solver is here as simple as possible: a complete decoupling is achieved by *applying* the partial assignments, and the sub-solver just gets the results. So each sub-instance is solved completely independently of the other, and the sub-solver only sees the sub-instance. For method A as well as for method B, the partial assignments contain everything: the decisions, the unit-propagation, the failed literals, the autarkies found (including pure literals).

On the implementation side, there are two simple data formats: either storing each partial assignment in its own file in DIMACS format (this is used for the experiments below), or creating an iCNF file[1], which here is basically just the concatenation of the instance and the partial assignments, put into one big file. Processing runs through the partial assignments, applies them to the original CNF, and calls the sub-solver on the sub-instance. Since only unsatisfiable instances are considered in this section, and the sub-instances are independent of each other, the order of the instances does not matter. All methods and all data are available in the OKLIBRARY, see [27]. The cutoff (the above parameter D) is determined ad hoc such that sub-instances only take *a few seconds* (this seems to be around the optimum, but with less overhead, as achieved by the system discussed in Section 2.5, one can partition further – the more cubes the better).

We report here only on two instance classes: determining unsatisfiability of van der Waerden (vdW) instances and palindromic vdW instances, using in both cases two colors, and thus the instances have a canonical translation into Boolean CNF. Such problems are explained (resp. introduced in the palindromic case) in [1], and they were also part of the SAT 2011 competition. The standard (Boolean) vdW problems are given by equations $\mathrm{vdW}(k_1,k_2) = n$, for natural numbers $k_1 \leq k_2 \leq n$, meaning that whenever $\{1,\ldots,n\}$ is partitioned into two parts, it holds that the first part contains an arithmetic progression (ap for short) of size k_1 or the second part contains an ap of size k_2 (and n is minimal with this property). This gives a CNF with n variables $v_1,\ldots,v_n$ and with two clause-sizes k_1,k_2, where the clauses of length k_1 are all the ap's of size k_1, as positive clauses, and the clauses of length k_2 are all the ap's of size k_2, as negative clauses. The *palindromic* (Boolean) vdW problems are given by equations $\mathrm{vdW}^{\mathrm{pd}}(k_1,k_2) = (n_1,n_2)$ $(n_1 < n_2)$, with a similar meaning, only that now only palindromic partitions are allowed, thus regarding the partition as a bit-string of length n, given by the values of $v_1,\ldots,v_n$, and requiring that $(v_1,\ldots,v_n) = (v_n,\ldots,v_1)$. By these equations, the number of variables is halved, replacing v_n with v_1 and so on, and shorter clauses are obtained. Subsumption elimination is performed on the instances. There are now two unsatisfiable problems, one using $\frac{n_1+1}{2}$ variables, with $n = n_1 + 1$ as the smallest n with an unsatisfiable problem, and one with $\frac{n_2+1}{2}$ variables, based on the smallest $n = n_2$ such that *all* $n' \geq n$ yield unsatisfiable problems. For standard vdW instances, lookahead solvers can perform better than conflict-driven solvers, while for palindromic vdW instances conflict-driven solvers are much better (here we are not speaking about cube-and-conquer, but about standard SAT solving). Method B for determining the cutoff was vastly superior (diminishing the variability of the sub-instances enormously), and is the only one considered here. MINISAT (version 2.2) performed very well as conquer-solver. All times are on a single core with about 2 GHz (parallelization has not been used), and the times for the cube-and-conquer approach are the total times, including all computations (writing each sub-instance to file etc.). All solvers mentioned below for comparison seem best performing (as ordinary SAT solvers, on the original (full) instances).

For $\mathrm{vdW}(3,15) = 218$ (yielding 13,362 clauses) the lookahead solver SATZ (version 215) needs about 20 hours, while with $D = 35$ (yielding 32,331 cubes) it is

[1] `http://users.ics.tkk.fi/swiering/icnf/`

solved in about 4 hours. The maximal time per job is 5 seconds, enabling trivial optimal parallelization with more than 2,000 processors (by just distributing the jobs for the subproblems to the first available processor). For $\mathrm{vdW}(4,8) = 146$ (yielding 4,930 clauses) PICOSAT (version 913) takes 8 hours. Setting $D = 20$ (yielding 65,270 cubes), it is solved in 4 hours, with maximal job-time of 22 seconds. PICOSAT for $\mathrm{vdW}(5,6) = 206$ was aborted after a week, while with $D = 20$ (yielding 91,001 cubes) it was solved in about one day. For $\mathrm{vdW}^{\mathrm{pd}}(3,25) = (586,607)$ (yielding 45,779 resp. 49,427 clauses), PRECOSAT (version 570) used in both cases about 13 days, while with $D = 45$ (yielding 9,120 resp. 13,462 cubes) the problems were solved in about 6.5 hours resp. 2 days. For $\mathrm{vdW}^{\mathrm{pd}}(4,12) = (387,394)$ (yielding 15,544 resp. 15,889 clauses) MINISAT (version 2.2) was aborted after 2 weeks, while setting $D = 30$ resp. $D = 34$ (yielding 132,131 resp. 147,237 cubes) solved the problems in 2 days resp. 8 hours. Finally, for $\mathrm{vdW}^{\mathrm{pd}}(5,8) = (312,323)$ (yielding 9,121 resp. 9,973 clauses), MINISAT used 3.5 days resp. 53 days, while setting $D = 20$ in both cases (yielding 22,482 resp. 87,667 cubes) solved it in 5 hours resp. 40 hours.

2.5 Creating Cubes: a General Methodology

This section shows how to modify a lookahead solver into a partitioning tool. First, we explain where to modify the code in Section 2.5.1. Second, we present an adaptive mechanism to cut off branches in Section 2.5.2. We conclude with some important heuristics in Section 2.5.3. The automatic partitioning provided here essentially is able to simulate the splitting characteristics from Section 2.4.

2.5.1 *General Framework*

The procedure CreateCubes, a modified lookahead solver for partitioning, shown in Algorithm 2.1, takes as input a CNF formula F and outputs two sets. The first set $\mathscr{A}$ is a disjunction of cubes for which each cube represents a set of assumptions that describes a cutoff branch in the DPLL tree. The cubes in $\mathscr{A}$ cover all subproblems of F that have not been refuted during the partition procedure. The second set $\mathscr{C}$ is a conjunction of clauses. Each of these (learnt) clauses is implied by F and represents a refuted branch in the DPLL tree. Hence the clauses in $\mathscr{C}$ can be added to F to obtain a logically equivalent formula $F' := F \cup \mathscr{C}$.

The recursive procedure has five inputs. Besides F, $\mathscr{A}$, and $\mathscr{C}$, it passes on the set of *decision literals* (denoted by φ_{dec}) and the set of *implied literals* (denoted φ_{imp}). Implied literals are assignments that were forced by BCP or some form of learning such as failed literal reasoning. Initially, CreateCubes is called with the input formula F and all the other parameters as empty sets.

Algorithm 2.1: The General Framework of the Procedure CreateCubes

input: CNF F, DNF $\mathscr{A}$, CNF $\mathscr{C}$, dec. lits. φ_{dec}, imp. lits. φ_{imp}

1 $\langle F, \varphi_{\text{imp}} \rangle :=$ LAsimplify_and_learn (F, φ_{dec}, φ_{imp});
2 **if** $\varphi_{\text{dec}} \cup \varphi_{\text{imp}}$ *falsify a clause in* F **then**
3   **return** $\langle \mathscr{A}, \mathscr{C} \cup \{\neg\varphi_{\text{dec}}\} \rangle$;
4 **if** *cutoff heuristic is triggered* **then**
5   **return** $\langle \mathscr{A} \cup \{\varphi_{\text{dec}}\}, \mathscr{C} \rangle$;
6 $l_{\text{dec}} :=$ LAdecide (F, φ_{dec}, φ_{imp});
7 $\langle \mathscr{A}, \mathscr{C} \rangle :=$ CreateCubes ($F, \mathscr{A}, \mathscr{C}, \varphi_{\text{dec}} \cup \{l_{\text{dec}}\}, \varphi_{\text{imp}}$) ;
8 **return** CreateCubes ($F, \mathscr{A}, \mathscr{C}, \varphi_{\text{dec}} \cup \{\neg l_{\text{dec}}\}, \varphi_{\text{imp}}$) ;

In line 1 of Algorithm 2.1, the method LAsimplify_and_learn is called. This method simplifies the formula by BCP and lookaheads, forcing some variables to certain truth values. All assigned variables are added to φ_{imp}. Additionally, it produces *local learnt clauses* which are added to F. In case the current assignment falsifies F then a conflict clause is learnt. This clause consists of the complements of the decisions and is added to $\mathscr{C}$ (line 3). Line 4 deals with cutting off branching which is further discussed in the next subsection. The procedure LAdecide on line 6 determines the next decision variable and preferred truth value based on lookaheads. There exists a vast body of work on these decision heuristics [26]. Section 2.5.3 offers the details of this procedure.

After CreateCubes is terminated, $\mathscr{A}$ and $\mathscr{C}$ are optimized. First, the clauses in $\mathscr{C}$ are reduced in size by applying self-subsumption resolution. For instance, returning to the example in Figure 2.1 with $(x_5 \vee x_2 \vee \neg x_3 \vee x_7), (x_5 \vee x_2 \vee \neg x_3 \vee \neg x_7) \in \mathscr{C}$, then the resolvent $(x_5 \vee x_2 \vee \neg x_3)$ replaces both antecedent clauses. When $\mathscr{C}$ is fully optimized, this set of conflict clauses is used to remove assumptions in $\mathscr{A}$. For instance, if $(\neg x_5 \wedge x_2 \wedge x_8 \wedge x_9) \in \mathscr{A}$, and $(x_5 \vee \neg x_2 \vee x_8) \in \mathscr{C}$, then x_8 is removed as an assumption because it will be forced by BCP after $\mathscr{C}$ is added to F.

2.5.2 Cutoff Heuristic

The heuristic that triggers the cutoff of a branch is of crucial importance in creating an effective partition. Ideally, this heuristic partitions the original problem into several subproblems such that 1) the runtimes to solve each of the subproblems are comparable and 2) the sum of these runtimes (at least) does not exceed the runtime of the original instance.

A simplified interpretation of the results discussed in Section 2.4 is that for some hard combinatorial problems both objectives can be achieved by cutting off a branch if a certain fraction (say 10%) of the variables are assigned. This measure is much easier to handle than the solution time for the sub-instances, which for the experiments reported in Section 2.4 was determined in an ad hoc manner. The total runtime to solve all subproblems was not just not bigger than the original runtime, but much smaller. So this metric is very useful for several small hard problems. However, for the larger industrial instances, the number of decisions appears to be also important to determine the hardness of a subproblem. Additionally, for these formulas sometimes a single decision assigns 10% of the variables, while for other formulas it requires over 100 decisions. In the former case the number of partitions becomes too small, while in the latter case the number of partitions becomes too large.

An alternative approach by Hyvärinen *et al.* [22] cuts off a branch after k decisions have been made (this was called method A in Section 2.4). The advantage of this approach is that one can clearly upper-bound the number of partitions in advance. However, branches with the same number of decisions are rarely equally hard to solve. It is often the case that assigning a decision literal x to $\mathbf{t}$ results in significantly more implied literals than assigning x to $\mathbf{f}$ or vice versa.

We combine both approaches by using the product of the number of decisions and the number of assigned variables, $|\varphi_{\text{dec}}| \cdot |\varphi_{\text{dec}} \cup \varphi_{\text{imp}}|$, as the cutoff metric. Furthermore, the refined procedure $\mathsf{CreateCubes}^*$, Algorithm 2.2, includes a dynamic cutoff mechanism. It implements the cutoff of a branch (with the cutoff heuristic discussed above) as shown in line 7 using a threshold parameter θ. Two lines update the value of θ. The first, the *increment rule* on line 1, raises the value by 5% without a condition. This rule aims to restore the value in case it was reduced too much. The second, the *decrement rule* on line 3, lowers the value by 30%. This rule tries to avoid two unfavorable situations described below.

First and most importantly, the value is decreased if the lookahead solver hits a conflict, meaning that the current node is a refuted branch. The rationale of this update is as follows. If the lookahead solver was able to show that the current node is conflicting, then probably a CDCL solver could have found the conflict faster. Additionally, if the CDCL solver had found the conflict, then it could have analyzed it and possibly computed a smaller reason for this conflict (than all decisions as computed by the lookahead solver). By lowering θ, the mechanism tries to cut off neighboring branches before a conflict emerges.

Secondly, the mechanism prevents the recursive procedure from going too deep into the DPLL tree. For most interesting instances, it appeared useful to decrease θ for all nodes with a depth larger than 20. In case one wants the mechanism to finish creating cubes within a few seconds, then the condition should be dependent on the size of the formula, such as $|\varphi_{\text{dec}}| + \log_2(|F|) > 30$.

Initially, θ should be large enough to ensure that the mechanism will cut off the tree at a reasonable depth. We used $\theta := 1000$ as initial value. Using a value which is a factor of 10 larger or smaller hardly influences the resulting partition. Using this initial value, θ will first be decreased before cutting off a branch.

Algorithm 2.2: The Procedure CreateCubes* with the Cutoff Mechanism

input : CNF F, DNF $\mathscr{A}$, CNF $\mathscr{C}$, dec. lits. φ_{dec}, imp. lits. φ_{imp}

1 $\theta := 1.05 \cdot \theta$;
2 $\langle F, \varphi_{\text{imp}} \rangle :=$ LAsimplify_and_learn $(F,\ \varphi_{\text{dec}},\ \varphi_{\text{imp}})$;
3 **if** $\varphi_{\text{dec}} \cup \varphi_{\text{imp}}$ *falsify a clause in F* ***or*** $|\varphi_{\text{dec}}| > 20$ **then**
4     $\theta := 0.7 \cdot \theta$;
5 **if** $\varphi_{\text{dec}} \cup \varphi_{\text{imp}}$ *falsify a clause in F* **then**
6   **return** $\langle \mathscr{A}, \mathscr{C} \cup \{\neg \varphi_{\text{dec}}\} \rangle$;
7 **if** $|\varphi_{\text{dec}}| \cdot |\varphi_{\text{dec}} \cup \varphi_{\text{imp}}| > \theta \cdot |\text{vars}(F)|$ **then**
8   **return** $\langle \mathscr{A} \cup \{\varphi_{\text{dec}}\}, \mathscr{C} \rangle$;
9 $l_{\text{dec}} :=$ LAdecide $(F,\ \varphi_{\text{dec}},\ \varphi_{\text{imp}})$;
10 $\langle \mathscr{A}, \mathscr{C} \rangle :=$ CreateCubes* $(F, \mathscr{A}, \mathscr{C}, \varphi_{\text{dec}} \cup \{l_{\text{dec}}\}, \varphi_{\text{imp}})$;
11 **return** CreateCubes* $(F, \mathscr{A}, \mathscr{C}, \varphi_{\text{dec}} \cup \{\neg l_{\text{dec}}\}, \varphi_{\text{imp}})$;

2.5.3 Heuristics for Splitting

Besides the development of the cutoff mechanism, the standard heuristics for lookahead solvers had to be tweaked in order to realize fast performance.

Decision heuristics

The default and costly lookahead evaluation heuristic (measurement) in most lookahead solvers is based on the clauses that are reduced but not satisfied during a lookahead. These clauses are weighted depending on their (new) length. In general, a clause of length k has a weight which is five times larger compared to a clause of length $k+1$. A cheaper heuristic counts the number of variables that are assigned during the lookahead.

For an example of both heuristics, consider the formula F below. Because the longest clauses have length 3, all "new" clauses have length 2, so no weights are required. Let $\text{eval}_{\text{cls}}(x_i)$ denote the clause-based heuristic that is the (weighted) sum of the reduced, not satisfied clauses, and let $\text{eval}_{\text{var}}(x_i)$ be the variable-based heuristic that is the number of assigned variables during the lookahead on $x_i = 1$. For example, $\text{eval}_{\text{var}}(\neg x_6) = 1$ and $\text{eval}_{\text{cls}}(\neg x_6) = 2$ because the lookahead on $x_6 = 0$ reduces two clauses from ternary to binary, and only x_6 is assigned. Notice that the values of the two heuristics are not necessarily related. $\text{eval}_{\text{cls}}(x_i)$ may be much smaller than $\text{eval}_{\text{var}}(x_i)$. For instance $\text{eval}_{\text{cls}}(\neg x_2) = 1$, while $\text{eval}_{\text{var}}(\neg x_2) = 4$.

$$F = (\neg x_1 \vee \neg x_3 \vee x_4) \wedge (\neg x_1 \vee \neg x_2 \vee \neg x_3) \wedge (\neg x_1 \vee x_2) \wedge (x_1 \vee x_3 \vee x_6) \wedge \\ (\neg x_1 \vee x_4 \vee \neg x_5) \wedge (x_1 \vee \neg x_6) \wedge (x_4 \vee x_5 \vee x_6) \wedge (x_5 \vee \neg x_6)$$

In general, lookahead solvers rank variables x_i by $\text{eval}(x_i) \cdot \text{eval}(\neg x_i)$. Ties are broken by $\text{eval}(x_i) + \text{eval}(\neg x_i)$. The decision heuristics select in each node of the DPLL tree the variable with the highest rank.

The default heuristic $\mathsf{eval}_{\mathrm{cls}}$ appeared to be quite effective on instances that had zero or few binary clauses. This is frequently the case for random and crafted instances used in the SAT competitions. However, we noticed that $\mathsf{eval}_{\mathrm{var}}$ was more effective on industrial instances. An advantage of $\mathsf{eval}_{\mathrm{var}}$ is that it does not require the eager data structures used in lookahead SAT solvers. Hence, this heuristic is relatively easy to implement in CDCL solvers.

Direction heuristics

Given a decision variable x, *direction heuristics* decide which branch (x to **t** or x to **f**) to explore first; see Section 5.3.2 in [18] for more information. Direction heuristics in lookahead solvers aim to improve performance on satisfiable formulas. Therefore, the solver prefers the branch that is most "likely" to be satisfiable. For methods to estimate such probabilities see Section 7.9 in [26], and see Subsection 4.6.2 in [33] for some discussions in the CSP context. As a cheap approximation one can take the less constrained branch first. This is the complementary strategy of the *first fail principle* [14] which is often used in constraint satisfaction. In case $\mathsf{eval}(x) < \mathsf{eval}(\neg x)$, x to **t** is explored first. Otherwise x to **f** is preferred. For a certain node with decision variable x, we refer to the branch with $\mathsf{eval}(x) < \mathsf{eval}(\neg x)$ as its *left branch*. The other branch we call its *right branch*.

The partition mechanism as described in Section 2.5.2 seems to be quite robust regarding the direction heuristics. The number of cubes and the average size of the cubes is hardly influenced by exploring the left or the right branch first. However, the order in which partitions are visited has a clear impact on performance related to the left and right branches, when considering how the *subproblems* are solved; see Section 2.6.1.

2.6 Solving Cubes

A CDCL solver deals with the second phase of the cube-and-conquer method. The solver takes as input the original formula F, optionally extended with the learnt clauses $\mathscr{C}$, and the set of assumption cubes $\mathscr{A}$. The latter is ordered based on some heuristic. For each cube $c \in \mathscr{A}$ based on this order, the CDCL solver solves $F \wedge c\,(\wedge \mathscr{C})$. First, we present how to solve the cubes sequentially in Section 2.6.1. Second, we discuss a parallel-solving approach in Section 2.6.2.

2.6.1 *Sequential Solving*

The sequential-solving procedure is rather straightforward and shown in Algorithm 2.3. Iteratively, a cube $c \in \mathscr{A}$ is selected (line 3) and assumed to be true, followed by solving the simplified formula (line 4). In case the result is satisfiable, the original formula is satisfiable and hence the procedure ends. After all cubes have been refuted, the formula is found to be unsatisfiable.

After refuting a cube, most CDCL solvers provide a technique to extract a subset of the cube that was required to prove unsatisfiability, known as AnalyzeFinal. It can be useful to add the clause – the complement of this subset – to the formula (line 6). Adding it can make refuting another cube easier. However, if $|\mathscr{A}|$ is much larger than $|F|$, the addition may significantly slow down performance.

Last but not least, we observed that removing some learnt clauses after refuting a cube can significantly improve the performance of cube-and-conquer. This can be explained by the intuition that the subproblems are relatively independent and hence the learnt clauses of one subproblem can hardly be reused for another subproblem. Removal of learnt clauses is realized by resetting the clause deletion policy after solving a cube (line 7). So the size of the clause database is reduced to its initial size and the least important clauses are kicked out.

Algorithm 2.3: The Pseudo-Code of SolveCubes Using the Partition

```
  input: CDCL solver S, CNF F, DNF 𝒜
1 S.Load (F);
2 while 𝒜 is not empty do
3     get a cube c from 𝒜 and remove c from 𝒜;
4     if S.SolveWithAssumptions (c) = satisfiable then
5         return satisfiable;
6     S.AnalyzeFinal () ;                              // optional
7     S.ResetClauseDeletionPolicy () ;
8 return unsatisfiable ;
```

Describing the cubes

In the partition procedure CreateCubes, the cube consists only of all decisions (φ_{dec}) from the root to the cutoff. Alternatively, one could describe a cube by all the assigned variables ($\varphi_{\text{dec}} \cup \varphi_{\text{imp}}$). The latter may include several assignments that a CDCL solver cannot reconstruct by BCP, for instance the failed literals. Recall that this approach is used is Section 2.4 and by Hyvärinen *et al.* [22, 23]. However, it seems that communicating implied variables to a CDCL solver does not improve runtime. Throughout our experiments, using cubes consisting of only decision literals resulted in stronger performance.

The order in which the decision literals are assumed into the CDCL solver influences the size of conflict clauses. The natural order – the order in which the decisions were made – appears to be the best alternative.

Ordering the cubes

During the experiments, we observed a relation between the time it requires to refute a cube and the number of right branches between the root and the cutoff of that cube: the more right branches (also known as *discrepancies*), the easier the corresponding subformula. On the other hand, for satisfiable formulas, cubes that cover a solution tend to have *few right branches*. Although we focused mostly on unsatisfiable formulas, we observed that for satisfiable benchmarks it pays off to solve the cubes with few right branches first. This strategy is known as *limited discrepancy search* [15].

There is also another reason for preferring this order, namely when solving cubes in parallel (see Section 2.6.2). In case CreateCubes produces an unbalanced tree, then frequently one or a few cubes will consume most of the computation costs to solve a formula. Therefore, one should solve the hard cubes first: a few cores attack these hard cubes, while others solve the easier ones. In contrast, if a hard cube needs to be solved at the end, all other cores may be idle as no unsolved cubes are left.

2.6.2 Solving Cubes in Parallel

A natural extension of the approach in the prior section is to consider solving the partitions in parallel. In existing work on parallel SAT solving [12] two main approaches are distinguishable. The first aims to partition the formula in an attempt to divide the total workload evenly over multiple cores, the second are so-called *portfolio* approaches [13]. Rather than partitioning the formula, *portfolio* systems run multiple solvers in parallel, each attempting to solve the same formula. The system finishes whenever the fastest solver finishes. Often such portfolios consist simply of multiple instances of the same CDCL solver. They can be configured such that each explores a different part of the search space — simply using different random seeds. Such parallel solvers mostly exploit the lack of robustness of SAT solvers, and can be surprisingly effective. Parallel SAT solvers of both types can be extended by exchanging learnt clauses.

In the solving phase of cube-and-conquer many partitions are independently solved and thus it can be easily parallelized. However, we treat this phase as one single incremental problem and use of incremental SAT. In [36] two different job assignment strategies for parallel incremental SAT were discussed and implemented in a tool called TARMO. That work was focused on Bounded Model Checking (BMC) but it can be seen as a general framework for parallel incremental SAT solving with clause sharing . The first strategy implemented is the *multijob* approach in which an idle core is assigned the first job that is not already assigned to any other core. When two cores are idle at the same time the job assignment order is undefined but it is guaranteed that no two cores ever work on the same job. The second strategy, called *multiconv*, is inspired by portfolio solvers, and it simply runs a conventional incremental SAT solver on all jobs on all cores. The latter can be effective for BMC, where jobs are difficult and job order is relevant. For cube-and-conquer however, we deal with an enormous number of jobs, most of which are very easy, which means there are no large deviations in single-job runtimes for the *multiconv* strategy to exploit. For this application *multijob* is a natural choice, although it is not ideal. If the partitioning is uneven, a small number of the jobs may be responsible for a large part of the runtime. Thus, many cores may end up sitting idle while waiting for a small number of cores with hard jobs to finish. In TARMO we experimented also with an extended strategy, *multijob+*, which is like *multijob* except that it will assign a job that is already being solved by some core to cores that would otherwise become idle. This modified strategy appeared beneficial for the performance of the cube-and-conquer solving phase.

Another feature of TARMO is its ability to share learnt clauses between solver threads. As discussed in [36] different settings are possible for the number of clauses shared. TARMO by default shares learnt clauses which lengths is below as it appears the most effective for this application.

After studying the parallelization of cube-and-conquer's solving phase using various versions of TARMO, a special purpose multithreaded version of the fast SAT solver LINGELING was created, which uses the basic *multijob* strategy. This special purpose solver called ILINGELING is faster than TARMO for this application, although it does not use clause sharing or the *multijob+* strategy yet.

2.7 Interleaving the Cube and Conquer Phases

Offline cube-and-conquer, i.e., performing the cube phase before the conquer phase, shows strong performance on several hard application benchmarks, beating both the lookahead and CDCL solvers that were used for the cube and conquer steps. However, on many other instances, either lookahead or CDCL outperforms cube-and-conquer. We observed that for benchmarks for which cube-and-conquer has relatively weak performance, two important assumptions do not hold in general.

First, in order for cube-and-conquer to perform well, lookahead heuristics must be able to split the search space into cubes that, combined, take less time for CDCL to solve. Otherwise, cube-and-conquer techniques are ineffective and CDCL would be the preferred solving technique. Second, lookahead must be able to cutoff cubes that are easy for CDCL to solve, and it should not cutoff cubes that are still hard for CDCL. When this assumption fails, the cutoff heuristic will perform badly, and the cube phase either generates too few cubes and leaves a potential performance gain unused, or generates too many cubes because cubes with fewer decisions are also easy for CDCL to solve.

2.7.1 Ineffective Lookahead Heuristics

To compare the performance of CDCL and cube-and-conquer, we ran both solver types[2] on all application benchmarks of SAT 2009. CDCL was able to solve 57 more benchmarks compared to cube-and-conquer within the timeout of 900 seconds (171 vs. 114). For some instances, the performance gap was huge (in favor of CDCL), in particular on satisfiable ones. This can be explained as follows. After a decision, the reduced formula might be harder (or at least not easier) than the original one. This may be caused by ineffective lookahead heuristics. In case a decision hardly reduces the search space, the conquer solver might need to solve two similar problems instead of one, thereby raising the computational costs. On satisfiable formulas this negative effect is expected to be more profound, since a single wrong decision might bring the solver into a part of the search space without solutions.

The main reason for this negative effect is that the key assumption underlying cube-and-conquer fails. This assumption expects that lookahead decision heuristics can select for a formula F a decision variable x in such a way that $F \cup \{x\}$ and $F \cup \{\neg x\}$ are easier to solve separately than F itself. It was shown that for several benchmarks this assumption holds [21]. However, the results above show that for many benchmarks in the SAT 2009 application suite this is not the case. For those, one would like to apply pure CDCL instead of cube-and-conquer.

Ineffective lookahead heuristics can be observed as follows. Given a formula F and a decision variable x, lookahead creates two branches $F \cup \{x\}$ and $F \cup \{\neg x\}$. The branch that reduces the formula the most is called the *right* branch, or a *discrepancy*. In case lookahead heuristics are effective, then with each decision, but especially each discrepancy, the formula becomes much simpler. Thus, after only a few discrepancies, lookahead (or CDCL) should be able to refute the branch. A cube that is reached through many discrepancies suggests that the lookahead heuristics have not been effective for that branch.

2.7.2 Concurrent Cube-and-Conquer

This section describes the *concurrent cube-and-conquer* (CCC) technique. We first describe CCC_∞ (CCC without cutting of branches), and extend it later by adding a cutoff heuristic for better resource utilization. CCC_∞ constructs a decision tree via the lookahead solver and simultaneously runs a CDCL solver on the newest node of this decision tree. Whenever the lookahead solver assigns a decision variable, the new literal is sent to the CDCL solver, which adds it as an assumption and restarts. This is repeated recursively until either solver proves unsatisfiability, which means that the cube is refuted and both solvers backtrack. Whereas (offline) cube-and-conquer cuts

---

[2] MINISAT 2.2 for CDCL; MINISAT 2.2 and MARCH_CC (cube phase) and IMINISAT 2.2 (conquer phase) for cube-and-conquer. All benchmarks were first preprocessed using LINGELING as suggested in [21]. We used the same version of LINGELING as in [21].

off branches explicitly using a heuristic, CCC_∞ cuts branches off implicitly when CDCL proves unsatisfiability before lookahead makes another decision.

Ideally, this approach is implemented within one solver. However, due to lack of appropriate data structures, current CDCL solvers only apply lookahead and other forms of preprocessing at the top level, and not under assumptions. For instance, tree-based lookahead [19] requires access to all binary clauses at all decision levels, which can only be accessed in a fast manner by using either full occurrence lists or three pointers for non-binary clauses. Both techniques are not easy to combine with data structures currently used in CDCL solvers.

On the other hand, lookahead solvers lack data structures for conflict analysis and learning, which is essential in CDCL solvers for allowing non-chronological backtracking and for cutting off repeated parts of the search. CC and CCC can be seen as two different ways of solving this dilemma by running both types of solvers separately, sequentially in CC and concurrently in CCC_∞.

CC was particularly useful if many cubes were generated, which means that CCC needs frequent synchronization. To limit the synchronization costs, CCC_∞ uses asynchronous message queues, where both solvers are peers. This architecture also makes it easy to integrate other solvers in the future.

The solvers in CCC_∞ communicate using two queues: the decision queue Q_{decision} and the result queue Q_{solved}. Whenever the lookahead solver assigns a decision variable, it pushes onto the queue the tuple $\langle$cube c_{id}, literal l_{dec}, *backtrackLevel*$\rangle$ comprising a uniquely allocated *id*, the decision literal, and the number of previously assigned decision variables (*backtrackLevel*). When the CDCL solver reads the new decision from the queue, it already knows all previous decision literals, and only needs to backtrack to the *backtrackLevel* and add l_{dec} as an assumption to start solving c_{id}. The *id* is used to identify the newly created cube.

If the CDCL solver proves unsatisfiability of a cube before it receives another decision, it pushes the c_{id} of the refuted cube to Q_{solved}. The solver then continues with the parent cube, by backtracking to the level where all but the last decision literal were assigned. When the lookahead solver reads the c_{id} from Q_{solved}, it backtracks to the level just above this cube's last decision variable and continues its search as if it proved unsatisfiability of the cube by itself.

The CDCL solver proves unsatisfiability of a cube if it encounters a complementary assignment when attempting to assign one of a cube's literals. This is not necessarily the last literal of the cube, so it may refute not only the cube corresponding to the latest decision read from Q_{decision}, but also one or more of its parent cubes. Therefore, it sends only c_{id} of the smallest cube which it refuted, which implies that the subcubes are also unsatisfiable.

To keep track of the cubes that are pending to be solved, both solvers keep the trail of decision literals (or assumptions for the CDCL solver) and the *id*s of the cubes up to and including each decision literal (or assumption). Whenever either solver proves unsatisfiability of the empty cube, or when it finds a satisfying assignment, the other solver is aborted.

It is possible that the lookahead solver already proved unsatisfiability of a cube when it receives the same result from the CDCL solver. The *id* is used to discard results on Q_{solved} for cubes that have already been closed. Similarly, it is possible that the lookahead solver makes a decision even though the CDCL solver already proved unsatisfiability of a parent of that cube. In that case the CDCL solver can discard the obsolete item on Q_{decision}.

2.7.3 Cubes on Demand

Running CDCL and lookahead in parallel as described in Section 2.7.2 promises to better combine the advantages of CDCL and lookahead than the original cube-and-conquer offline approach. It allows the search to switch between paradigms as soon as one becomes more effective. The original proposal however, only used two parallel solvers and could not make use of more than two processing units.

Inspired by this basic concurrent cube-and-conquer idea we developed the parallel SAT solver TREENGELING for multi-core computers, which not only combines the strengths of lookahead solving with CDCL solving, but also combines it with inprocessing. The main idea is to keep many instances of the base solver LINGELING in memory. Each instance, called a *node*, corresponds to an open branch in the lookahead search tree. The nodes are sorted by the number of remaining variables and preference is given to nodes with fewer variables. We use eight times more active nodes than there are processor cores available (including virtual cores due to hyper-threading) and as many inactive nodes as the system allows.

In a first phase the active nodes are simplified by all preprocessor algorithms available in LINGELING, then in a second phase they are searched with CDCL for a certain number of conflicts, limited by a conflict limit. In the last phase half of the larger active nodes are split using lookahead. This concludes one round. After each round finishes, closed nodes (the solver proved unsatisfiability of its branch) are removed and a new set of active nodes is determined by sorting all nodes w.r.t. the number of remaining variables. These three phases, i.e., simplification, search, and lookahead, are executed sequentially in order. But within each phase nodes are processed in parallel (using `pthreads`) on as many cores as available, scheduled by a working queue. If all nodes are removed or one node finds a solution, the solver stops.

It is important to realize that TREENGELING relies on heavy-duty simplification of each search node, which is only possible through copying (cloning) the solver and adding lookahead decisions as units, which in turn, however, prevents sharing learned clauses among cube-and-conquer nodes. This is the same trade-off made by other lookahead solvers though. It is only expected to perform well on instances where not much information can be shared among nodes. Lookahead bets on the quality of its global decision heuristic to split the search into "disconnected" parts.

For unsatisfiable instances, the implementation of TREENGELING is in essence deterministic, i.e., it always traverses the same search space and produces the same number of conflicts, as long as the maximum number of active nodes stays the same and the same memory limit is used. It is independent of the actual thread schedule, as determined by the operating system. The number of parallel threads during simplification, search, or lookahead does not influence the search. With more available cores, more threads can be run in parallel, without runtime penalty. However, in order to use more threads, more active nodes have to exist in parallel. The magic constant of eight times more active nodes than processor cores seems to work best.

The conflict limit is adapted dynamically. If more nodes are removed (found unsatisfiable during CDCL search) than added (through splitting), then the conflict limit is increased, otherwise it is decreased, both in a geometric way. The limit is increased more aggressively than decreased. Furthermore, there is a minimum $(1,000)$ and maximum $(100,000)$ conflict limit.

TREENGELING combines part of the infrastructure of PLINGELING with cube-and-conquer by running some additional parallel solver threads in portfolio manner, which export units to the worker cube-and-conquer threads, and import blocking clauses corresponding to closed branches. Since recent SAT competitions also featured certain random instances, on which local search solvers do fairly well in both the combinatorial and the application track, our local search solver YALSAT is run during inprocessing in these parallel solver threads, for a certain amount of time.

Note that TREENGELING, as well as PLINGELING, uses the same base library of LINGELING as used in its stand-alone sequential version, without modifications, except for registering call-backs, to support early termination and importing and exporting clauses.

Further, after forking the first parallel solver thread, the initial instance is preprocessed several times using the whole arsenal of preprocessing algorithms available in LINGELING. Then further parallel solver threads are started and the cube-and-conquer rounds are started.

As an example, consider a six-core machine with hyper-threading which has 12 virtual cores. Two threads are reserved for parallel solver threads running in a portfolio manner. The first one is started after parsing and copying the formula, and runs in parallel with preprocessing the original formula in 10 inprocessing rounds in another thread. After this initial phase the second portfolio solver is cloned from the preprocessed formula. Then the solver holding the preprocessed formula becomes the first root node. Since there are 11 virtual cores left, the maximum number of active nodes, which are simplified, searched, and split through lookahead in parallel is 88.

For large formulas, keeping many copies of LINGELING around needs a substantial amount of memory. Splitting is disabled if there is a risk of exhausting the available main memory. There is also a system limit on the maximum number of threads. This may be reached if too many copies of nodes are active. These issues will be addressed in future work.

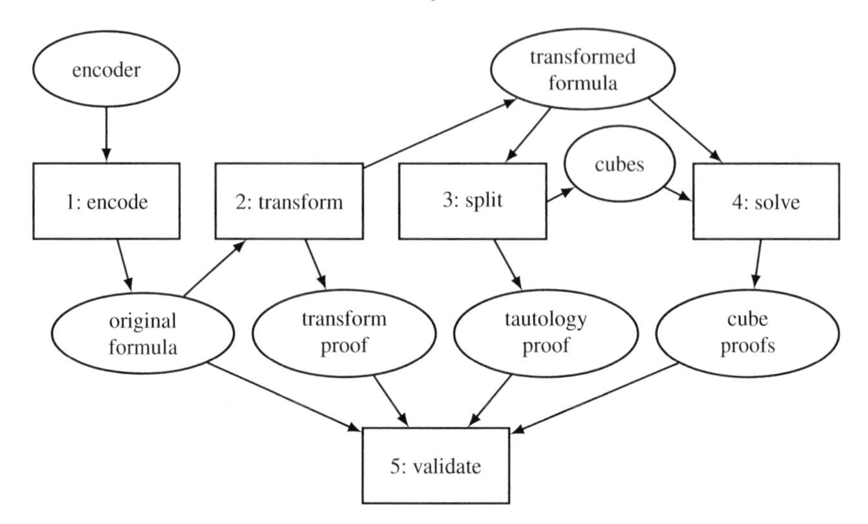

Fig. 2.2: Illustration of the framework to solve hard combinatorial problems. The phases are shown in the rectangular boxes, while the input and output files for these phases are shown in oval boxes

2.8 Proofs of Unsatisfiability

This section presents a framework for solving hard problems using cube-and-conquer as well as for producing and validating a proof of unsatisfiability. The framework consists of five phases: encode, transform, split, solve, and validate. The focus of the encode phase is to make sure that representation of the problem as a SAT instance is valid. The transform phase reformulates the problem to reduce the computation costs of the later phases. The split phase partitions the transformed formula into many, possibly millions of subproblems. The subproblems are tackled in the solve phase. The validation phase checks whether the proofs emitted in the prior phases are a valid refutation for the original formula. Figure 2.2 shows an illustration of the framework.

A *proof of unsatisfiability* (also called a *refutation*) for a formula F is a sequence of satisfiability-preserving transitions ending with some formula containing the empty clause. There are currently two prevalent types of unsatisfiability proofs: *resolution proofs* and *clausal proofs*. Both do not display the sequence of transformed formulas, but only list the axioms (from F) and the additions and (possibly) deletions. Several formats have been designed for resolution proofs [38, 10, 3] (which only add clauses), but they all share the same disadvantages. Resolution proofs are often enormous, and it is hard to express important techniques, such as conflict clause minimization, with resolution steps. Other techniques, such as bounded variable addition [30], cannot be polynomially simulated by resolution at all. Clausal proof formats [35, 34, 17] are syntactically similar; they involve a sequence of clauses that are claimed to be satisfiability-preserving, starting with the given formula. But now we might add clauses that are not logically implied, and we also might remove clauses (this is needed now in order to enable certain additions, which might depend on global conditions). The most popular format for clausal proofs is the DRAT ("Deletion Resolution Asymmetric Tautology") format [16].

Below we discuss the five phases of the framework in more detail.

Encode. The first phase of the framework focuses on making sure that the problem to be solved is correctly represented in SAT. In the second phase the representation will be optimized. The DRAT proof format can express all transformations.

Transform. The goal of the transformation phase is to massage the initial encoding so that the later phases may be executed more efficiently. A proof for the transformations is required to ensure that the changes are valid. Notice that a transformation that would be helpful for one later phase might be harmful for another phase. Selecting transformations is therefore typically a balance between different trade-offs. For example, bounded variable elimination [9] is a preprocessing technique that tends to speed up the solving phase. However, this technique is generally harmful for the splitting phase as it obscures the lookahead heuristics.

Split. Partitioning is crucial to solve hard combinatorial problems. Effective partitioning is based on global heuristics [21] – in contrast to the "local" heuristics used in CDCL solvers. The result of partitioning is a binary branching tree of which the leaf nodes represent a subproblem of the original problem. The subproblem is constructed by adding the conjunction of decisions that lead to the leaf as unit clauses. Figure 2.1 shows such a partitioning as a binary tree with seven leaf nodes (left) and the corresponding list of seven cubes (right). The cubes are shown in the iCNF format that is used for incremental solvers to guide their search.

Solve. The solving phase is the most straightforward part of the framework. It takes the transformed formula and cube files as input and produces a proof of unsatisfiability of the transformed formula. Two different approaches can be distinguished in general: one for "easy" problems and one for "hard" problems. A problem is considered easy when it can be solved in reasonable time, say within a day on a single core. In that case, a single cube file can be used and the incremental SAT solver will emit a single proof file. The more interesting case is when problems are hard and two levels of splitting are required, allowing parallel solving.

Validate. The last phase of the framework validates the results of the earlier phases. First, the encoding into SAT needs to be validated. This can be done by proving that the encoding tool is correct using a theorem prover. Alternatively, a small program can be implemented whose correctness can be checked manually. The second part consists of checking the three types of DRAT proofs produced in the earlier phases: the transformation, tautology, and the cube proofs. DRAT proofs can be merged easily by concatenating them. The required order for merging the proofs is: transformation proof, cube proofs, and tautology proof.

2.9 Experimental Evaluation

This section describes an evaluation of offline cube-and-conquer on application benchmarks from the SAT Competition in Section 2.9.1 and of a very hard problem in Ramsey Theory in Section 2.9.2.

2.9.1 Application Benchmarks

The experiments focus on the strength of cube-and-conquer on hard application benchmarks. For this chapter we used instances from the SAT 09 application category that were not solved during the competition (within the given timeout of 10,000 seconds) – the same set as used in [23]. We modified two existing SAT solvers according to the general method of cube-and-conquer. First, the lookahead SAT solver MARCH [32] was converted into a splitting tool called MARCH_CC. Second, the CDCL solver LINGELING was extended to deal with iCNF files. This version called ILINGELING also supports solving cubes in parallel. The sources of both tools are available on `http://fmv.jku.at/cnc/`.

Phase I of our cube-and-conquer implementation consists of A) simplifying the formula using the preprocessor of LINGELING (option -s) and B) calling MARCH_CC on the result. The cutoff mechanism in MARCH_CC is implemented as shown in Algorithm 2.2. Three benchmarks in the SAT 09 suite (9dlx* and sortnet*) remained too large after simplifying and caused memory problems for MARCH_CC. Therefore, we replaced $|\varphi_{\text{dec}}| > 20$ by $|\varphi_{\text{dec}}| > 10$ in the decrement rule for these instances. We used the cheap $\mathsf{eval}_{\text{var}}$ lookahead evaluation, because it resulted in improved performance compared to $\mathsf{eval}_{\text{cls}}$. The reported runtimes in Table 2.1 for phase I include both preprocessing and partitioning – the latter consuming most of the time. Notice that partitioning is based on lookahead. Hence, this part can relatively easily be parallelized. Since solving cubes requires more time than creating them, this optimization is left for future work. MARCH_CC outputs an iCNF file that concatenates the simplified formula and a line for each cube.

For phase II of cube-and-conquer, the iCNF file is provided to ILINGELING. We used a 12-core machine during this phase. On such a machine, ILINGELING starts 12 worker threads using separate LINGELING solvers. Idle threads ask for the first cube that has not been dealt with by another thread. After receiving a cube, LINGELING solves the reduced formula of the first phase with the cube as assumptions. After a cube is refuted, the clause database of the corresponding LINGELING is reduced as discussed in Section 2.6.1. A thread terminates either when a solution is found by one of the 12 solvers or when no new cube is available. ILINGELING terminates when all threads are terminated.

Table 2.1 shows the results of our cube-and-conquer implementation on hard SAT 2009 application instances. The experiments are run on a two 6-core AMD Opteron 2435 machine from 2009. This machine, part of a cluster, has 32 GB main memory and each job had a memory limit of 2.5 GB per core. Additionally it shows the results of three alternative solvers, which we obtained from [23]:

- PLINGELING 276, a multi-core portfolio solver using 12 cores [5].
- MANYSAT 1.5, a multi-core portfolio solver using 4 cores [13].
- PT-LEARN, an iterative partitioning solver with learning running on a grid [23].

The portfolio solvers PLINGELING and MANYSAT were run on exactly the same hardware as our implementation, while PT-LEARN was run on the M-grid environment consisting of nine clusters with CPUs from 2006 to 2009.

Table 2.1: Results on benchmarks of the SAT 2009 application suite that were not solved during that competition. S denotes satisfiable, U denotes unsatisfiable. Phase I uses LINGELING for preprocessing and MARCH_CC for partitioning. The column I shows the total time (in seconds) of both tools on a single core. Phase II uses ILINGELING to solve the cubes. Both the total time (sum of all threads) and the real time are listed. For the other solvers only the real time is provided (originating from [23]). – denotes that the timeout of 4 hours (14,400 seconds) was reached

Benchmark	S/U	number of cubes	I total	II total	II real	PLINGELING real	MANYSAT real	PT-LEARN real
9dlx_vliw_at_b_iq8	U	84	284	—	—	3256	2750	—
9dlx_vliw_at_b_iq9	U	40	314	—	—	5164	3731	—
AProVE07-25	U	98320	168	81513	6858	—	—	9967
dated-5-19-u	U	28547	478	5601	2538	4465	18080	2522
eq.atree.braun.12	U	86583	115	3218	269	—	—	4691
eq.atree.braun.13	U	83079	106	17546	1466	—	—	9972
gss-24-s100	S	339398	1853	14265	1191	2930	6575	3492
gss-26-s100	S	493870	1517	66489	5547	18173	—	10347
gus-md5-14	U	78488	649	—	—	—	—	13890
ndhf_xits_09_UNS	U	39351	128	—	—	—	—	9583
rbcl_xits_09_UNK	U	61653	210	132788	16900	—	—	9819
rpoc_xits_09_UNS	U	36733	255	104552	20665	—	—	8635
sortnet-8-ipc5-h19	S	583	271	48147	4023	2700	79010	4304
total-10-17-u	U	19773	948	5927	5561	3672	10755	4447
total-5-15-u	U	7865	192	—	—	—	—	18670

When we compare our approach with the two portfolio solvers PLINGELING and MANYSAT, then cube-and-conquer solves several more of these hard instances. Portfolio solvers are stronger on the three huge instances 9dlx* and sortnet*. A possible explanation could be that these instances must be “easy” relative to their size. Therefore, lookahead techniques cannot really help the CDCL solvers.

The PT-LEARN solver shows on most instances comparable performance to cube-and-conquer – although the latter is an order of magnitude faster on the eq.atree.braun* and gss* benchmarks. The comparison of the two solvers in Table 2.1 however is biased towards PT-LEARN: the experiments are run on similar hardware, but PT-LEARN runs up to 60 jobs at the same time, while cube-and-conquer runs at most 12 jobs. PT-LEARN suffers a bit from delays, while our solver runs on one machine. So, the presented results suggest that cube-and-conquer is actually the strongest solver on these hard application benchmarks.

Additional experiments suggest that our current implementation of cube-and-conquer is not optimal yet. For several instances, we observed improved real-time using fewer than 12 cores. For example, our 4-core cube-and-conquer experiments solved dated-5-19-u in 901 seconds. Also, total-10-17-u was solved in $2,632$ seconds using a single core. This time is almost half the 12-core real-time and faster than the other parallel SAT solvers. Notice that for both instances the real-time is relatively close to the total time, indicating that solving a certain cube requires most of the computational cost.

2.9.2 The Boolean Pythagorean Triples Problem

The Boolean Pythagorean Triples problem asks whether all bi-colorings of the positive integers result in a monochromatic solution of the Pythagorean equation $a^2 + b^2 = c^2$. We solved the Boolean Pythagorean Triples problem [20] by showing that there exists a bi-coloring of the numbers 1 to 7824 without a monochromatic Pythagorean Triple, but that no such coloring exists of the numbers 1 to 7825. To show that all bi-colorings of the numbers 1 to 7825 result in a monochromatic Pythagorean Triple, we constructed a formula F_{7825} that is satisfiable if and only if there exist a monochromatic-free Pythagorean Triple coloring. For each number $i \in \{1,\ldots,7825\}$, the formula has a Boolean variable x_i. Assigning x_i to true colors i red, while assigning x_i to false colors i blue. For each solution of $a^2 + b^2 = c^2$ with $a,b,c \leq 7825$, F_{7825} contains two clauses: a, b, or c is red $(x_a \vee x_b \vee x_c)$ and a, b, or c is blue $(\neg x_a \vee \neg x_b \vee \neg x_c)$.

The first step of solving F_{7825} consisted of splitting the formula into 10^6 subproblems [20] using MARCH_CC. Each of these subproblems, represented by a cube, was solved using cube-and-conquer by combing MARCH_CC and GLUCOSE 3.0. Figure 2.3 (left) shows a histogram of the size of the cube of the subproblems. Notice that the smallest cube has size 12 and the largest cubes have size 49. Hence the cutoff heuristics by MARCH_CC resulted in a highly unbalanced tree. Figure 2.3 (right) shows the time for the cube and conquer runtimes averaged per size of the cubes. The peak average of the cube runtime is around size 24, while the peak of the conquer runtime is around size 26. The cutoff heuristics of the cube solver for second-level splitting were based on the number of unassigned variables, $3,450$ variables to be precise.

A comparison between the cube, conquer, and validation runtimes is shown in Figure 2.4. The left scatter plot compares cube and conquer runtimes. It shows that within our experimental setup the cube computation is about twice as expensive compared to the conquer computation. The right scatter plot compares the validation and conquer runtimes. It shows that these times are very similar. Validation runtimes grow slightly faster compared to conquer runtimes. The average cube, conquer, and validation times for the 10^6 subproblems are 78.87, 47.52, and 60.62 seconds, respectively.

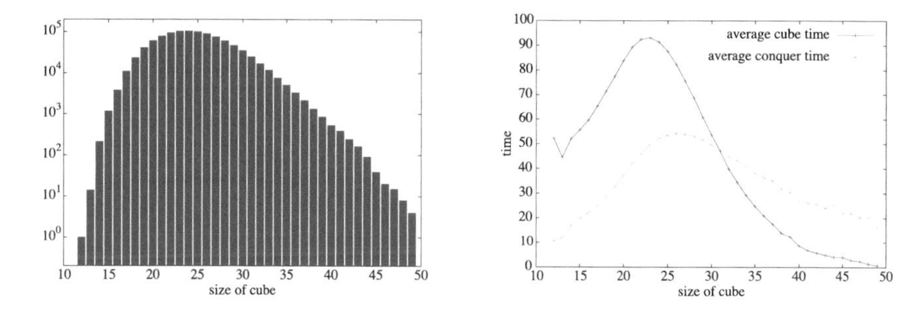

Fig. 2.3: Left, a histogram (logarithmic) of the cube size of the 10^6 subproblems. Right, average runtimes per size for the split (cube) and solve (conquer) phases

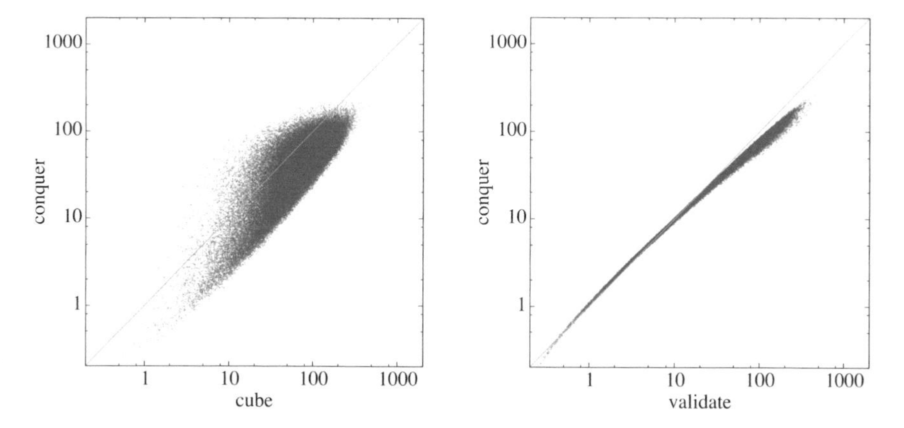

Fig. 2.4: Left, a scatter plot comparing the cube (split) and conquer (solve) time per subproblem. Right, a scatter plot comparing the validation and conquer time

Figure 2.5 compares the cube-and-conquer runtimes to solve the 10^6 subproblems with the runtimes of pure CDCL (using GLUCOSE 3.0) and pure lookahead (using MARCH_CC). The plot shows that cube-and-conquer clearly outperforms pure CDCL. Notice that no heuristics of GLUCOSE 3.0 were changed during any experiments for cube-and-conquer or pure CDCL. In particular, a variable decay of 0.8 was used throughout all experiments as this is the GLUCOSE default. However, we observed that a higher variable decay (between 0.95 and 0.99) would improve the performance of both cube-and-conquer and pure CDCL. We did not optimize GLUCOSE to keep it simple, and because the conquer part is already the cheapest phase of the framework (compared to split and validate); indeed, frequently speedups of two orders

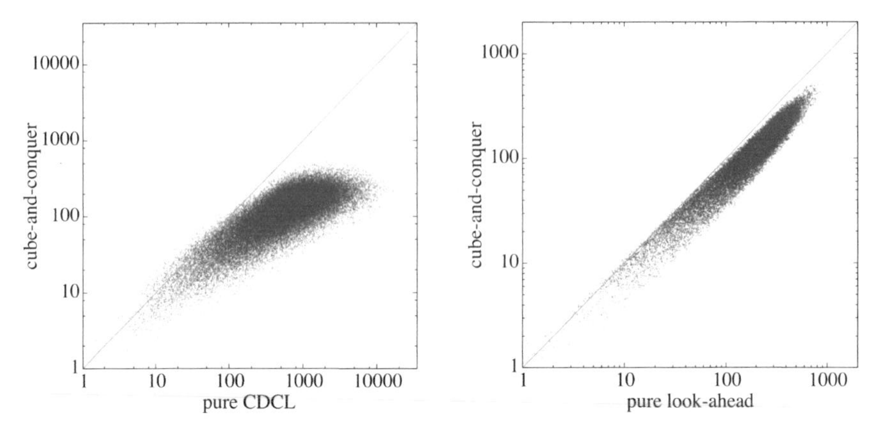

Fig. 2.5: Scatterplots comparing cube-and-conquer to pure CDCL (left) and pure lookahead (right) solving methods on the Pythagorean Triples subproblems

of magnitude could be achieved on the harder instances. Pure lookahead is also slower compared to cube-and-conquer, but the differences are smaller: on average cube-and-conquer is about twice as fast.

2.10 Conclusions

We presented the novel SAT-solving approach cube-and-conquer, which is a very powerful method to solve hard CNF formulas. Our approach combines sophisticated lookahead decision heuristics with the efficiency of CDCL solvers. Results on hard van der Waerden benchmarks using our basic method show reduced computational costs up to a factor of 20 compared to the fastest "pure" SAT solver. Using our cutoff mechanism, we were able to apply cube-and-conquer on hard application instances of the SAT competition. As a result, we outperform on most of these benchmarks the state-of-the-art parallel SAT solvers. Moreover, cube-and-conquer allowed us to solve the Boolean Pythagorean Triples problem. Interleaving the cube and conquer phases is a promising alternative to offline cube-and-conquer. The TREENGELING SAT solver, which is based on cubes on demand, has been very successful in recent SAT competitions.

Acknowledgements

The authors thank Siert Wieringa for his contributions to Section 2.6.2 and Peter van der Tak for his contributions to Section 2.7.2.

References

[1] Tanbir Ahmed, Oliver Kullmann, and Hunter Snevily. On the van der Waerden numbers $w(2;3,t)$. *Discrete Applied Mathematics*, 174:27–51, September 2014.
[2] Fahiem Bacchus. Enhancing Davis Putnam with extended binary clause reasoning. In *AAAI 2002*, pages 613–619, 2002.
[3] Armin Biere. PicoSAT essentials. *JSAT*, 4(2-4):75–97, 2008.
[4] Armin Biere. Bounded model checking. In Biere et al. [6], chapter 14, pages 455–481.
[5] Armin Biere. Lingeling, Plingeling, PicoSAT and PrecoSAT at SAT race 2010. 2010.
[6] Armin Biere, Marijn J.H. Heule, Hans van Maaren, and Toby Walsh, editors. *Handbook of Satisfiability*, volume 185 of *FAIA*. IOS Press, February 2009.
[7] Martin Davis, George Logemann, and Donald Loveland. A machine program for theorem-proving. *Commun. ACM*, 5(7):394–397, 1962.
[8] Olivier Dubois and Gilles Dequen. A backbone-search heuristic for efficient solving of hard 3-SAT formulae. In Bernhard Nebel, editor, *IJCAI*, pages 248–253. Morgan Kaufmann, 2001.
[9] Niklas Eén and Armin Biere. Effective preprocessing in SAT through variable and clause elimination. In Fahiem Bacchus and Toby Walsh, editors, *Theory and Applications of Satisfiability Testing, 8th International Conference, SAT 2005, St. Andrews, UK, June 19-23, 2005, Proceedings*, volume 3569 of *Lecture Notes in Computer Science*, pages 61–75. Springer, 2005.
[10] Niklas Eén and Niklas Sörensson. An extensible SAT-solver. In Enrico Giunchiglia and Armando Tacchella, editors, *SAT*, volume 2919 of *LNCS*, pages 502–518. Springer, 2003.
[11] Niklas Eén and Niklas Sörensson. Temporal induction by incremental SAT solving. *Electr. Notes Theor. Comput. Sci.*, 89(4):543–560, 2003.
[12] Youssef Hamadi. Conclusion to the special issue on parallel SAT solving. *JSAT*, 6(4):263, 2009.
[13] Youssef Hamadi, Saïd Jabbour, and Lakhdar Sais. ManySAT: a parallel SAT solver. *JSAT*, 6(4):245–262, 2009.
[14] Robert M. Haralick and Gordon L. Elliott. Increasing tree search efficiency for constraint satisfaction problems. *Artif. Intell.*, 14(3):263–313, 1980.
[15] William D. Harvey and Matthew L. Ginsberg. Limited discrepancy search. In *IJCAI 1995*, pages 607–613, 1995.
[16] Marijn J.H. Heule. The DRAT format and DRAT-trim checker. CoRR, abs/1610.06229, 2016. Source code available from: `https://github.com/marijnheule/drat-trim`.
[17] Marijn J.H. Heule, Warren A. Hunt, Jr, and Nathan Wetzler. Verifying refutations with Extended Resolution. In *CADE*, volume 7898 of *LNAI*, pages 345–359. Springer, 2013.
[18] Marijn J.H. Heule and Hans van Maaren. *Look-Ahead Based SAT Solvers*, chapter 5, pages 155–184. Volume 185 of Biere et al. [6], February 2009.

[19] Marijn J.H. Heule, Mark Dufour, Joris E. van Zwieten, and Hans van Maaren. March_eq: Implementing additional reasoning into an efficient look-ahead SAT solver. In Holger H. Hoos and David G. Mitchell, editors, *SAT (Selected Papers*, volume 3542 of *Lecture Notes in Computer Science*, pages 345–359. Springer, 2004.

[20] Marijn J.H. Heule, Oliver Kullmann, and Victor W. Marek. Solving and verifying the boolean Pythagorean Triples problem via Cube-and-Conquer. In Nadia Creignou and Daniel Le Berre, editors, *Theory and Applications of Satisfiability Testing - SAT 2016*, volume 9710 of *Lecture Notes in Computer Science*, pages 228–245. Springer, 2016.

[21] Marijn J.H. Heule, Oliver Kullmann, Siert Wieringa, and Armin Biere. Cube and conquer: Guiding CDCL SAT solvers by lookaheads. In Kerstin Eder, João Lourenço, and Onn Shehory, editors, *Hardware and Software: Verification and Testing (HVC 2011)*, volume 7261 of *Lecture Notes in Computer Science (LNCS)*, pages 50–65. Springer, 2012.

[22] Antti E. J. Hyvärinen, Tommi Junttila, and Ilkka Niemelä. Partitioning SAT instances for distributed solving. In *LPAR-17*, volume 6397 of *LNCS*, pages 372–386, 2010.

[23] Antti E. J. Hyvärinen, Tommi Junttila, and Ilkka Niemelä. Grid-based SAT solving with iterative partitioning and clause learning. In *CP 2011*, volume 6876 of *LNCS*, 2011.

[24] Hans Kleine Büning and Oliver Kullmann. *Minimal Unsatisfiability and Autarkies*, chapter 11, pages 339–401. Volume 185 of Biere et al. [6], February 2009.

[25] Oliver Kullmann. Investigating the behaviour of a SAT solver on random formulas. Technical Report CSR 23-2002, University of Wales Swansea, Computer Science Report Series, October 2002. 119 pages.

[26] Oliver Kullmann. *Fundaments of Branching Heuristics*, chapter 7, pages 205–244. Volume 185 of Biere et al. [6], February 2009.

[27] Oliver Kullmann. The `OKlibrary`: Introducing a "holistic" research platform for (generalised) SAT solving. *Studies in Logic*, 2(1):20–53, 2009.

[28] Oliver Kullmann. Green-Tao numbers and SAT. In Ofer Strichman and Stefan Szeider, editors, *SAT 2010*, volume 6175 of *LNCS*, pages 352–362. Springer, 2010.

[29] Chu Min Li and Anbulagan. Heuristics based on unit propagation for satisfiability problems. In *IJCAI (1)*, pages 366–371, 1997.

[30] Norbert Manthey, Marijn J.H. Heule, and Armin Biere. Automated reencoding of Boolean formulas. In *Proceedings of Haifa Verification Conference 2012*, 2012.

[31] Joao P. Marques-Silva, Ines Lynce, and Sharad Malik. *Conflict-Driven Clause Learning SAT Solvers*, chapter 4, pages 131–153. Volume 185 of Biere et al. [6], February 2009.

[32] Sid Mijnders, Boris de Wilde, and Marijn J.H. Heule. Symbiosis of search and heuristics for random 3-SAT. In David Mitchell and Eugenia Ternovska, editors, *LaSh 2010*, 2010.

[33] Peter van Beek. Backtracking search algorithms. In Francesca Rossi, Peter van Beek, and Toby Walsh, editors, *Handbook of Constraint Programming*, chapter 4, pages 85–134. 2006.
[34] Allen Van Gelder. Verifying RUP proofs of propositional unsatisfiability. In *ISAIM*, 2008.
[35] Nathan Wetzler, Marijn J.H. Heule, and Warren A. Hunt, Jr. DRAT-trim: Efficient checking and trimming using expressive clausal proofs. In Carsten Sinz and Uwe Egly, editors, *SAT 2014*, volume 8561 of *LNCS*, pages 422–429. Springer, 2014.
[36] Siert Wieringa, Matti Niemenmaa, and Keijo Heljanko. Tarmo: A framework for parallelized bounded model checking. In Lubos Brim and Jaco van de Pol, editors, *PDMC*, volume 14 of *EPTCS*, pages 62–76, 2009.
[37] Hantao Zhang. Combinatorial designs by SAT solvers. In Biere et al. [6], chapter 17, pages 533–568.
[38] Lintao Zhang and Sharad Malik. Validating SAT solvers using an independent resolution-based checker: Practical implementations and other applications. In *DATE*, pages 10880–10885, 2003.

Chapter 3
Parallel Maximum Satisfiability

Inês Lynce, Vasco Manquinho, and Ruben Martins

Abstract Developments in parallel Boolean Satisfiability (SAT) have motivated developments in parallel Maximum Satisfiability (MaxSAT), where MaxSAT is the optimization counterpart of SAT. Although many of the techniques implemented in parallel SAT can be extended to parallel MaxSAT, additional techniques are required to deal with the optimization part. This chapter provides an overview of the state of the art in parallel Maximum Satisfiability. The required background is first provided, namely the characteristics of the different MaxSAT algorithms. Solutions to parallel MaxSAT solving include portfolio approaches and search space splitting. Clause sharing is a key issue and so conditions for safe clause sharing are described. Deterministic parallel MaxSAT is another contribution in the field, having the additional challenge of synchronization strategies. Finally, future research directions are discussed.

3.1 Introduction

Maximum Satisfiability (MaxSAT) is an optimization version of Boolean Satisfiability (SAT) where the goal is to find an assignment to the problem variables such that the number of satisfied (unsatisfied) clauses is maximized (minimized) [44]. Many important application domains can be encoded as MaxSAT problems, such

Inês Lynce
INESC-ID / Instituto Superior Técnico, Universidade de Lisboa, Rua Alves Redol, 9, 1000-029 Lisboa, Portugal, e-mail: ines.lynce@tecnico.ulisboa.pt

Vasco Manquinho
INESC-ID / Instituto Superior Técnico, Universidade de Lisboa, Rua Alves Redol, 9, 1000-029 Lisboa, Portugal, e-mail: vmm@sat.inesc-id.pt

Ruben Martins
University of Texas at Austin, 2317 Speedway, M/S D9500, TX 78712-0233, e-mail: rmartins@cs.utexas.edu

Y. Hamadi und L. Sais (eds.), *Handbook of Parallel Constraint Reasoning*,
https://doi.org/10.1007/978-3-319-63516-3_3

Package	Dependencies	Conflicts
p_1	$\{p_2 \vee p_3\}$	$\{p_4\}$
p_2	$\{p_3\}$	$\{\}$
p_3	$\{p_2\}$	$\{p_4\}$
p_4	$\{p_2 \wedge p_3\}$	$\{\}$

Table 3.1: Example of a software package upgradeability problem

as software package upgrades [9], error localization in C code [41], debugging of hardware designs [22], haplotyping with pedigrees [34], course timetabling [1], and detection of Android malware [29].

For example, consider the software package upgradeability problem [49] where we have a set of software packages we want to install. Each package p_i has a set of dependencies and a set of conflicts. The dependencies denote packages which p_i depends on. Therefore, those packages must be installed for p_i to be installed. On the other hand, conflicts denote packages that cannot be installed for p_i to be installed. Table 3.1 shows an example of a software package upgradeability problem instance with four packages $\{p_1, p_2, p_3, p_4\}$. Each package has a set of dependencies and a set of conflicts. We can easily conclude that is not possible to install all packages of the problem presented in Table 3.1. Note that package p_4 requires package p_3 to be installed, but at the same time package p_3 has a conflict with package p_4.

Even though not all packages can be installed, the user may want to maximize the number of installed packages. This problem instance can be encoded as a partial MaxSAT problem. The encoding is a set of clauses to be split into a subset of hard clauses that *must be* satisfied and a subset of soft clauses that are *desired* to be satisfied. Consider the following hard clauses:

$$\begin{aligned} &[\neg p_1 \vee p_2 \vee p_3] \wedge [\neg p_1 \vee \neg p_4] \wedge [\neg p_2 \vee p_3] \wedge \\ &[\neg p_3 \vee p_2] \wedge [\neg p_3 \vee \neg p_4] \wedge [\neg p_4 \vee p_2] \wedge [\neg p_4 \vee p_3] \end{aligned} \tag{3.1}$$

These clauses correspond to the dependencies and conflicts between the different packages. For example, the clause $[\neg p_1 \vee p_2 \vee p_3]$ corresponds to the dependencies of package p_1, i.e., if p_1 is installed, then either p_2 or p_3 must also be installed. On the other hand, clause $[\neg p_1 \vee \neg p_4]$ corresponds to the conflicts of package p_1, i.e., if p_1 is installed then p_4 cannot be installed.

Since we want to maximize the number of installed packages, we include that information in the formula by using the following soft clauses:

$$(p_1) \wedge (p_2) \wedge (p_3) \wedge (p_4) \tag{3.2}$$

Therefore, in the resulting partial MaxSAT problem we have to satisfy all clauses in (3.1), while maximizing the number of satisfied clauses in (3.2). The assignment $\nu = \{p_1 = \mathit{true}, p_2 = \mathit{true}, p_3 = \mathit{true}, p_4 = \mathit{false}\}$ satisfies all clauses in Equation (3.1), while satisfying three out of four clauses in Equation (3.2).

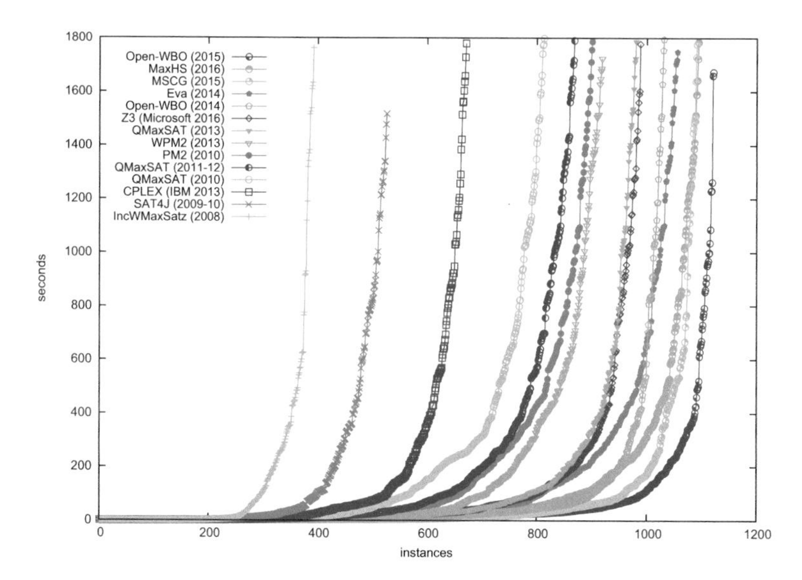

Fig. 3.1: Evolution of partial MaxSAT solvers

MaxSAT can indeed be used to effectively solve the software package upgradeability problem. For example, the widely used Eclipse platform[1] uses MaxSAT for managing the plugin's dependencies [18]. Improving MaxSAT algorithms will therefore result in more effective optimization solvers, which is expected to have a strong impact on several application areas.

MaxSAT solvers have been significantly improved over the last decade. Figures 3.1 and 3.2 show the evolution of partial MaxSAT and partial weighted MaxSAT solvers in the last decade. All experimental results for Figures 3.1 and 3.2 were obtained in the StarExec [80] cluster infrastructure on Intel(R) Xeon(R) E5-2609 processors (2.40 GHz) running Red Hat Enterprise Linux Workstation release 6.3 (Santiago) with a timeout of 1,800 seconds and a memory limit of 32 GB. We have included the best MaxSAT solvers from each year the MaxSAT evaluation was run[2] [55, 70, 25, 66, 68, 2, 3, 4, 18, 48, 5, 51, 75] together with CPLEX[3] and Z3.[4] CPLEX is a tool for solving linear optimization problems from IBM. We used the solver ILP [7], which converts a MaxSAT formula into a linear optimization problem and uses CPLEX as the back-end solver. Z3 is a Satisfiability Modulo Theory (SMT) solver and has recently been extended to support optimization [19]. We converted each MaxSAT formula into an equivalent SMT formula and used version 4.5.1 of Z3 to solve it. All

[1] `http://www.eclipse.org/`

[2] `http://www.maxsat.udl.cat/`

[3] `https://www-01.ibm.com/software/commerce/optimization/cplex-optimizer/`

[4] Version 4.5.1 available at `https://github.com/Z3Prover/z3`

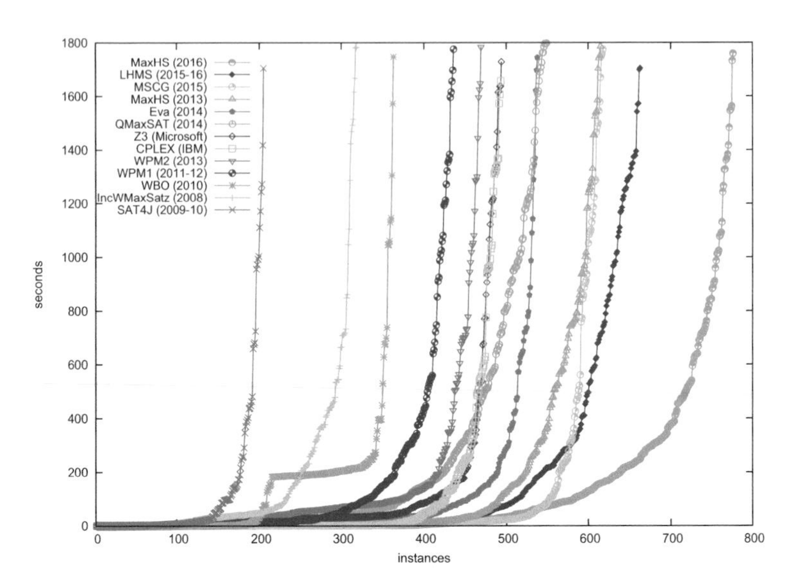

Fig. 3.2: Evolution of weighted partial MaxSAT solvers

solvers were evaluated on the 1,279 crafted and industrial partial benchmarks and the 961 weighted partial benchmarks from the MaxSAT Evaluation of 2016.[5]

Partial MaxSAT solvers can now solve 2.9× more benchmarks than the best solver in 2008. This remarkable improvement is due to a combination of new algorithmic ideas and the evolution of SAT solvers. Partial MaxSAT solvers are now better than other popular optimization tools such as CPLEX and Z3 for several Boolean optimization problems. Specifically, Open-WBO [64, 55] is able to solve 67% more benchmarks than CPLEX and 14% more benchmarks than Z3.

A similar picture can be seen for weighted partial MaxSAT solvers. MaxHS [24, 25] can solve 2.4× more benchmarks than the best solver in 2008. When compared to CPLEX and Z3, MaxHS can solve 57% more benchmarks than both of those tools.

The solvers included in the plots are all sequential solvers. However, exploiting new architectures is essential for the continued evolution of MaxSAT solvers. Nowadays, extra computational power is no longer coming from higher processor frequencies. On the other hand, multicore architectures and distributed systems are becoming predominant.

This chapter focuses on recent techniques that have been proposed for parallel MaxSAT solving and is organized as follows. Section 3.2 formally defines MaxSAT and provides a short overview of sequential MaxSAT solvers. Section 3.3 presents the two main approaches for parallel MaxSAT solving: (i) *portfolio*, and (ii) *search space splitting*. Section 3.4 presents a deterministic approach for parallel MaxSAT for multicore architectures and compares different synchronization mechanisms. Finally,

[5] `http://www.maxsat.udl.cat/16/index.html`

Section 3.5 concludes this chapter with ongoing challenges for parallel MaxSAT solvers and future research directions.

3.2 Maximum Satisfiability

A Boolean formula in conjunctive normal form (CNF) is a conjunction of clauses, where a clause is a disjunction of literals. A literal l is a Boolean variable x_i or its negation $\neg x_i$. An assignment is a correspondence between the Boolean variables in the formula and the truth values *true* and *false*. The assignment of variables is generalized to literals. A literal $l = x_i$ is said to be satisfied if x_i is assigned value *true* and unsatisfied if x_i is assigned value *false*. On the other hand, a literal $l = \neg x_i$ is satisfied if x_i is assigned value *false* and unsatisfied if x_i is assigned value *true*. A clause is satisfied if and only if at least one of its literals is satisfied. A formula ϕ is satisfied if all of its clauses are satisfied. The Boolean Satisfiability (SAT) problem can be defined as finding a satisfying assignment to a propositional formula ϕ or prove that such an assignment does not exist.

Consider the following CNF formula:

$$\phi = (x_1 \vee x_2 \vee x_3) \wedge (\neg x_1 \vee \neg x_2) \wedge (x_2 \vee \neg x_3) \tag{3.3}$$

Observe that ϕ is satisfiable, since there is at least one assignment ν that satisfies the formula. For example, $\nu = \{x_1 = true, x_2 = false, x_3 = false\}$ is a satisfying assignment to ϕ.

A CNF formula ϕ is unsatisfiable if there is no assignment that satisfies the formula. The following formula ϕ is unsatisfiable:

$$\phi = (x_1 \vee x_2 \vee x_3) \wedge (\neg x_1 \vee x_2) \wedge (x_2 \vee \neg x_3) \wedge (\neg x_2) \tag{3.4}$$

Given a CNF formula ϕ, the Maximum Satisfiability (MaxSAT) problem is to find an assignment to the formula variables such that the number of unsatisfied (satisfied) clauses in ϕ is minimized (maximized) [44]. Observe that maximizing the number of satisfied clauses is equivalent to minimizing the number of unsatisfied clauses. In the context of this chapter, we assume that the MaxSAT problem is a minimization problem.

Consider the CNF formula ϕ in (3.4). In this case, a MaxSAT optimal assignment to ϕ could be $\nu = \{x_1 = false, x_2 = false, x_3 = false\}$, since only one clause is unsatisfied.

MaxSAT has several variants such as partial MaxSAT, weighted MaxSAT, and weighted partial MaxSAT. In partial MaxSAT, some clauses in ϕ are considered hard while the others are considered soft. Let ϕ_h denote the set of hard clauses in ϕ, while ϕ_s denotes the set of soft clauses. The goal in partial MaxSAT is to find an assignment to the formula variables such that all hard clauses ϕ_h are satisfied, while minimizing the number of unsatisfied soft clauses in ϕ_s.

Consider the following CNF formula ϕ where hard clauses are enclosed within square brackets:

$$\phi = [x_1 \vee x_2 \vee x_3] \wedge (\neg x_1 \vee x_2) \wedge (x_2 \vee \neg x_3) \wedge (\neg x_2) \tag{3.5}$$

Observe that this formula is the same as in (3.4), but the first clause is labeled as hard ($\phi_h = \{[x_1 \vee x_2 \vee x_3]\}$), while the remaining clauses are labeled as soft. As a result, assigning *false* to all variables no longer produces an optimal assignment. Hence, one possible optimal assignment for the partial MaxSAT problem would be $\nu = \{x_1 = true, x_2 = false, x_3 = false\}$, where all clauses in ϕ_h are satisfied, while only one soft clause remains unsatisfied.

Finally, in the weighted versions of MaxSAT, each soft clause c_i is associated with an integer weight w_i such that $w_i \geq 1$. In this case, the goal is to find an assignment such that all hard clauses are satisfied and the total weight of unsatisfied soft clauses is minimized.

Consider the following weighted formula where an integer weight is associated with each soft clause:

$$\phi = [x_1 \vee x_2 \vee x_3] \wedge (\neg x_1 \vee x_2, 3) \wedge (x_2 \vee \neg x_3, 5) \wedge (\neg x_2, 2) \tag{3.6}$$

In this weighted MaxSAT formula, an optimal solution could be $\nu = \{x_1 = false, x_2 = true, x_3 = false\}$, where all hard clauses are satisfied and the sum of the weights of unsatisfied soft clauses is 2, i.e., the weight associated with assignment ν is only 2.

In a MaxSAT problem instance ϕ, if the set of hard clauses ϕ_h is not satisfiable, then ϕ is also unsatisfiable. For ease of explanation, we assume that ϕ_h is always satisfiable. Note that this can be tested by using a SAT solver on ϕ_h before calling any of the MaxSAT algorithms described in the chapter.

Cardinality constraints are a well-known generalization of propositional clauses. A cardinality constraint encodes that at most k out of n literals can be assigned to *true*, i.e., $\sum_{i=1}^{n} l_i \leq k$ where l_i is a literal. A generalization of cardinality constraints is pseudo-Boolean constraints where each literal can have a weight, i.e., $\sum_{i=1}^{n} w_i \cdot l_i \leq k$. In this case, the weighted sum of the literals assigned to *true* must be smaller than or equal to k.

Neither cardinality nor pseudo-Boolean constraints occur in MaxSAT formulations, but their use in MaxSAT algorithms is common [31, 50, 5, 38, 65]. However, in order to iteratively use a SAT solver, most MaxSAT algorithms encode cardinality constraints [16, 79, 12, 71] and pseudo-Boolean constraints into CNF [81, 28, 17, 40, 42].

3.2.1 Sequential MaxSAT Algorithms

In this section we briefly describe basic sequential MaxSAT algorithms. These algorithms are presented solely to introduce notions that can be used in the parallel framework. As a result, we refer to the vast literature on more advanced algorithms

Algorithm 3.1: Linear Search SAT-UNSAT Algorithm

Input: $\phi = \phi_h \cup \phi_s$
Output: optimal solution to ϕ

```
1  (φ_W, V_R, μ, ν_sol) ← (φ_h, ∅, +∞, ∅)
2  foreach c_i ∈ φ_s do
3      V_R ← V_R ∪ {r_i}                          // r_i is a new variable
4      c_R ← c_i ∪ {r_i}
5      φ_W ← φ_W ∪ {c_R}
6  while true do
7      (st, ν, φ_C) ← SAT(φ_W ∪ {CNF(Σ_{r_i ∈ V_R} w_i · r_i ≤ μ)})
8      if st = UNSAT then
9          return ν_sol
10     ν_sol ← ν                                  // save solution
11     μ ← (Σ_{r_i ∈ V_R} w_i · ν(r_i)) − 1        // update bound
```

and implementation details in order to build a state-of-the-art sequential MaxSAT algorithm.

Additionally, we focus on algorithms where a SAT solver is iteratively used in order to find an optimal solution to the MaxSAT problem. These algorithms have been shown to be more effective at tackling industrial benchmark instances. However, there is also a vast literature on branch-and-bound algorithms [44], more commonly used to solve randomly generated instances.

In the algorithms described in the next subsections, we assume that a SAT solver call SAT(ϕ) takes as input a CNF formula ϕ. The result of a call to a SAT solver is a triple (st, ν, ϕ_C), where st denotes the status of the formula: satisfiable (SAT) or unsatisfiable (UNSAT). If the solver returns SAT, then the assignment that satisfies ϕ is stored in ν. On the other hand, if the solver returns UNSAT, then ϕ_C contains an unsatisfiable formula that explains a reason for the unsatisfiability of ϕ.[6]

Finally, for ease of notation and for a better understanding of the algorithms, in the remaining sections we use set notation for CNF formulas and clauses. In particular, a CNF formula ϕ can be seen as a set of clauses and a clause c as a set of literals.

3.2.1.1 Linear Search Algorithms

One common approach for solving MaxSAT is to perform a search on the possible values of the weight assignments by iteratively calling a SAT solver. In these algorithms, a relaxation variable r_i is added to each soft clause c_i such that if the original soft clause c_i is unsatisfied, then r_i must be assigned to *true*. Hence, relaxation variables r_i represents whether the original soft clause c_i is satisfied (or not).

Algorithm 3.1 illustrates a linear search on the total weight of unsatisfied soft clauses. The algorithm maintains an *upper bound* μ on the weight of an optimal assignment. Observe that μ can be initialized with any value larger than the sum of

[6] A common approach to extract an unsatisfiable subformula is to use assumptions [26].

Algorithm 3.2: Linear Search UNSAT-SAT Algorithm

Input: $\phi = \phi_h \cup \phi_s$
Output: optimal solution to ϕ

```
1  (φW, VR, λ) ← (φh, ∅, 0)
2  foreach ci ∈ φs do
3      VR ← VR ∪ {ri}                         // ri is a new relaxation variable
4      cR ← ci ∪ {ri}
5      φW ← φW ∪ {cR}
6  while true do
7      (st, ν, φC) ← SAT(φW ∪ {CNF(Σ_{ri∈VR} wi · ri ≤ λ)}, ∅)
8      if st = SAT then
9          return ν                           // optimal assignment to φ
10     λ ← UpdateBound({wi : ri ∈ VR}, λ)
```

the weights of all soft clauses. The working formula ϕ_W initially contains all hard clauses from ϕ_h. Next, a relaxation variable r_i is added to each soft clause c_i and the resulting clause is added to the working formula (lines 2-5).

At each iteration of the algorithm, a SAT solver is called (line 7) on the working formula ϕ_W with an additional pseudo-Boolean constraint such that the total weight of the unsatisfied soft clauses must be smaller than the upper bound μ. Therefore, if the SAT solver call returns SAT, then ν contains an assignment with weight smaller than μ, thus improving on the previously found solution. As a result, ν is saved (line 10) and the upper bound μ is decreased (line 11). On the other hand, if the SAT solver returns UNSAT, then there is no better solution than the last one found and the algorithm ends (line 9).

Observe that the pseudo-Boolean constraint is encoded into CNF [81, 28, 17, 40, 42] in the SAT solver call (line 7). Otherwise, the SAT solver would have to be replaced by a pseudo-Boolean solver that is able to natively deal with these constraints [50, 18]. Moreover, if the MaxSAT formula is an instance of the partial MaxSAT problem where all soft clauses c_i have weight 1, then a cardinality constraint is used and encoded into CNF [16, 79, 12, 71].

Algorithm 3.2 also performs a linear search on the weight of unsatisfied soft clauses. However, in this case, the search maintains a *lower bound* λ on the weight of an optimal assignment. The lower bound λ is initialized at 0 and increases at each iteration until it reaches the optimal value.

The algorithm structure is very similar to Algorithm 3.1. The working formula is initialized in exactly the same way by adding a relaxation variable to each soft clause (lines 2-5). However, at each iteration, the SAT solver call checks whether there is a solution with weight λ (line 7). If that is the case, then an optimal solution was found and the algorithm ends (line 9). Otherwise, λ is increased to the next possible lower bound value using function `UpdateBound`. The `UpdateBound` function returns the smallest integer value υ such that $\upsilon > \lambda$ and $SubSetSum(\{w_i : r_i \in V_R\}, \upsilon)$ is *true* [6]. Function $SubSetSum(S, \upsilon)$ solves the well-known subset sum problem, i.e., it returns *true* if there is a subset S' of S such that the sum of the elements of S' equals

Algorithm 3.3: WMSU3 Algorithm

```
   Input: φ = φ_h ∪ φ_s
   Output: optimal solution to φ
 1 (φ_W, V_R, λ) ← (φ, ∅, 0)
 2 while true do
 3     (st, ν, φ_C) ← SAT(φ_W ∪ {CNF(Σ_{r_i∈V_R} w_i · r_i ≤ λ)})
 4     if st = SAT then
 5         return ν                          // optimal assignment to φ
 6     foreach c_i ∈ (φ_C ∩ φ_s) do
 7         V_R ← V_R ∪ {r_i}                 // r_i is a new variable
 8         c_R ← c_i ∪ {r_i}                 // c_i was not previously relaxed
 9         φ_W ← (φ_W \ {c_i}) ∪ {c_R}
10     λ ← UpdateBound({w_i : r_i ∈ V_R}, λ)
```

υ. Since the subset sum problem is NP-hard [32], a pseudo-polynomial algorithm based on dynamic programming is used. This allows us to skip over lower bound values that are not possible to attain, given the weights of the relaxed soft clauses in V_R [6]. Finally, notice that when the weight of all soft clauses is 1 (e.g., unweighted partial MaxSAT), then `UpdateBound` always increases λ by 1.

3.2.1.2 Unsatisfiability-Based Algorithms

In 2006, Fu and Malik [31] proposed the first MaxSAT algorithm that takes advantage of the ability of SAT solvers to be able to identify unsatisfiable subformulas (also known as an unsatisfiable core of a formula). Since then, many other MaxSAT algorithms have been proposed that also take advantage of this SAT solver feature.

As in Algorithm 3.2, the WMSU3 algorithm [52] also performs a lower bound search on the weight of the optimal solution. However, the WMSU3 algorithm takes advantage of the identification of unsatisfiable cores to delay the relaxation of soft clauses.

Algorithm 3.3 presents the pseudo-code of WMSU3, where the working formula is initialized with all clauses from ϕ (hard and soft). At each iteration, the SAT solver call verifies whether there is a solution with weight λ. Whenever the SAT solver returns UNSAT, ϕ_C contains an unsatisfiable subformula of ϕ_W. In that case, a new relaxation variable r_i is added to each soft clause $c_i \in \phi_C$ that has not already been relaxed (lines 6-9). Next, the lower bound λ is updated, as explained for Algorithm 3.2.

Observe that at each SAT solver call (line 3), the pseudo-Boolean constraint might not depend on the relaxation variables of *all* soft clauses. Since a clause is relaxed only when it appears in an unsatisfiable core, the pseudo-Boolean constraint is usually much smaller than the one in Algorithm 3.2. As a result, Algorithm 3.3 is much more effective. Finally, when the SAT solver call is satisfiable, then an optimal solution was found and the algorithm terminates (line 5).

Algorithm 3.4: Fu-Malik for Weighted MaxSAT Algorithm

Input: $\phi = \phi_h \cup \phi_s$
Output: optimal solution to ϕ
1 $(\phi_W, \lambda) \leftarrow (\phi, 0)$ // clauses in ϕ_s are marked as soft
2 **while** true **do**
3     $(\mathsf{st}, \nu, \phi_C) \leftarrow \mathtt{SAT}(\phi_W, \emptyset)$
4   **if** st = SAT **then**
5   **return** ν // optimal assignment to ϕ
6     $V_R \leftarrow \emptyset$
7     $m_C = \mathtt{min}\{\mathtt{weight}(c_i) \mid c_i \in \phi_C \wedge \mathtt{soft}(c_i)\}$
8   **foreach** $c_i \in \phi_C \wedge \mathtt{soft}(c_i)$ **do**
9         $V_R \leftarrow V_R \cup \{r\}$ // r is a new relaxation variable
10         $c_R \leftarrow c_i \cup \{r\}$ // c_R is marked as soft
11         $\mathtt{weight}(c_r) \leftarrow m_C$
12   **if** $\mathtt{weight}(c_i) > m_C$ **then**
13             $\mathtt{weight}(c_i) \leftarrow \mathtt{weight}(c_i) - m_C$
14             $\phi_W \leftarrow \phi_W \cup \{c_R\}$
15   **else**
16             $\phi_W \leftarrow (\phi_W \setminus \{c_i\}) \cup \{c_R\}$
17     $\phi_W \leftarrow \phi_W \cup \{\mathtt{CNF}(\sum_{r \in V_R} r \leq 1)\}$
18     $\lambda \leftarrow \lambda + m_C$

In the algorithms described so far, it is necessary to encode a pseudo-Boolean constraint at each iteration. However, the generalization of the original Fu-Malik [31] algorithm for weighted MaxSAT [50, 5] described in Algorithm 3.4 does not need pseudo-Boolean constraints. Algorithm 3.4 also performs a lower bound search on the optimal value of the MaxSAT problem. In this case, a constraint on the lower bound is not represented explicitly.

In Algorithm 3.4, the working formula is also initialized with all hard and soft clauses. When the working formula becomes satisfiable, then an optimal solution was found and the algorithm ends. On the other hand, while the working formula remains unsatisfiable, the SAT solver returns an unsatisfiable core ϕ_C. In this case, each soft clause in $c_i \in \phi_C$ is relaxed by creating a new relaxed clause c_r from c_i, extended with a new relaxation variable.

On line 7, the *weight of the core* m_C is the minimum weight of all soft clauses in ϕ_C. Each soft clause $c_i \in \phi_C$ with weight equal to m_C is removed and replaced with its relaxation c_r (line 16). Otherwise, its weight is decreased by m_C, thus resulting in a clause split, since the original weight is divided between c_i and c_r. Finally, at most one clause from the unsatisfiable core ϕ_C can be relaxed (line 17) and the lower bound is increased by m_C. Notice that no pseudo-Boolean constraints are involved in this algorithm. At each iteration, a new *AtMost1* cardinality constraint is encoded into the working formula [33, 8, 30, 74, 43, 22].

3.2.1.3 Other Algorithmic Solutions and Implementation Issues

In this chapter we have briefly introduced some algorithms for solving MaxSAT that can be easily used in the parallel setting. We have focused on algorithms that iteratively call a SAT solver, but other algorithmic solutions can be used instead.

A classic alternative is to use a specific branch and bound algorithm for MaxSAT. In these algorithms, two techniques are common at each node of the search tree: (1) the application of restrictive rules of MaxSAT inference [46, 21], and (2) the application of lower bound estimation procedures in order to prune the search [45, 10, 23, 47]. Furthermore, SAT-based techniques can also be used on hard clauses [11].

More recently, several new algorithms based on the identification of unsatisfiable cores have been proposed and an extended survey on this class of algorithms is available [65]. The organization of iterative and core-based algorithms for MaxSAT can be very diverse. For instance, some algorithms try to split the MaxSAT formula using different criteria [60, 3, 70]. Others take advantage of identifying disjoint unsatisfiable cores and perform binary search [38], while another approach is to integrate hitting-set minimization into MaxSAT solving [25], among many other diverse techniques [68].

In order to build a state-of-the-art MaxSAT solver, there are important implementation issues to be taken in consideration. First, the underlined SAT solver should be able to effectively handle assumptions [26]. In particular, for iterative and core-based algorithms, where the SAT solver is called successively, the usage of incremental SAT solving is crucial [56]. In this chapter, we focus mainly on the algorithmic techniques, but the integration with the SAT solver is essential to maximize the performance of the MaxSAT solver.

3.3 Parallel MaxSAT

In recent years, parallel SAT solvers have successfully exploited multicore and distributed architectures to speed up the performance of sequential SAT solvers.[7] When compared with SAT instances, MaxSAT instances tend to be more intricate [72]. When solving a MaxSAT instance, it is not sufficient to find an assignment that satisfies all clauses, but rather an assignment that satisfies all hard clauses and minimizes the sum of the weights of unsatisfied soft clauses. Hence, it comes as a natural step to develop parallel algorithms for MaxSAT.

Parallel MaxSAT solvers typically follow the architecture presented in Figure 3.3 and are based on two orthogonal approaches: (i) unsatisfiability-based algorithms that search on the lower bound of the optimal solution, i.e., that perform *lower bound search*, and (ii) linear search algorithms that search on the upper bound of the optimal solution, i.e., that perform *upper bound search*. A parallel search with these two orthogonal strategies results in a performance as good as the best strategy for

[7] We refer the interested reader to Chapter 1, Parallel Satisfiability for more details on parallel SAT solving.

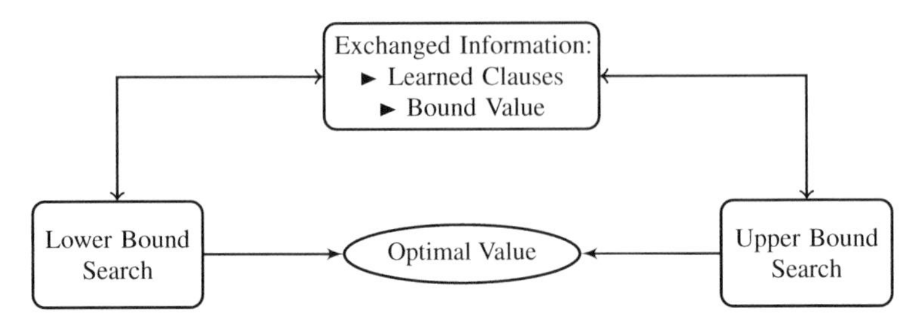

Fig. 3.3: Architecture of parallel MaxSAT solvers

each problem instance. However, if workers cooperate through clause sharing , it is possible to perform better than the best algorithm. Additionally, the two strategies can cooperate in finding the optimum value. If during the search the lower bound value provided by the unsatisfiability-based algorithms and the upper bound value provided by linear search algorithms become the same, it means that an optimal solution has been found. As a result, it is not necessary for any of the remaining workers to continue the search to prove optimality since their combined information already proves it.

The main differences between parallel MaxSAT solvers that follow the architecture depicted in Figure 3.3 are:

- *Resource allocation:* different strategies for splitting the available workers between lower bound search and upper bound search;
- *Information exchange:* different heuristics for clause sharing .

In this section, we present the most common ways of splitting the available resources and illustrate how parallel MaxSAT solvers exchange information to prune the search space and speed up the search.

3.3.1 Portfolio Approaches

A portfolio approach explores the parallelism given by different strategies on the same problem. Given n workers, a parallel portfolio MaxSAT solver will split the workers between lower bound search and upper bound search. However, if the algorithms that perform lower bound search are the same and the algorithms that perform upper bound search are also the same, then the gain from increasing the number of workers will be very small since all of them will be searching in a similar way. In order to diversify the search one may employ different strategies such as: (i) change the heuristics of the underlying SAT solver; (ii) use different algorithms on the lower bound search or upper bound search; (iii) change the encoding of cardinality and pseudo-Boolean constraints used in the MaxSAT algorithms.

The portfolio approaches for parallel MaxSAT solving presented in this section are closely related to the portfolio approaches for parallel SAT solving presented in

	PWBO-T4		PWBO-T8	
	Encoding	Search	Encoding	Search
Thread t_1	Commander	LB	Commander	LB
Thread t_2	Totalizer	LB	Totalizer	LB
Thread t_3	Sorters	UB	Ladder	LB
Thread t_4	PB	UB	Product	LB
Thread t_5	–	–	Sorters	UB
Thread t_6	–	–	PB	UB
Thread t_7	–	–	Sequential	UB
Thread t_8	–	–	Totalizer	UB

Table 3.2: Configuration of PWBO with four and eight threads

Chapter 1, Parallel Satisfiability. The main differences between these two approaches are: (i) parallel portfolio for MaxSAT uses two orthogonal algorithms, whereas parallel SAT solvers are usually based on the same algorithm; (ii) the diversification of the search in parallel SAT solving is usually done through different heuristics, while in parallel MaxSAT there are more available strategies to diversify the search.

An example of a portfolio solver for parallel MaxSAT that diversifies the search by using different cardinality encodings for each worker is given in Table 3.2. PWBO [57, 61] is a parallel MaxSAT solver for multicore architectures that uses different configurations for four and eight threads. To maintain a balance between lower and upper bound search PWBO uses the same number of threads for each kind of search, while diversifying the search through different encodings.

3.3.1.1 Parallel Unsatisfiability-Based Algorithms

Figure 3.4 illustrates parallel unsatisfiability-based algorithms. These algorithms work by iteratively identifying unsatisfiable cores and can use any unsatisfiability-based algorithms such as the ones presented in Section 3.2.1.2. While solving the formula, the parallel algorithm checks whether another worker has found a better lower bound value, i.e., if it has found an unsatisfiable core. If this is the case, then it imports the unsatisfiable core and relaxes it as if it had been found by this worker.

If a worker is not aware of a better lower bound value, then it continues the search process until it finds an unsatisfiable core or a solution to the formula. If it finds an unsatisfiable core, then it shares this unsatisfiable core with the remaining workers searching on the lower bound. Next, it relaxes the unsatisfiable core as previously described and continues the search on the new working formula. The procedure terminates when the working formula becomes satisfiable and the solver returns an optimal solution.

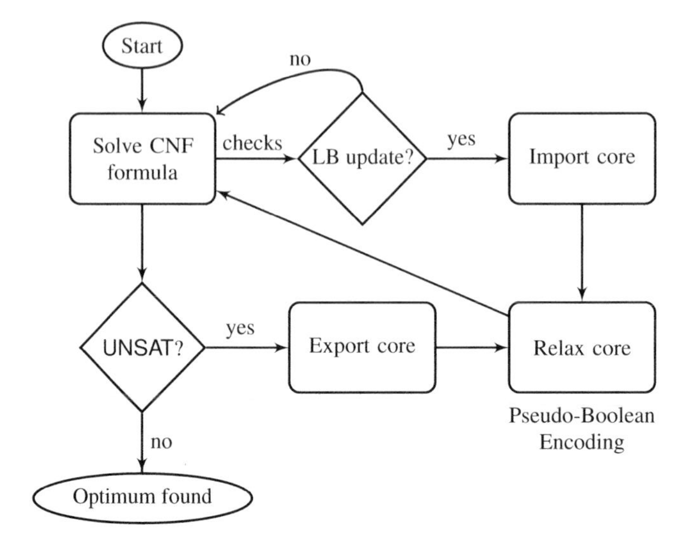

Fig. 3.4: Parallel unsatisfiability-based algorithms

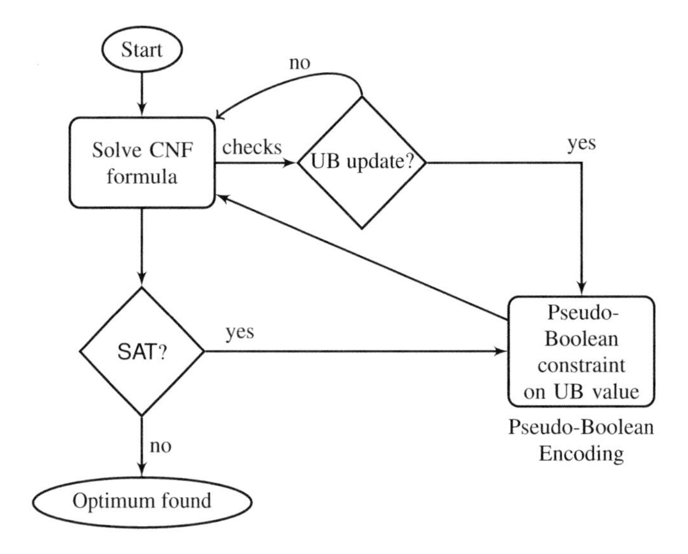

Fig. 3.5: Parallel linear search algorithms

3.3.1.2 Parallel Linear Search Algorithms

Figure 3.5 illustrates parallel linear search algorithms. These algorithms are based on Algorithm 3.1 presented in Section 3.2.1.1. Recall that the original MaxSAT formula ϕ is modified by adding a new relaxation variable r_i to each soft clause c_i from ϕ, resulting in an equivalent formula ϕ_W where one wants to minimize the number of relaxation variables assigned to *true*. In the parallel algorithm, whenever a new solution is found for ϕ_W, the upper bound value is updated and a new pseudo-Boolean constraint on the relaxation variables is added such that all solutions with a greater or equal value are excluded. During search, each algorithm checks whether there is a better upper bound value. If this is the case, it adds a pseudo-Boolean constraint considering the new upper bound value. Afterwards, it restarts the search on the constrained formula.

3.3.1.3 Implementation Issues

Note that there are a few implementation details not shown in Figures 3.4 and 3.5. In particular, in Figure 3.4 only one worker exports an unsatisfiable core for each lower bound value. Before exporting an unsatisfiable core, the respective worker checks whether its lower bound value is the greatest lower bound value among all workers. If this is the case, then it is safe to export the unsatisfiable core to the remaining workers. Otherwise, it discards its own unsatisfiable core and imports the unsatisfiable core that corresponds to the current lower bound value. Moreover, when a worker relaxes an unsatisfiable core, it updates its lower bound value. When a worker imports an unsatisfiable core from another worker, the relaxation procedure is the same as if this unsatisfiable core had been found by the importing worker. Therefore, each worker may preserve incrementality across iterations of the unsatisfiability-based algorithm by using the incremental schemes presented in the literature [55, 56].

In Figure 3.5, the encodings used by the linear search algorithms support *incremental strengthening* [12]. Since the upper bound value is always decreasing, the pseudo-Boolean constraint only needs to be encoded when the first upper bound value is found. In the following iterations, one can assign truth value *false* to some specific literals in the encoding such that it restricts the pseudo-Boolean constraint to the new upper bound value. Hence, all learned clauses from previous iterations remain valid and can therefore be kept.

3.3.2 Search Space Splitting

In portfolio solvers there is a race between different algorithms (or the same algorithm with different configurations) to reach a solution. In search-space-splitting solvers, the goal is to split the original problem such that each worker process has to deal with a smaller formula that is hopefully easier to solve. This section describes several procedures to split the search space when solving MaxSAT formulas.

3.3.2.1 Interval Splitting

As previously mentioned, when solving a MaxSAT formula, one can define an interval between a lower bound value and an upper bound value. Hence, a different approach for parallel MaxSAT is to split the search space by defining tentative bound values to narrow this interval.

Consider that n cores or machines are available. In this case, one worker can be used to search on the lower bound (using any unsatisfiability-based MaxSAT algorithm), one worker can be used to search on the upper bound (using a linear search algorithm), and the remaining $n-2$ workers can search considering different tentative bound values between a known lower bound and a known upper bound. The goal is to define tentative upper bound values that restrict the search space by enforcing a fixed upper bound value of the optimal solution. Since this fixed upper bound value is restricted to each worker, it is called *local upper bound value*. The search performed by each of these workers is called the *local upper bound search*. The iterative search on different local upper bound values leads to constant updates on the lower and upper bound values that will reduce the search space. Next, an example of this approach is described. Afterwards, a more detailed description of local upper bound search is presented.

Example 1. Consider a partial MaxSAT formula ϕ as input. For the input formula, one can easily find initial lower and upper bounds. Suppose the initial lower and upper bound values are 0 and 11, respectively. Moreover, consider also that the optimal solution is 3 and our goal is to find it using four workers: t_0, t_1, t_2, and t_3. Worker t_0 applies an unsatisfiability-based algorithm (i.e., searches on the lower bound of the optimum solution). This worker starts with a lower bound of 0 and will iteratively increase the lower bound until the optimum value is found.

Worker t_1 searches on the upper bound of the optimum solution. Hence, worker t_1 starts its search with upper bound value of 11. Workers t_2 and t_3 search on different local upper bound values. For example, workers t_2 and t_3 can start their search with local upper bound values of 3 and 7, respectively.

Suppose that worker t_2 finishes its computation and finds that the formula is unsatisfiable for an upper bound of 3. This means that there is no solution with values 0, 1 or 2. Therefore, the lower bound value can be updated to 3. Worker t_2 is now free to search on a greater local upper bound value, for example 5. In the meantime, worker t_3 finds a solution with value 6. Hence, the upper bound value can be updated to 6. Worker t_1 updates its upper bound value to 6 and worker t_3 is now free to search on a different local upper bound value, for example 4. Afterwards, consider that worker t_1 finds a solution with value 3. Again, the upper bound value can be updated to 3. Since the lower bound value is the same as the upper bound value, the optimum has been found and the search terminates.

Observe that this parallel search incorporates three types of algorithms: unsatisfiability-based, linear search, and local linear search. The unsatisfiability-based and linear search algorithms can be any two MaxSAT algorithms that follow these approaches. In what follows we describe the algorithm for parallel local linear search that is used by the remaining workers to perform local upper bound search.

Figure 3.6 illustrates parallel local linear search algorithms. Similarly to linear search algorithms, the original MaxSAT formula $\phi = \phi_h \cup \phi_s$ is modified by adding a new relaxation variable r_i to each soft clause c_i from ϕ_s, resulting in an equivalent

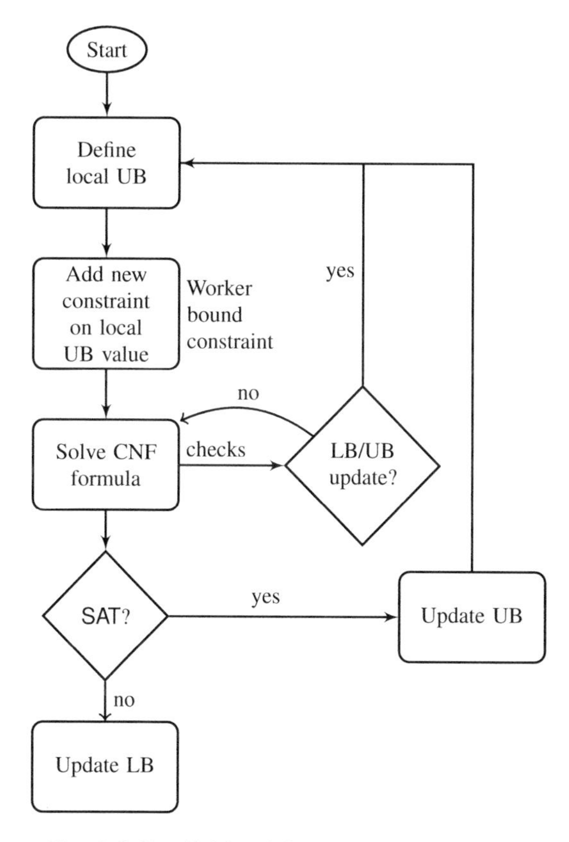

Fig. 3.6: Parallel local linear search algorithms

formulation ϕ_R where the goal is to minimize the weighted sum of relaxation variables assigned value 1.

These algorithms start by defining their local upper bound. Initially, the lower bound value is set to 0 and the upper bound value to the sum of the weights of soft clauses plus 1, i.e., $s = \sum_{c_i \in \phi_s} w_i + 1$. Therefore, considering k local workers, $t_1, \ldots, t_k$, a worker t_j will have an initial tentative upper bound value b_j of $j \times \lfloor (s+1)/(k+1) \rfloor$.

Next, worker t_j adds a constraint of the form $\sum r_i w_i \leq b_j - 1$ to exclude solutions with a value greater than or equal to b_j. Let this constraint be labeled the *worker bound constraint*. If an encoding to CNF is used, then all clauses that were created to encode this constraint will be labeled as worker bound constraints.

After adding a worker bound constraint, the algorithm starts the search. During the search, the algorithm checks whether another worker has found a lower bound that is greater than the current local upper bound or an upper bound that is smaller than the current local upper bound. If one of these cases occurs, then the algorithm will terminate its search and a new local upper bound is defined. Next, the search restarts using the new local upper bound value. If the algorithm is not informed that

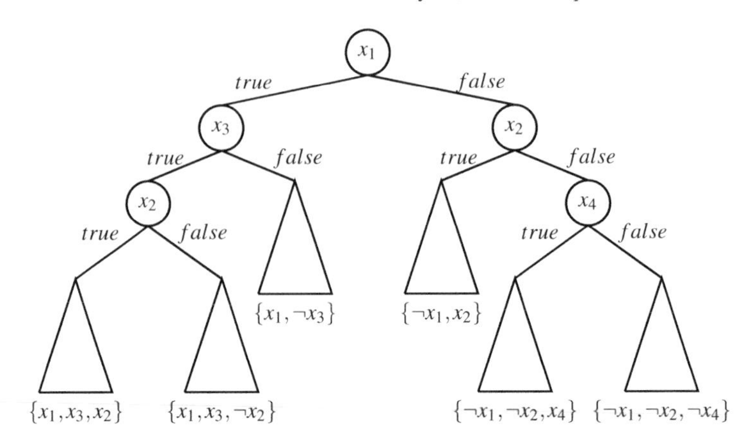

Fig. 3.7: Example of guiding path generation using search tree split

a better lower or upper bound value has been found, then it continues the search process until it finds a solution or proves that no solution exists for the current local upper bound value. If a solution is found, then the algorithm updates the upper bound value. Otherwise, if it proves that no solution exits, then the lower bound value is updated. In both cases, a new local upper bound value is set and the search restarts.

Although not shown in Figure 3.6, note that updates to the global lower and upper bounds can only take place when the new values improve the current ones. Observe also that when a worker is assigned a new tentative upper bound value, its value should cover the broadest range of yet untested bounds. More formally, the new local upper bounds should be chosen as follows. Let $B = \langle b_0, b_1, \ldots, b_{k-1}, b_k \rangle$ be a sorted list where b_0 corresponds to the lower bound and b_k corresponds to the upper bound, while the remaining b_j are the non-aborted worker local upper bounds. Let $[b_{m-1}, b_m]$, where $1 \leq m \leq k$, define an interval such that for all $1 \leq j \leq k$ we have $b_m - b_{m-1} \geq b_j - b_{j-1}$. In this case, the new upper bound of the aborted worker is $\lfloor (b_m + b_{m-1})/2 \rfloor$. The sorted list B is updated with the new value and this process is repeated for each aborted worker.

Example 2. Consider the following scenario. A MaxSAT formula ϕ is currently being solved by 4 worker processes. Worker t_0 is searching on the lower bound value of the optimal solution, and worker t_1 is searching on the upper bound value of the optimal solution. The current lower and upper bound values are 5 and 10, respectively. Worker t_2 is searching on a local upper bound with value 8 and worker t_3 should start computing with a new local upper bound. The sorted list B corresponds to $B = \langle 5, 8, 10 \rangle$. Note that the largest interval between two consecutive values in B is $[5, 8]$. Therefore, the new tentative upper bound value of t_3 should be $\lfloor (8+5)/2 \rfloor = 6$.

3.3.2.2 Guiding Paths

Another method of splitting the search space is to use guiding paths [77, 82]. In this case, one considers the search space as a binary tree, where each node corresponds to a variable and each of its edges corresponds to an assignment to that variable.

What the guiding paths approach does is to split the search tree into sub-trees and assign each of them to a distinct worker process. Figure 3.7 illustrates how the search tree could be split among 6 workers. Observe that the variable order for splitting does not have to be fixed, and the guiding paths do not need to have the same size (e.g., number of literals)

There are several issues regarding the generation of guiding paths, namely the variables chosen to split the search space and the length of the guiding path. The variables should be *important* in the formula such that the exploration of the remaining search spaces can be done effectively and in a balanced way. Moreover, the length of the guiding path should be large enough to effectively reduce the search space to be explored, but small enough so that just a small number of guiding paths should be tried.

There are many different strategies to generate guiding paths [76, 73, 39]. For instance, in order to identify relevant variables, some solvers make an initial SAT call that is aborted after a small number of conflicts is attained (or some other stopping criteria). Next, the variables with higher VSIDS [67] score are chosen to build the guiding paths [82].

Another issue is an uneven load balancing between the workers. It might be the case that some guiding paths result in very effective solver calls, while others will take much more time. As a result, some workers might become idle. In this case, solvers apply a work-stealing procedure [20] and a previously generated guiding path is extended with another variable, thus producing a new guiding path for the idle worker.

Yet another approach was proposed by Heule et al. [39] where a parallel SAT algorithm initially uses a *lookahead* solver to generate guiding paths in order to split the search tree. Lookahead solvers apply sophisticated reasoning at each branching step in order to guide the search more effectively. This approach was successfully adapted to solving several MaxSAT instances [69].

The parallel solver starts by generating a queue of guiding paths to be solved by the worker processes and an initial upper bound μ is defined. The guiding paths can be heuristically sorted (e.g. by the weight of the (un)satisfied soft clauses of the guiding path assignments) and given to available workers with the best upper bound computed thus far. Each worker can then apply any MaxSAT algorithm and returns the best solution found for the given path. If the newly found solution improves on the previous one, it is saved and the upper bound μ is updated.

Note that the number of guiding paths can be much larger than the number of workers. As a result, the remaining MaxSAT instances to be solved by each worker are smaller. Moreover, when a guiding path is solved, the worker can readily obtain the next one from the queue. Each time a guiding path is solved by a worker, it must be removed from the queue.

The MaxSAT instance is considered solved when the guiding path queue becomes empty. Additionally, in the context of MaxSAT, a generated guiding path may cause some of the soft clauses to be unsatisfied. If the number of unsatisfied soft clauses is greater than or equal to μ, then we know that no better solution will be found for that guiding path. In that case, the guiding path is discarded and removed from the queue. In fact, to prune guiding paths, one can also apply lower bounding procedures for MaxSAT commonly used in branch-and-bound algorithms [44].

3.3.2.3 Other Splitting Schemes and Implementation Issues

Besides interval splitting and guiding paths, there are other splitting schemes based on cut-and-join approaches. For instance, one can decompose the input problem into simpler subproblems that can be independently solved [78]. Next, the solutions to each subproblem can be joined and checked to build a global solution. In MaxSAT, several sequential algorithms already decompose the formula using different strategies, namely by using information from the weights of soft clauses [60, 3], finding disjoint unsatisfiable cores in binary search [38], or based on the structural analysis of the formula [70]. In these algorithms, one can decompose the formula to be solved in parallel and later joined in order to find an optimal solution for the MaxSAT formula.

As with the sequential solvers, parallel MaxSAT solvers can also be improved by a proper usage of current state-of-the-art SAT solvers, in particular by taking advantage of incremental SAT solver calls. In both approaches of interval splitting and guiding paths, each worker can be implemented in an incremental way, i.e., does not have to rebuild its working formula when a new interval or guiding path is provided. We refer the reader to the literature on the details of using a SAT solver incrementally to build state-of-the-art MaxSAT algorithms [55, 56].

An incremental scheme can be implemented at each worker by considering the guiding path literals as assumptions [27] in the SAT solver calls. Several incremental schemes for MaxSAT have already been proposed [55, 56] that can be easily extended to consider the guiding path literals [69]. Observe that considering these literals as assumptions in SAT solver calls, one can check whether the MaxSAT formula is unsatisfiable due to the guiding path literals. When the unsatisfiability does not depend on the guiding path, then the current upper bound is also a lower bound of the MaxSAT formula. As a result, the previously found solution is optimal and the solver can terminate, even if there are guiding paths in the queue.

A similar scheme can also be implemented for interval splitting. Here, the upper bound constraint can be encoded just once with the initial upper bound limit. Therefore, in the subsequent calls to the worker, the local upper bound will always be smaller than the initial bound. As a result, when the worker is testing a new upper bound, this limit can be encoded as assumptions in the SAT solver call by using incremental strengthening [12] or incremental weakening [55], depending on the new bound being tested.

3.3.3 Clause Sharing

Conflict-driven clause learning [53, 83] is crucial for the efficiency of modern SAT solvers. After detecting a conflict, i.e., a sequence of assignments that make a clause unsatisfiable, a new clause is learned to prevent the same conflict from occurring again in the subsequent search. The new clause results from the analysis of the implication graph which represents the dependencies between assignments. A more detailed explanation can be found in the literature [53, 83].

Clause learning is also essential to the efficiency of many modern MaxSAT solvers. In the context of parallel solving, *shared clauses* correspond to learned clauses that were exported by a worker and were given to other workers. The importing worker can then decide whether it incorporates the shared learned clause into its context or not. Sharing learned clauses helps to further prune the search space and boosts performance of parallel solvers.

3.3.3.1 Conditions for Safe Clause Sharing

Sharing clauses in parallel MaxSAT is not straightforward and poses additional challenges that need to be addressed. First, not all learned clauses can be shared among all workers since each worker may have a different working formula. For example, workers using unsatisfiability-based algorithms have relaxed the MaxSAT formula in a different way than workers using linear search algorithms and cannot share all learned clauses between them. Additionally, when using cardinality or pseudo-Boolean encodings, we also have to take into account the auxiliary variables used by those encodings. Therefore, each worker may contain variables not present in the other workers.

Workers that perform local upper bound search contain worker bound constraints.[8] Sharing conflict-driven learned clauses that are implied by worker bound constraints depends on the upper bound value of the worker. If an importing worker has an upper bound value smaller than or equal to the upper bound value of the exporting worker, then the import is safe. Otherwise, the import may be unsafe and the respective clauses are not shared. Therefore, it is necessary to define what is a local constraint and in which conditions it can be shared with other workers. Let the worker bound constraint be labeled as a *local constraint*. Let c be a conflict-driven learned clause and let ϕ_c be the set of constraints used in the implication graph to learn c. The new clause c is defined as a local constraint if at least one constraint in ϕ_c is a local constraint. The sharing procedure between different workers is as follows:

- Learned clauses that are not local constraints and do not have encoding variables can be safely shared between all workers;

[8] Workers that perform lower bound search on a lower bound value k can be seen as performing local upper bound search on the upper bound value k. Even though these workers do not have worker bound constraints, they have cardinality or pseudo-Boolean constraints that have the same effect as having a worker bound constraint.

- Learned clauses that are local constraints can only be shared from worker t_i to worker t_j if the local upper bound value of t_i is greater than or equal to the local upper bound value of t_j.

Example 3. Consider the scenario presented in Example 2 where we have four workers solving a MaxSAT formula. Worker t_0 is searching on the lower bound value 5, worker t_1 on the upper bound value 10, and workers t_2 and t_3 on local upper bound values 8 and 6, respectively. Learned local constraints from t_1 can be shared to all other workers since the local bound value of t_1 is larger than the local bound value of t_0, t_2 and t_3. However, learned local constraints from t_3 cannot be shared to t_2 and t_1 since the local bound of t_3 is smaller than the local bound of t_2 and t_1.

3.3.3.2 Clause-Sharing Heuristics

When sharing learned clauses, not all learned clauses can be shared since this would lead to an exponential blow up in memory and to many irrelevant clauses being shared. A clause is considered irrelevant if it never becomes unsatisfied or unit, which means that it does not help in pruning the search space. The problem of determining whether a shared clause will be useful in the future remains challenging and in practice heuristics are used to choose which learned clauses should be shared. Clause-sharing heuristics can be divided into the following three categories: (i) static, (ii) dynamic, and (iii) freezing. The static heuristics share learned clauses within a given cutoff, whereas the dynamic heuristics adjust this cutoff during the search. Alternatively, the freezing heuristics temporarily delay the incorporation of shared clauses until they are expected to be useful in the context of the importing worker.

Static Heuristics. The static heuristics are the most commonly used heuristics for clause sharing since they are simple but still efficient in practice. The following measures are used in these heuristics:

- *Size*: the clause size is given by the number of literals in a clause from an exporting worker. Small clauses are expected to be more useful than larger clauses. Clause size was originally used as a measure to select which learned clauses should be kept by the SAT solver [53, 54]. More recently, it has been adopted by parallel SAT solvers (e.g., [37]) and parallel MaxSAT solvers [61].
- *Literal Block Distance* (LBD) [15]: the literal block distance corresponds to the number of different decision levels involved in a clause to be shared from an exporting worker. The decision level of a literal denotes the depth of the decision tree at which the corresponding variable was assigned a value. Clauses with small LBD are considered to be more relevant.

Dynamic Heuristics. It has been observed that the size of learned clauses tends to increase over time. Consequently, in parallel solving, any static limit may lead to halting the clause-sharing process. Therefore, to continue sharing learned clauses it is necessary to dynamically increase the limit during search. In the context of parallel SAT solving, Hamadi et al. [36] proposed the following dynamic heuristic. At every k conflicts (corresponding to a period α) the number of shared learned clauses (s) is evaluated between each pair of workers $(t_i \rightarrow t_j)$ according to the following heuristic:

$$\lim_{t_i \rightarrow t_j}^{\alpha+1} = \begin{cases} \text{if s} < \text{m (sharing is small): } \lim_{t_i \rightarrow t_j}^{\alpha} + \text{quality}_{t_i \rightarrow t_j}^{\alpha} \times \frac{b}{\lim_{t_i \rightarrow t_j}^{\alpha}} \\ \text{if s} \geq \text{m (sharing is large): } \lim_{t_i \rightarrow t_j}^{\alpha} - (1 - \text{quality}_{t_i \rightarrow t_j}^{\alpha}) \times a \times \lim_{t_i \rightarrow t_j}^{\alpha} \end{cases}$$

where a and b are positive constants and the value of $\text{quality}_{t_i \rightarrow t_j}^{\alpha}$ corresponds to the quality of shared learned clauses that were exported from t_i and imported by t_j. If s is less than a given m, then the sharing in period k is considered to be small. Otherwise, if s is greater than or equal to m, then the sharing in period k is considered to be large.

A shared learned clause with n literals is said to have *quality* [36] if at least half of its literals are active when the learned clause is exported. A literal is *active* if its VSIDS heuristic [83] score is high in the exporting worker, i.e., if it is likely to be chosen as a decision variable by the exporting worker in the near future. The quality of sharing between each pair of workers $(t_i \rightarrow t_j)$ is given by the following heuristic:

$$\text{quality}_{t_i \rightarrow t_j}^{\alpha} = \frac{q}{s}$$

where q is the number of quality shared learned clauses and s is the total number of shared learned clauses in the period α.

If the quality of sharing is high then the increase (decrease) in the size limit of shared learned clauses will be larger (smaller). The idea behind this heuristic is that the information recently received from a worker t_i is qualitatively linked to the information to be received from the same worker t_i in the near future.

Freezing Heuristics. There are possible drawbacks to importing clauses shared by other workers. One drawback is that the newly imported clauses may be irrelevant in the context of the importing worker. Another possible drawback is that the exploration of the search space may be influenced in such a way that the search becomes more closely related to the exploration being performed by the worker from which the clauses originated. As a result, the diversification of the exploration of the search space is decreased by shifting the context of the current search in the importing worker.

The freezing heuristic addresses these issues by only incorporating shared clauses when they are expected to be useful in the near future. For that, the decision to incorporate new learned clauses shared by other workers must take into consideration the current search context where these clauses are to be integrated. As a result, these new clauses should improve the efficiency of the search being carried out as they do not imply a major change to the search context of the receiving worker.

Figure 3.8 illustrates the freezing procedure [59]. Each imported learned clause c is evaluated to determine whether it will be frozen or added to the working formula of the importing worker. If c is frozen then it will be reevaluated later. However, if c

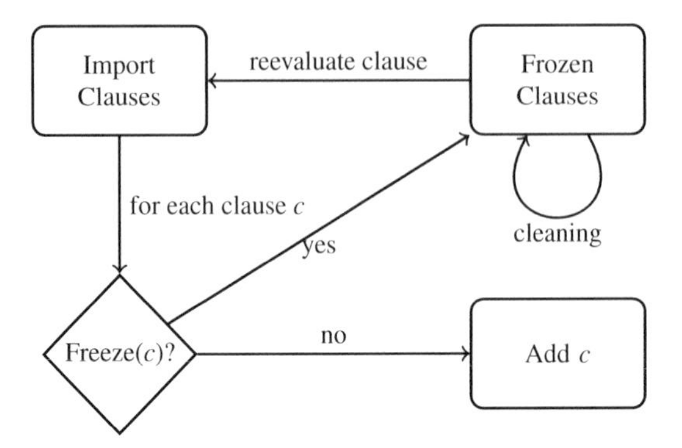

Fig. 3.8: Freezing procedure for sharing learned clauses

is assigned to the frozen state more than z times then it is permanently deleted. When evaluating c, our goal is to import clauses that are unsatisfied or that will become unit clauses in the near future. Next, the freezing heuristic is presented. According to the *status* of c in the importing worker (satisfied, unsatisfied, unit, or unresolved), it decides whether c should be frozen:

- c is *satisfied*: Let *clevel* denote the current decision level, $level(c)$ the highest decision level of the satisfied literals in c, $unassignedLits(c)$ the number of unassigned literals in c and $activeLits(c)$ the number of active literals in c. If $(clevel - level(c) \leq d)$ and $(unassignedLits(c) - activeLits(c) \leq e)$ (where d and e are constant values) then c is imported, otherwise it is frozen. A satisfied clause is expected to be useful in the near future assuming there is no need to backtrack significantly for the clause to become unit. It is also important that the number of unassigned literals is small, otherwise the clause may not become unit in the near future. Active literals are also taken into consideration since they will be assigned in the near future.
- c is *unsatisfied* or *unit*: c is always imported;
- c is *unresolved*: if $unassignedLits(c) - activeLits(c) \leq e$ then the clause is imported. Otherwise, it is frozen. Similarly to the satisfied case, if the number of unassigned literals is small then c is likely to be unit in the near future.

Freezing learned clauses was first proposed as a deletion strategy for learned clauses in SAT solving [14]. It was later extended for freezing shared clauses in the context of parallel MaxSAT [59] and parallel SAT [13].

3.3.3.3 Comparison Between Clause-Sharing Heuristics

The clause-sharing heuristics presented in this section have been compared against each other in the deterministic version of the portfolio-based parallel MaxSAT solver PWBO [63]. Since the search is deterministic, one can independently evaluate the gains coming from the use of different heuristics rather than the non-determinism of the solver.

This evaluation showed that sharing learned clauses does not significantly increase the number of solved instances. However, it does allow for a considerable reduction of the solving time. The freezing heuristic outperforms all other heuristics both in solving time and number of solved instances. On the other hand, the dynamic heuristics perform similarly to the static heuristics in terms of the number of solved instances but outperform them in terms of solving time. Even though PWBO only performs a limited form of clause sharing, it already shows a large impact of clause sharing on the solving time of a parallel MaxSAT solver. This opens future research directions to further improve clause sharing for parallel MaxSAT solving.

3.4 Deterministic Parallel MaxSAT

Despite being able to improve the performance of sequential MaxSAT solvers, current parallel MaxSAT solvers cannot be used in application domains that require reproducible results. For example, if we use a parallel MaxSAT solver in software verification [41, 22], different runs can report different bugs when verifying the same program. This behavior is unacceptable for the end user and restricts the use of parallel MaxSAT solvers for software verification applications.

The non-deterministic behavior of parallel MaxSAT solvers arises from the cooperation between workers. Sharing learned clauses and exchanging information on the lower and upper bounds can prune the search space and boost the performance of the parallel solver. However, this cooperation is also responsible for their non-deterministic behavior.

Hamadi et al. [35] proposed the first deterministic parallel SAT solver for multicore architectures. The deterministic solver only exchanges information between threads at fixed points during the search. These points are referred to as synchronization points and are defined based on the number of conflicts occurring during the search in each thread. Whenever a thread reaches a synchronization point (i.e., after detecting a given number of conflict assignments), it waits until the remaining threads reach that same point. Afterwards, when all threads have reached the synchronization point, they exchange learned clauses. This synchronization step guarantees the determinism of the cooperation between threads.

Synchronization points can also be applied to build a deterministic parallel MaxSAT solver [58, 63] for multicore architectures.[9] In what follows, the details of a deterministic parallel MaxSAT solver will be described.

[9] Synchronization points are not limited to multicore architectures and could also be used for building a deterministic parallel MaxSAT solver for distributed architectures.

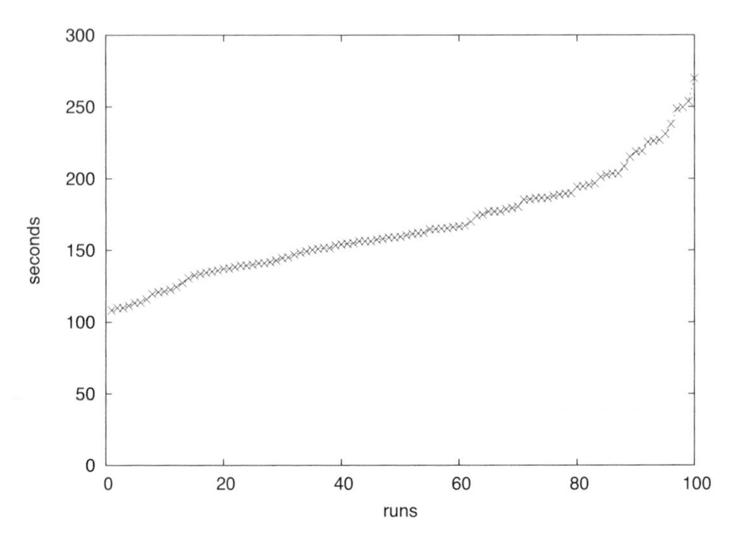

Fig. 3.9: Run time variation over 100 runs of the non-deterministic solver on the instance normalized-f20c10b_001_area_delay.wcnf (industrial category)

3.4.1 Motivation

The work on deterministic parallel MaxSAT solving can be motivated by the behavior illustrated in Figure 3.9. This figure shows the run time variation over 100 runs when running a non deterministic parallel MaxSAT solver with four threads on one instance of the partial MaxSAT industrial category of the MaxSAT Evaluation.[10] Run times vary between 108 seconds and 270 seconds, with an average run time of 166.41 seconds, a standard deviation of 35.59 seconds and a coefficient of variation [11] of 21.39%. Furthermore, 56 different optimal models were found during the 100 runs of the non-deterministic solver, showing that not only do the run times have a high variation but also the number of different models can be high.

For a more in-depth study of the variation of the non-deterministic solver, hundreds of partial MaxSAT instances of the industrial and crafted categories of the MaxSAT Evaluation were analyzed in detail [58, 63]. A generalized behavior was observed. The average coefficient of variation is around 21% and it is similar for both industrial and crafted benchmarks, although the average number of different models is lower in the crafted-benchmark set. Note that the value of the average coefficient of variation, despite being high, is similar to the variation of portfolio parallel SAT solvers on solving satisfiable instances [37].

As expected, these results support the idea that current state-of-the-art parallel solvers exhibit a high non-determinism on both running times and models found.

[10] `http://maxsat.ia.udl.cat/`

[11] The coefficient of variation is a normalized measure of dispersion and is given by $\Delta = \frac{\sigma}{\mu} \times 100$, where μ is the average time and σ is the standard deviation.

Therefore, if a parallel MaxSAT solver is to be widely used, it is necessary to build a deterministic version of the solver, as end users should be able to replicate the solver behavior for the same input.

3.4.2 Deterministic Solver

In this section, we present a deterministic parallel MaxSAT solver that ensures the reproducibility of results [58, 63]. Similarly to deterministic parallel SAT solving [35], synchronization points are used to guarantee the determinism of the cooperation between threads of the parallel MaxSAT solver.

The goal of the deterministic solver is to reproduce the same results on solving each problem instance by ensuring the following constraints: the solution reported by the solver and the search performed by each thread are always the same.

Figure 3.10 exemplifies an execution of the deterministic solver with four threads (although it can be easily generalized to any number of threads). In this example, threads t_1 and t_2 search on the lower bound value of the optimal solution, while threads t_3 and t_4 search on the upper bound value of the optimal solution. Each thread begins by performing its search as in the non-deterministic version [57, 61]. Every time a clause is learned, it is exported to the remaining threads. However, in the deterministic solver, learned clauses are only incorporated in other threads at synchronization points. This contrasts with the non-deterministic version where learned clauses can be imported on the fly.

When a thread that is searching on the lower bound finds a core, it stops the search and proceeds to the synchronization point. As can be seen in Figure 3.10, before reaching the synchronization point each thread exports the core that was found during the current period. A period corresponds to the search done between two consecutive synchronization points. Note that if a core has not been found in the last period, then only learned clauses are exported.

Remember that to each core is associated a cost that corresponds to an increase in the lower bound value and is used by the unsatisfiability-based algorithm to iteratively relax the MaxSAT formula [50]. Consider k threads performing lower bound search. At a synchronization point, all cores that were found in the last period are analyzed. Our goal is to import the core that corresponds to the largest increase in the lower bound value. If two threads find a core that corresponds to the same increase in the lower bound value, then the core with the smallest size is imported by all threads. If there are two cores with exactly the same size, then ties are broken by taking the thread identifiers in increasing order. Notice that, similarly to the non-deterministic version, all threads that are searching on the lower bound always have the same cores after a synchronization point.

Threads that are searching on the upper bound export their best solution and the corresponding upper bound value before reaching the synchronization point. At a synchronization point, each thread imports the smallest upper bound value among all

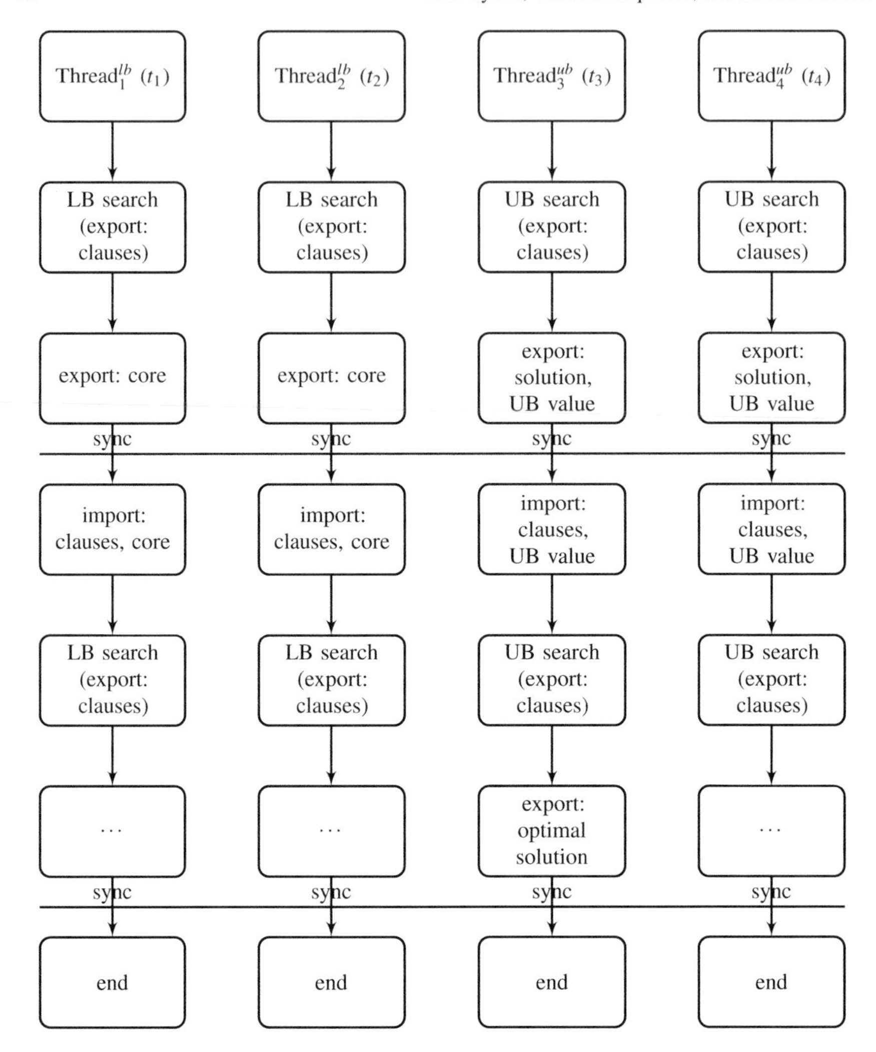

Fig. 3.10: Execution of the deterministic solver based on synchronization points

threads that have found a solution. As a result, all threads that are searching on the upper bound will have the same upper bound value after the synchronization point.

Learned clauses are also imported at synchronization points. Each thread imports the learned clauses that were exported by the remaining threads since the last synchronization point. Note that threads searching on the lower bound can also selectively import learned clauses from threads that are performing an upper bound search. The converse is also true [57]. In order to guarantee deterministic behavior, learned clauses must be imported in the same order. In our case, learned clauses are imported using ascending order with respect to the thread identifiers. Note that this is crucial to ensure that all procedures (such as unit propagation, conflict detection and clause learning, among others), maintain their behavior for different runs of the solver. Otherwise, the search space could be explored in a different order, thus resulting in non-deterministic behavior.

In addition, we must also guarantee the determinism of the solution reported. For a given problem instance, the variable assignments of the optimal solution that the solver outputs must always be the same for all runs of the solver. Every time a new solution is exported, it is only recorded if its corresponding value is smaller than the best value found so far. If the new solution has the same value as the current best value, then the thread identifier is used to decide whether the new solution is recorded or not. If the identifier of the exporting thread is smaller than the identifier of the thread where the previous solution was found, then the new solution is recorded. Otherwise, it is discarded. Finally, a thread stops when it proves optimality. However, the remaining threads are only terminated when their next synchronization point is reached. This is done to guarantee the determinism of the reported solution, since new optimal solutions may be found by the threads still running.

In the remainder of the section, we denote the synchronization scheme described above as *standard synchronization*. In this kind of synchronization, the threads that are searching on the lower bound reach a synchronization point every time a new core is found.

3.4.2.1 Standard Synchronization

As described in the previous section, the deterministic solver is based on the existence of synchronization points. However, a deterministic measure must be used to define synchronization points. Otherwise, the solver would remain non-deterministic. Hamadi et al. [35] proposed using the number of conflicts as a measure for defining the synchronization points. A simple strategy is to use a static number of conflicts to determine when a thread should reach a synchronization point. For example, each thread performs k conflicts before reaching the next synchronization point.

There is a trade-off between having a large number of synchronization points and how often the information is exchanged between threads. If k is too large, then the number of synchronization points is low but learned clauses and information regarding the bounds is exchanged less frequently. On the other hand, if k is too small then the number of synchronization points is higher which may degrade the performance of the solver. Note that other deterministic measures could be used instead of the number of conflicts, such as the number of propagations and the number of decisions.

An experimental evaluation [58, 63] has shown that the instantiation of k to $1,000$ conflicts achieves a good balance. The overall performance for other values (e.g., 100 and 10,000) was worse than when using $1,000$ conflicts. Moreover, the results for other measures were similar to using conflicts (e.g., variable assignments, decisions, etc.). More interestingly, the evaluation showed that the number of instances solved by the deterministic solver is comparable to the number of instances solved by the non-deterministic solver. On the other hand, the average idle time of each thread is high. Hence, it is expected that reducing the idle time will lead to an improvement in the performance of the deterministic solver. Next, different synchronization strategies are presented with the goal of reducing the idle time of the deterministic solver.

3.4.2.2 Period Synchronization

For problem instances with a large number of cores, standard synchronization may result in high idle times since at most one core can be found by each thread within each period. For example, consider a given instance having an optimal solution with value 100. Assume that the lower bound search is able to solve this instance very quickly by finding 100 cores, each with weight 1, whereas the upper bound search is unable to solve this instance. In the standard synchronization approach, if we set k to 1,000 conflicts then every time a core is found one has to wait for the threads searching on the upper bound to reach 1,000 conflicts. In this case, the lower bound threads would have to wait for the upper bound threads to reach more than 100,000 conflicts before finding a solution to the formula. This can require a prohibitive amount of time, which is a critical issue in the standard synchronization approach.

An alternative approach is to synchronize the threads that are searching on the lower bound only when they also reach the number of conflicts k that defines the length of the period. However, notice that in this case more than one core may be found between two synchronization points.

In this new approach, once all threads reach a synchronization point, the thread with the largest lower bound value is selected. (If two threads have the same lower bound value, then the thread with the smallest identifier is chosen.) In order to synchronize the lower bound threads, the chosen thread will export the cores that were found in the previous period to the remaining threads that are using an unsatisfiability-based approach.

In the remainder of the section we denote the kind of synchronization just described as *period synchronization* since the threads that are searching on the lower bound only stop at the end of each period. The main difference between standard and period synchronization is that in the standard synchronization at most one core is found by each thread during each period, whereas in the period synchronization it is possible to find a larger number of cores in a single period.

The main difference between the different synchronization techniques is shown by the instances where the lower bound search is crucial for finding the optimal solution. An experimental evaluation [58, 63] compared the standard synchronization with the period synchronization on instances that were solved optimally by the lower bound search on both deterministic solvers. For most of these instances, the period synchronization clearly outperforms the standard synchronization. Observe that if the optimal value is small then the standard synchronization may outperform the period synchronization. This is due to the fact that the idle time resulting from the synchronization for each core found is not significant. However, this case does not seem to be common and is restricted to only a few outliers.

3.4.2.3 Dynamic Synchronization

The main problem of a static strategy using a fixed number of conflicts k to define the length of a period is that different threads have different search behaviors and reach k conflicts at very different times. Therefore, using a static strategy may lead to high idle times at each synchronization point for the faster threads. This problem is further accentuated in parallel MaxSAT since the size of the formula can differ substantially between threads. For example, threads that search on the upper bound of the optimal solution and use CNF encodings to encode the constraint on the upper bound value may have a formula that is several times larger than working formulas in other threads.

An alternative approach that tries to minimize idle times is to use a dynamic strategy. In this case, each thread updates the number of conflicts that is required in order to reach the next synchronization point. Hamadi et al. [35] proposed to use the number of clauses learned in each thread for dynamically updating the necessary number of conflicts to reach the next synchronization point. A similar approach could also be used for parallel MaxSAT. Note that in deterministic parallel SAT all threads are initially run on the same formula. Therefore, all threads are initially working on a formula with the same size. However, as previously mentioned, this is not the case for our deterministic parallel MaxSAT solver. In fact, the size of the working formula can be substantially different in each thread. As a result, we propose to take into account the initial number of clauses and the number of learned clauses in each thread.

Let φ_i and ϕ_i denote the set of initial clauses and the set of learned clauses in the working formula of thread i, respectively. Consider that at synchronization point p, thread i has $|\varphi_i + \phi_i|$ clauses. Let m be the maximum number of clauses between all threads and k the number of conflicts to reach the first synchronization point. The number of conflicts that thread i needs to reach the next synchronization point is given by:

$$sync_i^{p+1} = \lceil k + (1 - \frac{|\varphi_i + \phi_i|}{m}) \times k \rceil$$

Threads that have more clauses will have smaller periods, whereas threads that have fewer clauses will have larger periods. The goal is to balance the number of conflicts required by each thread to reach the next synchronization point in an attempt to reduce the idle time of each thread.

An experimental evaluation [58, 63] has shown that the dynamic synchronization approach and the period synchronization solve the same number of instances. However, the average number of synchronization points decreases with the dynamic synchronization. Similarly, the percentage of idle time also decreases with the dynamic synchronization. The idle time was reduced significantly and the solving time improved.

3.4.3 Comparison Between Non-deterministic and Deterministic Solvers

A thorough experimental evaluation was performed to compare the non-deterministic version with the deterministic versions that use standard synchronization, period synchronization, and dynamic synchronization [58, 63]. All deterministic versions solved the same number of instances. On the other hand, the non-deterministic version was able to solve a few more instances than the deterministic versions. However, even though the non-deterministic solver is more efficient, the run times of the deterministic solver are comparable to the ones of the non-deterministic solver.

The comparison between the run times of the non-deterministic and dynamic deterministic solvers on industrial and crafted instances showed that the dynamic deterministic solver is clearly slower than the non-deterministic solver for instances that the non-deterministic solver solves in a few seconds. For these instances, it is expected that the overhead of synchronization will dominate the performance of the deterministic solver.

However, if we consider instances that required more time to be solved, then the performance of the deterministic solver is comparable to the performance of the non-deterministic solver. Moreover, for some instances the deterministic solver can actually outperform the non-deterministic solver. Observe that the two solvers perform different searches. Therefore, it is possible for the deterministic solver to outperform the non-deterministic solver. Nevertheless, this was not the case for the majority of the instances considered in the evaluation.

3.5 Research Directions

Parallel MaxSAT solvers [57, 61, 69] are taking their first steps towards using multicore and distributed architectures to speed up MaxSAT solving. However, there are still several challenges that need to be overcome in order for parallel MaxSAT solvers to become the staple of MaxSAT solving. Two of the most important challenges to be tackled in the near future are: (i) *scalability* and (ii) *limited clause sharing*.

3.5.1 Scalability

Current portfolio-based parallel MaxSAT solvers perform well for a small number of workers [57, 61] but lack scalability when the number of workers increases. One of the reasons for the lack of scalability is the difficulty of finding complementary approaches for a portfolio when the number of workers is large. Future research directions may involve diversifying the search through different MaxSAT algorithms for both lower bound search as well as upper bound search. Another research direction would be to build a hybrid version between the splitting and the portfolio approaches presented in Section 3.3. One could start with a splitting strategy and when the interval between the lower and upper bound values becomes small change to a portfolio approach. A hybrid approach may be more suitable for a large number of workers since it is possible to reduce the values of the optimal solution to a small interval.

Splitting approaches tend to be more scalable than portfolio-based approaches for parallel MaxSAT solving. Recently, there has been some work on distributed parallel MaxSAT that has shown scalability beyond 32 workers [69]. Pursuing this direction and improving the splitting strategy may further increase the scalability of parallel MaxSAT solvers.

Another alternative is to develop new approaches beyond splitting and portfolio. For example, partitioning techniques have been successfully used for improving sequential MaxSAT solvers [60, 62, 70]. These approaches are based on partitioning the formula into disjoint subformulas that can be solved independently and then merged to find a global optimal solution. Exploiting the inherently parallel structure of partitioning-based approaches is another research direction that may improve the scalability of parallel MaxSAT solvers.

3.5.2 Clause Sharing

Clause sharing can significantly improve the performance of parallel solvers. However, there is a trade-off between the amount of information exchanged and the benefit gained from this information. Current parallel MaxSAT solvers either use a limited form of clause sharing for multicore architectures [57, 61] or do not share learned clauses for distributed architectures [69].

To further improve the performance of MaxSAT solvers it is necessary to develop new clause-sharing heuristics that take into consideration distributed architectures with a large number of workers. The freezing heuristic presented in Section 3.3.3 is a first step towards controlling the amount of useful information that is exchanged between workers. However, for distributed architectures, the communication needs to be more sparse than for multicore architectures. Since most clause-sharing heuristics were proposed for multicore architectures with a small number of workers, it is necessary to improve clause-sharing heuristics for architectures with hundreds of workers. Clause sharing is intertwined with the scalability of parallel MaxSAT solvers, and its improvement will lead to more scalable parallel MaxSAT solvers.

Acknowledgments

This work was supported by national funds through Fundação para a Ciência e a Tecnologia (FCT) with reference UID/CEC/50021/2013.

References

[1] Achá, R.J.A., Nieuwenhuis, R.: Curriculum-based course timetabling with SAT and MaxSAT. Annals of Operations Research **218**(1), 71–91 (2014)

[2] An, X., Zhang, T., Fujita, H., Hasegawa, R.: QMaxSAT: A Partial Max-SAT Solver. Journal on Satisfiability, Boolean Modeling and Computation **8**, 95–100 (2012)

[3] Ansótegui, C., Bonet, M.L., Gabàs, J., Levy, J.: Improving WPM2 for (Weighted) Partial MaxSAT. In: Proc. International Conference on Principles and Practice of Constraint Programming, pp. 117–132. Springer (2013)

[4] Ansótegui, C., Bonet, M.L., Levy, J.: On Solving MaxSAT Through SAT. In: Proc. International Conference of the Catalan Association for Artificial Intelligence, pp. 284–292. IOS Press (2009)

[5] Ansótegui, C., Bonet, M.L., Levy, J.: Solving (Weighted) Partial MaxSAT through Satisfiability Testing. In: Proc. International Conference on Theory and Applications of Satisfiability Testing, pp. 427–440. Springer (2009)

[6] Ansótegui, C., Bonet, M.L., Levy, J.: A New Algorithm for Weighted Partial MaxSAT. In: Proc. AAAI Conference on Artificial Intelligence, pp. 3–8. AAAI Press (2010)

[7] Ansótegui, C., Gabàs, J.: Solving (Weighted) Partial MaxSAT with ILP. In: Proc. International Conference on Integration of AI and OR Techniques in Constraint Programming for Combinatorial Optimization Problems, pp. 403–409. Springer (2013)

[8] Ansótegui, C., Manyà, F.: Mapping problems with finite-domain variables into problems with Boolean variables. In: Proc. International Conference on Theory and Applications of Satisfiability Testing, pp. 1–15. Springer (2004)

[9] Argelich, J., Berre, D.L., Lynce, I., Marques-Silva, J., Rapicault, P.: Solving Linux Upgradeability Problems Using Boolean Optimization. In: Workshop on Logics for Component Configuration, pp. 11–22. Conference Proceedings (2010)

[10] Argelich, J., Li, C.M., Manyà, F.: An improved exact solver for Partial Max-SAT. In: Proc. of the International Conference on Nonconvex Programming: Local and Global Approaches, pp. 230–231. Conference Proceedings (2007)

[11] Argelich, J., Manyà, F.: Partial Max-SAT Solvers with Clause Learning. In: Proc. International Conference on Theory and Applications of Satisfiability Testing, pp. 28–40. Springer (2007)

[12] Asín, R., Nieuwenhuis, R., Oliveras, A., Rodríguez-Carbonell, E.: Cardinality Networks: a theoretical and empirical study. Constraints **16**(2), 195–221 (2011)

[13] Audemard, G., Hoessen, B., Jabbour, S., Lagniez, J.M., Piette, C.: Revisiting Clause Exchange in Parallel SAT Solving. In: Proc. International Conference on Theory and Applications of Satisfiability Testing, pp. 200–213. Springer (2012)
[14] Audemard, G., Lagniez, J.M., Mazure, B., Sais, L.: On Freezing and Reactivating Learnt Clauses. In: International Conference on Theory and Applications of Satisfiability Testing, pp. 188–200. Springer (2011)
[15] Audemard, G., Simon, L.: Predicting Learnt Clauses Quality in Modern SAT Solvers. In: Proc. International Joint Conferences on Artificial Intelligence, pp. 399–404. IJCAI/AAAI Press (2009)
[16] Bailleux, O., Boufkhad, Y.: Efficient CNF Encoding of Boolean Cardinality Constraints. In: Proc. International Conference on Principles and Practice of Constraint Programming, pp. 108–122. Springer (2003)
[17] Bailleux, O., Boufkhad, Y., Roussel, O.: New Encodings of Pseudo-Boolean Constraints into CNF. In: Proc. International Conference on Theory and Applications of Satisfiability Testing, pp. 181–194. Springer (2009)
[18] Berre, D.L., Parrain, A.: The Sat4j library, release 2.2. JSAT **7**(2-3), 59–6 (2010)
[19] Bjørner, N., Phan, A., Fleckenstein, L.: *ν*Z - An Optimizing SMT Solver. In: Proc. Tools and Algorithms for Construction and Analysis of Systems, pp. 194–199. Springer (2015)
[20] Böhm, M., Speckenmeyer, E.: A Fast Parallel SAT-Solver - Efficient Workload Balancing. Annals of Mathematics and Artificial Intelligence **17**, 381–400 (1996)
[21] Bonet, M.L., Levy, J., Manyà, F.: Resolution for Max-SAT. Artificial Intelligence **171**(8–9), 606–618 (2007)
[22] Chen, Y., Safarpour, S., Marques-Silva, J., Veneris, A.G.: Automated Design Debugging With Maximum Satisfiability. IEEE Transactions on CAD of Integrated Circuits and Systems **29**(11), 1804–1817 (2010)
[23] Darras, S., Dequen, G., Devendevill, L., Li, C.M.: On Inconsistent Clause-Subsets for Max-SAT Solving. In: Proc. International Conference on Principles and Practice of Constraint Programming, pp. 225–240. Springer (2007)
[24] Davies, J., Bacchus, F.: Solving MAXSAT by Solving a Sequence of Simpler SAT Instances. In: Proc. International Conference on Principles and Practice of Constraint Programming, pp. 225–239. Springer (2011)
[25] Davies, J., Bacchus, F.: Postponing optimization to speed up MAXSAT solving. In: Proc. International Conference on Principles and Practice of Constraint Programming, pp. 247–262. Springer (2013)
[26] Eén, N., Sörensson, N.: An extensible SAT-solver. In: Proc. International Conference on Theory and Applications of Satisfiability Testing, pp. 502–518. Springer (2003)
[27] Eén, N., Sörensson, N.: Temporal induction by incremental SAT solving. Electronic Notes in Theoretical Computer Science **89**(4), 543–560 (2003)
[28] Eén, N., Sörensson, N.: Translating pseudo-Boolean constraints into SAT. Journal on Satisfiability, Boolean Modeling and Computation **2**, 1–26 (2006)

[29] Feng, Y., Bastani, O., Martins, R., Dillig, I., Anand, S.: Automated Synthesis of Semantic Malware Signatures using Maximum Satisfiability. In: Network and Distributed System Security Symposium. The Internet Society (2017)
[30] Frisch, A.M., Peugniez, T.J., Doggett, A.J., Nightingale, P.: Solving Non-Boolean Satisfiability Problems with Stochastic Local Search: A Comparison of Encodings. Journal of Automated Reasoning **35**(1-3), 143–179 (2005)
[31] Fu, Z., Malik, S.: On solving the partial MAX-SAT problem. In: Proc. International Conference on Theory and Applications of Satisfiability Testing, pp. 252–265. Springer (2006)
[32] Garey, M.R., Johnson, D.S.: Computers and Intractability: A Guide to the Theory of NP-Completeness. W. H. Freeman (1979)
[33] Gent, I.P., Nightingale, P.: A new encoding of All Different into SAT. In: International Workshop on Modelling and Reformulating Constraint Satisfaction Problems. Conference Proceedings (2004)
[34] Graça, A., Lynce, I., Marques-Silva, J., Oliveira, A.L.: Efficient and Accurate Haplotype Inference by Combining Parsimony and Pedigree Information. In: Algebraic and Numeric Biology, pp. 38–56. Springer (2010)
[35] Hamadi, Y., Jabbour, S., Piette, C., Sais, L.: Deterministic Parallel DPLL. Journal on Satisfiability, Boolean Modeling and Computation **7**(4), 127–132 (2011)
[36] Hamadi, Y., Jabbour, S., Sais, L.: Control-Based Clause Sharing in Parallel SAT Solving. In: Proc. International Joint Conferences on Artificial Intelligence, pp. 499–504. IJCAI/AAAI Press (2009)
[37] Hamadi, Y., Jabbour, S., Sais, L.: ManySAT: a Parallel SAT Solver. Journal on Satisfiability, Boolean Modeling and Computation **6**(4), 245–262 (2009)
[38] Heras, F., Morgado, A., Marques-Silva, J.: Core-Guided Binary Search Algorithms for Maximum Satisfiability. In: Proc. AAAI Conference on Artificial Intelligence, pp. 36–41. AAAI Press (2011)
[39] Heule, M.J., Kullmann, O., Wieringa, S., Biere, A.: Cube and Conquer: Guiding CDCL SAT Solvers by Lookaheads. In: Hardware and Software: Verification and Testing, pp. 50–65. Springer (2012)
[40] Hölldobler, S., Manthey, N., Steinke, P.: A compact encoding of pseudo-Boolean constraints into SAT. In: KI 2013: Advances in Artificial Intelligence, pp. 107–118. Springer (2012)
[41] Jose, M., Majumdar, R.: Cause clue clauses: error localization using maximum satisfiability. In: Proc. Conference on Programming Language Design and Implementation, pp. 437–446. ACM Press (2011)
[42] Joshi, S., Martins, R., Manquinho, V.: Generalized Totalizer Encoding for Pseudo-Boolean Constraints. In: International Conference on Principles and Practice of Constraint Programming, pp. 200–209. Springer (2015)
[43] Klieber, W., Kwon, G.: Efficient CNF Encoding for Selecting 1 from N Objects. In: International Workshop on Constraints in Formal Verification. Conference Proceedings (2007)
[44] Li, C.M., Manyà, F.: MaxSAT, Hard and Soft Constraints. In: Handbook of Satisfiability, pp. 613–631. IOS Press (2009)

[45] Li, C.M., Manyà, F., Planes, J.: Exploiting unit propagation to compute lower bounds in branch and bound Max-SAT solvers. In: Proc. International Conference on Principles and Practice of Constraint Programming, pp. 403–414. Springer (2005)
[46] Li, C.M., Manyà, F., Planes, J.: New inference rules for Max-SAT. Journal of Artificial Intelligence Research **30**, 321–359 (2007)
[47] Lin, H., Su, K.: Exploiting inference rules to compute lower bounds for MAX-SAT solving. In: Proc. International Joint Conferences on Artificial Intelligence, pp. 2334–2339. IJCAI/AAAI Press (2007)
[48] Lin, H., Su, K., Li, C.M.: Within-problem Learning for Efficient Lower Bound Computation in Max-SAT Solving. In: Proc. AAAI Conference on Artificial Intelligence, pp. 351–356. AAAI Press (2008)
[49] Mancinelli, F., Boender, J., Cosmo, R.D., Vouillon, J., Durak, B., Leroy, X., Treinen, R.: Managing the Complexity of Large Free and Open Source Package-Based Software Distributions. In: Proc. International Conference on Automated Software Engineering, pp. 199–208. IEEE Computer Society Press (2006)
[50] Manquinho, V., Marques-Silva, J., Planes, J.: Algorithms for Weighted Boolean Optimization. In: Proc. International Conference on Theory and Applications of Satisfiability Testing, pp. 495–508. Springer (2009)
[51] Manquinho, V., Martins, R., Lynce, I.: Improving Unsatisfiability-Based Algorithms for Boolean Optimization. In: Proc. International Conference on Theory and Applications of Satisfiability Testing, pp. 181–193. Springer (2010)
[52] Marques-Silva, J., Planes, J.: On using unsatisfiability for solving maximum satisfiability. CoRR **abs/0712.1097** (2007). URL `http://arxiv.org/abs/0712.1097`
[53] Marques-Silva, J., Sakallah, K.: GRASP: A New Search Algorithm for Satisfiability. In: Proc. International Conference on Computer-Aided Design, pp. 220–227. IEEE Computer Society Press (1996)
[54] Marques-Silva, J., Sakallah, K.: GRASP: A Search Algorithm for Propositional Satisfiability. IEEE Transactions on Computers **48**(5), 506–521 (1999)
[55] Martins, R., Joshi, S., Manquinho, V., Lynce, I.: Incremental Cardinality Constraints for MaxSAT. In: Proc. International Conference on Principles and Practice of Constraint Programming, pp. 531–548. Springer (2014)
[56] Martins, R., Joshi, S., Manquinho, V., Lynce, I.: On Using Incremental Encodings in Unsatisfiability-based MaxSAT Solving. Journal on Satisfiability, Boolean Modeling and Computation **9**, 59–81 (2015)
[57] Martins, R., Manquinho, V., Lynce, I.: Exploiting Cardinality Encodings in Parallel Maximum Satisfiability. In: Proc. International Conference on Tools with Artificial Intelligence, pp. 313–320. IEEE Computer Society Press (2011)
[58] Martins, R., Manquinho, V., Lynce, I.: Clause Sharing in Deterministic Parallel Maximum Satisfiability. In: RCRA International Workshop on Experimental Evaluation of Algorithms for solving problems with combinatorial explosion. Conference Proceedings (2012)

[59] Martins, R., Manquinho, V., Lynce, I.: Clause Sharing in Parallel MaxSAT. In: Proc. Learning and Intelligent Optimization Conference, pp. 455–460. Springer (2012)
[60] Martins, R., Manquinho, V., Lynce, I.: On Partitioning for Maximum Satisfiability. In: Proc. European Conference on Artificial Intelligence, pp. 913–914. IOS Press (2012)
[61] Martins, R., Manquinho, V., Lynce, I.: Parallel Search for Maximum Satisfiability. AI Communications **25**(2), 75–95 (2012)
[62] Martins, R., Manquinho, V., Lynce, I.: Community-based Partitioning for MaxSAT Solving. In: Proc. International Conference on Theory and Applications of Satisfiability Testing, pp. 182–191. Springer (2013)
[63] Martins, R., Manquinho, V., Lynce, I.: Deterministic Parallel MaxSAT Solving. International Journal on Artificial Intelligence Tools **24**(3) (2015)
[64] Martins, R., Manquinho, V.M., Lynce, I.: Open-WBO: A Modular MaxSAT Solver. In: Proc. International Conference on Theory and Applications of Satisfiability Testing, pp. 438–445. Springer (2014)
[65] Morgado, A., Heras, F., Liffiton, M.H., Planes, J., Marques-Silva, J.: Iterative and core-guided MaxSAT solving: a survey and assessment. Constraints **18**(4), 478–534 (2013)
[66] Morgado, A., Ignatiev, A., Marques-Silva, J.: MSCG: Robust Core-Guided MaxSAT Solving. Journal on Satisfiability, Boolean Modeling and Computation **9**, 129–134 (2015)
[67] Moskewicz, M., Madigan, C., Zhao, Y., Zhang, L., Malik, S.: Chaff: Engineering an Efficient SAT Solver. In: Design Automation Conference, pp. 530–535. ACM (2001)
[68] Narodytska, N., Bacchus, F.: Maximum Satisfiability Using Core-Guided MaxSAT Resolution. In: Proc. AAAI Conference on Artificial Intelligence, pp. 2717–2723. AAAI Press (2014)
[69] Neves, M., Lynce, I., Manquinho, V.: DistMS: A Non-Portfolio Distributed Solver for Maximum Satisfiability. In: Proc. International Conference on Tools with Artificial Intelligence. IEEE Computer Society Press (2016)
[70] Neves, M., Martins, R., Janota, M., Lynce, I., Manquinho, V.M.: Exploiting resolution-based representations for MaxSAT solving. In: Proc. International Conference on Theory and Applications of Satisfiability Testing, pp. 272–286. Springer (2015)
[71] Ogawa, T., Liu, Y., Hasegawa, R., Koshimura, M., Fujita, H.: Modulo Based CNF Encoding of Cardinality Constraints and Its Application to MaxSAT Solvers. In: Proc. International Conference on Tools with Artificial Intelligence, pp. 9–17. IEEE Computer Society (2013)
[72] Papadimitriou, C.M.: Computational complexity. Addison-Wesley, Reading, Massachusetts (1994)
[73] Plaza, S., Kountanis, I., Andraus, Z., Bertacco, V., Mudge, T.: Advances and Insights into Parallel SAT Solving. In: Internacional Workshop on Logic & Synthesis, pp. 188–194. Conference Proceedings (2006)

[74] Prestwich, S.: Variable Dependency in Local Search: Prevention is Better than Cure. In: Proc. International Conference on Theory and Applications of Satisfiability Testing, pp. 107–120. Springer (2007)
[75] Saikko, P., Berg, J., Järvisalo, M.: LMHS: A SAT-IP Hybrid MaxSAT Solver. In: Proc. International Conference on Theory and Applications of Satisfiability Testing, pp. 539–546. Springer (2016)
[76] Schubert, T., Lewis, M., Becker, B.: PaMira - A Parallel SAT Solver with Knowledge Sharing. In: Workshop on Microprocessor Test and Verification, pp. 29–36. Conference Proceedings (2005)
[77] Schubert, T., Lewis, M., Becker, B.: PaMiraXT: Parallel SAT Solving with Threads and Message Passing. Journal on Satisfiability, Boolean Modeling and Computation **6**, 203–222 (2009)
[78] Singer, D., Monnet, A.: JaCk-SAT: A New Parallel Scheme to Solve the Satisfiability Problem (SAT) Based on Join-and-Check. In: Proc. Parallel Processing and Applied Mathematics, pp. 249–258. Springer (2008)
[79] Sinz, C.: Towards an Optimal CNF Encoding of Boolean Cardinality Constraints. In: Proc. International Conference on Principles and Practice of Constraint Programming, pp. 827–831. Springer (2005)
[80] Stump, A., Sutcliffe, G., Tinelli, C.: StarExec: A Cross-Community Infrastructure for Logic Solving. In: Proc. International Joint Conference on Automated Reasoning, pp. 367–373. Springer (2014)
[81] Warners, J.P.: A linear-time transformation of linear inequalities into conjunctive normal form. Information Processing Letters **68**(2), 63–69 (1998)
[82] Zhang, H., Bonacina, M.P., Hsiang, J.: PSATO: a Distributed Propositional Prover and Its Application to Quasigroup Problems. Journal of Symbolic Computation **21**, 543–560 (1996)
[83] Zhang, L., Madigan, C.F., Moskewicz, M.W., Malik, S.: Efficient Conflict Driven Learning in Boolean Satisfiability Solver. In: Proc. International Conference on Computer-Aided Design, pp. 279–285. IEEE Computer Society Press (2001)

Chapter 4
Parallel Solving of Quantified Boolean Formulas

Florian Lonsing and Martina Seidl

Abstract Quantified Boolean formulas (QBFs) extend propositional logic by universal and existential quantifiers over the propositional variables. In the same way as the satisfiability problem of propositional logic is the archetypical problem for NP, the satisfiability problem of QBFs is the archetypical problem for PSPACE. Hence, QBFs provide an attractive framework for encoding many applications from verification, artificial intelligence, and synthesis, thus motivating the quest for efficient solving technology. Already in the very early stages of QBF solving history, attempts have been made to parallelize the solving process, either by splitting the search space or by portfolio-based approaches. In this chapter, we review and compare approaches for solving QBFs in parallel.

4.1 Introduction

Since the late 1990s, there has been impressive progress in research on solving the propositional satisfiability problem (SAT) (see Chapter 1, Parallel Satisfiability). The boost in the performance of SAT solvers enabled routine applications of SAT to large-scale industrial problems [13, 19, 85]. In practice, nowadays SAT solvers are capable of solving formulas containing hundreds of thousands of variables and millions of clauses. This is in contrast to the computational intractability that follows from the NP-completeness of SAT.

Florian Lonsing
Institute of Information Systems, TU Wien,
Favoritenstr. 9-11, 1040 Wien, Austria, e-mail: florian.lonsing@tuwien.ac.at,
Supported by the Austrian Science Fund (FWF) under grant S11409-N23.

Martina Seidl
Institute for Formal Models and Verification, JKU Linz,
Altenbergerstr. 69, 4040 Linz, Austria, e-mail: martina.seidl@jku.at,
Supported by the Austrian Science Fund (FWF) under grant S11408-N23.

Y. Hamadi und L. Sais (eds.), *Handbook of Parallel Constraint Reasoning*,
https://doi.org/10.1007/978-3-319-63516-3_4

Motivated by the success story of SAT solving, problems from complexity classes beyond NP became the focus of intensive research.[1] The *polynomial hierarchy* [64, 82] is a theoretical framework to describe the complexity of problems beyond NP. Examples of problems in the polynomial hierarchy are conformant planning [75], problems related to answer set programming [27], or the computation of minimal unsatisfiable subformulas (MUSes) [54].

A natural extension of SAT is QSAT, the satisfiability problem of *quantified Boolean formulas* (QBFs) [32, 47]. In a nutshell, QBFs are propositional formulas that additionally may contain existential ($\exists$) and universal ($\forall$) quantifiers over the propositional variables. QBFs can be used to encode any problem in the polynomial hierarchy. For example, the QBFs

$$\forall x \exists y.((x \vee \neg y) \wedge (\neg x \vee y)) \tag{4.1}$$

and

$$\exists y \forall x.((x \vee \neg y) \wedge (\neg x \vee y)) \tag{4.2}$$

encode the equivalence of the variables x and y by the propositional CNF $((x \vee \neg y) \wedge (\neg x \vee y))$ under the quantifier prefixes $\forall x \exists y$ and $\exists y \forall x$, respectively. Intuitively, the QBF 4.1 asks whether for all possible assignments of variable x there exists an assignment of variable y such that the propositional CNF evaluates to true. In contrast to that, the QBF 4.2 asks whether there exists an assignment of y such that for all assignments of x the propositional CNF evaluates to true.

Like in propositional logic, the variables in a QBF are interpreted over the Boolean domain. Obviously, the QBF 4.1 is satisfiable since the assignment of the existential variable y can be selected depending on the assignment of the universal variable x in order to satisfy the CNF. The QBF 4.2 is unsatisfiable since it differs from the QBF 4.1 in the ordering of the variables in the quantifier prefix. Due to the ordering, in the QBF 4.2 the value of y is fixed for any value of x. Hence, in general the ordering of variables in the quantifier prefix impacts the satisfiability of a QBF.

When solving a propositional formula using a SAT solver, the solver can stop as soon as it finds an assignment to the variables which satisfies the formula. When solving a QBF, however, finding one assignment which satisfies its propositional part is not enough to show the satisfiability of the QBF. The presence of universal and existential quantifiers in a QBF and the ordering of variables in the quantifier prefix give rise to tree-shaped (counter)models for witnessing (un)satisfiability. These (counter)models represent the different choices of variable assignments that have to be made depending on the quantifier types and the ordering of variables. Figure 4.1 shows a model of the QBF 4.1 and a countermodel of the QBF 4.2.

In practice, tree-shaped models (countermodels) are represented as functions that provide strategies to assign the existential (universal) variables with respect to universal (existential) ones if the considered formula is satisfiable (unsatisfiable). For the QBF 4.1 the function $f_y(x) := x$ yields the strategy to assign variable y to the same value as the previously assigned variable x in order to satisfy the CNF

[1] Beyond NP research community website (June 2017): `http://beyondnp.org/`

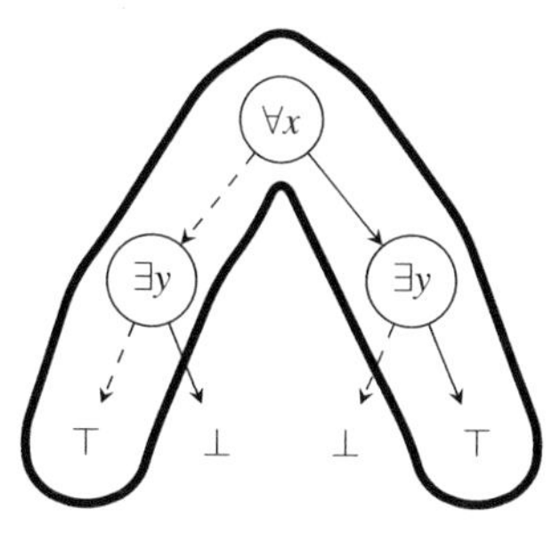

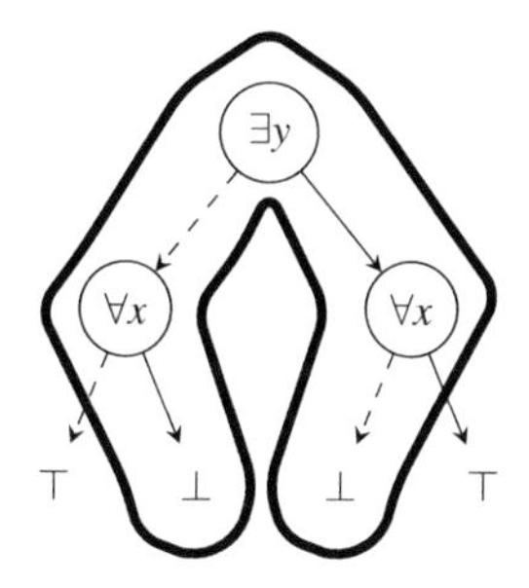

Fig. 4.1: Tree-shaped model (left) and countermodel (right) illustrating the satisfiability of the QBF 4.1 and the unsatisfiability of the QBF 4.2, respectively. (Counter)models are special subtrees of a formula's assignment tree. Dashed (solid) edges indicate that the variable in the source node is set to false (true). In the model every assignment along a path satisfies the propositional CNF $((x \vee \neg y) \wedge (\neg x \vee y))$ of the QBF, whereas in the countermodel all such assignments falsify the CNF

$((x \vee \neg y) \wedge (\neg x \vee y))$. Thus function $f_y(x)$ witnesses the satisfiability of the QBF 4.1. In a dual way, the function $f_x(y) := \neg y$ is a witness for the unsatisfiability of the QBF 4.2 as it represents a strategy to assign x to the opposite value of y in order to falsify the CNF.

As a consequence of extending propositional logic with quantifiers over the propositional variables to obtain the language of QBFs, the QBF satisfiability problem becomes PSPACE-complete [81]. The use of quantifiers allows for QBF encodings of problems that potentially are exponentially more succinct than the corresponding SAT encodings. Due to this property, QBFs are an attractive language for encoding and solving many practically relevant problems from domains such as, for example, formal verification [40], synthesis [14], or artificial intelligence [26, 75] (see [10] for a detailed survey). For solving such problems in practice, efficient QBF solvers are highly desirable.

Since the year 2000, there has been substantial progress in the development of efficient QBF solvers. Traditional QBF solvers can be classified into one of two dominant solving paradigms that have emerged: (1) search-based solving and (2) expansion-based solving.

Search-based QBF solvers implement a QBF-specific extension of the DPLL algorithm [18, 22] and of *conflict-driven clause learning (CDCL)* [66, 78, 79, 90]. CDCL is at the core of most modern SAT solvers. The QBF variant of CDCL, often referred to as QCDCL [35, 50, 91] implicitly searches for a tree-shaped (counter)model of the given QBF in the search space of all possible variable assignments. Thereby assignments encountered during the search that falsify the propositional part of the QBF, called *conflicts*, and assignments satisfying it, called *solutions*, are analysed. The analysis of conflicts and solutions allows the solver to learn new *clauses* (disjunctions of literals) and *cubes* (conjunctions of literals) for pruning the search space. QCDCL solvers may apply additional QBF-specific techniques such as duality-aware reasoning [36, 37] or the analysis of variable dependencies with respect to the quantifier structure of a QBF [57].

In contrast to SAT, where CDCL is almost the single dominant solving paradigm in practice, QCDCL-based QBF solving is complemented by *expansion-based QBF solving*. Expansion-based solvers rewrite a given QBF to a satisfiability-equivalent propositional formula by successively expanding the quantifiers [5, 12]. To counter the potential exponential blow-up of the formula size that may result from expansions, *counterexample-guided abstraction refinement (CEGAR)* [20] has proven to be powerful [42, 73]. With a CEGAR-based approach to expansion, the solver operates on an abstract representation of the formula and expands quantifiers lazily. This way, only those quantifiers are expanded which promise to be useful for solving the formula. Expansion has been found to be orthogonal to QCDCL from a proof complexity perspective [11, 43].

In SAT solving, typically expansion is not applied as a standalone approach to solving since the size of formulas to be solved is often prohibitive. Instead expansion is used as a pre- and inprocessing technique in a resource-bounded way [23, 45]. Inprocessing is an approach where preprocessing is dynamically interleaved with the search process in CDCL. Bounded expansion has also been applied successfully for QBF preprocessing [17].

Given the recently published literature on QBF solving, today the main focus of QBF solver development is still on sequential systems. In QBF solving there is no general consensus on which solving paradigm is superior in practice as the performance of solvers may be highly sensitive to the considered benchmarks (for example, see the results of related QBF evaluations and competitions [41, 60, 71]). The landscape of sequential QBF solving has changed and evolved as novel solving approaches have emerged, such as nested SAT solving [15], clause selection [44, 73] or the computation of functions that represent strategies of QBFs [72].

While parallelization is natural for most QBF-solving approaches, it also introduces additional complexity in solver engineering and development. Therefore it is not surprising that solver developers first focus on the implementation of stable and efficient sequential systems before facing the challenge of parallelization. Nevertheless, since the beginning of QBF solving in the early 2000s, several approaches have been investigated to parallelize QBF solvers and thus benefit from modern clusters and multicore processors. After a period of relatively little progress, the interest in parallel QBF solving has increased, which is reflected by the QBF competition *QBFEVAL'16* [71] held in 2016.[2] There, five different parallel solvers in six configurations participated, while in previous competitions the parallel-solving track had to be canceled due to the lack of participants.

In this chapter, we give an overview of previous and recent approaches to parallel QBF solving. To this end, we first review the necessary preliminaries related to QBFs and recapitulate relevant sequential-solving approaches. On this basis we first present general ideas of parallelization and then introduce and compare concrete approaches implemented in parallel QBF solvers. Finally, we conclude this chapter

[2] QBFEVAL'16 website: `http://www.qbflib.org/qbfeval16.php`

with a selection of challenges that have to be faced in order to make parallel QBF solving ready for applications in practice.

4.2 Background

In this section, we recapitulate syntax and semantics of QBFs and summarize the terminology used in the rest of this chapter.

The language $\mathscr{L}_{\mathscr{V}}$ of quantified Boolean formulas over a set of propositional variables $\mathscr{V}$ and truth constants $\top$ and $\bot$ is defined as the smallest set such that

1. if $x \in (\mathscr{V} \cup \{\top, \bot\})$ then $x \in \mathscr{L}_{\mathscr{V}}$;
2. if $\phi \in \mathscr{L}_{\mathscr{V}}$ then $\neg\phi \in \mathscr{L}_{\mathscr{V}}$;
3. if $\phi_1, \phi_2 \in \mathscr{L}_{\mathscr{V}}$ then $(\phi_1 \circ \phi_2) \in \mathscr{L}_{\mathscr{V}}$ where $\circ \in \{\vee, \wedge, \rightarrow, \leftrightarrow, \oplus\}$;
4. if $\phi \in \mathscr{L}_{\mathscr{V}}$ and $x \in \mathscr{V}$, then $(\mathsf{Q}x.\phi) \in \mathscr{L}_{\mathscr{V}}$ where $\mathsf{Q} \in \{\forall, \exists\}$.

If convenient and unambiguous, we omit parenthesis in QBFs $\phi \in \mathscr{L}_{\mathscr{V}}$. For a QBF $\mathsf{Q}x.\phi$, ϕ is the *scope* of the quantifier $\mathsf{Q}x$. A variable x is *free* in a QBF ϕ, if x does not occur in the scope of a quantifier $\mathsf{Q}x$ in ϕ. A QBF is *closed* if it does not contain any free variables. In the following, we consider only closed QBFs. Furthermore, we assume that for each $x \in \mathscr{V}$, a QBF contains at most one occurrence of $\mathsf{Q}x$. For $\exists x_1, \ldots, \exists x_n$ and $\forall y_1, \ldots, \forall y_n$ we also write $\exists X$ and $\forall Y$, respectively, where $X = \{x_1, \ldots, x_n\}$ and $Y = \{y_1, \ldots, y_n\}$. We define $\mathsf{var}(\phi) := \{x \mid \mathsf{Q}x \text{ occurs in } \phi, \mathsf{Q} \in \{\forall, \exists\}\}$.

A *literal* is a propositional variable $x \in \mathscr{V}$ or its negation $\neg x$. By $\neg l$ we denote the negation of literal l. Further, $\mathsf{var}(l) := x$ if $l = x$ or $l = \neg x$. A *clause* is a disjunction of literals. A *cube* is a conjunction of literals. A clause (cube) C is *tautological* (*contradictory*) if $\{x, \neg x\} \subseteq C$. A propositional formula is in *conjunctive normal form* (CNF) if it is a conjunction of clauses. A propositional formula is in *disjunctive normal form* (DNF) if it is a disjunction of cubes. When convenient, we interpret a formula in CNF (DNF) as a set of clauses (cubes) and clauses (cubes) as sets of literals.

A QBF ϕ is in *prenex conjunctive normal form* (PCNF) if it has the form $\Pi.\psi$ where $\Pi := \mathsf{Q}_1X_1, \ldots, \mathsf{Q}_nX_n$ is the prefix of ϕ and ψ is the matrix of ϕ. The matrix ψ is a propositional formula in CNF over the variables in Π. The variable sets X_i are pairwise disjoint and for $\mathsf{Q}_i \in \{\forall, \exists\}$, $\mathsf{Q}_i \neq \mathsf{Q}_{i+1}$. We define $\mathsf{var}(\Pi) := X_1 \cup \ldots \cup X_n$. The quantifier $\mathsf{quant}(\Pi, l)$ of literal l is Q_i if $\mathsf{var}(l) \in X_i$. Given literals l and k, then $l \leq_\Pi k$ if $\mathsf{quant}(\Pi, l) = \mathsf{Q}_i$ and $\mathsf{quant}(\Pi, k) = \mathsf{Q}_j$ and $i \leq j$. For example, QBFs 4.1 and 4.2 are in PCNF.

A QBF ϕ over variables $\mathscr{V}$ is in *negation normal form* (NNF) if (1) $\phi \in \mathscr{L}_{\mathscr{V}}$, (2) the negation symbol occurs only directly in front of variables or truth constants, and (3) the only binary connectives are conjunction ($\wedge$) and disjunction ($\vee$). Note that the NNF structure does not impose any restrictions on the positions of quantifiers.

A *partial assignment* of the variables $\mathsf{var}(\phi)$ of a QBF ϕ is a total mapping $A\colon \mathsf{var}(\phi) \mapsto \mathbb{B} \cup \{\mathsf{U}\}$, where $\mathbb{B} := \{\mathsf{T}, \mathsf{F}\}$ is the Boolean domain and U denotes that the assignment of a variable is undefined. A *full assignment* is a total mapping $A\colon \mathsf{var}(\phi) \mapsto \mathbb{B}$. Given an assignment A of QBF ϕ, we also write A as a set of literals $A = \{l_1, \ldots, l_n\}$ such that, for all $x \in \mathsf{var}(\phi)$, $x \in A$ if $A(x) = \mathsf{T}$, $\neg x \in A$ if $A(x) = \mathsf{F}$, and both $x \notin A$ and $\neg x \notin A$ if $A(x) = \mathsf{U}$. Then for any $l_i, l_j \in A$ with $i \neq j$, $\mathsf{var}(l_i) \neq \mathsf{var}(l_j)$.

For a QBF ϕ and an assignment A, $\phi[A]$ denotes the QBF ϕ *under* A which is obtained from ϕ as follows. For all $l \in A$ with $\mathsf{var}(l) = x$, the quantifier $\mathrm{Q}x$ is removed, any occurrence of x is replaced by $\top$ if $x \in A$ and by $\bot$ if $\neg x \in A$, followed by the usual simplifications of Boolean logic. For example, if $\phi := \Pi.\psi$ is in PCNF, then for all $l \in A$ any clause C with literal $l \in C$ is deleted, any occurrence of literal $\neg l$ is removed, and the variable $\mathsf{var}(l)$ of l and its quantifier $\mathtt{quant}(\Pi, l)$ are removed from the prefix. If $\phi[A]$ simplifies to $\top$ (written as $\phi[A] = \top$) then A is called a *satisfying assignment*. If $\phi[A]$ simplifies to $\bot$ (written as $\phi[A] = \bot$) then A is called a *falsifying assignment*.

An *assignment tree* of a QBF ϕ is a complete binary tree of depth $|\mathsf{var}(\phi)| + 1$ where the internal nodes of each level are associated with a variable of ϕ. The levels reflect the order of the quantifiers in the formula. The outgoing edges of an internal node labeled by variable x are associated with $\neg x$ and x, indicating that x is set to false and to true, respectively. A path from the root of the tree to a leaf represents a particular variable assignment. The leaves are labeled by the truth value of ϕ under the assignment of the respective path. Figure 4.1 shows examples of two assignment trees. The highlighted subtrees of the assignment trees represent a model and a countermodel, respectively.

The *semantics* of QBFs is defined recursively based on the syntactic structure as follows. The QBF $\phi := \top$ is satisfiable and the QBF $\phi := \bot$ is unsatisfiable. A QBF $\forall x.\phi$ is satisfiable iff $\phi[x]$ is satisfiable and $\phi[\neg x]$ is satisfiable. A QBF $\exists x.\phi$ is satisfiable iff $\phi[x]$ is satisfiable or $\phi[\neg x]$ is satisfiable. The Boolean connectives are interpreted according to standard semantics. Two QBFs ϕ and ϕ' are *satisfiability-equivalent* iff ϕ is satisfiable whenever ϕ' is satisfiable.

Example 1. The QBF 4.1 $\phi := \forall x \exists y.((x \vee \neg y) \wedge (\neg x \vee y))$ is satisfiable since both $\phi[\{\neg x\}] = \exists y.(\neg y)$ and $\phi[\{x\}] = \exists y.(y)$ are satisfiable. In contrast to that, the QBF 4.2 $\phi := \exists y \forall x.((x \vee \neg y) \wedge (\neg x \vee y))$ is unsatisfiable since neither $\phi[\{\neg y\}] = \forall x.(\neg x)$ nor $\phi[\{y\}] = \forall x.(x)$ is satisfiable.

In the following, we define the *Q-resolution calculus*, the formal framework of QBF solvers based on QCDCL. The calculus consists of rules that allow us to derive clauses and cubes from a given PCNF ϕ. The implementation of clause (cube) learning in QCDCL relies on the Q-resolution calculus.

Definition 1 (Q-Resolution Calculus [35, 48, 50, 91]). Let $\phi = \Pi.\psi$ be a formula in PCNF. The rules of the *Q-resolution calculus* are as follows:

$$\frac{C \cup \{l\}}{C} \quad \begin{array}{l} \text{if for all } x \in \mathsf{var}(\Pi)\colon \{x, \neg x\} \not\subseteq (C \cup \{l\}) \text{ and either} \\ (1)\ C \text{ is a clause, } \mathsf{quant}(\Pi, l) = \forall, \\ \quad l' <_\Pi l \text{ for all } l' \in C \text{ with } \mathsf{quant}(\Pi, l') = \exists \text{ or} \\ (2)\ C \text{ is a cube, } \mathsf{quant}(\Pi, l) = \exists, \\ \quad l' <_\Pi l \text{ for all } l' \in C \text{ with } \mathsf{quant}(\Pi, l') = \forall \end{array} \qquad (red)$$

$$\frac{C_1 \cup \{p\} \qquad C_2 \cup \{\neg p\}}{C_1 \cup C_2} \quad \begin{array}{l} \text{if for all } x \in \mathsf{var}(\Pi)\colon \{x, \neg x\} \not\subseteq (C_1 \cup C_2), \\ \neg p \notin C_1,\ p \notin C_2, \text{ and either} \\ (1)\ C_1, C_2 \text{ are clauses, } \mathsf{quant}(\Pi, p) = \exists \text{ or} \\ (2)\ C_1, C_2 \text{ are cubes, } \mathsf{quant}(\Pi, p) = \forall \end{array} \qquad (res)$$

$$\frac{}{C} \quad \begin{array}{l} A \text{ is an assignment, } \phi[A] = \top, \\ \text{and } C = (\bigwedge_{l \in A} l) \text{ is a cube} \end{array} \qquad (cu\text{-}init)$$

$$\frac{}{C} \quad \text{if for all } x \in \mathsf{var}(\Pi)\colon \{x, \neg x\} \not\subseteq C \text{ and } C \in \psi \text{ is a clause} \qquad (cl\text{-}init)$$

A QBF ϕ in PCNF is unsatisfiable (satisfiable) [35, 48, 50, 91] iff the empty clause (empty cube) $\emptyset$ is derivable from ϕ by applying the rules given in Def. 1. A derivation of the empty clause (cube) $\emptyset$ from ϕ starting with applications of the axiom rules *cl-init* (*cu-init*) is a *Q-resolution proof* of the unsatisfiability (satisfiability) of ϕ.

In the case of unsatisfiability, non-tautological clauses occurring in ϕ are selected by applications of axiom rule *cl-init*. In the case of satisfiability, cubes obtained from satisfying assignments are derived by applications of axiom rule *cu-init*.

The variants of rule *res* to resolve clauses or cubes, respectively, are similar to the resolution rule in propositional logic. In this chapter, we assume that the *pivot variable* p is existential (universal) when resolving clauses (cubes) by rule *res*. Furthermore, clauses (cubes) derived by *res* must not be tautological (contradictory). These restrictions define the most common variant of Q-resolution [48]. However, it has been shown that the restriction may be lifted, resulting in more powerful variants of Q-resolution [7, 29].

The main distinguishing feature between propositional resolution and Q-resolution is rule *red*, the *reduction* operation. *Universal (Existential) reduction* eliminates trailing universal (existential) literals from a non-tautological clause (non-contradictory cube) C with respect to the quantifier ordering. We write $\mathsf{UR}(C) = C'$ ($\mathsf{ER}(C) = C'$) to denote the clause (cube) C' resulting from clause (cube) C by universal (existential) reduction. For a PCNF $\phi = \Pi.\psi$, $\mathsf{UR}(\phi) = \Pi.(\bigwedge_{C \in \psi} \mathsf{UR}(C))$ is the PCNF resulting from universal reduction of every clause $C \in \psi$.

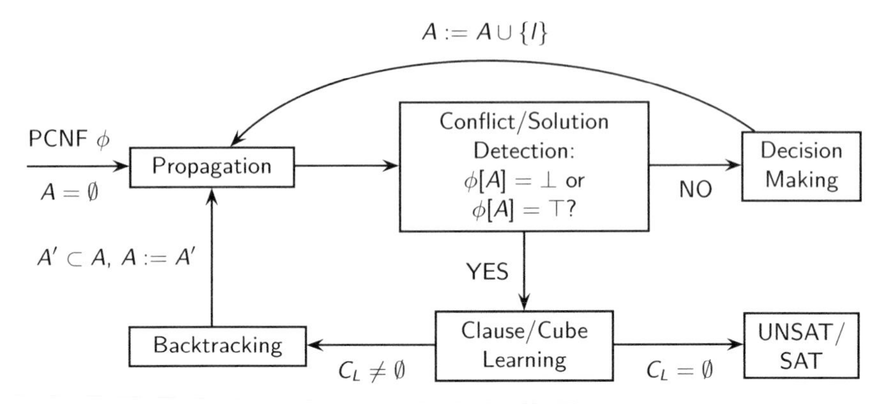

Fig. 4.2: Flowchart of QCDCL (adapted from [59]). Stages *propagation* and *conflict/-solution detection* are part of function `qbcp` in Algorithm 4.1, and stages *clause/cube learning* and *backtracking* are part of function `analyze`

4.3 Sequential Search-Based QBF Solving

Most parallel QBF solvers are based on the search-based QBF-solving paradigm. Therefore, we briefly recapitulate the core concepts and ideas behind search-based solvers.

Search-based QBF solving [18] lifts the DPLL algorithm [22] to QBF. Conflict-driven clause learning (CDCL) in SAT solving [66, 78, 79, 80, 90] extends DPLL by clause learning. Clauses are learned from conflicts to prune the search space during the search for a satisfying assignment. The QBF-specific variant of CDCL is usually called QCDCL [35, 50, 91]. In contrast to CDCL-based SAT solvers, QCDCL-based QBF solvers not only learn clauses from conflicts, but also cubes from solutions. Clauses and cubes are learned using the rules of the Q-resolution calculus. The pseudocode in Algorithm 4.1 and the flowchart in Figure 4.2 provide a high-level description of QCDCL.

From a high-level point of view, the basic building blocks of QCDCL such as propagation, decision making, learning, and backtracking are similar to CDCL. Given an input formula $\phi = \Pi.\psi$ in PCNF, the assignment tree of ϕ is traversed in a depth-first manner. QCDCL terminates if the empty clause (cube) is derived in clause (cube) learning, which shows the unsatisfiability (satisfiability) of ϕ.

Learned clauses and cubes are stored in separate sets ϕ_{CL} and ϕ_{CU}, respectively. In practice, the set of learned clauses ϕ_{CL} is added conjunctively to $\phi = \Pi.\psi$ to obtain the satisfiability-equivalent formula $\Pi.(\psi \wedge (\bigwedge_{C \in \phi_{CL}} C))$. In a similar way, the set of learned cubes is added disjunctively to $\phi = \Pi.\psi$ to obtain the satisfiability-equivalent formula $\Pi.(\psi \wedge (\bigvee_{C \in \phi_{CU}} C))$. Adding learned clauses (cubes) to ϕ preserves the satisfiability (unsatisfiability) of ϕ due to the soundness of the Q-resolution calculus.

Given the current assignment A (which is initially empty), *unit* and *pure literals* are detected and assigned during QBF-specific Boolean constraint propagation (called

QBCP, cf. function `qbcp` in Algorithm 4.1) [18, 31]. To this end, the PCNF $\phi[A]$ is considered, i.e., ϕ interpreted under A. Some literal l is unit in $\phi[A]$ if $\phi[A]$ contains a clause (l). Some literal l is pure in $\phi[A]$ if $\neg l$ does not occur in $\phi[A]$. Unit and pure literal detection is also applied to the learned clauses and cubes in sets ϕ_{CL} and ϕ_{CU}, respectively. While unit clause detection is similar to CDCL, in QBCP additionally universal reduction by rule *red* of the Q-resolution calculus is applied to the clauses in $\phi[A]$.

After the techniques in QBCP have been applied until saturation, in *conflict/solution detection* (part of function `qbcp`) it is checked whether $\phi[A] = \bot$ or $\phi[A] = \top$.

If neither $\phi[A] = \bot$ nor $\phi[A] = \top$ (line 5 in Algorithm 4.1) then A is extended by tentatively assigning some variable in *decision making* (function `assign_dec_var`). A SAT solver may assign any unassigned variable of the formula. However, this would not be sound in QCDCL. Only variables from the outermost, i.e., leftmost quantifier block of $\phi[A]$ may be assigned as decisions. As in SAT solving, it does not affect soundness whether a variable is first set to true or to false. After a variable has been assigned in decision making, propagation continues (function `qbcp`).

If $\phi[A] = \bot$ (line 7 in Algorithm 4.1) then ϕ (or ϕ_{CL}, respectively) contains a clause C for which $\mathsf{UR}(C[A]) = \emptyset$. This situation is called a *conflict*. Conflicts trigger clause learning, where a learned clause C_L is derived using the rules of the Q-resolution calculus. Thereby, C is successively resolved with *antecedent clauses* of unit literals identified during QBCP. The antecedent clause of a unit literal l is the clause in ϕ containing l that became unit in $\phi[A]$ during QBCP.

If $\phi[A] = \top$ (line 7 in Algorithm 4.1), then $\phi[A] = \emptyset$, i.e., ϕ reduces to the empty matrix under A, or ϕ_{CU} contains a cube C for which $\mathsf{ER}(C[A]) = \emptyset$. This situation is called a *solution*. A solution corresponds to a single path in the assignment tree of ϕ where the leaf is labeled with $\top$ (cf. Fig. 4.1). A SAT solver would terminate after a

Algorithm 4.1: Pseudocode of QCDCL

```
    Data: PCNF φ
    Result: True (false) if φ is satisfiable (unsatisfiable)
 1  Result R = UNDEF
 2  Assignment A = ∅
 3  while true do
        /* Simplify under A, propagation.                */
 4      (R,A) = qbcp(φ,A)
 5      if R == UNDET then
            /* Decision making.                          */
 6          A = assign_dec_var(φ,A)
 7      else
            /* Backtracking: R == UNSAT/SAT              */
 8          A' = analyze(R,A)
 9          if A' == INVALID then
10              return R
11          else
12              A = backtrack(A')
```

solution has been found. However, due to universally quantified variables, a QCDCL QBF solver in general must proceed and find further solutions. Solutions trigger cube learning, where a learned cube C_L is derived in a similar way to a learned clause. Cubes to be resolved on by rule *res* have to be derived by rule *cu-init* first.

Clause (cube) learning is part of function `analyze` in Algorithm 4.1. If the empty clause (cube) C_L is derived in clause (cube) learning ($C_L = \emptyset$ in Figure 4.2), then QCDCL terminates and reports the unsatisfiability (satisfiability) of the input PCNF ψ (line 10 in Algorithm 4.1).

Otherwise ($C_L \neq \emptyset$ in Figure 4.2), during backtracking the current assignment A is analyzed (line 8, function `analyze`) in order to retract a subassignment $A' \subseteq A$ of A (line 12). The subassignment A' is selected so that the learned clause (cube) becomes unit under the new assignment A that results from backtracking. Clauses (cubes) C_L having this property are called *asserting*. In QCDCL, typically only asserting clauses and cubes are learned. The run of QCDCL proceeds with the new assignment A resulting from backtracking.

Example 2 (Based on an example from [55]). Consider the PCNF $\phi = \Pi.\psi$ with prefix $\Pi = \exists z,z'\forall u\exists y$ and CNF

$$\psi = (u \vee \neg y) \wedge (\neg u \vee y) \wedge (z \vee u \vee \neg y) \wedge (z' \vee \neg u \vee y) \wedge (\neg z \vee \neg u \vee \neg y) \wedge (\neg z' \vee u \vee y)$$

Initially the current assignment A and the sets of learned clauses and cubes are empty. Propagation does not have any effect since ϕ does not contain unit literals (to keep the example simple, we do not carry out pure literal detection). Suppose that both z and z' are assigned *true* in decision making, i.e., $A := \{z,z'\}$, resulting in the PCNF $\phi[A] = \forall u\exists y.(u \vee \neg y) \wedge (\neg u \vee y) \wedge (\neg u \vee \neg y) \wedge (u \vee y)$. Again, $\phi[A]$ does not contain unit literals to be propagated. Hence, let A be extended by assignment $\{u\}$ in decision making, i.e., $A := \{z,z',u\}$, resulting in $\phi[A] = \exists y.(y) \wedge (\neg y)$. Suppose that variable y is assigned *true* by unit literal detection applied to $\phi[A]$, where $(\neg u \vee y) \in \phi$ is the antecedent clause of the derived assignment $\{y\}$. Clause $C_1 = (\neg z \vee \neg u \vee \neg y) \in \phi$ is falsified under $A := \{z,z',u,y\}$, i.e., $\mathsf{UR}(C_1[A]) = \emptyset$. In clause learning, C_1 is resolved with the antecedent clause $(\neg u \vee y)$ by pivot variable y, resulting in the asserting learned clause $C_{L,1} = (\neg z)$ after universal reduction.

Based on the result of `analyze`, the whole current assignment $A = \{z,z',u,y\}$ is retracted to the empty assignment $A = \emptyset$. Note that, in particular, all assignments of variables due to decision making are retracted, which corresponds to non-chronological backtracking. Since the learned clause $C_{L,1}$ is unit, i.e., a clause of size one, under the empty assignment A, propagation updates A to $A := \{\neg z\}$. Next, suppose that z' and u are assigned as decisions to obtain $A := \{\neg z, \neg z', \neg u\}$. Finally we get $A := \{\neg z, \neg z', \neg u, \neg y\}$ by unit literal detection. Every clause in ϕ is satisfied under A. In cube learning, the new cube $C_2 = (\neg z \wedge \neg z' \wedge \neg u \wedge \neg y)$ is derived using rule *cu-init* of the Q-resolution calculus. From C_2, the asserting learned cube $C_{L,2} = (\neg z \wedge \neg z' \wedge \neg u) = \mathsf{ER}(C_2)$ is derived by existential reduction (rule *red*).

After retracting $\{\neg u, \neg y\}$ from A to obtain $A := \{\neg z, \neg z'\}$ due to the result of `analyze`, $C_{L,2}$ becomes unit and hence A is extended to $A := \{\neg z, \neg z', u\}$, thus flipping the assignment of u. Cube $C_{L,2}$ is the antecedent cube of assignment $\{u\}$.

Finally $A := \{\neg z, \neg z', u, y\}$ by unit clause detection. Every clause in ϕ is satisfied under A. Cube $C_3 = (\neg z \wedge \neg z' \wedge u \wedge y)$ is derived by rule *cu-init* as before and further $C_4 = (\neg z \wedge \neg z' \wedge u) = \mathsf{ER}(C_3)$ by existential reduction of C_3. Q-resolution of the antecedent cube $C_{L,2}$ of assignment $\{u\}$ and C_4 using pivot variable u produces $C_5 = (\neg z \wedge \neg z')$. Finally, existential reduction of C_5 results in the empty cube, proving that ϕ is satisfiable.

4.4 Parallel QBF Solving at a Glance

In this section, we present approaches to parallel QBF solving that have been implemented in 11 different solvers summarized in Table 4.1. Before we discuss the individual solvers in detail in the next section, we first outline the basic ideas behind the approaches. Parallel QBF solving can be classified into *portfolio* approaches and approaches based on *search space splitting*.

A conceptually simple and straightforward way to solve a QBF in parallel is the use of a *portfolio* approach. Thereby, given a set of sequential solvers having different solving characteristics or one sequential solver in different configurations, the input formula is solved by running the solver instances in parallel on separate computing nodes. The nodes may be logically separated, like in threaded solvers, or physically separated like in distributed solvers.

Due to the orthogonality of QCDCL and expansion-based QBF solving that has been witnessed both in proof complexity [11, 43] and in experimental studies [60, 63], portfolio approaches appear to be a promising direction for the implementation of parallel QBF solvers. Since QBF solving by QCDCL and expansion has different characteristics depending on the input formula, a parallel QBF portfolio solver which combines these two solving paradigms can exploit the benefits of both approaches. However, in contrast to parallel SAT solving, where portfolio solvers are well studied and established (see Chapter 1, Parallel Satisfiability), few parallel portfolio QBF solvers have been presented.

We are aware of the following three parallel QBF solvers based on the portfolio approach. The solver HordeQBF [8] applies the HordeSAT framework [9] (cf. Chapter 1, Parallel Satisfiability) to QBF, allowing us to run different configurations of one QCDCL solver in a massively parallel manner. The solvers hiqqerfork and par-pd-depqbf are implemented as Linux shell scripts which run instances of sequential solvers in parallel processes. While hiqqerfork uses different configurations of one solver in the parallel processes, par-pd-depqbf uses two identical solver instances to solve different input formulas. To this end, par-pd-depqbf takes structured non-PCNF formulas ϕ in the QCIR format[3] as input and transforms both ϕ and its negation $\neg\phi$ into PCNF [30]. Then one process in par-pd-depqbf runs a

[3] QCIR format: `http://qbf.satisfiability.org/gallery/qcir-gallery14.pdf`

Algorithm 4.2: Splitting Algorithm for QBF Evaluation

```
   Data: QBF φ
   Result: True (false) if φ is satisfiable (unsatisfiable)
 1 begin
 2   switch φ do
 3     case ⊤ do
 4       return SAT
 5     case ⊥ do
 6       return UNSAT
 7     case ¬ψ do
 8       return NOT split(ψ)
 9     case ψ1 ∨ ψ2 do
10       return split(ψ1) OR split(ψ2)
11     case ψ1 ∧ ψ2 do
12       return split(ψ1) AND split(ψ2)
13     case ∃x.ψ do
14       return split(ψ[x/⊤]) OR split(ψ[x/⊥])
15     case ∀x.ψ do
16       return split(ψ[x/⊤]) AND split(ψ[x/⊥])
```

solver instance to solve the primal PCNF encoding of ϕ, and a second process runs an identical solver instance to solve the dual PCNF encoding of $\neg\phi$.

The most widely used approach to parallel QBF solving in terms of implemented solvers, however, is based on *search space splitting* by analyzing the formula structure as follows. Consider Algorithm 4.2 which shows a very basic recursive algorithm to evaluate a QBF of arbitrary syntactic structure. In fact, this algorithm is a direct translation of the QBF semantics given in Section 4.2 into pseudocode. The evaluation of a QBF is broken down into subproblems. The base cases of the evaluation are QBFs consisting of only a truth constant. Compound formulas containing operators such as negation, binary connectives, or quantifiers are evaluated depending on the respective semantics of the operators. That is, the result of evaluating a QBF depends on the results of evaluating its subformulas.

Algorithm 4.2 already illustrates the potential of parallel QBF solving. For example, if we want to solve the QBF $\forall x.\psi$, then we can solve $\psi[x]$ and $\psi[\neg x]$ in parallel processes and then combine the results according to the semantics of the universal quantifier. If either $\psi[x]$ or $\psi[\neg x]$ is found unsatisfiable in one process, then the other process can be stopped since the given QBF $\forall x.\psi$ has been proved unsatisfiable already. The situation is similar when solving a non-PCNF formula like $\psi_1 \vee \psi_2$. The subformulas ψ_1 and ψ_2 can be solved independently by two different processes—as soon as one of the subformulas is found to be satisfiable, the process evaluating the other subproblem can be stopped due to the semantics of the $\vee$ operator.

Based on the above observations related to Algorithm 4.2, an obvious way to parallelize QBF solving is to split the problem of evaluating the original formula into several subproblems, which are then distributed to the different client solvers. Either a sequential solver is called for each subproblem or the subproblem is split further.

Reconsider QBF $\phi = \exists z, z' \forall u \exists y.\psi$ from Example 2. The assignment tree of ϕ is shown in Fig. 4.3. Two processes could solve the subproblems $\phi[z]$ and $\phi[\neg z]$ independently and in parallel. Example 2 presented a sequential solver run in which the subproblem $\phi[z]$ was considered first, i.e., the variable z was first set to true in decision making. Only after undoing this decision in backtracking, the solver entered that part of the assignment tree that contains the model of ϕ (i.e., the left subtree in Fig. 4.3). If variable z were first set to the opposite value, i.e., false, then the extra work spent on evaluating the subproblem $\phi[z]$ would have been avoided altogether. It would not be necessary to wait for the solver to enter the part of the search space given by subproblem $\phi[\neg z]$, which contains the model. Moreover, if the two subproblems $\phi[\neg z]$ and $\phi[z]$ are solved in parallel, then the search can be stopped as soon as one subproblem witnesses the satisfiability of ϕ. If a subproblem, e.g., $\phi[\neg z]$ turns out to be too hard for a process to solve within certain resource limits, then it can be split again into further subproblems $\phi[\neg z, z']$ and $\phi[\neg z, \neg z']$, provided that the necessary computing resources are available. Again these subproblems can be solved independently of each other, and only the results of their evaluations need to be merged according to the semantics of the existential quantifier in $\exists z'$. Subproblems related to universal quantification are handled analogously.

As illustrated by Example 2, QCDCL solvers learn clauses and cubes from conflicts and solutions encountered during the search. When solving a QBF in parallel, these derived clauses and cubes potentially are helpful to other processes, even if a process had only a minor contribution to identifying a model or a countermodel for the given QBF. Therefore, sharing knowledge in terms of learned clauses and cubes with other processes is crucial in parallel QBF solving.

Research on parallel QBF solving has been focused on (1) the generation of subproblems, which are delegated to processes running on different computing nodes, and (2) knowledge sharing, i.e., the distribution of information derived by one process which is potentially useful for the others. Subproblem generation and knowledge sharing for parallel QBF solving are strongly inspired by the respective approaches to parallel SAT solving (see Chapter 1, Parallel Satisfiability). However, the SAT approaches cannot be ported to QBF in a straightforward way.

Generating subproblems (and hence also assembling the results returned by different processes) is complicated by the quantifier types of variables and by the order of the variables with respect to the quantifier structure of a QBF.

A variant of the *guiding path method* [88] as introduced for SAT solving has been found effective at generating subproblems in parallel QBF solving. With this method, a sequential solver instance in a separate computing node is provided with a set of *assumptions*. Assumptions are predefined variable assignments that the solver has to take into account in the solving process. This way, the subproblem that the solver has to solve is defined. Assumptions can also be understood as a special kind of decision variables the solver has to treat in a certain way. For example, the subproblems $\phi[\neg z]$ and $\phi[z]$ in Fig. 4.3 are defined by the sets $\{\neg z\}$ and $\{z\}$ of assumptions, respectively.

Typically, a master process generates sets of assumptions and distributes them to the solver instances running on the computing nodes. Based on the result of solving the subproblem, the solver may request further subproblems from the master. The

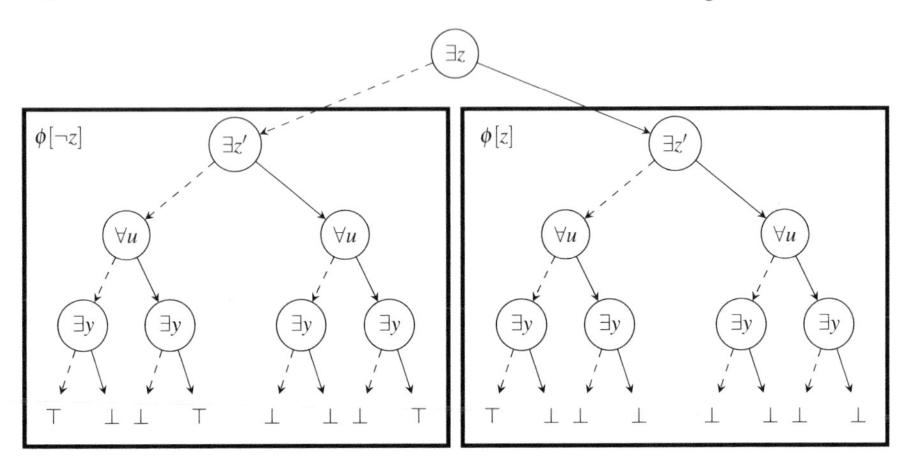

Fig. 4.3: Assignment tree of QBF ϕ from Example 2 and two subtrees of $\phi[\neg z]$ and $\phi[z]$

master combines the results of the subproblems depending on the quantifier types of the variables assigned in the set of assumptions (cf. the splitting algorithm in Algorithm 4.2). Due to the quantifier types and the ordering of variables in the quantifier prefix, the generation of subproblems and combination of results by the master process is more complicated than in the context of SAT solving. In this respect, care has to be taken to guarantee soundness and completeness of a parallel QBF solver. For instance, in Example 2 it would be unsound to generate subproblems by assumption $\{y\}$ only since y is not at the left end of the quantifier prefix of ϕ.

In contrast to SAT solvers, QCDCL solvers operating on a PCNF ϕ not only learn clauses from conflicts but also cubes from solutions. As illustrated by Example 2, initially the set of learned cubes is empty. Hence cubes have to be derived first by rule *cu-init* of the Q-resolution calculus based on satisfying assignments. Since a satisfying assignment of ϕ must satisfy every clause in ϕ, cubes derived by rule *cu-init* tend to be large and often contain a large number of the variables in ϕ. Therefore, sharing large cubes with other solver instances in a parallel setting is challenging not only because of their size but also since large cubes tend to have only a limited pruning effect on the search space.

QBF solvers that implement parallelization by the guiding path method are MPI-DepQBF, PAQuBE, PQSolve, and QMiraXT. They are distinguished by whether learning is supported or not, whether subproblems are generated by the master or by the client, and the way the result returned by the client solvers is represented. To summarize, the master process in a parallel QBF solver based on search space splitting carries out the following tasks (if supported by the respective concrete approach):

- administrate the currently distributed subproblems[4];
- maintain information about decision variables used to generate assumptions;
- request new subproblems from busy clients if there are idle clients;
- activate idle clients when new subproblems are available;
- manage information sharing among the clients;
- stop the clients if the given QBF has been solved.

In principle, the clients in a parallel solver based on search space splitting are responsible for the following tasks (if supported by the concrete approach):

- receive a subproblem (ideally in terms of assumptions);
- solve a subproblem and return the respective result to the master;
- share information with other clients;
- learn information from other clients;
- optionally generate subproblems, which are passed to the master (or to other clients);
- terminate if requested by the master.

The solvers pcaqe, PQUABS, and PQSAT also use syntactic properties of a formula to split the search space, but in a conceptually different manner to the guiding path method. These solvers are based on expansion. PQSAT extracts subformulas as subproblems, which may contain free variables. The clients processing the subproblems either eliminate the remaining quantifiers such that a propositional formula over these free variables is returned to the master, or they further split the subproblem. The solvers pcaqe and PQUABS extract a propositional formula for each quantifier block that is then used for evaluating the given QBF. Differently from the other solvers, which operate on formulas in PCNF, PQUABS operates on formulas in non-prenex form.

The rough classification of parallel QBF-solving paradigms presented above already illustrates the different approaches to leverage the power of modern computing systems. In the following section, we give a detailed description of individual parallel QBF solvers.

4.5 Parallel QBF-Solving Approaches

Table 4.1 summarizes and compares the parallel QBF-solving approaches that have been presented in the literature. The only approach not implemented is the one by Aspvall et al. [3], which is restricted to PCNFs with a maximum clause size of two and so far has been of theoretical interest only. For the other approaches implementations either are publicly available or at least experimental results have been published. Most parallel solvers are based on a sequential QBF solver such as DepQBF, QuBE, caqe, quabs, QSolve, and QSAT. Usually the sequential

[4] In PQSolve a client may distribute subproblems to other clients and hence becomes the master with respect to the particular subproblem.

Table 4.1: Comparison of parallel QBF solvers

parallel QBF solver	*base solver*	*QCDCL-based*	*portfolio*	*PCNF input format*	*information sharing*	*pre-/inprocessing*	*process management*	*QBFEVAL'16*	*publicly available*	*most recent paper*
Aspvall et al.	n.i.									1996 [3]
pcaqe	caqe[1]	×	×	✓	×	✓	Pthreads	✓	✓	–
hiqqerfork	DepQBF	✓	✓	✓	×	✓	`fork`	✓	–	–
HordeQBF	DepQBF[2]	✓	✓	✓	✓	✓	MPI	✓	✓	2016 [8]
MPIDepQBF	DepQBF	✓	×	✓	×	✓	MPI	✓	✓	2014 [46]
par-pd-depqbf	DepQBF	✓	✓	✓	×	✓	`fork`	✓	–	–
PAQuBE	QuBE	✓	×	✓	✓	✓	MPI	×	×	2011 [51]
PQSolve	QSolve	∼[4]	×	✓	×	✓	MPI	×	×	2000 [28]
PQSAT	QSAT	×	×	×	×	✓	MPI	×	×	2010 [67]
PQUABS	quabs	×	×	×	×	✓	Pthreads	×	✓	2016 [83]
QMiraXT	MiraXT[3]	✓	×	✓	✓	✓	Pthreads	×	✓	2009 [52]

✓ yes/supported × no/not supported – unpublished n.i. not implemented

[1] Picosat is the default SAT solver; also Minisat is supported

[2] any QCDCL solver could be used

[3] parallel SAT-solving framework

[4] DPLL-based

solvers are tightly integrated into the implementations of the clients. As the only exception, HordeQBF is based on a generic framework that allows integration of any QCDCL-based QBF solver supporting incremental solving and learning. QMiraXT implements its own QBF solver in order to be used with the MiraXT framework that was developed for parallel SAT solving. The majority of the parallel QBF solvers are based on QCDCL; only pcaqe, PQSAT, and PQUABS apply expansion-based techniques. Out of the QCDCL-based solvers, three support clause and cube sharing. All solvers apply certain simplification techniques either before the solving starts, i.e., as a preprocessing step, or dynamically as *inprocessing* [45] during solving, like DepQBF as used in HordeQBF. In the following, we discuss the individual solving approaches in detail.

Approach by Aspvall et al.

One of the first parallel approaches to QBF solving was presented by Aspvall et al. in 1996 [3]. The considered QBFs ϕ are in PCNF with a quantifier prefix having arbitrarily many quantifier alternations but with the restriction that clauses contain at most two literals. Formulas of this kind are also called *Q2CNF formulas*. In consequence, the satisfiability problem of Q2CNFs is not PSPACE-complete any more. Instead, the satisfiability of a Q2CNF ϕ can be decided by a sequential algorithm [4] in time $O(n+m)$, where n is the number of variables and m is the number of clauses in ϕ.

In principle, the approach by Aspvall et al. builds on the linear time sequential algorithm to solve Q2CNFs [4]. Let $G(\phi) := (V, E)$ be the directed *implication graph* of a Q2CNF formula $\phi = Q_1x_1 \dots Q_nx_n.\psi$ where the set $V = \{x_1, \dots, x_n, \neg x_1, \dots, \neg x_n\}$ of vertices is given by all possible literals $x_i \in \phi$, and for any clause $(l \vee k) \in \phi$ it holds that $(\neg l, k)$ and $(\neg k, l)$ are edges in E. A vertex of $G(\phi)$ is called existential (universal) if its associated variable is existentially (universally) quantified. Given a Q2CNF ϕ and the related graph $G(\phi)$, ϕ is satisfiable iff none of the following three conditions holds [4]:

1. Existential vertices l and $\neg l$ are in the same strongly connected component of $G(\phi)$.
2. A strongly connected component of $G(\phi)$ contains universal vertex l and existential vertex k with $k < l$.
3. There is a path between two universal vertices.

To test the satisfiability of a Q2CNF ϕ, first the transitive closure of $G(\phi)$ is represented as an adjacency matrix. Then the above conditions are checked in constant parallel time by assigning one processor to each pair of variables. Furthermore, Aspvall et al. present an algorithm to find models of satisfiable Q2CNFs: because of the restricted formula structure it is sufficient that the values of the existential variables are mapped either to truth constants or to one universal literal.

In the original publication [3] no implementation of the algorithm was reported, and we are not aware of any implementation published elsewhere. In practical QBF applications, encodings of problems typically have clauses of size bigger than two. Therefore it is unlikely that this approach will ever be implemented in a dedicated parallel Q2CNF solver. However, in the same way as the sequential version [4] of this algorithm is used to identify equivalent literals (e.g., in the preprocessor bloqqer [38]), also its parallel variant could be used for speeding up preprocessing.

Unpublished QBFEVAL'16 Participants: pcaqe, hiqqerfork, par-pd-depqbf

Three parallel solvers not formally published in the literature participated in the parallel track of QBFEVAL'16. These solvers are par-pd-depqbf, hiqqerfork, and pcaqe, which solved the largest number of formulas in the parallel track, i.e., 606, 598, and 585 formulas out of 825, respectively [71]. We briefly review these solvers in the following.

Both hiqqerfork and par-pd-depqbf may be considered to be portfolio-based solvers. The solver hiqqerfork is a portfolio solver in the classical sense, running different configurations of the sequential solver hiqqer. A short description of hiqqer can be found in [41]. The solver hiqqer uses modifications of the publicly available preprocessors bloqqer and qxbf before invoking the solver DepQBF.

The solver par-pd-depqbf is based on the insight that often it is not clear whether the *primal* or the *dual* encoding of a problem is preferable for a particular solver [30].

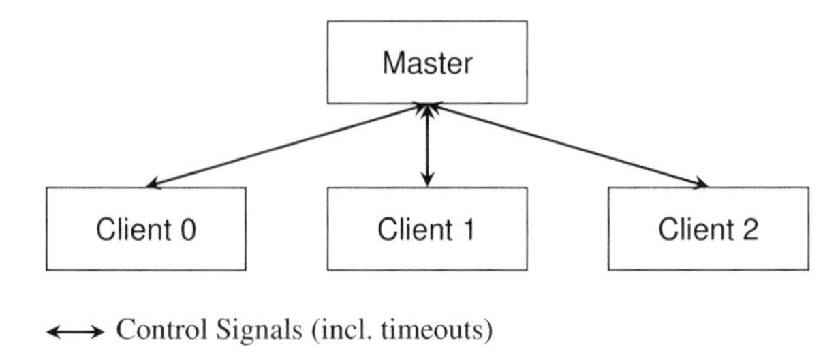

Fig. 4.4: Master-Client architecture of MPIDepQBF

The primal encoding represents the original problem, whereas the dual encoding represents its negation. The solver par-pd-depqbf runs exactly two identical instances of a sequential QBF solver in parallel. Given a structured non-PCNF formula ϕ in the QCIR format as input, one solver instance processes the primal encoding of ϕ as a PCNF, and the other solver instance processes the dual encoding of $\neg\phi$ as a PCNF. If either the primal or the dual version is solved, the whole solving process is stopped and the respective result is returned. The sequential back-end solver of par-pd-depqbf is DepQBF in combination with the preprocessor bloqqer. However, basically any QBF solver or preprocessor can be applied in par-pd-depqbf.

The solver pcaqe is a parallel version of the sequential solver caqe [73] which is based on a similar abstraction-based technique to that used in the solver PQUABS (see below). The solver pcaqe is part of the source code of caqe[5] and can be run with either Minisat or Picosat as back-end solver.

MPIDepQBF

The solver MPIDepQBF [46] relies on the sequential QCDCL-based solver DepQBF to solve any input formula ϕ in PCNF. To this end, ϕ is split into subproblems to be evaluated by client processes operating in parallel. The clients are independent of each other and do not exchange any information. However, information learned locally by a client is reused in different runs of that client. Keeping the information from run to run is realized by assumption-based reasoning. Assumptions are temporary (and partial) assignments of variables that define the formula to be solved by a client following the guiding path method.

In MPIDepQBF one dedicated master process coordinates an arbitrary number of clients via MPI (see Fig. 4.4). The sequential solver DepQBF applied by the clients provides an API similar to the APIs of most incremental SAT solvers [25, 68]. The API allows the solver to be provided with the formula to be solved by adding the respective variables, quantifiers, and clauses, and has functions to control the solving process. DepQBF was extended with assumption-based reasoning to integrate it into

[5] https://www.react.uni-saarland.de/tools/caqe/index.html

the framework implemented by MPIDepQBF. Apart from that, DepQBF was used out of the box without any changes.

Due to the use of assumptions, the clients are provided with the original formula ϕ to be solved in parallel only once. The master sends a set of assumptions to an idle client, which defines its subproblem to be solved, in addition to a timeout restricting the solving time. The client sends the result related to the subproblem back to the master (the result may be undefined if the solving process of the client timed out) and discards the set of assumptions. Then the master either generates a new subproblem in terms of new assumptions, or resends the previous subproblem to the client with an increased timeout. Information learned during a run of a client, e.g., like clauses and cubes, is not shared between the clients. However, assumption-based reasoning enables this information to be reused in different runs of the same client.

Given a PCNF $\phi := Q_1X_1 \dots Q_nX_n.\psi$, the master process in MPIDepQBF generates the subproblems to be solved by the clients as follows. First, the variables of each quantifier block in ϕ are sorted according to their respective number of variable occurrences. This heuristic ordering together with the quantifier ordering in the prefix of ϕ determines the order in which the variables will be assigned as assumptions to generate subproblems. Then a search tree is built in a similar way to assignment trees (cf. Fig. 4.3), which contains three types of nodes: sat, unsat, and open. Nodes of type sat and unsat represent solved subproblems whereas an open node corresponds to an unsolved subproblem and contains a variable assignment and a timeout.

Initially, the search tree is balanced and has n leaves which are of type open where n is the smallest power of 2 that is smaller than the total number of available clients. The result obtained from a client for a particular subproblem is incorporated into the search tree. For sat or unsat, the tree is simplified according to the quantifier rules in the splitting algorithm shown in Algorithm 4.2. If the result is a timeout, then the subproblem is either split further provided that additional clients are idle and hence waiting for work, or it is handed again to the same client with an increased timeout. If the tree is reduced to a single leaf node with sat or unsat then the formula is solved.

The master process is implemented in OCaml. Source code is available as part of the TOSS framework.[6] For simplifying the formula, the preprocessor bloqqer is used. MPIDepQBF is not limited to the use of DepQBF as a sequential back-end solver. In principle, any QBF solver supporting assumption-based reasoning can be integrated into MPIDepQBF. Further, the reuse of information learned locally within a run of a client has been found crucial for solving performance [46] but is not necessary for the basic workings of MPIDepQBF.

[6] http://toss.sourceforge.net/

HordeQBF

The solver HordeQBF [8] is based on the massively parallel SAT-solving framework HordeSAT,[7] which integrates sequential CDCL-based SAT solvers in a portfolio style [9]. HordeSAT features hierarchical parallelism on two levels. On the top level, several instances of HordeSAT are executed in parallel and communicate with each other via MPI. These are the master processes. On the bottom level, each master starts several *core CDCL solvers* as client processes in separate threads. Thus communication within a master is implemented via the shared-memory paradigm. The clients periodically put learned clauses in a pool which is managed by their respective master. The pool is stored in a shared-memory region, which enables sharing of learned clauses between clients at low communication overhead. Periodically the masters exchange the learned clauses in their respective pools via MPI. This way, clauses learned by a particular client in a certain master become available to all the other clients in the different masters. The runs of the clients are diversified by providing the core solvers with different parameter settings so that the solvers operate in different parts of the search space.

HordeQBF differs from HordeSAT only in the use of a sequential QCDCL QBF solver instead of a CDCL SAT solver. The communication framework as described above is unchanged. In order to integrate a QCDCL solver into HordeQBF to be used as a core solver in the clients, the solver has to implement an API that provides functions to achieve various tasks, for example:

- import the formula in the core solver;
- diversify the run of the core solver by parameter settings;
- start the core solver;
- import/export learned clauses;
- stop the search if the formula has been solved by any core solver.

Although QCDCL solvers learn cubes in addition to clauses, the HordeSAT framework does not have to be adapted to explicitly support sharing of cubes via a dedicated API function. Instead, learned clauses and cubes are treated as sets of literals which are augmented by a special marker literal. The marker literal indicates whether the literal set is supposed to be interpreted by a client as a clause or as a cube. The master processes communicating via MPI do not distinguish between clauses or cubes but only exchange literal sets provided by the clients. Depending on certain heuristics, clients may or may not import a shared clause or cube stored in the pool of their respective master.

In principle, HordeQBF can be combined with any QCDCL QBF solver that implements the HordeSAT API. In the first release [8], the search-based solver DepQBF version 5.0, which implements a dynamic variant of *blocked clause elimination (QBCE)* for learning smaller cubes [55], was integrated into the framework.

In HordeQBF the clients check whether new learned clauses or cubes are available in the pool after a *restart*. CDCL and QCDCL solvers periodically restart by retracting

[7] `http://baldur.iti.kit.edu/hordesat/`

the entire assignment and starting the search from scratch while keeping the learned clauses and cubes. In order to import learned clauses and cubes after a restart in DepQBF, its restart policy was modified such that it always fully retracts the assignment in a restart (cf. the original restart policy of DepQBF [56]). Learned clauses and cubes are imported, data structures are updated, and the search is resumed under the new constraints.

In order to diversify the different DepQBF instances, the master provides each solver instance with a random seed. Based on this random seed several (Q)CDCL-related parameters, such as the assignment cache [69], are randomly initialized. In consequence, the first value assigned to a decision variable is random. Further, parameters related to variable-activity scaling (see [30]), restarting parameters, and the percentage of learned clauses and cubes to be discarded periodically are set at random. Finally, various variants of dynamic QBCE and variants of different kinds of Q-resolution to learn new constraints are randomly turned on and turned off.

Experimental results with HordeQBF on application benchmarks showed superlinear average and median speedup on a cluster with up to 1024 processing cores [8].

PAQuBE

The QBF solver PAQuBE [51, 62] is a parallel version of QuBE [33], which pioneered QCDCL solving but currently is not being further developed. QuBE implements literal watching, conflict and solution analysis, and learning as well as advanced decision heuristics. Furthermore, QuBE uses the preprocessor SqueezBF [34] which considerably improves its performance. To integrate QuBE into the parallel architecture of PAQuBE, it was extended with assumption-based reasoning (like DepQBF was extended for the integration into MPIDepQBF). For conflict analysis and backjumping, assumptions require special treatment. Furthermore, literal watching had to be modified to correctly handle clauses and cubes obtained from other clients when backtracking.

Parallelization in PAQuBE is based on MPI and a master-client architecture as shown in Figure 4.5. One dedicated master controls $n-1$ sequential instances of QuBE. The master generates and distributes the subproblems and collects solutions using a specific variant of the guiding path method. Thereby, at any time all clients operate on subproblems rooted at variables from the same quantifier block of the given PCNF to be solved. Due to the scheduling policy, the master has to deal only with control signals but not with shared knowledge. In consequence, the master process spends most of the time sleeping. It only has to wake up when one client is idle and a new subproblem has to be requested from another client. Hence, the master does not need its own CPU. The existence of the master process is justified by the scheduling algorithm for the distribution of subproblems. Without a master process, it would be necessary for the clients to communicate among themselves to share subproblems, thus increasing the overall communication overhead.

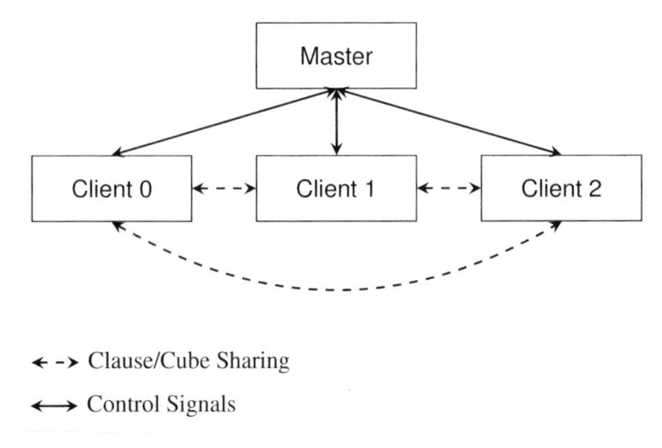

Fig. 4.5: Master-Client architecture of PAQuBE

To solve a formula by PAQuBE, first it is read by the clients. For simplifying the formula, the preprocessor SqueezBF is applied. One client informs the master about basic formula properties such as number of variables, number of clauses, and number of quantification levels. This information is necessary for scheduling the subproblems. Then one client starts to solve the preprocessed formula as it is without any assumptions. The other clients request a subproblem from the master, who forwards their requests to the busy client. For the assignment of subproblems, the SQLS algorithm introduced with the solver QMiraXT (see below for a description of this approach) is used. SQLS is a restricted, simplified variant of the scheduling algorithm of PQSolve. The master requests a subproblem with a root variable in the current quantifier block. If the asked client does not have such a problem, another client is asked. If no client can provide a subproblem of the requested form, the master moves to the next quantification level. This will continue until either all clients are waiting for new subproblems or until a subproblem with a topmost universally (existentially) quantified variable is found unsatisfiable (satisfiable).

PAQuBE realizes an advanced knowledge-sharing mechanism of learned clauses and cubes. The clients freely communicate with each other in order to share learned clauses and cubes derived while solving their subproblems. The master process is not involved in knowledge sharing. After a fixed number of decisions the clients check whether new messages either from the master or from some other client are available. At this time, also suitable learned clauses and cubes are shared with other clients. The clients have to share the clauses and cubes learned from their run as well as receive and learn clauses and cubes derived by other clients. In addition, cubes are compressed under the assumption that different cubes share many literals from the highest quantification levels. Therefore, the literals of a cube are sorted according to the prefix order and common parts of cubes are sent only once. As this knowledge exchange leads to a significant communication overhead, multiple clause- and cube-sharing strategies are implemented. In experiments it was shown that an adaptive method yielded the overall best results. In [53] the application of machine learning is suggested to control information sharing.

PQSAT

Da Mota et al. [67] presented a parallel architecture for QBF solving. In the following, we name this approach PQSAT because it is based on the sequential solver QSAT [70]. In contrast to most other systems, PQSAT does not require the formulas to be in PCNF. Instead it accepts arbitrarily structured formulas as input. Furthermore, the base solver QSAT used in PQSAT applies quantifier elimination rather than QCDCL. Thereby, quantified variables are successively eliminated from a given formula ϕ similarly to expansion.

PQSAT implements a parallel master-client architecture using MPI. The master reads the original QBF and splits it into several subproblems, which are distributed among the clients. For generating subproblems, the master analyses the syntactic structure of the formula in order to find subproblems which can be solved independently by the clients. For example, given the formula

$$\phi = \exists a \forall b.(((a \leftrightarrow b) \wedge (\forall c.(c \vee b))) \wedge (\exists d.(a \wedge \neg d)))$$

the subproblems $\phi_1 = \forall c.(c \vee b)$ and $\phi_2 = \exists d.(a \wedge \neg d)$ are extracted (cf. [21]). Note that variables b and a are free in ϕ_1 and ϕ_2, respectively. The task of the clients is to find propositional formulas over the free variables that are equivalent to the formulas in the subproblems by following the quantifier elimination approach implemented in QSAT [70]. For example, given a QBF $\Pi \exists x.(\psi_1 \wedge \psi_2)$ where ψ_1 does not contain any occurrence of x, the formula is rewritten to $\Pi.(\psi_1 \wedge \exists x.\psi_2)$ by minimizing the scope of $\exists x$. Then $\exists x.\psi_2$ is replaced by an equivalent formula without x. Universally quantified variables are eliminated in a similar manner. Quantifier elimination is repeated until a purely propositional formula is left. Then this propositional formula is passed to a SAT solver.

After subproblems have been assigned to the clients, the master waits for the respective results and assembles them in order to get the result of the full problem. As the subproblems may contain free variables, the clients must return an equivalent formula over these free variables without any quantifiers. The clients themselves may split their given subproblems into further subproblems if the given subproblem appears to be too difficult according to some syntactic measure of difficulty. If a client decides to split a subproblem, then it employs semantic splitting based on assignments to the free variables. The set of new subproblems is passed to the master node, who distributes them to other idle clients.

PQSolve

One of the first parallel QBF solvers was PQSolve, which was published in the year 2000 [28]. At that time, QBF-solving technology in general still was in its infancy. For example, neither learning as used in QCDCL-based QBF solvers nor expansion-based solving had been presented. Although PQSolve naturally lacks many techniques that are standard in modern solvers, it can be seen as a milestone in parallel QBF solving. PQSolve relies on QSolve as the base solver, which implements the DPLL algorithm for QBF with several then state-of-the-art heuristics and pruning techniques such as quantifier inversion, trivial truth, and trivial falsity [18, 74]. Thus PQSolve is an early distributed realization of DPLL for QBF.

The motivation for parallelizing QSolve stems from the common view of QBF solving as a two-person zero-sum game with complete information (cf. [76]). Thereby, the *universal player* assigns the universally quantified variables of a given QBF with the aim to falsify the formula, whereas the *existential player* assigns the existentially quantified variables in order to satisfy it. For the development of PQSolve, its authors applied techniques successfully used in parallel chess programs.

PQSolve implements a master-client architecture based on MPI where the role of master and client processes may change dynamically depending on the scheduling of subproblems and on the progress of the search. Furthermore, there may be more than one master process. This dynamic architecture of PQSolve is different from many other parallel QBF solvers and complicates the checking of termination conditions. To obtain a simpler design, solvers such as QMiraXT and PAQuBE implement a restricted variant of PQSolve's architecture and scheduling based on the SQLS algorithm.

PQSolve takes formulas in PCNF as input and works as follows: first one process is assigned to solve the input formula. All other processes are idle. If a process Q is idle then it sends a request for work to a random process P which is not idle. If the contacted busy process P has an unexplored part in its current search tree then it sends the respective formula to the requesting process Q similarly to the guiding path method. This way, P becomes the master of the client Q. The requesting client process Q now solves the formula and sends the result back to the master P. Then Q becomes idle again and the master-client relationship between P and Q is released. Process P incorporates the result into its search tree. If P has another open subproblem then it communicates that subproblem to the idle process Q. Otherwise, a request for work is sent to a random busy process. It may happen that a client's work on a subproblem becomes obsolete because of some pruning techniques applied in the master. In this case, the master informs the client to stop solving the respective subproblem.

Every process in PQSolve applies tests for trivial truth and trivial falsity. For the trivial truth check, only the existentially quantified variables are considered and all literals of universal variables are discarded from the PCNF. If the resulting propositional formula is satisfiable, then also the original PCNF is satisfiable. For the trivial falsity check, all variables are assumed to be existentially quantified. If the resulting propositional formula is unsatisfiable, then also the original PCNF is unsatisfiable. Trivial truth and falsity checks are simply realized with a SAT solver and can be done at any time during the search.

The subproblem handed over to a different process in PQSolve must be large enough to justify the communication overhead. The selection and scheduling of subproblems work as follows. Let $\{l_0, \ldots, l_m\}$ be the current assignment such that l_i was assigned before l_j if $i < j$. When receiving a request from another process, then the formula under assignment $\{l_0, \ldots, \neg l_i\}$ is passed to the other process such that $3 * |N(x_i) - P(x_i)| + i$ is minimal where $\mathsf{var}(l_i) = x_i$ and $P(x_i)$ is the number of positive occurrences of x_i and $N(x_i)$ is the number of negative occurrences of x_i.

To increase parallel efficiency, PQSolve implements *Helpful Master Scheduling*. A master process that has passed on a subproblem to a client has to wait for the result and thus stays idle after it has solved its own subproblem. In that case the master itself sends a request to the client, which in turn provides a subproblem (of its current one) to share the work.

To avoid irrelevant work, *Young Brothers Wait Scheduling* is applied. This approach tries to deal with the problem that when solving a formula under a certain assignment of some variable x, it is often not necessary to solve the formula under the dual assignment of x. In a parallel setting, situations of this kind result in a waste of work. Therefore, blocks of variables are considered. Only after the leftmost leaves of the subtrees obtained by setting the variables in a block have been fully evaluated are the subformulas related to the other subtrees passed to other processes.

PQUABS

The solver PQUABS [83] extends the sequential solver quabs [84], which processes formulas in prenex negation normal form (prenex NNF), by allowing input formulas to be in non-prenex NNF. That is, PQUABS is able to handle formulas with a tree-shaped quantifier structure in contrast to the linear quantifier structure of formulas in prenex NNF. For each maximal consecutive block of quantifiers of the same type, PQUABS builds a propositional abstraction of the input formula in a way that is similar to the approach implemented in caqe [73] and its parallel variant pcaqe. Thereby, the evaluation of a given QBF is broken down to evaluating a set of propositional abstractions. The abstraction of a quantifier block is linked to the abstractions of adjacent quantifier blocks in the syntactic structure of the formula via so-called interface literals. The interface literals express quantifier dependencies resulting from the ordering of quantifier blocks. The satisfiability of a subformula is communicated via assignments to the interface literals. There are two types of interface literals: one type to represent the assignments made by abstractions of outer quantifier blocks, and the other type to represent the assignments made by abstractions of inner quantifier blocks. A counterexample-guided abstraction refinement loop (CEGAR) is employed based on SAT solving to generate refined abstractions. Additionally, PQUABS analyses the quantifier structure of the given formula to avoid the use of interface literals whenever a subformula appears in the scope of only one quantifier block.

QMiraXT

The solver QMiraXT [52, 77] implements QCDCL combined with preprocessing. Unlike the other parallel QCDCL solvers (see Table 4.1), knowledge sharing is based on shared memory (see Figure 4.6) rather than message passing by MPI. QMiraXT is an extension of the parallel SAT solver MiraXT. While MiraXT and QMiraXT share a common architecture, the reasoning mechanisms of QMiraXT are adapted to QBF.

QMiraXT implements a decision heuristics similar to VSIDS [66], but takes the different quantification levels of the variables into account. That is, all variables of the current level have to be set before a variable of the next level is selected, similarly to QBF semantics. Two counters are used to keep track of positive and negative variable occurrences in the formula. When conflict clauses are added, these counters are increased. Further, they are periodically decreased to amplify the influence of more recent conflict clauses. From a set of existentially quantified variables, the variable that satisfies the largest number of clauses is chosen. A universally quantified variable is selected and assigned so that the number of implications by unit clauses that would result from the respective assignment is maximized.

QMiraXT eliminates unused variables and pure literals and performs substitution of equivalent literals. Then the complete solver Quantor [12] is applied as preprocessor. Quantor implements bounded variable elimination and universal variable expansion. Those techniques are applied until the formula reduces to a propositional formula. As the memory consumption of Quantor is not restricted, QMiraXT sets a memory limit (128 MB is reported in [52]) as well as a time limit of five seconds. Then the remaining QBF formula is processed in QCDCL style. This way, Quantor is applied in an incomplete manner as a preprocessor, what is very similar to the idea behind the preprocessor bloqqer.

The shared clause database (SCD) of QMiraXT contains every clause that is currently used by a client thread. Cubes are not stored because it was found [52] that in general they are too large, and storing them would slow down the performance of the solver. A clause is contained only once in the SCD and is marked as read-only. After a clause has been generated and added to the SCD it is available to all threads via shared-memory accesses. That is, unlike MPI-based communication as implemented in other parallel solvers, explicit exchange of shared clauses via messages is not required.

Clauses stored in the SCD may reside at any position in memory. To optimize memory accesses made by the threads, each thread maintains a *watched-literal reference list (WLRL)*. For every clause, the WLRL allows a thread to store two watched literals and an existentially quantified *cache literal* in its local memory. It has been shown that this caching policy optimizes memory accesses made by the threads.

QMiraXT has no controlling master process. Instead there is a *Master Control Object* (MCO), which coordinates the communication between the threads. The MCO is never directly involved in the communication. It stores messages on global events, for example, that the formula has been solved. The most important task of the MCO is the generation of subproblems. Subproblems are generated by the

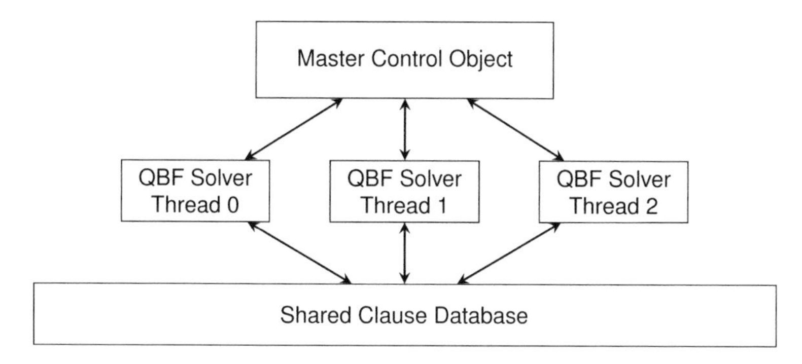

Fig. 4.6: Architecture of the solver QMiraXT [52]

guiding path method like in the solvers MPIDepQBF, PAQuBE, and PQSolve. To this end, the MCO provides the two functions `donateDecisionStack()` and `getDecisionStack()`, which both enforce the use of locks. Function `donateDecisionStack()` splits the *decision stack* of the current thread into two different decision stacks and provides another thread with one of them. The decision stack contains the decision variables in the ordering they were assigned in QCDCL. Function `getDecisionStack()` implements *single quantification level scheduling (SQLS)* [77]. SQLS is a restricted, simplified variant of the scheduling algorithm of PQSolve and is also employed by the solver PAQuBE. The splitting of the search space is done by the clients. Clients are allowed to split the search space on one quantification level. If no subproblems are available anymore, the threads block until either new subproblems have been provided by another thread or until all other threads terminate. If only one thread is running and all other decisions have been considered at the current decision level, it may use variables from the next quantification level. In this way, the clients manage the subproblem generation by themselves and no complicated management infrastructure is needed.

4.6 Challenges and Potential of Parallel QBF Solving

In the past and recent QBF solver landscape, the majority of the presented tools focus on sequential solving approaches. Thus the potential of modern computer architectures is currently not fully leveraged. In the following, we discuss challenges and opportunities that arise in the context of parallel QBF solving.

Preprocessing

In the context of sequential QBF solving, preprocessing has been shown to be substantially valuable to many state-of-the-art solvers [60]. All parallel approaches either perform some simplifications before solving, exploit the power of sequential preprocessors such as bloqqer, SqueezBF, or HQSpre [87], or apply the complete solver Quantor in a resource bounded way. In general, the goal of preprocessing is to simplify the input formula such that it becomes easier to solve. At the moment, however, no special parallel preprocessing techniques are applied. It would be a natural approach to apply expensive sequential preprocessing techniques in parallel. To this end, however, it has to be investigated whether preprocessing techniques that have been found beneficial in the context of sequential solving are also beneficial to parallel solving to the same extent. Furthermore, it may be necessary to tune sequential preprocessing techniques to parallel settings. For example, whereas in sequential solving the elimination of both variables and clauses from a formula is crucial, in parallel solving it may be more important to emphasize the removal of variables. Search space splitting is carried out based on the set of variables. Hence eliminating variables reduces the size of the search space and thus might simplify search space splitting.

Learning and Knowledge-Sharing Heuristics

All parallel solvers which support knowledge sharing (see Table 4.1) are based on QCDCL. In QCDCL, new learned clauses and cubes are derived using the Q-resolution calculus (Definition 1). The learned clauses can be shared with other threads or processes in order to prune the search space and thus speed up the overall search.

To limit the communication overhead that may result from sharing, suitable heuristics must be applied in order to select the clauses and cubes to be shared. For example, in parallel SAT solving, typical clause selection metrics are the length of a clause or the involvement of literals in conflicts.

While the quantifier structure of QBFs results in several restrictions that potentially limit the effectiveness of parallel solving techniques in general, at the same time it gives rise to additional selection criteria. Possible criteria are the number of universal (existential) literals in a clause (cube), or the quantification levels of literals of a clause (cube).

Effective selection criteria are particularly important when it comes to sharing learned cubes. In QCDCL learned cubes are first derived by rule *cu-init* of the Q-resolution calculus. Cubes derived this way tend to be large since their derivation relies on assignments that satisfy all clauses of the given PCNF. Due to the size of cubes, it may be costly to share large numbers of cubes. Furthermore, large cubes tend to prune only small parts of the search space. We see a lot of potential in the development of useful heuristics to decide on the benefit of sharing knowledge.

In general, cube learning in QCDCL may be a bottleneck also in sequential QBF solving. To mitigate the weaknesses of deriving only large cubes by rule *cu-init*, the Q-resolution calculus has been extended by additional axioms [59]. Derivations made by these additional axioms rely on the application of oracles to check the satisfiability of QBFs that arise during the solving process. In this respect, oracles implement resource-bounded procedures for QBF satisfiability checking. Cubes derived by the additional axioms are potentially smaller than cubes derived by rule *cu-init* in the traditional way. In parallel QBF solving based on QCDCL, there is considerable potential in parallelizing the calls of several oracles, which might implement orthogonal or incomplete solving techniques, for example. Since the cubes derived by such parallel oracle calls tend to be smaller than cubes derived by rule *cu-init*, sharing these cubes with other threads or processes in the solver would result in smaller communication overhead and better pruning of the search space.

Incremental Solving

An *incremental* QBF solver based on QCDCL [58, 61, 65] allows us to solve sequences $S := \langle \phi_0, \ldots, \phi_n \rangle$ of related PCNFs ϕ_i. Each PCNF ϕ_{i+1} is obtained from the previous PCNF ϕ_i by adding or deleting clauses, variables, or quantifiers. When solving a PCNF ϕ_i in S in an incremental way, the solver does not start from scratch. Instead, clauses and cubes learned when solving ϕ_i potentially can be kept and reused when solving the next PCNF ϕ_{i+1}. This way, the PCNFs ϕ_i might be solved faster than if each ϕ_i was solved independently and non-incrementally. For incremental solving, the solver must provide an API so that the same solver instance can be used to solve the PCNFs in S.

We are not aware of any approaches to parallelize incremental QBF solving. Hence the potential positive effects of combining the benefits of incremental and parallel solving are currently not leveraged. It might be possible to apply approaches from incremental and parallel SAT solving [86] also to QBF.

Expansion-Based Solving

Currently most parallel solvers implement search-based solving by QCDCL (see Table 4.1 above). However, recently expansion-based solving [5, 12] in combination with CEGAR [42, 43] has been shown to be powerful in solving many practically relevant classes of formulas.

Since expansion is orthogonal to QCDCL regarding proof complexity [11, 43], there is considerable potential in parallelizing solvers that employ CEGAR-based expansion. However, it has not been deeply investigated how to leverage the power of CEGAR approaches in parallel solving. For example, different processes could work on different abstractions of a formula at the same time and then share or synchronize counterexamples that they have found with respect to the different abstractions.

Duality-Aware Reasoning

In the context of QBF solving it is well known that reasoning on a propositional CNF introduces a bias towards the search for conflicts. A CNF is easily falsified by an assignment that falsifies at least one clause. Based on such falsifying assignments, in QCDCL learned clauses can be derived by rule *cl-init* of the Q-resolution calculus. Compared to falsifying assignments, it is more difficult to satisfy a CNF as all the clauses must be satisfied. Therefore, for the search for solutions, a formula in disjunctive normal form (DNF), i.e., a disjunction of cubes, would be better suited. A DNF is dual to a CNF in the sense that a DNF can be satisfied easily by satisfying at least one of its cubes.

To benefit from properties of both CNFs and DNFs, approaches have been presented that reason on a CNF and on a DNF representation of the given QBF at the same time. This way, propagations are performed on the CNF and on the DNF (e.g., [37, 49, 89]). However, so far these approaches have been realized systematically only in a sequential manner. The solver par-pd-depqbf is based on the observation of Van Gelder [30] who proposes to solve a formula in CNF as well as in DNF by calling two separate solver instances in parallel. However, in this approach there is neither communication nor knowledge sharing between the solver instances.

Proof Generation

The generation of proofs becomes more and more important for the practical applicability of QBF solvers. Proofs serve two purposes: on the one hand, they allow for the independent validation of the correctness of a solver's result by an efficient checker, and on the other hand they allow the extraction of *Skolem* and *Herbrand functions*. Skolem functions represent a *strategy* for the assignment of existential variables if a formula is satisfiable. Likewise, Herbrand functions represent a strategy for selecting the assignments of universal variables in unsatisfiable formulas (see also the informal presentation of these functions by means of the example in Fig. 4.1 in Section 4.1).

Strategies are crucial for practical applications of QBF solvers. For example, given a PCNF ϕ which models an instance of some problem to be solved, a solution to the problem instance can be computed from a strategy for ϕ.

Skolem and Herbrand functions can be efficiently extracted from *Q-resolution proofs* as produced by sequential QCDCL solvers [6]. However, in parallel QBF solving, currently none of the presented approaches supports the generation of proofs or strategies in terms of Skolem and Herbrand functions, respectively. However, for parallel solvers based on QCDCL a potential approach to proof generation would be to combine the respective proofs of the subproblems that have been solved by the different threads or processors. To this end, ideas from proof generation in parallel SAT solving [39] may also be applicable.

Testing and Debugging

One of the major challenges in developing a sequential QBF solver is a stable implementation, i.e., an implementation which does not crash and which returns correct results. In general, implementations of QBF solvers are more complex than implementations of SAT solvers due to the complexity of handling nested quantifiers that is present in QBFs. Furthermore, in order to achieve good solving performance, it is necessary to equip QBF solvers with advanced data structures and optimizations to prune the search space. At the same time, these optimizations may hinder the efficient implementation of advanced features such as proof generation and incremental solving.

For sequential solving, effective approaches to testing and debugging of solvers [16] exist. First, *fuzz testing* has proven itself to be powerful for finding problematic corner cases and conceptual errors in an implementation. A fuzz test generates random formulas according to predefined random models such that the formulas are not too hard to solve. The goal is to achieve a high testing throughput together with a uniform distribution of satisfiable and unsatisfiable instances.

Second, *delta debugging* is used to automatically simplify large formulas on which a solver exhibits incorrect behavior. To this end, clauses are successively removed from the formula and literals are removed from clauses so that the incorrect behavior of the solver is preserved. In the end, the result of delta debugging is a formula which is reasonably small so that the run of the solver can be inspected manually using traditional debugging techniques.

Third, *model-based testers* [1, 2] have been found particularly useful in testing the behavior of incremental solvers via the API provided for incremental use. While in fuzz testing and delta debugging solvers are considered as black boxes, a model-based testing environment comes with a tighter integration of the solver. Sequences of function calls of the solver's API are automatically generated and replayed in order to test solver behavior on the sequence. This approach also allows us to replay entire solver runs where certain bugs were triggered.

It is well known that testing and debugging of parallel solvers is far more complex than for sequential solvers. In parallel QBF solving, this problem is made worse by the higher complexity that is intrinsic to QBF solvers, compared to SAT solvers. For the development of robust parallel QBF solvers, it may be useful to combine the generation of proofs and strategies outlined above with approaches to automated testing and debugging.

4.7 Conclusion

Already in the very early years of QBF solving attempts were made to exploit the full computational power of modern computer architectures, ranging from multicore processors to huge clusters as found in modern cloud-based systems. However, compared to the advancements made in sequential QBF solving, a lot of the potential of parallelizing QBF solving has still not been exploited.

We have reviewed and classified parallel approaches to QBF solving that either were published in the literature or that participated in the parallel track of the QBF competition QBFEVAL'16 held in 2016. Overall, we identified 11 approaches; 10 of them are implemented. The implementations of five systems are publicly available.

Unfortunately, not all of the QBFEVAL'16 participants are among those solvers. As half of the systems are not available (anymore), we did not carry out an empirical evaluation. The parallel track of QBFEVAL'16 was not competitive due to the small number of participating systems. However, it is still remarkable that the track could be carried out, because in the previous editions of QBFEVAL it had to be canceled. This fact might be a first indicator of an upwards trend in parallel QBF solving. Given the high computational complexity of QBF solving in general, the large variety of sequential solvers and the power of modern computer architectures, we see considerable potential to speed up QBF solving by parallel approaches.

References

[1] Cyrille Artho, Armin Biere, and Martina Seidl. Model-Based Testing for Verification Back-Ends. In Margus Veanes and Luca Viganò, editors, *Proc. of the 7th Int. Conference on Tests and Proofs (TAP 2017)*, volume 7942 of *LNCS*, pages 39–55. Springer, 2013.

[2] Cyrille Artho, Martina Seidl, Quentin Gros, Eun-Hye Choi, Takashi Kitamura, Akira Mori, Rudolf Ramler, and Yoriyuki Yamagata. Model-Based Testing of Stateful APIs with Modbat. In Myra B. Cohen, Lars Grunske, and Michael Whalen, editors, *Proc. of the 30th Int. Conference on Automated Software Engineering (ASE 2015)*, pages 858–863. IEEE Computer Society, 2015.

[3] Bengt Aspvall, Christos Levcopoulos, Andrzej Lingas, and Robert Storlind. On 2-QBF Truth Testing in Parallel. *Information Processing Letters*, 57(2):89–93, 1996.

[4] Bengt Aspvall, Michael F. Plass, and Robert Endre Tarjan. A linear-time algorithm for testing the truth of certain quantified Boolean formulas. *Inf. Process. Lett.*, 8(3):121–123, 1979.

[5] Abdelwaheb Ayari and David A. Basin. QUBOS: Deciding Quantified Boolean Logic Using Propositional Satisfiability Solvers. In Mark Aagaard and John W. O'Leary, editors, *Proc. of the 4th Int. Conference on Formal Methods in Computer-Aided Design (FMCAD 2002)*, volume 2517 of *LNCS*, pages 187–201. Springer, 2002.

[6] Valeriy Balabanov and Jie-Hong R. Jiang. Unified QBF certification and its applications. *Formal Methods in System Design*, 41(1):45–65, 2012.

[7] Valeriy Balabanov, Jie-Hong Roland Jiang, Mikolás Janota, and Magdalena Widl. Efficient Extraction of QBF (Counter)models from Long-Distance Resolution Proofs. In Blai Bonet and Sven Koenig, editors, *Proc. of the 29th AAAI Conference on Artificial Intelligence (AAAI 2015)*, pages 3694–3701. AAAI Press, 2015.

[8] Tomas Balyo and Florian Lonsing. HordeQBF: A Modular and Massively Parallel QBF Solver. In Nadia Creignou and Daniel Le Berre, editors, *Proc. of the 19th Int. Conference on Theory and Applications of Satisfiability Testing (SAT 2016)*, volume 9710 of *LNCS*, pages 531–538. Springer, 2016.

[9] Tomas Balyo, Peter Sanders, and Carsten Sinz. HordeSat: A Massively Parallel Portfolio SAT Solver. In Marijn Heule and Sean Weaver, editors, *Proc. of the 18th Int. Conference on Theory and Applications of Satisfiability Testing (SAT 2015)*, volume 9340 of *LNCS*, pages 156–172. Springer, 2015.

[10] Marco Benedetti and Hratch Mangassarian. QBF-Based Formal Verification: Experience and Perspectives. *Journal on Satisfiability, Boolean Modeling and Computation*, 5(1-4):133–191, 2008.

[11] Olaf Beyersdorff, Leroy Chew, and Mikolás Janota. Proof Complexity of Resolution-based QBF Calculi. In Ernst W. Mayr and Nicolas Ollinger, editors, *Proc. of the 32nd Int. Symposium on Theoretical Aspects of Computer Science (STACS 2015)*, volume 30 of *LIPIcs*, pages 76–89. Schloss Dagstuhl - Leibniz-Zentrum fuer Informatik, 2015.

[12] Armin Biere. Resolve and Expand. In Holger H. Hoos and David G. Mitchell, editors, *Proc. of the 7th Int. Conference on Theory and Applications of Satisfiability Testing (SAT 2004)*, volume 3542 of *LNCS*, pages 59–70. Springer, 2004.

[13] Armin Biere, Marijn Heule, Hans van Maaren, and Toby Walsh, editors. *Handbook of Satisfiability*, volume 185 of *Frontiers in Artificial Intelligence and Applications*. IOS Press, 2009.

[14] Roderick Bloem, Robert Könighofer, and Martina Seidl. SAT-Based Synthesis Methods for Safety Specs. In Kenneth L. McMillan and Xavier Rival, editors, *Proc. of the 15th Int. Conference on Verification, Model Checking, and Abstract Interpretation (VMCAI 2014)*, volume 8318 of *LNCS*, pages 1–20. Springer, 2014.

[15] Bart Bogaerts, Tomi Janhunen, and Shahab Tasharrofi. Solving QBF instances with nested SAT solvers. In Adnan Darwiche, editor, *Proc. of the 2016 AAAI Workshop Beyond NP*, volume WS-16-05 of *AAAI Workshops*. AAAI Press, 2016.

[16] Robert Brummayer, Florian Lonsing, and Armin Biere. Automated testing and debugging of SAT and QBF solvers. In Ofer Strichman and Stefan Szeider, editors, *Proc. of the 13th Int. Conference on Theory and Applications of Satisfiability Testing (SAT 2010)*, volume 6175 of *LNCS*, pages 44–57. Springer, 2010.

[17] Uwe Bubeck and Hans Kleine Büning. Bounded Universal Expansion for Preprocessing QBF. In *Proc. of the 10th Int. Conference on Theory and Applications of Satisfiability Testing (SAT 2007)*, volume 4501 of *LNCS*, pages 244–257. Springer, 2007.

[18] Marco Cadoli, Andrea Giovanardi, and Marco Schaerf. An Algorithm to Evaluate Quantified Boolean Formulae. In Jack Mostow and Chuck Rich, editors, *Proc. of the 15th National Conference on Artificial Intelligence and 10th Innovative Applications of Artificial Intelligence Conference (AAAI/IAAI 1998)*, pages 262–267. AAAI Press / The MIT Press, 1998.

[19] Koen Claessen, Niklas Eén, Mary Sheeran, Niklas Sörensson, Alexey Voronov, and Knut Åkesson. SAT-Solving in Practice, with a Tutorial Example from Supervisory Control. *Discrete Event Dynamic Systems*, 19(4):495–524, 2009.

[20] Edmund M. Clarke, Orna Grumberg, Somesh Jha, Yuan Lu, and Helmut Veith. Counterexample-guided abstraction refinement for symbolic model checking. *Journal of the ACM*, 50(5):752–794, 2003.

[21] Benoit Da Mota. *Quantified Boolean formulae: formal processings and parallel computations.* Thesis, Université d'Angers, December 2010.

[22] Martin Davis, George Logemann, and Donald W. Loveland. A machine program for theorem-proving. *Communications of the ACM*, 5(7):394–397, 1962.

[23] Niklas Eén and Armin Biere. Effective preprocessing in SAT through variable and clause elimination. In Fahiem Bacchus and Toby Walsh, editors, *Proc. of the 8th Int. Conference on Theory and Applications of Satisfiability Testing (SAT 2005)*, volume 3569 of *LNCS*, pages 61–75. Springer, 2005.

[24] Niklas Eén and Niklas Sörensson. An Extensible SAT-solver. In Enrico Giunchiglia and Armando Tacchella, editors, *Proc. of the 9th Int. Conference on Theory and Applications of Satisfiability Testing (SAT 2006)*, volume 2919 of *LNCS*, pages 502–518. Springer, 2003.

[25] Niklas Eén and Niklas Sörensson. Temporal induction by incremental SAT solving. *Electr. Notes Theor. Comput. Sci.*, 89(4):543–560, 2003.

[26] Uwe Egly, Martin Kronegger, Florian Lonsing, and Andreas Pfandler. Conformant planning as a case study of incremental QBF solving. *Ann. Math. Artif. Intell.*, 80(1):21–45, 2017.

[27] Wolfgang Faber and Francesco Ricca. Solving hard ASP programs efficiently. In Chitta Baral, Gianluigi Greco, Nicola Leone, and Giorgio Terracina, editors, *Proc. of the 8th Int. Conference on Logic Programming and Nonmonotonic Reasoning (LPNMR 2005)*, volume 3662 of *LNCS*, pages 240–252. Springer, 2005.

[28] Rainer Feldmann, Burkhard Monien, and Stefan Schamberger. A Distributed Algorithm to Evaluate Quantified Boolean Formulae. In Henry A. Kautz and Bruce W. Porter, editors, *Proc. of the 17th Nat. Conference on Artificial Intelligence and 12th Conference on on Innovative Applications of Artificial Intelligence (AAA/IAAI 2000)*, pages 285–290. AAAI Press / The MIT Press, 2000.

[29] Allen Van Gelder. Contributions to the theory of practical quantified Boolean formula solving. In Michela Milano, editor, *Proc. of the 18th Int. Conference on Principles and Practice of Constraint Programming (CP 2012)*, volume 7514 of *LNCS*, pages 647–663. Springer, 2012.

[30] Allen Van Gelder. Primal and Dual Encoding from Applications into Quantified Boolean Formulas. In Christian Schulte, editor, *Proc. of the 19th Int. Conference on Principles and Practice of Constraint Programming (CP 2013)*, volume 8124 of *LNCS*, pages 694–707. Springer, 2013.

[31] Ian P. Gent, Enrico Giunchiglia, Massimo Narizzano, Andrew G. D. Rowley, and Armando Tacchella. Watched data structures for QBF solvers. In Enrico Giunchiglia and Armando Tacchella, editors, *Proc. of the 6th Int. Conference on Theory and Applications of Satisfiability Testing (SAT 2003)*, volume 2919 of *LNCS*, pages 25–36. Springer, 2003.

[32] Enrico Giunchiglia, Paolo Marin, and Massimo Narizzano. Reasoning with quantified Boolean formulas. In Armin Biere, Marijn Heule, Hans van Maaren, and Toby Walsh, editors, *Handbook of Satisfiability*, volume 185 of *Frontiers in Artificial Intelligence and Applications*, pages 761–780. IOS Press, 2009.
[33] Enrico Giunchiglia, Paolo Marin, and Massimo Narizzano. QuBE7.0. *Journal on Satisfiability, Boolean Modeling and Computation*, 7(2-3):83–88, 2010.
[34] Enrico Giunchiglia, Paolo Marin, and Massimo Narizzano. sQueezeBF: An Effective Preprocessor for QBFs Based on Equivalence Reasoning. In Ofer Strichman and Stefan Szeider, editors, *Proc. of the 13th Int. Conference on Theory and Applications of Satisfiability Testing (SAT 2010)*, volume 6175 of *LNCS*, pages 85–98. Springer, 2010.
[35] Enrico Giunchiglia, Massimo Narizzano, and Armando Tacchella. Clause/Term Resolution and Learning in the Evaluation of Quantified Boolean Formulas. *J. Artif. Intell. Res. (JAIR)*, 26:371–416, 2006.
[36] Alexandra Goultiaeva and Fahiem Bacchus. Recovering and Utilizing Partial Duality in QBF. In Matti Järvisalo and Allen Van Gelder, editors, *Proc. of the 16th Int. Conference on Theory and Applications of Satisfiability Testing (SAT 2013)*, volume 7962 of *LNCS*, pages 83–99. Springer, 2013.
[37] Alexandra Goultiaeva, Martina Seidl, and Armin Biere. Bridging the gap between dual propagation and CNF-based QBF solving. In Enrico Macii, editor, *Proc. of the Int. Conference on Design, Automation and Test in Europe (DATE 2013)*, pages 811–814. EDA Consortium / ACM DL, 2013.
[38] Marijn Heule, Matti Järvisalo, Florian Lonsing, Martina Seidl, and Armin Biere. Clause Elimination for SAT and QSAT. *J. Artif. Intell. Res. (JAIR)*, 53:127–168, 2015.
[39] Marijn J.H. Heule and Armin Biere. Compositional Propositional Proofs. In Martin Davis, Ansgar Fehnker, Annabelle McIver, and Andrei Voronkov, editors, *Proc. of the 20th Int. Conference on Logic for Programming, Artificial Intelligence, and Reasoning (LPAR-20)*, volume 9450 of *LNCS*, pages 444–459. Springer, 2015.
[40] Tamir Heyman, Dan Smith, Yogesh Mahajan, Lance Leong, and Husam Abu-Haimed. Dominant Controllability Check Using QBF-Solver and Netlist Optimizer. In Carsten Sinz and Uwe Egly, editors, *Proc. of the 17th Int. Conference on Theory and Applications of Satisfiability Testing (SAT 2014)*, volume 8561 of *LNCS*, pages 227–242. Springer, 2014.
[41] Mikolás Janota, Charles Jordan, Will Klieber, Florian Lonsing, Martina Seidl, and Allen Van Gelder. The QBF Gallery 2014: The QBF Competition at the FLoC Olympic Games. *Journal on Satisfiability, Boolean Modeling and Computation*, 9:187–206, 2016.
[42] Mikolás Janota, William Klieber, Joao Marques-Silva, and Edmund M. Clarke. Solving QBF with counterexample guided refinement. *Artif. Intell.*, 234:1–25, 2016.
[43] Mikolás Janota and Joao Marques-Silva. Expansion-based QBF solving versus Q-resolution. *Theor. Comput. Sci.*, 577:25–42, 2015.

[44] Mikolás Janota and Joao Marques-Silva. Solving QBF by Clause Selection. In Qiang Yang and Michael Wooldridge, editors, *Proc. of the 24th Int. Joint Conference on Artificial Intelligence (IJCAI 2015)*, pages 325–331. AAAI Press, 2015.

[45] Matti Järvisalo, Marijn Heule, and Armin Biere. Inprocessing Rules. In Bernhard Gramlich, Dale Miller, and Uli Sattler, editors, *Proc. of the 6th Int. Joint Conference on Automated Reasoning (IJCAR 2012)*, volume 7364 of *LNCS*, pages 355–370. Springer, 2012.

[46] Charles Jordan, Lukasz Kaiser, Florian Lonsing, and Martina Seidl. MPIDepQBF: Towards Parallel QBF Solving without Knowledge Sharing. In Carsten Sinz and Uwe Egly, editors, *Proc. of the 17th Int. Conference on Theory and Applications of Satisfiability Testing (SAT 2014)*, volume 8561 of *LNCS*, pages 430–437. Springer, 2014.

[47] Hans Kleine Büning and Uwe Bubeck. Theory of quantified Boolean formulas. In Armin Biere, Marijn Heule, Hans van Maaren, and Toby Walsh, editors, *Handbook of Satisfiability*, volume 185 of *Frontiers in Artificial Intelligence and Applications*, pages 735–760. IOS Press, 2009.

[48] Hans Kleine Büning, Marek Karpinski, and Andreas Flögel. Resolution for Quantified Boolean Formulas. *Inf. Comput.*, 117(1):12–18, 1995.

[49] William Klieber, Samir Sapra, Sicun Gao, and Edmund M. Clarke. A non-prenex, non-clausal QBF solver with game-state learning. In Ofer Strichman and Stefan Szeider, editors, *Proc. of the 13th Int. Conference on Theory and Applications of Satisfiability Testing (SAT 2010)*, volume 6175 of *LNCS*, pages 128–142. Springer, 2010.

[50] Reinhold Letz. Lemma and Model Caching in Decision Procedures for Quantified Boolean Formulas. In Uwe Egly and Christian G. Fermüller, editors, *Proc. of the Int. Conference on Automated Reasoning with Analytic Tableaux and Related Methods (TABLEAUX 2002)*, volume 2381 of *LNCS*, pages 160–175. Springer, 2002.

[51] Matthew Lewis, Tobias Schubert, Bernd Becker, Paolo Marin, Massimo Narizzano, and Enrico Giunchiglia. Parallel QBF Solving with Advanced Knowledge Sharing. *Fundamenta Informaticae*, 107(2-3):139–166, 2011.

[52] Matthew D.T. Lewis, Tobias Schubert, and Bernd Becker. QmiraXT - A Multithreaded QBF Solver. In Carsten Gremzow and Nico Moser, editors, *Methoden und Beschreibungssprachen zur Modellierung und Verifikation von Schaltungen und Systemen (MBMV)*, pages 7–16. Universitätsbibliothek Berlin, Germany, 2009.

[53] Tao Li and Nan-feng Xiao. Parallel solving model for quantified Boolean formula based on machine learning. *Journal of Central South University*, 20(11):3156–3165, 2013.

[54] Paolo Liberatore. Redundancy in logic I: CNF propositional formulae. *Artif. Intell.*, 163(2):203–232, 2005.

[55] Florian Lonsing, Fahiem Bacchus, Armin Biere, Uwe Egly, and Martina Seidl. Enhancing Search-Based QBF Solving by Dynamic Blocked Clause Elimination. In Martin Davis, Ansgar Fehnker, Annabelle McIver, and Andrei

Voronkov, editors, *Proc. of the 20th Int. Conference on Logic for Programming, Artificial Intelligence, and Reasoning (LPAR 2015)*, volume 9450 of *LNCS*, pages 418–433. Springer, 2015.

[56] Florian Lonsing and Armin Biere. DepQBF: A Dependency-Aware QBF Solver. *Journal on Satisfiability, Boolean Modeling and Computation*, 7(2-3):71–76, 2010.

[57] Florian Lonsing and Armin Biere. Integrating dependency schemes in search-based QBF solvers. In Ofer Strichman and Stefan Szeider, editors, *Proc. of the 13th Int. Conference on Theory and Applications of Satisfiability Testing (SAT 2010)*, volume 6175 of *LNCS*, pages 158–171. Springer, 2010.

[58] Florian Lonsing and Uwe Egly. Incremental QBF Solving. In Barry O'Sullivan, editor, *Proc. of the 20th Int. Conference on Principles and Practice of Constraint Programming (CP 2014)*, volume 8656 of *LNCS*, pages 514–530. Springer, 2014.

[59] Florian Lonsing, Uwe Egly, and Martina Seidl. Q-Resolution with Generalized Axioms. In Nadia Creignou and Daniel Le Berre, editors, *Proc. of the 19th Int. Conference on Theory and Applications of Satisfiability Testing (SAT 2016)*, volume 9710 of *LNCS*, pages 435–452. Springer, 2016.

[60] Florian Lonsing, Martina Seidl, and Allen Van Gelder. The QBF Gallery: Behind the scenes. *Artif. Intell.*, 237:92–114, 2016.

[61] Paolo Marin, Christian Miller, Matthew D.T. Lewis, and Bernd Becker. Verification of partial designs using incremental QBF solving. In Wolfgang Rosenstiel and Lothar Thiele, editors, *Proc. of the Design, Automation & Test in Europe Conference & Exhibition (DATE 2012)*, pages 623–628. IEEE, 2012.

[62] Paolo Marin, Massimo Narizzano, Enrico Giunchiglia, Matthew D.T. Lewis, Tobias Schubert, and Bernd Becker. Comparison of knowledge sharing strategies in a parallel QBF solver. In *Proc. of the Int. Conference on High Performance Computing & Simulation (HPCS 2009)*, pages 161–167. IEEE, 2009.

[63] Paolo Marin, Massimo Narizzano, Luca Pulina, Armando Tacchella, and Enrico Giunchiglia. Twelve Years of QBF Evaluations: QSAT Is PSPACE-Hard and It Shows. *Fundam. Inform.*, 149(1-2):133–158, 2016.

[64] Albert R. Meyer and Larry J. Stockmeyer. The Equivalence Problem for Regular Expressions with Squaring Requires Exponential Space. In *13th Annual Symposium on Switching and Automata Theory*, pages 125–129. IEEE Computer Society, 1972.

[65] Christian Miller, Paolo Marin, and Bernd Becker. Verification of partial designs using incremental QBF. *AI Commun.*, 28(2):283–307, 2015.

[66] Matthew W. Moskewicz, Conor F. Madigan, Ying Zhao, Lintao Zhang, and Sharad Malik. Chaff: Engineering an Efficient SAT Solver. In *Proc. of the 38th Design Automation Conference (DAC 2001)*, pages 530–535. ACM, 2001.

[67] Benoit Da Mota, Pascal Nicolas, and Igor Stéphan. A new parallel architecture for QBF tools. In *Proc. of the Int. Conference on High Performance Computing and Simulation (HPCS 2010)*, pages 324–330. IEEE, 2010.

[68] Alexander Nadel and Vadim Ryvchin. Efficient SAT Solving under Assumptions. In Alessandro Cimatti and Roberto Sebastiani, editors, *Proc. of the 15th*

Int. Conference on Theory and Applications of Satisfiability Testing (SAT 2012), volume 7317 of *LNCS*, pages 242–255. Springer, 2012.

[69] Knot Pipatsrisawat and Adnan Darwiche. A Lightweight Component Caching Scheme for Satisfiability Solvers. In João Marques-Silva and Karem A. Sakallah, editors, *Proc. of the 10th Int. Conference on Theory and Applications of Satisfiability Testing (SAT 2007)*, volume 4501 of *LNCS*, pages 294–299. Springer, 2007.

[70] David A. Plaisted, Armin Biere, and Yunshan Zhu. A satisfiability procedure for quantified Boolean formulae. *Discrete Applied Mathematics*, 130(2):291–328, 2003.

[71] Luca Pulina. The Ninth QBF Solvers Evaluation - Preliminary Report. In *Proc. of the 4th Int. Workshop on Quantified Boolean Formulas (QBF 2016)*, volume 1719, pages 1–13. CEUR Workshop Proceedings, 2016.

[72] Markus N. Rabe and Sanjit A. Seshia. Incremental Determinization. In Nadia Creignou and Daniel Le Berre, editors, *Proc. of the 19th Int. Conference on Theory and Applications of Satisfiability Testing (SAT 2016)*, volume 9710 of *LNCS*, pages 375–392. Springer, 2016.

[73] Markus N. Rabe and Leander Tentrup. CAQE: A certifying QBF solver. In Roope Kaivola and Thomas Wahl, editors, *Proc. of the Int. Conference on Formal Methods in Computer-Aided Design (FMCAD 2015)*, pages 136–143. IEEE, 2015.

[74] Jussi Rintanen. Improvements to the evaluation of quantified Boolean formulae. In Thomas Dean, editor, *Proc. of the 16th Int. Joint Conference on Artificial Intelligence (IJCAI 1999)*, pages 1192–1197. Morgan Kaufmann, 1999.

[75] Jussi Rintanen. Asymptotically Optimal Encodings of Conformant Planning in QBF. In *Proc. of the 22nd AAAI Conference on Artificial Intelligence (AAAI 2007)*, pages 1045–1050. AAAI Press, 2007.

[76] Thomas J. Schaefer. On the Complexity of Some Two-Person Perfect-Information Games. *J. Comput. Syst. Sci.*, 16(2):185–225, 1978.

[77] Tobias Schubert, Matthew D.T. Lewis, and Bernd Becker. PaMiraXT: Parallel SAT solving with threads and message passing. *Journal on Satisfiability, Boolean Modeling and Computation*, 6(4):203–222, 2009.

[78] João P. Marques Silva, Inês Lynce, and Sharad Malik. Conflict-driven clause learning SAT solvers. In Armin Biere, Marijn Heule, Hans van Maaren, and Toby Walsh, editors, *Handbook of Satisfiability*, volume 185 of *Frontiers in Artificial Intelligence and Applications*, pages 131–153. IOS Press, 2009.

[79] João P. Marques Silva and Karem A. Sakallah. GRASP - a new search algorithm for satisfiability. In *Proc. of the Int. Conference on Computer-Aided Design (ICCAD 1996)*, pages 220–227, 1996.

[80] João P. Marques Silva and Karem A. Sakallah. GRASP: A search algorithm for propositional satisfiability. *IEEE Trans. Computers*, 48(5):506–521, 1999.

[81] L. J. Stockmeyer and A. R. Meyer. Word problems requiring exponential time (preliminary report). In *Proc. of the 5th Annual ACM Symposium on Theory of Computing (STOC'73)*, pages 1–9, New York, NY, USA, 1973. ACM.

[82] Larry J. Stockmeyer. The Polynomial-Time Hierarchy. *Theor. Comput. Sci.*, 3(1):1–22, 1976.
[83] Leander Tentrup. Non-prenex QBF Solving Using Abstraction. In *Proc. of the 19th Int. Conference on Theory and Applications of Satisfiability Testing (SAT 2016)*, volume 9710 of *LNCS*, pages 393–401. Springer, 2016.
[84] Leander Tentrup. Solving QBF by abstraction. *CoRR*, abs/1604.06752, 2016.
[85] Yakir Vizel, Georg Weissenbacher, and Sharad Malik. Boolean satisfiability solvers and their applications in model checking. *Proceedings of the IEEE*, 103(11):2021–2035, 2015.
[86] Siert Wieringa and Keijo Heljanko. Asynchronous Multi-core Incremental SAT Solving. In Nir Piterman and Scott A. Smolka, editors, *Proc. of the 19th Int. Conference on Tools and Algorithms for the Construction and Analysis of Systems (TACAS 2013)*, volume 7795 of *LNCS*, pages 139–153. Springer, 2013.
[87] Ralf Wimmer, Sven Reimer, Paolo Marin, and Bernd Becker. HQSpre - An Effective Preprocessor for QBF and DQBF. In *Proc. of the 23rd Int. Conference on Tools and Algorithms for the Construction and Analysis of Systems (TACAS 2017)*, volume 10205 of *LNCS*, pages 373–390. Springer, 2017.
[88] Hantao Zhang, Maria Paola Bonacina, and Jieh Hsiang. PSATO: a Distributed Propositional Prover and its Application to Quasigroup Problems. *J. Symb. Comput.*, 21(4):543–560, 1996.
[89] Lintao Zhang. Solving QBF by Combining Conjunctive and Disjunctive Normal Forms. In *Proc. of the 21st Nat. Conference on Artificial Intelligence and the 8th Innov. Applications of Artificial Intelligence Conference (AAAI/IAAI 2006)*, pages 143–150. AAAI Press, 2006.
[90] Lintao Zhang, Conor F. Madigan, Matthew W. Moskewicz, and Sharad Malik. Efficient Conflict Driven Learning in Boolean Satisfiability Solver. In Rolf Ernst, editor, *Proc. of the Int. Conference on Computer-Aided Design (ICCAD 2001)*, pages 279–285. IEEE, 2001.
[91] Lintao Zhang and Sharad Malik. Conflict driven learning in a quantified Boolean Satisfiability solver. In Lawrence T. Pileggi and Andreas Kuehlmann, editors, *Proc. of the Int. Conference on Computer-Aided Design (ICCAD 2002)*, pages 442–449. ACM / IEEE Computer Society, 2002.

Chapter 5
Parallel Satisfiability Modulo Theories

Antti E.J. Hyvärinen and Christoph M. Wintersteiger

Abstract Satisfiability Modulo Theories (SMT) is an extension of the propositional satisfiability problem (SAT) to other, well-chosen (first-order) theories such as integers, reals, and bit-vectors. This approach currently enjoys much popularity, especially in the field of software verification, where SMT solvers have become the de facto standard tool for the discharge of verification conditions. The development of *parallel* SMT solvers is still in its infancy, but the first general paradigms have been established. This chapter provides an overview of the recent advances in this area, specifically algorithm portfolio, search-space partitioning, and problem decomposition techniques, and how they relate to each other in theory and practice.

5.1 Introduction

Satisfiability Modulo Theories (SMT) [5] is an initiative in the area of automated deduction that aims to foster development of techniques for satisfiability checking that go beyond solving purely Boolean SAT problems. The scope of SMT is first-order logic with particular, and well-chosen, background theories of industrial or academic interest. In contrast to general first-order logic, SMT does not require background theories to be finitely axiomatizable or even decidable and still allows us to compute results that are of practical interest efficiently.

The traditional application field of SMT solvers is software verification, where a restriction to particular background theories enabled the development of specialized decision procedures that perform particularly well in determining satisfiability. Frequently the software is modeled so that satisfying solutions correspond to bugs or other undesirable program behavior. Today, SMT solvers are applied in an increasing

Antti E.J. Hyvärinen
Università della Svizzera italiana, Lugano, Switzerland, e-mail: antti.hyvaerinen@usi.ch

Christoph M. Wintersteiger
Microsoft Research, e-mail: cwinter@microsoft.com

Y. Hamadi und L. Sais (eds.), *Handbook of Parallel Constraint Reasoning*,
https://doi.org/10.1007/978-3-319-63516-3_5

number of applications outside of the traditional areas, including computational biology (e.g. [84, 6]), chemistry (e.g., [32]), and material science (e.g., [31]). Current research also attempts to lift the approach of providing a carefully crafted set of background theories to other domains, for instance model checking [34], optimization [77], planning [40], and probabilistic inference [76] – all modulo theories.

For some (combinations of) background theories, the computational cost of SMT solving can be very high and, in terms of computational complexity, often greatly exceeds the cost of NP-complete problems such as Boolean SAT. It is therefore of general interest to study ways of improving the problem-solving performance by addition of more parallel computing power to the SMT solver.

SMT solvers are conceptually based on SAT solvers and certain parallelization techniques can be applied in very similar ways in both approaches. Perhaps surprisingly the intricate interaction with the theory solvers results in certain techniques that work in a predictable way in SAT solvers not maintaining similar behavior in SMT solving. In particular the techniques for partitioning search space turn out to be significantly different in SMT, the technique being very efficient when applied in the presence of some background theories and even detrimental in the presence of other background theories. It seems that the background theories require a very different set of trade-offs to be considered (e.g., [83, 65]).

In this chapter we address the challenges in parallelizing SMT solvers using both multi-core and cloud-based computing environments and a variety of parallelization approaches.

5.2 General Preliminaries

We rely on the basic definitions of Boolean variables, literals, clauses, etc. from Chapter 1, Parallel Satisfiability. Some terms used in the SMT research community and their publications may be confusing to the uninitiated reader, so we provide a brief description of the most important concepts here.

5.2.1 Theories

SMT (Satisfiability Modulo Theories) focuses on the satisfiability problem for first-order logic with particular, and well-chosen, background theories. Today, those theories are Booleans, arrays, bit-vectors, floating-point numbers, integers, real numbers, and uninterpreted symbols (equality and uninterpreted functions and sorts). From these theories, a number of fragments have been identified as academically and industrially important. Currently those fragments are as follows:

- core theory: Booleans and equalities,
- uninterpreted functions and sorts (UF),

- (infinite-size) arrays (including extensionality) (A),
- fixed-size bit-vectors (BV),
- floating-point numbers (FP),
- (non-)linear integer arithmetic (NIA, LIA),
- (non-)linear real arithmetic (NRA, LRA), and
- integer and real difference logic (IDL, RDL).

The abbreviations of those fragments are then combined to identify particular *logics*, where QF is used to indicate that a logic is quantifier-free. For instance, QF_AUFLIA is the quantifier-free theory of arrays, uninterpreted functions, and linear integer arithmetic. SMT solvers implement decision procedures for some or all of these logics and they are evaluated on a large set of community-contributed benchmarks available in the SMT library [5].

Note that SMT solvers do not necessarily implement specialized decision procedures for each logic. Instead, they employ *theory combination* strategies to craft decision procedures by combining more general *core theory solvers*. For instance, a general ‘arithmetic’ theory solver, usually implemented as a backtrackable variation of the Simplex algorithm [29], may be used for multiple logics involving some fragments of integer and real arithmetic.

5.2.2 The Underlying Conflict-Driven, Clause-Learning SAT Solver

The input to SMT solvers is usually a set of assertions (Boolean expressions, constraints), which may have a rich Boolean structure (the *skeleton*) that is not necessarily in any normal form. In practice it is often rewritten into Conjunctive Normal Form (CNF), such that existing SAT solver technology may be used to solve the skeleton. Any (partial or complete) model for the skeleton of the formula implies that some subset of theory literals needs to be solved.

Example 1. Consider the problem of solving

$$a = b \wedge (f(a) - f(b) = c) \wedge \neg(c \leq 0) ,$$

where $a, b, c \in \mathbb{N}$ are integers and $f : \mathbb{N} \to \mathbb{N}$ is an uninterpreted function with integer range and domain. First, we introduce new Boolean variables x_1, x_2, x_3 to obtain

$$x_1 \wedge x_2 \wedge \neg x_3 \;\wedge\; x_1 \equiv (a = b) \wedge x_2 \equiv (f(a) - f(b) = c) \wedge x_3 \equiv (c \leq 0) ,$$

where the first three conjuncts are purely Boolean. A model for this part of the formula is $x_1 = \mathit{true}, x_2 = \mathit{true}, x_3 = \mathit{false}$. This means that every theory solver will now have to solve a *conjunction of literals* instead of an arbitrary Boolean combination of literals. Here, these are

$$(a = b) \;\wedge\; (f(a) - f(b) = c) \;\wedge\; (c > 0) ,$$

which are solved independently by a solver for the theory of (linear) integers and a solver for uninterpreted functions. The theory solvers will determine that these constraints are unsatisfiable, and will return a concise *explanation* $\neg(x_1 \wedge x_2 \wedge \neg x_3) \equiv (\neg x_1 \vee \neg x_2 \vee x_3)$ that the Boolean solver may learn and use for further guiding the search.

There are of course many more details that can make a great difference in runtime performance in practice, but here it is enough to remember that existing SAT technology such as DPLL and CDCL solvers (see Chapter 1, Parallel Satisfiability) are immediately applicable to the Boolean skeleton of SMT formulas.

While SMT solvers learn skeleton clauses, they may also learn *lemmas* that involve theory-specific terms. These may be completely internal to a theory solver (e.g., in the form of caches), but they may also be exposed to the other theories involved. Conceptually, we can think of lemmas introducing new predicates (and thus new Boolean variables), or being new combinations of existing predicates or literals. Whether lemmas are theory-dependent or purely Boolean, heuristics similar to those in SAT solvers are employed to remove unnecessary clauses and lemmas periodically, and various simplification and minimization techniques are used to control memory usage.

5.2.3 Theory Combination

To combine theory solvers, the mechanism underlying many SMT solvers is the Nelson/Oppen theory combination framework [69]. The first step of this is to purify the formula into terms of single theories by introducing new variables for function application terms. As a result two theories only ever share uninterpreted symbols and constants. These sub-formulas are then solved independently and the theory solvers then exchange entailed equalities (syntactic and semantic).

Example 2. Suppose we need to solve the theory part of the two first conjuncts in Example 1:

$$(a=b) \wedge (f(a)-f(b)=0) .$$

The aim is to employ separate theory solvers for subsets of the constraints. Before that, we purify the constraints by introduction of new interface variables. Here, this results in

$$\underbrace{(a=b) \wedge (a=e_1) \wedge (b=e_2)}_{Integers} \wedge \underbrace{(f(e_1)-f(e_2)=e_3) \wedge (e_3=0)}_{Uninterpreted functions}$$

so that e_1, e_2, and e_3 are the interfaces between theories. Basic theory combination as defined by Nelson and Oppen now exchanges *all equalities* over interface variables implied by the current constraints. Note that these equalities are equalities between variables, not necessarily only between numerals. In this over-simplified example, because of $a = b$ we find the new equality $(e_1 = e_2)$, which the integer theory communicates to the UF theory solver. This can then derive $f(e_1) - f(e_2) \equiv f(e_1) - f(e_1)$ and thus $e_3 = 0$, satisfying all constraints.

Nelson and Oppen focused on equalities for the interface between theories, of which there may be an infinite number, and models themselves may be of infinite size, thus requiring theories to be 'stably infinite', and the theory solvers need to be able to perform case splits in non-convex theories. However, there have been extensions to deal with a larger set of theories since then, for instance by Tinelli and Zarba [80]. Variations with certain other desirable properties and better runtime performance include, for example, model-based theory combination [67], which, if possible, communicates fewer equalities, and delayed theory combination, which delays communication of some equalities to a later point in time [13, 17].

5.2.4 Interpolants

Let $v(\phi) = \{x_1, \ldots, x_n\}$ be the free variables of a first-order formula ϕ. Craig's interpolation theorem provides a way to characterize the relationship between two formulas when one implies the other.

Theorem 1 (Craig Interpolation [23]). *Let ϕ and ψ be first-order formulas. If $\phi \Rightarrow \psi$ then there exists an* Interpolant *I such that $\phi \Rightarrow I \wedge I \Rightarrow \psi$ and $v(I) \subseteq v(\phi) \cap v(\psi)$.*

Equivalently, there is an interpolant I such that $\phi \Rightarrow I \wedge I \Rightarrow \neg\psi$ whenever $\phi \wedge \psi$ is unsatisfiable, because $\phi \Rightarrow \neg\psi \equiv \neg(\phi \wedge \psi)$. Craig's theorem guarantees the existence of an interpolant, but does not provide an algorithm to compute one. Such algorithms are known for some logics, most importantly for propositional logic, and for some of them the relationship between multiple choices of interpolants are known. The most important interpolation algorithms for propositional logic can be expressed using the *labeled interpolation system* [27, 3], and includes the McMillan [66] and the Huang [48], Krajícek [61], Pudlák [73] (HKP) interpolants. It is, however, not necessary to understand the details of these algorithms in the remainder of this chapter.

It is worth noting that the Nelson/Oppen theory combination framework exchanges interface equalities, but their proof of soundness does in fact rely on the existence of Craig interpolants, which provides a more general view of theory combination, if all involved theories have interpolants and methods to compute them.

5.2.5 SMT Solvers

The software used for determining the satisfiability of formulas expressed in SMT are called SMT solvers. There are several SMT solver implementations with different strengths. The solvers offering most compelling support for the SMT language include CVC4 [4], Z3 [68], MathSAT [19], Yices [28]. Other solvers specializing in particular theories or problems include OpenSMT [53], MapleSTP, STP, Boolector [16], ABC [15], AProVE [35], iSAT [59], Minkeyrink, ProB [62], Q3B [58], raSAT [81], SMT-RAT [22], SMTInterpol [18], toysmt, Vampire [60], and veriT [12]. Many of these solvers take part in the annual SMT competition (for the latest edition see [21]).

5.3 Portfolios of SMT Solvers

Many branch-and-bound backtracking algorithms exhibit high variance in runtime when small alternations are introduced into the search process [79, 63, 72, 39]. Intuitively an 'unlucky' choice in the heuristic search can lead to a part of the branch-and-bound tree which is particularly difficult to solve. Sometimes such areas could have been avoided, had the search been performed in a slightly different order.

The SMT search can be seen as an instance of a branch-and-bound backtracking algorithm, and experiments confirm that the runtime of an SMT solver exhibits similar behavior. The range of the runtime depends on the instance being solved: for example the runtime of OpenSMT [53] for a fixed instance varies from twofold to several orders of magnitude. Two cumulative runtime distributions for benchmark instances from the SMT-LIB benchmark collection, showing the probability that an instance is solved in time less than a given t, are shown in Figure 5.1. The values are normalized to the minimum measured runtime of the respective instance to make the distributions comparable. Neither behavior is particularly unusual within the benchmark collection, but they represent very different behaviors. For one instance (purple, solid), the runtime ranges from 80 seconds to 400 seconds, resulting in roughly fivefold difference between the slowest and the fastest run. For the other (green, dashed), the runtime ranges from 0.2 seconds to 7.2 seconds, giving a much bigger, 36-fold difference. Both instances are unsatisfiable, and the difference between the two distributions means that the instances will benefit from parallelization approaches in very different ways.

The solving times of some SMT instances seem to obey a *heavy-tailed runtime distribution* [37]. Such distributions have a significant probability of producing 'outlier' samples, that is, a runtime which is far from the median. In practice, the distributions behave as if they had an infinite standard deviation or even an infinite mean. Since SMT solvers in particular in the quantifier-free cases reduce the problem to the satisfiability of a finite-sized propositional formula, the formulas have a finite search space. As the heuristic parameters are usually also finite, the distributions are, technically, finite as well. However, since the search space is in the worst case exponential in the size of the formula, the statistics can in practice be considered to be infinite for suitable formulas [38].

The small variations in the search can result, for example, from explanation generation inside the theory solvers or the process used for selecting decision literals. Most heuristics employ randomization to break ties, and often implement a form

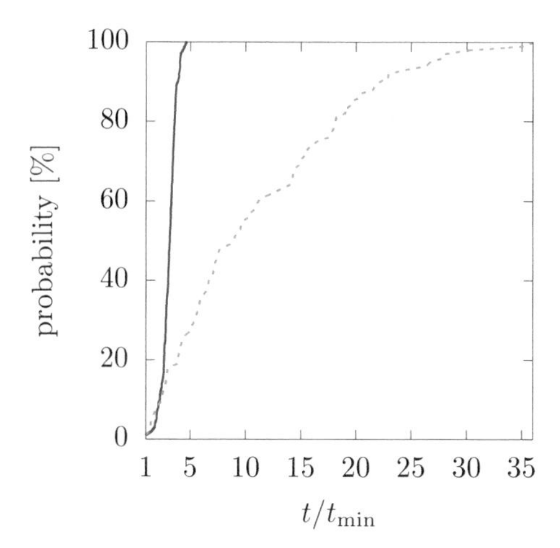

Fig. 5.1: Runtime distributions for two unsatisfiable formulas from the QF_UF category of SMT-LIB. The figure shows the cumulative solving probability with respect to the minimum measured solving time $t_{\min}$

of deliberate increase in the random behavior either by introducing a *heuristic equivalence parameter* [36] or by simply mixing the *random heuristic* together with a more context-dependent heuristic. Another source of randomness in runtimes can be obtained by using different algorithms for the core theory solvers, such as different variations of the Simplex algorithm, or using different pre-processing techniques that might be detrimental for some instances and very useful for others. Hence it is natural to express the runtime of a solver as a random variable and a related probability distribution.

Let T be the random variable describing the time required to solve a given formula ϕ with a (CDCL-based) SMT solver S randomized using, for example, some of the abovementioned approaches. The *cumulative runtime distribution* $q_T(t)$ gives the probability that $T \leq t$. We will use the cumulative distribution to express the expected time required to solve ϕ with S. By definition, this is

$$\mathbb{E}T = \int_0^\infty t q_T'(t) \mathrm{d}t \,, \tag{5.1}$$

where $q_T'(t)$ is the derivative of the cumulative distribution $q_T(t)$.

5.3.1 Parallel SMT Based on Algorithm Portfolios

The inherent randomness in SMT solver runtimes can be utilized in obtaining a natural parallelization approach. In such approaches the goal is to run in parallel several solvers with different heuristic parameters, such as restart and learning strategies, decision heuristics, or using different theory solvers, on the same formula and obtain the solution from the first solver determining the satisfiability. This *algorithm portfolio approach* [75, 49, 36] has been extensively studied in related areas [56, 57, 64, 71, 55, 33], and has recently proved surprisingly efficient in solving structured formulas [43, 44, 41, 7] in SAT as well as in SMT [83].

We first consider a simplified version of the problem where worker solvers communicate their success or failure in determining satisfiability to a master process. In Section 5.3.2 we extend this case by allowing the worker solvers to share clauses and lemmas with each other either through the master or directly. The plain algorithm portfolio approach is based on running several randomized SMT solvers in a distributed or parallel computing environment on a given formula, and obtaining the result from the first solver that finishes.

One effective approach is to simply introduce a small amount of randomness in the heuristic while keeping the search strategy of the solver otherwise unchanged. This provides an interesting setting for obtaining speed-up as it requires virtually no modification to the underlying solver. The results in, e.g., [83] also suggest that it compares favorably to many other portfolio-based approaches. In this case we are given a randomized solver and a formula such that the probability that the solver solves the instance within time t is $q_T(t)$. Assume now we are given n simultaneously running solvers. As the formula is solved if at least one of the solvers solves the formula within time t, the probability of solving within time t becomes

$$q_{T^n}(t) = 1 - (1 - q_T(t))^n \,. \tag{5.2}$$

Depending on the distribution $q_T(t)$, the expected runtime $\mathbb{E}T^n$ of the simple distribution approach can be be significantly lower than the expected runtime $\mathbb{E}T$ of a single solver.

5.3.2 Lemma Sharing in Portfolios

Since clauses and lemmas learned during solver execution are implied by the original problem, they may be shared freely between solvers, with the purpose of improving the performance of the receiving solver. The challenge with this approach is that the number of lemmas generated by an SMT solver is often very high and transferring all lemmas is too much overhead, so that it often has a detrimental effect on the overall performance. There are two approaches for avoiding this problem. One, taken in [83], is to place a strict limit on the number of literals in the transferred lemmas. The second, followed in [65], is to maintain a centralized database of lemmas from which the solvers receive a heuristically determined subset. The former allows a decentralized implementation, whereas the latter allows the use of more sophisticated heuristics. In both cases experiments show that the shared lemmas can improve solver performance significantly [83, 65].

5.3.3 Centralized Lemma Databases

Sharing of learned lemmas plays a central role in parallel SMT. The learned lemmas are transferred to a *lemma database*, where they are filtered using a *parallel lemma-sharing heuristic*, and then passed on to the running solvers. We first define some concepts that help to formalize the working of the database. Let S be a set of lemmas. The size of a lemma set $||S||$ is the total number of literals in S, that is, $||S|| = \sum_{C \in S} |C|$. Unit lemmas, i.e., lemmas consisting of a single literal, are handled specially in the process: they are always stored in the lemma database, and do not contribute to the size of the database.

Algorithm 5.1 shows a version of the CS-SDSMT algorithm and the related concepts. The lemma database, initialized on line 1, is denoted by LemmaDB, and is annotated with an index j to facilitate the representation of the results. The set U contains the unit lemmas that are already proven to be logical consequences of the input formula ϕ. The shorthand notation $\text{UP}(\phi) = \text{UP}(\phi, \emptyset)$ denotes computing the unit theory propagation closure of ϕ on the empty truth assignment (with no variables fixed to values).

The first part of the loop in lines 5–6 consists of submitting the formula, all unit lemmas, and a heuristically selected subset of LemmaDB of size at most *SubmSize* to the parallel computing environment so that the n computing resources are filled. The next phase is to receive the results in lines 8–14. The $\mathit{Receive}(i)$ function receives, from the resource i, a tuple consisting of the result of the computation, which can be Sat, Unsat, or Error (m/o), and a set L of learned lemmas. If the formula is found either satisfiable or unsatisfiable, the algorithm terminates. Otherwise the set of unit lemmas is updated using the learned lemmas on line 13 and the lemma database is updated on line 14, again using a heuristic function *Merge* and limiting the maximum size of the database to *MaxDBSize*.

The function *Merge* takes a central role in discussing lemma sharing. Firstly, the function acts as a heuristic for selecting learned lemmas, and secondly, it simplifies the learned lemmas using the set of literals U obtained by unit propagation. Two operations are involved in the simplification:

1. removing satisfied lemmas (lemmas C such that $C \cap U \neq \emptyset$), and
2. removing false literals $\neg l$ from lemmas so that for a given lemma C, the simplified lemma becomes $C' = \{l \in C \mid \neg l \notin U\}$.

Algorithm 5.1: The CS-SDSMT Algorithm

Input : Formula ϕ, n (number of computing elements), *MaxDBSize* (maximum database size), *SubmSize* (maximum submit size)

Output : Sat if ϕ is satisfiable, Unsat otherwise

```
1  LemmaDB^0 := ∅
2  U := UP(φ)
3  j := 0
4  while True do
5      for i := 1 to n do
6          Submit(φ ∪ U ∪ Choose(LemmaDB^j, SubmSize))
7          LemmaDB^{j+1} := LemmaDB^j
8          for i := 1 to n do
9              (result, L) := Receive(i)
10             if result is in {Sat, Unsat} then
11                 return result
12             else
13                 U := UP(φ ∪ U ∪ LemmaDB^{j+1} ∪ L)
14                 LemmaDB^{j+1} := Merge(U, LemmaDB^{j+1}, L, MaxDBSize)
15                 j := j + 1
```

5.3.4 Experiments on the Algorithmic Framework

It is interesting to contemplate the different types of heuristics that can be implemented both for *Choose* and *Merge*. This section studies the following four possibilities:

- $Choose_{123}$ only considers lemmas of length 1, 2, or 3. If the size of the resulting database is greater than the limit, the shorter lemmas are preferred. This type of approach is used in many portfolio solvers. For example, [7] only transfers lemmas of length 1 to other solvers, and [44] only lemmas that have at most eight literals.
- $Choose_{\text{len}}$ returns the shortest lemmas. This approach is more general than $Choose_{123}$, as it always returns lemmas even if the argument set contains only lemmas longer than some limit.
- $Choose_{\text{freq}}$ returns the most common learned lemmas. As the parallel search is allowed to overlap, it is not unlikely that the same lemma can be learned many times in different solvers.
- $Choose_{\text{rand}}$ returns a randomly selected set of lemmas.

5.3.5 Lemma Sharing and Partitioning

For certain instances partitioning of the search space and forcing the search to be performed on the partitions shows significant speed-up in the experiments. However,

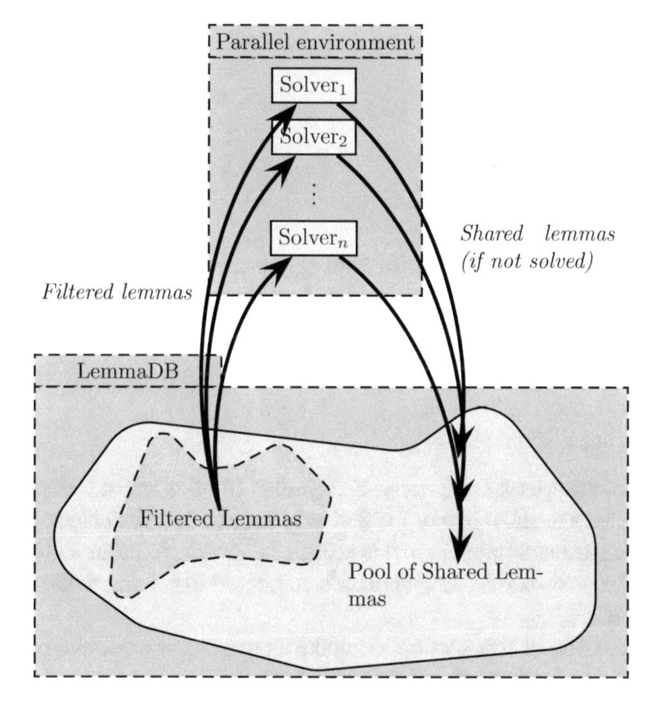

Fig. 5.2: The CS-SDSMT Process

partitioning the search space of the formula makes lemma sharing more complicated because of the constraints that result in the SMT solver learning lemmas that might not be logical consequences of other partitions.

5.4 Search-Space Partitioning in SMT

The algorithm portfolio approach described in Section 5.3 does not force the solvers to explore different search spaces on the formula, but instead relies on randomization in the heuristic to produce speed-ups. The idea in portfolios is that it is unlikely that two randomized solvers would be searching for the solution in a similar fashion.

A complementary approach is to use the divide-and-conquer paradigm to force the search performed by the parallel computing units not to overlap with each other. This can be achieved by constraining the original problem into a set of independently solvable *derived problems*, finding the solutions for them, and computing the final solution based on the results of the derived problems.

The constraints used for constructing the derived problems can be represented in SMT as conjunctions of clauses (lemmas). This section analyzes the effects of the partitioning approach on the expected time required to determine the satisfiability of a formula.

5.4.1 Plain Partitioning

Plain-partitioning is the straightforward approach where an SMT formula ϕ is divided into n derived formulas $\phi_1, \ldots, \phi_n$ that are solved in parallel with an SMT solver S. The derived formulas are obtained with a *partitioning function* and satisfy the following conditions (see Definition 1):

1. $\phi \equiv \phi_1 \vee \ldots \vee \phi_n$, and
2. $\phi_i \wedge \phi_j$ is unsatisfiable if $i \neq j$.

The unsatisfiability of all the derived formulas implies the unsatisfiability of ϕ, whereas it suffices to show one of the derived formulas is satisfiable to prove satisfiability of ϕ. Of particular interest in this section is how much faster a given formula is solved with the plain-partitioning approach compared to solving the formula directly with the solver S.

In the discussion of this section we make an assumption that the partitioning of the instance is done only once and the number of derived formulas is fixed. This is in contrast to many implementations of plain-partitioning where it is natural to use a form of load balancing where new derived formulas are constructed from formulas being solved as the satisfiability of previous formulas is determined. As a result, the number of derived formulas n is not fixed in these parallel solvers.

Despite such differences, an analysis of the plain-partitioning approach provides insight into practical parallel solving. The main result in this section is that the plain-partitioning approach is 'risky' in the following sense. Assume that for any cumulative probability distribution $q(t)$ there exists a formula ϕ_q such that the probability of solving ϕ_q with S in time less than or equal to t is $q(t)$. If the partitioning function is from a certain natural class described in Definition 1, and n is fixed and sufficiently large, there is always an unsatisfiable formula so that the expected runtime of the plain-partitioning approach will be higher than the expected runtime of the underlying solver S [52].

The approach is analyzed in a spirit similar to the analysis of the portfolio approach in Section 5.3. In particular, we will assume that given a formula, the time required to determine its satisfiability with a solver S is a random variable T with cumulative distribution $q_T(t)$. To simplify the discussion, we will assume for now that given a number $n \geq 2$, the partitioning function produces n derived instances which are all solved in parallel using n computing elements.

We will first introduce a model describing how a partitioning function affects the runtime distributions of the derived formulas. We assume that the solver S performs with the same probability a given search that takes time t_ϕ in the formula ϕ but, due to the partitioning constraints, a shorter time t_{ϕ_i} in the derived formulas ϕ_i. The efficiency $\varepsilon(n) = t_\phi / t_{\phi_i}$ of the partitioning function is assumed to depend only on the number n of derived formulas. This reasoning results in a model where, given a formula with the runtime distribution $q_T(t)$ on a solver S, the n derived formulas all have the distributions $q_T(\varepsilon(n)t)$.

The efficiency model that will be used in the proof is $\varepsilon(n) = n^\alpha$, where $0 \leq \alpha \leq 1$ is a constant depending on the partitioning function. This model can be motivated in two ways. Firstly, the efficiency satisfies the following natural properties:

1. $1 \leq \varepsilon(n) \leq n$,
2. $\varepsilon(n) \leq \varepsilon(n+1)$, and
3. $(\varepsilon(n))^p = \varepsilon(n^p)$ for all $p \in \mathbb{N}$,

The first condition states that the partitioning function should not make a particular search of S super-linearly faster or slow the search down. The second condition requires that the efficiency does not decrease as more derived formulas are created. The last condition states that if a partitioning function $P(\phi,n)$ is used to produce n^p derived formulas recursively, the resulting efficiency must equal the efficiency of $P(\phi,n^p)$ where the derived formulas are all generated at once.

Secondly, the model $\varepsilon(n) = n^\alpha$ can be derived from the following constructive application of partitioning. Assume there is a procedure for splitting the search space of an arbitrary formula ϕ following the runtime distribution $q_T(t)$ into a fixed number $n_0 \geq 2$ of derived formulas $\phi_1,\ldots,\phi_{n_0}$. Assume further that the derived formulas ϕ_i have runtime distributions $q_T(\beta t)$ where $1 \leq \beta \leq n_0$. Applying this procedure first to ϕ and then recursively to the derived formulas i times in total results in $n = n_0^i$ derived formulas with runtime distribution $q_T(\beta^i t)$. Hence the recursive application of the procedure results in a partitioning function $P(\phi,n)$ defined for values $n = n_0^i$ with efficiency β^i. Since $i = \log_{n_0} n$, we have

$$\beta^i = \beta^{\log_{n_0} n} = e^{\frac{\ln n}{\ln n_0}\ln\beta} = \left(e^{\ln n}\right)^{\frac{\ln\beta}{\ln n_0}} = n^{\frac{\ln\beta}{\ln n_0}} = n^\alpha ,$$

where $\alpha = \ln\beta/\ln n_0$. Alternative expressions for the efficiency include a linear model

$$\varepsilon'(n) = \max(\beta n, 1) ,$$

where $0 \leq \beta \leq 1$ is a constant. However, condition 3 does not hold for $\varepsilon'(n)$. For example, setting $\beta = 0.9$, $n = 2$, and $p = 2$ results in $(\varepsilon'(2))^2 = 3.24$, while $\varepsilon'(4) = 3.6$. We are now ready to define the partitioning function more precisely.

Definition 1. Given a formula ϕ with runtime distribution $q_T(t)$ on solver S and a partitioning factor $n \geq 2$, a *partitioning function* $P : (\phi,n) \mapsto (\Pi_1,\ldots,\Pi_n)$ is a function mapping the formula ϕ to n *partitioning constraints* $\Pi_1,\ldots,\Pi_n$. The partitioning constraints produce n *derived formulas* $\phi_i = \phi \wedge \Pi_i, 1 \leq i \leq n$. The derived formulas then satisfy the following two properties:

1. $\phi \equiv \phi_1 \vee \ldots \vee \phi_n$, and
2. $\phi_i \wedge \phi_j$ is unsatisfiable for all $i \neq j$.

The runtime distribution of each of the derived formulas on solver S is described by the probability distribution $q_T(\varepsilon(n)t)$, where

$$\varepsilon(n) = n^{\alpha}, 0 \leq \alpha \leq 1 \tag{5.3}$$

describes the *efficiency* of the partitioning function.

We will denote by $\mathbb{E}T^{n}_{\text{plain}(\alpha)}$ the expected time required to determine the satisfiability of ϕ with the plain-partitioning approach using a partitioning function with efficiency $\varepsilon(n) = n^{\alpha}$. A partitioning function is called *void* if $\alpha = 0$ and hence $\varepsilon(n) = 1$. In this case all the derived instances are as difficult to solve as the original formula. A partitioning function is called *ideal* if $\alpha = 1$, that is, $\varepsilon(n) = n$.

With these definitions, we are now ready to show that for non-ideal partitioning functions there are distributions where solving with plain partitioning is slower than solving with the underlying solver.

Proposition 1. *Let $P(\phi, n)$ be a partitioning function as in Definition 1, where $0 \leq \alpha < 1$, and S a SAT solver. Then for every n and every α there exists a distribution $q_n(t)$ such that if the solving of an unsatisfiable instance follows $q_n(t)$ on S, then the expected runtime $\mathbb{E}T$ of S is lower than the expected runtime*

$$\mathbb{E}T^{n}_{\text{plain}(\alpha)}$$

of the plain-partitioning approach.

Proof. The family of distributions $q_n(t)$ we will use in the proof is

$$q_n(t) = \begin{cases} 0 & \text{if } t < t_1, \\ 1 - \frac{1}{n} & \text{if } t_1 \leq t < t_2, \textit{and} \\ 1 & \text{if } t \geq t_2, \end{cases} \tag{5.4}$$

where $t_1 < t_2$. Thus the probabilities that the formula is solved by S in exactly time t_1 is $1 - 1/n$ and in time t_2 is $1/n$. The expected runtime for a formula following the distribution $q_n(t)$ on S is

$$\mathbb{E}T = (1 - \frac{1}{n})t_1 + \frac{1}{n}t_2 \,. \tag{5.5}$$

The expected runtime of the plain-partitioning approach using the partition function $\varepsilon(n) = n^{\alpha}$ can be derived by noting that all derived formulas need to be solved before the result can be determined. This means that either all solvers are 'lucky', and determine the unsatisfiability in time t_1/n^{α}, or at least one of the solvers runs for time t_2/n^{α}, which will then become the runtime of the approach. This results in

$$\mathbb{E}T^{n}_{\text{plain}(\alpha)} = \left(1 - \frac{1}{n}\right)^{n} \frac{t_1}{n^{\alpha}} + \left(1 - (1 - \frac{1}{n})^{n}\right) \frac{t_2}{n^{\alpha}} \,. \tag{5.6}$$

We claim that for every α, there are values for n, t_1, and t_2 such that $\mathbb{E}T < \mathbb{E}T^{n}_{\text{plain}(\alpha)}$. Dividing both sides of the resulting inequality by t_2 and setting $k = t_1/t_2$ results in

$$(1 - \frac{1}{n})k + \frac{1}{n} < \frac{(1 - \frac{1}{n})^{n}}{n^{\alpha}}k + \frac{1 - (1 - \frac{1}{n})^{n}}{n^{\alpha}} \,,$$

which we reorder to

$$k\left((1-\frac{1}{n})-\frac{(1-\frac{1}{n})^n}{n^\alpha}\right)<\frac{1-(1-\frac{1}{n})^n}{n^\alpha}-\frac{1}{n}.$$

Note that $(1-\frac{1}{n})>(1-\frac{1}{n})^n/n^\alpha$ when $n\geq 2$, and therefore the left-hand side of the inequality is positive and can be made arbitrarily small by setting k small. It remains to show that the right-hand side of the inequality is positive for sufficiently large n, i.e.,

$$\frac{n-(1-\frac{1}{n})^n n-n^\alpha}{n^{\alpha+1}}>0.$$

Since $n^{\alpha+1}$ is always positive, we may simplify this and factor n from the denominator, resulting in

$$1-(1-\frac{1}{n})^n-n^{\alpha-1}>0. \tag{5.7}$$

Noting that $\lim_{n\to\infty}(1-\frac{1}{n})^n=\frac{1}{e}\approx 0.3$, and that $\lim_{n\to\infty}1-n^{\alpha-1}=1$ if $\alpha<1$, we get the desired result, that is, for sufficiently large n, there are values t_1 and t_2 such that $t_1<t_2$ and $\mathbb{E}T<\mathbb{E}T^n_{\text{plain}(\alpha)}$.

The following example illustrates the performance of the plain-partitioning approach for distributions of type Equation (5.4).

Example 3. Assume there is a formula following the distribution $q_{20}(t)$ such that $t_1=1$ and $t_2=1000$, and a partition function $\varepsilon(n)=n^{0.7}$ for this formula. The expected runtime of the solver S, given by Equation (5.5), is $\mathbb{E}T\approx 50.95$, while the expected runtime of the plain-partitioning algorithm, from Equation (5.6), is $\mathbb{E}T^{20}_{\text{plain}(0.7)}\approx 78.84$. The scalability of the expected runtime $\mathbb{E}T^n_{\text{plain}(\alpha)}$ of the plain-partitioning approach is shown for the distribution $q_{20}(t)$ for different values of α in Figure 5.3.

Note that the proof does not hold if the partitioning function is ideal, since the left-hand side of Inequality (5.7) is negative if $\alpha=1$. In fact the condition that the partitioning function be ideal turns out to be sufficient to guarantee that the expected runtime of the plain-partitioning approach is never higher than the expected runtime of S, that is, $\mathbb{E}T\geq\mathbb{E}T^n_{\text{plain}(1)}$ for all n and T. To see this, we will first derive an expression for $\mathbb{E}T^n_{\text{plain}(\alpha)}$ for an arbitrary distribution $q_T(t)$ and an arbitrary partitioning function.

Let $q_T(t)$ be a runtime distribution of an unsatisfiable formula ϕ with a randomized SAT solver S, and $t_{\max}$ the maximum time required to solve ϕ with S (hence $q_T(t)=1$ if $t\geq t_{\max}$ and $q_T(t)<1$ otherwise). The n partitions have runtime distributions $q_T(\varepsilon(n)t)$ and since they all need to be shown unsatisfiable, the runtime distribution of the plain-partitioning approach is $q_T(\varepsilon(n)t)^n$. Hence by Equation (5.1) the expected runtime of the plain-partitioning approach is given by

$$\mathbb{E}T^n_{\text{plain}(\alpha)}=\int_0^{t_{\max}}t\frac{\mathrm{d}}{\mathrm{d}t}q_T(\varepsilon(n)t)^n\mathrm{d}t,$$

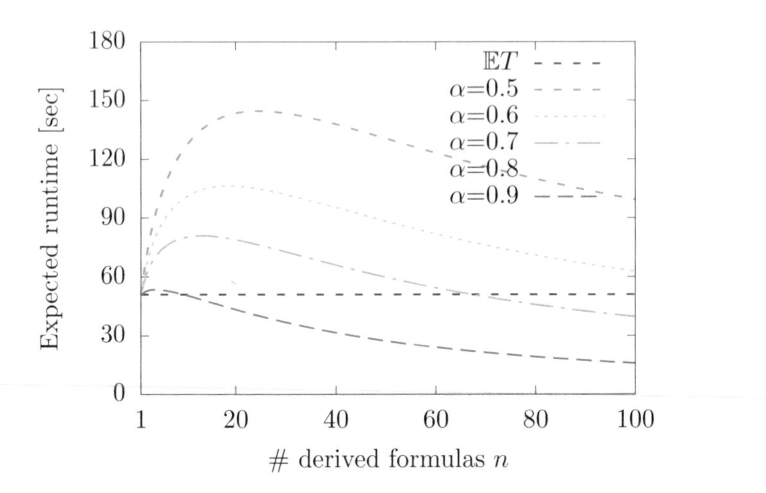

Fig. 5.3: The scalability of the plain-partitioning approach for the distribution $q_{20}(t)$ in Equation (5.4) where $t_1 = 1$ and $t_2 = 1,000$

where $\frac{\mathrm{d}}{\mathrm{d}t} q_T(\varepsilon(n)t)^n = n\varepsilon(n) q_T(\varepsilon(n)t)^{n-1} q_T'(\varepsilon(n)t)$ is the derivative of the distribution function. Substituting $\varepsilon(n)t = \tau$ above, the expected runtime can be written

$$\begin{aligned} \mathbb{E}T^n_{\text{plain}(\alpha)} &= \int_0^{t_{\max}} \frac{\tau}{\varepsilon(n)} n\varepsilon(n) q_T(\tau)^{n-1} q_T'(\tau) \frac{\mathrm{d}\tau}{\varepsilon(n)} \\ &= \int_0^{t_{\max}} \frac{n}{\varepsilon(n)} \tau q_T(\tau)^{n-1} q_T'(\tau) \mathrm{d}\tau \, . \end{aligned} \tag{5.8}$$

We can now state the following proposition that increasing the number of derived instances in ideal plain-partitioning does not result in increased expected runtime.

Proposition 2. *Let $n \geq 1$, $\varepsilon(n) = n^1 = n$ be the efficiency of an ideal partitioning function, and $q_T(t)$ be the runtime distribution of an unsatisfiable formula with a randomized solver. Then $\mathbb{E}T^n_{\text{plain}(1)} \geq \mathbb{E}T^{n+1}_{\text{plain}(1)}$.*

Proof. Substituting $\varepsilon(n) = n$ in Equation (5.8) results in

$$\mathbb{E}T^n_{\text{plain}(1)} = \int_0^{t_{\max}} \tau q_T(\tau)^{n-1} q_T'(\tau) \mathrm{d}\tau \, .$$

Since $q_T(\tau) \leq 1$ when $0 \leq \tau \leq t_{\max}$, we immediately have the desired result $\mathbb{E}T^n_{\text{plain}(1)} \geq \mathbb{E}T^{n+1}_{\text{plain}(1)}$.

Finally from Propositions 1 and 2 we get the main result concerning unsatisfiable instances.

Proposition 3. *The expected runtime of the plain-partitioning approach,*

$$\mathbb{E}T^n_{\text{plain}(\alpha)} \, ,$$

is guaranteed not to be higher than the expected runtime $\mathbb{E}T$ *of the underlying solver* S *if and only if the partitioning function is ideal, that is,* $\alpha = 1$.

It is a strong requirement that the efficiency of a partitioning function must be ideal in order to never increase the time required to solve a formula, and it would be tempting to draw the conclusion that this requirement is never met. The practical implications of the above negative result are not as dramatic. However, it is not completely impossible that unsatisfiable formulas have such pathological distributions, even when the solvers employ restart strategies known to eliminate this type of behavior [37]. Furthermore, it is not impossible for the partitioning function to provide even super-linear speed-ups if, for example, the partitioning constraints are related to the *back door set* [82] of the formula.

5.5 Decomposition

In some cases where partitioning is not applicable and where portfolios require infeasible amounts of memory in practice, a different approach is required. For instance, suppose the input formula is too large to fit into the local memory. In this case the problem must be *decomposed* into a series of smaller problems, which, in contrast to a partitioning, do not compose disjunctively.

Definition 2 (Decomposition). Let ϕ be a first-order formula in conjunctive normal form, i.e., $\phi = \phi_1 \wedge \cdots \wedge \phi_n$. A *decomposition* of ϕ into k sub-formulas is a set of formulas $\{\psi_1, \ldots, \psi_k\}$ such that

- each $\psi_i \subseteq \bigcup_{j \in J} \phi_j$, for some selection of indices J, and
- each ϕ_i is included in at least one ψ_i.

Note that the original problem ϕ is unsatisfiable if at least one of the ψ_i in the decomposition is unsatisfiable, but the converse is not a sufficient criterion for satisfiability, i.e., each ψ_i being satisfiable does not imply that ϕ is satisfiable.

In theory, this type of decomposition is 'ultimately lazy' in that it does not require us to extract any other type of semantic information embedded in the input problem, apart from the number of clauses. This enables us to decompose very large input problems efficiently; for instance, we can simply send a random selection of $\frac{1}{n}$th of the clauses to each of n processors or nodes, without ever inspecting their content.

The cost that we pay for this laziness is then in the reconciliation of (partial) models: suppose that we obtain (partial) models μ_i from independent SMT solver queries, one for each corresponding ψ_i, and let $v(\psi_i)$ be the variables in ψ_i. Two models μ_i, μ_j are trivially 'compatible' if the corresponding ψ_i, ψ_j do not share variables, i.e., when $v(\psi_i) \cap v(\psi_j) = \emptyset$. Models may however not be reconcilable if they assign different values to variables that occur in both ψ_i and ψ_j. This criterion is easy to check for simple models that map theory variables to numerals, but we should keep in mind that for more complex background theories this may amount to function equivalence checks that are not polynomial-time decidable, as is the case, for instance, for some problems involving arrays or quantifiers.

The reconciliation process is perhaps best understood by conceptual introduction of one additional sub-formula σ (and perhaps an associated SMT solver and/or processor or node) which we use to track models for the shared variables *only*. Initially, we pick an arbitrary assignment μ_σ to those shared variables, which we propagate to all ψ_i, after which the resulting, modified ψ_i do not share any variables anymore, which means that all ψ_i can now be solved in parallel without communication between the computing elements, while we know that, if all sub-formulas are found to be satisfiable, the conjunction of the corresponding models μ_i are extensions of μ_σ and thus their conjunction is, trivially, a model for ϕ.

Usually however, there will be at least one ψ_i which is not satisfiable under μ_σ. Thus, we have $\psi_i \Rightarrow \neg\sigma$ for at least one ψ_i, and $v(\psi_i) \cap v(\sigma) \subseteq v(\phi)$, where we expect $v(\psi_i) \cap v(\sigma)$ to be a small set in practice. This is thus, essentially, an interpolation problem.

Lemma 1. *Let* $\phi = \psi_1 \wedge \cdots \wedge \psi_k$, *let* μ_σ *be a model for the set of shared variables* $V = \bigcup_{(i,j)\in C} v(\psi_i) \cap v(\psi_j)$, *where* $C = \{(i,j) \mid 1 \leq i < j \leq k\}$. *If* I *is an interpolant for* $\psi_i \Rightarrow \neg\mu_\sigma$ *over* V, *i.e., we have* $\psi_i \Rightarrow I \wedge I \Rightarrow \neg\mu_\sigma$ *and* $v(I) \subseteq v(\psi_i) \cap v(\mu_\sigma)$, *then* $\phi \Rightarrow I$.

Proof. We have $\psi_i \Rightarrow I \wedge I \Rightarrow \neg\mu_\sigma$ by Craig's interpolation lemma. By construction we also have $\phi \Rightarrow \psi_i$; thus it follows that $\phi \Rightarrow I$ (and $\phi \Rightarrow \neg\mu_\sigma$ as in traditional learned clauses and lemmas).

Thus, the interpolant I is implied by ϕ, which means we may, conceptually, (conjunctively) add I to ϕ, to σ, or to its corresponding ψ_i, while preserving satisfiability, just as we would keep a learned clause or lemma (in a local or a shared lemma database). This enables us to formulate a sound and (relatively) complete reconciliation algorithm for decompositions as presented in Algorithm 5.2.

5.5.1 Experimental Evidence

While decomposition is necessary for very large formulas, there are some indications that the concept itself may help to improve performance on small, hard problems as well. So far, this has been shown to be the case for some propositional problems, for which multiple different interpolation techniques exist that can be compared. The experiments we present here are an example of this, on the small, but hard benchmark instances by Aloul et al. [2], which contain symmetries that can be broken by addition of symmetry-breaking predicates. Ideally, a good interpolation approach is able to detect such symmetries and it automatically breaks them by finding interpolants corresponding to symmetry-breaking predicates.

Figure 5.4 shows the runtime obtained for MiniSAT 1.14p [30] on each of Aloul et al.'s benchmarks, when decomposed into up to 50 sub-formulas (where '1' corresponds to no decomposition being performed). A time limit of 3,600 seconds

Algorithm 5.2: An Interpolation-based Reconciliation Algorithm

```
   Input   : Formula φ
   Output  : Sat if φ is satisfiable, Unsat otherwise
1  ψ_1, ..., ψ_k := decompose(φ)
2  σ := ∅
3  flag := true
4  while flag do
5      if σ is Unsat then
6          return Unsat
7      else
8          Let μ_σ be a model for σ
9      flag := false
10     foreach i in 1...k do
11         if ψ_i ∧ μ_σ is Unsat then
               /* ¬(ψ_i ∧ μ_σ) ⇔ ψ_i ⇒ ¬μ_σ is valid */
12             Let I be an interpolant for ψ_i ⇒ ¬μ_σ (over v(ψ_i) ∩ v(μ_σ))
13             σ := σ ∧ I
14             flag := true
15 return satisfiable
```

is enforced and all averages are computed with memory-out problems counted as $36,000 = 10 \times$ time limit. For representation of interpolants, both McMillan's and HKP interpolants, Reduced Boolean Circuits [1] are used and they are computed along a resolution proof found by MiniSAT. Since they are general expressions and not in conjunctive normal form yet, they are then added to the formula via Tseitin transformation (including introduction of new variables). To compute the runtime of a decomposition, all iterations of Algorithm 5.2 are executed *sequentially*, i.e., the runtime improvements we see in Figure 5.4 describe a completely sequential algorithm, yet for decompositions into about 10 or more sub-formulas, the average runtime over all benchmark problems is smaller than the average runtime of the unmodified MiniSAT. Of course, for an effective parallelization of this algorithm, it is necessary to implement a proper load-balancing strategy. More details of this evaluation are provided by Hamadi et al. [45].

5.5.2 Variations and Extensions

Clearly, keeping a single additional formula γ for assignments to shared variables is not ideal for all types of problems. For instance, there may be multiple non-overlapping sets of shared variables, such that γ itself may be decomposed into multiple parts trivially. Since all sub-formulas ψ_i as well as γ may grow (by addition of learned clauses and lemmas), a dynamic decomposition approach, which introduces new ψ_i on demand, may be a good choice in practice.

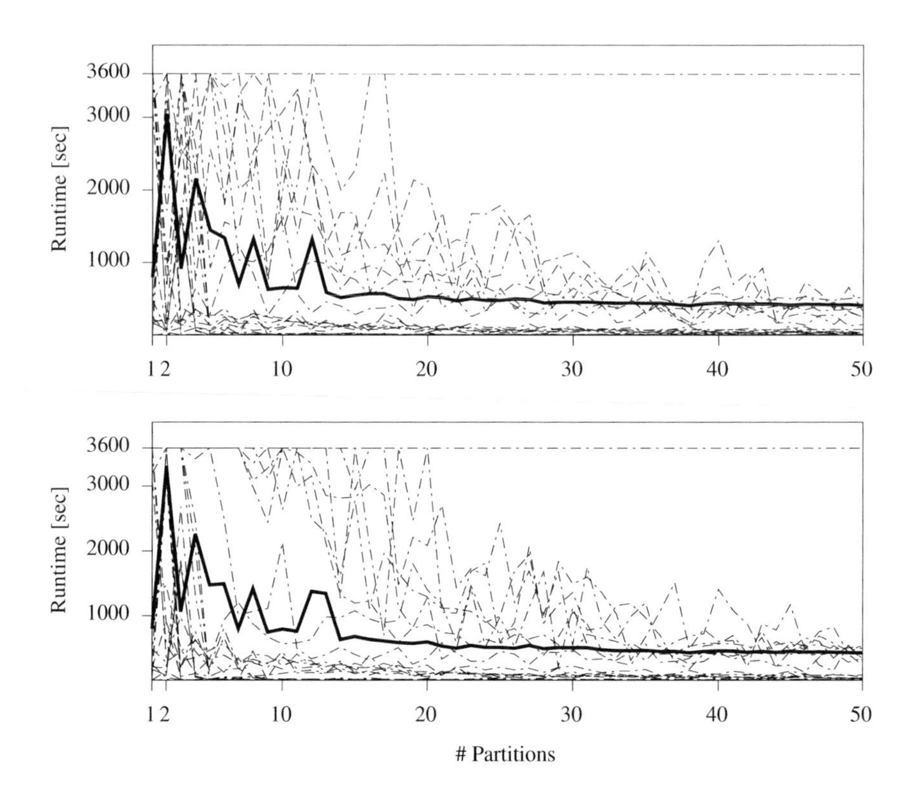

Fig. 5.4: Decompositions into up to 50 sub-formulas. Every line corresponds to one benchmark problem; their average is the bold line. Reconciliation via McMillan interpolants (top) and HKP interpolants (bottom)

Furthermore, in any given decomposition, it is not strictly necessary for all sub-formulas (and sub-solvers) to communicate with all others. Solvers only need to communicate with other solvers if their sub-formulas do in fact share variables, in which case they communicate satisfying assignments and interpolants to each other. This means that all sub-solvers may solve their problems independently, sharing their satisfying assignments with those solvers that solve problems involving shared variables. In return, and completely asynchronously, the recipients of satisfying assignments may respond with an interpolant that excludes at least one particular variable assignment, at a later time. The process terminates when all interpolants have been received and all satisfying assignments have been consumed, which indicates satisfiability of the whole problem; unsatisfiability is detected by one of the interpolants becoming false. In this fashion, it is possible to construct a completely asynchronous system in which no solver needs to have access to all of the data at any given time, while termination of the algorithm is determined by a distributed consensus algorithm (or variation thereof).

It is worth noting that Nelson/Oppen theory combination is essentially also a decomposition in the sense defined here, with reconciliation taking place via a particular kind of interpolant. A decomposition, however, does not require us to separate sub-formulas into sets of constraints belonging to the same theory. Instead, we may have *multiple* theory solvers of the *same* conceptual theory. For instance, we may have two solvers for the theory of linear integers, each of them solving only a subset of all (purified) linear integer constraints, exchanging interpolants between them.

5.6 Combinations of Parallelization Algorithms

It is useful to analyze the differences between the three parallelization approaches, portfolio, partitioning, and decomposing, discussed in the preceding sections. For instance Bonacina [9] constructs a taxonomy of the approaches based on this distinction, while Grama and Kumar [39] provides an overview of the different parallelization approaches for constraint solving using essentially the same distinction. Attempts at understanding the differences of in particular the portfolio and partitioning approaches include Bordeaux, Hamadi, and Samulowitz [11], and Bonacina [9].

A complementary approach to understanding the differences between the approaches is to try to combine their strengths. Some work towards this has been done in Bonacina [10]; Segre et al. [78] on a parallel SAT-solving approach called *nagging*; Hyvärinen, Junttila, and Niemelä [50] on the parallelization approach based on SAT solving through *scattering*; Dequen, Vander-Swalmen, and Karajecki [25], which implements a similar approach; Ohmura and Ueda [70], which implements a *safe-partitioning* approach on computing clusters for SAT solving; and Gebser et al. [33], which implements a plain-partitioning approach strengthened with a dedicated solver solving the original, unpartitioned formula.

In this section we go a step further, giving a generic framework for combining partitioning and portfolio called *parallelization trees* [54], which represents the instances of parallel algorithms as and/or trees. We give a more in-depth analysis of three of the approaches that we find particularly interesting; parts of this have previously been presented in [52] and [51].

5.6.1 The Parallelization Tree

A way to represent the combination of partitioning and portfolio in a unified framework is the *parallelization tree* abstract algorithmic framework. The idea is to provide a unified way of presenting and comparing different parallelization algorithms. The parallelization tree consists of two types of nodes: *and-nodes* and *or-nodes*. The tree is constructed using the following simple rules:

- The root and the leaves of the parallelization tree are and-nodes.

- Each and-node is associated with an SMT instance and, with the possible exception of the root of the parallelization tree, with one or more SMT solvers.
- All children of an and-node are or-nodes,
- All children of an or-node are and-nodes.

In a more formal treatment we adapt the partitioning function of Definition 1 to the construction of the parallelization tree through the operator $\mathrm{split}^k(n_1,\ldots,n_k,\phi)$.

Definition 3. The result of applying the operator split^k on an and-node ϕ is a tree rooted at the and-node ϕ with k children $o_1,\ldots,o_k$. Each child node o_i is an or-node and has as children the and-nodes $a^i_1,\ldots,a^i_{n_i}$. Finally, each and-node a^i_j is associated with the partition obtained by applying the (randomized) partitioning function P_{n_i} of Definition 1 on the formula ϕ.

The satisfiability of the instance is determined by the solvers and the tree structure as follows:

- The instance at the root of the parallelization tree is satisfiable if any instance among the and-nodes is shown satisfiable.
- A subtree rooted at an and-node is unsatisfiable if one of its children is unsatisfiable or at least one of the solvers associated with the and-node has shown the instance unsatisfiable.
- A tree rooted at an or-node is unsatisfiable if every tree rooted at its children is unsatisfiable.

We immediately obtain both the plain-partitioning and the portfolio approaches as instances of the parallelization tree approach:

- The *plain* partitioning approach $\mathrm{plain}(n,\phi)$ corresponds to the parallelization tree $\mathrm{split}^1(n,\phi)$ where each of the instances associated with the nodes $a^1_1,\ldots,a^1_n$ is solved with a single SMT solver.
- The *portfolio* approach $\mathrm{portf}(k,\phi)$ corresponds to the parallelization tree consisting of the root associated with the instance ϕ and using k SMT solvers to solve the instance.

However, in addition to these two algorithms the parallelization tree approach allows us to easily define other, less trivial algorithms from the literature:

- The *safe*-partitioning approach $\mathrm{safe}(n,s,\phi)$ corresponds to the parallelization tree $\mathrm{split}^1(n,\phi)$ and solving each of the instances $a^1_1,\ldots,a^1_n$ with s SMT solvers.
- The *repeated*-partitioning approach $\mathrm{rep}(n,k,\phi)$ corresponds to the parallelization tree $\mathrm{split}^k(n,\ldots,n,\phi)$ where each instance associated with the nodes

$$a^1_1,\ldots,a^1_n,\ldots,a^k_1,\ldots,a^k_n$$

  is solved with one SMT solver.
- The *iterative*-partitioning approach $\mathrm{iter}(k,\phi)$ corresponds to the infinite parallelization tree where every instance associated with an and-node is solved with a single SMT solver and every and-node associated with an instance ϕ_a has

the single or-child and and-grandchildren constructed by applying the operator $\text{split}^1(n, \phi_a)$.

Figure 5.5 illustrates the corresponding parallelization trees and the solver assignments. When clear from the context, we omit the formula ϕ as well as the other parameters from the partitioning approach.

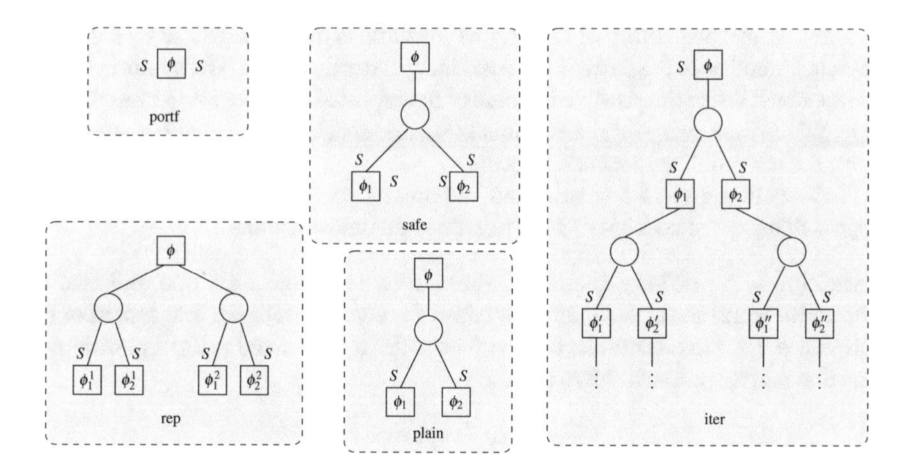

Fig. 5.5: Example parallelization trees (clockwise from the top left): $\text{portf}(2, \phi)$, $\text{safe}(2, 2, \phi)$, $\text{iter}(2, \phi)$, $\text{plain}(2, \phi)$, and $\text{rep}(2, 2, \phi)$. The and-nodes are drawn with boxes, and the or-nodes with circles. The SMT solvers are indicated with the symbol S (Figure adapted from [54])

Concrete SMT instantiations of the parallelization tree include the CVC4 and Z3 SMT solvers, which implement a portfolio, and PBoolector [74], which implements an iterative-partitioning approach. The OpenSMT2 solver [53] implements the full parallelization tree framework.

5.6.2 *Iterative Partitioning with Partition Trees*

The result of Proposition 3, showing that the plain-partitioning approach is 'vulnerable' to certain distributions of unsatisfiable formulas, raises the question whether there are other solving techniques that use a partitioning function but are immune to the increased expected runtimes in all unsatisfiable cases. Given an unsatisfiable formula, the challenge in plain-partitioning is that the number of formulas needed to show unsatisfiability increases as more derived formulas are produced.

A trivial solution to this problem is to attempt to solve both the formula ϕ and the derived formulas using $n+1$ computing elements. This solution corresponds to solving the formula with the plain-partitioning approach and the underlying solver S in parallel, and guarantees that the expected runtime of the approach will be at most as high as the expected runtime of S. However, by Proposition 3, it is possible that the runtime of the plain-partitioning approach increases as more resources are used, which adversely affects the behavior of the proposed solution.

The *iterative-partitioning approach* [50], is based on a hierarchical partitioning of formulas into increasingly constrained derived formulas, which are organized as a tree. The satisfiability of the original formula is then determined by solving a sufficient number of the derived formulas independently with S. The intuition behind the approach is that the possible increase of the expected runtime due to Proposition 3 is avoided since every time a formula is partitioned, there is an added attempt to solve the unpartitioned formula directly.

This section gives a formalization and an analysis of the iterative-partitioning approach using the concept of a *partition tree* defined as follows.

Definition 4. A *partition tree* $\mathscr{T}_\phi$ of a formula ϕ is a finite n-ary tree rooted at v_0. The nodes v_i are associated with constraints: the constraints of the root consist of the formula ϕ and the constraints of the other nodes are obtained using a partitioning function on their parents. More precisely,

$$Constr(v_0) := \phi \ ,$$

and given a node v_i, its children $v_{i,1},\ldots,v_{i,n}$, and a rooted path $v_0,\ldots,v_i$ in the partition tree, the partitioning constraints of the child nodes are

$$Constr(v_{i,k}) := \Pi_k \text{ where } \Pi_k \in P(Constr(v_0) \wedge \ldots \wedge Constr(v_i), n) \ .$$

Finally, each node v_i represents the derived formula

$$\phi_{v_i} := Constr(v_0) \wedge \ldots \wedge Constr(v_i) \ .$$

In the iterative-partitioning approach a partition tree $\mathscr{T}_\phi$ is constructed in breadth-first order and the solving of each derived formula ϕ_{v_i} is attempted in parallel with a solver S until the satisfiability of ϕ is determined. The satisfiability of a node v_i is determined either by solving ϕ_{v_i} with S, or determining the satisfiability of all the child nodes $v_{i,1},\ldots,v_{i,n}$.

The iterative-partitioning approach guarantees that its expected runtime does not increase as more computing elements are introduced, even if the partitioning function is void. We will show this for partition trees $\mathscr{T}_\phi^k$, where all rooted paths to the leaves are of length k. As is conventional, we say that the height of $\mathscr{T}_\phi^k$ is k.

Proposition 4. *Let ϕ be an unsatisfiable formula, $\mathscr{T}_\phi^k$ and $\mathscr{T}_\phi^m$ be two partition trees of height k and m, respectively, constructed with a void partition function, and $k < m$. Then the expected runtime of the partition tree approach using $\mathscr{T}_\phi^m$ is less than or equal to the expected runtime of the partition tree approach using $\mathscr{T}_\phi^k$.*

Proof. We show by induction on the height of the partition tree that the probability that ϕ is solved within time t cannot decrease, from which the claim follows. Let $q(t)$ be the probability that ϕ is solved sequentially within time t, $q'(t)$ be its derivative at t, and let $q_i(t)$ denote the probability that ϕ is solved within time t using a partition tree $\mathscr{T}^i_\phi$ of height i. Then the probability $q_0(t) = q(t)$. The probability that the formula is solved within time t with the partition tree approach using a tree of height one is $q_1(t) = \int_0^t (q'(\tau) + (1-q(\tau))nq'(\tau)q(\tau)^{n-1})\mathrm{d}\tau$, that is, the integral of the sum of probability $q'(\tau)\mathrm{d}\tau$ that the formula is solved in the root of the tree at time τ, and the probability that the formula has not been solved in the root, has been solved by all children but one by time τ, and is solved at time τ in the last child. A direct calculation shows that $q_1(t) \geq q_0(t)$. Assume now that $q_k(t) \geq q_{k-1}(t)$ for all $t \geq 0$. As previously, $q_{k+1}(t) = \int_0^t (q'(\tau) + (1-q(\tau))nq'_k(\tau)q_k(\tau)^{n-1})\mathrm{d}\tau = q(t) + q_k(t)^n - \int_0^t q(\tau)nq'_k(\tau)q_k(\tau)^{n-1}\mathrm{d}\tau$. Integration by parts on the negative term results in $q_{k+1}(t) = q(t) + q_k(t)^n - q_k(t)^n q(t) + \int_0^t q_k(\tau)^n q'(\tau)\mathrm{d}\tau = q(t) + (1-q(t))q_k(t)^n + \int_0^t q_k(\tau)^n q'(\tau)\mathrm{d}\tau$. By the induction hypothesis $q_{k+1}(t) \geq q(t) + (1-q(t))q_{k-1}(t)^n + \int_0^t q_{k-1}(\tau)^n q'(\tau)\mathrm{d}\tau = q_k(t)$.

In practice the construction of the tree is not atomic, but the nodes of the tree are expanded at different times in breadth-first order. As the construction of the tree is not immediate, the tree expansion can use information obtained from earlier solving attempts. The straightforward way to use this information, as in Example 4 (and first published in [50]), is not to expand a subtree rooted at a formula shown unsatisfiable. This example further illustrates the use of iterative partitioning and the related partition tree.

Example 4. Figure 5.6 illustrates how the partition tree approach runs in an environment with $m = 8$ parallel resources. The left tree shows the initial setup, and the right tree shows how the solving has proceeded after one of the SAT solvers terminates in a memory out and three of the solvers return unsatisfiable for their respective formulas. In both trees the shaded area indicates the set of formulas currently being solved. The formulas shown unsatisfiable are labeled with Unsat and the formula that has exceeded its resource limit is labeled with Error (m/o) on the right-hand side tree. There is no need to solve $v_{0,1,1}$ once $v_{0,1,1,1}$ and $v_{0,1,1,2}$ are shown unsatisfiable.

5.6.3 *Safe and Repeated Partitioning*

Another approach to avoiding the increase of expected runtime in solving unsatisfiable instances is to combine the plain-partitioning approach with randomization. This way the inherent randomness in runtimes of SAT and SMT solvers and the reduction in search space provided by the partitioning function can be used simultaneously to improve performance. We present two such composite approaches:

- *Safe partitioning* uses the partitioning function to derive formulas each of which is solved with the portfolio approach; and

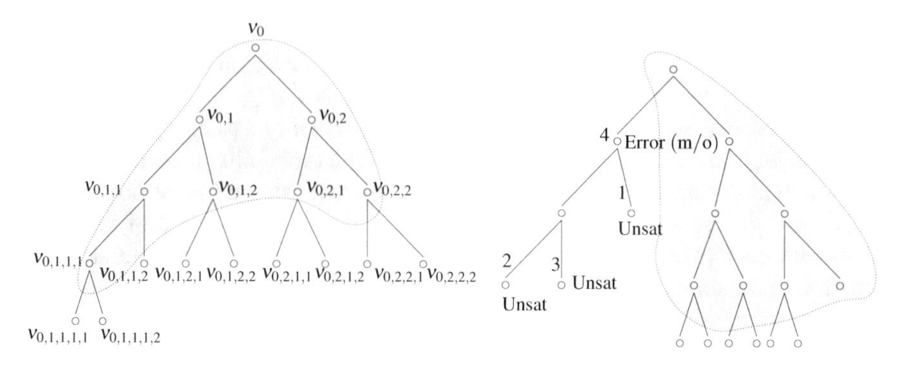

Fig. 5.6: Illustration of the partition tree approach. The shaded area represents jobs running simultaneously, the numbers indicate the order in which the jobs terminate, and the solid lines represent the edges of the tree

- *Repeated partitioning* produces several sets of derived formulas with a partitioning function, and solves these sets in parallel using one solver per derived instance.

The use of safe partitioning has been suggested in [70, 33], whereas the repeated-partitioning approach is closely related to *hard restarts* in guiding-path-based approaches (e.g., [33]). Here, we analyze a setting where n^2 resources are used so that in safe partitioning the partitioning function results in n partitions that are solved using n solvers each. In repeated partitioning the partitioning function is repeated n times for the same formula, resulting again in n^2 formulas.

Safe partitioning applies a partitioning function $P(\phi,n) = (\Pi_1,\ldots,\Pi_n)$, and solves each derived formula $\phi \wedge \Pi_i$, $1 \leq i \leq n$, with a portfolio of n solvers. It suffices then to show each derived instance unsatisfiable with one solver. Intuitively this approach improves performance because derived formulas should be easier to solve than the original formula, and, assuming the solving times of the derived formulas obey a non-trivial random distribution, the portfolio approach results in lower runtimes for the derived formulas. The repeated-partitioning approach, on the other hand, consists of applying a family of partitioning functions $P^j(\phi,n) = (\Pi_1^j,\ldots,\Pi_n^j)$, $1 \leq j \leq n$, and solving each derived formula $\phi \wedge \Pi_i^j$, $1 \leq i \leq n$, $1 \leq j \leq n$ with a solver S. To show a formula unsatisfiable it suffices to show unsatisfiable any set of derived formulas $\phi \wedge \Pi_1^k,\ldots,\phi \wedge \Pi_n^k$ for a fixed k. This approach is expected to provide speed-ups as the derived formulas are easier to solve than the original formula, but also because it is possible that one of the partitioning functions P^j performs better than some other partitioning function.

Based on the definition we can immediately give the runtime distributions of the two composite approaches using Equations (5.2) and (5.8) for simple distribution and plain-partitioning. The cumulative runtime distribution for safe partitioning of unsatisfiable formulas $q_{T_{\mathrm{safe}}}(t)$ is given by substituting $q_T(t)$ in (5.2) by (5.8), yielding

$$q_{T_{\text{safe}}}(t) = (1-(1-q(\varepsilon(n)t))^n)^n\,, \tag{5.9}$$

and the repeated partitioning by substituting $q_T(t)$ in (5.8) by (5.2), resulting in

$$q_{T_{\text{rep}}}(t) = 1-(1-q(\varepsilon(n)t)^n)^n\,. \tag{5.10}$$

From Equations (5.9) and (5.10) it follows that the expected runtime of the repeated partitioning is always at least the expected runtime of the safe partitioning, independent of the partitioning function or number of computing elements (n).

Proposition 5. *Let $q_T(t)$ be the runtime distribution of an unsatisfiable formula. Then $\mathbb{E}T_{\text{safe}} \leq \mathbb{E}T_{\text{rep}}$.*

Proof. Since $\mathbb{E}T_{\text{safe}} \leq \mathbb{E}T_{\text{rep}}$ if $q_{T_{\text{safe}}}(t) \geq q_{T_{\text{rep}}}(t)$ for all $0 \leq t \leq t_{\max}$, it suffices to show that that $q_{T_{\text{rep}}}(t) = 1-(1-q(\varepsilon(n)t)^n)^n \leq q_{T_{\text{safe}}}(t) = (1-(1-q(\varepsilon(n)t))^n)^n$. Substituting $q(\varepsilon(n)t) = x$, this is equivalent to $1-(1-x^n)^n \leq (1-(1-x)^n)^n$ for $0 \leq x \leq 1$. We will show this by showing $f(x) = (1-(1-x)^n)^n - 1 + (1-x^n)^n \geq 0$ for $0 \leq x \leq 1$. First note that $f(x) = f(1-x)$ is symmetric with respect to $x = 1/2$. Since $f(0) = 0$, it suffices to show that $f(x)$ is increasing when $0 \leq x \leq 1/2$, that is, $\mathrm{d}/\mathrm{d}x\,(f(x)) \geq 0$, whenever $0 \leq x \leq 1/2$. By computing the derivative, we have

$$\begin{aligned}
\frac{\mathrm{d}}{\mathrm{d}x}f(x) &= n^2\,(1-(1-x)^n)^{n-1}\,(1-x)^{n-1} - n^2(1-x^n)^{n-1}x^{n-1}\\
&= n^2\left((1-(1-x)^n)^{n-1}\,(1-x)^{n-1} - (1-x^n)^{n-1}x^{n-1}\right)\,.
\end{aligned}$$

Since n^2 is positive, it suffices to confirm that the parenthesized expression is positive. By rearranging the terms, we get

$$\begin{aligned}
&\left((1-(1-x)^n)^{n-1}\,(1-x)^{n-1} - (1-x^n)^{n-1}x^{n-1}\right) =\\
&\left((1-x)-(1-x)^{n+1}\right)^{n-1} - \left(x - x^{n+1}\right)^{n-1}\,.
\end{aligned}$$

The expression above is positive, since the distance between $(1-x)$ and $(1-x)^{n+1}$ is greater than or equal to the distance between x and x^{n+1} whenever $0 \leq x \leq 1/2$, as can be verified by confirming that the claim holds for $n = 1$ and noting that $\mathrm{d}/\mathrm{d}n((1-x)-(1-x)^{n+1}) = -(1-x)^{n+1}\log(1-x) \geq \mathrm{d}/\mathrm{d}n(x-x^{n+1}) = -x^{n+1}\log x$. The latter can be shown using induction by noting that $-(1-x)^n\log(1-x) \geq -x^n\log x$ for $n = 2$, assuming that the claim holds for $n = k$ and noting that in this case also $-(1-x)^{k+1}\log(1-x) \geq -x^{k+1}\log x$, since $(1-x) \geq x$ when $0 \leq x \leq 1/2$.

By Proposition 5, the cumulative runtime distribution of safe-partitioning is less than that of repeated partitioning for all t and unsatisfiable instances. We now show that, when the number of resources is fixed to N, there are distributions of unsatisfiable instances for which the expected runtime of the safe partitioning approach is greater than the expected runtime of a single solver. From this it follows that whenever partitioning is performed, it is possible that the expected performance of the approach is worse than that of a single solver. An example is a two-step distribution where the probability of solving the instance exactly at time t_1 is p and the probability of solving the instance exactly at time t_2 is $(1-p)$. The expected runtime of a single solver for this type of instance is $\mathbb{E}T = pt_1 + (1-p)t_2$. The safe partitioning approach with void partitioning function has the expected runtime $\mathbb{E}T^N_{\mathrm{safe}(0)} = \left(1-(1-p)^N\right)t_1 + \left(1-\left(1-(1-p)^N\right)^N\right)t_2$. For example, if $p = 0.01$, $t_1 = 1$, $t_2 = 1{,}000{,}000$ and $N = 2$, the expected runtime of the safe-partitioning approach is approximately 1% higher than the expected runtime of a single solver.

Conversely, if the formula to be solved is satisfiable, we have the following.

Proposition 6. $\mathbb{E}T_{\mathrm{safe}} = \mathbb{E}T_{\mathrm{rep}}$ *for satisfiable instances.*

Proof. If instead r of the k partitions are satisfiable, the expected runtime of safe partitioning becomes

$$D^k_{\mathrm{part}}(D^k_{\mathrm{portf}}(q(t))) = 1-(1-(1-(1-q(\varepsilon(k)t))^k))^r = 1-(1-q(\varepsilon(k)t))^{kr} ,$$

and the expected runtime for repeated partitioning equally becomes

$$D^k_{\mathrm{portf}}(D^k_{\mathrm{part}}(q(t))) = 1-(1-(1-(1-q(\varepsilon(k)t))^r))^k = 1-(1-q(\varepsilon(k)t))^{rk} .$$

These types of step probability functions turn out to be interesting to compare the working of different partitioning approaches on extreme cases. In Figure 5.7 we show the behavior of some of the partitioning algorithms discussed in this chapter when the number of parallel computing elements is increased. The distribution used in the simulation is defined by

$$q_T(t) = \begin{cases} 0 & \text{for } 0 \le t < 1, \\ 0.8 & \text{for } 1 \le t < 1{,}000{,}000, \text{ and} \\ 1 & \text{for } t \ge 1{,}000{,}000. \end{cases} \tag{5.11}$$

5.6.4 Constructing Partitions

As seen from the preceding analytical discussion, the quality of the partitioning function is critical for performance, and, in case of plain, safe, and repeated partitioning, avoiding an increase in expected runtime is too. The partitioning functions considered here introduce new constraints, represented as clauses, to a formula. We consider two types of partitioning functions: the *DPLL-based partitioning* producing only unit clauses and the *scattering-based partitioning*, which produces longer clauses. Heuristics for constructing the constraints are used for increasing the likelihood of obtaining partitions that result in low runtime. All implementations of the partitioning functions are built on a CDCL SAT solver underlying the SMT solver.

The first partitioning function discussed here uses the unit propagation lookahead (see, e.g., [46]). The goal is to use as decision literals the literals that result in the highest number of unit propagations.

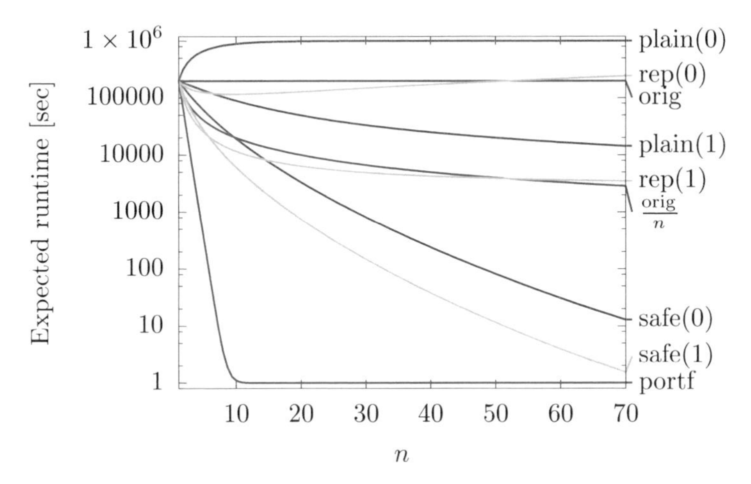

Fig. 5.7: The expected runtimes of different approaches on an (artificial) unsatisfiable instance having the distribution described in Equation (5.11)

Computing the full lookahead for a formula ϕ is worst-case quadratic in the number of variables in ϕ. To guarantee scalability the implementations only study a subset of promising literals of ϕ and use several optimizations in the computation. The *lookahead DPLL* partitioning function implements such optimizations to produce evenly sized derived formulas and uses both theory and Boolean propagation to determine the heuristic value of the variables. Given a formula ϕ, promising literals l are studied by computing the number of literals in the unit propagation closure $\mathrm{UP}(\phi, l)$ and $\mathrm{UP}(\phi, \neg l)$. As the number of literals in $\mathrm{UP}(\phi, l)$ might differ dramatically compared to $\mathrm{UP}(\phi, \neg l)$, the implementation scores literals based on the minimum of these two numbers. Once a heuristically good literal has been selected, the corresponding two derived formulas $\phi \wedge \mathrm{UP}(\phi, l)$ and $\phi \wedge \mathrm{UP}(\phi, \neg l)$ are recursively handled in a similar way. The binary tree up to the depth n constructed this way can be interpreted as consisting of 2^n derived formulas covering all potential satisfying truth assignments of ϕ, and the idea in DPLL-based partitioning is to return exactly these formulas as the derived formulas.

It is interesting to study partitioning functions producing more general constraints. The derived formulas in DPLL-based partitioning are of the form $\phi \wedge l_1 \wedge \ldots \wedge l_n$, but there is no need to limit partitioning functions to producing only constraints of unit clauses. Scattering-based partitioning produces both unit and longer clauses as the constraints. The idea is to first run the SMT solver for a fixed time to tune the heuristic of the solver. If the satisfiability of the formula is not determined in this time, the solver restarts, and starts to produce derived formulas. The first derived formula is produced by making the decisions $l_1^1, \ldots, l_{d_1}^1$, and producing the formula $\phi \wedge l_1^1 \wedge \ldots \wedge l_{d_1}^1$ as in DPLL-based partitioning. Then, instead of selecting the next branch of the search tree, the negation of the literals is added as a clause to ϕ. The solver restarts again, makes new decisions $l_1^2, \ldots, l_{d_2}^2$, and produces the formula

$\phi \wedge (\neg l_1^1 \vee \ldots \vee \neg l_{d_1}^n) \wedge l_1^2 \wedge \ldots \wedge l_{d_2}^2$. The process is continued until a sufficient number of derived formulas are produced. The idea leads to a partitioning function producing the derived formula ϕ_i such that

$$\phi_i = \begin{cases} \phi \wedge (l_1^1) \wedge \ldots \wedge (l_{d_1}^1) & \text{if } i = 1, \\ \phi \wedge (\neg l_1^1 \vee \ldots \vee \neg l_{d_1}^1) \wedge \\ \quad \wedge \ldots \wedge (\neg l_1^{i-1} \vee \ldots \vee \neg l_{d_{i-1}}^{i-1}) \wedge \\ \quad (l_1^i) \wedge \ldots \wedge (l_{d_i}^i) & \text{if } 1 < i < n, \\ \phi \wedge (\neg l_1^1 \vee \ldots \vee \neg l_{d_1}^1) \wedge \ldots \wedge \\ \quad \wedge (\neg l_1^{n-1} \vee \ldots \vee \neg l_{d_{n-1}}^{n-1}) & \text{if } i = n. \end{cases} \tag{5.12}$$

Essentially the derived formulas consist of the original formula ϕ, a conjunction of unit clauses $(l_1) \wedge \ldots \wedge (l_d)$, and clauses representing negations of the previously selected unit clauses. In order for the derived formulas to be of roughly equal size, the number of new unit clauses, denoted by d_i, should not in general be the same in all derived formulas. The selection of the number d_i is motivated so that the expected runtime of each derived formula should be t/n, where t is the expected runtime of the original formula and n is the total number of derived instances produced by the partitioning function. Hence the goal fraction r_i of the runtime for the derived formula ϕ_i can be obtained from the equality

$$\frac{t}{n} = (t - (i-1)\frac{t}{n})r_i \,,$$

where $(i-1)\frac{t}{n}$ is the runtime already contributed to the derived formulas $\phi_1, \ldots, \phi_{i-1}$. Solving the above for r_i results in

$$r_i = \frac{1}{n - i + 1} \,. \tag{5.13}$$

Here, we assume that conjoining a literal with a formula halves the expected runtime of the formula, and therefore the number d_i is chosen to be the integer minimizing the difference

$$\Delta = |r_i - 2^{-d_i}| \,. \tag{5.14}$$

Example 5. Let ϕ be a propositional formula and P a partitioning function producing three partitions. From Equation (5.13), the first fraction of the search space should be $r_1 = 1/3$. The value $d_1 = 2$ minimizes Δ in Equation (5.14); the first derived formula becomes, by Equation (5.12), $\phi_1 = \phi \wedge (l_1^1) \wedge (l_2^1)$. Similarly, $r_2 = 1/2$ and the value $d_2 = 1$ minimizes Δ; the second derived formula becomes $\phi_2 = \phi \wedge (\neg l_1^1 \vee \neg l_2^1) \wedge (l_1^2)$. The final derived formula then becomes $\phi_3 = \phi \wedge (\neg l_1^1 \vee \neg l_2^1) \wedge (\neg l_1^2)$.

The approach for choosing values for d_i using the model in Equation (5.13) is not the only choice we have. The following example illustrates how the scattering approach can 'simulate' a DPLL-based partitioning.

Example 6. Let ϕ be a formula. Our target will be to build a partitioning function producing four derived formulas. Let the first derived formula be $\phi_1 = \phi \wedge (l_1) \wedge (l_2)$. Setting $d_2 = 1$ we may choose $\phi_2 = \phi \wedge (\neg l_1 \vee \neg l_2) \wedge (l_1)$ as the second derived formula. Since $\mathrm{UP}((\neg l_1 \vee \neg l_2) \wedge (l_1)) = \{l_1, \neg l_2\}$, the solving of ϕ_2 will proceed exactly as if the second derived formula had been $\phi_2 = \phi \wedge (l_1) \wedge (\neg l_2)$, corresponding to the DPLL-based partitioning. Similarly it is possible to choose $d_3 = 1$ in Equation (5.12) and $\phi_3 = \phi \wedge (\neg l_1 \vee \neg l_2) \wedge (\neg l_1) \wedge (l_3)$ resulting in the search corresponding to the formula $\phi \wedge \neg l_1 \wedge l_3$, derived from the DPLL-based-partitioning, and finally $\phi_4 = \phi \wedge (\neg l_1 \vee \neg l_2) \wedge (\neg l_1) \wedge (\neg l_3)$.

The approach presented in the above example generalizes to higher numbers of derived formulas. Let $S_n = (d_1, \ldots, d_n)$ denote the sequence producing n derived instances as in Example 6. Let $S_i = (d_1, \ldots, d_i)$ and $T_j = (e_1, \ldots, e_j)$ be two such sequences. We denote by $S_n + 1$ the sequence $(d_1 + 1, \ldots, d_n + 1)$ and by $(S_i) \cdot (T_j)$ the concatenation of the two sequences $(d_1, \ldots, d_i, e_1, \ldots, e_j)$. The scattering-based partitioning function simulates the DPLL-based partitioning function producing $n = 2^k, k \geq 0$ derived instances by using a fixed variable ordering and the sequence S_n defined recursively as $S_1 = S_{k^0} = (0)$ and $S_{2^k} = (S_{2^{k-1}} + 1) \cdot (S_{2^{k-1}})$.

5.7 Further topics

Theory Solver Parallelization: Two core algorithms of an SMT theory solver are the congruence closure algorithm based on the *E-graph* data structure [26] and an incremental implementation of the Simplex algorithm [29]. For most non-incremental SMT problems most of the runtime of an SMT solver is spent on the theory solvers. This presents an interesting challenge for parallelization of the theory solvers, since while the total time spent in theories is high, each individual call to the solver is usually very short. However, the performance of a theory solver parallelization is of course highly dependent on the type and behavior of the theory that it decides. While this type of parallelization is often used for very specific problems in many areas, their application in SMT with multiple theories being combined is not very common yet. However, this area is just starting to be explored, for instance by Hadarean et al. [42] who use lazy and eager bit-vector solvers in parallel.

Parallelization of Incremental SMT: Software and hardware verification using symbolic model-checking is undeniably the most important application driving the development of SMT solvers today. Many successful symbolic model-checking approaches reduce the verification problem to repeated, related queries to an SMT solver [14]. This frequently results in a very large set of (relatively) simple queries. Thus, the problem of parallelization across multiple, but related, problems is of very high importance as well.

References

[1] Abdulla, P.A., Bjesse, P., Eén, N.: Symbolic reachability analysis based on SAT-solvers. In: S. Graf, M.I. Schwartzbach (eds.) Tools and Algorithms for Construction and Analysis of Systems, 6th International Conference, TACAS 2000, Held as Part of the European Joint Conferences on the Theory and Practice of Software, ETAPS 2000, Berlin, Germany, March 25 - April 2, 2000, Proceedings, *Lecture Notes in Computer Science*, vol. 1785, pp. 411–425. Springer (2000)

[2] Aloul, F.A., Ramani, A., Markov, I.L., Sakallah, K.A.: Solving difficult SAT instances in the presence of symmetry. In: Proceedings of the 39th Design Automation Conference, DAC 2002, New Orleans, LA, USA, June 10-14, 2002, pp. 731–736. ACM (2002)

[3] Alt, L., Fedyukovich, G., Hyvärinen, A.E.J., Sharygina, N.: A proof-sensitive approach for small propositional interpolants. In: A. Gurfinkel, S.A. Seshia (eds.) Verified Software: Theories, Tools, and Experiments - 7th International Conference, VSTTE 2015, San Francisco, CA, USA, July 18-19, 2015. Revised Selected Papers, *Lecture Notes in Computer Science*, vol. 9593, pp. 1–18. Springer (2015)

[4] Barrett, C., Conway, C.L., Deters, M., Hadarean, L., Jovanovic, D., King, T., Reynolds, A., Tinelli, C.: CVC4. In: G. Gopalakrishnan, S. Qadeer (eds.) Computer Aided Verification - 23rd International Conference, CAV 2011, Snowbird, UT, USA, July 14-20, 2011. Proceedings, *Lecture Notes in Computer Science*, vol. 6806, pp. 171–177. Springer (2011)

[5] Barrett, C., Fontaine, P., Tinelli, C.: The Satisfiability Modulo Theories Library (SMT-LIB). `www.SMT-LIB.org` (2016)

[6] Benque, D., Bourton, S., Cockerton, C., Cook, B., Fisher, J., Ishtiaq, S., Piterman, N., Taylor, A.S., Vardi, M.Y.: BMA: Visual tool for modeling and analyzing biological networks. In: P. Madhusudan, S.A. Seshia (eds.) Computer Aided Verification - 24th International Conference, CAV 2012, Berkeley, CA, USA, July 7-13, 2012 Proceedings, *Lecture Notes in Computer Science*, vol. 7358, pp. 686–692. Springer (2012)

[7] Biere, A.: Lingeling, Plingeling, PicoSAT and PrecoSAT at SAT Race 2010. Technical Report 10/1, Institute for Formal Models and Verification, Johannes Kepler University (2010)

[8] Biere, A., Bloem, R. (eds.): Computer Aided Verification - 26th International Conference, CAV 2014, Held as Part of the Vienna Summer of Logic, VSL 2014, Vienna, Austria, July 18-22, 2014. Proceedings, *Lecture Notes in Computer Science*, vol. 8559. Springer (2014)

[9] Bonacina, M.P.: A taxonomy of parallel strategies for deduction. Annals of Mathematics and Artificial Intelligence **29**(1–4), 223–257 (2000)

[10] Bonacina, M.P.: Combination of distributed search and multi-search in Peers-mcd.d. In: Proceedings of the 1st International Joint Conference on Automated Reasoning (IJCAR 2001), *Lecture Notes in Artificial Intelligence*, vol. 2083, pp. 448–452. Springer (2001)

[11] Bordeaux, L., Hamadi, Y., Samulowitz, H.: Experiments with massively parallel constraint solving. In: Proceedings of the 21st International Joint Conference on Artificial Intelligence (IJCAI 2009), pp. 443–448 (2009)
[12] Bouton, T., Oliveira, D.C.B.D., Déharbe, D., Fontaine, P.: veriT: An open, trustable and efficient SMT-solver. In: R.A. Schmidt (ed.) Automated Deduction - CADE-22, 22nd International Conference on Automated Deduction, Montreal, Canada, August 2-7, 2009. Proceedings, *Lecture Notes in Computer Science*, vol. 5663, pp. 151–156. Springer (2009)
[13] Bozzano, M., Bruttomesso, R., Cimatti, A., Junttila, T.A., Ranise, S., van Rossum, P., Sebastiani, R.: Efficient satisfiability modulo theories via delayed theory combination. In: K. Etessami, S.K. Rajamani (eds.) Computer Aided Verification, 17th International Conference, CAV 2005, Edinburgh, Scotland, UK, July 6-10, 2005, Proceedings, *Lecture Notes in Computer Science*, vol. 3576, pp. 335–349. Springer (2005)
[14] Bradley, A.R.: SAT-based model checking without unrolling. In: R. Jhala, D.A. Schmidt (eds.) Verification, Model Checking, and Abstract Interpretation - 12th International Conference, VMCAI 2011, Austin, TX, USA, January 23-25, 2011. Proceedings, *Lecture Notes in Computer Science*, vol. 6538, pp. 70–87. Springer (2011)
[15] Brayton, R.K., Mishchenko, A.: ABC: an academic industrial-strength verification tool. In: T. Touili, B. Cook, P.B. Jackson (eds.) Computer Aided Verification, 22nd International Conference, CAV 2010, Edinburgh, UK, July 15-19, 2010. Proceedings, *Lecture Notes in Computer Science*, vol. 6174, pp. 24–40. Springer (2010)
[16] Brummayer, R., Biere, A.: Boolector: An efficient SMT solver for bit-vectors and arrays. In: S. Kowalewski, A. Philippou (eds.) Tools and Algorithms for the Construction and Analysis of Systems, 15th International Conference, TACAS 2009, Held as Part of the Joint European Conferences on Theory and Practice of Software, ETAPS 2009, York, UK, March 22-29, 2009. Proceedings, *Lecture Notes in Computer Science*, vol. 5505, pp. 174–177. Springer (2009)
[17] Bruttomesso, R., Cimatti, A., Franzén, A., Griggio, A., Sebastiani, R.: Delayed theory combination vs. Nelson-Oppen for satisfiability modulo theories: a comparative analysis. Ann. Math. Artif. Intell. **55**(1-2), 63–99 (2009)
[18] Christ, J., Hoenicke, J., Nutz, A.: SMTInterpol: An interpolating SMT solver. In: A.F. Donaldson, D. Parker (eds.) Model Checking Software - 19th International Workshop, SPIN 2012, Oxford, UK, July 23-24, 2012. Proceedings, *Lecture Notes in Computer Science*, vol. 7385, pp. 248–254. Springer (2012)
[19] Cimatti, A., Griggio, A., Schaafsma, B.J., Sebastiani, R.: The MathSAT5 SMT solver. In: N. Piterman, S.A. Smolka (eds.) Tools and Algorithms for the Construction and Analysis of Systems - 19th International Conference, TACAS 2013, Held as Part of the European Joint Conferences on Theory and Practice of Software, ETAPS 2013, Rome, Italy, March 16-24, 2013. Proceedings, *Lecture Notes in Computer Science*, vol. 7795, pp. 93–107. Springer (2013)
[20] Cimatti, A., Sebastiani, R. (eds.): Theory and Applications of Satisfiability Testing - SAT 2012 - 15th International Conference, Trento, Italy, June 17-20,

2012. Proceedings, *Lecture Notes in Computer Science*, vol. 7317. Springer (2012)
[21] Conchon, S., Déharbe, D., Heizmann, M., Weber., T.: 11th International Satisfiability Modulo Theories Competition (SMT-COMP 2016). `http://smtcomp.sourceforge.net/2016/` (2016)
[22] Corzilius, F., Kremer, G., Junges, S., Schupp, S., Ábrahám, E.: SMT-RAT: an open source C++ toolbox for strategic and parallel SMT solving. In: Heule and Weaver [47], pp. 360–368
[23] Craig, W.: Linear reasoning. a new form of the Herbrand-Gentzen theorem. J. Symb. Log. **22**(3), 250–268 (1957)
[24] Creignou, N., Berre, D.L. (eds.): Theory and Applications of Satisfiability Testing - SAT 2016 - 19th International Conference, Bordeaux, France, July 5-8, 2016, Proceedings, *Lecture Notes in Computer Science*, vol. 9710. Springer (2016)
[25] Dequen, G., Vander-Swalmen, P., Krajecki, M.: Toward easy parallel SAT solving. In: Proceedings of the 21st IEEE International Concerence on Tools with Artificial Intelligence (ICTAI 2009), pp. 425–432. IEEE Press (2009)
[26] Detlefs, D., Nelson, G., Saxe, J.B.: Simplify: a theorem prover for program checking. Journal of the ACM **52**(3), 365–473 (2005)
[27] D'Silva, V., Kroening, D., Purandare, M., Weissenbacher, G.: Interpolant strength. In: G. Barthe, M.V. Hermenegildo (eds.) Verification, Model Checking, and Abstract Interpretation, 11th International Conference, VMCAI 2010, Madrid, Spain, January 17-19, 2010. Proceedings, *Lecture Notes in Computer Science*, vol. 5944, pp. 129–145. Springer (2010)
[28] Dutertre, B.: Yices 2.2. In: Biere and Bloem [8], pp. 737–744
[29] Dutertre, B., de Moura, L.M.: A fast linear-arithmetic solver for DPLL(T). In: T. Ball, R.B. Jones (eds.) Computer Aided Verification, 18th International Conference, CAV 2006, Seattle, WA, USA, August 17-20, 2006, Proceedings, *Lecture Notes in Computer Science*, vol. 4144, pp. 81–94. Springer (2006)
[30] Eén, N., Sörensson, N.: An extensible SAT-solver. In: E. Giunchiglia, A. Tacchella (eds.) Theory and Applications of Satisfiability Testing, 6th International Conference, SAT 2003. Santa Margherita Ligure, Italy, May 5-8, 2003 Selected Revised Papers, *Lecture Notes in Computer Science*, vol. 2919, pp. 502–518. Springer (2003)
[31] Ermon, S., LeBras, R., Gomes, C.P., Selman, B., van Dover, R.B.: SMT-aided combinatorial materials discovery. In: Cimatti and Sebastiani [20], pp. 172–185
[32] Fagerberg, R., Flamm, C., Merkle, D., Peters, P.: Exploring chemistry using SMT. In: M. Milano (ed.) Principles and Practice of Constraint Programming - 18th International Conference, CP 2012, Québec City, QC, Canada, October 8-12, 2012. Proceedings, *Lecture Notes in Computer Science*, vol. 7514, pp. 900–915. Springer (2012)
[33] Gebser, M., Kaminski, R., Kaufmann, B., Schaub, T., Schnor, B.: Cluster-based ASP solving with Claspar. In: Proceedings of the 11th International Conference on Logic Programming and Nonmonotonic Reasoning (LPNMR 2011), *Lecture Notes in Computer Science*, vol. 6645, pp. 364–369. Springer (2011)

[34] Ghilardi, S., Ranise, S.: MCMT: A model checker modulo theories. In: J. Giesl, R. Hähnle (eds.) Automated Reasoning, 5th International Joint Conference, IJCAR 2010, Edinburgh, UK, July 16-19, 2010. Proceedings, *Lecture Notes in Computer Science*, vol. 6173, pp. 22–29. Springer (2010)
[35] Giesl, J., Brockschmidt, M., Emmes, F., Frohn, F., Fuhs, C., Otto, C., Plücker, M., Schneider-Kamp, P., Ströder, T., Swiderski, S., Thiemann, R.: Proving termination of programs automatically with AProVE. In: S. Demri, D. Kapur, C. Weidenbach (eds.) Automated Reasoning - 7th International Joint Conference, IJCAR 2014, Held as Part of the Vienna Summer of Logic, VSL 2014, Vienna, Austria, July 19-22, 2014. Proceedings, *Lecture Notes in Computer Science*, vol. 8562, pp. 184–191. Springer (2014)
[36] Gomes, C.P., Selman, B.: Algorithm portfolios. Artificial Intelligence **126**(1–2), 43–62 (2001)
[37] Gomes, C.P., Selman, B., Crato, N., Kautz, H.: Heavy-tailed phenomena in satisfiability and constraint satisfaction problems. Journal of Automated Reasoning **24**(1/2), 67–100 (2000)
[38] Gomes, C.P., Selman, B., Kautz, H.: Boosting combinatorial search through randomization. In: Proceedings of the 15th National Conference on Artificial Intelligence (AAAI 1998), pp. 431–437. AAAI Press (1998)
[39] Grama, A., Kumar, V.: State of the art in parallel search techniques for discrete optimization problems. IEEE Transactions on Knowledge and Data Engineering **11**(1), 28–34 (1999)
[40] Gregory, P., Long, D., Fox, M., Beck, J.C.: Planning modulo theories: Extending the planning paradigm. In: L. McCluskey, B.C. Williams, J.R. Silva, B. Bonet (eds.) Proceedings of the Twenty-Second International Conference on Automated Planning and Scheduling, ICAPS 2012, Atibaia, São Paulo, Brazil, June 25-19, 2012. AAAI (2012)
[41] Guo, L., Hamadi, Y., Jabbour, S., Sais, L.: Diversification and intensification in parallel SAT solving. In: 16th International Conference on Principles and Practice of Constraint Programming (CP 2010), *Lecture Notes in Computer Science*, vol. 6308, pp. 252 – 265. Springer (2010)
[42] Hadarean, L., Bansal, K., Jovanovic, D., Barrett, C., Tinelli, C.: A tale of two solvers: Eager and lazy approaches to bit-vectors. In: Biere and Bloem [8], pp. 680–695
[43] Hamadi, Y., Jabbour, S., Sais, L.: Control-based clause sharing in parallel SAT solving. In: Proceedings of the 21st International Joint Conference on Artificial Intelligence (IJCAI 2009), pp. 499–504 (2009)
[44] Hamadi, Y., Jabbour, S., Sais, L.: ManySAT: a parallel SAT solver. Journal on Satisfiability, Boolean Modeling and Computation **6**(4), 245 – 262 (2009)
[45] Hamadi, Y., Marques-Silva, J., Wintersteiger, C.M.: Lazy decomposition for distributed decision procedures. In: J. Barnat, K. Heljanko (eds.) Proceedings 10th International Workshop on Parallel and Distributed Methods in verifiCation, PDMC 2011, Snowbird, Utah, USA, July 14, 2011., *EPTCS*, vol. 72, pp. 43–54 (2011)

[46] Heule, M., van Maaren, H.: Look-ahead based SAT solvers. In: A. Biere, M. Heule, H. van Maaren, T. Walsh (eds.) Handbook of Satisfiability, *Frontiers in Artificial Intelligence and Applications*, vol. 185, pp. 155–184. IOS Press (2009)

[47] Heule, M., Weaver, S. (eds.): Theory and Applications of Satisfiability Testing - SAT 2015 - 18th International Conference, Austin, TX, USA, September 24-27, 2015, Proceedings, *Lecture Notes in Computer Science*, vol. 9340. Springer (2015)

[48] Huang, G.: Constructing Craig interpolation formulas. In: D. Du, M. Li (eds.) Computing and Combinatorics, First Annual International Conference, COCOON '95, Xi'an, China, August 24-26, 1995, Proceedings, *Lecture Notes in Computer Science*, vol. 959, pp. 181–190. Springer (1995)

[49] Huberman, B.A., Lukose, R.M., Hogg, T.: An economics approach to hard computational problems. Science **275**(5296), 51–54 (1997)

[50] Hyvärinen, A.E.J., Junttila, T., Niemelä, I.: A distribution method for solving SAT in grids. In: Proceedings of the 9th International Conference on Theory and Applications of Satisfiability Testing (SAT 2006), *Lecture Notes in Computer Science*, vol. 4121, pp. 430–435. Springer (2006)

[51] Hyvärinen, A.E.J., Junttila, T.A., Niemelä, I.: Partitioning search spaces of a randomized search. Fundam. Inform. **107**(2-3), 289–311 (2011)

[52] Hyvärinen, A.E.J., Manthey, N.: Designing scalable parallel SAT solvers. In: Cimatti and Sebastiani [20], pp. 214–227

[53] Hyvärinen, A.E.J., Marescotti, M., Alt, L., Sharygina, N.: OpenSMT2: An SMT solver for multi-core and cloud computing. In: Creignou and Berre [24], pp. 547–553

[54] Hyvärinen, A.E.J., Marescotti, M., Sharygina, N.: Search-space partitioning for parallelizing SMT solvers. In: Heule and Weaver [47], pp. 369–386

[55] Inoue, K., Soh, T., Ueda, S., Sasaura, Y., Banbara, M., Tamura, N.: A competitive and cooperative approach to propositional satisfiability. Discrete Applied Mathematics **154**(16), 2291–2306 (2006)

[56] Janakiram, V.K., Agrawal, D.P., Mehrotra, R.: A randomized parallel backtracking algorithm. IEEE Transactions on Computers **37**(12), 1665–1676 (1988)

[57] Janakiram, V.K., Gehringer, E.F., Agrawal, D.P., Mehrotra, R.: A randomized parallel branch-and-bound algorithm. International Journal of Parallel Programming **17**(3), 277 – 301 (1988)

[58] Jonás, M., Strejcek, J.: Solving quantified bit-vector formulas using binary decision diagrams. In: Creignou and Berre [24], pp. 267–283

[59] Kalinnik, N., Ábrahám, E., Schubert, T., Wimmer, R., Becker, B.: Exploiting different strategies for the parallelization of an SMT solver. In: M. Dietrich (ed.) Methoden und Beschreibungssprachen zur Modellierung und Verifikation von Schaltungen und Systemen (MBMV), Dresden, Germany, February 22-24, 2010, pp. 97–106. Fraunhofer Verlag (2010)

[60] Kovács, L., Voronkov, A.: First-order theorem proving and Vampire. In: N. Sharygina, H. Veith (eds.) Computer Aided Verification - 25th International Conference, CAV 2013, Saint Petersburg, Russia, July 13-19, 2013.

Proceedings, *Lecture Notes in Computer Science*, vol. 8044, pp. 1–35. Springer (2013)
[61] Krajícek, J.: Interpolation theorems, lower bounds for proof systems, and independence results for bounded arithmetic. J. Symb. Log. **62**(2), 457–486 (1997)
[62] Krings, S., Bendisposto, J., Leuschel, M.: From failure to proof: The prob disprover for B and Event-B. In: R. Calinescu, B. Rumpe (eds.) Software Engineering and Formal Methods - 13th International Conference, SEFM 2015, York, UK, September 7-11, 2015. Proceedings, *Lecture Notes in Computer Science*, vol. 9276, pp. 199–214. Springer (2015)
[63] Li, G.J., Wah, B.W.: Computational efficiency of parallel combinatorial OR-Tree searches. IEEE Transactions on Software Engineering **16**(1), 13–31 (1990)
[64] Luby, M., Ertel, W.: Optimal parallelization of Las Vegas algorithms. In: Proceedings of the 11th Annual Symposium on Theoretical Aspects of Computer Science (STACS 1994), *Lecture Notes in Computer Science*, vol. 775, pp. 463–474. Springer (1994)
[65] Marescotti, M., Hyvärinen, A.E.J., Sharygina, N.: Clause sharing and partitioning for cloud-based SMT solving. In: C. Artho, A. Legay, D. Peled (eds.) Automated Technology for Verification and Analysis - 14th International Symposium, ATVA 2016, Chiba, Japan, October 17-20, 2016, Proceedings, *Lecture Notes in Computer Science*, vol. 9938, pp. 428–443 (2016)
[66] McMillan, K.L.: Interpolation and SAT-based model checking. In: W.A. Hunt Jr., F. Somenzi (eds.) Computer Aided Verification, 15th International Conference, CAV 2003, Boulder, CO, USA, July 8-12, 2003, Proceedings, *Lecture Notes in Computer Science*, vol. 2725, pp. 1–13. Springer (2003)
[67] de Moura, L.M., Bjørner, N.: Model-based theory combination. Electr. Notes Theor. Comput. Sci. **198**(2), 37–49 (2008)
[68] de Moura, L.M., Bjørner, N.: Z3: an efficient SMT solver. In: C.R. Ramakrishnan, J. Rehof (eds.) Tools and Algorithms for the Construction and Analysis of Systems, 14th International Conference, TACAS 2008, Held as Part of the Joint European Conferences on Theory and Practice of Software, ETAPS 2008, Budapest, Hungary, March 29-April 6, 2008. Proceedings, *Lecture Notes in Computer Science*, vol. 4963, pp. 337–340. Springer (2008)
[69] Nelson, G., Oppen, D.C.: Simplification by cooperating decision procedures. ACM Trans. Program. Lang. Syst. **1**(2), 245–257 (1979)
[70] Ohmura, K., Ueda, K.: c-sat: A parallel SAT solver for clusters. In: Proceedings of the 12th International Conference on Theory and Applications of Satisfiability Testing (SAT 2009), *Lecture Notes in Computer Science*, vol. 5584, pp. 524–537. Springer (2009)
[71] Petrik, M., Zilberstein, S.: Learning parallel portfolios of algorithms. Annals of Mathematics and Artificial Intelligence **48**(1-2), 85–106 (2006)
[72] Prestwich, S., Mudambi, S.: Improved branch and bound in constraint logic programming. In: Proceedings of the 1st International Conference on Principles and Practice of Constraint Programming (CP 1995), *Lecture Notes in Computer Science*, vol. 976, pp. 534–548. Springer (1995)

[73] Pudlák, P.: Lower bounds for resolution and cutting plane proofs and monotone computations. J. Symb. Log. **62**(3), 981–998 (1997)

[74] Reisenberger, C.: PBoolector: a parallel SMT solver for QF_BV by combining bit-blasting with look-ahead. Master's thesis, Johannes Kepler Universität Linz, Linz, Austria (2014)

[75] Rice, J.R.: The algorithm selection problem. Advances in Computers **15**, 65 – 118 (1976)

[76] de Salvo Braz, R., O'Reilly, C., Gogate, V., Dechter, R.: Probabilistic inference modulo theories. In: S. Kambhampati (ed.) Proceedings of the Twenty-Fifth International Joint Conference on Artificial Intelligence, IJCAI 2016, New York, NY, USA, 9-15 July 2016, pp. 3591–3599. IJCAI/AAAI Press (2016)

[77] Sebastiani, R., Trentin, P.: OptiMathSAT: A tool for optimization modulo theories. In: D. Kroening, C.S. Pasareanu (eds.) Computer Aided Verification - 27th International Conference, CAV 2015, San Francisco, CA, USA, July 18-24, 2015, Proceedings, Part I, *Lecture Notes in Computer Science*, vol. 9206, pp. 447–454. Springer (2015)

[78] Segre, A.M., Forman, S.L., Resta, G., Wildenberg, A.: Nagging: A scalable fault-tolerant paradigm for distributed search. Artificial Intelligence **140**(1/2), 71–106 (2002)

[79] Speckenmeyer, E., Monien, B., Vornberger, O.: Superlinear speedup for parallel backtracking. In: Proceedings of the 1st international conference on Supercomputing (SC 1987), *Lecture Notes in Computer Science*, vol. 297, pp. 985–993. Springer (1988)

[80] Tinelli, C., Zarba, C.G.: Combining nonstably infinite theories. J. Autom. Reasoning **34**(3), 209–238 (2005)

[81] Tung, V.X., Khanh, T.V., Ogawa, M.: raSAT: An SMT solver for polynomial constraints. In: N. Olivetti, A. Tiwari (eds.) Automated Reasoning - 8th International Joint Conference, IJCAR 2016, Coimbra, Portugal, June 27 - July 2, 2016, Proceedings, *Lecture Notes in Computer Science*, vol. 9706, pp. 228–237. Springer (2016)

[82] Williams, R., Gomes, C.P., Selman, B.: Backdoors to typical case complexity. In: Proceedings of the 18th International Joint Conference on Artificial Intelligence (IJCAI 2003), pp. 1173–1178. Morgan Kaufmann (2003)

[83] Wintersteiger, C.M., Hamadi, Y., de Moura, L.: A concurrent portfolio approach to SMT solving. In: A. Bouajjani, O. Maler (eds.) Computer Aided Verification, 21st International Conference, CAV 2009, Grenoble, France, June 26 - July 2, 2009. Proceedings, *Lecture Notes in Computer Science*, vol. 5643, pp. 715–720. Springer (2009)

[84] Yordanov, B., Wintersteiger, C.M., Hamadi, Y., Kugler, H.: SMT-based analysis of biological computation. In: G. Brat, N. Rungta, A. Venet (eds.) NASA Formal Methods, 5th International Symposium, NFM 2013, Moffett Field, CA, USA, May 14-16, 2013. Proceedings, *Lecture Notes in Computer Science*, vol. 7871, pp. 78–92. Springer (2013)

Chapter 6
Parallel Theorem Proving

Maria Paola Bonacina

Abstract This chapter surveys the research in parallel or distributed strategies for mechanical theorem proving in first-order logic, and explores some of its connections with the research in the parallelization of decision procedures for satisfiability in propositional logic (SAT). We clarify the key role played by the *Clause-Diffusion methodology for distributed deduction* in moving parallel reasoning from the *parallelization of inferences* to the *parallelization of search*, which is the dominating paradigm today. Since the quest for parallel first-order proof procedures has not been pursued recently, we endeavour to relate lessons learned from investigations of parallel theorem proving and parallel SAT-solving with novel advances in theorem proving, such as SGGS (*Semantically-Guided Goal-Sensitive reasoning*), a method that lifts the CDCL (Conflict-Driven Clause Learning) procedure for SAT to first-order logic.

6.1 Introduction

Research on parallel theorem proving, meaning automatic theorem proving (ATP) in first-order logic, began in the mid and late 1980's, flourished in the 1990's, and came pretty much to a halt in the early 2000's [192, 50, 210, 85, 37]. Research on parallel satisfiability solving, meaning satisfiability in propositional logic (SAT), began in the early 1990's and is still actively pursued today (cf. [163, 114, 1, 158] and the chapters on 1, Parallel Satisfiability and 2, Cube and Conquer for Satisfiability). It is probably unknown to most authors active in parallel SAT-solving that Hantao Zhang began work on his parallel SAT solver PSATO [227, 228], that pioneered the divide-and-conquer approach to parallel SAT-solving, after learning about the *Clause-Diffusion*

Maria Paola Bonacina
Dipartimento di Informatica, Università degli Studi di Verona, Strada Le Grazie 15, I-37134 Verona, Italy, EU. e-mail: mariapaola.bonacina@univr.it

Y. Hamadi und L. Sais (eds.), *Handbook of Parallel Constraint Reasoning*,
https://doi.org/10.1007/978-3-319-63516-3_6

methodology for distributed deduction [27, 48, 49, 51] and its implementation in the Aquarius theorem prover [46, 27, 47, 52].

In previous work, we surveyed parallel theorem proving twice, first at the height of the interest in this topic [27, 50], and then when the involvement of the scientific community with this fascinating subject was already decreasing [36, 37]. In this chapter we revisit parallel theorem proving, in the light of advances in theorem proving itself and in parallel SAT-solving, with the aim of providing the readers with material to reflect about the connections between:

- Past work in parallel theorem proving and selected contemporary approaches to theorem proving;
- Past work in parallel theorem proving and later work in parallel SAT-solving;
- Selected approaches to parallel SAT-solving and potentially new leads for parallel theorem proving.

Since the time of the latest investigations in parallel theorem proving, the field has witnessed a significant growth of the paradigm of *model-based reasoning* [44], where a theorem-proving method is *model-based*, if the state of a derivation contains a representation of a candidate partial model that unfolds with the derivation. Traditional model-based first-order methods include *subgoal-reduction strategies* based on *model elimination* and *model-elimination tableaux*, whose parallelization received considerable attention [194, 7, 75, 63, 144, 170, 209, 104, 219]. It is therefore an interesting question to ask what we may learn from past work on parallel theorem proving towards the parallelization of more recent or new approaches to model-based theorem proving.

The growth of model-based reasoning has various motivations. One motivation is the relevance of models for applications. For instance, an assignment to program variables is a model from a logical point of view [40]. Thus, models become "moles" to exercise program paths in testing or examples for example-driven synthesis, and a reasoner that generates models supports automated test generation and program synthesis [82, 141]. Another motivation is inspiration from the practical successes of solvers for propositional satisfiability (cf. [157] and the chapter on 1, Parallel Satisfiability) and satisfiability modulo theories (SMT) (cf. [83] and the chapter on 5, Parallel Satisfiability Modulo Theories), that are model-based because built on the CDCL (Conflict-Driven Clause Learning) procedure [161, 162, 171, 160], which is inherently model-based. Therefore, another spontaneous question is what we may learn from past work on both parallel theorem proving and parallel SAT-solving towards the parallelization of recent model-based first-order methods. Examples of the latter include DPLL($\Gamma + \mathcal{T}$) [81, 56, 57], that integrates an *ordering-based strategy* in an SMT solver, and SGGS, for *Semantically-Guided Goal-Sensitive reasoning* [60, 59, 61, 62], that generalizes CDCL to first-order logic.

To this end we reconsider and expand our analyses of the parallelization of theorem proving [50, 37], covering *subgoal-reduction strategies*, *ordering-based strategies*, and *instance-based strategies*. Ordering-based strategies are based on *ordered resolution* and *ordered paramodulation/superposition*. They represent the state of the art for first-order logic with equality and are implemented in top-notch theorem-provers such as SPASS [218], E [190], and Vampire [137]. Although they are not model-based, their connection with SAT or SMT solvers is a current research topic [43, 57, 182]. Instance-based strategies integrate instance generation at the first-order level with deciding satisfiability at the propositional or ground level, by a SAT or SMT solver [125, 134]. They are model-based, or at least model-driven, in as much as instance generation is geared to exclude the models proposed by the solver.

We illustrate a selection of methods already covered in our previous surveys [50, 37], to make this chapter self-contained, give the reader a direct impression of those methods, and have material for discussion. For example, ROO [154, 155, 156] illustrates the approach to parallelization by parallel inferences in shared memory that was the state of the art before Clause-Diffusion, and Team-Work [84, 9, 88, 10, 89, 92, 91, 87, 197] is a forerunner of the *portfolio approach*.

Then we summarize twelve years of research on the *Clause-Diffusion methodology* [27, 48, 49, 51] to parallelize ordering-based strategies, including *Modified Clause-Diffusion* [28, 29], and the Clause-Diffusion provers *Aquarius* [46, 27, 47, 52], *Peers* [58, 51], and *Peers-mcd* [30, 31, 33, 38]. We reflect on the impact of Clause-Diffusion on subsequent research: Clause-Diffusion played a central role, because it was the first approach to move from the parallelization of inferences to the parallelization of search, so that it can be considered a forerunner of all parallel or distributed search methods in both theorem proving and satisfiability.

In the final discussion we draw connections between parallel theorem proving and parallel SAT-solving (e.g., [107, 112, 113, 123, 187, 111, 116]), and we discuss ideas for future work in parallel theorem proving, especially parallel model-based theorem proving.

This chapter is organized in three parts: Section 6.2 provides a parallelization-oriented survey of theorem-proving strategies; Section 6.3 presents the approaches to parallel theorem proving; and Section 6.4 contains the final discussion.

It is our contention that much of the research in parallel and distributed theorem proving was simply ahead of its time, with respect to both theorem proving and parallel or distributed computing, and we hope that this chapter will contribute to maintain its intellectual legacy alive and fruitful for future research.

6.2 Theorem Proving Strategies

We begin our analysis of the parallelizability of theorem proving with a classification of theorem-proving strategies in three categories: *subgoal-reduction based*, *ordering-based*, and *instance-based* strategies. Ordering-based strategies include *expansion-oriented* and *contraction-based* strategies. In this section we survey these classes of theorem-proving strategies, covering inference system, search plan, proof generation, redundancy control, and use of models. The presence of *backward contraction* inferences, the *size* of the database of clauses generated and kept by the strategy, and the degree of *dynamicity* of the database of clauses, are among the relevant issues for parallelization.

6.2.1 Subgoal-Reduction Strategies

We use φ and ψ for clauses, σ for most general unifiers, L and L' for literals, C and D for disjunctions of literals, $\simeq$ for equality, and $\Box$ for the empty clause, which represents a contradiction.

In theorem proving *subgoal reduction* stems from *ordered linear resolution* [138]: at each step the strategy resolves the current goal clause $\varphi_i = L \vee C$ with an input clause $\psi = L' \vee D$, such that $L\sigma = \neg L'\sigma$. The next goal clause φ_{i+1} is the resolvent $(D \vee C)\sigma$, where L, seen as a *subgoal* to be solved, has been replaced by or *reduced to* a new bunch of subgoals D. Such a procedure is called *linear*, because at every resolution step one of the parents is the previous resolvent; it is called *linear input*, if, in addition, the side clause ψ is always an input clause. The *ordered* attribute refers to the requirement that the literals in goal clauses be resolved away in a fixed pre-defined order, determined by the *literal-selection rule* of the strategy, also known as AND-*rule*. A typical example is to select literals in left-to-right order. Using only input clauses as side clauses is sufficient for problems made of Horn clauses, or clauses with at most one positive literal [119]. For full first-order logic, also ancestor goal clauses have to be considered for side clauses, so that ψ may be a φ_j with $j < i$.

Model elimination (ME) can be described as a variant of ordered linear resolution [153, 71, 54], where L is saved in φ_{i+1} as a *boxed*, or *framed*, literal $[L]$ (*A-literal* in the original ME terminology), so that φ_{i+1} has the form $(D \vee [L] \vee C)\sigma$. The resulting inference rule is called *ME-extension*. In this manner, resolution with an ancestor goal clause can be replaced by *ME-reduction*, that reduces a goal clause $L' \vee D \vee [L] \vee C$ to $(D \vee [L] \vee C)\sigma$ when $L\sigma = \neg L'\sigma$. Thus, ME is a linear input strategy for full first-order logic.

Independent of resolution, the concept of ME is to prove that the input set S of clauses is unsatisfiable by eliminating all possible candidate models [152]. In order to satisfy a set S of first-order clauses, it is necessary to satisfy all its clauses. In order to satisfy a clause, it is necessary to satisfy all its ground instances. In order to satisfy a ground instance, it is necessary to satisfy one of its literals. If the current goal clause $\varphi_i = L \vee C$ and an input clause $\psi = L' \vee D$ are such that $L\sigma = \neg L'\sigma$, no model can satisfy $\varphi_i\sigma$ and $\psi\sigma$ by satisfying $L\sigma$ and $L'\sigma$. The next goal clause $\varphi_{i+1} = (D \vee [L] \vee C)\sigma$ generated by ME-extension says precisely this: the literal $L\sigma$ is framed to denote that it has been added to the current candidate model, so that $\varphi_i\sigma$ is satisfied; since a model that satisfies $L\sigma$ cannot satisfy $L'\sigma$, some other literal in $D\sigma$ must be satisfied to satisfy $\psi\sigma$. In this sense, the literals of $D\sigma$ are *subgoals* of $L\sigma$. An ME-reduction step that reduces a goal clause $L' \vee D \vee [L] \vee C$ to $(D \vee [L] \vee C)\sigma$, when $L\sigma = \neg L'\sigma$, reckons that $L'\sigma$ cannot be satisfied in a model that contains L.

ME-tableaux make this model elimination process perspicuous by building a tableau, that is a tree, whose nodes are labeled by literals, and whose branches represent candidate models [144, 19, 25, 20, 142, 146, 39]. A branch is *closed* if it contains two complementary literals, and it is *open* otherwise. An open branch represents a candidate model, whereas a closed branch represents an eliminated model. A tableau is *closed* if all its branches are, which means that all candidate models have been eliminated. If $L\sigma = \neg L'\sigma$ for the leaf L of an open branch and an input clause $\psi = L' \vee D$, ME-extension extends the branch with children labeled by the literals of ψ, applies σ to the tableaux, and closes the branch with the complementary literals $L\sigma$ and $L'\sigma$. If a branch contains literals L and L' such that $L\sigma = \neg L'\sigma$, ME-reduction applies σ to the tableau and closes the branch.

A *subgoal-reduction derivation* can be described in the form

$$(S;\varphi_0) \vdash (S;\varphi_1) \vdash \ldots (S;\varphi_i) \vdash \ldots$$

where S is the input set of clauses, $\varphi_0 \in S$ is the input clause designated as *initial goal*, and $\varphi_1, \ldots, \varphi_i, \ldots$ is the succession of proceeding goal clauses. The initial goal clause $\varphi_0 = L_1 \vee \ldots \vee L_k$ yields an *initial tableau* where the root has k children labeled by $L_1, \ldots, L_k$. Note that the literals $L_1, \ldots, L_k$ may share variables, which means that the branches of the tableau may share variables, which is why substitutions apply to the entire tableau. The literals in the current goal clause φ_i label the leaves of the open branches of the tableau, and the framed literals in φ_i label the inner nodes of the tableau.

A derivation is a *refutation* if $\varphi_k = \Box$ for some k, $k > 0$, or, equivalently, if the tableau is closed. A subgoal-reduction strategy is *refutationally complete* if, whenever S is unsatisfiable and $S \setminus \{\varphi_0\}$ is satisfiable, there exists a refutation of S by the strategy starting with $(S;\varphi_0)$. If S is unsatisfiable and $S \setminus \{\varphi_0\}$ is satisfiable, it is the addition of φ_0 that makes the set unsatisfiable, and this is why φ_0 is the *initial goal clause*. Since all generated clauses descend from the initial goal clause, subgoal-reduction strategies are *goal-sensitive*. An unsatisfiable S may contain more than one clause φ_0 such that S is unsatisfiable and $S \setminus \{\varphi_0\}$ is satisfiable, so that there may be a choice of initial goal clause.

Given a refutation with $\varphi_k = \Box$, the comb shaped resolution tree formed by the sequence of goal clauses $\varphi_0, \ldots, \varphi_{k-1}, \Box$ and their companion side clauses is the generated *proof*. The linear shape of the generated proof reveals the linear nature of the strategy. In ME-tableaux, the closed tableau represents the proof.

In subgoal-reduction strategies, *redundancy* appears in the form of repeated subgoals or subgoal instances, and it is countered by techniques called *C-reduction* [195], *caching* [8, 54], *regressive merging* [215], *folding-up* [143, 103] or *success substitutions* [146], and *tabling* or *memoing* [217]. C-reduction, caching, and regressive merging are used in model elimination, folding-up and success substitutions in tableaux, tabling and memoing in declarative programming. In essence, all these techniques descend from the idea of *lemmatization* or *lemmaizing* [152, 8, 54]. Adopting tableaux parlance, if the strategy manages to close a sub-tableau whose root is labeled with literal L, it means that no model of the set S of clauses contains L, that is, $S \models \neg L$. Thus, $\neg L$ can be learned as a lemma, and applied to resolve away, or close, any future subgoal L' such that $L\sigma = L'\sigma$ [143, 103, 39]. [1] Lemmatization causes the database of clauses to grow, if the lemmas are added to S, but a common charac-

[1] If the sub-tableau is closed using *ME*-reductions with ancestors $L_1, \ldots, L_n$ of L, no model with $L_1, \ldots, L_n$ contains L; the lemma is $\neg L \vee \neg L_1 \vee \ldots \vee \neg L_n$; and $\neg L$ can be applied as a unit lemma only below $L_1, \ldots, L_n$ (cf. Section 2.5 of [39]).

teristic of caching techniques is to store the information on the learned lemma $\neg L$, or dually, on the solved subgoal L, without bothering S. For example, in folding-up, the information is stored in the tableau, at the node labeled with L.

Subgoal-reduction strategies use *depth-first search* (DFS) to search for a proof, with *backtracking* to get out of the dead-end represented by a φ_i to which no inference applies (e.g., its leftmost literal can be neither ME-extended nor ME-reduced). A specific DFS plan is characterized by a literal-selection rule and a *clause-selection rule*, also known as OR-*rule*, that determines the order in which the input clauses are tried. A typical example is to try clauses in top-down order, that is, in the order they are written in the input file. Backtracking undoes the latest inference and substitution application to enable the strategy to try a different inference. For completeness, DFS is enriched with *iterative deepening* [133, 206] on the number of resolution or ME-extension inferences. Thus, subgoal-reduction strategies develop and keep in memory *one proof attempt at a time*, and switch to another one by backtracking.

Prolog Technology Theorem Proving (PTTP) is a major paradigm for subgoal-reduction strategies [202, 203, 205]. PTTP implements ME on top of the *Warren Abstract Machine* (WAM), the virtual machine designed for Prolog [216]. The linear input nature of ME is crucial, because Prolog uses a variant of ordered linear input resolution. The WAM is a stack machine, with goal clauses stored on the stack, and input clauses compiled as a Prolog program. A stack machine implements DFS naturally. For theorem proving, PTTP adds iterative deepening and the *occur check* in unification: when computing a most general unifier σ, the substitution σ cannot include a pair $x \leftarrow t$, where x occurs in t. Unification in Prolog omits this check for the sake of performance, and because Prolog, at least in its basic formulation, is a relational language with a limited use of function symbols, so that the likelihood of a pair $x \leftarrow t$, where x occurs in t, is deemed low.

6.2.2 Ordering-Based Strategies

Ordering-based strategies have two kinds of inference rules. *Expansion inference rules* generate and add clauses:

$$\frac{S}{S'} \quad S \subset S'$$

where $S \subset S'$ says that the existing set S of clauses is being expanded by adding something. An expansion inference rule is *sound*, if whatever is added is a logical consequence of what pre-existed, that is, if $S' \subseteq Th(S)$, where $Th(S) = \{\varphi : S \models \varphi\}$. Examples include *binary resolution* and *factoring* [185], that add a *binary resolvent* or a *factor*, *paramodulation* [183, 174], that adds a *paramodulant*, *superposition* [132, 120, 12], which is ordered paramodulation where the literal paramodulated into is an equality, *equational factoring* [13], and *reflection*, which is resolution with $x \simeq x$. Together these inference rules build equality into resolution [121, 186, 13, 53, 173].

Contraction inference rules delete clauses or replace them by smaller ones:

$$\frac{S}{\overline{\overline{S'}}} \quad S \not\subseteq S' \quad S' \prec_{mul} S$$

where $S \not\subseteq S'$ tells that something has been deleted; $S' \prec_{mul} S$ says that S' is smaller than S in the *multiset extension* [96] of a *well-founded* ordering $\prec$ on terms, literals, and clauses [93]; and the double inference line [41] emphasizes the diversity of contraction with respect to the traditional notion of inference. Soundness for contraction is called *adequacy* [41]: a contraction inference rule is *adequate*, if whatever is deleted is a logical consequence of what is kept, that is, if $S \subseteq Th(S')$. Since $S \subseteq Th(S')$ implies $Th(S) \subseteq Th(S')$, soundness for contraction is also called *monotonicity* [35], meaning monotonicity of inferences with respect to theoremhood. Examples of contraction inference rules include *tautology deletion*, *subsumption* [185, 189], *clausal simplification*, which is a combination of unit resolution and subsumption, *demodulation* [222] or *simplification* (i.e., simplification by an equation) [132, 120, 12], *functional subsumption* (i.e., subsumption between equations) [120], *purity deletion* [80, 54], and *conditional simplification* [42]. Repeated simplification is called *normalization* or reduction to *normal form*, meaning a form that cannot be rewritten further, and normalization can be viewed as a single contraction step. The normal form of a clause φ is denoted $\varphi\downarrow$.

These strategies are called *ordering-based*, because they use the ordering $\prec$ to define contraction rules and restrict expansion rules: resolution is *ordered resolution*, factoring is *ordered factoring*, paramodulation is *ordered paramodulation*, and superposition is natively ordering-restricted. This means that only maximal literals are resolved upon, factorized, paramodulated or superposed into and from, and only maximal sides of equations are paramodulated or superposed into and from, where maximality is tested in the clause instantiated by the most general unifier of the inference step [120, 12, 121, 186, 13, 53, 173]. Ordering-based strategies search for the proof by *best-first search* and do not need backtracking. Best-first search is implemented by the *given-clause algorithm* that we shall cover in Section 6.3.2 because it is relevant to parallelization approaches. Several presentations, surveys, and systematizations of ordering-based theorem proving and its orderings are available [177, 94, 178, 98, 97, 15, 172, 41, 149, 179].

Ordering-based strategies are *not model-based*: in ordering-based inference systems models remain implicit, and come to the forefront only in the proofs of refutational completeness. For example, a proof technique uses *transfinite semantic trees* to survey models and show that the inference system excludes them all [121]. Another proof technique is based on *saturation*. A set of clauses is *saturated*, if any inference with premises in the set is redundant (cf. Section 6.2.2.2 for redundancy). Refutational completeness is established by showing that a saturated set of clauses that does not contain $\Box$ is satisfiable [13].

However, ordering-based strategies may use a *fixed* interpretation for *semantic guidance*, as exemplified in *semantic resolution* [198], *hyperresolution* [184], and *resolution with set of support* [221] (cf. Sections 2.6 in [35] and 2.1 in [44]).

Hyperresolution resolves a clause $L_1 \vee \ldots \vee L_q \vee C$, called *nucleus*, and clauses $L'_1 \vee D_1, \ldots, L'_q \vee D_q$, with $q \geq 1$, called *electrons*, such that $L_i\sigma = \neg L'_i\sigma$ for all i, $1 \leq i \leq q$, to generate the *hyperresolvent* $(D_1 \vee \ldots \vee D_q \vee C)\sigma$ [184]. *Positive hyperresolution* assumes a fixed Herbrand interpretation I that contains all negative literals, and generates only clauses that are *false* in I, namely clauses whose literals are all positive. Such a clause is called *positive*. The electrons are required to be positive clauses, and $L_1 \vee \ldots \vee L_q$ are required to be all and only negative literals in the nucleus. Thus, positive electrons are used to resolve away all negative literals in the nucleus to get a positive hyperresolvent. *Negative hyperresolution* is defined dually with all signs exchanged.

In a *strategy with set of support*, all clauses issued from the negation of the conjecture are considered *goal clauses*. The input set S of clauses is subdivided into the set of support SOS, that contains the goal clauses, and its complement $T = S \setminus SOS$. T is assumed to be satisfiable (e.g., it contains the axioms of a theory), so that if S is unsatisfiable, the unsatisfiability is caused by SOS. Every expansion inference is required to be *supported*, meaning that at least one parent is in SOS. The generated clauses are added to SOS, and since they all descend from goal clauses, the strategy is *goal-sensitive*. A strategy with set of support is compatible with contraction (tautology deletion, subsumption, clausal simplification), provided also clauses generated by backward contraction are inserted in SOS [54]. Since T does not get expanded, a strategy with set of support is complete for problems with equality only if T is saturated. The interplay of *parallelism and semantic guidance* in ordering-based strategies has not been explored thus far, as we shall discuss in Section 6.4.

6.2.2.1 Expansion-Oriented Strategies

The distinction between *expansion-oriented* and *contraction-based* strategies towards analyzing parallelism [50] depends on the distinction between *forward* and *backward* contraction. In *forward contraction*, a newly generated clause φ, called *raw clause* [27, 50], is deleted or normalized into $\varphi\downarrow$ by previously existing clauses. In *backward contraction*, such a $\varphi\downarrow$ is applied to contract previously existing clauses. Expansion-oriented strategies apply at most forward contraction. Accordingly, an *expansion-oriented derivation* has the form

$$(S_0;N_0) \vdash (S_1;N_1) \vdash \ldots (S_i;N_i) \vdash \ldots$$

where S_i is the set of clauses in the database of clauses, and N_i is the set of *raw clauses*. Every clause in S_i has an *identifier*, typically a positive integer generated progressively by the prover, and is ready to be used as premise. Raw clauses are clauses that were just generated and still need to undergo forward contraction. Initially, $S_0 = S$ is the input set of clauses and $N_0 = \emptyset$. Expansion takes premises in S_i and adds raw clauses to N_{i+1}. Forward contraction deletes clauses in N_i and adds their non-trivial normal forms to S_{i+1}. It follows that $S_0 \subseteq S_1 \subseteq \ldots \subseteq S_i \subseteq \ldots$, that is, for an expansion-oriented strategy the database of clauses is *monotonically increasing*.

A derivation is a *refutation* if $\Box \in S_k$ for some k, $k > 0$. A strategy is *refutationally complete* if, whenever S is unsatisfiable, there exists a refutation of S by the strategy. Ordering-based strategies develop *multiple proof attempts* that remain *implicit* in the set of clauses S_i. Only when $\Box \in S_k$, the strategy reconstructs the generated *proof* in the form of the *ancestor-tree* of $\Box$ [29], denoted $\Pi(\Box)$. The reconstruction starts from $\Box$ and proceeds backward until it reaches the input clauses. For instance, if φ is a resolvent of φ_1 and φ_2, $\Pi(\varphi)$ has root labeled φ and subtrees $\Pi(\varphi_1)$ and $\Pi(\varphi_2)$. If φ is generated as the normal form of a pre-existing clause ψ with respect to equations $\varphi_1,\ldots,\varphi_n$, $\Pi(\varphi)$ has root labeled φ and subtrees $\Pi(\varphi_1),\ldots,\Pi(\varphi_n)$ and $\Pi(\psi)$. Every clause has its own variables, and *variants*, that is, clauses that are identical up to variable renaming, are treated as distinct clauses. Therefore, no clause is generated twice, and $\Pi(\varphi)$ is uniquely defined given φ. Since a clause may be used as premise more than once, $\Pi(\varphi)$ is an ancestor-tree if we allow the same clause to label more than one node, an *ancestor-graph* otherwise.

At the time of our first analysis of parallel theorem proving [27, 50], it was already understood that backward contraction is indispensable for theorem proving by ordering-based strategies, especially in the presence of equality. Thus, the class of expansion-oriented strategies was introduced to cover parallel resolution-based theorem-proving methods without backward contraction, mostly for first-order logic without equality [150, 151, 72, 76, 127] or propositional logic [100], and parallelizations [196, 214, 115, 68, 69] of the *Buchberger algorithm* [64] to compute Gröbner bases of ideals generated by sets of polynomials. Buchberger algorithm is a *completion procedure* like *Knuth-Bendix completion* [132], with an expansion inference rule similar to superposition and a contraction rule similar to simplification [65]. However, Buchberger algorithm is guaranteed to converge with or without backward contraction, that can be delayed to a post-processing phase. On the other hand, validity in equational theories, first-order logic, and first-order logic with equality is a semi-decidable problem, so that theorem-proving methods are only *semi-decision procedures*, and backward contraction is crucial in practice to find a proof and terminate. Since expansion-oriented theorem-proving strategies today have mostly pedagogical and historical interest, and Buchberger algorithm is not a first-order theorem-proving strategy, we refer the reader interested in their parallelization to our previous surveys [50, 37].

6.2.2.2 Contraction-Based Strategies

Contraction-based strategies apply both forward and backward contraction eagerly and as much as possible. A *contraction-based derivation* has the form

$$(S_0;N_0;R_0) \vdash (S_1;N_1;R_1) \vdash \ldots (S_i;N_i;R_i) \vdash \ldots$$

where S_i is the set of clauses in the database of clauses, those with an identifier and ready to be used as premises; N_i is the set of *raw clauses*, that is, clauses just generated and still to be subject to forward contraction; and R_i is the set of clauses deleted by backward contraction. Initially, $S_0 = S$ is the input set of clauses and $N_0 = R_0 = \emptyset$. Expansion takes premises in S_i and adds raw clauses to N_{i+1}. Forward contraction deletes clauses in N_i and adds their non-trivial normal forms to S_{i+1}. Backward contraction detects which clauses in S_i can be contracted, moves them to N_{i+1}, and also copies them in R_{i+1}. In this way, backward and forward contraction are implemented by the same operations, and clauses generated by backward contraction are treated in the same way as clauses generated by expansion. The copy in the R component is made for the purpose of proof reconstruction. The database S_i of clauses may either expand or shrink, and therefore it is *non-monotonic*.

Forward contraction applies to a clause before it is established in the database; it can be seen as part of the process that leads to install a new clause in the database. With backward contraction, every clause in the database may be subject to contraction. Thus, the notion of *persistent clauses* becomes relevant: a clause is *persistent*, if it is never deleted after it has entered S_i at some stage i, $i \geq 0$. The set of persistent clauses, called the *limit* of the derivation, is defined as $S_\infty = \bigcup_{i \geq 0} \bigcap_{j \geq i} S_j$.

The notions of *refutation*, *refutational completeness*, and *proof reconstruction* are the same as for expansion-oriented strategies. Assume $\Box \in S_k$: while an expansion-oriented strategy finds in S_k all ancestors of $\Box$, as clauses deleted by forward contraction are not premises of other steps, a contraction-based strategy reconstructs the proof from $S_k \uplus R_k$. Indeed, clauses deleted by backward contraction may be ancestors of $\Box$, because they may have been used as premises before being deleted, and therefore they may be parents of other clauses. Proof reconstruction is the reason for the R_i component.

If the N_i and R_i components are omitted, the derivation has the form

$$S_0 \vdash S_1 \vdash \dots S_i \vdash \dots$$

where $S_0 = S$ is the input set, and at every step S_{i+1} is derived from S_i by an inference that can be either an expansion or a contraction inference.

A key monotonicity property of contraction-based derivations is $\rho(S_0) \subseteq \rho(S_1) \subseteq \dots \subseteq \rho(S_i) \subseteq \dots$, where $\rho(S)$ is the set of clauses that are *redundant* in S. This monotonicity property means that if a clause is redundant at a certain stage of the derivation, it will be redundant at all subsequent stages (cf. Lemma 2.6.4 in [27]), a principle later popularized by the slogan *"once redundant, always redundant"* [41]. A clause is redundant in S if adding it or removing it from S neither improves nor worsens minimal proofs, where improving means making smaller, and worsening means making larger, with respect to a well-founded *proof ordering* (cf. [53, 41] and Chapter 2 of [27]). Clauses that are not redundant are called *irredundant*. Clauses deleted by contraction rules are redundant, and so are clauses whose generation is prevented by the ordering-based restrictions of expansion rules.

The notion of redundancy is extended from clauses to inferences: an inference is *redundant* if it uses or generates redundant clauses, and *irredundant* otherwise. In turn, redundancy is connected with *fairness*: intuitively, the two concepts are dual, because redundancy aims at capturing which inferences can be ignored, and fairness aims at capturing which inferences must be considered to find a proof. Refutational completeness of the inference system ensures that if the input set S is unsatisfiable, then there exist refutations. Fairness is the complementary property: if refutations exist, a fair derivation is guaranteed to be a refutation. Similar to redundancy, also fairness is defined based on proof orderings: whenever a minimal proof of the target theorem is reducible by inferences, it is reduced eventually [27, 53, 41]. In practice, a derivation is *fair*, if all irredundant inferences are considered eventually. A search plan is *fair*, if it generates a fair derivation for all inputs. The combination of a *refutationally complete inference system* and a *fair search plan* yields a *complete theorem-proving strategy*.

Contraction-based strategies feature a search plan that prioritizes contraction over expansion, in order to ensure that redundant clauses are deleted prior to being selected as expansion premises. Such a search plan is called *simplification-first* [45], *contraction-first* [50], or *eager-contraction* [29] search plan.

6.2.3 Instance-Based Strategies

All first-order clausal theorem-proving strategies can be seen as ways to implement *Herbrand's theorem*, which says that a set S of first-order clauses is unsatisfiable if and only if there exists a finite set of ground instances of clauses of S that is unsatisfiable [71]. The semi-decidability of first-order theorem-proving descends from this theorem. *Instance-based strategies* represent the theorem-proving paradigm most directly inspired by Herbrand's theorem. The basic idea is to generate ground instances of input clauses, and test them for propositional unsatisfiability. The first such procedure was *Gilmore method* [71], followed by SATCHMO [159], and *hyperlinking* [140], the latter at the beginning of the renewed interest for the Davis-Putnam-Logemann-Loveland (DPLL) procedure [80, 79, 71] for propositional satisfiability.

A clause $L_1 \vee \ldots \vee L_q$, called *nucleus*, and clauses $L'_1 \vee D_1, \ldots, L'_q \vee D_q$, with $q \geq 1$, called *electrons*, such that $L_i\sigma = \neg L'_i\sigma$ for all i, $1 \leq i \leq q$, form a *hyperlink*. Hyperlinking generates the instance of the nucleus $(L_1 \vee \ldots \vee L_q)\sigma$. An instance generated by hyperlinking is termed *hyperinstance*. Variants of a same clause may be used in a hyperlink, and all literals of the nucleus are linked, since the purpose is not to generate a hyperresolvent (cf. Section 6.2.2), but to instantiate the nucleus as much as possible. Since only instances are generated, all contraction is forward contraction, limited to unit subsumption and clausal simplification, because unrestricted subsumption would delete all instances and defeat the purpose of the strategy. An *instance-based derivation* has the form

$$(S_0; F_0) \vdash (S_1; F_1) \vdash \ldots (S_i; F_i) \vdash \ldots$$

where S_i is the set of clauses in the database of clauses, those with an identifier and ready to be used as premises, and F_i is the set of *generated instances*. Initially, $S_0 = S$ is the input set of clauses and $F_0 = \emptyset$. Instance generation takes premises in S_i and adds new clauses to F_{i+1}. Forward contraction deletes clauses in F_i and adds their non-trivial normal forms to F_{i+1}. If all hyperlinks in S_i have been considered, and contraction has been applied, all clauses in $S_i \cup F_i$ are made ground by replacing all variables by a new constant. A SAT solver is applied to the resulting ground set: if it is unsatisfiable, the procedure halts successfully; otherwise, the next phase of hyperlinking starts with state $(S_{i+1}; F_{i+1})$ where $S_{i+1} = S_i \cup F_i$ and $F_{i+1} = \emptyset$. It follows that $S_0 \subseteq S_1 \subseteq \ldots \subseteq S_i \subseteq \ldots$, that is, the database of clauses is *monotonically increasing*.

While early instance-based strategies had a generate-and-test flavor, proceeding ones, such as *CLINS-S* [73, 74], *ordered semantic hyperlinking* (OSHL) [181, 180], *Inst-Gen* [105, 106, 136, 135], as well as methods that hybridize instance generation and tableaux, such as the *disconnection calculus* [26, 145, 147, 148] and *hypertableaux* [16, 24], progressively emphasized *model-driven* instance generation, putting model building in the driver's seat. The model-building component of the procedure maintains a candidate model. The instance-generation component generates ground instances that are *false* in the model in order to exclude it. The model-building component updates the model to satisfy those ground instances, and the game continues until a contradiction arises.

In summary, for subgoal-reduction strategies the database of clauses is *fixed* and equal to the input set, hence relatively *small*; for expansion-oriented and instance-based strategies it is *large* and *monotonically increasing*; for contraction-based strategies it is *large* and *non-monotonic*. Since expansion-oriented strategies today have mostly pedagogical and historical interest, from now on we use ordering-based strategies to mean contraction-based strategies.

6.3 Parallelization of Theorem Proving

We distinguish three types of parallelism for deduction: *fine-grain parallelism* or *parallelism at the term/literal level*, *medium-grain parallelism* or *parallelism at the clause level*, and *coarse-grain parallelism* or *parallelism at the search level*.

In parallelism at the term/literal level, the parallelization affects operations *below the inference level*, as in *parallel rewriting*, where parallel rewrite steps together make a normalization inference; or *below the clause level*, as in AND-*parallelism*, where alternative inferences apply in parallel to different literals of a clause. In parallelism at the clause level, the parallelization affects operations *at the inference level*, so that parallelism at the clause level means *parallel inferences*. The possibility of *conflicts* between parallel inferences, and the impact of backward contraction on their incidence, emerge as key issues for fine and medium-grain parallelism. This discovery [27, 50] led to the move *from parallelism at the clause level to parallelism at the search level* pioneered by Clause-Diffusion.

In parallelism at the search level, the parallelization affects entire derivations, as multiple processes search in parallel for a proof, so that parallelism at the search level means *parallel search* and involves *communication* among the processes. Parallel search yields *multi-search*, where the parallel processes employ different search plans, *distributed search*, where the search space is subdivided among the processes, and their combination.

6.3.1 Parallelism at the Term or Literal Level

The classification of types of parallelism is based on the granularity of data accessed in parallel, leading to a distinction among *fine-grain*, *medium-grain*, and *coarse-grain* parallelism. Intuitively, the finer the granularity, the higher the possibility that parallel processes incur into *conflicts*. For inferences, *fine-grain parallelism* means having parallel processes access in parallel distinct *terms* or *literals* of a clause, so that fine-grain parallelism is *parallelism at the term or literal level.*

6.3.1.1 Parallelism at the Literal Level for Subgoal-Reduction Strategies

For subgoal-reduction strategies, fine-grain parallelism is AND-*parallelism*, where parallel processes access and reduce in parallel distinct literals of a goal clause. However, literals of the same clause may *share variables*, so that the parallel processes may be *in conflict*, in the sense that they need to instantiate the same variables by different unifiers.

For example, assume that the goal clause φ contains literals $\neg P(x)$ and $\neg Q(x,y)$, where P and Q are predicate symbols, and x and y are variable symbols. The two literals share the variable x. Let S include the clauses $\psi_1 = P(a) \vee C$ and $\psi_2 = Q(f(z),z) \vee D$, where a is a constant symbol, f is a function symbol, and z is a variable symbol. A process that resolves upon $\neg P(x)$ and $P(a)$ and a process that resolves upon $\neg Q(x,y)$ and $Q(f(z),z)$ are in conflict, because the first one needs to apply the substitution $x \leftarrow a$ and the second one needs to apply the substitution $x \leftarrow f(z)$. For this reason, already early provers parallelizing subgoal-reduction strategies, such as PARTHEO [194], METEOR [7], and Parthenon [75, 63], avoided AND-parallelism.

6.3.1.2 Parallelism at the Term Level for Ordering-Based Strategies

For ordering-based strategies, fine-grain parallelism is *parallel term rewriting*, where a term t is rewritten by applying in parallel multiple rewrite rules, or equations applied according to the ordering $\succ$. Given two equations that apply to a term t, it is well-known that there are three cases [132].

The first possibility is that the two equations rewrite t at *disjoint positions*. For example, the equations $i(i(x)) \simeq x$ and $f(x,y) \simeq f(y,x)$ match disjoint positions of the term $h(i(i(a)), f(a,b))$, where f, h, and i are function symbols, and a and b are constant symbols. The two steps can be applied in parallel, yielding $h(a, f(b,a))$, under an ordering $\succ$ where $a \succ b$.

The second possibility is that the two equations have a *variable overlap*. For example, the equations (1) $h(x,x) \simeq x$ and (2) $f(y,b) \simeq y$ overlap at a variable position in $f(h(a,a),b)$, because $h(x,x)$ matches with $h(a,a)$, $f(y,b)$ matches with $f(h(a,a),b)$, and the position of $h(a,a)$ corresponds to that of the variable y in $f(y,b)$. The two equations can be applied in any order, because the two rewriting sequences $f(h(a,a),b) \rightarrow_{(1)} f(a,b) \rightarrow_{(2)} a$ and $f(h(a,a),b) \rightarrow_{(2)} h(a,a) \rightarrow_{(1)} a$ yield the same result.

The third possibility is that the two equations *overlap* at a *non-variable* position. For example, the equations (1) $f(z,e) \simeq z$ and (2) $f(l(x,y),y) \simeq x$, where e is another constant symbol and l another function symbol, overlap at a non-variable position in $f(l(a,e),e)$, as both match the whole term. It is impossible to apply both equations, because the first one rewrites the term to $l(a,e)$ and the second one to a, as shown in the peak $l(a,e) \leftarrow_{(1)} f(l(a,e),e) \rightarrow_{(2)} a$. The two rewriting steps are *in conflict*, as they aim at replacing the same term by different terms.

An *overlap* tout court is a non-variable overlap: two equations *overlap*, if the left-hand side of one unifies with a *non-variable* subterm of the other. An overlap is a pre-condition to apply superposition. In the above example, the left hand sides $f(z,e)$ and $f(l(x,y),y)$ of the equations (1) $f(z,e) \simeq z$ and (2) $f(l(x,y),y) \simeq x$ overlap as they unify with most general unifier $\{y \leftarrow e, z \leftarrow l(x,e)\}$. Indeed, superposition generates from the two equations the new equation $l(x,e) \simeq x$, closing the peak $l(x,e) \leftarrow_{(1)} f(l(x,e),e) \rightarrow_{(2)} x$, of which the above peak is an instance. A sufficient and necessary condition to avoid conflicts is to exclude the non-variable overlap case by requiring the equations to be *non-overlapping*, which means that neither left-hand side unifies with a *non-variable subterm* of the other.

Historically, *parallel rewriting* [108, 109, 95, 129] allows parallel processes to apply in parallel equations that match the term at disjoint positions, while *concurrent rewriting* [131, 3, 4] allows them to apply equations that match at disjoint positions or have a variable overlap.

In equational declarative languages for specification or programming, equations are required to be *regular*, that is, non-overlapping and *left-linear*. The latter property says that no variable occurs repeated in the left-hand side. Regularity suffices to ensure uniqueness of normal forms, which means that the set of equations defines a functional program, in the sense that the output $t\downarrow$ is unique for a given input term t to be reduced [118]. Thus, the study of parallel or concurrent term rewriting was motivated primarily by the quest for fast implementations of interpreters of equational declarative languages [108, 109, 95, 129].

In theorem proving it is impossible to restrict the attention to non-overlapping equations, since this would mean barring superposition, which is the main expansion inference rule to deduce equations from equations. The same consideration applies to *Knuth-Bendix completion*, where superposition first appeared [132]. Nevertheless, the possibility of implementing Knuth-Bendix completion on top of parallel [66] or concurrent [130] rewriting, the latter only in the ground case, was explored. If all equations are ground, superposition collapses to simplification, and all operations of completion are done by rewriting. In the non-ground case, superposition is done sequentially, and only normalization can take advantage of parallel rewriting.

6.3.2 Parallelism at the Clause Level

Medium-grain parallelism for inferences means having parallel processes access in parallel distinct *clauses*, and perform one or more inferences with those clauses as premises. Thus, medium-grain parallelism is *parallelism at the clause level.*

6.3.2.1 Parallelism at the Clause Level for Subgoal-Reduction and Instance-Based Strategies

For subgoal-reduction strategies parallelism at the clause level is OR-*parallelism*, where parallel processes access in parallel distinct goal clauses, and resolve them with as many side clauses generating new goal clauses. This means trying in parallel the proof attempts that a sequential strategy tries in sequence by going from one to the next via backtracking. Each goal clause is seen as a *task* (φ, j, k), where φ is a goal clause, j is the number of ME-extension steps used to generate φ, and k is the limit of iterative deepening. The task consists of reducing φ to $\Box$ by applying at most $k-j$ ME-extension steps. When a new goal clause φ_{i+1} is generated from a goal clause φ_i, a new task $(\varphi_{i+1}, j+1, k)$ is generated from task (φ_i, j, k). A task (φ, j, k) is active only if $j < k$.

Assume that there are n processes, all with initial limit k for iterative deepening. As soon as there are n active tasks, all processes may be active. A way of initializing the derivation is to have a sequential preprocessing phase where one process proceeds sequentially, generating at least n tasks. Then, a *parallel subgoal-reduction derivation*, with parallelism at the clause level, has the form

$$(S; G_0) \vdash (S; G_1) \vdash \ldots (S; G_i) \vdash \ldots$$

where the G_i component represents the set of active tasks.

Each process maintains a *queue* of its active tasks and the distribution of tasks among the processes is realized by *task stealing*. When the queue of a process is empty, that process steals active tasks from the queues of others. Task stealing is implemented by representing a task (φ, j, k) by an encoding of the WAM operations (cf. Section 6.2.1) that generate (φ, j, k) from the input set of clauses. When there are no more active tasks, the search restarts with the initial goal and a higher limit of iterative deepening. Communication of tasks is achieved by message passing in

PARTHEO [194], in shared memory in Parthenon [75, 63], and either way in METEOR [7].

For instance-based strategies, parallelism at the clause level means having multiple parallel processes picking different clauses as nuclei and generate in parallel all their hyperinstances [223]. However, a most natural way to parallelize hyperlinking [140] is to execute in parallel the instance generation and satisfiability testing phases [223]. This may be an example where parallelization contributed to improve the underlying theorem-proving method, as the notion of doing in parallel instance generation and satisfiability testing may have given ammunition to the design of instance-based strategies with a tighter integration between model building and instance generation (cf. Section 6.2.3).

6.3.2.2 Parallelism at the Clause Level for Ordering-Based Strategies

ROO [154, 155, 156] is the paradigmatic example of parallelism at the clause level for ordering-based strategies. ROO is a parallelization of up to version 2.2 of the OTTER *theorem prover* [164, 165, 166, 169]. The idea of ROO is to parallelize the *given-clause algorithm* at the heart of OTTER. This algorithm was later adopted by most theorem provers implementing ordering-based strategies, such as SPASS [218], E [190] and its predecessor DISCOUNT [10, 90], Vampire [137], Gandalf [211], WALDMEISTER [117], and Zipperposition [77]. In the given-clause algorithm the database of clauses is organized as two lists of clauses, that we call `already-selected` and `to-be-selected` [35, 41]. In OTTER, these lists were named originally `axioms` and `set-of-support`, abbreviated `sos`; in later versions `axioms` was renamed `usable`. In E these lists are called `active` and `passive`.

The standard initialization is to start with all input clauses in `to-be-selected` and empty `already-selected`. For a strategy with set of support (cf. Section 6.2.2), one starts with `already-selected` containing the clauses of T and `to-be-selected` those of *SOS*, which explains the original names of the two lists in OTTER.

The given-clause algorithm prescribes to perform a loop, until either a refutation is found or the list `to-be-selected` becomes empty. In the latter case the input set of clauses is recognized to be satisfiable. For a first-order theorem prover, termination typically occurs either with a refutation or when the prover hits a predefined time or space threshold. At every iteration of the loop, the prover selects from `to-be-selected` the *best* clause according to a heuristic evaluation function [5, 166, 88, 33, 169, 191]. This clause is the *given clause*. Thus, the given-clause algorithm realizes a *best-first search*. The prover performs all applicable expansion inferences having as premises the given clause and clauses in `already-selected` and moves the given clause from `to-be-selected` to `already-selected`. Every raw clause φ thus generated is subject to forward contraction, so that it is either deleted or reduced to a normal form $\varphi\downarrow$ (where $\varphi\downarrow$ and φ may be identical), which gets an identifier and is appended to `to-be-selected`.

For backward contraction, the prover detects which previously existing clause ψ can be contracted by such a $\varphi\downarrow$ and subjects every such ψ to forward contraction. Any resulting $\psi\downarrow$ gets a *new identifier*, is added to `to-be-selected`, and will try in turn to backward-contract other clauses. In the case of backward contraction it cannot be that ψ and $\psi\downarrow$ are identical, because ψ was found reducible to begin with. The backward contraction phase terminates when the set of clauses S is such that $\rho(S) = \emptyset$, which is guaranteed to occur eventually thanks to the well-foundedness of the ordering $\succ$.

The OTTER version [169] of the given-clause algorithm applies backward contraction to both lists, so that the set S such that $\rho(S) = \emptyset$ at the end of every iteration of the loop is the union of `already-selected` and `to-be-selected`. The E version [188, 190], tried first in DISCOUNT [10, 90], applies backward contraction only to `already-selected`, so that the set S such that $\rho(S) = \emptyset$ at the end of every iteration of the loop contains the clauses in `already-selected`. If a clause in `already-selected` is backward contracted, its descendants in `to-be-selected` are deleted.

The idea of the E versionis that it is not necessary to keep `to-be-selected` fully reduced, since clauses in `to-be-selected` are not used as premises of expansion inferences. Since `to-be-selected` is allowed to contain redundant clauses, the given clause is subject to forward contraction as soon as extracted from `to-be-selected` and prior to be used as expansion premise. Most contemporary provers implementing ordering-based strategies feature both the OTTER and E versions of the given-clause algorithm.

At the time of ROO, only the OTTER version of the given-clause algorithm existed. The concept of ROO is to store the lists `already-selected` and `to-be-selected` in shared memory, and let n parallel processes pick n given clauses and perform in parallel the ensuing expansion and forward contraction inferences. The expansion and forward contraction phases for a given clause is called *task A*. The parallel processes are not allowed to append the clauses thus generated to `to-be-selected`, because that could cause conflicts in accessing the shared list, and more importantly because the clauses generated in parallel are not guaranteed to be irredundant. Indeed, if N_1 and N_2 are the sets of clauses generated by parallel processes p_1 and p_2, the clauses in N_1 are not forward contracted with respect to those in N_2 and vice versa. ROO features an additional list, termed `K-list`, and lets the parallel processes append their new clauses to the `K-list`. A single process performs contraction within the `K-list` and then transfers all clauses from the `K-list` to `to-be-selected`. This activity is called *task B*.

All the more, the parallel processes are not allowed to do backward contraction, in order to avoid conflicts in deleting or rewriting clauses in `already-selected` and `to-be-selected` in shared memory. They are only allowed to test for backward contraction: if a parallel process discovers that one of its newly generated clauses can backward-contract a clause ψ in `already-selected` or `to-be-selected`, it adds the identifier of ψ to a list named `to-be-deleted`. The single process in charge of task B then proceeds to backward-contract every such ψ. All processes follow the same schedule: execute task B, if either `K-list` or `to-be-deleted` is not empty and no other process is doing task B; execute task A otherwise.

Thus, Roo has to do backward contraction sequentially, as only one process is allowed to execute task B at any given time. Since an ordering-based prover typically spends most of its time doing contraction, and especially backward contraction, Roo incurs in a problem identified as the *backward-contraction bottleneck* [27, 50], which manifests itself as follows: the single process executing task B lags behind, `K-list` and `to-be-deleted` grow too long, and the other processes remain idle waiting for clauses to reach `to-be-selected` and become available as given clauses.

The backward-contraction bottleneck affects also the application of the *transition-based approach to parallel programming* [224] to *Knuth-Bendix completion* [225]. The considered version of Knuth-Bendix completion is the original one [132], that only handles rewrite rules and fails if it generates an equation that cannot be oriented by the ordering into a rewrite rule. For theorem proving, *unfailing* or *ordered* completion [120, 12], that handles also equations, and therefore does not fail, supersedes the original Knuth-Bendix completion [132]. Nevertheless, the transition-based parallelization of Knuth-Bendix completion [225] is relevant to our analysis as another instance of parallelism at the clause level. It performs parallel inferences in shared memory, with locks and critical regions to prevent conflicts between inference steps that involve the same rewrite rules. Backward contraction causes a *write-bottleneck* as all the backward-contraction inferences ask write-access to shared memory. It is plausible that the difficulty with backward contraction suggested applying the transition-based approach to Buchberger algorithm [68, 69] instead, since in Buchberger algorithm backward contraction is not as crucial (cf. Section 6.2.2.1).

Because of the backward-contraction bottleneck, parallelism at the clause level was largely abandoned, and an approach à la Roo was never tried in combination with the E version of the given-clause algorithm.

6.3.3 The Rise of Parallel Search

The above analysis of parallelism at the term/literal and clause levels reveals that a key element to understand whether and how theorem proving can be parallelized is an abstract analysis of the *conflicts* between parallel inferences [27, 50]. The analysis is abstract in the sense of not being tied to a memory model or an implementation.

Two expansion inferences read their premises and generate and add their consequences. If they add their consequences to a shared data structure, some access control is required, but two expansion inferences are not in conflict in an essential way, because they only read their premises. On the other hand, contraction inferences delete or rewrite one of their premises, and therefore determine three types of conflicts:

1. *Write-write conflict between contraction inferences*: two contraction steps aim at rewriting the same clause φ;

2. *Write-read conflict between contraction inferences*: a contraction step aims at rewriting a clause φ that another contraction step aims at using as premise to contract some other clause ψ;
3. *Write-read conflict between contraction and expansion inferences*: a contraction step aims at rewriting a clause φ that an expansion step aims at using as premise to generate other clauses.

Conflicts of Type (1) are exemplified by the conflicts in parallel rewriting (cf. Section 6.3.1). Conflicts of Types (2) and (3) are due to backward contraction, because a raw clause is not used as premise of another inference while it is subject to forward contraction. A conflict of Type (2) is harmless: the *once redundant always redundant* principle ensures that no matter which step commits, the other clause, whether φ or ψ, will still be reducible [27]. Conflicts of Type (3) are the most problematic, because φ is redundant, a clause generated by φ will also be redundant, and therefore the contraction step should have priority.

Subgoal-reduction strategies have a static database of clauses given by the input set and no contraction. The absence of contraction means *no conflicts* among inferences. A static, relatively small, database of clauses represents *read-only* data that can be kept in shared memory, and even compiled as done for declarative programs. Instance-based strategies have a monotonically increasing database of clauses given by the input set plus instances, and only forward contraction. The absence of backward contraction means *no conflicts* among inferences. This explains why approaches to parallelize subgoal-reduction and instance-based strategies adopted parallelism at the clause level (cf. Section 6.3.2.1).

The situation is different for ordering-based strategies: the database of generated clauses is large, and non-monotonic, in fact *highly dynamic*, due to backward contraction, which causes conflicts among inferences. There is *no read-only data*, as any clause can be rewritten by proceeding ones. Contraction is essential for equational reasoning: indeed subgoal-reduction and instance-based strategies as described thus far are for first-order logic, and ordering-based strategies for first-order logic with equality. This analysis motivates resorting to *coarse-grain parallelism* for ordering-based strategies for first-order logic with equality.

Coarse-grain parallelism in deduction is *parallelism at the search level* or *parallel search*: multiple processes $p_0, \ldots, p_{n-1}$ search in parallel for a proof, each developing *its own derivation* and maintaining *its own database of clauses*. It is sufficient that one of the processes succeeds, and as soon as that happens, all may halt. Since each process has its own database of clauses, the issue of conflicts disappears, and especially the *backward-contraction bottleneck* cannot arise. While parallelism at the term/literal or clause levels aim at speeding up a given search, parallelism at the search level aims at finding a proof sooner by generating new searches, by searching in different ways.

The counterpart of allowing every process to build its own database of clauses is the redundancy of having the same clauses in all or some of these databases. However, this duplication is not considered redundancy in parallel search, as long as the clauses are not redundant in the logical sense (cf. Section 6.2.2.2). Moreover, as long as contraction inferences are adequate, having φ in the database of a process and $\varphi\downarrow$ in the database of another process is not a correctness issue, unlike in other distributed applications, where the lack of *agreement* or *coherence* may affect correctness.

A general issue with parallelism at the search level is to *differentiate* the searches conducted in parallel by the processes. Intuitively, it is wasteful to have different processes visit the same search space, performing the same inferences in the same order. On the other hand, it is unavoidable that their searches have something in common, given that they are all solving the same problem. The idea is *to minimize the overlap* of the searches performed by the parallel processes [27, 50, 51, 30, 31, 32, 34, 38].

One approach to this issue is to differentiate the processes by letting them execute *different search plans on the same data*. The dual approach is to differentiate them by letting them execute *the same search plan on different data*. A way to differentiate data is to *subdivide clauses and inferences* among the processes, in order to subdivide the work to be executed. In parallel theorem proving this distinction was presented first as *competition versus cooperation* [192, 210, 85]. The analogue in parallel SAT-solving is the dichotomy between the *portfolio approach* and the *divide-and-conquer approach*. However, these terminologies suggest that the two parallelization principles may not coexist, while several methods explore their combination. We distinguish between *multi-search* and *distributed search* [37, 38].

Both multi-search and distributed search approaches feature *communication* among the parallel deductive processes. In the case of distributed search the rationale for communication is obvious: since the space to be searched is subdivided, communication is needed for completeness. However, also in the case of multi-search communication is necessary, otherwise multi-search reduces to running independent experiments in parallel.

For *subgoal-reduction strategies*, parallel search is typically *multi-search*, because the database of clauses is small and static. On the other hand, the large database of generated and kept clauses of ordering-based strategies suggests distributed search, and the notion of subdividing the search space by subdividing clauses and inferences. Since ordering-based strategies also offers a variety of search plans, *both multi-search and distributed search* have been applied to *ordering-based strategies*. In the rest of this section, we cover multi-search for subgoal-reduction strategies, multi-search for ordering-based strategies, and distributed search for ordering-based strategies.

6.3.4 Multi-search

A *multi-search method* is a parallel search method where the parallel deductive processes apply different search plans to search for a solution. As a way to differentiate the searches further, multi-search may also allow the processes to employ different inference systems. In multi-search with *homogeneous systems*, the deductive processes have different search plans and the same inference system. In multi-search approaches with *heterogeneous systems*, the deductive processes differ in the inference system or in both inference system and search plan.

6.3.4.1 Multi-search for Subgoal-Reduction Strategies

For subgoal-reduction strategies, ways of differentiating the search plans include assigning to the parallel processes different literal-selection rules [207], different clause-selection rules, different limits for iterative deepening, different choices of initial goal clause, or any combination thereof. These possibilities have been explored in the successors of PARTHEO [194], namely SETHEO, E-SETHEO, SPTHEO, CPTHEO and P-SETHEO [144, 170, 209, 104, 219].

A *multi-search subgoal-reduction derivation* with n processes $p_0, \ldots, p_{n-1}$ takes the form

$$(S;G_0^j) \vdash (S;G_1^j) \vdash \ldots (S;G_i^j) \vdash \ldots$$

where S is the input set of clauses, and G_i^j is the set of active tasks at process p_j, $0 \leq j \leq n-1$, and stage i, $i \geq 0$ (cf. Section 6.3.2.1 for the notion of task). The processes may communicate tasks, so that each process may have a set of active tasks as an effect of communication. If the processes start with different limits $k_0, \ldots, k_{n-1}$ for iterative deepening, a process may have in its set active tasks with different limits, such as (φ, n, k) and (φ', n', k'): if $k < k'$, task (φ, n, k) must be given higher priority by the process, in order to preserve completeness.

An example of heterogeneous system is HPDS [208], with three deductive processes and a *deduction controller*. The three deductive processes execute *guided linear deduction* (GLD), which is similar to model elimination (cf. Section 6.2.1), hyperresolution (HR) (cf. Section 6.2.2), and *unit-resulting resolution* (UR) [204], respectively. The latter inference rule resolves a clause $L_1 \vee \ldots \vee L_q \vee L_{q+1}$, called *nucleus*, and unit clauses $L_1', \ldots, L_q'$, with $q \geq 1$, called *electrons*, such that $L_i\sigma = \neg L_i'\sigma$ for all i, $1 \leq i \leq q$, to generate the unit clause $L_{q+1}\sigma$. If the L_{q+1} literal is allowed to be absent, UR resolution is allowed to generate $\Box$. As UR resolution alone does not form a refutationally complete inference system, its purpose is to accelerate the generation of unit clauses for other inference rules. For example, UR resolution is used to generate unit lemmas for a PTTP prover [204].

HPDS implements this concept in a parallel setting. Every process is endowed with forward and backward subsumption, employs a DFS plan with iterative deepening, and sends the clauses it generates, including subsumed clauses tagged as such, to the deduction controller. The deduction controller forwards to the GLD and HR processes the unit clauses generated by the UR process, and feeds the latter with the clauses generated by the other two. It may also forward to the HR process clauses generated by GLD, but not vice versa, so that the GLD process only receives unit lemmas. Furthermore, the deduction controller gives every process information on clauses subsumed by the other processes.

Another instance of multi-search with heterogeneous systems is CPTHEO [104], built on top of the model elimination prover SETHEO [144]. SETHEO is equipped with a resolution-based prover preprocessor, named Delta [193], with the idea of generating in advance, by resolution, clauses that could be useful as lemmas for the subgoal-reduction derivation. CPTHEO replaces preprocessing by cooperation in a parallel setting: it launches SETHEO and Delta in parallel, and lets SETHEO use the clauses generated by Delta as lemmas, according to different communication schemes.

For instance, SETHEO sends to Delta goal clauses from tasks that SETHEO cannot solve in the current limit of iterative deepening. Delta responds with lemmas that resolve with those goal clauses. SETHEO restarts with its next round of iterative deepening and a database of clauses enriched with the received lemmas.

Alternatively, SETHEO sends to Delta the literals labeling the open leaves in its current tableau. Delta replies by sending lemmas including *similar* literals of opposite sign. Similarity is measured according to various heuristic criteria [103]. In either scheme Delta ranks its generated clauses and selects the best to be sent as lemmas. The ranking is based on clause size, with small size deemed preferable, size of $\Pi(\varphi)$ for clause φ, with large size deemed preferable, assuming that a clause that required more inferences to be generated is more precious, and similarity-based criteria [103].

6.3.4.2 Multi-search for Ordering-Based Strategies

Multi-search for ordering-based strategies was introduced with the *Team-Work method* [84, 9, 88, 10, 89, 91, 92, 87, 197]. Team-Work is devised for purely equational problems, but its concept applies just as well to first-order logic with equality. The Team-Work method provides for n deductive processes $p_0,\ldots,p_{n-1}$, one of which also plays the role of *supervisor*. All processes start with the same input problem, the same inference system, a time period, again the same for all processes, but different search plans. For instance, in the context of the given-clause algorithm, this may mean different evaluation functions to select the given clause [5, 166, 88, 33, 169, 191]. Every process develops its own derivation and builds its own database of clauses independently. When the allotted time period expires, every process evaluates its current database of clauses, based on a set of heuristic measures, the same for all processes. For example, the number of generated clauses may indicate how productive a process has been, while the number of deleted clauses may suggest whether the process has generated some very effective simplifiers or subsumers.

Then, every process sends to the supervisor its scores according to the heuristic measures. The supervisor picks a winner, the one with the best scores, and broadcasts its identity, say p_j. The winner p_j becomes the supervisor for the next round, and all the other processes send to p_j their best clauses according to other heuristic criteria, relative to individual clauses [5, 88], rather than the whole database. For example, an equation that has simplified many other clauses may be deemed precious. The new supervisor p_j broadcasts its database, enriched with these best clauses received from the others. In this manner, all deductive processes restart with the best database generated thus far and augmented with selected good clauses from the other derivations.

A *multi-search ordering-based derivation* with n processes $p_0,\ldots,p_{n-1}$ has the form

$$S_0^j \vdash S_1^j \vdash \ldots S_i^j \vdash \ldots$$

where S_i^j is the database of clauses at process p_j, $0 \leq j \leq n-1$, and stage i, $i \geq 0$. Initially, $S_0^0 = S_0^1 = \ldots = S_0^{n-1} = S$ is the input set of clauses. Such a derivation is a *refutation* if $\Box \in S_i^j$ for some i and j. A *Team-Work derivation* is a multi-search ordering-based derivation characterized by a series $\mathscr{A} = i_0 < i_1 < \ldots < i_k < i_{k+1} < \ldots$ of special stages, where $i_0 = 0$, and for all $i \in \mathscr{A}$, $S_i^0 = S_i^1 = \ldots = S_i^{n-1}$: the stages in $\mathscr{A}$ are those where all the processes restart with the same database.

Fairness of a multi-search derivation does not require that all search plans be fair. In the context of Team-Work, it is sufficient that at least one of the search plans is fair, and that a database produced by a fair search plan is selected as the winner infinitely often [9].

Starting at least with OTTER [164, 165, 166, 169], automatic theorem provers have many options and parameters that can be set for each problem. A multi-search à la Team-Work adds even more, including the set of heuristics to evaluate databases, the set of heuristics to evaluate clauses, and the time period. One may also program the prover to vary selected parameters during a derivation. For example, the time period may increase over time, so that the processes cooperate a lot at the beginning and behave more independently later, or vice versa. The sequential basis for the implementation of Team-Work is the DISCOUNT theorem prover [10, 90], meaning that every p_j executes an instance of DISCOUNT.

The purpose of Team-Work is to *interleave* and *combine* different search plans. The periodic restart from a common database lets a process apply a search plan to a database generated by another search plan, realizing the interleaving. The mechanism whereby the database of the winner is enriched with clauses deemed good by other processes provides the combination. Since different search plans may generate clauses in different orders, their interleaving and combination may enable one of the processes to discover a proof sooner than any of the search plans would allow if applied sequentially. The downsides include the delays imposed by the periodic synchronizations, and the risk that the heuristics are misleading, so that discarding the databases with lower scores will make the search longer rather than shorter.

Ingredients of Team-Work appeared also in multi-search approaches with heterogeneous systems. For example, *requirement-based cooperative theorem proving* [102] prescribes to run SPASS and DISCOUNT in parallel. The two provers communicate by *expansion requests* and *contraction requests*. In an *expansion request*, a prover sends to the other a clause φ, and the receiver replies by sending all resolvents between φ and the clauses in its `already-selected` list. In a *contraction request*, a prover sends to the other a clause φ, and the receiver replies by sending all its clauses that contract φ.

The TECHS system [86] is even more heterogeneous, as it runs in parallel SPASS, DISCOUNT, and SETHEO, thereby mixing contraction-based and subgoal-reduction strategies, a feature that recalls HPDS and CPTHEO (cf. Section 6.3.4.1). In TECHS, SPASS and DISCOUNT exchange equations, while SPASS and SETHEO exchange lemmas, from SPASS to SETHEO, and subgoals, from SETHEO to SPASS. These heterogeneous systems share with Team-Work the notion of heuristic selection of good clauses to be shared. For example, short clauses are deemed good, so that unit clauses and especially unit equations are the best.

The legacy of the Team-Work method is threefold. First, the notion of interleaving search plans migrated into the design of search plans for sequential theorem provers: the prover is programmed to execute a search plan for a fixed interval of time, then another one for the next interval, and so on. This feature is available, for instance, in Vampire [137]. This development is rather natural as interleaving is a standard way to simulate a parallel computation by a sequential computation. In the theory of parallel computing, a parallel computation that can be sequentially simulated by interleaving is not regarded as truly concurrent, although we are not aware of results on sequential derivations simulating multi-search derivations. Second, the notion of letting a process send to another one its best clauses is connected with *learning*, in the sense of learning the results of other derivations starting from the same problem [90]. This concept is generalized to learning from proofs of similar problems, as in the approaches that apply machine learning and big data technologies to theorem proving [99, 213]. Third, Team-Work can be considered a forerunner of the *portfolio approach* to parallel SAT-solving (cf. Section 6.4.1 in this chapter and the chapter on 1, Parallel Satisfiability).

6.3.5 Distributed Search

A *distributed-search method* is a parallel-search method where the search space is *subdivided* among the parallel deductive processes, in order to subdivide the work to be performed, and possibly reach a solution sooner. As a way to differentiate the searches further, distributed search may also allow the processes to apply different search plans, leading to methods with *both distributed-search and multi-search*.

In general, subdividing the work may mean subdividing data, as in *data-driven* parallelism, or subdividing operations, as in *operation-driven* parallelism. In theorem proving, there are typically few inference rules and a huge number of generated clauses, and therefore the subdivision and the parallelism are naturally data-driven. However, the subdivision is designed knowing which inferences need to be applied to the clauses, so that the two aspects are intertwined. This also means that distributed search is usually coupled with *homogeneous systems*, where all deductive processes feature the same inference system, although in principle it could be combined also with *heterogeneous systems*, where the deductive processes employ different inference systems.

6.3.5.1 Distributed Search for Ordering-Based Strategies

Distributed search for ordering-based strategies was introduced with the *Clause-Diffusion methodology* [27, 48, 49, 51], implemented in the *Aquarius* [46, 27, 47, 52] and *Peers* [58, 51] provers, and then investigated through *Modified Clause-Diffusion* [28, 29], the *Peers-mcd* [30, 31, 33, 38] prover, and a formal analysis of distributed search for contraction-based proof search [32, 34]. To the best of our knowledge,

Clause-Diffusion was the first parallel-search method for automatic first-order theorem proving, and many of the elements of the analysis of parallelism for deduction (cf. Section 6.3.3) were discovered with and around Clause-Diffusion and its developments. The reason for calling it a *methodology* is that Clause-Diffusion came since the start with a choice of solutions for several issues. In this presentation we cover all issues and the most mature and most successful solutions, hence Modified Clause-Diffusion, referring the interested readers to the original articles for other possibilities.

6.3.5.2 The Basic Clause-Diffusion Mechanisms

Clause-Diffusion provides for n deductive processes $p_0, \ldots, p_{n-1}$, that are all *peers*. In a Clause-Diffusion prover, n is a parameter set by the user. All processes start with the same input problem, inference system, and search plan, although different search plans may be assigned. Every process develops its own derivation and builds its own database of clauses independently. The processes are *asynchronous*, as the only synchronization occurs when one sends all others a halting message. This happens, for example, when one of the processes finds a proof.

Clause-Diffusion subdivides the search space by subdividing clauses, so that *every clause is owned by a process*. A *distributed-search ordering-based derivation*, or *distributed derivation* for short, has the form

$$(O_0;NO_0)^j \vdash (O_1;NO_1)^j \vdash \ldots (O_i;NO_i)^j \vdash \ldots$$

where for every process p_j, $0 \leq j \leq n-1$, and stage i, $i \geq 0$, $S_i^j = O_i^j \uplus NO_i^j$ is the *local database* of clauses at p_j; O_i^j is the set of clauses *owned* by p_j; NO_i^j is the set of clauses *not owned* by p_j; and $\bigcup_{j=0}^{n-1} S_i^j$ represents the *global database* at stage i. Initially, $S_0^0 = S_0^1 = \ldots = S_0^{n-1} = S$ is the input set of clauses. In the early Clause-Diffusion terminology owned clauses are termed *residents* and the others *visitors* or *visiting clauses* [27, 51]. A distributed derivation is a *refutation* if $\Box \in S_i^j$ for some i and j.

Since every clause is owned by a process, for every stage i, $i \geq 0$, we have $\bigcup_{j=0}^{n-1} O_i^j = \bigcup_{j=0}^{n-1} S_i^j$. This also means that every clause $\varphi \in NO_i^j$ is owned by some p_k, with $k \neq j$, so that $\varphi \in O_l^k$ for some $l \geq 0$. Furthermore, under the customary assumptions that every clause has its own variables, and variants are distinct clauses, every clause is owned by *only one* process, so that $O_i^j \cap O_i^k = \emptyset$ for all $i \geq 0$ and $0 \leq j \neq k \leq n-1$.

Assume that a clause ψ is generated by process p_j, and that its normal form after forward contraction $\varphi = \psi\downarrow$ is not trivial, so that φ is kept. *Regardless of whether ψ was generated by expansion or backward contraction*, process p_j assigns φ to some p_k according to an *allocation criterion*. The number k becomes part of the *identifier* of φ: for example, if φ is the m-th clause generated and kept by p_j, its identifier includes the fields $\langle k, m, j\rangle$. These three components suffice to identify a clause *uniquely* across all processes, so that the identifier of a clause is a *global* attribute.

If $k = j$, p_j adds φ to O^j; if $k \neq j$, p_j adds φ to NO^j. Either way, p_j applies φ to backward-contract clauses in S^j, and broadcasts it as an *inference message* $\langle \varphi, k, m, j \rangle$ to all other processes. This broadcasting mechanism is the reason for the name Clause-Diffusion, as clauses are *diffused*. These messages are called *inference messages*, because received clauses will be used for inferences.

Any other process p_q, $q \neq j$, upon receiving the inference message $\langle \varphi, k, m, j \rangle$ applies forward contraction to the received clause φ. If φ is deleted by forward contraction no other operation is needed. Otherwise, let $\varphi \downarrow$ be the normal form of φ with respect to S^q, where $\varphi \downarrow$ and φ may be identical. If $k = q$, p_q adds $\varphi \downarrow$ to O^q; if $k \neq q$, p_q adds $\varphi \downarrow$ to NO^q. Either way, p_q applies $\varphi \downarrow$ to backward-contract clauses in S^q.

6.3.5.3 The Subdivision of Clauses in Clause-Diffusion

Allocation criteria to subdivide clauses play an important role in differentiating the searches and limiting their *overlap* [32, 34]. The intuition is that different searches, and searches that differ from a sequential one, may enable one of the processes to find a proof sooner. A simple option is that each process assigns clauses according to a *round-robin schedule*, called *alternate-fit* [27, 51] or *rotate* [31]: p_j assigns φ to p_k for $k = (q+1)$ *mod* n, if p_j assigned the previous clause to p_q.

In the *half-alternate-fit* criterion [27, 51], p_j assigns every other clause to itself and in a round-robin manner otherwise. Let p_{q_1} and p_{q_2} be the two most recently used destinations; if $q_1 = j$, p_j assigns φ to p_k for $k = (q_2 + 1)$ *mod* n; if $q_1 \neq j$, p_j assigns φ to itself.

Alternatively, every process p_j may estimate the work-load of each process as measured by the number of generated clauses, a criterion named *best-fit* [27, 51] or *select-min* [58]. Clearly, p_j knows exactly how many clauses it generated thus far. For all other processes $p_q, q \neq j$, p_j may consider the latest inference message $\langle \psi, k, m, q \rangle$ received from p_q and take m as an estimate of the number of clauses generated by p_q. Then p_j assigns the next φ to the process with the smallest estimated work-load. However such a criterion may lead the processes to assign too many clauses to others, since a process may under-estimate the work-load of others but not its own. Therefore, this criterion may be corrected by letting each process assign a fixed percentage of clauses to itself as in the *half-alternate-fit* criterion.

A different approach is to determine the owner of a clause *based on properties of the clause itself*. For example, assume that every symbol in the signature has an associated *weight*. The sum of the weights of the symbols occurring in a clause is the weight of the clause. This is a feature that the Clause-Diffusion provers inherit from OTTER, where it is used for *deletion by weight*, a contraction rule that allows the prover to delete all clauses whose weight is above a certain threshold [164, 165, 166, 169]. Such a rule is not adequate (cf. Section 6.2.2), but it may be useful in practice. A simple weight-based allocation criterion is to assign clause φ to

p_k, where $k = w \ mod \ n$ and w is φ's weight. This criterion was called *syntax* in the *Peers* prover [58, 51].

The next step is to use *information from the ancestor-graph* $\Pi(\varphi)$ (cf. Section 6.2.2.1) in order to allocate φ. Since theorem provers save anyway the data to generate $\Pi(\varphi)$ for every kept clause φ in order to be able to build $\Pi(\Box)$, storing this information is no additional burden. This concept is achieved by the *ancestor-graph oriented* (AGO) allocation criteria [31]. The general idea is to use information from the finite portion of the search space that has been generated to assign clauses to processes and therefore induce a subdivision of the search space that lies ahead.

The AGO criterion *parents* determines φ's owner by applying a function f to the identifiers of φ's parents. As the function f may vary, this is actually a family of criteria. If φ was generated from premises ψ_1 and ψ_2 by a binary expansion inference rules, such as resolution, paramodulation, or superposition, its parents are ψ_1 and ψ_2. If φ is a factor of ψ, its parent is ψ. If φ was obtained by normalizing ψ during backward contraction, ψ is considered as the sole parent. Since f is a function, clauses that have the same parent(s) are assigned to the same process. The intuition is that clauses that have the same parents are spatially close in the search space, and therefore should belong to the same process. If they were assigned to different processes, the effect could be to bring those different processes to be active in the same region of the search space, increasing their overlap.

The AGO criterion *majority* considers all ancestors of clause φ, that is, all clauses that occur in its ancestor-graph $\Pi(\varphi)$. It assigns to every process p_j a number of votes equal to the number of clauses in $\Pi(\varphi)$ owned by p_j. The process, say p_k, that gets the most votes owns φ. Ties are broken arbitrarily. The idea is that a process that owns the most ancestors of φ is already most active in the region of the search space where φ is, and therefore should get φ as well. Assigning φ to another process, say p_q, with $q \neq k$, could increase the overlap between p_k and p_q.

It remains what to do with input clauses. One process, say p_0, reads the input file and handles input clauses like raw clauses. Most allocation criteria listed above apply regardless of whether the clause was read or generated. The select-min criterion does not apply to input clauses, because at the beginning the processes have no workload: therefore, select-min assigns input clauses in round-robin fashion. The AGO criteria do not apply to input clauses, because input clauses do not have ancestors. The parents criterion assigns all input clauses to p_0. The majority criterion cannot proceed in this manner, because otherwise all clauses would belong to p_0, as p_0 would always have the majority of ancestors. This does not happen with the parents criterion, because the function f applies to the entire identifiers of parents, not only to the owners. Thus, also the majority criterion assigns input clauses in round-robin style.

6.3.5.4 The Subdivision of Inferences in Clause-Diffusion

In Clause-Diffusion the ownership of clauses induces a *subdivision of expansion inferences* as follows. Assume that p_j is about resolving clauses $\varphi = L \vee C$ and

$\psi = \neg L' \vee D$, such that $L\sigma = L'\sigma$. Clause-Diffusion allows p_j to proceed with the inference if and only if p_j *owns* ψ, that is, *the parent with the negative literal resolved upon*. Similarly, assume that p_j is about paramodulating or superposing clause $\varphi = l \simeq r \vee C$ into clause $\psi = L[s] \vee D$, such that $s\sigma = l\sigma$. Clause-Diffusion allows p_j to proceed with the inference if and only if p_j *owns* ψ, that is, *the clause paramodulated or superposed into*. When paramodulating φ into ψ, the prover needs to consider all non-variable subterms of ψ and only l and r in φ. In other words, there is more work connected with the clause paramodulated into. For superposition, that is, paramodulation into equalities, a prover needs to test for both superposition of $\varphi = l \simeq r \vee C$ into $\psi = s \simeq t \vee D$ and superposition of ψ into φ. The owner of ψ will superpose φ into ψ and the owner of φ will superpose ψ into φ. For factoring, p_j is allowed to generate the factors of ψ if and only if it *owns* ψ. For hyperresolution and unit-resulting resolution, p_j is allowed to proceed if and only if it *owns the nucleus* of the inference step.

As far as contraction inferences are concerned, there is *no subdivision of forward-contraction inferences*, as every process p_j applies all the clauses in its current local database S^j to try to delete or reduce a raw clause it has generated. There is also *no subdivision of backward-contraction inferences that delete clauses*, such as subsumption, functional subsumption, or tautology elimination (cf. Section 6.2.2). Every process p_j is allowed to use any clause in S^j to delete any other clause in S^j by such an inference rule, regardless of ownership.

On the other hand, ownership is used to *subdivide backward-contraction inferences that generate new clauses*, such as clausal simplification and equational simplification or normalization. Assume that process p_j detects that clause $\varphi \in S^j$ can be backward-simplified by some other clause $\psi \in S^j$. If p_j owns φ, p_j is allowed to generate $\varphi\downarrow$. If p_j does not own φ, p_j is allowed to delete φ, but it is not allowed to generate $\varphi\downarrow$. Whoever owns φ will generate $\varphi\downarrow$, give it a *new identifier*, and broadcast it as inference message.

6.3.5.5 Distributed Global Contraction, Distributed Fairness, and Distributed Proof Reconstruction

Clause-Diffusion led to formulate and solve three general issues in distributed search for ordering-based strategies: *distributed fairness* [27, 48, 51, 29], *distributed proof reconstruction* [29], and *distributed global contraction* [27, 51, 29].

Distributed fairness, that is, fairness of a distributed derivation, is guaranteed by two conditions. First, each process must be *locally fair*, which means it considers eventually all irredundant inferences. Second, all persistent irredundant clauses must be broadcast eventually. Clause-Diffusion fulfills the second condition eagerly, by broadcasting kept clauses right after forward contraction. The reason for this eager choice is the second property, namely *distributed proof reconstruction*.

Proof reconstruction requires to save the clauses deleted by backward contraction (cf. Section 6.2.2.2). Thus, the *distributed derivation* takes the form

$$(O_0;NO_0;R_0)^j \vdash (O_1;NO_1;R_1)^j \vdash \ldots (O_i;NO_i;R_i)^j \vdash \ldots$$

where for every process p_j, $0 \leq j \leq n-1$, and stage i, $i \geq 0$, $S_i^j = O_i^j \uplus NO_i^j$ is the database of clauses at process p_j and stage i, partitioned into owned (O_i^j) and not owned (NO_i^j) clauses, while R_i^j is the set of clauses that p_j deleted by backward contraction. *Distributed proof reconstruction* means that if $\Box \in S_i^k$, process p_k can reconstruct $\Pi(\Box)$ by consulting only $S_i^k \uplus R_i^k$. In order to guarantee distributed proof reconstruction, it is not sufficient that all persistent irredundant clauses be broadcast eventually, since clauses deleted by backward contraction, that are redundant and not persistent, may be needed to reconstruct the proof. A stronger, and sufficient, condition is that all clauses ever used as premises are broadcast. This is why Clause-Diffusion lets every process broadcast a clause φ after φ emerges from forward contraction, that is, as soon as φ is ready to be used as premise [29].

The problem of *distributed global contraction* is to ensure that notwithstanding the subdivision of inferences among the parallel processes, if φ is globally redundant at some stage i, φ is recognized redundant eventually by every process. Formally, if $\varphi \in \rho(\bigcup_{j=0}^{n-1} S_i^j)$ at some stage i, then for all processes p_j, $0 \leq j \leq n-1$, there exists a stage l, $l \geq i$, such that $\varphi \in \rho(S_l^j)$. Assume that $\varphi \in \bigcup_{j=0}^{n-1} S_i^j$, and $\varphi \in \rho(\bigcup_{j=0}^{n-1} S_i^j)$, because there is a $\psi \in \bigcup_{j=0}^{n-1} S_i^j$ such that ψ can delete φ by contraction. By the broadcasting mechanism of Clause-Diffusion, the two clauses are guaranteed to meet at every process, so that global redundancy becomes local redundancy. By fairness, every process eventually applies ψ to delete φ, so that global contraction becomes local contraction, unless the derivation succeeds sooner. Furthermore, by the subdivision of backward simplification, distributed global contraction is achieved while avoiding both the redundancy of letting all processes generate $\varphi\downarrow$ and the redundancy of preventing all processes but the owner from deleting φ.

In summary, Clause-Diffusion is a methodology to transform a sequential ordering-based theorem-proving strategy into a distributed one, in the sense that each parallel process executes the sequential strategy, modified with subdivision of labor and communication according to Clause-Diffusion. If the requirements for distributed fairness are fulfilled, a complete sequential strategy yields a complete distributed strategy.

6.3.5.6 The Clause-Diffusion Provers

At the implementation level, Clause-Diffusion is a methodology to transform a sequential ordering-based theorem prover into a distributed one, and indeed all Clause-Diffusion provers have a pre-existing sequential code base.

The first Clause-Diffusion prototype is *Aquarius* [46, 27, 47, 52]. Aquarius is the parallelization of OTTER 2.2 [165], using PCN for communication by message passing [101, 70]. Aquarius implements the *rotate* allocation criterion, with variants such as letting every process p_j own the factors of φ if p_j owns φ, or even allowing every process to own all input clauses. The latter trick violates the principle that every clause is owned by *only one* process, and it was tried only to watch its effect in experiments, especially when the input clauses include the axioms of some theory. Since Otter implements *unfailing* or *ordered* completion [120, 12], Aquarius offers also a Clause-Diffusion parallelization of ordered completion. Aquarius features also multi-search, since its options enable the user to shut off the subdivision of clauses, so that every process assigns all its generated clauses to itself, and attach different search plans to the processes. For a Clause-Diffusion prover that uses the given-clause algorithm, different search plans may mean different evaluation functions to select the given clause [164, 5, 165, 166, 88, 33, 169, 191].

While Aquarius, like OTTER, handles first-order logic with equality, the subsequent Clause-Diffusion provers focus on equational logic. A reason for this choice is that a motivation for exploring distributed search is to avoid the backward-contraction bottleneck, and backward contraction is crucial to solve equational problems.

The second Clause-Diffusion prototype is *Peers* [58, 51], whose name, chosen by Bill McCune, emphasizes that the deductive processes in Clause-Diffusion are peers. Peers is the parallelization of code from the *Otter Parts Store* (OPS), for theorem proving in equational theories possibly with *associative-commutative* (AC) function symbols, using p4 for communication by message passing [67]. If paramodulation is done modulo AC [175], there are generally so many AC-paramodulants that generating all AC-paramodulants between the given equation and all those in `already-selected` is too much for an iteration of the given-clause loop (cf. Section 6.3.2.2). Therefore, Peers employs a variant of the given-clause algorithm, called *pairs algorithm*: in every iteration of the loop the prover selects *a pair of equations* and performs all expansion inferences from the equations in the pair, provided at least one of them comes from `to-be-selected`. The evaluation function to select the best clause as given clause is replaced by an evaluation function that selects the best pair of equations.

Peers implements the *rotate*, *syntax*, and *select-min* allocation criteria, with variants such as allowing every process p_j to own $\varphi\downarrow$, if p_j owns φ and $\varphi\downarrow$ is generated by backward contraction. Assume that the input set S is satisfiable. In principle, a theorem-proving strategy may not terminate, because it is a semi-decision procedure. In practice, a theorem prover terminates on a satisfiable input, because either it generates a finite saturated set (cf. Section 6.2.2), or, more likely, because it reaches a predefined threshold on running time or memory space. For the first kind of situation, Peers implements the *Dijkstra-Pnueli global termination detection algorithm* [212] to recognize that all processes are idle. For the second kind of situation, a process p_k that reached a threshold broadcasts a message to inform all others that it quits the search. Clause-Diffusion allows $p_0,\dots,p_{k-1},p_{k+1},\dots,p_{n-1}$ to continue, but in Peers and its successors, for simplicity, such a message from p_k is a halting message.

The third Clause-Diffusion prototype is *Peers-mcd*, thus named because it implements Modified Clause-Diffusion. The first version, called *Peers-mcd.a* [29], is obtained by modifying Peers to execute Modified Clause-Diffusion, still using code from OPS as sequential base and p4 for message passing.

The second version, dubbed *Peers-mcd.b* [30], is the parallelization, according to Modified Clause-Diffusion, of version 0.9 of the EQP prover [167] for equational theories possibly with *associative-commutative* (AC) function symbols. In addition to ordered paramodulation or superposition (cf. Section 6.2.2), EQP features *blocking* [199, 139, 11, 128] and *basic paramodulation* [14]. Blocking prevents a paramodulation step whose most general unifier contains at least a pair $x \leftarrow t$ where t is reducible. Basic paramodulation stipulates that a term is *basic*, if it is not introduced by a substitution, and restricts paramodulation and simplification to apply only to basic terms. The restriction to simplification is not implemented in EQP, renouncing refutational completeness. In terms of search plan, EQP features both given-clause algorithm and pairs algorithm. Peers-mcd.b and its successors inherit all these features, and adopt the Message Passing Interface (MPI) and its implementation mpich for message passing [110].

Peers-mcd.b is the first Clause-Diffusion prover to implement the AGO allocation criteria (cf. Section 6.3.5.3). The EQP prover made history by proving mechanically that *Robbins algebras are Boolean* [168, 78], a conjecture remained open in mathematics since 1933 and considered a challenge in automatic theorem proving since 1990 [220]. Thanks to the AGO allocation criteria, Peers-mcd.b exhibited *super-linear speedup* on several problems, including two lemmas representing two thirds of the proof of the Robbins theorem [30, 31], and the *Levi commutator problem in group theory* [33].

The following version of Peers-mcd is *Peers-mcd.c*, that features version 0.9d of EQP as sequential base. Peers-mcd.c maintains the super-linear speedup in the first two lemmas that form the proof of the Robbins theorem, and adds an almost linear speedup in the third lemma [37].

Peers-mcd.d [38] still has EQP0.9d as sequential base. It differs from all previous versions of Peers-mcd, because it offers distributed search, multi-search, and their combination. It can run in one of three modes: (1) *pure distributed-search mode:* the search space is subdivided among the processes; all processes execute the same search plan; (2) *pure multi-search mode:* the search space is not subdivided; every process executes a different search plan; and (3) *hybrid mode:* the search space is subdivided, and the processes execute different search plans.

A first way to differentiate the search plans in Peers-mcd.d is to have half the processes execute the given-clause algorithm and the other half execute the pairs algorithm. Another way is to let the processes employ different evaluation functions to select the given clause or pair of equations. The two ways may also be combined, if the number of processes is sufficiently high.

Peers-mcd.d implements three heuristic evaluation functions to select given clauses based on their similarity with the target theorem [5, 88, 38]. If multi-search with similarity-based heuristics is selected, process p_k executes the given-clause algorithm with the first heuristic function if $k \ mod \ 3 = 0$, with the second heuristic function if $k \ mod \ 3 = 1$, and with the third heuristic function if $k \ mod \ 3 = 2$. These heuristics do not apply to the pairs algorithm.

Peers-mcd.d also turns the `pick-given-ratio` parameter into a way of differentiating searches in multi-search. This parameter appeared first in OTTER [164, 165, 166, 169] and has been adopted by most ordering-based theorem provers [191]. It allows the prover to mix best-first search and breadth-first search: if the parameter `pick-given-ratio` has value x, the given-clause/pair algorithm picks the oldest, rather than the best, equation/pair once every $x+1$ choices. In other words, it picks the best according to the heuristic evaluation function x times, then the oldest, and then it repeats. Peers-mcd.d lets each process use a different value of `pick-given-ratio`: if multi-search with different ratios is selected, process p_k resets its `pick-given-ratio` to $x+k$.

Prior to the Robbins theorem, another challenge problem for automatic theorem provers were the *Moufang identities in alternative rings* [6]. Alternative rings are rings where the product is not associative. The first automated proofs of these identities by a sequential prover involve several ingredients [6], including inference rules that *build the cancellation laws* in the inference system [122]. Peers-mcd.d proves the Moufang identities in alternative rings *without cancellation laws* and exhibiting several instances of *super-linear speedup* with respect to EQP0.9d [38]. This finding suggests that parallel search can even compensate for a weaker inference system. These results are obtained in pure distributed-search mode or hybrid mode, whereas multi-search alone shows no speedup at all. The best performances arise in hybrid mode. Thus, distributed search is necessary to conquer these problems, and the addition of multi-search improves the outcome further.

In summary, super-linear speedup by Clause-Diffusion is possible, precisely because parallel search, and all the more distributed search, does not mean executing in parallel the same steps of the sequential search, but generating a *different search*, that may visit the search space in a different way. The analysis of the experiments shows that whenever there is a super-linear speedup, the Clause-Diffusion prover *generates fewer clauses* than the sequential prover, *retains a higher percentage* of them, and generates a *different proof* [31, 38]. Generating fewer clauses and retaining more of them suggest better focus and less redundancy. Thus, the interpretation of the experiments is that an effective subdivision of the search space prevents the processes from overlapping too much, reduces the amount of redundancy, and allows the winning process to focus on a proof sooner. Since the proof is often not unique, these differences also reflect in a different proof being found. Note that different proof does not necessarily mean shorter proof: in theorem proving a shorter proof may require a longer run. The observation of super-linear speedup also indicates that the sequential search plan is not optimal for the problem, which is not surprising, given the generic and still largely syntactic nature of most heuristics in theorem proving.

While generating a different search may yield a faster proof, up to the point of a super-linear speedup, it also means that *scalability* may be irregular. Precisely because the point is not to use more computers to do the same steps, there is no guarantee that the performance improves regularly with the number of processes. For example, it may happen that the performance scales well with up to six processes, and becomes worse with seven or eight. A pattern of this type suggests that the problem may not be hard enough to justify more computing power beyond a certain point, so that subdividing the search space further is counterproductive.

In other cases, the performance oscillates: two processes do better than one, but four do worse than two, and six speed up again; or, neither four nor six improve, but seven or eight do. In these instances, an explanation is that the subdivision of the search space in Clause-Diffusion depends on the number of processes, as it is done by dynamic allocation of generated clauses during the derivation. Assume that we have two processes p_0 and p_1. When we add a third process p_2, the portions of the search space assigned to p_0 and p_1 change with respect to what they were with two processes. The three searches developed by p_0, p_1, and p_2, differ from those developed by p_0 and p_1 when running as two processes. Since the result depends on the subdivision of the search, it may happen that two processes do better than four on a certain combination of problem and strategy. However, combining distributed search and multi-search may smooth these oscillations improving scalability [38].

6.4 Discussion

In this section first we draw connections between parallel theorem proving and parallel satisfiability solving. The readers will find more by reading this chapter together with those on 1, Parallel Satisfiability and 2, Cube and Conquer for Satisfiability. Then, we discuss future directions for research in parallelization of theorem proving in the light of advances in first-order model-based reasoning [44].

6.4.1 Parallel Theorem Proving and Parallel Satisfiability

The idea of subdivision of the search space in Clause-Diffusion influenced the design of the parallel SAT solver PSATO [227, 228], which is considered a forerunner of the *divide-and-conquer approach* to parallel SAT-solving. More generally, research in parallel SAT-solving inherited from research in parallel theorem proving the focus on *parallel search*. In addition, inferences and data in propositional logic are simpler than in first-order logic, so that there is no room for parallelism below or at the inference level. The concepts of *distributed search* and *multi-search* apply with the same meaning also in parallel SAT-solving, corresponding to the *divide-and-conquer* and *portfolio* approaches, respectively.

PSATO is a *distributed-search* parallelization of SATO [226, 229], that implements the DPLL procedure [80, 79, 71] for propositional satisfiability. The original Davis-Putnam (DP) procedure [80] is for first-order logic, and features propositional, or ground, resolution. The Davis-Putnam-Logemann-Loveland (DPLL) procedure [79] replaces propositional resolution with *splitting*, seen as breaking disjunctions apart by *case analysis*, to avoid the growth of clauses and the non-determinism of resolution. Splitting is understood also as *guessing*, or *deciding*, the truth value of a propositional variable, in order *to search for a model* of the given set of clauses. Thus, DPLL is a *model-based* procedure, where all operations are centered around a candidate partial model, called *context*, represented by a sequence, or *trail*, of literals.

A PSATO derivation features $n+1$ processes, with one master process that subdivides the work, and n client processes each searching for a model by executing SATO. The key idea is to subdivide the search space by using *guiding paths*. The notion of guiding path is inspired by the view of the search space of a SAT problem as the tree of recursive calls of the DPLL procedure. In this tree a node has typically two outgoing arcs, one labeled L and the other labeled $\neg L$, where L is a literal occurring in the input problem. The two arcs correspond to the two cases of the case-splitting on L (either L is *true* or L is *false*), and lead to the two ensuing recursive calls, one where L is asserted and one where $\neg L$ is asserted.

A guiding path is a path in this tree; it is represented as a sequence of pairs $\langle(L_1,\delta_1),(L_2,\delta_2),\ldots,(L_k,\delta_k)\rangle$, where, for $1 \leq i \leq k$, the L_i's are the literals labeling the path; $\delta_i = 1$, if L_i is a first child; and $\delta_i = 0$, if L_i is a second child. A node labeled $(L,1)$ is *open*, because L is still to be flipped; a node labeled $(L,0)$ is *closed*, because L has been already flipped. A *job* is given by a pair (S,P), where S is the input set of clauses and P is a guiding path. Given a path $P = \langle(L_1,0),(L_2,0),\ldots,(L_i,1),\ldots,(L_k,\delta_k)\rangle$, where i is the smallest index for which $\delta_i = 1$, two new *disjoint* paths are generated by splitting on L_i, yielding $P_1 = \langle(L_1,0),(L_2,0),\ldots,(\neg L_i,0)\rangle$ and $P_2 = \langle(L_1,0),(L_2,0),\ldots,(L_i,0),\ldots,(L_k,\delta_k)\rangle$.

In PSATO, the master process is responsible for preparing the jobs and assigning a job and a time limit to each client process. Every client will return either `sat` with a model of S; or `unsat`, meaning that its assigned subtree contains no model; or a guiding path, representing the search remaining when the time is up. The subtrees assigned to the clients are *disjoint* portions of a finite search space, so that the subdivision has *no overlap* by definition. In contrast, in first-order theorem proving the search space is infinite, its representation is far more complex [55, 34, 39], and a strategy may at most try to *limit the overlap* of the searches by heuristic subdivision criteria as done in Clause-Diffusion (cf. Section 6.3.5.3).

The transition from the DPLL to the CDCL (Conflict-Driven Clause Learning) procedure [161, 162, 171, 160] is a game changer in parallel SAT-solving like in sequential SAT-solving. CDCL means *conflict-driven SAT*: when the current candidate model falsifies a clause, called *conflict clause*, this conflict is *explained* by a heuristically controlled series of resolution steps, where every resolvent is also a conflict clause. A resolvent is *learned*, and the candidate partial model is repaired in such a way to remove the conflict, by satisfying the learned clause and backjumping as far away as possible from the conflict.

Learning a conflict clause is a form of *lemmatization*, as every resolvent is a lemma, a logical consequence of the input set of clauses. All learned clauses are former conflict clauses. Similar to other situations (cf. Section 6.2.1), a purpose of learning lemmas is to avoid repetitions: in CDCL it prevents the procedure from falling repeatedly in the same conflicts. In this sense, learning clauses is a way of pruning the search space.

The CDCL procedure involves several ingredients, in addition to conflict-driven clause learning and backjumping. *Activity-based decision heuristics* select the literal for the next decision by counting how many times a literal appear in learned clauses and favoring *most active* literals [230].

Clausal propagation consists of detecting *conflict clauses* and *implied literals*. A *conflict clause* is a clause whose literals are all *false* in the current candidate model. A literal is *implied* if it is the only unassigned literal of a clause: such a literal must be added to the trail in order to satisfy the clause, which is the *justification* of the implied literal. In the *two watched literals* scheme for clausal propagation [230, 126], it is sufficient to watch two non-*false* (i.e., either *true* or unassigned) literals per clause in order to detect conflict clauses and implied literals. Indeed, a conflict clause has zero non-*false* literals, and a justification has one non-*false* literal, so that a clause with two is neither a conflict clause nor a justification.

The possibility of periodically *restarting* the search with an empty trail and a set of clauses augmented with learned clauses may serve the purpose of compacting the trail or changing dynamically the order with which literals are picked for decision.

From the point of view of our analysis of parallelization of reasoning, *clause learning* is a key difference between DPLL and CDCL. Parallelizing DPLL can be seen as analogous to parallelizing tableau-based *subgoal-reduction strategies*: the database of clauses is *fixed*, equal to the input set, and the strategy searches for a model by exploring a tree that represents a survey of all possible interpretations. On the other hand, parallelizing CDCL can be seen as analogous to parallelizing *expansion-oriented strategies*, as the database of clauses *grows* due to *learning*. In CDCL learned clauses can be deleted based on heuristics (e.g., delete the oldest, or the least involved in resolution). These deletions can be considered a kind of forward contraction, while there is no analogy with backward contraction, since, for example, input clauses are not subject to deletion.

For CDCL, the definition of guiding path is updated to abandon the reference to the search space of a recursive DPLL procedure: a guiding path is simply a sequence of literals, and a node labeled L is *open*, if L is a decided literal, *closed*, if L is an implied literal [187]. Also, the notion of guiding path is replaced by that of *cube* [116]. Logically speaking, a cube is a conjunction, or a set, of literals. In practice, cubes are typically much longer than guiding paths [116].

In keeping with the model-based character of the CDCL procedure, a cube can be understood as an *assignment* that assigns *true* to the literals in the cube. Then, the SAT problem is generalized to the *satisfiability modulo assignment* (SMA) problem, defined as the problem of deciding the satisfiability of S with respect to an assignment J to some of the literals in S. If J is empty, SMA reduces to SAT, while an intermediate state of a SAT search is an SMA instance, since during the search a SAT solver maintains a partial candidate model represented by an assignment of truth values to propositional variables. Approaches to parallel SAT-solving by distributed search such as PaMiraXT [187] and *cube and conquer* [116] (cf. the dedicated chapter), attack a SAT problem with input set S, by having n processes $p_0, \ldots, p_{n-1}$ working in parallel on n SMA instances with input set S and initial assignments $J_0, \ldots J_{n-1}$, each containing a distinct cube.

Approaches to parallel SAT-solving by *multi-search* assign to the processes $p_0, \ldots, p_{n-1}$ different search plans, as in ManySAT [113]. Similar to ATP systems, also SAT solvers have many options and parameters that define the search plan and whose variation may serve the purpose of differentiating the searches. For example, the p_j's may employ different heuristics to pick the next literal for decision, or different heuristics to determine when to restart. Another way to differentiate the searches is to use *randomization* as in CL-SDSAT [123]: a randomized SAT solver makes a certain percentage of its decisions at random, starting from a given randomized seed, rather than based on a heuristic. Then, the p_j's may use different percentages or different seeds.

Activity-based decision heuristics and restart heuristics tend to *intensify* the search of a process, meaning that the process focuses on a certain region of the search space. In parallel search, this phenomenon may be useful to *reduce the overlap* between the processes, if each p_j focuses on a different region [112, 111].

In both distributed-search and multi-search parallel SAT-solving methods, the processes may *communicate learned clauses* [112, 113, 187, 123]. A learned clause φ is not sent to a process whose initial cube satisfies φ: indeed, in a model-based strategy a satisfied clause is *redundant* [62]. Upon receiving a learned clause, a process needs to determine its two watched literals for clausal propagation.

Since learned clauses are generated resolvents, communication of learned clauses in parallel SAT-solving reminds one of Clause-Diffusion (cf. Section 6.3.5.2). The possibility of applying heuristics to select for broadcasting only useful learned clauses is in the spirit of the Team-Work method (cf. Section 6.3.4.2). A typical heuristic is to broadcast learned clauses whose *size* is below a certain threshold. This is similar to what happens in Clause-Diffusion with *deletion by weight*: a clause whose weight is above the threshold gets deleted by forward contraction and therefore it is not broadcast. This kind of heuristic can be made dynamic by varying the threshold during the search [112]. In propositional logic the size of a clause is the number of its literals. In a SAT solver the size of a clause is the number of its non-*false* literals with respect to the current candidate model. Thus, a clause may have different sizes under different cubes. Therefore, whether a learned clause is communicated depends on the given cube, as suggested in PMSAT [107].

Since a purpose of learning conflict clauses is to prune the search space, receiving from process p_k a learned conflict clause may help process p_j prune its search space. This is analogous to what happens in parallel search for ordering-based strategies, where receiving from process p_k a good simplifier may help process p_j prune its search space. On the other hand, communication is a cost in both contexts.

In parallel SAT-solving, the communication of learned clauses may be at odd with having low overlap or no overlap: if the processes delve into remote regions of the search space, sharing learned clauses may become useless [112]. In parallel search for theorem proving it is much harder to avoid overlapping searches, and therefore this issue does not arise. The observation of this phenomenon in parallel SAT-solving leads to the notion of subdividing the processes into groups [111]. Processes within a group cooperate, by sharing information such as learned clauses. Each group is devoted to search a different region of the search space, by letting all processes in the group start with the same cube, which is distinct from the cubes given to all other groups.

6.4.2 Parallelism and First-Order Model-Based Reasoning

Motivations for renewing the quest for parallel first-order theorem-proving methods are not different from those for injecting parallelism in SAT solvers: problems from applications get bigger and bigger; it is hard to improve sequential performance; and parallel hardware is available. In addition, the ATP problem is harder (only semi-decidable) and still far less understood than the SAT problem. The research of new approaches to ATP is certainly not over, and there are also approaches that are not new but never or barely considered for parallelization.

The investigation of ways to combine semantics and parallelism in theorem proving is still largely an open problem. *Semantically-guided strategies* assume a *fixed* interpretation for semantic guidance. Among ordering-based strategies, a basic paradigm is that of *semantic resolution*, with *hyperresolution* and *resolution with set of support* as special cases (cf. Section 6.2.2). Among instance-based strategies, *ordered semantic hyperlinking* (OSHL) enriches hyperlinking with semantic guidance (cf. Section 6.2.3). A natural idea is to devise *multi-search* methods where the processes employ *different guiding interpretations* for semantic resolution or OSHL. A simple example is to have two parallel processes, one using positive and the other negative hyperresolution.

Another possibility is to design a method that combines distributed search as in Clause-Diffusion (cf. Section 6.3.5) with a multi-search scheme where the processes adopt different guiding interpretations. While Clause-Diffusion is a general paradigm, it targets especially contraction-based strategies for equational theories and first-order logic with equality (cf. Section 6.2.2.2). Thus, the challenge is to combine multi-search with different guiding interpretations with distributed search for a logic including equality.

Model-based strategies build a candidate partial model and declare unsatisfiability when a contradiction arises, showing that no candidate can be completed in a model of the input set of clauses. Beside model elimination (ME) and ME-tableaux strategies (cf. Sections 6.2.1, 6.3.1.1, 6.3.2.1, 6.3.4.1), there are other classes of strategies that aim at being model-based for first-order logic and have not been considered for parallelization. This is the case for most model-oriented instance-based strategies and hybrid strategies that combine instance generation with tableaux (cf. Section 6.2.3 and Section 7.3 of [39]), as well as for the *model evolution calculus* that lifts the DPLL procedure to first-order logic [17, 18, 22, 23, 21].

Another example are the methods that integrate an ordering-based strategy for first-order logic with equality with a model-building method. In early approaches the two engines were loosely coupled and the model-building method was a model finder enumerating small models [200]. In later approaches the integration is tight and the model finder is replaced with a CDCL-based SAT [182] or SMT solver [57]. A straightforward approach to parallelization is to have two parallel processes, one executing the first-order strategy and one executing the solver. More ambitious schemes could devote multiple processes to both kinds of reasoning, parallelizing, in the sense of parallel search, both ordering-based strategy and solver. Such schemes could combine approaches to parallel search for SAT solvers (cf. Section 6.4.1 and the chapters on 1, Parallel Satisfiability and 2, Cube and Conquer for Satisfiability), SMT solvers (cf. the chapter on 5, Parallel Satisfiability Modulo Theories), and ordering-based first-order provers (cf. Sections 6.3.4.2 and 6.3.5).

SGGS (*Semantically-Guided Goal-Sensitive reasoning*) generalizes CDCL to first-order logic, and is both model-based and semantically-guided [60, 59, 61, 62]. Other approaches to generalizing CDCL include DPLL($\mathscr{S}\mathscr{X}$) [176] and NRCL [2] for effectively propositional logic, and *conflict resolution* [201, 124] for first-order logic. SGGS searches for a model of the input set S of clauses, starting from a given *initial Herbrand interpretation* I, and building interpretations $I[\Gamma_1]$, $I[\Gamma_2]$, $I[\Gamma_3]\ldots$, represented by *SGGS clause sequences* $\Gamma_1, \Gamma_2, \Gamma_3 \ldots$. An SGGS clause sequence is a sequence of constrained clauses with selected literals. An SGGS-derivation has the form $\Gamma_0 \vdash \Gamma_1 \vdash \Gamma_2 \vdash \Gamma_3 \vdash \ldots$, where Γ_0 is empty and $I[\Gamma_0] = I$. The current SGGS clause sequence corresponds to the current trail in CDCL. The main SGGS activities correspond to those of CDCL as follows.

The SGGS analogue of CDCL decision is *selection* of a literal in any clause added to the current SGGS clause sequence Γ. Selected literals differentiate $I[\Gamma]$ from Γ. SGGS is possibly the first method that features *clausal propagation at the first-order level*. Clausal propagation in SGGS relies on the concepts of *uniform falsity* and *dependence*. A literal is *uniformly false* in an interpretation, if all its ground instances are *false* in that interpretation. For I, a literal is I-true if it is *true* in I, and I-false if it is uniformly false in I. SGGS requires that all literals in an SGGS clause sequence are either I-true or I-false. This invariant ensures that all ground instances of a literal in the sequence are in harmony with respect to I. A literal L *depends* on a selected literal M, if M precedes L in Γ, and all ground instances of L appear negated among the ground instances of M that M contributes to $I[\Gamma]$, so that M's selection makes L *uniformly false* in $I[\Gamma]$.

Most SGGS concepts and activities are defined *modulo semantic guidance* by I, because the system endeavours to make $I[\Gamma]$ different from I, since $I \not\models S$ (if $I \models S$, the problem is solved). For example, it is the I-false selected literals in Γ that differentiate $I[\Gamma]$ from I. Similarly, it is the dependence of I-true literals on I-false selected literals that is recorded by *assignments*; and it is I-all-true clauses, or clauses whose literals are all I-true, that are *conflict clauses* or *justifications* of *implied literal*. When all literals of an I-all-true clause are assigned, it means that in an attempt to diversify $I[\Gamma]$ from I to satisfy other clauses, the system made that I-all-true clause uniformly false in $I[\Gamma]$. When all literals of an I-all-true clause but one are assigned, the non-assigned one must be selected, and it is an implied literal, as *all* its ground instances must be *true* in $I[\Gamma]$ to satisfy the clause.

The SGGS inference system includes *SGGS-extension*, *SGGS-splitting*, *SGGS-resolution*, *SGGS-move*, and *SGGS-deletion*. SGGS-extension is an *instance generation* mechanism. SGGS-extension extends the sequence Γ and the candidate model $I[\Gamma]$, by adding to Γ an instance of an input clause which covers ground instances not satisfied by $I[\Gamma]$. The clause is instantiated in a way that enforces the invariant whereby all literals in Γ are either I-true or I-false.

SGGS-splitting has nothing to do with DPLL splitting. SGGS-splitting of a clause φ by a clause ψ replaces φ by a *partition*, where all ground instances that a specified literal in φ has in common with ψ's selected literal are confined to one clause of the partition. This enables SGGS-resolution or SGGS-deletion to remove such intersections between literals, eliminating duplications or contradictions in the representation of the candidate model. *SGGS-resolution* is a restricted form of first-order resolution, where an implied literal in a justification resolves away a literal that depends on it: for this reason it uses *matching* rather than unification, and allows the resolvent to *replace* the parent that is not a justification. *SGGS-deletion* removes *disposable* clauses, that are redundant, because satisfied by the interpretation induced by the clauses on their left in Γ. In a model-based approach a satisfied clause is redundant.

If SGGS-extension adds a clause in conflict with $I[\Gamma]$, the first-order CDCL mechanism of SGGS applies. It comprises *explanation* and *solving* inferences. If the conflict clause includes I-false literals, SGGS-resolution *explains* the conflict by resolving away those I-false literals with implied literals in Γ. An SGGS-extension adding such a clause makes sure that this is possible by applying an appropriate substitution. The explanation inferences yield either $\Box$ or an I-all-true conflict clause, which is then subject to the solving inferences.

If the conflict clause does not include I-false literals, only the solving inferences are applied: the conflict clause is *moved* to the left of the clause which its selected literal is assigned to. This *SGGS-move* solves the conflict by *flipping* the truth value in $I[\Gamma]$ of *all* ground instances of this selected literal. It corresponds to backjumping in CDCL. The moved clause is *learned* in the sense that it becomes the justification of its selected literal. Prior to the move, splitting inferences may apply to make the selected literal of the clause to be moved so precise, that the move will indeed flip the truth value of all its ground instances. Every SGGS-extension with a conflict clause is followed by the explanation and solving inference that solve the conflict.

Because of the novelty of SGGS, its parallelization is a research goal for the long term. Since SGGS is semantically guided by the initial interpretation I, the notion of a parallel search with multiple SGGS processes, each using a different I for semantic guidance, applies here too. Similar to hyperresolution, the simplest example is to have two parallel SGGS processes, one using an I where all negative literals are *true*, and the other using an I where all positive literals are *true*. Most excitingly, SGGS opens the possibility of lifting to the first-order level the ideas for distributed search (e.g., cubes) or multi-search put forth for CDCL.

References

[1] Martin Aigner, Armin Biere, Christoph M. Kirsch, Aina Niemetz, and Mathias Preiner. Analysis of portfolio-style parallel SAT solving on current multi-core architectures. In Daniel Le Berre and Allen Van Gelder, editors, *Notes of the Fourth Workshop on Pragmatics of SAT (POS), Sixteenth International Conference on Theory and Applications of Satisfiability Testing (SAT)*, pages 28–40, 2013.

[2] Gábor Alagi and Christoph Weidenbach. NRCL – a model building approach to the Bernays-Schönfinkel fragment. In Carsten Lutz and Silvio Ranise, editors, *Proceedings of the Tenth International Symposium on Frontiers of Combining Systems (FroCoS)*, volume 9322 of *Lecture Notes in Artificial Intelligence*, pages 69–84. Springer, 2015.
[3] Iliès Alouini. Concurrent garbage collector for concurrent rewriting. In Jieh Hsiang, editor, *Proceedings of the Sixth International Conference on Rewriting Techniques and Applications (RTA)*, volume 914 of *Lecture Notes in Computer Science*, pages 132–146. Springer, 1995.
[4] Iliès Alouini. *Étude et mise en oeuvre de la réecriture conditionnelle concurrente sur des machines parallèles à mémoire distribuée*. PhD thesis, Université Henri Poincaré Nancy 1, May 1997.
[5] Siva Anantharaman and Nirina Andrianarivelo. Heuristical criteria in refutational theorem proving. In Alfonso Miola, editor, *Proceedings of the First International Symposium on Design and Implementation of Symbolic Computation Systems (DISCO)*, volume 429 of *Lecture Notes in Computer Science*, pages 184–193. Springer, 1990.
[6] Siva Anantharaman and Jieh Hsiang. Automated proofs of the Moufang identities in alternative rings. *Journal of Automated Reasoning*, 6(1):76–109, 1990.
[7] Owen L. Astrachan and Donald W. Loveland. METEORS: high performance theorem provers using model elimination. In Robert S. Boyer, editor, *Automated Reasoning: Essays in Honor of Woody Bledsoe*, pages 31–60. Kluwer Academic Publishers, The Netherlands, 1991.
[8] Owen L. Astrachan and Mark E. Stickel. Caching and lemmaizing in model elimination theorem provers. In Deepak Kapur, editor, *Proceedings of the Eleventh International Conference on Automated Deduction (CADE)*, volume 607 of *Lecture Notes in Artificial Intelligence*, pages 224–238. Springer, 1992.
[9] Jürgen Avenhaus and Jörg Denzinger. Distributing equational theorem proving. In Claude Kirchner, editor, *Proceedings of the Fifth International Conference on Rewriting Techniques and Applications (RTA)*, volume 690 of *Lecture Notes in Computer Science*, pages 62–76. Springer, 1993.
[10] Jürgen Avenhaus, Jörg Denzinger, and Matthias Fuchs. DISCOUNT: a system for distributed equational deduction. In Jieh Hsiang, editor, *Proceedings of the Sixth International Conference on Rewriting Techniques and Applications (RTA)*, volume 914 of *Lecture Notes in Computer Science*, pages 397–402. Springer, 1995.
[11] Leo Bachmair and Nachum Dershowitz. Critical pair criteria for completion. *Journal of Symbolic Computation*, 6(1):1–18, 1988.
[12] Leo Bachmair, Nachum Dershowitz, and David A. Plaisted. Completion without failure. In Hassam Aït-Kaci and Maurice Nivat, editors, *Resolution of Equations in Algebraic Structures*, volume II: Rewriting Techniques, pages 1–30. Academic Press, Cambridge, England, 1989.

[13] Leo Bachmair and Harald Ganzinger. Rewrite-based equational theorem proving with selection and simplification. *Journal of Logic and Computation*, 4(3):217–247, 1994.
[14] Leo Bachmair, Harald Ganzinger, Christopher Lynch, and Wayne Snyder. Basic paramodulation. *Information and Computation*, 121(2):172–192, 1995.
[15] Leo Bachmair, Harald Ganzinger, David McAllester, and Christopher A. Lynch. Resolution theorem proving. In John Alan Robinson and Andrei Voronkov, editors, *Handbook of Automated Reasoning*, volume 1, chapter 2, pages 535–610. Elsevier, Amsterdam, The Netherlands, 2001.
[16] Peter Baumgartner. Hyper tableaux – the next generation. In Harrie de Swart, editor, *Proceedings of the Seventh International Conference on Automated Reasoning with Analytic Tableaux and Related Methods (TABLEAUX)*, volume 1397 of *Lecture Notes in Artificial Intelligence*, pages 60–76. Springer, 1998.
[17] Peter Baumgartner, Alexander Fuchs, and Cesare Tinelli. Implementing the model evolution calculus. *International Journal on Artificial Intelligence Tools*, 15(1):21–52, 2006.
[18] Peter Baumgartner, Alexander Fuchs, and Cesare Tinelli. Lemma learning in the model evolution calculus. In Miki Hermann and Andrei Voronkov, editors, *Proceedings of the Thirteenth Conference on Logic, Programming and Automated Reasoning (LPAR)*, volume 4246 of *Lecture Notes in Artificial Intelligence*, pages 572–586. Springer, 2006.
[19] Peter Baumgartner and Ulrich Furbach. Consolution as a framework for comparing calculi. *Journal of Symbolic Computation*, 16(5):445–477, 1993.
[20] Peter Baumgartner and Ulrich Furbach. Variants of clausal tableaux. In Wolfgang Bibel and Peter H. Schmitt, editors, *Automated Deduction - A Basis for Applications*, volume I: Foundations - Calculi and Methods, chapter 3, pages 73–102. Kluwer Academic Publishers, The Netherlands, 1998.
[21] Peter Baumgartner, Björn Pelzer, and Cesare Tinelli. Model evolution calculus with equality - revised and implemented. *Journal of Symbolic Computation*, 47(9):1011–1045, 2012.
[22] Peter Baumgartner and Cesare Tinelli. The model evolution calculus as a first-order DPLL method. *Artificial Intelligence*, 172(4/5):591–632, 2008.
[23] Peter Baumgartner and Uwe Waldmann. Superposition and model evolution combined. In Renate Schmidt, editor, *Proceedings of the Twenty-Second International Conference on Automated Deduction (CADE)*, volume 5663 of *Lecture Notes in Artificial Intelligence*, pages 17–34. Springer, 2009.
[24] Markus Bender, Björn Pelzer, and Claudia Schon. E-KRHyper 1.4: extensions for unique names and description logic. In Maria Paola Bonacina, editor, *Proceedings of the Twenty-Fourth International Conference on Automated Deduction (CADE)*, volume 7898 of *Lecture Notes in Artificial Intelligence*, pages 126–134. Springer, 2013.
[25] Wolfgang Bibel and Elmer Eder. Methods and calculi for deduction. In Dov M. Gabbay, Christopher J. Hogger, and John Alan Robinson, editors, *Handbook of Logic in Artificial Intelligence and Logic Programming*, volume I: Logical Foundations, pages 68–183. Oxford University Press, Oxford, England, 1993.

[26] Jean-Paul Billon. The disconnection method. In Pierangelo Miglioli, Ugo Moscato, Daniele Mundici, and Mario Ornaghi, editors, *Proceedings of the Fifth International Conference on Automated Reasoning with Analytic Tableaux and Related Methods (TABLEAUX)*, volume 1071 of *Lecture Notes in Artificial Intelligence*, pages 110–126. Springer, 1996.

[27] Maria Paola Bonacina. *Distributed automated deduction.* PhD thesis, Department of Computer Science, State University of New York at Stony Brook, December 1992.

[28] Maria Paola Bonacina. On the reconstruction of proofs in distributed theorem proving with contraction: a modified Clause-Diffusion method. In Hoon Hong, editor, *Proceedings of the First International Symposium on Parallel Symbolic Computation (PASCO)*, volume 5 of *Lecture Notes Series in Computing*, pages 22–33. World Scientific, 1994.

[29] Maria Paola Bonacina. On the reconstruction of proofs in distributed theorem proving: a modified Clause-Diffusion method. *Journal of Symbolic Computation*, 21(4–6):507–522, 1996.

[30] Maria Paola Bonacina. The Clause-Diffusion theorem prover Peers-mcd. In William W. McCune, editor, *Proceedings of the Fourteenth International Conference on Automated Deduction (CADE)*, volume 1249 of *Lecture Notes in Artificial Intelligence*, pages 53–56. Springer, 1997.

[31] Maria Paola Bonacina. Experiments with subdivision of search in distributed theorem proving. In Markus Hitz and Erich Kaltofen, editors, *Proceedings of the Second International Symposium on Parallel Symbolic Computation (PASCO)*, pages 88–100. ACM Press, 1997.

[32] Maria Paola Bonacina. Analysis of distributed-search contraction-based strategies. In Jürgen Dix, Luis Fariñas del Cerro, and Ulrich Furbach, editors, *Proceedings of the Sixth European Workshop on Logics in Artificial Intelligence (JELIA)*, volume 1489 of *Lecture Notes in Artificial Intelligence*, pages 107–121. Springer, 1998.

[33] Maria Paola Bonacina. Mechanical proofs of the Levi commutator problem. In Peter Baumgartner et al., editor, *Notes of the Workshop on Problem Solving Methodologies with Automated Deduction, Fifteenth International Conference on Automated Deduction (CADE)*, pages 1–10, 1998.

[34] Maria Paola Bonacina. A model and a first analysis of distributed-search contraction-based strategies. *Annals of Mathematics and Artificial Intelligence*, 27(1–4):149–199, 1999.

[35] Maria Paola Bonacina. A taxonomy of theorem-proving strategies. In Michael J. Wooldridge and Manuela Veloso, editors, *Artificial Intelligence Today - Recent Trends and Developments*, volume 1600 of *Lecture Notes in Artificial Intelligence*, pages 43–84. Springer, Berlin, Germany, 1999.

[36] Maria Paola Bonacina. Ten years of parallel theorem proving: a perspective. In Bernhard Gramlich, Hélène Kirchner, and Frank Pfenning, editors, *Notes of the Third Workshop on Strategies in Automated Deduction (STRATEGIES), Second Federated Logic Conference (FLoC)*, pages 3–15, 1999.

[37] Maria Paola Bonacina. A taxonomy of parallel strategies for deduction. *Annals of Mathematics and Artificial Intelligence*, 29(1–4):223–257, 2000.
[38] Maria Paola Bonacina. Combination of distributed search and multi-search in Peers-mcd.d. In Rajeev P. Gore, Alexander Leitsch, and Tobias Nipkow, editors, *Proceedings of the First International Joint Conference on Automated Reasoning (IJCAR)*, volume 2083 of *Lecture Notes in Artificial Intelligence*, pages 448–452. Springer, 2001.
[39] Maria Paola Bonacina. Towards a unified model of search in theorem proving: subgoal-reduction strategies. *Journal of Symbolic Computation*, 39(2):209–255, 2005.
[40] Maria Paola Bonacina. On theorem proving for program checking – Historical perspective and recent developments. In Maribel Fernandez, editor, *Proceedings of the Twelfth International Symposium on Principles and Practice of Declarative Programming (PPDP)*, pages 1–11. ACM Press, 2010.
[41] Maria Paola Bonacina and Nachum Dershowitz. Abstract canonical inference. *ACM Transactions on Computational Logic*, 8(1):180–208, 2007.
[42] Maria Paola Bonacina and Nachum Dershowitz. Canonical ground Horn theories. In Andrei Voronkov and Christoph Weidenbach, editors, *Programming Logics: Essays in Memory of Harald Ganzinger*, volume 7797 of *Lecture Notes in Artificial Intelligence*, pages 35–71. Springer, 2013.
[43] Maria Paola Bonacina and Mnacho Echenim. Theory decision by decomposition. *Journal of Symbolic Computation*, 45(2):229–260, 2010.
[44] Maria Paola Bonacina, Ulrich Furbach, and Viorica Sofronie-Stokkermans. On first-order model-based reasoning. In Narciso Martí-Oliet, Peter Olveczky, and Carolyn Talcott, editors, *Logic, Rewriting, and Concurrency: Essays Dedicated to José Meseguer*, volume 9200 of *Lecture Notes in Computer Science*, pages 181–204. Springer, Berlin, Germany, 2015.
[45] Maria Paola Bonacina and Jieh Hsiang. High performance simplification-based automated deduction. In *Transactions of the Ninth U.S. Army Conference on Applied Mathematics and Computing*, number 92-1, pages 321–335. Army Research Office, 1991.
[46] Maria Paola Bonacina and Jieh Hsiang. A system for distributed simplification-based theorem proving. In Bertrand Fronhöfer and Graham Wrightson, editors, *Proceedings of the First International Workshop on Parallelization in Inference Systems (December 1990)*, volume 590 of *Lecture Notes in Artificial Intelligence*, pages 370–370. Springer, Berlin, Germany, 1992.
[47] Maria Paola Bonacina and Jieh Hsiang. Distributed deduction by Clause-Diffusion: the Aquarius prover. In Alfonso Miola, editor, *Proceedings of the Third International Symposium on Design and Implementation of Symbolic Computation Systems (DISCO)*, volume 722 of *Lecture Notes in Computer Science*, pages 272–287. Springer, 1993.
[48] Maria Paola Bonacina and Jieh Hsiang. On fairness in distributed deduction. In Patrice Enjalbert, Alain Finkel, and Klaus W. Wagner, editors, *Proceedings of the Tenth Annual Symposium on Theoretical Aspects of Computer Science*

(STACS), volume 665 of *Lecture Notes in Computer Science*, pages 141–152. Springer, 1993.
[49] Maria Paola Bonacina and Jieh Hsiang. On subsumption in distributed derivations. *Journal of Automated Reasoning*, 12:225–240, 1994.
[50] Maria Paola Bonacina and Jieh Hsiang. Parallelization of deduction strategies: an analytical study. *Journal of Automated Reasoning*, 13:1–33, 1994.
[51] Maria Paola Bonacina and Jieh Hsiang. The Clause-Diffusion methodology for distributed deduction. *Fundamenta Informaticae*, 24(1–2):177–207, 1995.
[52] Maria Paola Bonacina and Jieh Hsiang. Distributed deduction by Clause-Diffusion: distributed contraction and the Aquarius prover. *Journal of Symbolic Computation*, 19:245–267, 1995.
[53] Maria Paola Bonacina and Jieh Hsiang. Towards a foundation of completion procedures as semidecision procedures. *Theoretical Computer Science*, 146:199–242, 1995.
[54] Maria Paola Bonacina and Jieh Hsiang. On semantic resolution with lemmaizing and contraction and a formal treatment of caching. *New Generation Computing*, 16(2):163–200, 1998.
[55] Maria Paola Bonacina and Jieh Hsiang. On the modelling of search in theorem proving – towards a theory of strategy analysis. *Information and Computation*, 147:171–208, 1998.
[56] Maria Paola Bonacina, Christopher A. Lynch, and Leonardo de Moura. On deciding satisfiability by DPLL($\Gamma + \mathscr{T}$) and unsound theorem proving. In Renate Schmidt, editor, *Proceedings of the Twenty-second International Conference on Automated Deduction (CADE)*, volume 5663 of *Lecture Notes in Artificial Intelligence*, pages 35–50. Springer, 2009.
[57] Maria Paola Bonacina, Christopher A. Lynch, and Leonardo de Moura. On deciding satisfiability by theorem proving with speculative inferences. *Journal of Automated Reasoning*, 47(2):161–189, 2011.
[58] Maria Paola Bonacina and William W. McCune. Distributed theorem proving by Peers. In Alan Bundy, editor, *Proceedings of the Twelfth International Conference on Automated Deduction (CADE)*, volume 814 of *Lecture Notes in Artificial Intelligence*, pages 841–845. Springer, 1994.
[59] Maria Paola Bonacina and David A. Plaisted. Constraint manipulation in SGGS. In Temur Kutsia and Christophe Ringeissen, editors, *Proceedings of the Twenty-Eighth Workshop on Unification (UNIF), Sixth Federated Logic Conference (FLoC)*, Technical Reports of the Research Institute for Symbolic Computation, pages 47–54. Johannes Kepler Universität, 2014. Available at `http://vsl2014.at/meetings/UNIF-index.html`.
[60] Maria Paola Bonacina and David A. Plaisted. SGGS theorem proving: an exposition. In Stephan Schulz, Leonardo De Moura, and Boris Konev, editors, *Proceedings of the Fourth Workshop on Practical Aspects in Automated Reasoning (PAAR), Sixth Federated Logic Conference (FLoC), July 2014*, volume 31 of *EasyChair Proceedings in Computing (EPiC)*, pages 25–38, 2015.

[61] Maria Paola Bonacina and David A. Plaisted. Semantically-guided goal-sensitive reasoning: model representation. *Journal of Automated Reasoning*, 56(2):113–141, 2016.

[62] Maria Paola Bonacina and David A. Plaisted. Semantically-guided goal-sensitive reasoning: inference system and completeness. *Journal of Automated Reasoning*, 56(2):165–218, 2016.

[63] Soumitra Bose, Edmund M. Clarke, David E. Long, and Spiro Michaylov. Parthenon: A parallel theorem prover for non-Horn clauses. *Journal of Automated Reasoning*, 8(2):153–182, 1992.

[64] Bruno Buchberger. *An algorithm for finding a basis for the residue class ring of a zero-dimensional polynomial ideal (in German)*. PhD thesis, Department of Mathematics, Universität Innsbruck, 1965.

[65] Bruno Buchberger. History and basic features of the critical-pair/completion procedure. *Journal of Symbolic Computation*, 3:3–38, 1987.

[66] Reinhard Bündgen, Manfred Göbel, and Wolfgang Küchlin. Strategy-compliant multi-threaded term completion. *Journal of Symbolic Computation*, 21(4–6):475–506, 1996.

[67] Ralph M. Butler and Ewing L. Lusk. User's guide to the p4 programming system. Technical Report 92/17, Mathematics and Computer Science Division, Argonne National Laboratory, Argonne, Illinois, October 1992.

[68] Soumen Chakrabarti and Katherine A. Yelick. Implementing an irregular application on a distributed memory multiprocessor. In *Proceedings of the Fourth ACM SIGPLAN Symposium on Principles and Practice of Parallel Programming*, pages 169–178, 1993.

[69] Soumen Chakrabarti and Katherine A. Yelick. On the correctness of a distributed memory Gröbner basis algorithm. In Claude Kirchner, editor, *Proceedings of the Fifth International Conference on Rewriting Techniques and Applications (RTA)*, volume 690 of *Lecture Notes in Computer Science*, pages 77–91. Springer, 1993.

[70] K. Many Chandy and Stephen Taylor. *An Introduction to Parallel Programming*. Jones and Bartlett, Burlington, Massachusetts, 1991.

[71] Chin-Liang Chang and Richard Char-Tung Lee. *Symbolic Logic and Mechanical Theorem Proving*. Academic Press, Cambridge, England, 1973.

[72] P. Daniel Cheng and J. Y. Juang. A parallel resolution procedure based on connection graph. In *Proceedings of the Sixth Annual Conference of the American Association for Artificial Intelligence (AAAI)*, pages 13–17, 1987.

[73] Heng Chu and David A. Plaisted. Model finding in semantically guided instance-based theorem proving. *Fundamenta Informaticae*, 21(3):221–235, 1994.

[74] Heng Chu and David A. Plaisted. CLINS-S: a semantically guided first-order theorem prover. *Journal of Automated Reasoning*, 18(2):183–188, 1997.

[75] Edmund M. Clarke, David E. Long, Spiro Michaylov, Stephen A. Schwab, Jean-Philippe Vidal, and Shinji Kimura. Parallel symbolic computation algorithms. Technical Report CMU-CS-90-182, School of Computer Science, Carnegie Mellon University, Pittsburgh, Pennsylvania, October 1990.

[76] Susan E. Conry, Douglas J. MacIntosh, and Robert A. Meyer. DARES: a Distributed Automated REasoning System. In *Proceedings of the Eleventh Annual Conference of the American Association for Artificial Intelligence (AAAI)*, pages 78–85, 1990.
[77] Simon Cruanes. *Extending superposition with integer arithmetic, structural induction, and beyond*. PhD thesis, École Polytechnique, Université Paris-Saclay, September 2015.
[78] Bernd I. Dahn. Robbins algebras are Boolean: a revision of McCune's computer-generated solution of Robbins problem. *Journal of Algebra*, 208:526–532, 1998.
[79] Martin Davis, George Logemann, and Donald Loveland. A machine program for theorem-proving. *Communications of the ACM*, 5(7):394–397, 1962.
[80] Martin Davis and Hilary Putnam. A computing procedure for quantification theory. *Journal of the ACM*, 7:201–215, 1960.
[81] Leonardo de Moura and Nikolaj Bjørner. Engineering DPLL(T) + saturation. In Alessandro Armando, Peter Baumgartner, and Gilles Dowek, editors, *Proceedings of the Fourth International Conference on Automated Reasoning (IJCAR)*, volume 5195 of *Lecture Notes in Artificial Intelligence*, pages 475–490. Springer, 2008.
[82] Leonardo de Moura and Nikolaj Bjørner. Bugs, moles and skeletons: Symbolic reasoning for software development. In Jürgen Giesl and Reiner Hähnle, editors, *Proceedings of the Fifth International Conference on Automated Reasoning (IJCAR)*, volume 6173 of *Lecture Notes in Artificial Intelligence*, pages 400–411. Springer, 2010.
[83] Leonardo de Moura and Nikolaj Bjørner. Satisfiability modulo theories: introduction and applications. *Communications of the ACM*, 54(9):69–77, 2011.
[84] Jörg Denzinger. *Team-Work: a method to design distributed knowledge based theorem provers*. PhD thesis, Department of Computer Science, Universität Kaiserslautern, 1993.
[85] Jörg Denzinger and Bernd Ingo Dahn. Cooperating theorem provers. In Wolfgang Bibel and Peter H. Schmitt, editors, *Automated Deduction – A Basis for Applications*, volume II: Systems and Implementation, chapter 14, pages 383–416. Kluwer Academic Publishers, Amsterdam, The Netherlands, 1998.
[86] Jörg Denzinger and Dirk Fuchs. Cooperation of heterogeneous provers. In Thomas Dean, editor, *Proceedings of the Sixeenth International Joint Conference on Artificial Intelligence (IJCAI)*, pages 10–15. Morgan Kaufmann Publishers, 1999.
[87] Jörg Denzinger, Marc Fuchs, and Matthias Fuchs. High performance ATP systems by combining several AI methods. In Martha E. Pollack, editor, *Proceedings of the Fifteenth International Joint Conference on Artificial Intelligence (IJCAI)*, pages 102–107. Morgan Kaufmann Publishers, 1997.
[88] Jörg Denzinger and Matthias Fuchs. Goal-oriented equational theorem proving using Team-Work. In Bernhard Nebel and Leonie Dreschler-Fischer, editors, *Proceedings of the Eighteenth German Conference on Artificial Intelligence*

(KI), volume 861 of *Lecture Notes in Artificial Intelligence*, pages 343–354. Springer, 1994.
[89] Jörg Denzinger and Martin Kronenburg. Planning for distributed theorem proving: the Team-Work approach. In Steffen Hölldobler, editor, *Proceedings of the Twentieth German Conference on Artificial Intelligence (KI)*, volume 1137 of *Lecture Notes in Artificial Intelligence*, pages 43–56. Springer, 1996.
[90] Jörg Denzinger, Martin Kronenburg, and Stephan Schulz. DISCOUNT: a distributed and learning equational prover. *Journal of Automated Reasoning*, 18(2):189–198, 1997.
[91] Jörg Denzinger and Jürgen Lind. TWlib: a library for distributed search applications. In Chu-Sing Yang, editor, *Proceedings of the International Conference on Artificial Intelligence, International Computer Symposium (ICS)*, pages 101–108. National Sun-Yat Sen University, 1996.
[92] Jörg Denzinger and Stephan Schulz. Recording and analyzing knowledge-based distributed deduction processes. *Journal of Symbolic Computation*, 21(4–6):523–541, 1996.
[93] Nachum Dershowitz. Orderings for term-rewriting systems. *Theoretical Computer Science*, 17(3):279–301, 1982.
[94] Nachum Dershowitz and Jean-Pierre Jouannaud. Rewrite systems. In Jan van Leeuwen, editor, *Handbook of Theoretical Computer Science*, volume B, pages 243–320. Elsevier, Amsterdam, The Netherlands, 1990.
[95] Nachum Dershowitz and Naomi Lindenstrauss. An abstract concurrent machine for rewriting. In Hélène Kirchner and W. Wechler, editors, *Proceedings of the Second International Conference on Algebraic and Logic Programming (ALP)*, volume 463 of *Lecture Notes in Computer Science*, pages 318–331. Springer, 1990.
[96] Nachum Dershowitz and Zohar Manna. Proving termination with multiset orderings. *Communications of the ACM*, 22(8):465–476, 1979.
[97] Nachum Dershowitz and David A. Plaisted. Rewriting. In John Alan Robinson and Andrei Voronkov, editors, *Handbook of Automated Reasoning*, volume 1, chapter 9, pages 535–610. Elsevier, Amsterdam, The Netherlands, 2001.
[98] Norbert Eisinger and Hans Jürgen Ohlbach. Deduction systems based on resolution. In Dov M. Gabbay, Christopher J. Hogger, and John Alan Robinson, editors, *Handbook of Logic in Artificial Intelligence and Logic Programming*, volume I: Logical Foundations, pages 184–273. Oxford University Press, Oxford, England, 1993.
[99] Zachary Ernst and Seth Kurtenbach. Toward a procedure for data mining proofs. In Maria Paola Bonacina and Mark E. Stickel, editors, *Automated Reasoning and Mathematics: Essays in Memory of William W. McCune*, volume 7788 of *Lecture Notes in Artificial Intelligence*, pages 229–239. Springer, 2013.
[100] Michael Fisher. An alternative approach to concurrent theorem proving. In James Geller, Hiroaki Kitano, and Christian B. Suttner, editors, *Parallel Processing for Artificial Intelligence 3*, pages 209–230. Elsevier, Amsterdam, The Netherlands, 1997.

[101] Ian Foster and Steve Tuecke. Parallel programming with PCN. Technical Report 91/32, Mathematics and Computer Science Division, Argonne National Laboratory, Argonne, Illinois, December 1991.

[102] Dirk Fuchs. Requirement-based cooperative theorem proving. In Jürgen Dix, Luis Fariñas del Cerro, and Ulrich Furbach, editors, *Proceedings of the Sixth Joint European Workshop on Logic in Artificial Intelligence (JELIA)*, volume 1489 of *Lecture Notes in Artificial Intelligence*, pages 139–153. Springer, 1998.

[103] Marc Fuchs. Controlled use of clausal lemmas in connection tableau calculi. *Journal of Symbolic Computation*, 29(2):299–341, 2000.

[104] Marc Fuchs and Andreas Wolf. Cooperation in model elimination: CPTHEO. In Claude Kirchner and Hélène Kirchner, editors, *Proceedings of the Fifteenth International Conference on Automated Deduction (CADE)*, volume 1421 of *Lecture Notes in Artificial Intelligence*, pages 42–46. Springer, 1998.

[105] Harald Ganzinger and Konstantin Korovin. New directions in instantiation-based theorem proving. In *Proceedings of the Eighteenth IEEE Symposium on Logic in Computer Science (LICS)*, pages 55–64. IEEE Computer Society Press, 2003.

[106] Harald Ganzinger and Konstantin Korovin. Theory instantiation. In Miki Hermann and Andrei Voronkov, editors, *Proceedings of the Thirteenth Conference on Logic, Programming and Automated Reasoning (LPAR)*, volume 4246 of *Lecture Notes in Artificial Intelligence*, pages 497–511. Springer, 2006.

[107] Luís Gil, Paulo F. Flores, and Luis Miguel Silveira. PMSat: a parallel version of Minisat. *Journal on Satisfiability, Boolean Modeling and Computation*, 6:71–98, 2008.

[108] Joseph A. Goguen, Sany Leinwand, José Meseguer, and Timothy Winkler. The rewrite rule machine 1988. Technical Report PRG-76, Oxford University Computing Laboratory, Oxford, England, August 1989.

[109] Joseph A. Goguen, José Meseguer, Sany Leinwand, Timothy Winkler, and Hitoshi Aida. The rewrite rule machine. Technical Report SRI-CSL-89-6, Computer Science Laboratory, SRI International, Menlo Park, California, March 1989.

[110] William Gropp, Ewing Lusk, and Anthony Skjellum. *Using MPI: Portable Parallel Programming with the Message Passing Interface*. MIT Press, Cambridge, Massachusetts, 1994.

[111] Long Guo, Youssef Hamadi, Said Jabbour, and Lakhdar Sais. Diversification and intensification in parallel SAT solving. In Dave Cohen, editor, *Proceedings of the Sixteenth International Conference on Principles and Practice of Constraint Programming (CP)*, volume 6308 of *Lecture Notes in Computer Science*, pages 252–265. Springer, 2010.

[112] Youssef Hamadi, Said Jabbour, and Lakhdar Sais. Control-based clause sharing in parallel SAT solving. In Craig Boutilier, editor, *Proceedings of the Twenty-First International Joint Conference on Artificial Intelligence (IJCAI)*, pages 409–504. AAAI Press, 2009.

[113] Youssef Hamadi, Said Jabbour, and Lakhdar Sais. ManySAT: a parallel SAT solver. *Journal on Satisfiability, Boolean Modeling and Computation*, 6:245–262, 2009.
[114] Youssef Hamadi and Christoph M. Wintersteiger. Seven challenges in parallel SAT solving. *AI Magazine*, 34(2):99–106, 2013.
[115] D. J. Hawley. A Buchberger algorithm for distributed memory multi-processors. In Hans P. Zima, editor, *Proceedings of the First International Conference of the Austrian Center for Parallel Computation (ACPC)*, volume 591 of *Lecture Notes in Computer Science*. Springer, 1991.
[116] Marijn Heule, Oliver Kullmann, Siert Wieringa, and Armin Biere. Cube and conquer: guiding CDCL SAT solvers by lookaheads. In Kerstin Eder, João Lourenço, and Onn M. Shehory, editors, *Proceedings of the Seventh International Haifa Verification Conference (HVC)*, volume 7261 of *Lecture Notes in Computer Science*, pages 50–65. Springer, 2012.
[117] Thomas Hillenbrand. Citius, altius, fortius: lessons learned from the theorem prover WALDMEISTER. In Ingo Dahn and Laurent Vigneron, editors, *Proceedings of the Fourth International Workshop On First-Order Theorem Proving (FTP)*, volume 86 of *Electronic Notes in Theoretical Computer Science*. Elsevier, 2003.
[118] Christoph M. Hoffmann and Michael J. O'Donnell. Programming with equations. *ACM Transactions on Programming Languages and Systems*, 4(1):83–112, 1982.
[119] Alfred Horn. On sentences which are true in direct unions of algebras. *Journal of Symbolic Logic*, 16:14–21, 1951.
[120] Jieh Hsiang and Michaël Rusinowitch. On word problems in equational theories. In Thomas Ottman, editor, *Proceedings of the Fourteenth International Colloquium on Automta, Languages, and Programming (ICALP)*, volume 267 of *Lecture Notes in Computer Science*, pages 54–71. Springer, 1987.
[121] Jieh Hsiang and Michaël Rusinowitch. Proving refutational completeness of theorem proving strategies: the transfinite semantic tree method. *Journal of the ACM*, 38(3):559–587, 1991.
[122] Jieh Hsiang, Michaël Rusinowitch, and Ko Sakai. Complete inference rules for the cancellation laws. In John McDermott, editor, *Proceedings of the Tenth International Joint Conference on Artificial Intelligence (IJCAI)*, pages 990–992. Morgan Kaufmann Publishers, 1987.
[123] Antti E. J. Hyvärinen, Tommi Junttila, and Ilka Niemelä. Incorporating clause learning in grid-based randomized SAT solving. *Journal on Satisfiability, Boolean Modeling and Computation*, 6:223–244, 2009.
[124] Daniyar Itegulov, John Slaney, and Bruno Woltzenlogel Paleo. Scavenger 0.1: a theorem prover based on conflict resolution. In Leonardo de Moura, editor, *Proceedings of the Twenty-Sixth Conference on Automated Deduction (CADE)*, volume 10395 of *Lecture Notes in Artificial Intelligence*, pp. 344–356, Springer, 2017.
[125] Swen Jacobs and Uwe Waldmann. Comparing instance generation methods for automated reasoning. *Journal of Automated Reasoning*, 38:57–78, 2007.

[126] Himanshu Jain. *Verification using satisfiability checking, predicate abstraction and Craig interpolation*. PhD thesis, School of Computer Science, Carnegie Mellon University, September 2008.

[127] Anita Jindal, Ross Overbeek, and Waldo C. Kabat. Exploitation of parallel processing for implementing high-performance deduction systems. *Journal of Automated Reasoning*, 8:23–38, 1992.

[128] Deepak Kapur, David Musser, and Paliath Narendran. Only prime superposition need be considered in the Knuth-Bendix completion procedure. *Journal of Symbolic Computation*, 6:19–36, 1988.

[129] Owen Kaser, Shaunak Pawagi, C. R. Ramakrishnan, I. V. Ramakrishnan, and R. C. Sekar. Fast parallel implementations of lazy languages – the EQUALS experience. In John L. White, editor, *Proceedings of the ACM Conference on LISP and Functional Programming*, pages 335–344. ACM Press, 1992.

[130] Claude Kirchner, Christopher Lynch, and Christelle Scharff. Fine-grained concurrent completion. In Harald Ganzinger, editor, *Proceedings of the Seventh International Conference on Rewriting Techniques and Applications (RTA)*, volume 1103 of *Lecture Notes in Computer Science*, pages 3–17. Springer, 1996.

[131] Claude Kirchner and Patrick Viry. Implementing parallel rewriting. In Bertrand Fronhöfer and Graham Wrightson, editors, *Proceedings of the First International Workshop on Parallelization in Inference Systems (December 1990)*, volume 590 of *Lecture Notes in Artificial Intelligence*, pages 123–138. Springer, Berlin, Germany, 1992.

[132] Donald E. Knuth and Peter B. Bendix. Simple word problems in universal algebras. In John Leech, editor, *Proceedings of the Conference on Computational Problems in Abstract Algebras*, pages 263–298. Pergamon Press, Oxford, England, 1970.

[133] Richard E. Korf. Depth-first iterative deepening: an optimal admissible tree search. *Artificial Intelligence*, 27(1):97–109, 1985.

[134] Konstantin Korovin. An invitation to instantiation-based reasoning: from theory to practice. In Renate Schmidt, editor, *Proceedings of the Twenty-Second International Conference on Automated Deduction (CADE)*, volume 5663 of *Lecture Notes in Artificial Intelligence*, pages 163–166. Springer, 2009.

[135] Konstantin Korovin. Inst-Gen: a modular approach to instantiation-based automated reasoning. In Andrei Voronkov and Christoph Weidenbach, editors, *Programming Logics: Essays in Memory of Harald Ganzinger*, volume 7797 of *Lecture Notes in Artificial Intelligence*, pages 239–270. Springer, 2013.

[136] Konstantin Korovin and Christoph Sticksel. iProver-Eq: An instantiation-based theorem prover with equality. In Jürgen Giesl and Reiner Hähnle, editors, *Proceedings of the Fifth International Conference on Automated Reasoning (IJCAR)*, volume 6173 of *Lecture Notes in Artificial Intelligence*, pages 196–202. Springer, 2010.

[137] Laura Kovàcs and Andrei Voronkov. First order theorem proving and Vampire. In Natasha Sharygina and Helmut Veith, editors, *Proceedings of the Twenty-*

Fifth International Conference on Computer-Aided Verification (CAV), volume 8044 of *Lecture Notes in Computer Science*, pages 1–35. Springer, 2013.

[138] Robert Kowalski and Donald Kuehner. Linear resolution with selection function. *Artificial Intelligence*, 2:227–260, 1971.

[139] Dallas S. Lankford and A. M. Ballantyne. The refutation completeness of blocked permutative narrowing and resolution. In William H. Joyner Jr., editor, *Proceedings of the Fourth Conference on Automated Deduction (CADE)*, pages 168–174, 1979. Available at `http://www.cadeinc.org/`.

[140] Shie-Jue Lee and David A. Plaisted. Eliminating duplication with the hyperlinking strategy. *Journal of Automated Reasoning*, 9:25–42, 1992.

[141] K. Rustan M. Leino and Aleksandar Milicevic. Program extrapolation with Jennisys. In *Proceedings of the Twenty-Seventh Conference on Object-Oriented Programming, Systems, Languages, and Applications (OOPSLA)*, pages 411–430. ACM, 2012.

[142] Reinhold Letz. Clausal tableaux. In Wolfgang Bibel and Peter H. Schmitt, editors, *Automated Deduction - A Basis for Applications*, volume I: Foundations - Calculi and Methods, chapter 2, pages 43–72. Kluwer Academic Publishers, Amsterdam, The Netherlands, 1998.

[143] Reinhold Letz, Klaus Mayr, and Christian Goller. Controlled integration of the cut rule into connection tableau calculi. *Journal of Automated Reasoning*, 13(3):297–338, 1994.

[144] Reinhold Letz, Johann Schumann, Stephan Bayerl, and Wolfgang Bibel. SETHEO: a high performance theorem prover. *Journal of Automated Reasoning*, 8(2):183–212, 1992.

[145] Reinhold Letz and Gernot Stenz. DCTP - a disconnection calculus theorem prover. In Rajeev P. Goré, Alexander Leitsch, and Tobias Nipkow, editors, *Proceedings of the First International Joint Conference on Automated Reasoning (IJCAR)*, volume 2083 of *Lecture Notes in Artificial Intelligence*, pages 381–385. Springer, 2001.

[146] Reinhold Letz and Gernot Stenz. Model elimination and connection tableau procedures. In John Alan Robinson and Andrei Voronkov, editors, *Handbook of Automated Reasoning*, chapter 28, pages 2015–2114. Elsevier, Amsterdam, The Netherlands, 2001.

[147] Reinhold Letz and Gernot Stenz. Proof and model generation with disconnection tableaux. In Robert Nieuwenhuis and Andrei Voronkov, editors, *Proceedings of the Eighth International Conference on Logic, Programming and Automated Reasoning (LPAR)*, volume 2250 of *Lecture Notes in Artificial Intelligence*, pages 142–156. Springer, 2001.

[148] Reinhold Letz and Gernot Stenz. Integration of equality reasoning into the disconnection calculus. In Uwe Egly and Christian G. Fermüller, editors, *Proceedings of the Fifteenth International Conference on Analytic Tableaux and Related Methods (TABLEAUX)*, volume 2381 of *Lecture Notes in Artificial Intelligence*, pages 176–190. Springer, 2002.

[149] Vladimir Lifschitz, Leora Morgenstern, and David A. Plaisted. Knowledge representation and classical logic. In Frank van Harmelen, Vladimir Lifschitz,

and Bruce Porter, editors, *Handbook of Knowledge Representation*, volume 1, pages 3–88. Elsevier, Amsterdam, The Netherlands, 2008.

[150] Rasiah Loganantharaj. *Theoretical and implementational aspects of parallel link resolution in connection graphs*. PhD thesis, Department of Computer Science, Colorado State University, 1985.

[151] Rasiah Loganantharaj and Robert A. Müller. Parallel theorem proving with connection graphs. In Jörg Siekmann, editor, *Proceedings of the Eighth International Conference on Automated Deduction (CADE)*, volume 230 of *Lecture Notes in Computer Science*, pages 337–352. Springer, 1986.

[152] Donald W. Loveland. A simplified format for the model elimination procedure. *Journal of the ACM*, 16(3):349–363, 1969.

[153] Donald W. Loveland. A unifying view of some linear Herbrand procedures. *Journal of the ACM*, 19(2):366–384, 1972.

[154] Ewing L. Lusk and William W. McCune. Experiments with ROO: a parallel automated deduction system. In Bertrand Fronhöfer and Graham Wrightson, editors, *Proceedings of the First International Workshop on Parallelization in Inference Systems (December 1990)*, volume 590 of *Lecture Notes in Artificial Intelligence*, pages 139–162. Springer, Berlin, Germany, 1992.

[155] Ewing L. Lusk, William W. McCune, and John K. Slaney. Parallel closure-based automated reasoning. In Bertrand Fronhöfer and Graham Wrightson, editors, *Proceedings of the First International Workshop on Parallelization in Inference Systems (December 1990)*, volume 590 of *Lecture Notes in Artificial Intelligence*, pages 347–347. Springer, Berlin, Germany, 1992.

[156] Ewing L. Lusk, William W. McCune, and John K. Slaney. ROO: a parallel theorem prover. In Deepak Kapur, editor, *Proceedings of the Eleventh International Conference on Automated Deduction (CADE)*, volume 607 of *Lecture Notes in Artificial Intelligence*, pages 731–734. Springer, 1992.

[157] Sharad Malik and Lintao Zhang. Boolean satisfiability: from theoretical hardness to practical success. *Communications of the ACM*, 52(8):76–82, 2009.

[158] Norbert Manthey. Towards next generation sequential and parallel SAT solvers. *Constraints*, 20(4):504–505, 2015.

[159] Rainer Manthey and François Bry. SATCHMO: a theorem prover implemented in Prolog. In Ewing Lusk and Ross Overbeek, editors, *Proceedings of the Ninth International Conference on Automated Deduction (CADE)*, volume 310 of *Lecture Notes in Computer Science*, pages 415–434. Springer, 1988.

[160] João P. Marques Silva, Inês Lynce, and Sharad Malik. Conflict-driven clause learning SAT solvers. In Armin Biere, Marjin Heule, Hans Van Maaren, and Toby Walsh, editors, *Handbook of Satisfiability*, volume 185 of *Frontiers in Artificial Intelligence and Applications*, chapter 4, pages 131–153. IOS Press, Amsterdam, The Netherlands, 2009.

[161] João P. Marques-Silva and Karem A. Sakallah. GRASP: A new search algorithm for satisfiability. In *Proceedings of the International Conference on Computer-Aided Design (ICCAD)*, pages 220–227, 1997.

[162] João P. Marques Silva and Karem A. Sakallah. GRASP: A search algorithm for propositional satisfiability. *IEEE Transactions on Computers*, 48(5):506–521, 1999.
[163] Ruben Martins, Vasco M. Manquinho, and Inês Lynce. An overview of parallel SAT solving. *Constraints*, 17(3):304–347, 2012.
[164] William W. McCune. OTTER 2.0 users guide. Technical Report 90/9, Mathematics and Computer Science Division, Argonne National Laboratory, Argonne, Illinois, March 1990.
[165] William W. McCune. What's new in OTTER 2.2. Technical Report TM-153, Mathematics and Computer Science Division, Argonne National Laboratory, Argonne, Illinois, July 1991.
[166] William W. McCune. OTTER 3.0 reference manual and guide. Technical Report 94/6, Mathematics and Computer Science Division, Argonne National Laboratory, Argonne, Illinois, January 1994. Revised August 1995.
[167] William W. McCune. 33 Basic test problems: a practical evaluation of some paramodulation strategies. In Robert Veroff, editor, *Automated Reasoning and its Applications: Essays in Honor of Larry Wos*, pages 71–114. MIT Press, Cambridge, Massachusetts, 1997.
[168] William W. McCune. Solution of the Robbins problem. *Journal of Automated Reasoning*, 19(3):263–276, 1997.
[169] William W. McCune. OTTER 3.3 reference manual. Technical Report TM-263, Mathematics and Computer Science Division, Argonne National Laboratory, Argonne, Illinois, August 2003.
[170] Max Moser, Ortrun Ibens, Reinhold Letz, Joachim Steinbach, Christoph Goller, Johann Schumann, and Klaus Mayr. The model elimination provers SETHEO and E-SETHEO. *Journal of Automated Reasoning*, 18(2):237–246, 1997.
[171] Matthew W. Moskewicz, Conor F. Madigan, Ying Zhao, Lintao Zhang, and Sharad Malik. Chaff: Engineering an efficient SAT solver. In David Blaauw and Luciano Lavagno, editors, *Proceedings of the Thirty-Ninth Design Automation Conference (DAC)*, pages 530–535, 2001.
[172] Robert Nieuwenhuis and Albert Rubio. Paramodulation-based theorem proving. In John Alan Robinson and Andrei Voronkov, editors, *Handbook of Automated Reasoning*, volume 1, chapter 7, pages 371–443. Elsevier, Amsterdam, The Netherlands, 2001.
[173] Robert Niewenhuis and A. Rubio. Theorem proving with ordering and equality constrained clauses. *Journal of Symbolic Computation*, 19(4):321–351, 1995.
[174] Gerald E. Peterson. A technique for establishing completeness results in theorem proving with equality. *SIAM Journal of Computing*, 12(1):82–100, 1983.
[175] Gerald E. Peterson and Mark E. Stickel. Complete sets of reductions for some equational theories. *Journal of the ACM*, 28(2):233–264, 1981.
[176] Ruzica Piskac, Leonardo de Moura, and Nikolaj Bjørner. Deciding effectively propositional logic using DPLL and substitution sets. *Journal of Automated Reasoning*, 44(4):401–424, 2010.

[177] David A. Plaisted. Mechanical theorem proving. In Ranan B. Banerji, editor, *Formal Techniques in Artificial Intelligence*, pages 269–320. Elsevier, Amsterdam, The Netherlands, 1990.
[178] David A. Plaisted. Equational reasoning and term rewriting systems. In Dov M. Gabbay, Christopher J. Hogger, and John Alan Robinson, editors, *Handbook of Logic in Artificial Intelligence and Logic Programming*, volume I: Logical Foundations, pages 273–364. Oxford University Press, Oxford, England, 1993.
[179] David A. Plaisted. Automated theorem proving. *Wiley Interdisciplinary Reviews: Cognitive Science*, 5(2):115–128, 2014.
[180] David A. Plaisted and Swaha Miller. The relative power of semantics and unification. In Andrei Voronkov and Christoph Weidenbach, editors, *Programming Logics: Essays in Memory of Harald Ganzinger*, volume 7797 of *Lecture Notes in Artificial Intelligence*, pages 317–344. Springer, 2013.
[181] David A. Plaisted and Yunshan Zhu. Ordered semantic hyper linking. *Journal of Automated Reasoning*, 25:167–217, 2000.
[182] Giles Reger, Martin Suda, and Andrei Voronkov. Playing with AVATAR. In Amy P. Felty and Aart Middeldorp, editors, *Proceedings of the Twenty-Fifth International Conference on Automated Deduction (CADE)*, volume 9195 of *Lecture Notes in Artificial Intelligence*, pages 399–415. Springer, 2015.
[183] George A. Robinson and Larry Wos. Paramodulation and theorem-proving in first-order theories with equality. In Donald Michie and Bernard Meltzer, editors, *Machine Intelligence*, volume 4, pages 135–150. Edinburgh University Press, Edinburgh, Scotland, 1969.
[184] John Alan Robinson. Automatic deduction with hyper-resolution. *International Journal of Computer Mathematics*, 1:227–234, 1965.
[185] John Alan Robinson. A machine oriented logic based on the resolution principle. *Journal of the ACM*, 12(1):23–41, 1965.
[186] Michaël Rusinowitch. Theorem-proving with resolution and superposition. *Journal of Symbolic Computation*, 11(1 & 2):21–50, 1991.
[187] Tobias Schubert, Matthew Lewis, and Bernd Becker. PaMiraXT: parallel SAT solving with threads and message passing. *Journal on Satisfiability, Boolean Modeling and Computation*, 6:203–222, 2009.
[188] Stephan Schulz. E – A brainiac theorem prover. *Journal of AI Communications*, 15(2–3):111–126, 2002.
[189] Stephan Schulz. Simple and efficient clause subsumption with feature vector indexing. In Maria Paola Bonacina and Mark E. Stickel, editors, *Automated Reasoning and Mathematics: Essays in Memory of William W. McCune*, volume 7788 of *Lecture Notes in Artificial Intelligence*, pages 45–67. Springer, 2013.
[190] Stephan Schulz. System description: E 1.8. In Ken McMillan, Aart Middeldorp, and Andrei Voronkov, editors, *Proceedings of the Nineteenth International Conference on Logic, Programming and Automated Reasoning (LPAR)*, volume 8312 of *Lecture Notes in Artificial Intelligence*, pages 735–743. Springer, 2013.

[191] Stephan Schulz and Martin Möhrmann. Performance of clause selection heuristics for saturation-based theorem proving. In Nicola Olivetti and Ashish Tiwari, editors, *Proceedings of the Eighth International Conference on Automated Reasoning (IJCAR)*, volume 9706 of *Lecture Notes in Artificial Intelligence*, pages 330–345. Springer, 2016.
[192] Johan Schumann. Parallel theorem provers – an overview. In Bertrand Fronhöfer and Graham Wrightson, editors, *Proceedings of the First International Workshop on Parallelization in Inference Systems (December 1990)*, volume 590 of *Lecture Notes in Artificial Intelligence*, pages 26–50. Springer, Berlin, Germany, 1992.
[193] Johann Schumann. Delta: a bottom-up pre-processor for top-down theorem provers. In Alan Bundy, editor, *Proceedings of the Twelfth International Conference on Automated Deduction (CADE)*, volume 814 of *Lecture Notes in Artificial Intelligence*, pages 774–777. Springer, 1994.
[194] Johann Schumann and Reinhold Letz. PARTHEO: a high-performance parallel theorem prover. In Mark E. Stickel, editor, *Proceedings of the Tenth International Conference on Automated Deduction (CADE)*, volume 449 of *Lecture Notes in Artificial Intelligence*, pages 28–39. Springer, 1990.
[195] Robert E. Shostak. Refutation graphs. *Artificial Intelligence*, 7:51–64, 1976.
[196] Kurt Siegl. Gröbner bases computation in STRAND: a case study for concurrent symbolic computation in logic programming languages (Master thesis). Technical Report 90-54.0, Research Institute for Symbolic Computation (RISC), Linz, Austria, November 1990.
[197] Carsten Sinz, Jörg Denzinger, Jürgen Avenhaus, and Wolfgang Küchlin. Combining parallel and distributed search in automated equational deduction. In *Proceedings of the Fourth International Conference on Parallel Processing and Applied Mathematics (PPAM) – Revised Papers*, pages 819–832, 2001.
[198] James R. Slagle. Automatic theorem proving with renamable and semantic resolution. *Journal of the ACM*, 14(4):687–697, 1967.
[199] James R. Slagle. Automated theorem proving for theories with simplifiers, commutativity, and associativity. *Journal of the ACM*, 21:622–642, 1974.
[200] John Slaney, Ewing Lusk, and William W. McCune. SCOTT: Semantically constrained Otter. In Alan Bundy, editor, *Proceedings of the Twelfth International Conference on Automated Deduction (CADE)*, volume 814 of *Lecture Notes in Artificial Intelligence*, pages 764–768. Springer, 1994.
[201] John Slaney and Bruno Woltzenlogel Paleo. Conflict resolution: a first-order resolution calculus with decision literals and conflict-driven clause learning. *Journal of Automated Reasoning*, in press:1–27, 2017.
[202] Mark E. Stickel. A Prolog technology theorem prover. *New Generation Computing*, 2(4):371–383, 1984.
[203] Mark E. Stickel. A Prolog technology theorem prover: implementation by an extended Prolog compiler. *Journal of Automated Reasoning*, 4:353–380, 1988.

[204] Mark E. Stickel. PTTP and linked inference. In Robert S. Boyer, editor, *Automated Reasoning: Essays in Honor of Woody Bledsoe*, pages 283–296. Kluwer Academic Publishers, Amsterdam, The Netherlands, 1991.

[205] Mark E. Stickel. A Prolog technology theorem prover: new exposition and implementation in Prolog. *Theoretical Computer Science*, 104:109–128, 1992.

[206] Mark E. Stickel and W. Mabry Tyson. An analysis of consecutively bounded depth-first search with applications in automated deduction. In *Proceedings of the Ninth International Joint Conference on Artificial Intelligence (IJCAI)*, pages 1073–1075. Morgan Kaufmann Publishers, 1985.

[207] David Sturgill and Alberto Maria Segre. Nagging: a distributed, adversarial search-pruning technique applied to first-order inference. *Journal of Automated Reasoning*, 19(3):347–376, 1997.

[208] Geoff Sutcliffe. A heterogeneous parallel deduction system. In Ryuzo Hasegawa and Mark E. Stickel, editors, *Proceedings of the FGCS Workshop on Automated Deduction: Logic Programming and Parallel Computing Approaches*, pages 5–13, 1992.

[209] Christian B. Suttner. SPTHEO: a parallel theorem prover. *Journal of Automated Reasoning*, 18(2):253–258, 1997.

[210] Christian B. Suttner and Johann Schumann. Parallel automated theorem proving. In Laveen N. Kanal, Vipin Kumar, Hiroaki Kitano, and Christian B. Suttner, editors, *Parallel Processing for Artificial Intelligence*. Elsevier, Amsterdam, The Netherlands, 1994.

[211] Tanel Tammet. Gandalf. *Journal of Automated Reasoning*, 18(2):199–204, 1997.

[212] Stephen Taylor. *Parallel Logic Programming Techniques*. Prentice Hall, Upper Saddle River, New Jersey, 1989.

[213] Josef Urban and Jirí Vyskocil. Theorem proving in large formal mathematics as an emerging AI field. In Maria Paola Bonacina and Mark E. Stickel, editors, *Automated Reasoning and Mathematics: Essays in Memory of William W. McCune*, volume 7788 of *Lecture Notes in Artificial Intelligence*, pages 240–257. Springer, 2013.

[214] Jean-Philippe Vidal. The computation of Gröbner bases on a shared memory multiprocessor. In Alfonso Miola, editor, *Proceedings of the First International Symposium on Design and Implementation of Symbolic Computation Systems (DISCO)*, volume 429 of *Lecture Notes in Computer Science*, pages 81–90. Springer, 1990.

[215] Kevin Wallace and Graham Wrightson. Regressive merging in model elimination tableau-based theorem provers. *Journal of the IGPL*, 3(6):921–937, 1995.

[216] David H. D. Warren. An abstract Prolog instruction set. Technical Report 309, Artificial Intelligence Center, SRI International, Menlo Park, California, October 1983.

[217] David S. Warren. Memoing for logic programs. *Communications of the ACM*, 35(3):94–111, 1992.

[218] Christoph Weidenbach, Dylana Dimova, Arnaud Fietzke, Rohit Kumar, Martin Suda, and Patrick Wischnewski. SPASS version 3.5. In Renate Schmidt, editor, *Proceedings of the Twenty-Second International Conference on Automated Deduction (CADE)*, volume 5663 of *Lecture Notes in Artificial Intelligence*, pages 140–145. Springer, 2009.

[219] Andreas Wolf. P-SETHEO: strategy parallelism in automated theorem proving. In Harrie de Swart, editor, *Proceedings of the Seventh International Conference on Automated Reasoning with Analytic Tableaux and Related Methods (TABLEAUX)*, volume 1397 of *Lecture Notes in Artificial Intelligence*, pages 320–324. Springer, 1998.

[220] Larry Wos. Searching for open questions. *Newsletter of the Association for Automated Reasoning*, 15, May 1990.

[221] Larry Wos, Daniel F. Carson, and George A. Robinson. Efficiency and completeness of the set of support strategy in theorem proving. *Journal of the ACM*, 12:536–541, 1965.

[222] Larry Wos, George A. Robinson, Daniel F. Carson, and Leon Shalla. The concept of demodulation in theorem proving. *Journal of the ACM*, 14(4):698–709, 1967.

[223] Chih-Hung Wu and Shie-Jue Lee. Parallelization of a hyper-linking based theorem prover. *Journal of Automated Reasoning*, 26(1):67–106, 2001.

[224] Katherine A. Yelick. *Using abstraction in explicitly parallel programs*. PhD thesis, Laboratory for Computer Science, Massachusetts Institute of Technology, July 1991.

[225] Katherine A. Yelick and Steven J. Garland. A parallel completion procedure for term rewriting systems. In Deepak Kapur, editor, *Proceedings of the Eleventh International Conference on Automated Deduction (CADE)*, volume 607 of *Lecture Notes in Artificial Intelligence*, pages 109–123. Springer, 1992.

[226] Hantao Zhang. SATO: an efficient propositional prover. In William W. McCune, editor, *Proceedings of the Fourteenth International Conference on Automated Deduction (CADE)*, volume 1249 of *Lecture Notes in Artificial Intelligence*, pages 272–275. Springer, 1997.

[227] Hantao Zhang and Maria Paola Bonacina. Cumulating search in a distributed computing environment: a case study in parallel satisfiability. In Hoon Hong, editor, *Proceedings of the First International Symposium on Parallel Symbolic Computation (PASCO)*, volume 5 of *Lecture Notes Series in Computing*, pages 422–431. World Scientific, 1994.

[228] Hantao Zhang, Maria Paola Bonacina, and Jieh Hsiang. PSATO: a distributed propositional prover and its application to quasigroup problems. *Journal of Symbolic Computation*, 21(4–6):543–560, 1996.

[229] Hantao Zhang and Mark E. Stickel. Implementing the Davis-Putnam method. *Journal of Automated Reasoning*, 24(1–2):277–296, 2000.

[230] Lintao Zhang and Sharad Malik. The quest for efficient Boolean satisfiability solvers. In Andrei Voronkov, editor, *Proceedings of the Eighteenth International Conference on Automated Deduction (CADE)*, volume 2392 of *Lecture Notes in Artificial Intelligence*, pages 295–313. Springer, 2002.

Chapter 7
Parallel Answer Set Programming

Agostino Dovier, Andrea Formisano, and Enrico Pontelli

Abstract *Answer Set Programming (ASP)* has become, in recent years, the paradigm of choice for the logic programming community and for a wide variety of application domains. Thanks to its declarative nature, ASP offers excellent opportunities for performance improvements through transparent exploitation of parallelism. This Chapter provides a survey on the main techniques and approaches in the literature to enable exploitation of parallelism in the execution of Answer Set Programming solvers. The survey explores the approaches along two orthogonal dimensions. The first dimension considers the different levels of complexity and features of the underlying language, ranging from propositional Datalog/definite Horn clauses to full ASP. The second dimension, instead, explores the different levels of granularity of exploitation of parallelism, ranging from fine grain parallelism, exploited using general-purpose graphical processing units, to very large grain parallelism exploited on distributed platforms.

7.1 Introduction

The paradigm of *logic programming* can be traced back to the late 1960s and early 1970s, and the seminal work of researchers such as McCarthy, Robinson, Hayes, and Kowalski—especially in their efforts to argue for the declarative front (opposed

Agostino Dovier
University of Udine, Dept. of Mathematics, Computer Science, and Physics
e-mail: agostino.dovier@uniud.it

Andrea Formisano
University of Perugia, Dept. of Mathematics and Computer Science
e-mail: andrea.formisano@unipg.it

Enrico Pontelli
New Mexico State University, Dept. of Computer Science
e-mail: epontell@cs.nmsu.edu

Y. Hamadi und L. Sais (eds.), *Handbook of Parallel Constraint Reasoning*,
https://doi.org/10.1007/978-3-319-63516-3_7

to the procedural front promoted by researchers like Papert and Minsky) in the debate on knowledge representation that permeated the development of Artificial Intelligence in the 1960s. In the early 1970s, Hayes and Kowalski [37, 43] proposed a unifying solution to this debate, through the exploration of SLD-Resolution, which laid the foundation for the ability to use logic as a programming language, thus providing a knowledge representation framework that provides both a declarative and a procedural reading [44].

The first outcome of this line of work was the *Prolog* programming language, which for decades served as the cornerstone of the logic programming community. Prolog is a Turing complete programming language, with a declarative model-theoretic semantics as well as a top-down procedural semantics. The success of Prolog was driven by its declarative nature paired with sophisticated compiled implementation models—e.g., based on the abstract machine originally designed by D.H.D. Warren [85]. Some of the initial promises of the logic programming paradigm were not completely satisfied by the language Prolog. Some of the issues that plagued Prolog include its inability to provide competitive performance in application domains for which logic programming offered ideal modeling solutions—in particular, database applications and combinatorial problems, the gap between the declarative semantics and the procedural behavior (due to compromises in the implementation), and the limitations of the language in capturing non-monotonic forms of knowledge (e.g., as found in commonsense reasoning).

These issues prompted the development of alternative logic programming languages, more suitable to address some of these needs. The database community promoted the creation of the language Datalog [11], which provides a clean declarative semantics and is supported by highly efficient bottom-up execution models, but at the price of severe syntactic restrictions limiting the expressive power of the language. The hybridization of logic programming with constraint programming [45] allowed the speeding up of the logic programming approaches to constraint satisfaction and optimization problems but did not address the knowledge representation issues.

In [31] Gelfond and Lifschitz proposed a novel semantics, the *stable model semantics*, for handling negation in logic programming, offering an alternative way of looking at non-monotonic reasoning through the lenses of logic programming [4]. The stable model semantics is elegant and, compared to the other semantics proposed for logic programming with negation (e.g., the well-founded semantics), is more suitable to capture uncertainty and defeasible reasoning. Under simple syntactic restrictions, it was proved that establishing the existence of a stable model is NP-complete [61]. This suggested a new, semantic-based, programming paradigm with exactly the expressive power of the NP class [64, 60]. In these proposals, the notion of *answer set* became a synonym of stable model, and programming in those settings was termed *Answer Set Programming (ASP).* ASP represents, to a large extent, the holy grail of logic programming, providing an elegant and purely declarative programming framework, matched with efficient solvers. ASP has gained strong popularity in recent years, supported by a research emphasis on the practical applications of the paradigm and the development of efficient and highly competitive solvers.

ASP solvers are typically composed of two main modules (see Figure 7.1): (1) A preprocessing module that transforms a program into its ground, equivalent version, and (2) A solving method that alternates non-deterministic choices and deterministic inferences to construct the answer sets of the program. One of the first solvers proposed that launched the success of the ASP solver is SMODELS, combined with its grounder LPARSE [77]. A number of solvers have appeared over the years (e.g., the DLV system, which was the first system to support ASP programs with disjunctive heads [49]). The current state of the art is represented by the ASP solver CLASP (with its grounder GRINGO) which implements learning capabilities typical of SAT solvers [27].

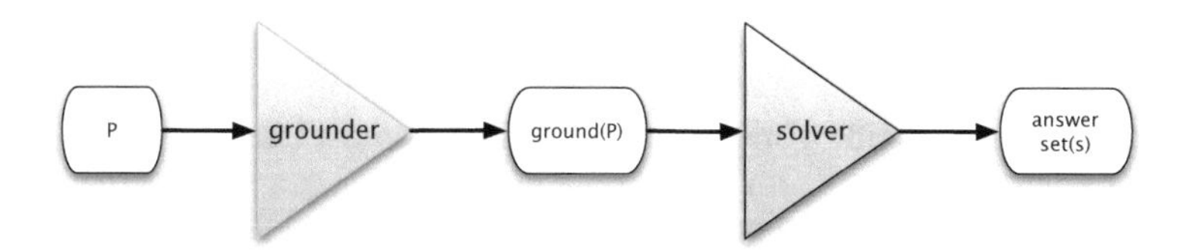

Fig. 7.1: Classical ASP solving pipeline

Declarative languages offer unprecedented opportunities for the use of parallelism to speed up execution. A declarative language, being not procedural, removes the need to perform operations in a strict order and reduces the number of dependencies among operations, thus opening the doors for concurrent execution. The potential for transparent exploitation of parallelism in logic programming emerged almost immediately with the birth of the paradigm [69]. The literature on parallel execution of Prolog is rich and has extensively explored many of the issues related to exploitation of parallelism in a top-down goal-oriented computation model (e.g., [36]). On the other hand, the literature on exploitation of parallelism in bottom-up execution models of logic programming, as found in Datalog and in ASP, has been more sparse. Yet, the fully declarative nature of ASP and Datalog offers even greater opportunities for parallelism than Prolog (which has a more "procedural" operational semantics).

This Chapter provides, to the best of our knowledge, the first comprehensive survey concerning the transparent exploitation of parallelism in ASP (and restricted versions of the paradigm, such as Datalog). In particular, we first discuss the main techniques used to parallelize the grounding phase (Section 7.3). We then explore how parallelism can be applied to parallelize the search process that underlies the construction of answer sets. We first consider some classical approaches for parallelizing Datalog (Section 7.4) and then focus on approaches that have been used to parallelize the search for an answer set of a general ASP program (Section 7.5). Some results on using GPUs for parallelizing ASP search are also reported, as well as the possibility of using the Map-Reduce framework for parallelizing Datalog and some of the components of the ASP computation—thus, opening the doors for parallelization of very large ASP programs (e.g., programs operating on extensive knowledge bases).

This Chapter has some connections with Chapter 1, Parallel Satisfiability in this book (SAT solving and propositional ASP solving have some common parts such as clause learning and a DPLL core), with Chapter 9, Parallel Constraint Programming (ASP can be used for solving constraint satisfaction problems), and with Chapter 15, Selection and Configuration of Parallel Portfolios (Section 7.5.4 deals with portfolio techniques for ASP).

7.2 Background

In this section, we provide a brief overview of the theoretical foundations of answer set programming. We start with the description of the syntax and semantics of both Datalog (definite logic programs) and full ASP (normal logic programs). We also provide a brief overview of the main algorithms that underlie the design of most common solvers for these logic programming languages.

7.2.1 Definite Logic Programming

The content of this section represents a brief review of the syntax and semantics of traditional logic programming—the interested reader is referred to [54, 30].

Let us consider the signature $\Sigma = \langle \mathscr{F}, \mathscr{X}, \mathscr{P} \rangle$ for a logic language, where

- $\mathscr{F}$ is a denumerable collection of function symbols; each symbol $f \in \mathscr{F}$ is associated with an arity $ar(f) \geq 0$.
- $\mathscr{X}$ is a denumerable collection of variables.
- $\mathscr{P}$ is a denumerable collection of predicate symbols; each symbol $p \in \mathscr{P}$ is associated with an arity $ar(p) \geq 0$.

The concepts of term and atomic formula (atom) are defined following the traditional structure as in traditional first order logic.

A term t in Σ is a syntactic structure defined recursively as follows:

- a variable $x \in \mathscr{X}$ is a term.
- If $f \in \mathscr{F}$, $ar(f) = n$, and $t_1, \ldots, t_n$ are terms, then $f(t_1, \ldots, t_n)$ is a term.

We denote by $F(\Sigma)$ the set of all terms of Σ; we denote by $F(\mathscr{F})$ the set of all terms that do not contain any element from $\mathscr{X}$. An atomic formula (atom) A in Σ is a syntactic structure defined as follows: if $t_1, \ldots, t_n$ are terms, $p \in \mathscr{P}$, and $ar(p) = n$, then $p(t_1, \ldots, t_n)$ is an atomic formula. The set of all atoms is denoted by $\Pi(\Sigma)$; the set $\Pi(\mathscr{F})$ denotes all atoms that do not contain elements from $\mathscr{X}$. Given an atom A, we denote by $pred(A) \in \mathscr{P}$ the predicate used to construct the atom A.

Definite logic programs assemble atoms to compose restricted forms of implications—satisfying the structure of *Horn clauses*. A *clause* (or *rule*) is of the form

$$head \leftarrow \underbrace{b_1,\ldots,b_m}_{Body} \tag{7.1}$$

where $head, b_1,\ldots,b_m$ are atoms and $m \geq 0$. If $m = 0$, then the clause is referred to as a *fact*—and syntactically we omit everything to the right of the *head*.

A *definite logic program* P is a collection of clauses. A term (atom, clause, program) is said to be *ground* if it does not contain any variable from $\mathscr{X}$. We will often refer to the *ground instances* of a term (atom, clause, program) α as the set of all entities α' obtained by consistently replacing each variable with an element from $F(\mathscr{F})$. Given any syntactic entity α (e.g., a term, an atom, a rule, a program), $\mathsf{ground}(\alpha)$ denotes the set of all ground instances of α.

The semantics of definite logic programs is given through Herbrand models. An interpretation I is a subset of $\Pi(\mathscr{F})$. A ground atom A is true in I if $A \in I$ (denoted by $I \models A$), false otherwise (denoted by $I \not\models A$). An interpretation I satisfies a ground rule $head \leftarrow b_1,\ldots,b_m$ if either $I \models head$ or there exists $i \in \{1,\ldots,m\}$ such that $I \not\models b_i$. An interpretation I is a model of a definite logic program P if I satisfies each rule in $\mathsf{ground}(P)$. We will often be interested in comparing interpretations based on how many atoms are made true—this can be realized by simply comparing the interpretations using the $\subseteq$ relation. Note that each definite logic program P has a unique minimal model M_P which is also the set of logical consequences of the "logical theory" P [54].

The minimal model of a definite logic program P can be computed in a bottom-up fashion. Let us define the *immediate consequence operator* of a program P as

$$T_P(I) = \{head \mid (head \leftarrow b_1,\ldots,b_m) \in \mathsf{ground}(P), \{b_1,\ldots,b_m\} \subseteq I\} \tag{7.2}$$

The computation of the minimal model is a polynomial process (w.r.t. the size of $\mathsf{ground}(P)$) and can be summarized as in Algorithm 7.1.

Algorithm 7.1: Naive Computation of the Least Model

```
1 procedure LEASTMODEL(P)
2 I ← ∅
3 repeat
4 |   I' ← I
5 |   I ← T_P(I')
6 until I = I'
```

7.2.2 Normal Logic Programs and Answer Set Programming

While definite logic programs allow us to express monotonic behavior, the language of logic programs needs to be extended to allow the expression of non-monotonic reasoning (e.g., the ability to withdraw consequences as new facts are added to the program). A *normal clause* has the form

$$head \leftarrow b_1,\ldots,b_m, not\ c_1,\ldots,not\ c_n \tag{7.3}$$

where $head, b_1,\ldots,b_m,c_1,\ldots,c_n$ are atoms. Given a normal clause r, we will write $head(r) = head$, $pos(r) = \{b_1,\ldots,b_m\}$, and $neg(r) = \{c_1,\ldots,c_n\}$. We also denote by $\varphi_{pos}(r)$ the conjunction $b_1 \wedge \cdots \wedge b_m$ and by $\varphi_{neg}(r)$ the conjunction $\neg c_1 \wedge \cdots \wedge \neg c_n$. A *normal logic program* is a collection of normal clauses.

Due to the presence of negated literals in clause bodies, the logical semantics is inadequate to deal with this formalism. There is no longer the notion of a unique minimal model that characterizes the logical consequences. If we consider, for instance, the program $P = \{p \leftarrow not\ q\}$,[1] it admits two (minimal) models: $\{p\}$ and $\{q\}$. Let us observe that (1) the intersection of the two models is $\emptyset$ which is not a logical model of P, and (2) there are no reasons in the program for believing in q and so the latter model looks "weaker" than the former.

The extension of the T_P operator (eq. (7.2)) to normal logic programs

$$T_P(I) = \left\{ head \;\middle|\; \begin{array}{l} (head \leftarrow b_1,\ldots,b_m, not\ c_1,\ldots,not\ c_n) \in \mathsf{ground}(P) \wedge \\ \{b_1,\ldots,b_m\} \subseteq I \wedge \{c_1,\ldots,c_n\} \cap I = \emptyset \end{array} \right\} \tag{7.4}$$

is not monotone. In the above example $T_P(\emptyset) = \{p\}$ while $T_P(\{q\}) = \emptyset$.

The semantics of a normal logic program is given in terms of *answer sets* [31]. Given a normal program P and an interpretation I, we define the *reduct* of P w.r.t. I (denoted by P^I) as the set of rules

$$P^I = \left\{ head \leftarrow b_1,\ldots,b_m \;\middle|\; \begin{array}{l} (head \leftarrow b_1,\ldots,b_m, not\ c_1,\ldots,not\ c_n) \in \mathsf{ground}(P) \wedge \\ \{c_1,\ldots,c_n\} \cap I = \emptyset \end{array} \right\} \tag{7.5}$$

Note that the reduct is a definite logic program. An interpretation I is an *answer set* if I is the least model of P^I. Note that a program may have zero, one, or multiple answer sets; deciding whether a ground normal logic program P admits a stable model is NP-complete [14]. A normal logic program under the answer set semantics is referred to as an *answer set program*.

A number of algorithms for computing answer sets of a program have been proposed. Proposed systems range from implementations that rely on translation of ASP into other paradigms (e.g., [51, 50]), to ad hoc solvers, e.g., relying on variations of the traditional *Davis-Putnam-Logemann-Loveland (DPLL)* procedure [15, 16] or on nogood propagation (see Section 7.2.4.1 and [27]).

---

[1] Let us observe that $p \leftarrow not\ q$ is logically equivalent to $p \vee q$.

Algorithm 7.2: Basic SMODELS Procedure

```
1  procedure compute(P: Program; S: Interpretation)
2  S′ ← expand(P,S);
3  if ¬consistent(S′) then
4  │ return False;
5  if complete(S′) then
6  │ return S′;
7  ℓ ← select_atom(P,S′);
8  S′ ← compute(P,S′ ∪ {ℓ});
9  if S′ ≠ False then
10 │ return compute(P,S′ ∪ {¬ℓ});
11 else
12 │ return S′;
```

Let us consider, for example, a simplified version of the procedure that underlies the SMODELS system. Algorithm 7.2 describes the overall structure of the computation. The recursive function progressively builds an answer set by adding one new literal at each recursive call.[2] The procedure `expand`(P,S) (line 2) deterministically expands the interpretation by adding literals whose truth value is uniquely determined by the program P and the partial interpretation S (for instance, if q and r are in S and the clause $p \leftarrow q,r$ is in P, then p is added to S'). The steps performed in `expand` guarantee that the interpretation converges towards an answer set (and not only to a generic logical model). If an inconsistency is reached (line 4—namely if both ℓ and $\neg\ell$ are in S for some atom ℓ), then an incorrect choice has been performed and backtracking is started. If each atom is represented in S' (line 6), then the procedure has determined an answer set. Otherwise, the procedure `select_atom` chooses an atom ℓ that does not appear (positively or negatively) in S'. The algorithm explores the two alternatives of adding either ℓ (line 8) or $\neg\ell$ (line 10) to S' and continue the construction of the answer set. The structure of this procedure, as well as similar procedures that are at the heart of several ASP solvers, is derived from the traditional Davis-Putnam-Logemann-Loveland procedure [16, 15].

Constraints of the form $\perp \leftarrow b_1,\ldots,b_m, not\ c_1,\ldots, not\ c_n$ are common in ASP programming. In this case $\perp$ is superfluous, since any constraint can be replaced by a normal clause (e.g., $q \leftarrow not\ q, b_1,\ldots,b_m, not\ c_1,\ldots, not\ c_n$) that has the same stable models, provided a new predicate (q in the example) is used for each rewriting.

During the grounding stage, each non-ground clause is replaced by a set of clauses in which every variable is replaced by the constant symbols in the program. In order to simplify this process, every variable in the clause should be limited by some atom with a "simple" definition.

Given a program P, let us define the (complete) dependency graph $\mathscr{G}^{+,-}(P) = \langle V,E\rangle$ as follows [5, 7]: V is the set of predicate symbols defined in P and a labeled edge $\langle p,q,\ell\rangle$ belongs to E if and only if there is a clause $p(\cdots) \leftarrow \cdots q(\cdots)\cdots$ in P.

[2] For the sake of this procedure, we consider interpretations (stored in the variable S) that contain both positive and negative literals.

The edge is labeled $+$ (resp., $-$) if q occurs positively (resp., negatively). Notice that an edge may be labeled by both $+$ and $-$.

A predicate p is a *domain* predicate if every path in $\mathscr{G}^{+,-}(P)$ starting from p does not contain cycles involving edges labeled by $-$. A clause is *strongly range restricted* if every variable in it occurs also as an argument of an atom built with a domain predicate occurring positively in its body. A program is *strongly range restricted* if every clause in it is strongly range restricted.

7.2.3 Datalog

Datalog [11] is a fragment of logic programming widely used in deductive databases. The syntax of its basic version is the same as that of definite logic programs, with the additional restriction that only function symbols of arity 0 (i.e., constant symbols) can be used and the following additional restriction (*safety*) is required: for each rule $head \leftarrow body$ each variable in *head* should appear in at least one atom in *body*.

The semantics of Datalog is simply represented by the minimal model semantics described earlier—the semantics of a Datalog program P is simply given by the minimal model M_P. The computation of the semantics of a Datalog program P is typically realized using a form of *bottom-up* computation. The procedure in Algorithm 7.1 can be implemented using standard relational algebra operators (e.g., selection, projection, join); absence of non-constant function symbols and the safety condition ensure finiteness of the process.

Negation is allowed in extended versions of Datalog. In this case, typically, a notion of stratification (e.g., [75]) is required, which guarantees the existence of a unique minimal model (computed by iterating Algorithm 7.1 for the different strata of the program).

7.2.4 Alternative ASP Computation Models

7.2.4.1 Program Completion

An alternative viable solving technique for ASP relies on the theoretical connections that exist between the stable models of a program P and the minimal models of the *completion* of P [12]. Without loss of generality, let us consider the case of ground programs only.

The *positive dependency graph* $\mathscr{G}(P) = \langle V, E \rangle$ is the graph where V is the set of ground atoms occurring in P, while $(A, B) \in E$ if and only if there is a ground rule $r \in P$ such that $A = head(r)$ and $B \in pos(r)$ (basically it is the restriction of $\mathscr{G}^{+,-}(P)$ to edges labeled $+$ for a propositional program). A loop in $\mathscr{G}(P)$ is any set of atoms $L \subseteq V$ inducing a non-trivial strongly connected component in $\mathscr{G}(P)$. A program P is said to be *tight* (resp., *non-tight*) if there are no (resp., there are) loops in $\mathscr{G}(P)$.

The completion P_{cc} of a program P is defined as the following formula:

$$P_{cc} = \bigwedge_{a \in atom(P)} \left(a \leftrightarrow \bigvee_{r \in P, head(r)=a} (\varphi_{pos}(r) \wedge \varphi_{neg}(r)) \right)$$

It is well known [29, 21] that the stable models of a tight program P coincide with the minimal models of P_{cc}. To obtain an analogous result holding for non-tight programs, one has to consider an additional class of *loop formulae*. Given a loop L in $\mathscr{G}(P)$, the corresponding loop formula is defined as:

$$\varphi_L = \bigvee_{a \in L} a \rightarrow \bigwedge_{r \in ER(L)} (\varphi_{pos}(r) \wedge \varphi_{neg}(r))$$

where $ER(L) = \{r \mid r \in P \wedge head(r) \in L \wedge pos(r) \cap L = \emptyset\}$. Intuitively, the loop formulae ensure that any atom that is *part of a loop and true* has to be made true by a rule that is not involved in the loop

It can be shown that, if P_Λ is the set of all loop formulae of a program P, then the stable models of P are exactly the minimal models of $P_{cc} \wedge P_\Lambda$. Since a SAT solver can be used to determine the models of the formula $P_{cc} \wedge P_\Lambda$, this enables the design of ASP solvers based on state-of-the-art SAT solvers. This option has been pursued, for instance, in solvers such as ASSAT and CMODELS [51, 32].

7.2.4.2 Conflict-Driven Search

Let us consider the basic techniques that are employed in the implementation of the CLASP solver, which uses a *conflict-driven* search strategy for stable model building (see [27] for a detailed treatment). The basic idea consists of translating the completion of a program P into a collection of *nogoods*, whose solutions (see below) correspond to stable models of P. The search for the solutions proceeds by executing a DPLL-like procedure.

More specifically, the technique describes both assignments σ and nogoods as sets of *signed atoms*—i.e., entities of the form Tp or Fp, denoting that p has been assigned `true` or `false`, respectively. Given an assignment σ, let $\sigma^T = \{p : Tp \in \sigma\}$ and $\sigma^F = \{p : Fp \in \sigma\}$. An assignment σ requires that, for each atom p, $\{Tp, Fp\} \not\subseteq \sigma$. A *total* assignment σ is such that, for every atom p, $\{Tp, Fp\} \cap \sigma \neq \emptyset$. Given a (possibly partial) assignment σ and a nogood δ, we say that δ is *violated* if $\delta \subseteq \sigma$. An assignment σ is a *solution* for a set of nogoods Δ if no $\delta \in \Delta$ is violated by σ.

Given a program P, we distinguish between two types of nogoods: the *completion nogoods*, which are derived from the completion of P, and the *loop nogoods*, which are derived from the loop formulae of P. If σ is an assignment for a program P, then σ^T is a stable model of P if and only if σ is a solution of the set of all completion and loop nogoods.

The CLASP system [27] explores a search space composed of all interpretations for the atoms in P, organized as a binary tree. The successful construction of a branch in the tree corresponds to the identification of an answer set of the program. If a (possibly partial) assignment violates a nogood, then backjumping procedures are used to backtrack to the node in the tree that caused the failure. The tree construction and the backjumping procedures in CLASP are implemented in such a way as to guarantee that, if a branch is successfully constructed, then the outcome will be an answer set of the program. CLASP's search is guided by nogoods. During deterministic propagation phases (*unit propagation*) nogoods are used to determine additional needed assignments. For example, given a nogood δ and a partial assignment σ such that $\delta \setminus \sigma = \{Fp\}$ (resp., $\delta \setminus \sigma = \{Tp\}$), then we can infer the need to add Tp (resp., Fp) to σ in order to avoid violation of δ.

CLASP exploits statically generated completion nogoods, while it dynamically introduces loop nogoods when they are needed to rule out unsupported models.

7.2.4.3 ASP Computation

We report briefly here a *computation-based* characterization of answer sets. It is based on an incremental construction process, where the choices are performed at the level of which rules are actually applied to extend the partial answer set.

An *ASP Computation* [53] of a program P is a sequence of interpretations $I_0 = \emptyset, I_1, I_2, \ldots$ satisfying the following conditions:

- $I_i \subseteq I_{i+1}$ for all $i \geq 0$ (*Persistence of Beliefs*)
- $I_\infty = \bigcup_{i=0}^{\infty} I_i$ is such that $T_P(I_\infty) = I_\infty$ (*Convergence*)[3]
- $I_{i+1} \subseteq T_P(I_i)$ for all $i \geq 0$ (*Revision*)
- if $a \in I_{i+1} \setminus I_i$ then there is a rule $a \leftarrow body$ in P such that I_j is a model of *body* for each $j \geq i$ (*Persistence of Reason*).

I_0 can be the empty set or, more generally, a set of atoms that are logical consequences of P. We say that a computation $I_0, I_1, \ldots$ converges to I if $I = \bigcup_{i=0}^{\infty} I_i$. The results in [53] prove that, given a ground program P, an interpretation I is an answer set of P if and only if there exists an ASP computation that converges to I. I is the set of atoms that are "true" in the answer set. This technique will be further discussed in Section 7.5.2.2 while describing a parallel GPU-based conflict-driven ASP-solver.

[3] T_P is defined in Equation (7.4).

7.3 Parallelizing the Grounding Phase

7.3.1 Introduction

As anticipated in the introduction, the ASP-solving process is usually composed of two phases, performed by two different modules of the ASP solver (see Figure 7.1). During the first phase a *grounder* is in charge of replacing each non-ground rule with the complete set of its ground instances. The grounding $ground(P)$ of a program P is such that the stable models of $ground(P)$ are exaclty the stable models of P. In general, this is achieved by uniformly replacing all the variables occurring in each rule with elements of the Herbrand universe of P (typically, the constants occurring in P). Then, $\mathsf{ground}(P)$ is processed by the solver module, which computes the solutions of such a ground program. Note that $\mathsf{ground}(P)$ might be of exponential size with respect to the size of the given program P [14].

Almost all available solvers follow this process, with some differences in the way the two modules are integrated. For instance, in the cases of CLINGO [28] and DLV [49], the grounder and the solver are tightly coupled to form a single tool, while in other approaches two distinct tools are available to carry out the two tasks. This is the case of GRINGO+CLASP and LPARSE+SMODELS [27, 77]. Modern grounders employ several techniques and optimizations in order to reduce, as much as possible, the size of the output program, while preserving its equivalence to P, with the hope that smaller ground programs could potentially lead to greater efficiency during the solving step.

In order to generate small propositional programs, the instantiation proceeds in a bottom-up fashion, starting from the facts included in the program and following the dependencies encoded by the program rules. Modern grounders exploit structural information of the input program and combine smart techniques for query evaluation originally developed in the field of deductive databases and briefly discussed in Section 7.4.

Dependencies among atoms of a program P are described through the (positive) dependency graph $\mathcal{G}(P)$ (Section 7.2.4.1). P is partitioned into *modules* consisting of the strongly connected components (SCCs) of $\mathcal{G}(P)$; the dependencies among predicates encoded by the rules induce a partial order among the modules. The grounder proceeds by processing the modules of P following a topological order of the SCCs of $\mathcal{G}(P)$. In this way, when a module M has to be grounded, all needed data (i.e., the ground instances upon which the rules in M depend) are available. Moreover, only the ground atoms that can potentially be derived (during the solving step) by rules in M are considered. This reduces the size of the resulting ground program and limits the combinatorial explosion that might otherwise occur if the full collection of constants were blindly used.

Specific treatment is adopted for those modules containing recursive rules—recall that, while the SCCs of $\mathcal{G}(P)$ form a directed acyclic graph, each module is not necessarily a stratified subprogram of P. In these cases, a fix point technique is applied, locally to the recursive module. This technique is derived from well-known algorithms designed for Datalog (e.g., the naive algorithm in Algorithm 7.1, semi-naive algorithms [82]). In order to efficiently explore all possible alternative instantiations, backjumping techniques are used and optimal strategies, similar to those developed for database query optimization, are employed to decide the order in which body atoms have to be instantiated.

7.3.2 Naive Parallel Grounding

A first investigation of the parallelization of grounding has been conducted in [6]. In that study, the authors propose a distributed implementation of the grounder LPARSE as part of a full-blown distributed ASP solver. The approach exploits the property of strong range restrictedness of LPARSE programs (see Section 7.2.2), and the presence of domain predicates, in order to statically partition the program rules. In principle, each rule can be grounded by a different processor. This basic idea is intuitively captured by Algorithm 7.3. The target architecture is a Beowulf cluster. The system is organized as a master-slave structure, where the master agent[4] computes the program partition and delegates the grounding of each component to different slaves. In line 4 the task is split among the available processors (i is the generic "slave" processor identifier) and in line 6 the master collects the data obtained in the variable GP. Load balancing can be heuristically controlled by assigning weights to rules—essentially, by computing an estimate of the number of expected ground instances of each rule.

Algorithm 7.3: Parallel Grounding on Beowulf Cluster (from [6])

1 **Procedure** PARALLELLPARSE (P: Program)
2 $\mathsf{GP} \leftarrow \{a \mid a \text{ is a ground instance of a domain predicate}\}$;
3 $P \leftarrow P \setminus \mathsf{GP}$;
4 **for each** $r^i \in P$ **— in parallel do**
5 $\quad r^i_g \leftarrow \textsc{GroundRule}(r^i)$;
6 $\mathsf{GP} \leftarrow \mathsf{GP} \cup \bigcup_i r^i_g$

7.3.3 Multi-level Parallel Grounding

A systematic study has been undertaken concerning the parallelization of the DLV instantiator on multicore/multiprocessor systems, adopting an SMP (Symmetric Multi-Processing) architecture, where concurrent threads communicate through a shared memory. This research started with [10] and evolved over the years into a

[4] Throughout the Chapter, we use the terms *agent* and *processor* with the same meaning, i.e., to identify a unit capable of performing concurrent computation.

Algorithm 7.4: Component Level Parallelism

```
1 Procedure ComponentsInstantiator (P: Program, 𝒢(P): DependencyGraph, GP:
  GroundProgram)
2 S ← {a | a is a fact in P} /* S is a set of ground atoms */
3 GP ← ∅
4 𝒞 ← {C | C is an SCC of 𝒢(P)}
5 while 𝒞 ≠ ∅ do /* until all SCCs have been processed */
6     𝒞' ← {C | C ∈ 𝒞 is an SCC of 𝒢(P) without incoming edges};
7     for each C^i ∈ 𝒞' —in parallel do /* spawn a t_i for each C^i */
8         RulesInstantiator(P,C^i,S,𝒢(P),𝒞,GP) /* mod. 𝒞 and GP */
9     thread_join(t) /* wait for (some) thread termination */
```

manifold strategy described in [68]. The approach identifies three levels of parallelism in the grounding process and, for each of them, it proposes different techniques to take advantage of the underlying multi-threaded system.

The first level of parallelism, called the *components level* in [10], exploits a partitioning of the given program P into modules, according to the dependencies among the SCCs of $\mathcal{G}(P)$. The modules must be processed in topological order, but the grounding concerning independent modules can be performed in parallel. Algorithm 7.4 shows a possible high-level description of this phase. In particular, in line 8 a new thread, executing the procedure RulesInstantiator, is spawned as soon as a module is ready for grounding. This happens when all the modules it depends on have been grounded. The synchronization step in line 9 ensures that the procedure ComponentsInstantiator looks for groundable modules only when a thread completes its task.

In the second level of parallelism, called the *rules level*, the rules of each single module M are instantiated in parallel. Two kind of rules have to be considered. *Recursive rules* are those defining a predicate p also occurring positively in the body of some rule in M. The remaining rules are called *exit rules*. The latter are grounded first, by spawning a sufficient number of threads, each one concurrently executing the procedure SingleRuleInstantiator—see Algorithm 7.5, line 5. Observe that the parent thread (the one executing ComponentsInstantiator) waits for the termination of all the children threads at the barrier in lines 6–7. Hence each exit rule is processed, once, by a different thread. The calls to SingleRuleInstantiator concurrently update the set $\mathcal{NS}$ of newly generated ground atoms (i.e., the heads of the newly generated ground rules).

The treatment of recursive rules is slightly more complex. In particular, the *semi-naïve* algorithm of Datalog [82], is used to evaluate a fix point for the grounding of these rules. Starting from the situation resulting from the grounding of exit rules (the value of set $\mathcal{NS}$ in line 9), concurrent threads repeatedly instantiate all recursive rules until no new instance is obtained (lines 8–16). A synchronization barrier paces this iteration, ensuring that in each iteration each thread always operates on the portion of the atoms' extension added during the previous iteration. Note that, at the end of the procedure the SCC C is removed from the collection of SCCs still to be processed; this affects the condition in line 5 of Algorithm 7.4 and ensures its termination.

Algorithm 7.5: Rule Level Parallelism (adapted from [68])

```
1  procedure RULESINSTANTIATOR (P: Program, C: SCC, S: SetOfAtoms, 𝒢(P):
   DependencyGraph, 𝒞: SetOfSCCs, GP: GroundProgram)
2  ΔS ← ∅ /* ΔS is a set of ground atoms                              */
3  𝒩S ← ∅ /* 𝒩S is a set of ground atoms                              */
4  for each r^i ∈ Exit(C,P) —in parallel do /* spawn a t_i for each r^i  */
5  └ SINGLERULEINSTANTIATOR(r^i,S,ΔS,𝒩S,GP) /* modify 𝒩S,ΔS,GP     */
6  for each r^i ∈ Exit(C,P) do /* synchronization barrier            */
7  └ thread_join(t_i) /* wait for all threads                         */
8  repeat/* process recursive rules in C                              */
9  │  ΔS ← 𝒩S
10 │  𝒩S ← ∅
11 │  for each r^j ∈ Recursive(C,P) — in parallel do /* spawn t_j for r^j   */
12 │  └ SINGLERULEINSTANTIATOR(r^j,S,ΔS,𝒩S,GP) /* modify 𝒩S,GP       */
13 │  for each r^j ∈ Recursive(C,P) do /* synchronization barrier      */
14 │  └ thread_join(t_j) /* wait for all threads                      */
15 │  S ← S ∪ ΔS
16 until 𝒩S = ∅
17 𝒞 ← 𝒞 \ C /* the SCC C has been processed                        */
```

As far as rule-level parallelism is concerned, one can observe that the smallest task a thread can perform consists of the grounding of one rule. In other words, all instances of a single rule must be generated by the same thread. Clearly, this strategy does not take into account the different structure and complexity of the various rules—specifically, the number of obtainable instances and the hardness of their instantiation process. In order to optimize load balancing and improve granularity control (the regulation of the amount of work assigned to each thread), a third level of parallelism is introduced. The basic idea consists of splitting the grounding of a *hard* rule r among different concurrent threads. This is achieved by selecting one of the body atoms of r, partitioning its extension, and constraining the action of each thread to one portion of the partition. This involves the solution of a number of sub-problems.

1. First, one has to estimate the work needed to ground a rule (essentially, the cardinality of the outcome) by applying well-known query optimization techniques developed in relational database theory (see, for instance, [82]). More specifically the grounding of the conjunction of those body atoms that share variables can be seen as a natural join operation among relations (the extensions of the atoms). Consequently, the size of the outcome is estimated by considering the size of extensions of the body atoms and the *selectivity* of each variable. Essentially, for each join-variable, one considers the number of distinct values it might assume w.r.t. the extensions of the body atoms. When the estimated size of the grounding of a rule exceeds a threshold, the rule is classified as *hard*,

and the computation of its grounding is split between two or more threads. Note that the number of splits is a parameter that may be heuristically controlled. On the other hand, in order to obtain a better load balance, a set of easy rules might be assigned to the same thread (for simplicity, this option is not shown in Algorithms 7.5 and 7.6).

2. Second, once a hard rule is identified, one has to choose the body atom to be split (line 3 in Algorithm 7.6). This choice is also made by taking into account the estimates of the size of the extensions of the body atoms and by determining the best order for the computation of the join (see [48] for further details). At this point, the partition of the split atom is computed (line 4 in Algorithm 7.6), and a pool of threads is spawned accordingly (line 5).

The entire process is dynamic, in the sense that each time a rule is grounded (for recursive rules this might happen many times), the estimations are evaluated with respect to the currently known atoms' extensions. This differentiates the approach from the techniques developed, for instance, in parallel Datalog evaluation, where rule assignment to processors is usually determined statically, for example by applying hashing functions, or fixing other parameters such as the number of splits, the split atoms, etc. A further optimization concerns the lifetime of threads. Instead of spawning new threads whenever needed, [68] suggests the creation of a global pool of threads. Whenever a task has to be executed, it is assigned to one of the free threads in the pool. Similarly, when a thread completes its task, it is inserted back into the pool and will wait for the next available task. This limits the overhead due to the creation and termination of threads.

Algorithm 7.6: Single-Rule Level Parallelism (adapted from [68])

1 **procedure** SINGLERULEINSTANTIATOR (r: Rule, S: SetOfAtoms, ΔS: SetOfAtoms, $\mathscr{N}S$: SetOfAtoms, GP: GroundProgram)
2 $s \leftarrow numberOfSplits(r, S, \Delta S)$ `/* heuristically evaluate optimal` s `*/`
3 $L \leftarrow selectSplitLiteral(r, s)$ `/* heuris. select a literal to split */`
4 $Splits \leftarrow$ SPLITEXTENSION$(L, s, S, \Delta S)$ `/* split the extension for` L `*/`
5 **for each** $sp^i \in Splits$ — **in parallel do** `/* spawn a` t_i `for each split */`
6   INSTANTIATERULE$(r, L, sp^i, S, \Delta S, \mathscr{N}S, \text{GP})$
7 **for each** $sp^i \in Splits$ **do** `/* synchronization barrier */`
8     $thread\_join(t_i)$ `/* wait for all threads */`

We conclude this section by mentioning an interesting approach described in [57]. In this case, the authors propose a portfolio-like framework to perform the grounding phase. The focus is not on parallelism per se, but on the design of an automated strategy for selecting the best grounder for each given input program. The motivating idea is that different grounders, as well as different settings in their configuration options, may offer significantly different performance, both in terms of the time spent for grounding a program and in the size of the outcome. The proposed framework exploits machine-learning techniques to classify input programs w.r.t. a number of easy-to-evaluate heuristic features. The most promising grounding engine (and its controlling options setting) is selected and run. The system is aimed at selecting a single executor for the grounding phase. Nevertheless, the ideas in [57] might be developed to design a *portfolio parallel grounder*, where different grounders are used to process different portions of the same input program. (See Section 7.5.4 for details of portfolio approaches applied to the solving phase.)

7.4 Parallelizing the Inference Phase I: Parallel Datalog

Several works (e.g., [86, 87, 24, 25, 90]) explored the parallelization of Datalog, especially on distributed architectures. These approaches deal with Datalog in its simple version, without negation (although the results could be applied to stratified Datalog programs with negation), thus they parallelize the computation of M_P (see Algorithm 7.1), assuming that the processors do not have a shared memory. Communication between computing units is instead allowed through explicit message passing. Recent approaches using shared memory are also briefly discussed.

The principle underlying the approaches [86, 87, 24, 25] is the *partition of the ground rules* of the Datalog program among k processors $\mathsf{U}_1,\dots,\mathsf{U}_k$. This is also known as *rule distribution*. Let us consider a Datalog program P and let r be a clause of P. We identify in r a set of distinct variables $disc(r)$ that appear in the body of r—referred to as the *discriminating set*. Let us define also a hash function $h_r : F(\mathscr{F})^{|disc(r)|} \mapsto \{1,\dots,k\}$; this function partitions the possible assignments of values to the variables in $disc(r)$ among the k processors.

For each predicate $q \in \mathscr{P}$ and for each pair $i, j \in \{1,\dots,k\}$ we introduce the following new predicates:

- q^i_{in}, describing instances of q entering processor U_i,
- q^i_{out}, describing instances of q produced by processor U_i, and
- q^i_j, describing instances of q produced by processor U_i and to be sent to processor U_j.

Given an atom A based on predicate q, we will denote by A^i_{in} (resp. A^i_{out}, A^i_j) the atom obtained by replacing the predicate q with q^i_{in} (resp. q^i_{out}, q^i_j). The program P is transformed as follows:

- The rule r of the form $head \leftarrow b_1,\dots,b_m$ is replaced by the rule

  $$head^i_{out} \leftarrow b^i_{1,in},\dots,b^i_{m,in}, h_r(disc(r)) = i$$

  Let us observe that the last atom is an equality atom; equality is allowed in Datalog, and as soon as the variables are replaced by values, it is simply a literal true/false property. This rule allows us to generate in processor U_i the relevant instances of rule r;
- For every recursive atom A in the body of r and for every processor U_j, generate the communication rule

$$A^i_j \leftarrow A^i_{out}, h_r(disc(r)) = j$$

used to send instances of A generated by processor U_i to the relevant processor U_j;

- For each recursive predicate p in the program, let A be an atom with predicate p and fresh new variables as arguments; then generate the receiving rule $A^i_{in} \leftarrow A^i_j$.
- For each recursive atom A, the collection rule is defined as $A \leftarrow A^i_{out}$.

While we described a general case, in practice the partitioning process can be focused on specific rules, e.g., selected rules that are part of a loop in the dependency graph $\mathscr{G}(P)$. Moreover, some of these approaches work also in the absence of data transmission between processors (e.g., [86, 87]) if the program is *decomposable* (a property, however, which is in general undecidable).

The method has also been modified to operate on multicore shared-memory machines—by exploring hash functions to partition computation of relations that guarantee the avoidance of locks [89]. Significant speedups can be obtained by using as little synchronization as possible during the program evaluation. The shared memory available in GPUs has been exploited in [63], where the authors distribute the load of computing the model between the various GPU threads that can access and modify the data in the shared memory. Particular care is taken in parallelizing the natural join operation that underlies the naive or semi-naive bottom-up computation of Datalog. Monotonicity of pure Datalog is crucial for the correctness of the shared-memory approach.

In [90] the authors show that a simple rule partition schema might require a lot of unnecessary message passing and propose to combine a rule partition schema with a new *data partition* parallel schema.

7.5 Parallelizing the Inference Phase II: Parallel ASP

7.5.1 Parallelizing the Search Process

7.5.1.1 General Idea and Seminal Work

The majority of the solvers used for ASP are based on a search process—just as exemplified in the skeleton SMODELS algorithm (Algorithm 7.2). The non-deterministic process incrementally constructs an answer set by alternating deterministic expansions of the interpretation (e.g., the `expand` step shown in Section 7.2.2) with non-deterministic choices (e.g., the addition of a yet undetermined literal to the interpretation—see the `select_atom` operation used in the SMODELS structure in Section 7.2.2). The ASP computation can, thus, be visualized as the construction/-exploration of a binary search tree (see, e.g., Figure 7.2), where the internal nodes correspond to the non-deterministic choices performed (e.g., the addition of an atom, positively or negatively, to the interpretation being constructed), while the segment of a branch between two consecutive internal nodes correspond to the deterministic phase of the construction of the answer set (e.g., the `expand` operation). It is possible to parallelize the computation by distributing the construction of different branches of the search tree between different processors (Figure 7.2). Effectively, this leads to different processors concurrently constructing/exploring different parts of the search tree (Figure 7.3).

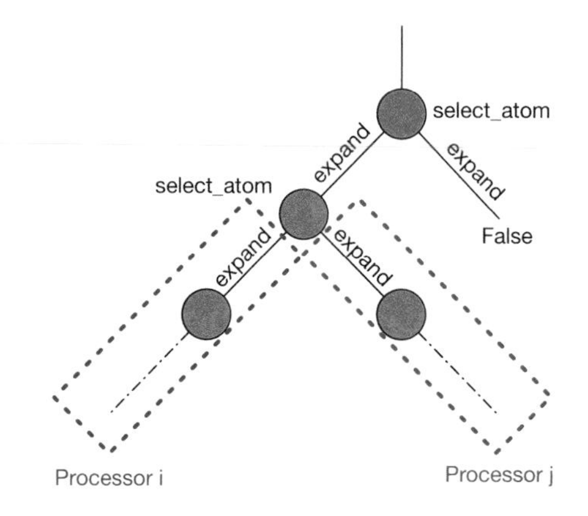

Fig. 7.2: Search tree and search parallelism

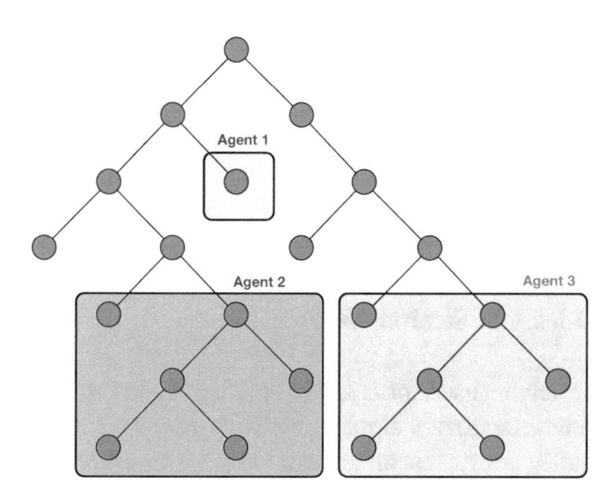

Fig. 7.3: Search parallelism

In principle, the different branches of the search tree correspond to different assignments of truth values to the atoms in the language—and they will thus potentially lead to distinct answer sets. This facilitates the exploitation of parallelism, as the parallel computations are independent and they can potentially be carried out, once started, without any additional communication. In practice, most state-of-the-art ASP solvers collect knowledge from each branch of the search tree being explored, to improve pruning of unnecessary branches in the rest of the search tree. A typical technique used is *clause learning* [8], which allows the learning of *nogoods* (sets of literals that can never concurrently hold true in any answer set—see Section 7.2.4.2). These techniques, if extended to the case of search parallelism, will require communication about concurrent computations—e.g., to exchange learned nogoods.

The idea of exploiting search parallelism was originally presented by El-Khatib and Pontelli [20, 70] and by Finkel et al. [22]. These two concurrent developments followed very similar directions, with the difference that the work in [20, 70] developed on a shared-memory platform, using Posix threads, while [22] relied on a distributed architecture, implemented using PVM [78].

The overall structure of a typical ASP computation for search parallelism is summarized in Algorithm 7.7. The parallel execution is conducted by a finite set of computing units (e.g., threads or processes)—all proposed models do not account for the dynamic creation/removal of computing units during execution to avoid additional overheads. The initial step (line 3) statically assigns a subtree of the search tree to the processor. Each processor performs a standard computation on the locally assigned subtree (line 5) until the entire subtree has been completely explored. The only novelty is represented by the need to occasionally respond to requests to share work with other processors (line 6)—i.e., allow other processors to explore parts of the local subtree. Upon completion of the exploration of the local subtree, the processor will attempt to communicate with other processors to get access to other unexplored subtrees (line 19).

While the overall idea of search parallelism is simple, its actual realization has to address some critical challenges; the two main challenges are: (1) how to move a processor to a different part of the search tree (referred to as *task sharing* and (2) how to locate which part of the subtree one processor should explore next (referred to as *scheduling*). These two issues are addressed in the following subsections.

7.5.1.2 Techniques for Task Sharing

The goal of the task-sharing phase is to "relocate" a processor to a different part of the search tree (Figure 7.4). This is typically necessary when a processor has exhausted a subtree and needs to access unexplored alternatives (which may be left behind by another processor). For example, in Figure 7.4, agent x has no alternatives left to explore, while agent y has at least one choice point (the *open node*) with unexplored alternatives. Intuitively, task sharing requires the data structures owned by one processor (e.g., agent x) to be modified to reflect the structure of the branch of the search tree currently being explored by another processor (e.g., agent y). Regardless of the approach used to achieve task sharing, we recognize two processors: the *receiver*, which is the processor that is moving to a different part of the search tree,

Algorithm 7.7: Overall Structure of a Parallel Search ASP Computation

```
1  procedure COMPUTE (P: Program)
2  S ← expand(P,∅)
3  (Branch,ℓ) ← Select_Private_Node(P,S) /* Initial partition          */
4  while (¬Termination_Detection()) do
5      while (¬Completed_Local_Task(Branch)) do
6          if (Need_to_Schedule()) then /* respond to requests         */
7              Branch ← Scheduling(Branch)
8          Branch ← Branch+[ℓ]; S ← expand(P,S∪{ℓ})
9          if (¬consistent(S)) then /* backtrack                       */
10             (ℓ,S,Branch) ← backtrack(P,S,Branch)
11         else if (complete(S)) then
12             Output (S);
13             if (¬Complete_Local_Task(Branch)) then
14                 (ℓ,S,Branch) ← backtrack(P,S,Branch)
15         else
16             atom ← select_atom(P,S);
17             CHOICE: ℓ ← atom OR ℓ ← ¬atom /* choice point          */
18         Branch ← Branch+[ℓ]
19     (ℓ,S,Branch) ← Look_for_Work() /* seek work from others         */
```

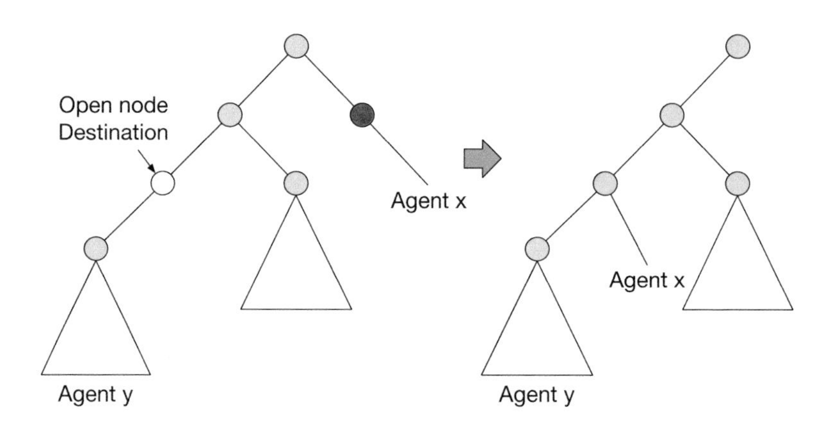

Fig. 7.4: Intuition about task sharing

and the *sender*, which is the processor that is offering local unexplored alternatives to the *receiver*.

Addressing this problem is not trivial—especially when the processors do not have access to any shared memory—e.g., the data structures representing the branch of agent y are not directly accessible by agent x. The problem resembles a similar problem explored in other domains where search parallelism has been considered—e.g., the *binding environment* problem extensively discussed in the parallel Prolog literature [36, 73, 71].

Even though a wide variety of design options have been explored (see, e.g., [72]), two options have emerged as the most promising—i.e., *copying* and *recomputation*.

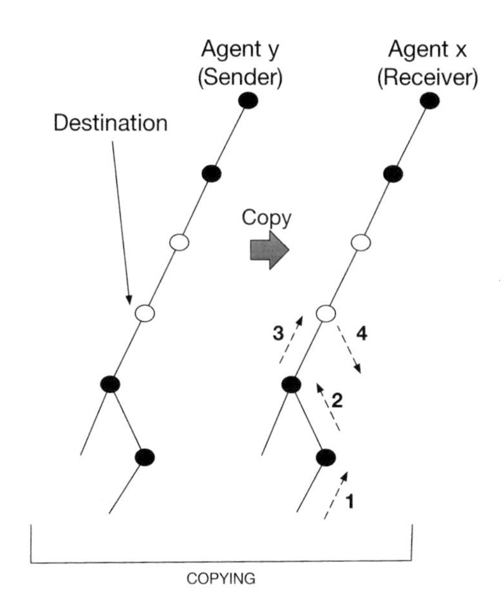

Fig. 7.5: Task sharing using copying

Copying: the intuition behind copying is simple: the *sender* creates a duplicate of the data structures it owns (representing its own branch of the search tree) and transfers this copy to the *receiver*. The receiver installs the data structures received and simply starts backtracking to locate the first unexplored alternative (see Figure 7.5 on the right). This approach is simple and apparently suitable for distributed-memory approaches. The challenge lies in the fact that search trees are potentially very large, and branches are often represented (in most state-of-the-art sequential implementations) using complex data structures (e.g., multiple arrays of cross-linked records [65, 27]). This makes copying a very expensive operation, requiring either very large chunks of memory (e.g., the entire memory image of a processor) to be transferred or a significant amount of time to be invested to extract the minimum necessary components to be transferred (e.g., determine the "difference" between the branches of the two processors involved in the task sharing). This is, for example, different from the case of Prolog, where the branch of the search tree can be efficiently described by a collection of stacks, allowing the parallel implementation to efficiently perform *incremental copying* [36].

Recomputation: the symmetrical approach consists of allowing the receiver to *reconstruct* the branch of the sender. In order to make the recomputation possible, the *sender* needs to provide the receiver with sufficient information to properly reconstruct the branch. The minimum amount of information needed is represented by the literals that have been "guessed" by the `CHOICE` operation (line 17 of Algorithm 7.7) along the branch of the *sender*—indeed, the design of the `expand` operation is such that, given the sequence of literals $\ell_1,\ldots,\ell_n$ that have been chosen to construct a branch, and given

$$
\begin{aligned}
E_0 &= \emptyset \\
E_i &= \mathtt{expand}(P, \{\ell_i\} \cup E_{i-1}) \;\; i \geq 1
\end{aligned}
$$

we have that

$$
\mathtt{expand}(P, \{\ell_1,\ldots,\ell_n\}) = \bigcup_{i=0}^{n} E_i
$$

As such, the only piece of information necessary for the receiver is the set of literals chosen in the construction of the branch—a piece of information that is typically easy to collect, as most sequential implementations maintain these choices in a stack to facilitate backtracking. Two versions of the recomputation approach can be envisioned. One approach combines recomputation with backtracking (see Figure 7.6(left))—by having the *receiver* backtrack to the nearest common ancestor in the search tree between the *sender* and the *receiver*, and perform a recomputation from that point on. The alternative is to allow the *receiver* to backtrack to the root of the search tree (e.g., the first `expand` in line 2) and reconstruct the entire branch (see Figure 7.6(right)). The advantage of the first approach is the potential need to reconstruct only a segment of the branch of the *sender*. The main disadvantage is the need for the *sender* and *receiver* to communicate in order to determine their nearest common ancestor; this is trivial in the case of a shared-memory implementation, but might require non-trivial exchanges in the case of distributed-memory implementations [46, 72]. The second approach benefits from the simplicity of positioning the *receiver* on the root of the tree (which can be realized with a simple memory-copying operation), not requiring any communication apart from the exchange of the set of chosen literals; the disadvantage is the need to recompute the entire branch (which could be potentially very long).

The three methods have shown distinct performance on different benchmarks (e.g., Figure 7.7), demonstrating that dynamic selection of the task-sharing scheme is necessary, adapting the task-sharing scheme to the structure of the computation. Further variations of these methods have been discussed in [72].

7.5.1.3 Scheduling and Load Balancing

The use of task-sharing is necessary to allow a finite number of processors to cooperate in constructing and exploring the search tree underlying an ASP computation. While the task sharing technique provides the mechanism to allow one processor to move from one part of the search tree to another, the open question is how to determine the parameters of the sharing operation—i.e., who is the *sender*, who is the *receiver*, what is the destination point, and when a sharing operation should be performed. Different design dimensions can be explored.

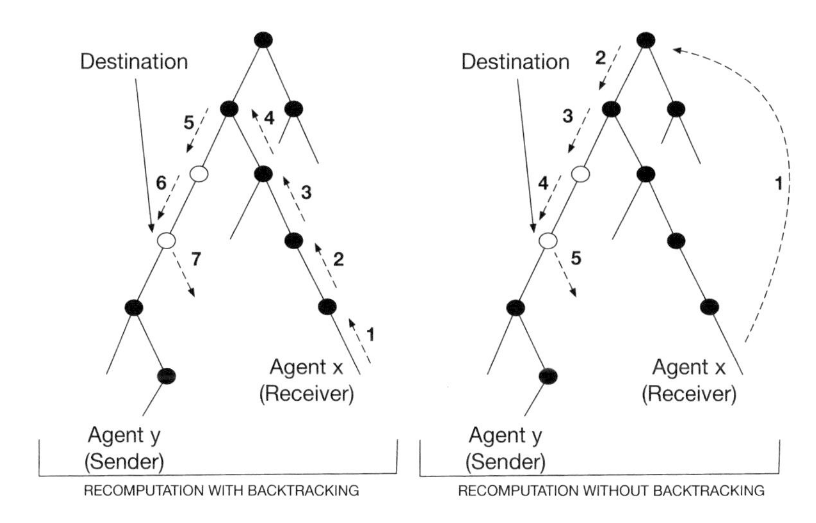

Fig. 7.6: Recomputation-based task sharing

Scheduling Symmetry: A common scheduling design is the use of an *asymmetric* scheduler, often referred to as a *centralized* scheduler. In this strategy, a distinction is introduced between agents that are in charge of leading the scheduling efforts (*masters*) and agents in charge of task execution (*slaves* or *workers*). Typically, a single master is used. The master is in charge of keeping track of available tasks for parallel execution (i.e., unexplored alternatives in the search tree), while the *workers* are standard ASP solvers in charge of exploring assigned parts of the search tree. Whenever a worker completes the exploration of the assigned search space, it requests a new unexplored alternative from the master and restarts the computation. The advantage is simplicity of communication and implementation, at the price of a potential communication bottleneck.

The opposite design alternative is represented by *symmetric scheduling*. In this model, all agents play the roles of both worker and master. Whenever one agent completes the exploration of a part of the search space, it can acquire unexplored alternatives from any of the other agents. The benefit of this model is the avoidance of bottlenecks, since the distribution of unexplored tasks is spread among all the agents. On the other hand, the lack of a reference master may require agents to perform multiple communication acts before being able to locate a valid unexplored alternative.

Scheduling Initiation: another important design decision concerns which processor initiates the task-sharing activity. Traditional scheduling designs make use of *receiver-initiated* models, where agents completing their assigned tasks seek new tasks to explore. An alternative design is *sender-initiated* scheduling, where agents that carry excessive unexplored alternatives volunteer their transfer to agents who are idle or have fewer open alternatives.

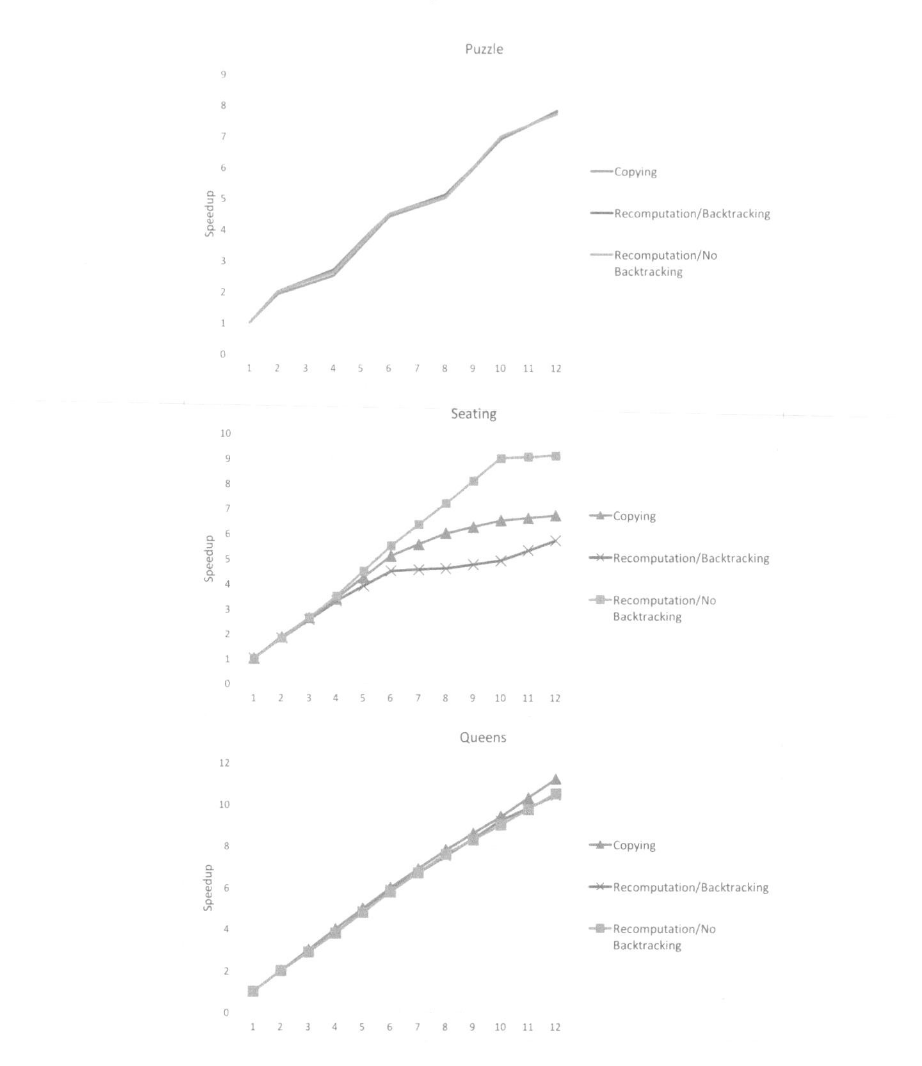

Fig. 7.7: Speedups using copying and recomputation with and without backtracking

Location Policy: looking at the process of task sharing as moving an agent from one position in the search tree to another, scheduling strategies may either select a random position or attempt to move to the "closest" unexplored alternative in the tree—where the distance is measured in terms of the cost of task sharing (e.g., backtracking+recomputation). The former method is frequently used in symmetric scheduling, while distance-based methods are used in (a) shared-memory implementations, where it is feasible to share among agents their respective positions in the tree, and (b) implementations based on asymmetric scheduling, where the master can maintain a map of the positions of the agents.

These distinct options have been compared in several experimental systems (e.g., [46, 47, 72]). While significant performance differences can be observed depending on the type of benchmark and the type of implementation (e.g., shared-memory vs. distributed-memory systems), a consistent observation is the dominance of symmetric scheduling methods over asymmetric methods, due to the complexity of communication, especially in search trees that have a mixed combination of long and short branches. Figures 7.8 and 7.9 show a comparison of speedups for selected benchmarks between symmetric and asymmetric scheduling methods.

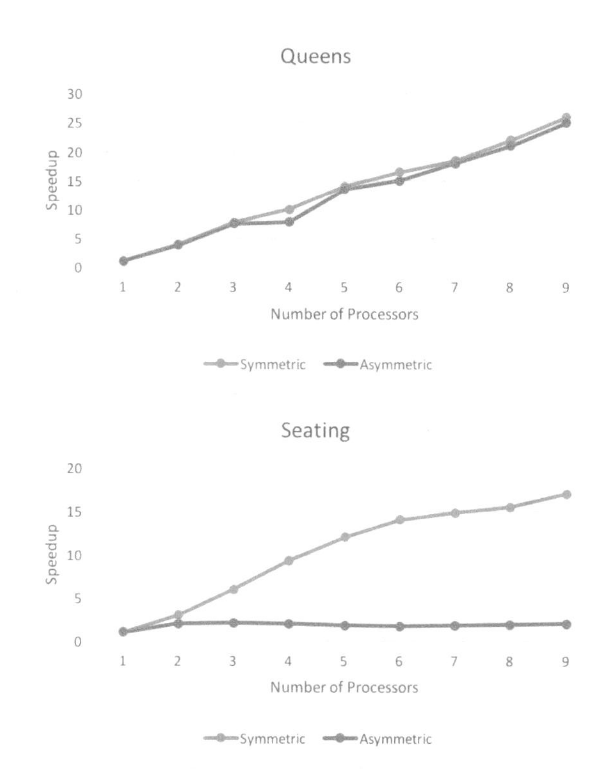

Fig. 7.8: Symmetric vs. asymmetric scheduling

7.5.1.4 Parallelizing Lookahead

Typical algorithms for the computation of answer sets have been enriched with a large number of optimizations, including a variety of techniques that have been drawn from the field of satisfiability testing (e.g., [8]). These optimizations may offer additional opportunities for parallelization.

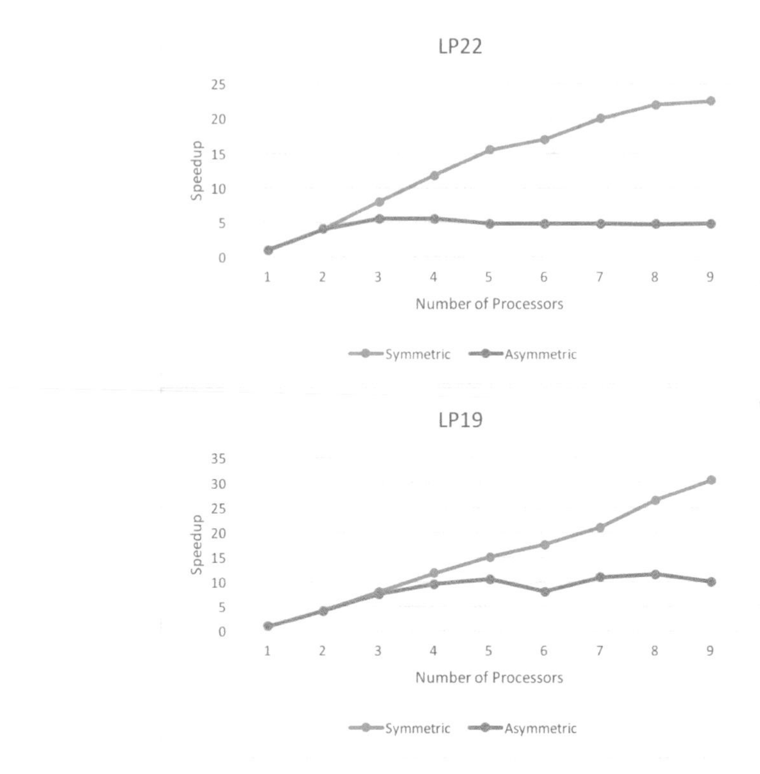

Fig. 7.9: Symmetric vs. asymmetric scheduling

Let us examine one such opportunity. Starting with the development of one of the earliest ASP systems, SMODELS [66], the *lookahead* technique [38] has been used to reduce the search space. Lookahead builds on two important principles of the `expand` operation (see Section 7.2.2): (1) the operation is deterministic and very efficient; (2) the operation can determine inconsistencies. The typical SMODELS process makes non-deterministic choices, selecting an atom that does not appear in the partially constructed answer set and non-deterministically adding it as a positive or negative literal. Lookahead performs a preliminary test to guide such selection, as summarized in Algorithm 7.8.

Algorithm 7.8: Naive Lookahead

```
1  procedure LOOKAHEAD(P: Program, S: Interpretation)
2  for each A ∈ Π(𝓕) \ {B | B ∈ S ∨ ¬B ∈ S} do
3      S1 ← expand(P, S ∪ {A})
4      S2 ← expand(P, S ∪ {¬A})
5      if ( ¬consistent(S1) ∧ ¬consistent(S2) ) then
6          return False
7      else if ( ¬consistent(S1) ) then
8          S ← S ∪ {¬A}
9      else if ( ¬consistent(S2) ) then
10         S ← S ∪ {A}
11     else
12         return A
```

Algorithm 7.9: Parallel Lookahead

```
1  procedure PARALLEL LOOKAHEAD(Id: Processor; P: Program, S:
   Interpretation)
2  for each (A ∈ Π(𝓕) \ {B | B ∈ S ∨ ¬B ∈ S} ∧ π(A) == Id) do
3      S1 ← expand(P, S ∪ {A})
4      S2 ← expand(P, S ∪ {¬A})
5      if ( ¬consistent(S1) ∧ ¬consistent(S2) ) then
6          Signal(Termination)
7      else if ( ¬consistent(S1) ) then
8          S ← S ∪ {¬A}
9          Broadcast(¬A)
10     else if ( ¬consistent(S2) ) then
11         S ← S ∪ {A}
12         Broadcast(A)
13     else
14         Signal(Success)
15         return A
16     Gather(S′)
17     S ← S ∪ S′
```

An obvious source of parallelism is in the for each loop of line 2. While the code in Algorithm 7.8 introduces a dependence among the iterations (due to the incremental growth of the interpretation S), these additions are all deterministic. The lookahead computation can be parallelized by distributing the iterations of the loop among different processes/threads. A sample parallel structure is summarized in Algorithm 7.9. An implementation of this structure has been presented in [6]. The set of atoms is partitioned among available processors by the function π (line 2). If any processor detects inconsistencies, a termination signal is issued (line 6).

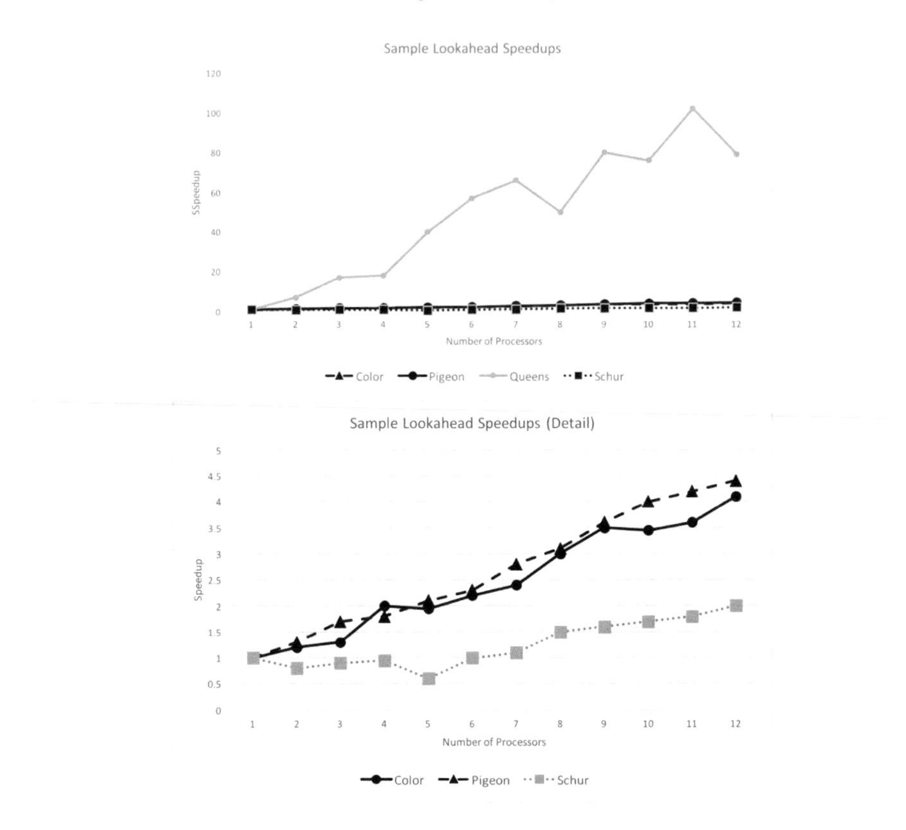

Fig. 7.10: Sample speedups from parallel lookahead

Elements that can be deterministically added to the interpretation are broadcasted to all processors (lines 9 and 12) and collected at each iteration (line 16). Figure 7.10 summarizes the speedups observed from parallelization of lookahead on some sample benchmarks.

7.5.2 GPU-Based Parallelism

Graphical Processing Units (GPUs) are highly parallel structured devices, originally developed to support efficient computer graphics and rapid image processing. In recent years, the use of such powerful multicore systems has become pervasive in general-purpose applications that are not directly related to computer graphics, but demand massive computational power. Vendors such as AMD and NVIDIA provide dedicated APIs and development environments, and promote language extensions such as *OpenCL* [42] and *CUDA (Computing Unified Device Architecture)* [67] to support the development of GPU-based applications.

As concerns the hardware, a GPU consists of hundreds or even thousands of identical computing units (*cores*) and provides access to both on-chip memory (used for registers and shared memory) and off-chip memory (used for cache and global memory). The underlying conceptual model for parallelism is *Single-Instruction Multiple-Thread (SIMT),* where the same instruction is executed by different threads that run on cores, while data and operands may differ from thread to thread.

A CUDA program includes parts meant for execution on the CPU (referred to as the *host*) and parts meant for parallel execution on the GPU (referred to as the *device*). Typically, the host code transfers data to the device memory, starts parallel computations on the device, and retrieves the results from device memory.

The API supports interaction, synchronization, and communication between host and device. Each device computation is described as a collection of concurrent threads, each executing the same device function (called a *kernel*, in CUDA terminology). These threads are hierarchically organized in *blocks* of threads and *grids* of blocks. Device global memory is accessible by all threads, whereas threads of the same block may access high-throughput on-chip *shared memory*.

7.5.2.1 GPU-Based Datalog Solving

The shared memory available in GPUs is exploited in [63] where the authors distribute the load of computing the model of a Datalog program between the various threads, which can access and modify the data in the shared memory. Particular care is given to the parallelization of the join operation.

The bottom-up semantics of a Datalog program can be reduced to a computation that makes use of the relational algebra operators *select*, *join*, and *projection*. The computation order is driven by the dependency graph, and fixpoint procedures are employed in case of recursive predicate definition.

The host preprocesses the program, converting each rule into an internal, numerical, representation; the host decides which kind of relational algebra operators are needed for each rule, while their executions are delegated to the device.

In more detail, *select* is implemented using three different device function executions. The first one marks the rows of a matrix that satisfy the selection predicate, the second one performs a prefix sum on the marks to determine the size of the results buffer and the location where each GPU thread must write the results, and the last device function writes the results.

The *projection* operator simply moves the elements of each required column to a different location.

The authors implement *single join*, *multijoin*, and *selfjoin* operations according to the number of variables involved. Let us focus here on the single join only, where the authors adapted the *Indexed Nested Loop Join* implementation. They build an array for each of the two columns to be joined, sort one of them, create a Cache Sensitive Search (CSS) Tree [74] for the sorted column, search the tree to determine the join positions, perform a first join to determine the size of the result, and finally perform a second join to write the result. CSS Trees can be built in parallel and are addressed arithmetically (rather than with pointers), thus making them well suited for GPUs. Empirical evaluations show excellent speedups, up to 200 times for some of the benchmarks, using a GPU with 512 cores.

In [40] the authors study the problem of parallelizing backtracking methods using GPUs. Backtracking methods deal with irregular accesses to the problem instance, with different memory accesses in different nodes of the same level, in a search space that is possibly of exponential size and shaped as an unbalanced tree. All these issues clash with the expected "regularity" of the SIMT parallelism model of GPUs. The authors apply their method to the *maximal clique enumeration (MCE)* problem, parallelizing a backtracking algorithm either on a GPU or using multicore processors. The implementation exploits coarse-grained parallelism to handle the construction of the tree, and fine-grained parallelism to execute tasks in each node. They test the two implementations on four different datasets, two of them coming from real-world problems (from biology and climate). The results of the comparison between the two kinds of parallelism do not identify a clear winner, but in both cases (multi-threaded implementations on standard multicore platforms and GPU-level parallelization) there is a significant speedup w.r.t. the sequential approach. The coarse-grain parallelization of backtracking algorithms can be easily ported to a GPU architecture—by simply implementing the process of exploring a subtree as a kernel, executed by multiple threads on the device, and by properly partitioning the global memory of the GPU. However, the resulting performance is highly dependent on how balanced the search tree is. Fine-grain parallelization of activities in a single node potentially offers good speedup, but usually this is obtained by designing ad hoc GPU-oriented algorithms, specifically devised to fully benefit from SIMT parallelism.

A similar empirical study is the subject of [13], where the authors show the possibilities and the limits of parallelizing the unit propagation procedure (done in each node of the search tree) and backtracking search (coarse grain). While the former seems to always benefit from GPU parallelism, the latter proved to be effective only for problems of size falling within a given range—not too small, otherwise the transfer time would not be justified, and not too large, otherwise the GPU parallel computation will exceed memory capacity. The experimental study reported in [13] shows that excellent speedups can be achieved by using GPU parallelism as a subroutine of the SAT computation, performed when the number of remaining unknown variables is close to 70.

7.5.2.2 GPU-Based Conflict-Driven ASP Solving

Overall Design: The approaches to the parallelization of ASP described earlier turn out to be hardly applicable in the case of GPUs and do not ensure adequate scalability of the solution. This is because GPUs' model of parallelism is profoundly different. Indeed, GPUs are designed to operate with a very large number of lightweight

Algorithm 7.10: `GPU-ASP-Computation`

```
1  procedure GPU-ASP-COMPUTATION(Δ: SetOfNogoods, P: GroundProgram)
2  current_dl ← 1 /* initial decision level                              */
3  A ← ∅ /* initial assignment is empty                                  */
4  (A,Violation) ← InitialPropagation(A,Δ) /* GPU parallelism            */
5  if Violation then
6      return no answer set
7  else
8      while true do
9          (Δ_A,Violation) ← NoGoodCheckAndPropagate(A,Δ) /* GPU par. */
           A ← A ∪ Δ_A;
10         if Violation ∧ (current_dl = 1) then
11             return no answer set
12         else if Violation then
13             (current_dl,Λ) ← ConflictAnalysis(Δ,A) /* GPU par.        */
14             Δ ← Δ ∪ Λ /* add learned nogoods                          */
15             A ← A \ {p̄ ∈ A | current_dl < dl(p̄)}
16         if (A is not total) then
17             (p̄,Selected) ← Selection(Δ,A)/* GPU parallelism           */
18             if Selected then
19                 current_dl ← current_dl + 1
20                 dl(p̄) ← current_dl
21                 A ← A ∪ {p̄}/* extend the assignment                   */
22             else
23                 A ← A ∪ {Fp : p is unassigned}
24         else
25             return A^T ∩ atom(P) /* stable model found                */
```

threads, homogeneously operating in a tightly synchronous fashion. Consequently, device code has to maximize fair load balancing among threads and avoid *divergence* in threads' computations (namely, threads of the same block which concurrently execute different instructions). Moreover, to fully benefit from the complex memory architecture of GPUs, one has to prefer specific regular patterns in accessing each type of device memory. Conversely, the presence of irregular memory accesses or thread divergence often cause the serialization of the kernel executions, yielding poor overall parallel performance.

A viable strategy to fully exploit GPUs' computational power consists of parallelizing the execution of those "easy" activities (i.e., those presenting polynomial time complexity), such as nogood checking, unit propagation, conflict analysis, and learning, that are basic components of the entire ASP solver. In this frame of mind, the approach proposed in [18, 19] builds on the notion of *ASP computation* combined with conflict analysis and learning techniques derived from those adopted in [27] for the CLASP solver. To the best of our knowledge, this is the only investigation ever proposed focused on the parallelization of ASP solving on GPUs.

Once the host has transferred the input nogoods to GPU global memory, the computation develops on the device as shown in Algorithm 7.10. The overall structure of the *GPU-ASP-Computation* procedure is the conventional structure of a conflict-driven ASP solver [27], except for the adopted selection heuristics, which implements the ASP computation (Section 7.2.4.3), and for the parallelization of all the functions involved. In particular, as usual for conflict-driven solvers, the procedure maintains a (partial) truth assignment, which is repeatedly extended by alternating decision steps and unit propagation phases. During each decision step, heuristics are used to select an unassigned atom and to assign it a truth value. To track taken decisions, at each decision step a *decision level* is incremented. During the propagation phase, further truth values for unassigned atoms can be derived from the current assignment, using program rules/nogoods. All these assignments occur at the *current decision level*.

Whenever a failure is encountered, namely a rule/nogood cannot be satisfied by the current partial assignment, a conflict analysis step is run to detect the decision step causing the failure and its decision level, say ℓ. Moreover, (at least) one new nogood is derived by applying standard clause-learning techniques [8]. At this point a backjumping step restores the state of the partial assignment corresponding to the decision made at level $\ell - 1$, the current decision level is updated accordingly, and the search continues. The newly introduced nogoods prevent the solver from repeating failing decisions. The computation ends as soon as the assignment becomes complete or no backjump is possible (because a failure occurs already before taking the first decision, namely, at the top decision level).

Implementation Details: Let us describe in more detail the main steps of Algorithm 7.10, where Δ is the set of nogoods (initialized with those computed from the program P), A is the partial truth assignment, and *current_dl* is the current decision level.

Since the set Δ may include some (input) unitary nogoods, a preliminary parallel computation partially initializes A by propagating them. This is done by the procedure *InitialPropagation* (line 4), which runs a grid of threads with one thread for each unitary nogood. Hence, all current unitary nogoods are processed in parallel. The sign of the literal in each unitary nogood is analyzed and the corresponding entry of A is set accordingly. If one thread finds such an atom already assigned in an inconsistent way, a `Violation` flag is set to true and the computation ends (line 5).

The procedure *NoGoodCheckAndPropagate* (line 9) performs unit propagation and, at the same time, checks nogoods for violation, w.r.t. A.

As mentioned earlier, to better exploit the SIMT parallelism and maximize the number of concurrently active threads, in each device computation the workload has to be distributed among the threads as uniformly as possible. To this aim, the set of nogoods is partitioned depending on their size and *NoGoodCheckAndPropagate* is organized as different steps, each one implemented by a different kernel grid. Each kernel grid deals with all nogoods having the same size.

The procedure *NoGoodCheckAndPropagate* iterates by repeatedly running such grids of kernels, which behave as follows. In each iteration the computation is driven by the atoms that have been assigned a truth value in the previous iteration. The motivation is that only these atoms may trigger either a conflict or a propagation. In particular, in the first iteration, the procedure relies on the atoms that have been assigned by the *InitialPropagation* procedure. Similarly, the atoms assigned in an iteration will possibly affect the following step. Thus, each grid of kernels involves a number of blocks that is equal to the number of atoms assigned during the previous iteration. The threads in each block process nogoods that share the same assigned atom. This strategy in distributing the load among grids (depending on the size of the nogoods) and among threads (depending on the assigned atoms) ensures that all threads of the same block perform similar amounts of work, minimizing thread divergence and irregularity in memory accesses. Specific data structures are used in order to efficiently determine, after each iteration and for each assigned atom, which are the input nogoods to be considered in the next iteration.

A similar approach is adopted to process those nogoods that are learned at run-time through the conflict analysis step (see below). They are partitioned depending on their size and processed by different grids, accordingly.

A further optimization, which is applied when processing nogoods of large size, relies on a standard technique based on *watched literals* [8]. In this manner, each thread may reduce the number of nogoods to be inspected by checking the watched literals and acting accordingly.

Notice that, in each iteration of *NoGoodCheckAndPropagate*, all significant nogoods are processed in parallel. On one hand, this implies that all possible propagations are performed as soon as possible. On the other hand, several violations/conflicts might be detected as soon as they are triggered by propagations. The iterations end when at least one conflict is generated or no more propagations are possible. The variables Δ_A and `Violation` are set accordingly. Then, the partial assignment A is updated by adding the set of newly assigned atoms Δ_A (line 9). In case of failure at the top level, the procedure ends (line 11). If a failure occurs at a deeper level (line 12), a conflict analysis phase is executed (see below). Otherwise, the procedure proceeds by executing a new decision step and increasing the current decision level (lines 16–23), unless A is complete (line 25).

The *Selection* procedure (line 17) determines the atoms to be decided. The selection is performed by heuristically ranking all potential candidates. (Well-known heuristics such as the Jeroslow-Wang and its variants [41], counting heuristics [8], as well as considering the "activity" of atoms [33] are exploited.) Such ranking is evaluated on the device, by processing in parallel all atoms that are potentially selectable in accordance with the development of an ASP-computation (cf., Section 7.2.4.3). The selected atom ($\overline{p}$ in lines 16–23), is assigned its decision level $dl(\overline{p})$. In case no selectable atom exists, A is completed by setting false all remaining unassigned atoms (line 23). The formal properties of the ASP computation ensure that A is a stable model [53].

The *ConflictAnalysis* procedure is used to resolve the conflicts previously detected by *NoGoodCheckAndPropagate*. For each conflict, it identifies a decision level (and the corresponding decided atom $\overline{p}$) the computation should backjump to, in order to remove the violation. It should be mentioned that the approach adopted in [18, 19] for implementing conflict analysis and learning adapts techniques originally developed for sequential solvers [8, 27]. These techniques derive a new nogood by performing a linear sequence of resolution steps, that resolve the violated nogood against input nogoods. This process is essentially sequential and hard to parallelize. Hence, this is the part of the GPU-based solver that least exploits the SIMT parallelism. Some initial investigations of the design of natively GPU-oriented parallel algorithms for learning have been made in [23]. Nevertheless, at the time of writing, the *ConflictAnalysis* procedure described in [19] is a parallelization of the algorithm described in [8, 27], and adopted for the CLASP solver to identify a unique implication point (UIP [62]) and to determine the lower decision level among those causing the detected conflicts. Notice that *ConflictAnalysis* can analyze all the conflicts in parallel. Moreover, for each conflict, a block of threads performs the sequence of resolution steps deriving one new nogood. In each step, the atoms composing intermediate resolvents are processed in parallel, equally partitioned among the threads. The procedure detects the best level to backjump to, performs backjumping, and returns a new set of nogoods Δ.

Notice that the solver described by Algorithm 7.10 is somehow simplified with respect to the concrete implementation reported on in [19], which encompasses also well-known techniques such as *restart* and *forgetting* [8] to speedup the search.

7.5.3 Moving Towards Large-Scale Architectures

7.5.3.1 The Map-Reduce Programming Model

The increasing availability of large data sets—including large knowledge bases—has pushed the development of frameworks and programming models that are suitable for processing large amounts of data. In particular, emphasis has been placed on the development of programming models that move computations to the data, to avoid large data transfers.

The *Map-Reduce* framework, originally introduced by Google [17], is a programming framework designed to operate over a distributed file system (e.g., Google File System). The underlying distributed file system automatically handles the partitioning of files into blocks (e.g., 128MB in the Hadoop File System [3]) and their distribution across data servers, to ensure scalability and fault-tolerance.

Figure 7.11 summarizes the structure of a typical Map-Reduce application. The workflow depicted in the figure is fixed, and the programmer's job consists only of developing the *Map* and the *Reduce* tasks. The workflow may include a large number of instances of both the Map and Reduce tasks—typically executed on the same servers that provide the distributed file system.

The tasks are designed to operate on data organized in $\langle$*key, value*$\rangle$ pairs. In particular, the Map tasks act as "partitioners", aimed at producing collections of $\langle$*key, value*$\rangle$ pairs from blocks of data (parts of the input file), while the Reduce tasks collect each key with the list of all values associated with the key, to produce the final result. Note that the `combiners`, in charge of assembling the output of the Map tasks according to the keys, are built-in in the Map-Reduce framework.

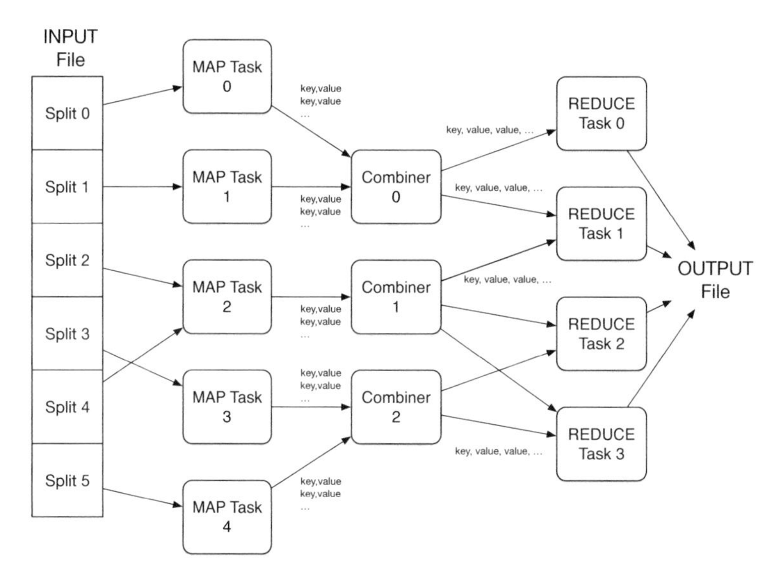

Fig. 7.11: Structure of a Map-Reduce application

7.5.3.2 Datalog and Map-Reduce

Datalog has been the focus of parallelization on large-scale distributed platforms using the Map-Reduce model; this focus is not surprising: Datalog has been a platform of choice for advanced database applications, and Map-Reduce enables the use of Datalog over very large-scale datasets.

The core of the Datalog bottom-up computation is the determination that a logic programming rule is applicable with respect to an interpretation. If we view the collection of facts with the same predicate in an interpretation as a database relation, then the application of a rule can be viewed as a combination of relational algebra operations. For example, a clause of the form $p(Y) \leftarrow q(a,X), r(X,Y)$ contributes to the relation p according to the relational algebra expression $\pi_Y(\sigma_{q_1=a}(q) \bowtie r)$.

The core of the application of a clause is the execution of natural joins among the relations in the body of the clause. The literature [1] has explored the Map-Reduce implementation of natural joins. Let us start by considering the simpler case of binary relations and clause bodies composed of two atoms—i.e., clauses of the form $p(X,Z) \leftarrow q(X,Y), r(Y,Z)$. Such a join can be realized with a Map-Reduce model organized as follows:

- Each Map task receives facts of either relation; each fact $q(a,b)$ is mapped to a key-value pair of the type $(b,(a,q))$ while each fact $r(b,c)$ is mapped to the pair $(b,(c,r))$. Thus, the key is the value of common argument between the two facts.
- Each Reduce task receives elements of the form $(key, List)$ and produces facts of the form $p(a,c)$ for each $(a,q) \in List$ and $(c,r) \in List$.

This can extended to the case of multiple elements in the body of a clause, by creating complex keys that capture the values of the variables used to connect facts in the join; let us consider the case of a rule $p(X,Y) \leftarrow q(X,Z), r(Z,T), s(T,Y)$; a Map-Reduce scheme can be realized as follows:

- Each Map task translates facts $q(a,b)$ to pairs $((b,k),(a,q))$ for each legal term k; facts of the form $s(c,d)$ are mapped to pairs $((k,c),(d,s))$; facts of the form $r(b,c)$ are mapped to pairs $((b,c),r)$.
- Each Reduce task receives elements of the form $((b,c),List)$ and produces facts $p(a,d)$ for each $(a,q) \in List$, $(d,s) \in List$ such that $r \in List$.

Generalizations to more complex cases are straightforward; for example, if the first body element in the clause above is replaced by $q(v,X,Z)$, then the Map task will produce the pair $((b,k),(a,q))$ only for facts of the form $q(v,a,b)$. The model can be extended trivially to implement a single iteration of the Datalog semi-naive algorithm. In order to provide the entire computation of the least Herbrand model, it becomes necessary to iterate the Map-Reduce phase until a fixpoint is determined; efficient schemes to support iterations of Map-Reduce pipelines (e.g., HaLoop [9]) have been investigated and successfully applied to the case of Datalog [2].

7.5.3.3 Towards ASP: Well-Founded Semantics and Map-Reduce

As discussed in the introductory Section 7.2.2, the addition of negation as failure to definite clause programs (hence to pure Datalog) has the consequence of possibly leading to multiple minimal models. Nevertheless, there are large classes of programs with negation that still have a single Herbrand model. The most popular syntactic restriction is *stratification* [75, 4]. A logic program with negation P is stratified if there exists a level mapping function $p : \mathscr{P} \to \mathbb{N}$ such that, for every rule $head \leftarrow b_1,\ldots,b_n, not\ c_1,\ldots, not\ c_m$ in P, we have that $p(pred(head)) \geq p(pred(b_i))$ and $p(pred(head)) > p(pred(c_j))$ for $1 \leq i \leq n$ and $1 \leq j \leq m$.[5] Each stratified program is guaranteed to have a unique least Herbrand model. For simplicity, we will assume that the levels in the program are assigned consecutively from 0 to a maximum level ℓ. The computation model used for stratified programs is a simple extension of that used for Datalog, as illustrated in Algorithm 7.11.

Algorithm 7.11: Stratified Datalog Computation

```
1 Procedure STRATIFIED(P)
2 I ← ∅
3 for Level ← 0 to ℓ do
4 |  P' ← {ground(r) ∈ P | p(pred(head(r))) = Level}^I
5 |_ I' ← least_Herbrand_model(P' ∪ I)
```

[5] This is equivalent to stating that the graph $\mathscr{G}^{+,-}(P)$, defined in Section 7.2.2, does not have cycles with negative edges.

The extension of the Map-Reduce Datalog computation to the case of stratified programs requires two components:

- The individual Map-Reduce pipeline described for applying one clause, through multi-way joins, has to be extended to include negated literals.
- An additional loop has to be added around the Map-Reduce pipeline that corresponds to the loop in lines 3–5 of Algorithm 7.11.

While the second item is simple, the first item requires work [79]. While positive atoms are solved using natural joins, negated elements in a clause require the use of anti-joins. Let us illustrate the modified Map-Reduce pipeline with an example. Let us consider the clause $p(X) \leftarrow q(X), not\ r(X)$; then

- *Map:* for each atom $q(a)$ received, the pair $(a,+)$ is returned; for each atom $r(b)$ received, the pair $(b,-)$ is generated.
- *Reduce:* for each element $(key, List)$ received, the task produces the result $p(key)$ if $+ \in List$ and $- \notin List$.

An analogous method can be used to deal with non-stratified programs under the *well-founded semantics* [84]. This semantics determines a unique minimal three-valued model and can be computed as explained below. If at the end of the computation all the atoms are assigned (to true or to false) then it is also a stable model. The core of the computation is a variant of the immediate consequence operator: given a ground program P

$$T_{P,J}(I) = \{head(r) \mid r \in P, pos(r) \subseteq I, neg(r) \cap J = \emptyset\}$$

where $pos(r)$ denotes the atoms that appear non-negated in the body of rule r, while $neg(r)$ are those atoms that appear negated in the body of r. Intuitively, given sets of true atoms I and false atoms J, the operator determines what rule heads can be immediately derived using such knowledge. Let us denote by $lfp(T)$ the least fixpoint of the operator T. The well-founded semantics of a program can be defined according to the following rules (P^+ indicates the definite rules in P):

$$\begin{array}{ll} K_0 = lfp(T_{P^+,\emptyset}) & U_0 = lfp(T_{P,K_0}) \\ K_{i+1} = lfp(T_{P,U_i}) & U_{i+1} = lfp(T_{P,K_{i+1}}) \end{array}$$

Given a fixpoint (K,U) of this sequence, the well-founded model of P is composed of the atoms in K (true elements) and all the elements not in U (false elements); the remaining atoms are undefined.

A Map-Reduce model for the computation of the well-founded model follows the same scheme shown earlier—since the computation of $T_{P,J}$—except that there are two iterative Map-Reduce pipelines (see Figure 7.12). Experimental results show great potential for scalability [81].

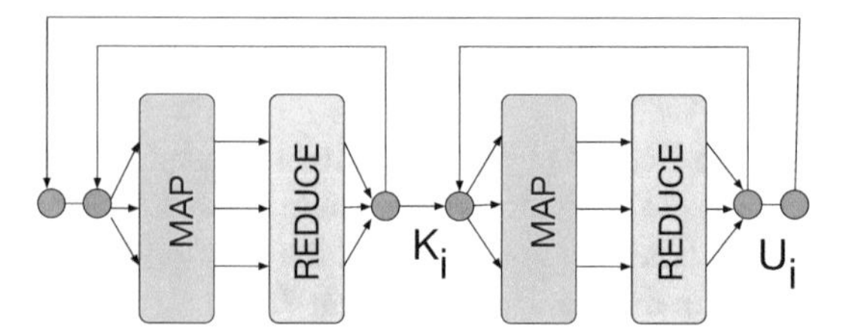

Fig. 7.12: Nested Map-Reduce pipelines for the well-founded semantics

7.5.3.4 Other Relevant Applications of Map-Reduce

Analogous principles to those discussed before have been applied to develop Map-Reduce pipelines for other logics. The authors of [80] provide a Map-Reduce pipeline for a defeasible logic; under the assumption of stratified defeasible theories, the computation is analogous to the case of stratified Datalog, with the exception that the processing within each level of the stratification requires two Map-Reduce pipelines, one to determine consequences and one to perform defeasible reasoning.

The work in [83] illustrates the mapping of RDFS and OWL reasoning to logic programming rules and the use of Map-Reduce pipelines for the computation, with particular emphasis on the optimization of the encoding to take advantage of the specific types of rules derived from RDFS and OWL encoding.

7.5.4 Portfolio Approaches for ASP

Portfolio is a meta-search technique that can easily benefit from parallelism; it applies a family of different solvers with the aim to exploit the best of them on a specific instance. As is common practice in the literature on portfolio techniques, we consider in the same way the application of completely different solvers or of the same solver used with different search parameters. Constraint solving in general, and Boolean constraint solving in particular, are highly sensitive to parameter tuning. Given an ASP program *P*, and a set *A* of solvers, portfolio methods use actual input data to set up the most promising parameter assignments for a complete run. In Figure 7.13, we describe the basic portfolio scheme adopted by CLASP—in the CLASPFOLIO system [26]—which has been inspired by SATzilla [88].

Given a program *P* a (usually partial) run of the *A* solvers on *P* is executed and halted when a timeout is reached. This part can be completely parallelized. At the end of the computation a set of *features* is analyzed and according to the values of these features the most promising solver is selected for running the complete computation. Also in this case a parallel architecture would allow us to run (independently) more than one solver on the complete input in the same time. Chapter 15, Selection and Configuration of Parallel Portfolios of this book extensively covers this topic; we just report here the main portfolio approaches for ASP solving.

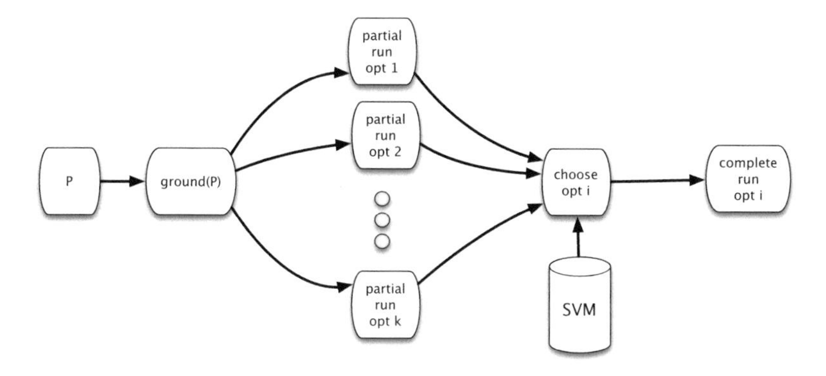

Fig. 7.13: A simple portfolio scheme. Partial executions with different search parameters can run in parallel. A parallel architecture can allow us to execute more than one complete run in parallel

Following [52], let us assume we have a set A of solvers, a set I of instances (programs) and a function $\ell : A \times I \longrightarrow \mathbb{R}$ to be minimized (e.g., running time). A selector S is a function $S : I \rightarrow A$ aimed at minimizing $\ell(S(P), P)$ with $P \in I$. Portfolio methods adopt a machine-learning technology to define the selector function S using a set T of training instances, where a set F of features is considered. A selector trained on a set of test instances is evaluated on a (disjoint) set of instances I by comparing $\sum_{P \in I} \ell(S(P), P)$.

In the case of CLASPFOLIO [26] starting from the ground program, a lightweight version of CLASP is used to extract the features F that are mapped to a score for each configuration in the portfolio. Data-related features (e.g., number of constraints, number of variables, etc.) and Search-related features (e.g., average backjump length, length of learned clauses, etc.) are used by the selector to retrieve the most promising configuration. The selector is based on an SVM trained using the Support Vector Regression technique.

In [76] the authors elaborate on the previous approaches showing also the potential of the portfolio techniques for finding a sequence of parameter settings to be switched in the computation. They also test local search methods for adjusting solver configurations.

In [58] the authors presents ME-ASP, a multi-engine approach for ASP. Six multinomial classification methods to train the selector are compared: Aggregation Pheromone density-based pattern Classification (APC), Decision rules (FURIA), Decision trees (J48), Multinomial Logistic Regression (MLR), Nearest-neighbor (NN), and Support Vector Machine (SVM). Training is done on a set of easy-to-compute features split into four families: Problem size features, Balance features, "Proximity to horn" features, and ASP peculiar features. In [59] ME-ASP is extended with a policy adaptation. The multi-engine nature of the approach makes it suitable for parallelism.

In [39] the authors present CLASPFOLIO 2, which extends the rigid architecture of Figure 7.13 to provide "a modular and open architecture that allows for integrating several different approaches and techniques", including different feature generators, several approaches to solver selection, variable solver portfolios, and solver-schedule-based pre-solving techniques. In [52] the same authors (plus Frank Hutter) present AUTOFOLIO, partially based on CLASPFOLIO 2, a general approach for automatically determining a strong algorithm selection method for a particular dataset, by using algorithm configuration to search through a flexible design space of algorithm selection methods.

7.6 Discussion and Conclusions

In this Chapter, we provided a brief overview of the principal approaches that have been investigated to enable the parallel computation of ASP (and related logic programming languages, like Datalog). A clear message that can be derived from the presentation is that highly declarative logic programming languages, like ASP, offer outstanding opportunities for the transparent exploitation of parallelism. The presence of search-based execution models and the relative lack of control dependencies (as present in traditional imperative programming) facilitate the mapping of the execution models to parallel processes. A broad adoption of parallelism in ASP will enhance scalability and applicability of the paradigm to real-world problems. Techniques such as mapping of search parallelism on multicore threads are now mature and can be realized in a robust and sustainable manner.

Research in the field of parallel ASP is still growing and moving in even more promising directions. State-of-the-art ASP engines such as CLASP maintain support for parallel execution, thus ensuring that the evolution of sequential techniques does not impair the exploitation of parallelism.

The most promising research direction, at the time of the development of this Chapter, is represented by the exploitation of very coarse-grained parallelism over distributed architectures (e.g., using Map-Reduce and similar frameworks). All state-of-the-art ASP solvers still rely on ground-then-solve, thus forcing ASP programmers to contend with potential explosion in the size of the program as a result of grounding. Large-scale parallelism provides an elegant avenue to mitigate the impact of large grounding. The recent development of programming frameworks for handling very large graphs (e.g., [34, 56, 55]) is opening up new opportunities for parallelization of ASP, whose computation can be reduced to graph transformations—as preliminarily explored in [35].

The effort on GPU-level parallelism, on the other hand, is still in its infancy. While the works discussed in this Chapter have highlighted the potential of this type of parallelism in ASP, the engineering of robust and scalable techniques still requires further research. Furthermore, existing studies have focused on the use of the CUDA framework, which is specific only to NVIDIA graphics cards. The impact of porting the techniques to a more general GPU programming framework (such as OpenCL) and to non-NVIDIA architectures is an interesting open line of research.

Acknowledgements The research pursued by the authors on the topics of this Chapter has been partially supported by NSF grants CBET-1401639, HRD-1345232, CNS-1337884, and DGE-0947465, by INdAM GNCS 2014–2017 grants, by PRID ENCASE, and by YASMIN (R.d.B.-UniPG2016/17) and FCRPG.2016.0105.021 projects.

References

[1] Afrati FN, Ullman JD (2010) Optimizing joins in a map-reduce environment. In: Proc. of 13th International Conference on Extending Database Technology, pp 99–110
[2] Afrati FN, Borkar VR, Carey MJ, Polyzotis N, Ullman JD (2011) Map-reduce extensions and recursive queries. In: 14th International Conference on Extending Database Technology, pp 1–8
[3] Apache Software Foundation (2016) Apache Hadoop. `http://hadoop.apache.org`
[4] Apt K, Bol R (1994) Logic Programming and Negation: A Survey. Journal of Logic Programming 19/20:9–71
[5] Apt K, Blair H, Walker A (1989) Towards a Theory of Declarative Knowledge. In: Minker J (ed) Foundations of Deductive Databases and Logic Programming, Morgan Kaufmann
[6] Balduccini M, Pontelli E, Elkhatib O, Le H (2005) Issues in Parallel Execution of Non-Monotonic Reasoning Systems. Parallel Computing 31(6):608–647
[7] Baral C (2003) Knowledge representation, reasoning and declarative problem solving. Cambridge University Press
[8] Biere A, Heule M, Van Maaren H, Walsh T (eds) (2009) Handbook of Satisfiability. IOS Press
[9] Bu Y, Howe B, Balazinska M, Ernst M (2010) Haloop: efficient iterative data processing on large clusters. In: Very Large Data Bases (VLDB) Conference, ACM, pp 285–296
[10] Calimeri F, Perri S, Ricca F (2008) Experimenting with parallelism for the instantiation of ASP programs. Journal of Algorithms 63(1-3):34–54
[11] Ceri S, Gottlob G, Tanca L (1990) Logic Programming and Databases. Springer
[12] Clark K (1978) Negation as failure. In: Gallaire H, Minker J (eds) Logic and Data Bases, Plenum
[13] Dal Palù A, Dovier A, Formisano A, Pontelli E (2015) CUD@SAT: SAT solving on GPUs. J Exp Theor Artif Intell 27(3):293–316
[14] Dantsin E, Eiter T, Gottlob G, Voronkov A (2001) Complexity and expressive power of logic programming. ACM Comput Surv 33(3):374–425
[15] Davis M, Putnam H (1960) A Computing Procedure for Quantification Theory. Journal of the ACM 7:201–215

[16] Davis M, Logemann G, Loveland D (1962) A machine program for theorem proving. Communications of the ACM 5(7):394–397
[17] Dean J, Ghemawat S (2004) MapReduce: Simplified Data Processing on Large Clusters. Tech. rep., Google, Inc.
[18] Dovier A, Formisano A, Pontelli E, Vella F (2015) Parallel Execution of the ASP Computation - an Investigation on GPUs. In: Proceedings of the Technical Communications of the 31st International Conference on Logic Programming, CEUR-WS.org, no. 1433 in CEUR Workshop Proceedings
[19] Dovier A, Formisano A, Pontelli E, Vella F (2016) A GPU implementation of the ASP computation. In: Gavanelli M, Reppy JH (eds) Practical Aspects of Declarative Languages - 18th International Symposium, PADL 2016. Proceedings, Springer, Lecture Notes in Computer Science, vol 9585, pp 30–47
[20] El-Khatib O, Pontelli E (2000) Parallel Evaluation of Answer Sets Programs Preliminary Results. In: Workshop on Parallelism and Implementation of Logic Programming
[21] Fages F (1994) Consistency of Clark's completion and existence of stable models. Methods of Logic in Computer Science 1(1):51–60
[22] Finkel R, Marek V, Moore N, Truszczyński M (2001) Computing Stable Models in Parallel. In: Provetti A, Tran S (eds) Proceedings of the AAAI Spring Symposium on Answer Set Programming, AAAI/MIT Press, Cambridge, MA, pp 72–75
[23] Formisano A, Vella F (2014) On multiple learning schemata in conflict driven solvers. In: Bistarelli S, Formisano A (eds) Proceedings of the 15th Italian Conference on Theoretical Computer Science, CEUR-WS.org, CEUR Workshop Proceedings, vol 1231, pp 133–146
[24] Ganguly S, Silberschatz A, Tsur S (1990) A Framework for the Parallel Processing of Datalog Queries. In: Garcia-Molina H, Jagadish H (eds) Proceedings of ACM SIGMOD Conference on Management of Data, ACM Press, New York, pp 143–152
[25] Ganguly S, Silberschatz A, Tsur S (1992) Parallel Bottom-Up Processing of Datalog Queries. Journal of Logic Programming 14(1-2):101–126
[26] Gebser M, Kaminski R, Kaufmann B, Schaub T, Schneider MT, Ziller S (2011) A portfolio solver for answer set programming: Preliminary report. In: Delgrande JP, Faber W (eds) Logic Programming and Nonmonotonic Reasoning - 11th International Conference, LPNMR 2011, Vancouver, Canada, May 16-19, 2011. Proceedings, Springer, Lecture Notes in Computer Science, vol 6645, pp 352–357
[27] Gebser M, Kaminski R, Kaufmann B, Schaub T (2012) Answer Set Solving in Practice. Morgan and Claypool Publishers
[28] Gebser M, Kaminski R, Kaufmann B, Schaub T (2014) Clingo = ASP + control: Preliminary report. CoRR abs/1405.3694
[29] Gelfond M (2007) Answer sets. In: Handbook of Knowledge Representation. Chapter 7, Elsevier

[30] Gelfond M, Kahl Y (2014) Knowledge Representation, Reasoning, and the Design of Intelligent Agents The Answer-Set Programming Approach. Cambridge University Press
[31] Gelfond M, Lifschitz V (1988) The Stable Model Semantics for Logic Programs. In: International Symposium on Logic Programming, MIT Press, pp 1070–1080
[32] Giunchiglia E, Lierler Y, Maratea M (2006) Answer set programming based on propositional satisfiability. J Autom Reasoning 36(4):345–377
[33] Goldberg E, Novikov Y (2007) BerkMin: A fast and robust SAT-solver. Discrete Applied Mathematics 155(12):1549–1561
[34] Gonzalez JE, Xin RS, Dave A, Crankshaw D, Franklin MJ, Stoica I (2014) GraphX: Graph Processing in a Distributed Dataflow Framework. In: Proceedings of the 11th USENIX Symposium on Operating Systems Design and Implementation, USENIX
[35] Grossi G, Marchi M, Pontelli E, Provetti A (2008) Experimental Analysis of Graph-based Answer Set Computation over Parallel and Distributed Architectures. Journal of Logic and Computation 19(4):697–715
[36] Gupta G, Pontelli E, Carlsson M, Hermenegildo M, Ali K (2001) Parallel Execution of Prolog Programs: a Survey. ACM Transactions on Programming Languages and Systems 23(4):472–602
[37] Hayes PJ, Kowalski RA (1969) Semantic trees in automatic theorem proving. Machine Intelligence 4:87–101
[38] Heule M, van Maaren H (2009) Look-ahead Based SAT Solvers. In: Handbook of Satisfiability, IOS Press, chap 5, pp 155–184
[39] Hoos H, Lindauer MT, Schaub T (2014) claspfolio 2: Advances in algorithm selection for answer set programming. TPLP 14(4-5):569–585
[40] Jenkins J, Arkatkar I, Owens JD, Choudhary AN, Samatova NF (2011) Lessons Learned from Exploring the Backtracking Paradigm on the GPU. In: Proc. of Euro-Par 2011, Springer Verlag, pp 425–437
[41] Jeroslow RG, Wang J (1990) Solving propositional satisfiability problems. Ann Math Artif Intell 1:167–187
[42] Khronos Group Inc (2015) OpenCL: The open standard for parallel programming of heterogeneous systems. `http://www.khronos.org`
[43] Kowalski RA (1970) Search strategies for theorem-proving. Machine Intelligence 5:181–201
[44] Kowalski RA (1974) Predicate Logic as a Programming Language. In: Proceedings IFIPS, pp 569–574
[45] Lassez J, Jaffar J (1987) Constraint logic programming. In: Proc. 14th ACM POPL
[46] Le H, Pontelli E (2005) An Investigation of Sharing Strategies for Answer Set Solvers and SAT Solvers. In: Euro-Par, Springer Verlag, pp 750–760
[47] Le H, Pontelli E (2007) Dynamic Scheduling in Parallel Answer Set Programming Solvers. In: High Performance Computing Symposium, ACM Press, pp 367–374

[48] Leone N, Perri S, Scarcello F (2001) Improving ASP instantiators by join-ordering methods. In: Logic Programming and Non-Monotonic Reasoning, Springer Verlag, pp 280–294
[49] Leone N, Pfeifer G, Faber W, Eiter T, Gottlob G, Perri S, Scarcello F (2006) The DLV system for knowledge representation and reasoning. ACM Trans Comput Log 7(3):499–562
[50] Lierler Y, Maratea M (2004) Cmodels-2: SAT-based Answer Set Solver Enhanced to Non-tight Programs. In: Lifschitz V, Niemelä I (eds) Proceedings of the 7th International Conference on Logic Programming and NonMonotonic Reasoning Conference (LPNMR'04), Springer Verlag, vol 2923, pp 346–350
[51] Lin F, Zhao Y (2004) ASSAT: Computing Answer Sets of a Logic Program by SAT Solvers. Artificial Intelligence 157(1):115–137
[52] Lindauer MT, Hoos HH, Hutter F, Schaub T (2015) Autofolio: An automatically configured algorithm selector. J Artif Intell Res (JAIR) 53:745–778
[53] Liu L, Pontelli E, Son TC, Truszczyński M (2010) Logic programs with abstract constraint atoms: The role of computations. Artificial Intelligence 174(3-4):295–315
[54] Lloyd J (1987) Foundations of Logic Programming. Springer-Verlag, Heidelberg
[55] Low Y, Bickson D, Gonzalez J, Guestrin C, Kyrola A, Hellerstein JM (2012) Distributed GraphLab: a framework for machine learning and data mining in the cloud. Journal of the Proceedings of the VLDB Endowment 5(8):716–727
[56] Malewicz G, Austern MH, Bik AJC, Dehnert JC, Horn I, Leiser N, Czajkowski G (2010) Pregel: a system for large-scale graph processing. In: Proceedings of the 2010 ACM SIGMOD International Conference on Management of data, ACM Press
[57] Maratea M, Pulina L, Ricca F (2013) Automated selection of grounding algorithm in answer set programming. In: Baldoni M, Baroglio C, Boella G, Micalizio R (eds) AI*IA 2013: Advances in Artificial Intelligence - XIIIth International Conference of the Italian Association for Artificial Intelligence, Turin, Italy, December 4-6, 2013. Proceedings, Springer, Lecture Notes in Computer Science, vol 8249, pp 73–84
[58] Maratea M, Pulina L, Ricca F (2014) A multi-engine approach to answer-set programming. TPLP 14(6):841–868
[59] Maratea M, Pulina L, Ricca F (2015) Multi-engine ASP solving with policy adaptation. J Log Comput 25(6):1285–1306
[60] Marek V, Truszczyński M (1999) Stable models and an alternative logic programming paradigm. In: The Logic Programming Paradigm, Springer Verlag, pp 375–398
[61] Marek W, Truszczyński M (1991) Autoepistemic Logic. Journal of the ACM 38(3):588–619
[62] Marques Silva JP, Sakallah KA (1999) GRASP: A search algorithm for propositional satisfiability. IEEE Transactions on Computers 48(5):506–521
[63] Martinez-Angeles CA, de Castro Dutra I, Costa VS, Buenabad-Chávez J (2014) A Datalog engine for GPUs. In: Hanus M, Rocha R (eds) Declarative Program-

ming and Knowledge Management - Declarative Programming Days, KDPD 2013, Unifying INAP, WFLP, and WLP, Kiel, Germany, September 11-13, 2013, Revised Selected Papers, Springer, Lecture Notes in Computer Science, vol 8439, pp 152–168
[64] Niemelä I (1999) Logic Programs with Stable Model Semantics as a Constraint Programming Paradigm. Annals of Mathematics and AI 25
[65] Niemelä I, Simons P (1996) Efficient Implementation of the Well-founded and Stable Model Semantics. In: Joint International Conference and Symposium on Logic Programming, MIT Press, pp 289–303
[66] Niemelä I, Simons P (1997) Smodels - An Implementation of the Stable Model and Well-Founded Semantics for Normal LP. In: Logic Programming and Non-monotonic Reasoning, Springer Verlag, pp 421–430
[67] NVIDIA Corporation (2015) NVIDIA CUDA Zone. `https://developer.nvidia.com/cuda-zone`
[68] Perri S, Ricca F, Sirianni M (2013) Parallel instantiation of ASP programs: techniques and experiments. Theory and Practice of Logic Programming 13(2):253–278
[69] Pollard GH (1981) Parallel execution of Horn clause programs. PhD thesis, Imperial College, London, Dept. of Computing
[70] Pontelli E, El-Khatib O (2001) Exploiting Vertical Parallelism from Answer Set Programs. In: AAAI Spring Symposium on Answer Set Programming: Towards Efficient and Scalable Knowledge Representation and Reasoning
[71] Pontelli E, Ranjan D, Dal Palù A (2002) An Optimal Data Structure to Handle Dynamic Environments in Non-Deterministic Computations. Computer Languages 28(2):181–201
[72] Pontelli E, Le H, Son T (2010) An Investigation in Parallel Execution of Answer Set Programs on Distributed Memory Platforms. Computer Languages, Systems and Structures 36(2):158–202
[73] Ranjan D, Pontelli E, Gupta G (1999) On the Complexity of Or-Parallelism. New Generation Computing 17(3):285–308
[74] Rao J, Ross KA (1999) Cache conscious indexing for decision-support in main memory. In: Atkinson MP, Orlowska ME, Valduriez P, Zdonik SB, Brodie ML (eds) VLDB'99, Proceedings of 25th International Conference on Very Large Data Bases, September 7-10, 1999, Edinburgh, Scotland, UK, Morgan Kaufmann, pp 78–89
[75] Shepherdson J (1989) Negation in Logic Programming. In: Minker J (ed) Foundations of Deductive Databases and Logic Programming, Morgan Kaufmann
[76] Silverthorn B, Lierler Y, Schneider M (2012) Surviving solver sensitivity: An ASP practitioner's guide. In: Dovier A, Costa VS (eds) Technical Communications of the 28th International Conference on Logic Programming, ICLP 2012, September 4-8, 2012, Budapest, Hungary, Schloss Dagstuhl - Leibniz-Zentrum fuer Informatik, LIPIcs, vol 17, pp 164–175
[77] Simons P, Niemelä I, Soininen T (2002) Extending and implementing the stable model semantics. Artificial Intelligence 138(1-2):181–234

[78] Sunderam V (1990) PVM: a framework for parallel distributed computing. Concurrency: Practice & Experience 2(4)
[79] Tachmazidis I, Antoniou G (2013) Computing the Stratified Semantics of Logic Programs over Big Data through Mass Parallelization. In: Theory, Practice, and Applications of Rules on the Web - 7th International Symposium, RuleML 2013
[80] Tachmazidis I, Antoniou G, Flouris G, Kotoulas S, McCluskey L (2012) Large-scale Parallel Stratified Defeasible Reasoning. In: Proceedings of the European Conference on Artificial Intelligence (ECAI), IOS Press, pp 738–743
[81] Tachmazidis I, Antoniou G, Faber W (2014) Efficient Computation of the Well-Founded Semantics over Big Data. Theory and Practice of Logic Programming 14(4-5):445–459
[82] Ullman JD (1988) Principles of Database and Knowledge-Base Systems. Computer Science Press, Maryland
[83] Urbani J, Kotoulas S, Maassen J, van Harmelen F, Bal H (2012) WebPIE: A Web-Scale Parallel Inference Engine using MapReduce. Journal of Web Semantics 10:59–75
[84] Van Gelder A, Ross K, Schlipf J (1991) The Well-Founded Semantics for General Logic Programs. Journal of the ACM 38(3):620–650
[85] Warren DHD (1980) Logic programming and compiler writing. Software – Practice and Experience 10(2):97–125
[86] Wolfson O (1988) Sharing the load of logic-program evaluation. In: Jajodia S, Kim W, Silberschatz A (eds) Proceedings of the International Symposium on Databases in Parallel and Distributed Systems, Austin, Texas, USA, December 5-7, 1988, IEEE Computer Society, pp 46–55
[87] Wolfson O, Silberschatz A (1988) Distributed Processing of Logic Programs. In: Boral H, Larson P (eds) Proceedings of the SIGMOD International Conference on Management of Data, ACM, ACM Press, New York, pp 329–336
[88] Xu L, Hutter F, Hoos HH, Leyton-Brown K (2008) Satzilla: Portfolio-based algorithm selection for SAT. J Artif Intell Res (JAIR) 32:565–606
[89] Yang M, Shkapsky A, Zaniolo C (2015) Parallel bottom-up evaluation of logic programs: Deals on shared-memory multicore machines. In: De Vos M, Eiter T, Lierler Y, Toni F (eds) Proceedings of the Technical Communications of the 31st International Conference on Logic Programming (ICLP) 2015, CEUR-WS.org, CEUR Workshop Proceedings, vol 1433
[90] Zhang W, Wang K, Chau SC (1995) Data Partition and Parallel Evaluation of Datalog Programs. IEEE Transactions on Knowledge and Data Engineering 7:163–176

Chapter 8
Parallel Solvers for Mixed Integer Linear Optimization

Ted Ralphs, Yuji Shinano, Timo Berthold, and Thorsten Koch

Abstract In this chapter, we provide an overview of the current state of the art with respect to solution of mixed integer linear optimization problems (MILPs) in parallel. Sequential algorithms for solving MILPs have improved substantially in the last two decades and commercial MILP solvers are now considered effective off-the-shelf tools for optimization. Although concerted development of parallel MILP solvers has been underway since the 1990s, the impact of improvements in sequential solution algorithms has been much greater than that which came from the application of parallel computing technologies. As a result, parallelization efforts have met with only relatively modest success. In addition, improvements to the underlying *sequential* solution technologies have actually been somewhat detrimental with respect to the goal of creating scalable *parallel* algorithms. This has made efforts at parallelization an even greater challenge in recent years. With the pervasiveness of multi-core CPUs, current state-of-the-art MILP solvers have now all been parallelized and research on parallelization is once again gaining traction. We summarize the current state-of-the-art and describe how existing parallel MILP solvers can be classified according to various properties of the underlying algorithm.

Ted Ralphs
Lehigh University, Bethlehem, PA, USA, e-mail: ted@lehigh.edu

Yuji Shinano
Zuse Institute Berlin, Takustraße 7, 14195 Berlin, Germany, e-mail: shinano@zib.de

Timo Berthold
Fair Isaac Germany GmbH, Germany, Takustraße 7, 14195 Berlin, Germany, e-mail: timoberthold@fico.com

Thorsten Koch
Zuse Institute Berlin, Takustraße 7, 14195 Berlin, Germany, e-mail: koch@zib.de

Y. Hamadi und L. Sais (eds.), *Handbook of Parallel Constraint Reasoning*,
https://doi.org/10.1007/978-3-319-63516-3_8

8.1 Introduction

This chapter addresses the solution of *mixed integer linear optimization problems* (MILPs) on parallel computing architectures. A MILP is a problem of the following general form:

$$\min_{x \in \phi} c^\top x \tag{MILP}$$

where the set

$$\phi = \{x \in \mathbb{R}^n \mid Ax \leq b, l \leq x \leq u, x_j \in \mathbb{Z}, \text{ for all } j \in I\}$$

is the *feasible region*, described by a given matrix $A \in \mathbb{Q}^{m \times n}$; vectors $b \in \mathbb{Q}^m$ and $c, l, u \in \mathbb{Q}^n$; and a subset $I \subseteq \{1, \ldots, n\}$ indicating which variables are required to have integer values. Members of ϕ are called *solutions* and are assignments of values to the *variables*. The polyhedron

$$\P = \{x \in \mathbb{R}^n \mid Ax \leq b, l \leq x \leq u\}$$

is the feasible region of a *linear optimization problem* (LP)

$$\min_{x \in \P} c^\top x \tag{LP}$$

known as the *LP relaxation*. The class of problems that can be expressed in this form is quite broad and many optimization problems arising in practice can be modeled as MILPs (see, e.g., [66]). As a language for describing optimization problems, MILP (and mathematical optimization, more generally) has proven to be flexible, expressive, and powerful.

All state-of-the-art solvers for MILP employ one of many existing variants of the well-known branch-and-bound algorithm of [56]. This class of algorithm searches a dynamically constructed tree (known as the *search tree*), following the general scheme of the generic tree search algorithm specified in Algorithm 8.1. Sequential al-

Algorithm 8.1: A Generic Tree Search Algorithm

```
1 Add root r to a priority queue Q.
2 while Q is not empty do
3     Remove a node i from Q.
4     Process the node i.
5     Apply pruning rules.
6     if Node i can be pruned then
7         Prune (discard) node i.
8     else
9         Create successors of node i (branch) by applying a successor function, and add
          the successors to Q.
```

gorithms for solving MILPs vary broadly based on their methods of processing nodes (line 4); their strategies for the order of processing nodes (*search strategy*, line 3);

their rules for pruning nodes (line 5); and their strategies for creating successors (*branching strategy*, line 9). We provide some details on how each of these steps is typically managed in Section 8.2. In the case of a parallel algorithm, some additional strategic factors come into play, such as how the search is managed (now in parallel), what information is shared as global knowledge, and the specific mechanism for sharing this knowledge on a particular target computer architecture.

Tree search algorithms appear to be naturally parallelizable, and soon after the advent of networked computing researchers began to experiment with parallelizing branch and bound. In [40], Gendron and Crainic chronicle the early history, dating the first experiments to somewhere in the 1970s. It did not take long after the first large-scale systems became available for it to be realized that good parallel performance is often difficult to achieve. In the case of MILPs, this is particularly true (and becoming more so over time) for reasons that we summarize in Section 8.3.

Despite the enormity of the existing literature on solving search problems in parallel, the case of MILP appears to present unique challenges. Although some progress has been made in the more than two decades in which parallel algorithms for MILP have been seriously developed, it is fair to say that many challenges remain. The difficulty comes from a few intertwining sources. For one, the most capable sequential solvers are commercial software and therefore only available to members of the research community as black boxes with which parallel frameworks can interact in limited ways. Even if this were not the case, it is also generally true that more sophisticated solvers are inherently more difficult to parallelize in a scalable fashion because of their greater exploitation of global information sharing and increasing emphasis on operations that limit the size of the search tree (and hence limit opportunities for parallelization), among other things. Despite an appearance to the contrary, state-of-the-art sequential algorithms for solving MILPs depend strongly on the order in which the nodes of the tree are processed, and advanced techniques for determining this order are in part responsible for the dramatic improvements that have been observed in sequential solution algorithms [59]. It is difficult to replicate this ordering in parallel without centralized control mechanisms, which themselves introduce inefficiencies. Moreover, sequential algorithms heavily exploit the warm-starting capabilities of the underlying LP solver in that they can usually process the child of a node shoe parent has just been processed, a phenomena which is more difficult to exploit in combination with the load-balancing techniques described in Section 8.3.2.5. Finally, most research has focused on parallelization of less sophisticated sequential algorithms, which both limits the sophistication of the overall algorithm (and hence limits the degree to which challenging open problems can be tackled) and inherently involves different challenges to achieving scalability.

The limits to scalability are by now well understood, but the degree to which these limits can be overcome is unclear and will unfortunately remain a moving target. Scalability involves the interplay of a changing landscape of computer architectures and evolving sequential algorithms, neither of which is being developed with the goal of making this (or any other) particular class of algorithms efficient. The days in which we could expect to see exponential improvements to sequential algorithms are gone, so the future will likely bring an increasing emphasis on the development of algorithms that effectively exploit parallel architectures.

In the remainder of this chapter, we survey the current landscape with respect to parallel solution of MILPs. By now, most existing solvers have been parallelized in some fashion, while in addition, there are also a few frameworks and approaches for parallelizing solvers that are only available as black boxes. We begin by briefly surveying the current state of the art in sequential solution algorithms in Section 8.2. In Section 8.3, we provide an overview of issues faced in parallelizing these algorithms and the design decisions that must be taken during development. In Section 8.4, we review the approaches taken by a number of existing solvers. Finally, in Section 8.5, we discuss the tricky issue of how to measure performance of a solver before concluding in Section 8.6.

8.2 Sequential Algorithms

8.2.1 Basic Components

As we have already mentioned, most modern solvers employ sophisticated variants of the well known branch-and-bound algorithm of [56]. The basic approach is straightforward, yet effective: the feasible region of the original optimization problem is partitioned to obtain smaller *subproblems*, each of which is recursively solved using the same algorithm. This recursive process can be interpreted naturally as a tree search algorithm and visualized as the exploration of a *search tree*. The goal of the search is to implicitly enumerate all potential assignments of the values of the integer variables. The power of the algorithm comes from the bounds that are calculated during the processing of nodes in the search tree and are used to truncate the recursive partitioning process in order to avoid what would eventually amount to the costly complete enumeration of all solutions.

An updated version of Algorithm 8.1 that reflects the specific way in which a tree search is managed according to the basic principles of branch and bound is shown in Algorithm 8.2. In employing the metaphor of this algorithm as the exploration of a search tree, we associate each subproblem with a *node* in the search tree and describe the tree itself through parent-child relationships, beginning with the root node. Thus, each subproblem has both a *parent* and zero or more *children*. Subproblems with no children are called *terminal* or *leaf* nodes. In Algorithm 8.2, the set Q is the set of terminal subproblems of the search tree as it exists at the end of each iteration of the algorithm. We next describe the specific elements of this algorithm in more detail.

Bounding. The processing of a subproblem in line 4 of Algorithm 8.2 consists of the computation of updated upper and lower bounds on the value of an optimal solution to the subproblem.

The lower bound is calculated with the help of a relaxation, which is constructed so as to be easy to solve. The lower-bounding scheme of most MILP solvers is

Algorithm 8.2: A Generic Branch-and-Bound Algorithm

1 Add root optimization problem r to a priority queue Q. Set global upper bound $U \leftarrow \infty$ and global lower bound $L \leftarrow -\infty$
2 **while** $L < U$ **do**
3 Remove the highest priority subproblem i from Q.
4 **Bound** the subproblem i to obtain (updated) final upper bound $U(i)$ and (updated) final lower bound $L(i)$.
5 Set $U \leftarrow \texttt{min}\{U(i), U\}$.
6 **if** $L(i) < U$ **then**
7 **Branch** to create child subproblems $i_1, \ldots, i_k$ of subproblem i with
- upper bounds $U(i_1), \ldots, U(i_k)$ (initialized to ∞ by default); and
- initial lower bounds $L(i_1), \ldots, L(i_k)$ (initialized to $L(i)$ by default)
by partitioning the feasible region of subproblem i.
8 Add $i_1, \ldots, i_k$ to Q.
9 Set $L \leftarrow \texttt{min}_{i \in Q} L(i)$.

based on solution of a strengthened version of the LP relaxation (LP), which can be solved efficiently, i.e., in time polynomial in the size of the input data [51, 67]. The strengthening is done using techniques stemming from another, more involved procedure for solving MILPs known as the *cutting-plane method* [41] that was developed prior to the introduction of the branch-and-bound algorithm. The basic idea of the cutting-plane method is to iteratively solve and strengthen the LP relaxation of an MILP. To do so, in each iteration, one or more *valid inequalities* are added to the LP relaxation. These inequalities have to fulfill two requirements:

1. they are violated by the computed optimal solution to the LP relaxation, and
2. they are satisfied by all "improving solutions" (those with objective values smaller than U) in the feasible set ϕ.

Since they "cut off" the observed optimal solution to LP relaxation, such inequalities are called *cutting planes* or *cuts*. For an overview of cutting plane algorithms for MILP, see, e.g., [63, 99].

An upper bound, on the other hand, results when either the solution to the relaxation is also a member of ϕ or a solution feasible for the subproblem is found using an auxiliary method, such as a *primal heuristic* designed for that purpose (see Section 8.2.2 below). Note that when the solution to the relaxation is in ϕ, the pruning rule applies, since in that case, the upper and lower bounds for the subproblem are equal. For an overview of primal heuristics for MILP, see, e.g., [32, 8].

The bounds on individual subproblems are aggregated to obtain global upper and lower bounds (lines 9 and 5), which are used to avoid the complete enumeration of all (usually exponentially many) potential assignments of values to the integer variables. If a subproblem's lower bound is greater than or equal to the global upper bound (this includes the case in which the subproblem is *infeasible*, e.g., has no feasible solution), that subproblem can be pruned (line 6). The difference between the upper and lower bounds at a given point in the algorithm is referred to as the *optimality gap* and can be expressed as either an *absolute gap* (the difference itself) or a *relative gap* (the ratio of the difference to the lower or upper bound). The progress of the algorithm is sometimes expressed in terms of how the gap decreases over time, with a final gap of zero representing the successful completion of the algorithm.

Branching. The task of *branching* in line 7 of Algorithm 8.2 is to successively partition the feasible region ϕ into regions defining smaller *subproblems* until the individual subproblems can either be solved explicitly or it can be proven that their feasible region cannot contain an optimal solution. This partitioning is done using logical disjunctions that must be satisfied by all feasible solutions. A solution with least cost among all those found by solving the subproblems yields the global optimum.

The *branching strategy* or *branching method* is an algorithm for determining the particular disjunction to be used for partitioning the current subproblem once the processing is done. As with the addition of cutting places, we generally require that these disjunctions be violated by the solution to the relaxation solved to obtain the lower bound. Note that part of the task of branching is to determine initial bounds for each of the created subproblems. By default, the initial bound of the child subproblem is set equal to the final bound for that of its *parent*. However, some more sophisticated branching strategies involve the calculation of more accurate initial bounds for each of the children, which can be used instead. For an overview of branching strategies for MILP, see, e.g., [3, 11].

Search. Finally, the *search strategy* is the scheme for prioritizing the subproblems and determining which to process next on line 3. The scheme typically involves a sophisticated strategy designed to accelerate the convergence of the global upper and lower bounds. This generally involves a careful balance of *diving*, in which one prioritizes nodes deep in the tree at which we are likely to be able to easily discover *some* feasible solution, and *best bound*, in which one prioritizes nodes whose feasible regions are likely to contain *high-quality* solutions (though extracting those solutions may be more difficult).

Generally speaking, diving emphasizes improvement in the upper bound, while best bound emphasizes improvement in the lower bound. Once the upper and lower bounds are equal, the solution process terminates. We mentioned in Section 8.1 that the rate of convergence depends strongly on an effective search order. The fact that it is extremely difficult to replicate the same ordering in the parallel algorithm that would have been observed in a sequential one is one of the major impediments to achieving scalability. For an overview on search strategies for MILP, see [59] and [3].

8.2.2 Advanced Procedures

In addition to the fundamental procedures of *branching* and *bounding*, there are a number of auxiliary subroutines that enhance the performance of the basic algorithm. The most important such subroutines used in MILP solvers are those for preprocessing, primal heuristics, and conflict analysis, each of which we explain here briefly.

Primal Heuristics. These are algorithms that try to find feasible solutions of good quality for a given optimization problem within a reasonably short amount of time. There is typically no guarantee that they will find any solution, let alone an optimal one. General-purpose heuristics for MILP are often able to find solutions with objective values close to the global lower bound (and thus with provably high quality) for a wide range of problems; they have become a fundamental ingredient of state-of-the-art MILP solvers [9]. Primal heuristics have a significant relevance as supplementary procedures since the early knowledge of a high-quality feasible solution helps to prune the search tree by bounding and enhances the effectiveness of certain procedures that strengthen the problem formulation.

Preprocessing. These are procedures to transform the given problem instance into an equivalent instance that is (hopefully) easier to solve. The task of preprocessing is threefold: first, it reduces the size of the problem by removing irrelevant information, such as redundant constraints or fixed variables. Second, it strengthens the LP relaxation of the model by exploiting integrality information, e.g., to tighten the bounds of the variables or to improve coefficients in the constraints. Third, it extracts information from the model that can be used later to improve the effectiveness of branching and the generation of cutting planes. For an overview on preprocessing for MILP, see, e.g., [2, 39, 61].

Conflict analysis. This analysis is performed to analyze and track the logical implications whenever a subproblem is found to be infeasible. To track these implications, a so-called *conflict graph* is constructed, which gets updated whenever a subproblem is deemed to be infeasible. This graph represents the logic of how the combination of constraints enforcing the partitioning in the branching step led to the infeasibility, which makes it possible to prohibit the same combination from arising in other subproblems. More precisely, the conflict graph is a directed acyclic graph in which the vertices represent bound changes of variables and the arcs correspond to bound changes implied by logical deductions, so-called propagation steps. In addition to the inner vertices, which represent the bound changes from domain propagation, the graph features source vertices for the bound changes that correspond to branching decisions and an artificial target vertex representing infeasibility. Then, each cut in the graph that separates the branching decisions from the artificial infeasibility vertex gives rise to a valid conflict constraint. For an overview of conflict analysis for MILP, see, e.g., [1, 98].

Figure 8.1 illustrates the connections between the main algorithmic components of an MILP solver. The development of MILP solvers started long before parallel computing became a mainstream topic. Among the first commercial mathematical optimization software was IBM's MPS/360 and its predecessor MPSX, which were introduced in the 1960s. Interestingly, the MPS input data format, designed to work with the punch-card input system of early computers, is still the most widely supported format for state-of-the-art MILP solvers half a century later.

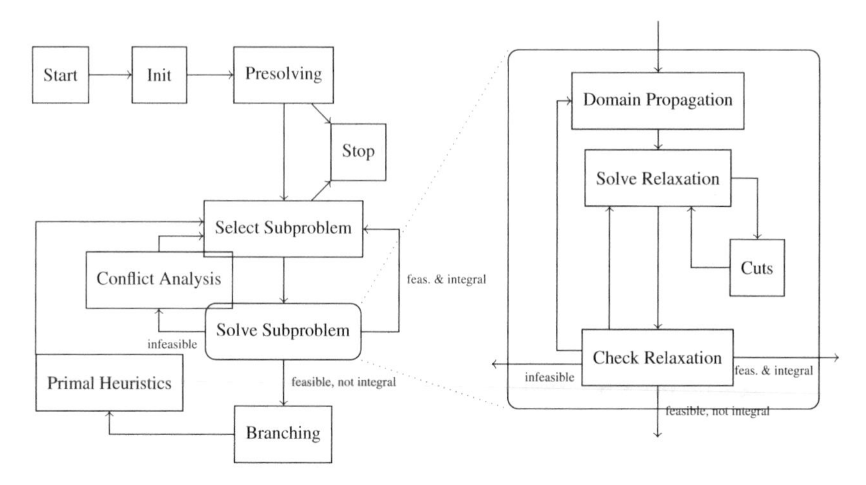

Fig. 8.1: Flowchart of the main solving loop of a typical MILP solver

Today, there is a wide variety of commercial software available for solving MILPs, including Xpress [31], Gurobi [44], and CPLEX [18]. All of them providing a deterministic parallelization for shared-memory systems. There are also several academic, noncommercial alternatives, such as CBC [35], Glpk [62], Lpsolve [30], SYMPHONY [75], DIP [74], and SCIP [81]. Some of these feature parallel algorithms, some do not. Every two years, Robert Fourer publishes a list of currently available codes in the field of linear and integer programming, the 2015 edition being the latest at the time of writing this chapter [36].

8.3 Parallel Algorithms

In this section, we discuss a variety of issues that arise in designing, implementing, and measuring the performance of parallel algorithms for solving MILPs. The material here is a high-level overview and thus intentionally avoids some details that are unnecessary in describing general concepts. Parallel algorithms can be assessed with respect to both *correctness* and *effectiveness*. Generally speaking, correctness is a mathematical property of an algorithm that must be proven independent of a particular implementation, though details of the implementation and even properties of the underlying hardware may sometimes matter in proving that a particular algorithm will terminate successfully without entering a race condition or other related error caused by the parallelization itself. We don't directly address correctness here, since correctness of this class of parallel algorithms typically follows easily from correctness of an associated sequential algorithm (with some minor exceptions that we'll point out). We focus instead on how the design of a parallel algorithm informs its *effectiveness*, a general term denoting how well the algorithm performs according to some specified measure.

Naturally, it is only once an algorithm has been implemented and deployed on a particular *computational platform* that we can assess its effectiveness. The end goal of the design and implementation process is what we call a *solution platform*: the combination of an algorithm and a *computational platform* consisting of a particular architecture, communication network, storage system, OS, compiler tool chain, and other related auxiliary components. We take a rather simplistic view of the details of the computational platform, assuming that it consists of a collection of *processors* capable of executing sequential programs and an interconnection network that allows any pair of processors to communicate. The details of *how* the processors communicate and the latency involved in such communication are vitally important in practice, but beyond the scope of this overview. For our purposes, the processors can be co-located on a single *central processing unit* (CPU) and communicate through memory or be located on physically distinct computers, with communication occurring over a network. We discuss more details about architectures in Section 8.3.3.1, but we also recommend [6] and [45] to the interested reader for more in-depth discussion and technical definitions.

In general, a complete parallel algorithm can be viewed as a collection of procedures and control mechanisms that can be divided roughly into two groups. Some are primarily geared towards governing the parallelization strategy: these include mechanisms for moving data between processors (see Section 8.3.2.4) and mechanisms for determining what each processor should be doing (see Section 8.3.2.5). Others are primarily methods for performing one of the subtasks associated with a standard sequential algorithm listed in Section 8.2 (branching, bounding, generating valid inequalities, etc.). It is the control mechanism governing the parallelization strategy that we generally refer to as the *parallel algorithm*. The remaining methods form the *underlying sequential algorithm*, since this latter collection of methods is generally sufficient to specify a complete sequential algorithm. Because most state-of-the-art sequential algorithms are based on some variant of tree search, a particular parallelization strategy can usually be used in tandem with any number of underlying sequential algorithms. We explore this idea in more detail in Section 8.3.2.1.

8.3.1 Scalability and Performance

By its very nature, tree search appears to be highly parallelizable—each time the successor function is applied, new tasks are generated that can be executed in parallel. By splitting the nodes in the set Q among multiple processors, it would seem that the search can be accomplished much faster. Unfortunately, there are a number of reasons why a naive implementation of this idea has proven not to work well, especially in the case of MILPs. At a high level, this is largely due to a few important factors.

- Most modern solvers spend a disproportionate amount of time processing the shallowest nodes in the tree (particularly the root node) because this generally results in a smaller overall tree, which generally leads to smaller running times in the case of a sequential algorithm. However, it is generally not until these shallowest nodes are processed that all computing resources can be effectively utilized.

- The tree being searched is constructed dynamically and its shape cannot be easily predicted [70, 17]. On top of that, state-of-the-art MILP solvers generate a highly unbalanced tree in general. This makes it difficult to divide the node queue into subsets of nodes whose processing (including the time to explore the entire resulting subtrees) will require approximately the same amount of effort.
- The order in which the nodes are considered can make a dramatic difference to the efficiency of the overall algorithm and enforcing the same ordering in a parallel algorithm as would be observed in a highly tuned sequential algorithm without a loss of efficiency is extremely difficult.
- Most state-of-the-art solvers generate a wide range of information while searching one part of the tree that, if available, could inform the search in another part of the tree.
- The combination of the above two points means that it's generally desirable/necessary to move large amounts of data dynamically during the algorithm, but this is costly, depending on the architecture on which the algorithm is being executed.

It is not difficult to see that there are several important tradeoffs at play. On the one hand, we would like to effectively share data in order to improve algorithmic efficiency. On the other hand, we must also limit communication to some extent or the cost of communication will overwhelm the cost of the algorithm itself. There is also a tradeoff between the time spent exploring the shallowest part of the tree and how quickly the algorithm is able to fully utilize all available processing power.

8.3.1.1 Scalability

Generally speaking, *scalability* is the degree to which a solver platform is able to take advantage of the availability of increased computing resources. In this chapter, we focus mainly on the ability to take advantage of more processors, but the increased resources could also take the form of additional memory, among other things. Scalability is a property of an entire solution platform, since scalability is inherently affected by properties of the underlying computational platform. Most parallel algorithms are designed to exhibit good performance on a particular such platform, but we do not attempt to survey the properties of platforms affecting scalability in this article and rather focus mainly on the properties of the algorithms themselves. For a good general introduction to computational platforms, see [45] and [6] (also see Section 8.3.3.1 for a brief summary). In the remainder of this section, we briefly survey issues related to scalability of algorithms.

Phases. The concept of algorithm phases is an important one in parallelizing tree search algorithms. Conceptually, a parallel algorithm consists of three phases. The *ramp-up phase* is the period during which work is initially partitioned and allocated to the available processors. In our current setting, this phase can be defined loosely as lasting until all processors have been assigned at least one task. The second phase is the *primary phase*, during which the algorithm operates in steady state. This is followed by the *ramp-down* phase, during which termination procedures are executed and final results are tabulated and reported. Defining when the ramp-down phase begins is slightly problematic, but we define it here as the earliest time at which one of the processors becomes permanently idle.

In tree search, a naive parallelization considers a "task" to be the processing of a single subproblem in the search tree. The division of the algorithm into phases is to highlight the fact that when parallelized naively, a tree search algorithm cannot take full advantage of available resources during the ramp-up and ramp-down portions of the algorithm without changing the granularity of the tasks (see Section 8.3.2.2). For some variants of the branch-and-bound algorithm and for instances with large size or particular structure, the processing of the root subproblems and immediate descendants can be very time-consuming relative to the time to search deeper parts of the tree, which results in lengthy ramp-up and ramp-down periods. This can make good scalability difficult to achieve.

Sequential Algorithm. Although the underlying sequential algorithm is often viewed essentially as a black box from the viewpoint of the parallel algorithm, it is not possible in general to parallelize a sequential algorithm without inherently affecting how it operates. Most obviously, the parallel algorithm generally reorders the execution of tasks and thus changes the search strategy. However, other aspects of the parallel algorithm's approach can affect the operation of the sequential parts in important ways. The parallel algorithm may even seek to control the sequential algorithm to make the combination more scalable. The degree to which the parallel algorithm can control the sequential algorithm and the degree to which it has access to information inside the "black box" has important implications.

Overhead. Scalability of parallel MILP solvers is often assessed by measuring the amount of *parallel overhead* introduced by the parallelization scheme. Roughly speaking, parallel overhead is the work done in parallel that would not have been performed in the sequential algorithm. It can be broken down into the following general categories (see [53]).

- *Communication overhead*: Computation time spent sending and receiving information, including time spent inserting information into the send buffer and reading it from the receive buffer. This is to be differentiated from time spent waiting for access to information or for data to be transferred from a remote location.
- *Idle time (ramp-up/ramp-down)*: Time spent waiting for initial tasks to be allocated or waiting for termination at the end of the algorithm. The ramp-up phase includes inherently sequential parts of the algorithm, such as time spent reading in the problem, processing the root node, etc., but also the time until enough B&B nodes are created to utilize all available processors. The ramp-up and ramp-down time is highly influenced by the shape of the search tree. If the tree

is "well balanced" and "wide" (versus deep), then both ramp-up and ramp-down time will be minimized.

- *Idle time (latency/contention/starvation)*: Time spent waiting for data to be moved from where it is currently stored to where it is needed. This can include time waiting to access local memory due to contention with other threads, time spent waiting for a response from a remote thread either due to inherent latency or because the remote thread is performing other tasks and cannot respond, and even time spent waiting for memory to be allocated to allow for the storage of locally generated data.
- *Performance of redundant work*: Time spent performing work (other than communication overhead) that would not have been performed in the sequential algorithm. This includes the evaluation of nodes that would not have been evaluated with fewer threads. The primary reason for the occurrence of redundant work is differences in the order in which the search tree nodes are explored. In general, one can expect that the performance of redundant work will increase when parallelizing the computation, since information that would have been used to avoid the enumeration of certain parts of the search space may not yet have been available (locally) at the time the enumeration took place in the parallel algorithm. However, it is entirely possible that the parallel algorithm will explore fewer nodes in some cases.

Effective parallelization is about controlling overhead, but how best to do this can vary depending on a wide range of factors and there is no "one-size-fits-all" solution. Properties of the search algorithm itself, properties of the instances to be solved, and properties of the architecture on which the algorithm will be deployed all play a role in determining what approach should be taken. This is one of the reasons why we see such a wide variety of approaches when we consider solvers in the wild. Evidence of this is presented in Section 8.4.

8.3.1.2 Performance

It is important to point out that scalability, also called *parallel performance*, is *not* the same as overall "performance." As we have already noted, parallelization of more sophisticated sequential algorithms is inherently more difficult in terms of achieving scalability than parallelization of a less sophisticated algorithm. Moreover, it could be the case that a solver platform exploiting a more sophisticated underlying sequential algorithm, although not as scalable when parallelized, would nevertheless outperform a more scalable but less sophisticated solver platform on the usual metrics used to assess overall effectiveness and solution power, such as wall-clock solution time or ability to solve difficult instances. The fact that there is a performance tradeoff between scalability and overall performance is one of the fundamental difficulties in determining how to measure performance and in assessing progress in this challenging field of research. We discuss measures of performance in much greater detail in Section 8.5.

8.3.2 Properties

In this section, we provide some means by which to classify the existing algorithms. This is certainly not the first attempt to do such classification. Already in 1994, [40] provided a very thorough classification for parallel branch-and-bound algorithms. An updated survey was later published in [19]. By now, much has changed and it is no longer possible to provide a classification of existing solvers in the traditional sense of a partitioning into subsets taking similar approaches. Modern solvers vary along many different axes in a complex design space. No partition of algorithms into subsets based on fundamental properties will likely be satisfactory, or if so, the subsets will have cardinality one. We therefore simply list basic properties of existing algorithms that we refer to in Section 8.4 when describing existing algorithms and software.

8.3.2.1 Abstraction and Integration

The separation between the parallel and sequential parts of the algorithm discussed in the previous part need not be as clean as we have indicated in our discussion in Section 8.3.1.1. In some cases, there is a tight integration of parallel and sequential parts into a single monolithic whole. In other cases, the parallel algorithm is completely separated from the underlying sequential algorithm or even encapsulated as a *framework*, whose overall approach does not depend on the details of the sequential algorithm. To make the clean separation of the parallel algorithm and the underlying sequential algorithm possible, the parallel algorithm must have a high level of *abstraction* and a low level of *integration* with the sequential solver. The concept of abstraction is similar to that which forms the basic philosophy underlying object-oriented programming languages, but we distinguish here between two different types of abstraction: *algorithmic abstraction* and *interface abstraction*.

To exemplify what we mean by *algorithmic abstraction*, consider Algorithm 8.2 once again. This tree search algorithm is stated at a high level of algorithmic abstraction because the internal details of how the bounding, branching, and prioritization methods work are not specified and the algorithm does not depend on these details. Naturally, any concrete implementation of this algorithm must specify these elements, but the algorithm as stated constitutes an *abstract framework* for branch and bound that is agnostic with respect to the internal implementation of these algorithmic elements.

Of course, there is some interaction between the unspecified elements of the underlying branch-and-bound algorithm and the abstract framework. These constitute the algorithmic *interface*, by which the framework interacts with the components of the underlying algorithm through the production of outputs in a specified form. For example, the bounding method on line 4 in Algorithm 8.2 is required to produce a lower bound that the framework may then use later on line 3 to determine the search order or on line 6 to prune nodes. The framework does not need to know *how* the bound is produced, it only requires that the underlying sequential methods produce it *somehow*. This may be done, for example, through a function call to the underlying sequential algorithm that encapsulates the processing of a single node in the search tree.

The flexibility and level of access to internal parts of the sequential algorithm that are present in a particular sequential solver's API determine the degree to which the parallel solver is able to control various aspects of the underlying sequential algorithm. Lack of ability to execute certain internal procedures independent of the sequential algorithm may limit the options for parallelization. In some cases, the only access to the sequential solver's functionality is by limited execution of the sequential solution algorithm, as a "black box," for solution of subproblems associated with nodes in the search tree. In other cases, fine-grained access to individual procedures, such as those for bounding and branching of individual nodes in the search tree, is available.

A related concept is that of *implementational abstraction*, which is a more practical measure of the degree to which the *implementation* (in the form of source code) of a parallel search algorithm is customized to work in tandem with a particular *implementation* of an underlying sequential solver. This has primarily to do with whether or not the interface between the parallel and sequential parts is well defined and "generic." In general, frameworks intended to work with multiple underlying sequential solvers need to have high degrees of both implementational and algorithmic abstraction. When there is a tight integration between the parallel and sequential parts (especially when they are produced as a monolithic whole), the level of implementational abstraction is often low, but this does not mean that the level of algorithmic abstraction is low.

Implementations with a low degree of abstraction depend on internal interfaces that may either not be well defined or depend on the passing of data that only a solver with a particular internal implementation can produce. Implementations with a high degree of abstraction interact with the underlying sequential solver only through well defined interfaces that only require the passing of data that almost any sequential algorithm should be able to produce. How those interfaces work and at what points in the sequential algorithm the solver is required to pass information depends on the level of granularity of the solver.

With an abstract framework, such as the one specified in Algorithm 8.2, parallelization becomes much easier, since we can conceptually parallelize the abstraction without dependence on the unspecified internals of the various algorithmic elements. A simple way of parallelizing Algorithm 8.2 is just to maintain set Q and global bounds L and U as global data, but allow multiple subproblems to be simultaneously bounded (this is one variant of the Master-Worker coordination scheme described in Algorithms 8.5 and 8.6). In other words, each processor independently executes the same processing loop, but all update the same global bounds, access the same common global queue (Q) of subproblems at the top of the loop, and insert into the same global queue any newly produced subproblems. As a practical matter, this requires the avoidance of conflicts in writing and reading global data, which is one primary source of overhead, but as long as this can be achieved, the convergence of the algorithm in parallel follows from the same principles that guarantee convergence of the sequential algorithm.

8.3.2.2 Granularity

One way in which we can differentiate algorithmic approaches is by their *granularity*. Roughly speaking, the granularity refers to what is considered to be the atomic unit of work in the algorithm. We consider four primary levels of granularity, but of course, there are many gradations and most algorithms either fall somewhere between these levels or may even take different approaches in different phases of the algorithm. Ranked from coarse to fine, they are *tree parallelism*, *subtree parallelism*, *node parallelism*, and *subnode parallelism*.

Tree parallelism. Several trees can be searched in parallel using different strategies and the knowledge generated when building one tree can be used for the construction of other trees. In other words, several tree searches look for solutions in the same state space at the same time. Each tree search algorithm can take a different approach, e.g., different successor function, search strategy, etc. For example, Pekny [71] developed a parallel tree search algorithm for solving the Traveling Salesman Problem in which the trees being built differed only in the successor functions (branching mechanism). A brute-force strategy for parallelization is to concurrently start single-threaded solves of slightly perturbed (but mathematically equivalent) copies of the same MILP problem. One implementation of this is the racing scheme described in Algorithm 8.3

Subtree parallelism. Multiple subtrees of the same overall search tree may be explored simultaneously, but independently (without sharing information). This can be accomplished, for example, by passing a single subproblem to a sequential solver and executing the sequential algorithm for some fixed amount of time. Most of the more advanced algorithms described later in this chapter use some form of subtree parallelism to avoid the overhead associated with more fine-grained approaches.

Node parallelism. A single search tree can be searched in parallel by executing an algorithm similar to the sequential one, but processing multiple nodes simultaneously, with a centralized control mechanism of some sort and more information sharing. A straightforward implementation would utilize a master process to coordinate the search, distribute nodes from the queue, and collect results. This is the most widely used type of parallelism for tree search algorithms and is reflected in the Master-Worker coordination scheme described in Algorithms 8.5 and 8.6.

Subnode parallelism. The parallelism may be introduced by performing the processing of a single node in parallel. This can be done effectively with decomposition-based algorithms, such as column generation, for example.

- *Strong branching*: A natural candidate for parallelization is the so-called strong branching approach to selecting a branching disjunction, in which some pre-computations are done to evaluate the potential effectiveness of each candidate.

Similarly, probing or domain propagation techniques, which also involve precomputations to simplify a given subproblem and improve the efficiency of later bounding computations, might also be parallelizable [42].
- *Solution of LPs*: All solution algorithms for solving MILPs rely fundamentally on the ability to quickly solve sequences of LPs, usually in the context of deriving a bound for a given subproblem. The barrier algorithm for solving the initial LP relaxation in the root node, always one of the first steps in any algorithm for solving a MILP, can be parallelized effectively. Recently, there has also been progress in the development of a parallel implementation of the dual simplex algorithm, the most commonly used algorithm for solving the LP relaxations in non-root nodes, due to its superior warm-starting capabilities (see [47]).
- *Heuristics*: Primal heuristics can often be executed independently of the search itself. Hence, they can be used to employ processor idle time to best effect. Furthermore, multiple expensive heuristics, such as large neighborhood search algorithms, can be run concurrently or can themselves be parallelized.
- *Cutting*: During bounding of individual search tree nodes, the relaxations are strengthened by executing additional sub-procedures, called *separation algorithms*. Multiple such algorithms can be executed simultaneously to achieve faster strengthening.
- *Decomposition*: If a problem naturally decomposes due to block structure in the matrix A, the problem can be solved as a sequence of completely independent subproblems. Even when the matrix A does not have perfect block structure (it only decomposes after removing some rows or columns), decomposition methods, such as Dantzig-Wolfe [21], can be employed, which enable parallelization of the bounding procedure.

8.3.2.3 Adaptivity

Another general property of algorithms is the degree to which they are *adaptive*. All algorithms must be adaptive to some extent. For example, most algorithms switch strategies from phase to phase, e.g., initially pursuing tree-level parallelism for exploring shallow parts of multiple trees during the ramp-up phase, then switching to subtree or node parallelism during the primary phase. In general, an algorithm for solution of MILPs could be changed adaptively as the solution process evolves based on many different possible aspects of the underlying algorithm, such as the global upper and lower bounds, the distribution of subproblems across processors, the overall "quality" of the remaining work (see Section 8.3.2.5), and others. When, how, and which part of algorithms are adapted at run-time is a crucial aspect of performance.

8.3.2.4 Knowledge Sharing

Knowledge is the data generated during the course of a search, such as solutions, valid inequalities, conflict information, bounds, and other "globally valid" data generated as a by-product of the search in one part of the tree that may inform the search in another. Such knowledge can be used to guide the search, e.g., by determining the order in which to process available nodes. *Global knowledge* is knowledge that is valid for all subproblems in the search tree. *Local knowledge* is knowledge that is only valid with respect to certain subproblems and their descendants. The general categories of knowledge to be shared in solving MILPs includes the following.

- *Bounds*: Global upper bounds, as well as the bounds of individual nodes in the search tree can be shared.
- *Nodes*: Descriptions of the search tree nodes themselves are data that can (and usually must) be shared.
- *Solutions*: Feasible solutions from one part of the tree may help with computations in another part.
- *Pseudocost estimates*: So-called *pseudocost estimates* are used to predict the effectiveness of branching disjunctions based on historical statistics.
- *Valid inequalities*: Valid inequalities used to strengthen relaxations in one part of the tree may be useful in strengthening relaxations in another part of the tree.

With knowledge about the progress of the search, the processes participating in the search are able to make better decisions. When knowledge about global bounds, for example, is shared among processes, it is easier to avoid the performance of redundant work and the parallel search can be executed in a fashion more similar to its sequential counterpart. If all processes had complete knowledge of all global information, then no redundant work should be performed at all, in principle.

Despite the obvious benefits of knowledge sharing, there are several challenges associated with it. Most obviously, increased sharing of knowledge can have significant impact on parallel overhead. Communication overhead and a possible increase in idle time due to the creation of communication bottlenecks are among the inherent costs of sharing knowledge. An increase in redundant work, on the other hand, is the main cost of *not* sharing knowledge. This highlights a fundamental tradeoff: sharing of knowledge reduces the performance of redundant work and can improve the effectiveness of the underlying sequential algorithms, but these benefits come at a price. The goal is to strike the proper balance in this tradeoff. Trienekens and de Bruin [92] give a detailed description of the issues involved in knowledge generation and sharing.

8.3.2.5 Load Balancing

Load balancing is the method by which information describing the units of work to be done is moved from place to place as needed to keep all processors busy with useful work. It is characterized by how often it is done, what information must be moved (mainly a product of what are considered to be the atomic tasks), and whether it is done statically (just once at the beginning) or dynamically (throughout the algorithm).

Static load balancing. The first task of any parallel algorithm is to create enough units of work to ensure all processors can be employed in useful work as quickly as possible. The goal is to make the initial work distribution as even as possible in order to avoid having to either rebalance later or absorb the overhead resulting from the idling of processors whose work is finished quickly.

In a pure static load-balancing scheme, each processor works independently on its assigned tasks following the initial distribution, reporting final results at the end. If the initial work distribution is uneven, there is no recourse. Such a scheme requires very little communication, but can result in the performance of a vast amount of redundant work, as well as a large amount of idle time due to the fact that the tasks assigned to one processor may be much less time-consuming than the tasks assigned to another processor. Because it is extremely challenging to predict the difficulty of any given subproblem, static load balancing on its own is usually not very effective, though a well-designed static load-balancing scheme has proven effective [79] in some cases.

In an early paper, Henrich [46] summarized four categories of initialization methods and describes the advantages and disadvantages of each. We repeat these here to give a flavor of these methods.

- *Root initialization*: When applying this method, one process processes the root of the search tree and creates children according to a branching scheme that may or may not be the same as the one used during the primary phase. These children of the root are then distributed to other processors and the process is continued iteratively until all processors are busy. Root initialization is the most common approach due to the fact that it is easy to implement. For example, one can use the same branching scheme and mechanisms for distributing nodes to idle processors that are used during the primary phase of the algorithm. The method is most effective when the time to process a node is short and the number of children created during branching is large, so that all processors are engaged in work as early as possible. A major shortcoming of root initialization is that when the processing times are not short or when the number of available processors is large, many of the processes are *idle* while waiting to receive their allocation of nodes.
- *Enumerative initialization*: This method broadcasts the root node to all processes, which then perform an initial tree search according to the sequential algorithm. When the number of leaf nodes on each processor is at least the number of processes, processes can stop expanding. The ith process then keeps the ith node and deletes the rest. In this method, all processes are working from the very beginning and no communication is required. On the other hand, there is redundant work because each process is initially doing an identical task. Note also that it is crucial that all processors generate identical search trees (this

requires determinism of the sequential algorithm, see Section 8.3.2.7). This method has been successfully implemented in PEBBL [29].
- *Selective initialization*: This method starts with the broadcasting of the root node to each process. Each process then generates one single path from the root (without generating any others). The method requires little communication, but requires a sophisticated scheme to ensure processes work on distinct paths. It also requires determinism as mentioned above.
- *Direct initialization*: This method does not build up the search tree explicitly. Instead, each process directly creates a node at a certain depth in the search tree. The number of nodes at this depth should be no less than the number of processes. This method requires little computation and communication, but it works only if the structure of the search tree is known in advance.

More sophisticated schemes have been implemented recently that build on these basic ideas, such as the *racing ramp-up* of the Ubiquity Generator (UG) framework [87, 94] (described in Section 8.3.3.3) or the *spiral* and *two-level root initialization* schemes of the CHiPPS framework [105, 100].

Dynamic Load Balancing. Although static load balancing is effective in certain cases, the uneven processor utilization, processor speed, memory size, and the change of tree shape due to the pruning of subtrees can make the workloads of processes gradually become unbalanced, especially in distributed computing environments. This necessitates dynamic load balancing, which involves reallocating workload among the processes during the execution of a parallel program.

Dynamic load balancing has been studied in many computational contexts and the literature abounds with surveys of various techniques that have been implemented ([23] and [97] are but two thorough and relevant such surveys). Load balancing in general parallel tree search generally addresses the need to distribute the quantity of work evenly among the processors (or at least to keep all processors busy). The challenge of tree search for solving optimization problems is that not all work in the current queue has the same priority. When work of higher priority is produced (new nodes are generated through branching), this work should preempt the currently existing lower-priority work. Unfortunately, the distribution of high-priority work can (and is very likely to) become uneven through the natural course of the algorithm. Even when all processors are busy, some of the work being done may not be useful (it will turn out to be redundant in hindsight). For this reason, load balancing needs to consider both *quality* and *quantity* of work when redistributing workload. The definitions of quality and quantify of work follow.

Definition 1. The *quality of work* is a numeric value to measure the possibility that the region to be explored (represented by a node or a set of nodes) contains *high quality* solutions.

Definition 2. The *quantity of work* is a numeric value to measure the *amount* of work. For instance, this could be the estimated number of nodes to be processed in a subtree.

The relative quality (and also quantity) of existing work may change as global information is received. Furthermore, because it is far more efficient to process a node in the search tree directly following the processing of its parent on the same processor due to the ability to warm start the computation, one must be careful not to be overly aggressive in balancing the load.

For these reasons, load balancing in optimization applications can be more challenging than in many general tree search applications. This is certainly not a new observation in the context of parallel optimization and was discovered early during the development of parallel algorithms for branch and bound [26, 27].

In general tree search, a vast number of methods have been proposed to dynamically balance workloads, many of which have been focused on tree-based computations [54, 69, 78, 80, 89]. Despite the wide array of work on this topic, most schemes, even when described as very general, have been targeted at particular problem classes or particular architectures and are best suited for these use cases. Few truly general-purpose schemes have been proposed for the simple reason that requirements vary dramatically for different applications and algorithmic approaches.

Broadly, load-balancing strategies can be categorized based on the degree of centralization of the mechanism and based on whether the balancing is initiated by a separate manager process, by the sender, or by the receiver. Centralized schemes involving one or more processors that act as managers, tracking workload and directing transfers, are generally referred to as *work-sharing* schemes, whereas schemes in which transfers are initiated by individual processors with either surplus or deficit workload are referred to as *work-stealing*.

Kumar, *et al.* [54] studied a number of early dynamic load-balancing schemes, such as *asynchronous round robin*, *nearest neighbor*, and *random polling*, and performed extensive scalability testing. We briefly describe these schemes here to provide a flavor of how such schemes work.

- In an asynchronous round robin scheme, each process maintains an independent variable *target*, which is the identification of the process to ask for work, i.e., whenever a process needs more work, it sends a request to the process identified by the value of the target. The value of the target is incremented each time the process requests work. Assuming P is the number of processes, the value of the target on process i is initially set to $(i+1)$ modulo P. A process can request work independently of other processes. However, it is possible that many requests may be sent to processes that do not have enough work to share.
- The nearest neighbor scheme assigns each process a set of neighbors. Once a process needs more work, it sends a request to its immediate neighbors. This scheme ensures locality of communication for requests and work transfer. A disadvantage is that the workload may take a long time to be balanced globally.
- In the random-polling scheme, a process sends a request to a randomly selected process when it needs work. All processors have the same probability of being selected. Although it is very simple, random polling is quite effective in some applications [54].

There are many important tradeoffs encapsulated in the load-balancing scheme. The work units themselves can be considered to be "knowledge" of a certain type that can be shared either more or less aggressively. More aggressive balancing will result in less idling and less redundant work, but also increased overhead, much as in the case of sharing of other types of knowledge. Modern solvers generally employ more sophisticated, adaptive schemes and these will be described in Section 8.4.1.

8.3.2.6 Synchronization and Coordination

Synchronization is a requirement that at some step in a parallel algorithm, a set of processes needs to proceed simultaneously from a known state for reasons of either correctness or efficiency. One common purpose of synchronization is to enforce *determinism* (see Section 8.3.2.7), but the need for synchronization arises in many applications for a variety of reasons. It is a natural requirement when parallelizing an algorithm that was originally designed to be sequential and sequential algorithms often aggregate results obtained in a distributed fashion at intermediate points.

Synchronization can be achieved either by using a *barrier* or by some kind of counting scheme [96]. A barrier is a mechanism that prevents processes from continuing past a specified point in a parallel program until certain other processes reach this point. The substantial downside of introducing synchronization is that some processes might reach the synchronization point much more quickly than others and will waste time in waiting for other processes to reach the same state. From the standpoint of parallel scalability, synchronization introduces overhead and overhead leads to a loss of performance. Therefore, synchronization is to be avoided if possible.

The most obvious need for synchronization arises from the need to compute accurate global bounds. It is evident that if Algorithm 8.2 were to be parallelized straightforwardly in an asynchronous fashion, computing accurate global bounds would not be possible. This is because the global lower bound requires minimizing over the terminal nodes (those without children) in the current search tree, but the set Q doesn't always contain all terminal nodes. In the sequential algorithm, this problem doesn't arise because the bound is only calculated at a time when we *know* that set Q *does* contain all terminal nodes. In parallel, however, some form of synchronization is required.

Fortunately, the maintenance of accurate global lower and upper bounds is not a strict requirement in parallel branch and bound. An upper bound is only needed for applying pruning rules and certain other advanced functionality, but any provably valid upper bound will suffice. It is not necessary (though it is generally desirable) for a global upper bound to be accurately computed and known to all processes. Likewise, knowledge of the global lower bound is not necessary either except possibly as a measure of progress. Application of pruning rules only requires the lower bound computed for an individual subproblem to be accurate.

A related, though different, phenomenon is the introduction of *coordination*, which may introduce communication bottlenecks due to the requirement that an algorithm *coordinate* actions. Coordination does not always require strict synchronization, but it may introduce points at which one process must wait for a reply from another one. Load balancing is an obvious example of coordination, but there are many other possibilities. In general, coordination leads to overhead. At the same time, coordination can also improve the performance of the underlying sequential algorithm, so this introduces a fundamental tradeoff and a balance that must be struck. We discuss coordination schemes used in modern solvers in Section 8.3.3.3.

The appropriate amount of synchronization and coordination has to do with the relative cost of communication on the computational platform on which the algorithm is to be deployed. An asynchronous execution mode is appropriate for algorithms running on networks of workstations and computational grids, where synchronization is hard to achieve. A major issue with asynchronous algorithms is that there is typically no process that has accurate information about the overall state of the search. This can cause difficulties in effectively balancing workload and detecting termination. Load-balancing and termination detection schemes should have the ability to deal with the issue of innaccurate information.

8.3.2.7 Determinism

A deterministic solver platform is one that is guaranteed to perform the same operations and to produce the same end result when provided with the same input. Although all valid parallel algorithms should produce a valid result in the end, it is often the case that more than one valid result is possible (due to alternative optimal solutions) and even if not, the intermediate computations done to achieve the result can vary substantially. Small variations in the timing of the discovery of intermediate feasible solutions may lead to small differences in the global upper bounds that get applied in the pruning step of an algorithm, leading to a different search tree being produced. Even tiny differences in the execution path in the beginning stages of a parallel algorithm can lead to very large differences in the overall execution and can even cause changes in running time of an order of magnitude or more. Non-determinism is even easily possible in the case of a sequential algorithm. For example, some algorithms depend on the generation of random numbers. It's also possible for tie-breaking to occur differently on different computational platforms, depending on how memory is allocated.

It should be clear then that ensuring an algorithm executes deterministically, even in the sequential case, requires careful attention to detail. In the parallel case, this is not only difficult, but inevitably requires some kind of synchronization and control of when inter-process communication happens. Thus, determinism also leads to a degradation in performance.

Further, it is not entirely clear what determinism *means* in the parallel case. In the sequential case, determinism means performing exactly the same atomic operations in exacctly the same sequence (putting aside possible reordering of operations by hyper-threading and such). In the parallel case, we generally cannot hope to ensure that individual atomic operations are performed in precisely the same order (*strong deterministic parallelism*, but rather must make a similar requirement that allows for different ordering at a low level with order preserved at a higher level (*weak deterministic parallelism* [68]. The precise definition is implementation-dependent, but one might require for example, that exactly the same search tree be explored with exactly the same set of subproblems and exactly the same relaxations solved at each step to produce exactly the same sequence of bounds and thus exactly the same end result. Accomplishing this requires not only synchronization, but also awareness that computations may vary on different processing elements on a single computational platform.

8.3.3 Implementation

In this section, we discuss some of the algorithmic issues that arise mainly in the implementation phase when the conceptual algorithm is translated into a computer program, compiled on a given computational platform, and deployed.

8.3.3.1 Platform

Although it is possible for algorithms to be conceived independently of the computational platform on which they are to be run, design must, to some extent, be informed by one or more target platforms. We have so far discussed a number of important tradeoffs, such as the fact that the sharing of knowledge may both *reduce* the total amount of computation time spent executing the parts of the algorithm associated with the underlying sequential algorithm while *increasing* the parallel overhead. We have the same sort of tradeoff with regards to the frequency of load balancing. Determining the appropriate compromise between the two sides of such tradeoffs is only possible in light of a particular platform. It is only with specific knowledge of the platform that particular hardware parameters, such as the communication latency, can be known.

Naturally, the development of a solver is time-consuming and one would therefore like to ensure effectiveness across as wide a variety of platforms as possible. To a certain extent, this can be done through parameterization. We specify certain parameters, such as the frequency of load balancing, and then tune them at run-time, based on collected statistics. This works fine for relatively small variations occurring between platforms, but for larger variations, design decisions must be made that inherently limit the algorithm to certain classes of platform.

Communication Network. Perhaps the most important property of the platform that needs to be taken into account is the nature of the underlying communication network. A full treatment of the properties of communication networks is well beyond the scope of this article. Roughly speaking, we can characterize the network by its *latency* and its *topology*. The former is a measure of how long it takes data to travel between specific pairs of processors and the latter concerns the general physical layout of the network. These two are connected in that the expected latency between a pair of processors has largely to do with their relative physical locations.

We generally divide platform into two broad categories: *shared memory* and *distributed memory*, though the distinction is not as clear on modern architectures as it was historically. In a shared-memory architecture, all processors have access to a common memory and data does not need to be physically moved in order for different cores to make use of it. In such an architecture, latency is, in principle, negligible. In distributed-memory architectures, on the other hand, processors have physical memories and one must account for the non-negligible time it takes to move data from the memory of one processor to the memory of another.

In modern practice, one finds that these distinctions are rather blurred at best. With respect to shared-memory architectures, modern CPUs have multiple computing *cores*, which, although they share a physical address space, may or may not share an actual physical memory. Further, a single computing device may have multiple multi-core CPUs that communicate over a bus. There may be significant differences in the latency passing data between pairs of cores on the same CPU versus on different CPUs, although from the point of view of the communication abstraction at the level of the programming interface, these things appear identical. To make matters worse, even access to different data in the same memory from the same individual cores may have different latencies due to the existence of a complex memory hierarchy in which the kernel attempts to cache data that is predicted to be needed in the near future. This exacerbates the imbalance in the time to access the same data from different cores. All of this may be more or less invisible to the programmer except inasmuch as there are ways to indirectly influence the communication patterns. This topic is also beyond the scope of the present article. For more information, see [45].

Shared-memory architectures also present the very real issue of *memory contention*, in which multiple processors attempt to access the same physical memory simultaneously or to use the same physical channel for retrieving such data. Contention results in longer memory access times than would otherwise be expected and may also necessitate the use of *locks*. A *lock* restricts access to certain memory to only a single physical processor until that lock is released. The use of locks is necessary in some situations to prevent run-time error conditions, but also contributes to contention and leads to effective increases in the time required to access data.

All modern CPUs have multiple cores and most modern computers have multiple such CPUs. A distributed-memory architecture generally consists of multiple computers connected by a communication network. Because each of these computers is a miniature parallel computer in its own right, "pure" distributed architectures have essentially ceased to exist. For efficiency, all algorithms that target distributed-memory architectures must thus account for the massive differences in the time to communicate data between cores on the same physical computer versus cores on different physical computers.

Taking account of the major differences in performance between the different possible types of computers discussed above can lead to major differences in the high-level design of parallel algorithms, as we discuss below in Section 8.3.3.3.

Communication Protocol. Aside from the many variations in architecture that we have just discussed, algorithm design may also be informed by the chosen *communication protocol*, the mechanism by which data is transferred from one processor to another. Protocols are usually closely associated with some underlying physical transfer mechanism. What we refer to as the "communication protocol" here is the *interface* used by the programmer to cause the data transfer programmatically.

Very broadly, there are two categories of communications protocols: *threads* and *message passing*. The threads model is generally associated with shared-memory architectures and involves communication that is performed by passing data through local memory. Message passing is generally associated with distributed-memory architectures and involves communication that is performed by passing data across a network. These two categories of protocol are in turn associated with the two basic ways of achieving parallelism programmatically: multi-threaded computation versus multi-process computation. A *process* presents the execution of a single computer program with its own private memory address space. A process can spawn multiple *threads*, all of which may execute sequential procedures simultaneously with access to the same memory.

The importance of both the communication protocol and the associated programming model and architecture is both that it can influence the efficiency of data transfer and, merely by restrictions in the interface itself, limit the algorithmic options available. The most prevalent communication protocols at the time of writing are

- Threads
  - OpenMP (Open Multi-Processing): an interface standard for providing instructions in source code that allow for the semi-automatic parallelization of *multi-threaded* applications by OpenMP-aware compilers using communication through shared memory.
  - pthreads (POSIX Threads): an interface standard for allowing communication between individual threads of a single process running on a shared-memory computer.
- Message Passing
  - MPI (Message Passing Interface): an interface standard and a set of associated libraries for allowing separate processes running either on the same computer or on remotely located computers to communicate with each other either through shared memory or over an associated communication network (the precise conduit depends on the details of the implementation of the MPI library used).
  - PVM (Parallel Virtual Machine): A much older and no longer commonly used message-passing protocol similar to MPI.

Programming Language. As with communication protocols, the choice of programming language can also heavily influence both the options for parallelization and the design of a given algorithm. Many newer high-level languages (e.g., Go) include parallel constructs directly in the language. Older, low-level languages, such as C, do not include direct support for parallelization. On the other hand, low-level languages tend to be preferred for implementation of numerical algorithms for well-known efficiency reasons.

8.3.3.2 Frameworks and Solvers

In Section 8.3.2.1, we described the level of abstraction of an algorithm as a fundamental property. At the level of an actual implementation, an abstract parallel algorithm, in which the parallel algorithm does not depend on the details of the associated sequential algorithm except through well defined interfaces, can be implemented completely independently of the sequential solver and can even be made to work with multiple sequential solvers. The implementation of a parallel algorithm in such a way as to enable any sequential algorithm (with possible slight modification to conform to the interface) to work within the scheme specified by the parallel algorithm is called a *framework*. The combination of a parallel framework and (the implementation of) a particular sequential algorithm make a *solver* (which when deployed on a particular platform becomes a *solver platform*). Naturally, it is not necessary for a solver to utilize a framework in order to execute in parallel. In some cases, the solver includes its own parallel algorithm in a tightly integrated package. We provide examples of both solvers and frameworks in Section 8.4 below.

8.3.3.3 Coordination Mechanisms

In this section, we review what we generally refer to as *coordination mechanisms*. This is a general term for the overall way in which the parallel algorithm controls execution, including both static and dynamic load balancing and the interaction with the underlying sequential solver. We review here the most common existing mechanisms employed by the solvers and frameworks discussed in Section 8.4.1.

It should be highlighted that in all the approaches reviewed below that involve dynamic load balancing, the adaptive tuning of granularity is a centrally important concept. Generally speaking, the atomic unit of work that we consider is the exploration of a subtree using the strategy of the underlying sequential algorithm, but with a work limit imposed. This work limit is generally imposed in the form of a limit on the execution time or a limit on the number of nodes enumerated. Depending on the specific details of how the parallel and sequential solvers interact (through a restrictive API or by direct internal function calls with access to the solver's internals), the work limit may be imposed in different ways.

Parallel Racing. Despite the decades of effort that have resulted in increasingly sophisticated sequential solution platforms, current state-of-the-art MILP solvers still have a high *performance variability*, which means that the impact on performance of seemingly performance neutral changes in the input or in minor implementation details of the solver can sometimes result in large variations in solution time and other performance measures. As the most striking example of this kind of variation, simply permuting the rows or columns of the constraint matrix, which yields an identical instance from a mathematical standpoint, can cause even the most sophisticated solvers to vary wildly [52]. It is impossible to predict, for a particular instance, what perturbations to the model or to the solver's run-time parameters will minimize running time.

Given this situation, one of the most straightforward ways to parallelize an existing sequential algorithm is simply to exploit this performance variability by executing the same sequential algorithms either with different parameter settings or with different permuted instances of the same MILP (or both) in parallel. We call this approach *parallel racing*, since it executes a simple race among the solver instances with no communication. The computation is terminated when the first solver finishes, and a winner is declared. The idea dates back to the early days of the development of parallel branch-and-bound algorithms [71, 64, 48] and has recently been shown to be surprisingly effective in some cases [33].

This approach has some obvious advantages. It is simple to implement and can be used easily with any sequential solver or combination of different sequential solvers and thus can capitalize on all available sequential solver technology. It requires no coordination of the solver instances and thus reduces parallel overhead to nearly zero. The cost of this simplicity, of course, is that the algorithm may perform a potentially vast amount of redundant work. By passing *some* small amount of data between the solvers, the basic procedure can easily be improved. Algorithm 8.3 shows a simplified racing-type algorithm in which global upper bounds are communicated during computation. Naturally, it would be possible to communicate other information as well, but there is a tradeoff between the communication overhead and the improvement that comes from communicating such data. See Section 8.4.2 for a description of the UG framework that provides an implementation of this approach. Commercial solvers CPLEX and Gurobi can also execute in this manner (see Section 8.4.1). There is a study specialized for this paradigm [15].

Algorithm 8.3: Basic Racing Algorithm

Input : Set of N different MILP solvers, N processors indexed by $S = \{1, \ldots, N\}$, and a MILP instance to be solved
Output : An optimal solution

1. terminated ← **false**.
2. Spawn N parallel processes with solver i solving the MILP instance on processor i, $\forall i \in S$.
3. **while** terminated = **false**. **do**
4.   (i, tag) ← Wait for message from solver processes. `// Returns source SOLVER identifier and message tag`
5.   **if** tag = incumbentValue **then**
6. $\forall j \in S \setminus \{i\}$: Send the incumbent value to solver j.
7.   **else**
8.   terminated ← **true** `// tag = optimalSolution.`
9. Output optimal solution.

Pure Static Load Balancing. Another simple approach to parallelization is to use a static load-balancing scheme (see Section 8.3.2.5) to generate and distribute a set of subproblems that can then be solved independently in parallel. This approach has advantages similar to those of the parallel-racing approach—no coordination is needed after the initial subproblem generation and distribution phase and thus parallel overhead is near zero. As with parallel racing, it can be used to parallelize almost any sequential solver. It can therefore take advantage of the state-of-the-art performance of sequential solvers. Finally, the amount of redundant work is minimized.

The scheme is, however, vitally dependent on the ability to predict a priori the difficulty of the subproblems being generated in order to balance the load. This problem of predicting difficulty is notoriously difficult except for problems with certain special structure and lies at the core of the difficulty of parallelizing. If the predictions made are not accurate, then some solvers will end up solving their assigned subproblem(s) well before others and this will result in a potentially large amount of idle time for the assigned processors, introducing large overhead.

As with racing strategies, this idea dates back to the early days of the development of parallel branch-and-bound algorithms (see, e.g., [57]). A very recent implementation of it, the so-called SelfSplit approach, is described in [34]. As in the parallel-racing case, the basic scheme can be improved by allowing *some* communication between the solvers, at the cost of increased complexity and a small amount of overhead. Algorithm 8.4 provides the basic outline of an algorithm similar to that of [34].

Algorithm 8.4: Static Load-Balancing Algorithm

Input : Single MILP solver, set of N processors $i \in S = \{1, \dots, N\}$, and a MILP instance to be solved

Output : An optimal solution

1 Spawn N identical processes solving the MILP instance on processors 1 to N with a fixed limit of L nodes.
2 **forall** $i \in S$ *IN PARALLEL* **do**
3   **forall** *leaf nodes j in the partial search tree* **do**
4   compute a score for the difficulty of node j.
5   Sort the nodes by decreasing scores.
6   Assign a color c between 1 and N to all nodes, in round robin.
7   Discard all nodes with color $c \neq i$ from the branch-and-bound tree.
8   Enumerate the remaining parts of the search tree.
9   Output solution x_i^* for solver i.
10 Output $x^* = \text{argmin}_{i \in S} c^\top x_i^*$.

Master-Worker. The *Master-Worker paradigm* is a well-known and widely used paradigm for many parallel tasks. The basic scheme involves a single *Master* process that coordinates the efforts of a set of *Workers*. In most cases, the role of the Master is to balance the load as effectively as possible, though it may possibly play other roles as well. In its straightforward implementation, the Master may become a communication bottleneck as the amount of communication with Workers may increase linearly with the number of Workers, eventually becoming a bottleneck. Naturally, there are approaches to combat this, such as having the Master request Workers to send information (such as the description of a workload) directly to each other rather than via the Master. The granularity of the workload can also be dynamically changed so that Workers do more work independently and intervals between communication with the Master are increased, decreasing the communication load on the Master.

This scheme, while not quite as simple as the previously described ones, is still rather simplified—all coordination decisions are made in a single sequential process. The potential advantage of this scheme is that the Master maintains a complete (though inevitably somewhat outdated) picture of the state of the entire procedure. As long as the Master's global view remains accurate, this allows the search order in the parallel algorithm to replicate, to a large extent, the search order that would be observed in sequential mode. The downside, of course, is that for this fine-grained global view to be maintained accurately requires a high communication frequency with Workers. There must inevitably be a point at which the Master becomes a communication bottleneck. Here again, there is an obvious tradeoff at play. Less data being shared will result in less effective coordination decisions in the long run, most likely resulting in redundant work being done. More data being shared results in higher overhead.

Despite the simplicity and the potential downsides, this approach has been used successfully to solve difficult open instances (`a1c1s1`, `roll3000`, `timtab2`) from MIPLIB 2003 in a large computational grid. A MILP instance was decomposed carefully using CPLEX, and the generated subproblems were distributed across a computational grid. Solution of the subproblems continued until predefined termination criteria, such as a time limit, were met. Subproblems not yet solved were decomposed using CPLEX again, and the newly generated subproblems were again distributed on a computational grid [14]. FATCOP [16] is a solver developed to perform this process automatically.

Algorithms 8.5 and 8.6 show a simplified Master-Worker-based parallel algorithm for solution of MILPs. After sending a subproblem to a Worker, there is no communication between the Master and the Worker. If a Worker solves a received subproblem, an optimal solution for the subproblem is returned; otherwise, the terminal nodes of the search tree for the subproblem (a new collection of subproblems) are returned after the Worker performs. The Master coordinates distribution of the global collection of subproblems to Workers. In order to keep the number of subproblems as small as possible, the search strategy is the so-called *depth-first* strategy, which processes the deepest node in the search tree first and minimizes the generation of new terminal nodes. All subproblems managed by the Master are treated independently, which makes fault tolerance easy to handle when computing resources are being added and removed at random, as on a computational grid. In the algorithm, N could be changed dynamically.

Supervisor-Worker. In contrast to the Master-Worker paradigm, the idea of Supervisor-Worker is that the Supervisor functions only to coordinate workload, but does not

Algorithm 8.5: Master (Master-Worker)

Input : Single MILP solver, set of N processors $i \in S = \{1,\ldots,N\}$ and an MILP instance to be solved
Output : An optimal solution

```
1  Spawn N Workers with the MILP solver on processors 1 to N.
2  x* ← NULL.
3  I ← S. // Idle processors
4  A ← ∅. // Busy processors
5  Q ← {0}. // Queue of indices of subproblems for processing,
      0 is the index of the root problem
6  R ← ∅. // Subproblems currently being processed
7  while Q ≠ ∅ and R ≠ ∅ do
8      while I ≠ ∅ and Q ≠ ∅ do
9          i ∈ I, I ← I \ {i}, A ← A ∪ {i}.
10         j ∈ Q, Q ← Q \ {j}, R ← R ∪ {(i, j)}.
11         Send subproblem j and best solution to processor i.
12     (i, tag) ← Wait for message. // Returns processor identifier and
           message tag
13     if tag = optimalSolutionFound or tag = solutionFound then
14         Receive solution x̂ from processor i.
15         if x* = NULL or cᵀx̂ < cᵀx* then
16             x* ← x̂.
17     Receive list of candidate subproblems generated by processor i and add them to Q.
18     R ← R \ {(i, j)}.
19     A ← A \ {i}, I ← I ∪ {i}.
20 Output x*.
```

Algorithm 8.6: Worker (Master-Worker)

Input : A subproblem and an incumbent solution
Output : A termination code, improved solution (if found), and a list of candidate subproblems

```
1  Set initial global upper bound based on the incumbent solution.
2  Set termination criteria and the other parameters, such as search strategy etc.
3  Execute sequential solution algorithm until termination criteria are reached.
4  if algorithm solves subproblem to optimality then
5      tag ← optimalSolutionFound.
6  else
7      if algorithm found feasible solution then
8          tag ← solutionFound.
9      else
10         tag ← noSolutionFound.
11 Send candidate subproblems, any solution found, and tag to Master.
```

actually store the data associated with the search tree. The terminal nodes of the search tree in the Workers are collected on demand and a set of subproblems in the Supervisor works as a buffer to ensure subproblems are available to idle Workers as needed. This coordination scheme has been successful in solving open instances from both MIPLIB2003 and MIPLIB2010 by using the UG framework described in Section 8.4.2 and the underlying sequential solver SCIP on a large supercomputer. This coordination scheme is also used in the CPLEX distributed MILP solver mentioned in Section 8.4.1. Algorithms 8.7 and 8.8 show a parallel algorithm with a simplified Supervisor-Worker coordination scheme similar to the one used in UG.

In the Supervisor-Worker approach in UG, the load balancing is accomplished mainly by toggling the collection mode flag in the Worker. Turning collecting mode on results in additional "high-quality" subproblems being sent to the Supervisor, which can then be distributed to Workers. Naturally, the method of selecting which Worker to collect from is crucial to the effectiveness of the approach. Some additional keys to avoiding having the Supervisor become a communication bottleneck are:

- Frequency of status updates can be controlled depending on the number of Workers.
- The maximum number of Workers in collection mode is capped and the Workers are carefully chosen in a dynamic fashion.

Naturally, there is a tradeoff between the frequency of communication and the number of Workers in collection mode and the degree to which the parallel search order replicates the sequential one. As the number of processors is scaled up, this tradeoff must be carefully navigated.

Multiple-Master-Worker and Master-Hub-Worker. An alternative approach to ensuring that the Master doesn't become a bottleneck is to either create additional Master processes (Multiple-Master-Worker) or even to create a layer of "middle management" (Master-Hub-Worker). In both schemes, Workers are grouped into collectives called Hubs, each of which has its own Hub Master. In creating this management hierarchy, the hope is that the Hub Masters can effectively balance the workload within the collective for some time before having to coordinate with other Hubs through the Master to do higher-level global balancing. Naturally, more levels can be added and experiments with schemes such as having a dynamic number of levels have appeared in the literature [106].

The CHiPPS framework, described in Section 8.4.2, uses the Master-Hub-Worker paradigm, whereas the PEBBL framework, described in 8.4.2, uses a Multiple Master approach. This scheme can be extended to allow for even more layers in a hierarchical load-balancing scheme [49]. A basic scheme similar to the one in CHiPPS is described in Algorithms 8.9–8.11. The keys to ensuring the effectiveness of this framework are:

- The number of clusters (Hub Masters) and thereby the cluster size can be dynamically controlled.
- The frequency of status updates between Workers and Hubs, as well as Hubs and Masters can be fixed or automatically adjusted adaptively (the default).

Algorithm 8.7: Supervisor (Supervisor-Worker)

Input : Single MILP solver, set of N processors $i \in S = \{1,\dots,N\}$ and a MILP instance to be solved

Output : An optimal solution

```
1  Spawn N Workers with the MILP solver on processors 1 to N.
2  collectMode ← false.
3  x* ← NULL.
4  I ← N\{1}. // Idle processors
5  A ← {1}. // Busy processors
6  Q ← ∅. // Queue of indices of subproblems for processing, 0
       is the index of the root problem
7  R ← {(1,0)}. // Subproblems currently being processed, 0 is
       the index of the root problem
8  Send the root problem to processor 1.
9  while Q ≠ ∅ and R ≠ ∅ do
10     (i,tag) ← Wait for message. // Returns processor identifier and
           message tag
11     if tag = solutionFound then
12         Receive solution x̂ from processor i if x* = NULL or cᵀx̂ < cᵀx* then
13             x* ← x̂.
14     else
15         if tag = subproblem then
16             Receive a subproblem indexed by k from processor i.
17             Q ← Q ∪ {k}.
18         else
19             if tag = terminated then
20                 R ← R\{(i,j)}. // j is the index of the terminated
                       subproblem
21                 A ← A\{i}, I ← I ∪ {i}.
22             else
23                 if tag = status then
24                     if collectMode = true then
25                         if there are enough heavy subproblems in Q then
                               // heavy subproblem is a subproblem
                                  which is expected to generate a
                                  large subtree
26                             Send message with tag = stopCollecting to processors
                                 in collecting mode.
27                             collectMode ← false
28                     else
                           // collectMode = false
29                         if there are not enough heavy subproblems in Q then
30                             Select processors which have heavy subproblems.
31                             Send message with tag = startCollecting to the
                                 selected processors.
32                             collectMode ← true.
33     while I ≠ ∅ and Q ≠ ∅ do
34         i ∈ I, I ← I\{i}, A ← A ∪ {i}.
35         subproblem j ∈ Q, Q ← Q\{j}, R ← R ∪ {(i,j)}.
36         Send subproblem j and x* to processor i.
37 ∀i ∈ S : Send message with tag = termination to processor i.
38 Output x*.
```

Algorithm 8.8: Worker (Supervisor-Worker)

```
  Input   : A MILP solver and an original MILP instance to be solved
 1 collectMode ← false.
 2 terminate ← false.
 3 while terminate = false do
 4     (i, tag) ← Wait for message from Supervisor. // Returns Supervisor
           identifier 0 and message tag
 5     if tag = subproblem then
 6         Receive subproblem and solution from Supervisor.
 7         Solve the subproblem, periodically communicating with supervisor as follows
               - Send message with tag solutionFound anytime a new solution is discovered.
               - Periodically send message with tag status to report current lower bound
               for this subproblem.
               - When messages with tag startCollecting or stopCollecting are received,
               toggle collectMode.
               - When collectMode = true, periodically send message with
               tag subproblem containing best candidate subproblem.
 8         Send a message with tag = terminated.
 9     else
10         if tag = termination then
11             terminate ← true.
```

- The frequency of inter-cluster and intra-cluster load balancing can be fixed or automatically adjusted adaptively (the default).
- The granularity of the work unit can also be fixed or automatically adjusted adaptively (default).

As with Supervisor-Worker, there is a clear tradeoff in adjusting these parameters between communication overhead and the ability to replicate the sequential search order.

Self Coordination. Recently, a completely decentralized approach to parallel branch-and-bound was introduced and implemented in PIPS-SBB [65], a distributed-memory parallel solver for Stochastic Mixed Integer Programs (SMIPs). Parallel PIPS-SBB features a lightweight mechanism for redistributing the most promising nodes among all the parallel processors without the need for a centralized load coordinator. This alternative scheme seeks to keep the load in balance without formally introducing any notion of a separate process to coordinate the load. In order to accomplish this, a synchronization point must be added, potentially introducing an alternative source of overhead. This scheme is untested with regard to solving generic MILPs, so it is unclear how to assess this tradeoff. Nevertheless, we introduce the basic scheme here in Algorithm 8.12. Instead of point-to-point communications, parallel processors exchange subproblems via all-to-all collective MPI asynchronous communications, allowing rebalance of the computational load using a single communication step. Parallel processors proceed to solve subproblems until the problem has been solved to optimality.

Algorithm 8.9: Master (Master-Hub-Worker)

```
Input  : Single MILP solver, set of N processors i ∈ S = {1,...,N}, number of hubs H,
         and a MILP instance to be solved
Output : An optimal solution
1  Spawn N Processes with the MILP solver on processors 1 to N
   // Process 1 is the master, processes 1 to H are hubs
      (master is also hub), all processes also function as
      workers.
2  x* ← NULL.
3  L_i ← −∞, W_i ← 0, 2 ≤ i ≤ H. // Best bound and workload of cluster i
4  Do initial static load balancing. // Either 2-level root initialization
      or spiral
5  while ∃i, W_i > 0 do
6      while timeSinceLastBalanceCheck < masterBalancePeriod do
7          (i, tag) ← Check for messages. // Returns processor identifier
              and message tag or NULL
8          if tag = solutionFound then
9              Receive newSolution from processor i.
10             if x* = NULL or c^T x̂ < c^T x* then
11                 x* ← x̂.
12         else
13             if tag = hubStatusUpdate then
14                 Update W_i, L_i.
15             else
16                 Process message as hub or worker (see Algorithms 8.10 and 8.11).
17         Do a unit of work. // As worker
18         Update timeSinceLastBalanceCheck
19     if W_i < workloadThreshold or L_i > boundThreshold then
20         Balance cluster loads.
21 ∀i ∈ S : Send message with tag = termination to processor i.
22 Output x*.
```

8.4 Software

In this section, we summarize software architectures of existing software for solving MILPs in parallel. The development of parallel software for solving MILPs has a long history by now and many solvers and frameworks have preceded the ones listed here. In addition to the ones listed below, previous efforts include ABACUS [50], PPBB-LIB [93], FATCOP [16], PARINO [58], MW [43], BoB [7], Bob++ [24], PUBB [88], PUBB2 [86], ParaLEX [84], ZRAM [12], and MallBa [5].

We divide this section into two subsections. In the first, we describe solvers that have embedded, generally tightly integrated parallelization schemes. In the second, we describe frameworks that can be used in tandem with multiple underlying sequential solvers.

Algorithm 8.10: Hub Master (Master-Hub-Worker)

```
Input    : Process index k, set S_k of workers assigned to cluster
1  L_i ← −∞, W_i ← 0, i ∈ S_k. // Best bound and workload of worker i
2  terminate ← false.
3  Participate in initial static load balancing.
4  while terminate = false do
5      while timeSinceLastBalanceCheck < hubBalancePeriod and
        timeSinceLastHubReport < hubReportPeriod do
6          (i, tag) ← Check for message. // Returns processor identifier
               and message tag or NULL
7          if tag = masterRequestsBalance or tag = workerRequestsBalance then
8              Identify donors and notify them of need to donate.
9          else
10             if tag = workerStatusUpdate then
11                 Update W_i, L_i.
12             else
13                 if tag = terminate then
14                     terminate ← true.
15                 else
16                     Process message as worker (see Algorithm 8.11).
17         Do a unit of work and request balance if necessary. // As worker
18         Incorporate worker status into hub status.
19         Increment timeSinceLastBalanceCheck, timeSinceLastHubReport.
20     if ∃i, W_i < workloadThreshold or L_i > boundThreshold then
21         Balance load of workers.
22     if timeSinceLastHubReport ≥ hubReportPeriod then
23         Send hub status to master.
```

8.4.1 Solvers

BLIS and DisCO. BLIS [103, 105] is an open-source parallel MILP solver that is part of the CHiPPS hierarchy to be described in section 8.4.2 below. DisCO [13] is a recent re-implementation and generalization of BLIS that supports the solution of mixed integer second-order conic optimization problems.

CBC. CBC [35] is an open-source solver originally developed by IBM. It employs a simple thread-based Master-Worker scheme, which is a straightforward parallelization of its sequential algorithm. Nodes are handed off by the master thread to idle workers one at a time and the results collected, with all global data stored centrally. The sequential algorithm is itself quite sophisticated and this simple approach to parallelization is quite effective at a small scale, since it mirrors the algorithmic approach taken by the underlying sequential solver. CBC has a deterministic execution mode in which the parallelization is at the subtree level. In this mode, each thread works on an entire subtree, with the amount of work fixed deterministically. After all threads complete their unit of work, there is a synchronization point, after which computation continues from this deterministic state.

Algorithm 8.11: Worker (Master-Hub-Worker)

```
   Input   : Process index k
 1 L_k ← −∞, W_k ← 0. // Best bound and workload of this worker
 2 terminate ← false.
 3 Participate in initial static load balancing.
 4 while terminate = false do
 5     while timeSinceLastWorkerReport < workerReportPeriod do
 6         (i, tag) ← Check for message. // Returns processor identifier
               and message tag or NULL
 7         if tag = hubRequestsDonation then
 8             Identify subtree to donate or split current tree.
 9         else
10             if tag = subTree then
11                 Receive donated subtree.
12             else
13                 if tag = terminate then
14                     terminate ← true.
15         Do a unit of work on best locally available subtree, send improved solution (if
             found), and request more work if necessary.
16         Increment timeSinceLastWorkerReport.
17     if timeSinceLastHubReport ≥ hubReportPeriod then
18         Send worker status to hub.
```

Algorithm 8.12: Self Coordination Algorithm

```
   Input   : Single MILP solver, set of N processors i ∈ S = {1,...,N} and an MILP
             instance to be solved
   Output  : An optimal solution
 1 Spawn N identical processes solving the MILP instance on processors 1 to N.
 2 forall i ∈ S IN PARALLEL do
 3     Add the root problem to priority queue Q_1.
 4     Set upper bound U^i ← ∞ and lower bound L^i ← −∞ on processor i.
 5     while min_{i∈S}{L^i} < min_{i∈S}{U^i} do
 6         if Load imbalance exists or synchronization point is reached then
 7             Exchange best solutions, set U^i ← min_{1≤i≤N}{U^i}.
 8             Determine the top M candidate subproblems from ∪_{i∈S} Q_i and redistribute
                 them among all processors in a round robin fashion.
 9             if termination conditions are met then
10                 return
11         Remove subproblem s from Q_i.
12         Process subproblem s, update U^i, L^i.
13         if s not fathomed then
14             Branch to create children of s and add them to Q^i.
```

DIP. DIP (Decomposition in Integer Programming) [74, 37] is a decomposition-based solver that takes a different approach to parallelism than any of the others listed so far. DIP was built on the ALPS tree search framework [104], although it does not currently take advantage of the built-in ability of ALPS to parallelize the tree search at the subtree level. Rather, it parallelizes the bounding process of individual search tree nodes (subnode parallelism). This can be done in a number of different ways. First, it can utilize an interior point-based LP solver to solve the LPs that arise. More importantly, however, since it can recursively use a MILP solver for solving the *column-generation subproblem* that must be solved during the bounding process, this step can itself be parallelized by using one of the other solvers listed in this section. Furthermore, when there is block structure present (see [95]), the solution of the subproblem can itself be decomposed into independent subproblems that can then also be solved in parallel. These two strategies may even be hybridized.

FICO Xpress-Optimizer. The internal parallelization of the FICO Xpress-Optimizer is based on a general task scheduler that is independent of the concrete MILP-solving application. It can handle the execution of interdependent tasks in a deterministic fashion. A core aspect of its design is the capability to handle *asymmetric tasks* that might have different levels of complexity. It is not only possible to have, e.g., cutting, heuristics, and exploration of the branch-and-bound tree parallelized individually, but to run tasks of each type at the same time.

The parallel design of Xpress avoids fixed synchronization points. At the time when a task is created, it gets a *deterministic stamp*. The task may only use information which is itself tagged with a smaller stamp. In this way, the task uses only a subset of the information that would be available if a synchronization had been triggered when the task was created. The idea is that the potential performance loss from using slightly "outdated" information will be easily made up for by the performance gain from dropping the need for regular complete synchronization. When information is collected, all data that are transferred back to global data receive a deterministic stamp. All tasks that have a stamp which is greater than this, will be allowed to use that information.

Concerning synchronization, the incumbent solution is always shared globally and as soon as possible. Apart from solutions, Xpress shares branching statistics and selected cuts. For both kinds of information, each task has a local variant that contains more and potentially newer information, which is combined with the global information. For the ramp-up, a root-initialization-like scheme is used.

In principle, tasks could be anything; in practice, they either refer to subtrees, more particularly individual dives, or to the execution of expensive heuristics. In that sense, Xpress employs a node parallelization scheme, with a slight flavor of subtree parallelization since the nodes explored within local search heuristics are not necessarily distinct from the nodes in the main tree search or within other heuristics. For the main tree search, however, it holds that no node is explored twice. Subnode parallelization is exclusively used at the root node, mainly for solving the global LP relaxation by parallel barrier and/or parallel dual simplex, subordinately for parallel heuristics and cutting.

A problem with deterministic parallelization approaches that use fixed synchronization points is that they do not scale well on large numbers of cores. One decision made as a consequence is to break with the one-to-one association between threads and tasks to be performed. As a consequence, more tasks, typically by a factor between two and four, are maintained than there are threads available. By doing so, Xpress aims at immediately having new tasks available for a thread when it completes a previous task, without needing to wait for other threads to synchronize. As a consequence, the load balancer might need to dynamically put certain tasks on hold when they require information from a task that is currently not being executed and exchange them with the task lagging behind. For more details on the parallelization of Xpress, see [10].

As an important consequence, by breaking the link between threads and tasks, it is possible to make the solution path independent of the number of threads used—it only depends on the maximum number of tasks that may exist at the same time. Moreover, the parallelization of Xpress is not only deterministic, but Xpress as a whole is also platform-independent, meaning that the solver takes exactly the same solution path independent of whether the underlying machine is a Mac, Windows, or Linux system and what brand of CPUs is used.

SYMPHONY. SYMPHONY [75, 77, 73, 22] was originally developed in the early 1990s as a framework that was intended to be customized by the addition of user-defined subroutines for generation of valid inequalities and other functionality (known today as *callbacks*). It did not initially have an execution mode as a generic MILP solver, but this capability was added later by leveraging libraries for I/O and generation of valid inequalities provided by the COIN-OR project [60].

SYMPHONY was originally designed to run on distributed memory platforms and was later modified to run on shared-memory platforms. It is implemented mainly in pure C and in its distributed execution mode, it uses the message-passing protocol PVM for communication. Generally speaking, it employs a Master-Worker coordination mechanism with node-level parallelism (the unit of work is a single node), though it has a variety of execution modes, some of which enable sub-node parallelism and parallelize some auxiliary processes that involve knowledge sharing.

SYMPHONY's functionality is divided into five modules that are designed to execute independently in parallel or in various bundled combinations.

- *Master*: This module contains functions that perform problem initialization and I/O. The primary reason for a separate master module is fault tolerance, as this module is not heavily tasked once the computation has begun.
- *Tree Manager* (TM): The TM controls the execution of the algorithm by maintaining a complete description of the search tree and deciding which candidate node should be chosen as the next to be processed at each step of the algorithm.
- *Node Processor* (NP): The NP modules perform basic node processing to calculate bounds and also perform the branching operation.
- *Cut Generator* (CG): The CG modules generate valid inequalities used to strengthen the relaxations solved by the NP modules. Multiple CG modules can be executed in parallel in tandem with NP modules.

- *Cut Pool* (CP): The CP modules store previously generated inequalities and act as auxiliary cut generators. It is possible to have multiple cut pools for different parts of the tree and even to store locally valid inequalities in them.

It is possible to combine the modules in various ways, such as either

- combining the NP module with the CG module to obtain one single sequential module that performs both functions or
- combining the CP, TM, and Master modules into a single module maintaining all global information.

After processing each node and making a branching decision, the NP module queries the TM module as to what to do next: retain one of the child nodes just generated and continue "diving" or wait for a new node to be sent. This approach minimizes redundant computation by ensuring that all NP modules are processing high-quality nodes, but increases communication overhead substantially. Scalability is limited by the TM's ability to handle incoming requests from the NP modules.

SYMPHONY's data structures are designed to ensure that all data that needs to be stored and communicated is represented as compactly as possible. All data in the tree is stored using a differencing scheme in which only the differences between a child node and parent node are stored. Descriptions of valid inequalities are only stored once and referred to elsewhere by index. In this way, parallel overhead is reduced as much as possible.

SYMPHONY also has a shared-memory parallel mode implemented using the OpenMP protocol to create a multi-threaded program that functions in roughly the same fashion as the distributed parallel version but with all communication through memory rather than over the network. The scalability issues with the shared-memory version are similar to those of the distributed version.

SYMPHONY has been used to develop a number of custom solvers for combinatorial problems, such as the vehicle routing problem [72].

Other Commercial Sovers. CPLEX [18] and Gurobi [44] are commercial solvers that also have parallel execution modes. However, not much public information is available on the approach to parallelization that these solvers take.

8.4.2 Frameworks

BCP. BCP [55, 76] is a framework for implementing parallel branch, cut, and price algorithms. It was initially developed as a re-implementation and generalization of SYMPHONY (which was also a framework at that time) in the C++ language using the more modern MPI message-passing protocol. It has extensive support for implementing column generation algorithms, whereas SYMPHONY's support for such algorithms was not developed. Its basic modular design is similar to that of SYMPHONY, however, and is described above. It employs the Master-Worker coordination scheme with node parallelism in a fashion similar to SYMPHONY, with a complete description of the tree maintained centrally and all decisions about search order made centrally. Its limitations from a scalability standpoint are also similar to SYMPHONY's.

CHiPPS. CHiPPS [101, 102, 103, 105, 104] is a generic framework for performing parallel tree search, but with particular support for branch-and-bound-based algorithms for optimization. CHiPPS is implemented in C++ and uses MPI as its communications protocol. The coordination mechanism is a Master-Hub-Worker scheme with subtree parallelism. The unit of work performed by a Worker is the exploration of an entire subtree until some specific criteria are met (time limit, node limit, etc.). These criteria can be dynamically adjusted to limit overhead.

The base layer of CHiPPS is ALPS, which is an abstract implementation of parallel tree search. To develop an algorithm using ALPS, the user must provide implementations of the node-processing method and the branching method of the tree search algorithm to be implemented, as well as providing classes for storing descriptions of the problem data and the data required to describe a node in the search tree.

ALPS is optimized for "data-intensive" tree search algorithms in which the amount of data required to describe a single subproblem in the search tree may be large and in which additional types of knowledge also might be shared. ALPS has an extensible mechanism for defining new types of knowledge and a general mechanism for storing such knowledge in auxiliary pools and sharing it between processors. Each processor has a *knowledge broker* responsible for routing all communication. All that's required to convert a sequential algorithm to a parallel one is to replace the serial knowledge broker with the parallel one. No other part of the implementation depends on the communication protocol or even whether the algorithm is to be executed in parallel.

The BiCePs layer, built on top of ALPS, provides support for implementing relaxation-based branch-and-bound algorithms for solving optimization problems. It provides an abstract notion of modeling "objects," collections of which can be used to describe subproblems. Subtrees, in turn, are described using a compact differencing scheme in which nodes are described in terms of differences between parent and child.

CHiPPS employs a unique load-balancing scheme in which entire subtrees are shifted directly from one worker to another to balance the load instead of individual nodes. A subtree can be seen as a collection of related nodes, which can be stored more efficiently if kept together as a single unit. By load balancing in this way, we hope to minimize both communication overhead and storage overhead. The overall mechanism is a hierarchical coordination scheme with several static balancing options; sender-initiated balancing (if necessary); and periodic intra- and inter-cluster dynamic load balancing, as described earlier in Algorithms 8.9–8.11.

CHiPPS has been used to develop three MILP solvers to date: DIP, BLIS (the third layer of the CHiPPS hierarchy), and DisCO (a generalization of BLIS to support solution of mixed integer conic optimization problems). It has also been used to develop MibS, a solver for mixed integer bilevel optimization problems [91].

PEBBL. The developerfpment of the Parallel Enumeration and Branch-and-Bound Library (PEBBL) [28] was sponsored by Sandia National Laboratories and has been ongoing for close to two decades. Its purpose was to support the solution of optimization problems arising in applications of interest to that laboratory. The PEBBL project itself resulted from the splitting of the parallel MILP solver PICO into an abstract framework for implementing parallel branch-and-bound algorithms and the parts of PICO specific to the solution of MILPs. PICO is now an application layer built on top of the base layer PEBBL. PEBBL uses a multiple-master-worker coordination mechanism with a sophisticated load-balancing scheme described in detail in [28] to achieving scalability.

UG. The core idea behind UG was to make it possible to utilize a powerful state-of-the-art MILP solver as the underlying sequential solver while still achieving good parallel performance. Development was started in 2001 using a general parallel branch-and-bound software framework PUBB2 [85]. After recognizing how difficult it is to use a powerful black-box solver with a general parallel branch-and-bound framework in order to improve overall solver performance, development was begun on ParaLEX [84], which was specialized for the CPLEX solver on a distributed-memory computing environment. ParaLEX was redesigned in 2008 [83], after which the idea to have a general software framework to exploit state-of-the-art MILP solvers was conceived.

Ubiquity Generator (UG) framework [87, 94] is a generic framework to parallelize an existing state-of-the-art branch-and-bound-based solver, which is referred to as the *base solver*, from "outside." UG is composed of a collection of base C++ classes, which define interfaces that can be customized for any base solvers (MILP/MINLP solvers) and allow descriptions of subproblems and solutions to translated into a solver-independent form. Additionally, there are base classes that define interfaces for different message-passing protocols. Implementations of ramp-up, dynamic load balancing, and check-pointing and restarting mechanisms are available as a generic functionality. The branch-and-bound tree is maintained as a collection of subtrees by the base solvers, while UG only extracts and manages a small number of subproblems (typically represented by variable bound changes) from the base solvers for load balancing.

The basic concept of UG is thus to abstract from a base solver and parallelization library and to provide a framework that can be used, in principle, to parallelize any powerful state-of-the-art base solver on any computational environment (shared or distributed memory, multithreading or massively parallel). For a particular base solver, only the interface to UG in the form of specializations of base classes, as provided by UG, needs to be implemented. Similarly, for a particular parallelization library (e.g., MPI), a specialization of an abstract UG class is necessary.

The message-passing functions used in UG are limited as much as possible and are wrapped within the base class. Therefore, adding support for an additional parallelization library should be easy. The most-used libraries for implementing distributed parallel programs are MPI implementations. The virtual functions in the base class provided by UG can be mapped straightforwardly onto corresponding MPI functions. Pthreads is a popular library that is used to make multi-threaded programs and the UG specialization for pthreads uses a simple message queue implementation, which has been developed as a part of the UG code.

From the UG framework point of view, a particular instantiated parallel solver is referred to as

ug [a specific solver name, a specific parallelization library name].

Here, the specific parallelization library is used to realize the message-passing-based communications. Solvers have been developed for the non-commercial SCIP solver (ParaSCIP (= ug [SCIP, MPI]), FiberSCIP (= ug [SCIP, Pthreads])) and the commercial Xpress solver (ParaXpress (= ug [Xpress, MPI]), FiberXpress (= ug [Xpress, Pthreads])). UG has also been used to parallelize the PIPS-SBB solver for two-stage stochastic programming problems (ug [PIPS-SBB,MPI]).

UG employs a Supervisor-Worker coordination mechanism with subtree-level parallelism (the unit of work is a subtree). One of the most important characteristics of UG is that it makes algorithmic changes to the base solver, such as multiple preprocessing, and performs very adaptive algorithms, such as racing ramp-up. These features make it difficult to measure the performance of an instantiated parallel solver. However, from a solvability point of view, instantiated parallel solvers are among the most successful ones. ParaSCIP successfully solved 14 previously unsolved instances from MIPLIB2003 and MIPLIB2010 as of writing this document.

UG has been developed mainly in concert with SCIP. Therefore, ug [SCIP,*] is the most mature and has user-customizable libraries. By using these libraries with the plug-in architecture of SCIP, a customized parallel solver can be developed with minimal effort. One of the successful results of using this development mechanism is the SCIP-Jack solver for Steiner Tree Problems and its variants. ug [SCIP-Jack, MPI] solved three open instances from the SteinLib[90] benchmark set [38]. The largest-scale computation conducted with ParaSCIP is up to 80,000 cores on TITAN at Oak Ridge National Laboratory[82].

8.5 Performance Measurement

Performance measurement presents exceedingly difficult challenges when it comes to parallel MILP solvers. As we mentioned in Section 8.3.1.1, performance of parallel MILP solvers is often assessed by measuring the amount of *parallel overhead* introduced by the parallelization scheme. The direct measurement of such overhead is problematic, so parallel overhead is often measured indirectly. The most common way of doing this involves measuring the *efficiency*, which is an intuitive and simple measure that focuses on the effect of using more cores, assumed to be the bottleneck resource, to perform a fixed computational task (e.g., solve a given optimization problem). The efficiency of a parallel program running on N processors is

$$E_N := (T_0/T_N)/N$$

with T_0 being the sequential running time and T_N being the parallel running time with N threads. Generally speaking, the efficiency attempts to measure the fraction of work done by the parallel algorithm that could be considered "useful." An algorithm that scales perfectly on a given system would have an efficiency of $E_N = 1$ for all N. A related measure is the *speed-up*, which is simply

$$S_N := NE_N$$

Although this way of measuring performance seems reasonable, one faces many problems with it in practice. We outline these problems in the sections below before discussing alternatives.

8.5.1 Performance Variability

Sequential Algorithms. Modern MILP solvers employ complex algorithms. Many algorithmic decisions are made heuristically, such as, e.g., when and how often primal heuristics are called, which disjunction to select for branching, or how many cutting planes should be generated/added. Furthermore, the results of certain operations are not necessarily unique. For example, the root LP might have several different optimal solutions and which one is selected will be influenced by the breaking of ties during algorithmic decision-making. How these ties are broken may end up being determined by any number of factors, including details of the hardware on which the algorithm is run. Small differences in algorithm parameters, how ties are broken, and other details, can result in enormous variations in the course of the algorithm. The tree generated by the algorithm, even when the algorithm itself is deterministic for a given input, might vary greatly in both structure and size, depending on such things as the order of constraints or variables in the model or slight numerical differences introduced by different CPU types.

Some instances are more vulnerable to variation than others. One way to test whether a particular instance is prone to performance variability in regard to a particular solver is to solve the instance multiple times, each with a different permutation of the rows and columns of the constraint matrix (and other associated input data). Such permutations create a problem that is mathematically equivalent to the original one, but for which the running time of a given algorithm might vary dramatically (see, e.g., [20, 52]).

The computer architecture might have an influence also. As mentioned, different CPUs might introduce slight numerical differences. NUMA architectures where the assignment of processes to cores is done by the operating system can easily have performance variations of 10%. In general, when doing benchmarking it is useful to bind processes to cores if possible to limit such variation.

To complicate matters, modern CPUs may change their clock frequency depending on the number of cores currently employed. A CPU might run much faster single-threaded and otherwise empty then it does when all cores are fully loaded. Switching this behavior off decreases variability, but on the other hand, real-world performance might be quite different.

As a result of all this performance variability, the number of instances that one would need to include in a test set and the number of experiments one would need to do in order to ensure that an observed, e.g., 5%, performance increase is actually statistically significant is rather high [4].

Parallel Algorithms. On top of the issues noted above, when an algorithm is executed in parallel, additional sources of variability are introduced, including possible non-determinism of the algorithm itself, due to the unpredictability of the order in which operations occur. As we mentioned in Section 8.3.2.7, it is usually possible to implement parallel algorithms in a deterministic mode. However, this will inevitably worsen performance, due to the required introduction of synchronization points, and if the goal is to assess the non-deterministic variant of the algorithm, then enforcing determinism does not make sense. Due to timing issues and the reasons mentioned above, the number of branch-and-bound nodes generated to solve a particular instance might vary strongly depending on the number of cores used (see [53] for more details).

Measuring Running Time. Finally, we briefly mention that the measurement of running time is itself problematic in the case of a parallel algorithm. In the sequential case, one typically measures the running time of an algorithm not using wall-clock time (the actual real time elapsed), but rather the amount of CPU time taken by the process. The CPU time and the wall-clock time can differ if other processes are running on the computer, which they often are, especially if testing is done on a platform shared with others. In the parallel case, however, one needs to rely on wall-clock time measurement because the running time must include idle time, which may not be properly measured if one only includes CPU time. It is the total elapsed running time of the parallel algorithm that matters, but as we have already noted, wall-clock time measurements are inherently more variable due to the possible influence of external processes and other extraneous factors.

8.5.2 Comparisons

It is usual in the literature to compare alternative algorithms for solving the same problem on the basis of some objective measures of performance, and this is a primary reason for measuring such performance. We have so far motivated why performance measurement is problematic due to inherent variability. This is a general phenomenon that affects all parallel algorithms. In the case of comparing parallel MILP solvers, however, there are additional problematic factors. In general, the main goal of algorithmic research in MILP solvers is to reduce the number of branch-and-bound nodes required to solve an instance. At present, it is not unusual for an instance to be solved in the root node or within fewer than a few hundred nodes. Unfortunately, as we mentioned earlier, reductions in the size of the search tree have a negative impact on scalability. This means that differences in the underlying sequential algorithm can impact our assessment of scalability of parallel algorithms, although these differences may be tangential to the differences we are actually attempting to observe (differences due to the approach to parallelization). In the extreme, one could imagine comparing a parallel algorithm employing an underlying sequential solver capable of solving most instances in a given benchmark set by enumerating only a handful of nodes against one that requires thousands of nodes. The former parallel algorithm will likely be more effective (faster) overall, while the latter is more likely to be scalable. Separating the effects of the underlying sequential solver from the approach taken by the parallel algorithm itself is difficult at best. This is further highlighted by the difficulties encountered in selecting a proper test set.

8.5.3 Instance Selection

Due to well-established data formats collections of widely used data-sets, e.g., the MIPLIB2010 [52], are available and are generally recommended to be used for comparisons. However, it is important to realize that properties of individual instances can limit the scalability that it will be possible to achieve, independent of a given solver's approach to parallelization. Instances that can be solved by most solvers in a small number of nodes or for which the LP relaxation in the root node is extremely difficult to solve will not scale with any current solver. It is therefore important to select instances that are suitable for parallel testing.

- Instances should produce a tree suitably large and broad enough that parallelization is both necessary and effective. Unfortunately, this property depends very much on the effectiveness of the underlying sequential solver. An instance may be suitable in this regard with respect to one solver and not with respect to another. In addition, the size of the tree is not fixed and may vary based on random factors, as noted above.
- If one wants to use efficiency as a performance measure, it is important that instances be solvable with one processor (or at least a small number of processors), since this is used as a baseline for assessing the amount of parallel overhead. Unfortunately, instances that can be solved in a reasonable amount of time on a single processor may not be difficult enough with a large number of processors to be interesting and may not be suitable with respect to the first criterion.

This makes standard benchmark sets of only limited use in testing parallel performance, at least insofar as we limit ourselves to parallelizing the tree search itself. Naturally, subnode parallelism could be employed in the case of small trees, but this approach has so far not been pursued very vigorously.

It should be noted that the performance of a solver on any single instances has very little meaning. After the release of MIPLIB 2010 the overall geometric mean performance of CPLEX, Gurobi, and Xpress was nearly equal, while the performance on individual instances varied by a factor of up to 1,500.

8.5.4 Alternative Performance Measures

Although efficiency is the most commonly employed measure of performance, we have motivated above why it might be slightly problematic in the case of measuring the performance of parallel MILP solvers. [53] suggests alternative measures based on separation of the overall running time into two factors: the number of search tree nodes required to be processed (the size of the search tree) and the throughput rate (the number of search tree nodes processed per second per core). Variation in the former can be mainly attributed to the performance of redundant work due to differences in the search order and a possible lack of global knowledge of the upper bound. Variation in the latter is mainly due to other sources of overhead, such as idle time. Whereas the former measure is subject to the effects of algorithmic variabilities described earlier, the latter is not. If both the size of the tree and the throughput rate remain constant as the number of cores is increased, then the result will be an efficiency of one (ideal). Otherwise, either the size of the tree must have increased (i.e., redundant work is being performed) or the throughput rate has dropped due to increased overhead. By considering these two kinds of statistics together, along with any other fine-grained measurements we can obtain (direct measurement of various sources of overhead, such as idle time from blocking), we obtain a more nuanced picture of performance(e.g. computational results in [87]).

Overall it can be said that benchmarking parallel MILP solvers is a very difficult topic. To get meaningful results, one needs well-defined settings, a large number of suitable instances, the ability to execute a large number of experiments, and a clear understanding regarding the factors influencing the results.

8.5.5 Summary Measures

Finally, we mention that a few guidelines have been established regarding how to summarize performance over an entire benchmark set. If the results over several instances are to be combined, it has been observed that it is better done using the geometric mean or the shifted geometric mean, as opposed to the arithmetic mean. Experience has shown that the latter is often dominated by a few instances in a given test set. When comparing two or more solvers special care has to be given to the question of how to deal with instances that can be solved by only a subset of the solvers. Choosing the time limit will have an substantial impact on the overall result. The same applies to the difficult question of how to deal with wrong results in a useful way. On the other hand, only selecting those instances for comparison which can be solved by all solvers is certainly a disadvantage to those solvers that are superior in this regard. Alternatives to single summary statistics, such as performance profiles [25], should also be considered.

8.6 Concluding Remarks

In this chapter, we have provided an overview of the main challenges involved in parallelizing solution methods for MILP solvers. We have also surveyed the current state of the art in terms of available software implementations. Although tremendous effort has been directed towards the development of scalable parallel algorithms, the tension between scalability and overall effectiveness is ever present and strategies for parallelization must constantly evolve in order to effectively exploit improvements in sequential solvers. Replicating the algorithmic schemes of sequential solvers in parallel continues to pose significant challenges. Nevertheless, substantial progress has been observed and more is expected as technology continues to evolve. New frontiers, such as the exploitation of GPUs, will continue to pose interesting research questions for years to come.

Acknowledgements This work has been supported by the Research Campus Modal (*Mathematical Optimization and Data Analysis Laboratories*) funded by the Federal Ministry of Education and Research (BMBF Grant 05M14ZAM), by the DFG SFB/Transregio 154, and by Lehigh University. All responsibility for the content is assumed by the authors.

References

[1] Achterberg, T.: Conflict analysis in mixed integer programming. Discrete Optimization **4**(1), 4–20 (2007). Special issue: Mixed Integer Programming
[2] Achterberg, T., Bixby, R.E., Gu, Z., Rothberg, E., Weninger, D.: Presolve reductions in mixed integer programming. ZIB-Report 16-44, Zuse Institute Berlin, (2016)
[3] Achterberg, T., Koch, T., Martin, A.: MIPLIB 2003. ORL **34**(4), 1–12 (2006)
[4] Achterberg, T., Wunderling, R.: Mixed integer programming: Analyzing 12 years of progress. In: M. Jünger, G. Reinelt (eds.) Facets of Combinatorial Optimization: Festschrift for Martin Grötschel, pp. 449–481. Springer Berlin Heidelberg (2013)
[5] Alba, E., Almeida, F., Blesa, M., Cabeza, J., Cotta, C., Díaz, M., Dorta, I., Gabarró, J., León, C., Luna, J., Moreno, L., Pablos, C., Petit, J., Rojas, A., Xhafa, F.: Mallba: A library of skeletons for combinatorial optimisation. In: B. Monien, R. Feldmann (eds.) Euro-Par 2002 Parallel Processing: 8th International Euro-Par Conference, Paderborn, Germany, August 27–30, 2002 Proceedings, pp. 927–932. Springer Berlin Heidelberg (2002). DOI 10.1007/3-540-45706-2_132
[6] Barney, B.: Introduction to Parallel Computing. `https://computing.llnl.gov/tutorials/parallel_comp/`
[7] Bénichou, M., Cung, V.D., Dowaji, S., Cun, B.L., Mautor, T., Roucairol, C.: Building a parallel branch and bound library. In: Solving Combinatorial

Optimization Problems in Parallel, Lecture Notes in Computer Science **1054**, pp. 201–231. Springer, Berlin (1996)

[8] Berthold, T.: Primal heuristics for mixed integer programs. Diploma thesis, Technische Universität Berlin (2006)

[9] Berthold, T.: Heuristic algorithms in global MINLP solvers. Ph.D. thesis, Technische Universität Berlin (2014)

[10] Berthold, T., Farmer, J., Heinz, S., Perregaard, M.: Parallelization of the FICO Xpress-Optimizer. In: G.M. Greuel, T. Koch, P. Paule, A. Sommese (eds.) Mathematical Software – ICMS 2016, pp. 251–258. Springer International Publishing (2016). DOI 10.1007/978-3-319-42432-3_31

[11] Berthold, T., Salvagnin, D.: Cloud branching. In: C. Gomes, M. Sellmann (eds.) Integration of AI and OR Techniques in Constraint Programming for Combinatorial Optimization Problems, *Lecture Notes in Computer Science*, vol. 7874, pp. 28–43. Springer Berlin Heidelberg (2013)

[12] Brüngger, A., Marzetta, A., Fukuda, K., Nievergelt, J.: The parallel search bench ZRAM and its applications. Annals of Operations Research **90**(0), 45–63 (1999). DOI 10.1023/A:1018972901171

[13] Bulut, A., Ralphs, T.K.: Disco version 0.95 (2017). DOI 10.5281/zenodo.237107

[14] Bussieck, M.R., Ferris, M.C., Meeraus, A.: Grid-enabled optimization with GAMS. IJoC **21**(3), 349–362 (2009). DOI 10.1287/ijoc.1090.0340

[15] Carvajal, R., Ahmed, S., Nemhauser, G., Furman, K., Goel, V., Shao, Y.: Using diversification, communication and parallelism to solve mixed-integer linear programs. Operations Research Letters **42**(2), 186–189 (2014). DOI 10.1016/j.orl.2013.12.012

[16] Chen, Q., Ferris, M.C., Linderoth, J.: Fatcop 2.0: Advanced features in an opportunistic mixed integer programming solver. Annals of Operations Research **103**(1), 17–32 (2001). DOI 10.1023/A:1012982400848. URL `http://dx.doi.org/10.1023/A:1012982400848`

[17] Cornuéjols, G., Karamanov, M., Li, Y.: Early estimates of the size of branch-and-bound trees. INFORMS J. on Computing **18**(1), 86–96 (2006). DOI 10.1287/ijoc.1040.0107

[18] IBM ILOG CPLEX Optimizer. `http://www-01.ibm.com/software/integration/optimization/cplex-optimizer/`

[19] Crainic, T., Le Cun, B., Roucairol, C.: Parallel branch-and-bound algorithms. In: E. Talbi (ed.) Parallel Combinatorial Optimization, pp. 1–28. Wiley, New York (2006)

[20] Danna, E.: Performance variability in mixed integer programming (2008). Presentation, Workshop on Mixed Integer Programming (MIP 2008), Columbia University, New York. `http://coral.ie.lehigh.edu/~jeff/mip-2008/talks/danna.pdf`

[21] Dantzig, G.B., Wolfe, P.: Decomposition principle for linear programs. Operations Research **8**(1), 101–111 (1960)

[22] DeNegre, S., Ralphs, T.K.: A branch-and-cut algorithm for bilevel integer programming. In: Proceedings of the Eleventh INFORMS Comput-

ing Society Meeting, pp. 65–78 (2009). DOI 10.1007/978-0-387-88843-9_4. URL http://coral.ie.lehigh.edu/~ted/files/papers/BILEVEL08.pdf

[23] Dinan, J., Olivier, S., Sabin, G., Prins, J., Sadayappan, P., Tseng, C.W.: Dynamic load balancing of unbalanced computations using message passing. In: 2007 IEEE International Parallel and Distributed Processing Symposium, pp. 1–8 (2007). DOI 10.1109/IPDPS.2007.370581

[24] Djerrah, A., Cun, B.L., Cung, V.D., Roucairol, C.: Bob++: Framework for solving optimization problems with branch-and-bound methods. In: 2006 15th IEEE International Conference on High Performance Distributed Computing, pp. 369–370 (2006). DOI 10.1109/HPDC.2006.1652188

[25] Dolan, E.D., Moré, J.J.: Benchmarking optimization software with performance profiles. Mathematical Programming **91**(2), 201–213 (2002). DOI 10.1007/s101070100263. URL http://dx.doi.org/10.1007/s101070100263

[26] Eckstein, J.: Control strategies for parallel mixed integer branch and bound. In: Proceedings of the 1994 conference on Supercomputing, pp. 41–48. IEEE Computer Society Press (1994)

[27] Eckstein, J.: Distributed versus centralized storage and control for parallel branch and bound: Mixed integer programming on the CM-5. Comput. Optim. Appl. **7**(2), 199–220 (1997). URL http://dx.doi.org/10.1023/A:1008699010646

[28] Eckstein, J., Hart, W.E., Phillips, C.A.: Pebbl: an object-oriented framework for scalable parallel branch and bound. Mathematical Programming Computation **7**(4), 429–469 (2015). DOI 10.1007/s12532-015-0087-1. URL http://dx.doi.org/10.1007/s12532-015-0087-1

[29] Eckstein, J., Phillips, C.A., Hart, W.E.: PEBBL 1.0 user guide (2007)

[30] Eikland, K., Notebaert, P.: lp_solve 5.5.2. http://lpsolve.sourceforge.net

[31] FICO Xpress-Optimizer. http://www.fico.com/en/Products/DMTools/xpress-overview/Pages/Xpress-Optimizer.aspx

[32] Fischetti, M., Lodi, A.: Heuristics in mixed integer programming. In: J.J. Cochran, L.A. Cox, P. Keskinocak, J.P. Kharoufeh, J.C. Smith (eds.) Wiley Encyclopedia of Operations Research and Management Science. John Wiley & Sons, Inc. (2010). Online publication

[33] Fischetti, M., Lodi, A., Monaci, M., Salvagnin, D., Tramontani, A.: Improving branch-and-cut performance by random sampling. Mathematical Programming Computation **8**(1), 113–132 (2016)

[34] Fischetti, M., Monaci, M., Salvagnin, D.: Self-splitting of workload in parallel computation. In: H. Simonis (ed.) Integration of AI and OR Techniques in Constraint Programming: 11th International Conference, CPAIOR 2014. Proceedings, pp. 394–404. Springer International Publishing (2014). DOI 10.1007/978-3-319-07046-9_28

[35] Forrest, J.: CBC MIP solver. http://www.coin-or.org/Cbc

[36] Fourer, R.: Linear programming: Software survey. OR/MS Today **42**(3) (2015)
[37] Galati, M.V., Ralphs, T.K., Wang, J.: Computational experience with generic decomposition using the DIP framework. In: Proceedings of RAMP 2012 (2012). URL http://coral.ie.lehigh.edu/~ted/files/papers/RAMP12.pdf
[38] Gamrath, G., Koch, T., Maher, S.J., Rehfeldt, D., Shinano, Y.: SCIP-Jack—a solver for STP and variants with parallelization extensions. Mathematical Programming Computation **9**(2), 231–296 (2017)
[39] Gamrath, G., Koch, T., Martin, A., Miltenberger, M., Weninger, D.: Progress in presolving for mixed integer programming. Mathematical Programming Computation **7**(4), 367–398 (2015)
[40] Gendron, B., Crainic, T.G.: Parallel branch-and-branch algorithms: Survey and synthesis. Operations Research **42**(6), 1042–1066 (1994). DOI 10.1287/opre.42.6.1042. URL http://dx.doi.org/10.1287/opre.42.6.1042
[41] Gomory, R.E.: Outline of an algorithm for integer solutions to linear programs. Bulletin of the American Mathematical Society **64**(5), 275–278 (1958)
[42] Gottwald, R.L., Maher, S.J., Shinano, Y.: Distributed domain propagation. ZIB-Report 16-71, Zuse Institute Berlin, (2016)
[43] Goux, J.P., Kulkarni, S., Linderoth, J., Yoder, M.: An enabling framework for master-worker applications on the computational grid. In: Proceedings the Ninth International Symposium on High-Performance Distributed Computing, pp. 43–50 (2000). DOI 10.1109/HPDC.2000.868633
[44] Gurobi Optimizer. http://www.gurobi.com/
[45] Hager, G., Wellein, G.: Introduction to High Performance Computing for Scientists and Engineers. CRC Press, Inc., Boca Raton, FL, USA (2010)
[46] Henrich, D.: Initialization of parallel branch-and-bound algorithms. In: Second International Workshop on Parallel Processing for Artificial Intelligence(PPAI-93) (1993)
[47] Huangfu, Q., Hall, J.: Parallelizing the dual revised simplex method. Tech. rep., arXiv preprint arXiv:1503.01889 (2015)
[48] Janakiram, V.K., Gehringer, E.F., Agrawal, D.P., Mehrotra, R.: A randomized parallel branch-and-bound algorithm. International Journal of Parallel Programming **17**(3), 277–301 (1988). DOI 10.1007/BF02427853
[49] Jeannot, E., Mercier, G., Tessier, F.: Topology and affinity aware hierarchical and distributed load-balancing in Charm++. In: Proceedings of the First Workshop on Optimization of Communication in HPC, COM-HPC '16, pp. 63–72. IEEE Press, Piscataway, NJ, USA (2016). DOI 10.1109/COM-HPC.2016.12
[50] Jünger, M., Thienel, S.: Introduction to ABACUS—a branch-and-cut system. Operations Research Letters **22**, 83–95 (1998)
[51] Khachiyan, L.G.: A polynomial algorithm in linear programming. Doklady Akademii Nauk SSSR **244**(5), 1093–1096 (1979). English translation in Soviet Math. Dokl. 20(1):191–194, 1979
[52] Koch, T., Achterberg, T., Andersen, E., Bastert, O., Berthold, T., Bixby, R.E., Danna, E., Gamrath, G., Gleixner, A.M., Heinz, S., Lodi, A., Mittelmann, H.,

Ralphs, T., Salvagnin, D., Steffy, D.E., Wolter, K.: MIPLIB 2010. Math. Prog. Comp. **3**, 103–163 (2011)
[53] Koch, T., Ralphs, T., Shinano, Y.: Could we use a million cores to solve an integer program? Mathematical Methods of Operations Research **76**(1), 67–93 (2012). DOI 10.1007/s00186-012-0390-9. URL `http://dx.doi.org/10.1007/s00186-012-0390-9`
[54] Kumar, V., Grama, A.Y., Vempaty, N.R.: Scalable load balancing techniques for parallel computers. Journal of Parallel and Distributed Computing **22**(1), 60–79 (1994)
[55] Ladányi, L.: BCP: Branch-cut-price framework (2000). URL `https://projects.coin-or.org/Bcp`
[56] Land, A.H., Doig, A.G.: An automatic method of solving discrete programming problems. Econometrica **28**(3), 497–520 (1960)
[57] Laursen, P.S.: Can parallel branch and bound without communication be effective? SIAM Journal on Optimization **4**, 288–296 (1994)
[58] Linderoth, J.: Topics in parallel integer optimization. Ph.D. thesis, School of Industrial and Systems Engineering, Georgia Institute of Technology, Atlanta, GA (1998)
[59] Linderoth, J.T., Savelsbergh, M.: A computational study of search strategies for mixed integer programming. INFORMS Journal on Computing **11**, 173–187 (1998)
[60] Lougee-Heimer, R.: The common optimization interface for operations research. IBM Journal of Research and Development **47**(1), 57–66 (2003)
[61] Mahajan, A.: Presolving mixed-integer linear programs. In: J.J. Cochran, L.A. Cox, P. Keskinocak, J.P. Kharoufeh, J.C. Smith (eds.) Wiley Encyclopedia of Operations Research and Management Science. John Wiley & Sons, Inc. (2010). DOI 10.1002/9780470400531.eorms0437. Online publication
[62] Makhorin, A.: the GNU linear programming kit. `http://www.gnu.org/software/glpk`
[63] Marchand, H., Martin, A., Weismantel, R., Wolsey, L.: Cutting planes in integer and mixed integer programming. Discrete Applied Mathematics **123**(1), 397–446 (2002)
[64] Miller, D., Pekny, J.: Results from a parallel branch and bound algorithm for the asymmetric traveling salesman problem. Operations Research Letters **8**(3), 129–135 (1989). DOI http://dx.doi.org/10.1016/0167-6377(89)90038-2
[65] Munguia, L.M., Oxberry, G., Rajan, D.: PIBS-SBB: A parallel distributed-memory branch-and-bound algorithm for stochastic mixed-integer programs. In: 2016 IEEE International Parallel and Distributed Processing Symposium Workshops (IPDPSW), pp. 730–739 (2016). DOI 10.1109/IPDPSW.2016.159
[66] Nemhauser, G.L., Wolsey, L.A.: Integer and combinatorial optimization. Wiley (1988)
[67] Nesterov, Y., Nemirovski, A.: Interior-Point Polynomial Algorithms in Convex Programming. Studies in Applied and Numerical Mathematics. Society for Industrial and Applied Mathematics (1994)

[68] Olszewski, M., Ansel, J., Amarasinghe, S.: Kendo: efficient deterministic multithreading in software. ACM SIGPLAN Notices **44**(3), 97–108 (2009). DOI 10.1145/1508284.1508256

[69] Osman, A., Ammar, H.: Dynamic load balancing strategies for parallel computers. URL `http://citeseer.nj.nec.com/osman02dynamic.html`

[70] Ozaltin, O.Y., Hunsaker, B., Schaefer, A.J.: Predicting the solution time of branch-and-bound algorithms for mixed-integer programs. INFORMS J. on Computing **23**(3), 392–403 (2011). DOI 10.1287/ijoc.1100.0405

[71] Pekny, J.F.: Exact parallel algorithms for some members of the traveling salesman problem family. Ph.D. thesis, Carnegie-Mellon University, Pittsburgh, PA, USA (1989)

[72] Ralphs, T.K.: Parallel branch and cut for capacitated vehicle routing. Parallel Computing **29**, 607–629 (2003). DOI 10.1016/S0167-8191(03)00045-0. URL `http://coral.ie.lehigh.edu/~ted/files/papers/PVRP.pdf`

[73] Ralphs, T.K.: Parallel branch and cut. In: E. Talbi (ed.) Parallel Combinatorial Optimization, pp. 53–101. Wiley, New York (2006). URL `http://coral.ie.lehigh.edu/~ted/files/papers/PBandC.pdf`

[74] Ralphs, T.K., Galati, M.V., Wang, J.: Dip version 0.92 (2017). DOI 10.5281/zenodo.246087

[75] Ralphs, T.K., Guzelsoy, M., Mahajan, A.: Symphony version 5.6 (2017). DOI 10.5281/zenodo.237456

[76] Ralphs, T.K., Ladányi, L.: COIN/BCP user's manual. Tech. rep., COR@L Laboratory, Lehigh University (2001). URL `http://coral.ie.lehigh.edu/~ted/files/papers/BCP-Manual.pdf`

[77] Ralphs, T.K., Ladányi, L., Saltzman, M.J.: Parallel branch, cut, and price for large-scale discrete optimization. Mathematical Programming **98**, 253–280 (2003). DOI 10.1007/s10107-003-0404-8. URL `http://coral.ie.lehigh.edu/~ted/files/papers/PBCP.pdf`

[78] Sanders, P.: A detailed analysis of random polling dynamic load balancing. In: International Symposium on Parallel Architectures Algorithms and Networks, pp. 382–389 (1994)

[79] Sanders, P.: Randomized static load balancing for tree-shaped computations. In: Workshop on Parallel Processing, pp. 58–69 (1994)

[80] Sanders, P.: Tree shaped computations as a model for parallel applications. In: ALV'98 Workshop on application based load balancing, pp. 123–132 (1998)

[81] SCIP: Solving Constraint Integer Programs. `http://scip.zib.de/`

[82] Shinano, Y., Achterberg, T., Berthold, T., Heinz, S., Koch, T., Winkler, M.: Solving open MIP instances with ParaSCIP on supercomputers using up to 80,000 cores. In: 2016 IEEE International Parallel and Distributed Processing Symposium (IPDPS), pp. 770–779. IEEE Computer Society, Los Alamitos, CA, USA (2016)

[83] Shinano, Y., Achterberg, T., Fujie, T.: A dynamic load balancing mechanism for new ParaLEX. In: Proceedings of ICPADS 2008, pp. 455–462 (2008)

[84] Shinano, Y., Fujie, T.: ParaLEX: A parallel extension for the CPLEX mixed integer optimizer. In: F. Cappello, T. Herault, J. Dongarra (eds.) Recent Advances in Parallel Virtual Machine and Message Passing Interface. Proceedings, pp. 97–106. Springer Berlin Heidelberg (2007). DOI 10.1007/978-3-540-75416-9_19
[85] Shinano, Y., Fujie, T., Kounoike, Y.: Effectiveness of parallelizing the ILOG-CPLEX mixed integer optimizer in the PUBB2 framework. In: H. Kosch, L. Böszörményi, H. Hellwagner (eds.) Euro-Par 2003 Parallel Processing: Proceedings, pp. 451–460. Springer Berlin Heidelberg (2003). DOI 10.1007/978-3-540-45209-6_67
[86] Shinano, Y., Fujie, T., Kounoike, Y.: Pubb2: A redesigned object-oriented software tool for implementing parallel and distributed branch-and-bound algorithms. In: Proceedings of ISTEAD International Conference: Parallel and Distributed Computing and Systems, pp. 639–647 (2003)
[87] Shinano, Y., Heinz, S., Vigerske, S., Winkler, M.: FiberSCIP – a shared memory parallelization of SCIP. INFORMS Journal on Computing, Published online 2017, https://doi.org/10.1287/ijoc.2017.0762
[88] Shinano, Y., Higaki, M., Hirabayashi, R.: A generalized utility for parallel branch and bound algorithms. In: Proceedings of the Seventh IEEE Symposium on Parallel and Distributed Processing, pp. 392–401 (1995). DOI 10.1109/SPDP.1995.530710
[89] Sinha, A., Kalé, L.V.: A load balancing strategy for prioritized execution of tasks. In: Seventh International Parallel Processing Symposium, pp. 230–237. Newport Beach, CA. (1993)
[90] SteinLib Testdata Library. `http://steinlib.zib.de/steinlib.php`
[91] Tahernejad, S., Ralphs, T., DeNegre, S.: A branch-and-cut algorithm for mixed integer bilevel linear optimization problems and its implementation. Tech. rep., COR@L Laboratory Technical Report 16T-015-R3, Lehigh University (2016)
[92] Trienekens, H.W.J.M., de Bruin, A.: Towards a taxonomy of parallel branch and bound algorithms. Tech. Rep. EUR-CS-92-01, Department of Computer Science, Erasmus University (1992)
[93] Tschoke, S., Polzer, T.: Portable parallel branch and bound library (2008). `http://www.cs.uni-paderborn.de/cs/ag-monien/SOFTWARE/PPBB/ppbblib.html`
[94] UG: Ubiquity Generator framework. `http://ug.zib.de/`
[95] Wang, J., Ralphs, T.K.: Computational experience with hypergraph-based methods for automatic decomposition in discrete optimization. In: Proceedings of the Conference on Constraint Programming, Artificial Intelligence, and Operations Research, pp. 394–402 (2013). DOI 10.1007/978-3-642-38171-3
[96] Wilkinson, B., Allen, M.: Parallel Programming: Techniques and Applications Using Networked Workstations and Parallel Computers. Prentice-Hall, Inc, New Jersey, USA (1999)

[97] Willebeek-LeMair, M.H., Reeves, A.P.: Strategies for dynamic load balancing on highly parallel computers. IEEE Transactions on Parallel and Distributed Systems **4**, 979–993 (1993). DOI 10.1109/71.243526

[98] Witzig, J., Berthold, T., Heinz, S.: Experiments with conflict analysis in mixed integer programming. ZIB-Report 16-63, Zuse Institute Berlin, (2016)

[99] Wolter, K.: Implementation of Cutting Plane Separators for Mixed Integer Programs. Master's thesis, Technische Universität Berlin (2006)

[100] Xu, Y.: Scalable algorithms for parallel tree search. Ph.D. thesis, Department of Industrial and Systems Engineering, Lehigh University, Bethlehem, PA, USA (2007)

[101] Xu, Y., Ralphs, T.K., Ladányi, L., Saltzman, M.: Alps version 1.5 (2016). DOI 10.5281/zenodo.245971

[102] Xu, Y., Ralphs, T.K., Ladányi, L., Saltzman, M.: Biceps version 0.94 (2017). DOI 10.5281/zenodo.245652

[103] Xu, Y., Ralphs, T.K., Ladányi, L., Saltzman, M.: Blis version 0.94 (2017). DOI 10.5281/zenodo.246079

[104] Xu, Y., Ralphs, T.K., Ladányi, L., Saltzman, M.J.: Alps: A framework for implementing parallel search algorithms. In: The Proceedings of the Ninth INFORMS Computing Society Conference, pp. 319–334 (2005). DOI 10.1007/0-387-23529-9_21. URL `http://coral.ie.lehigh.edu/~ted/files/papers/ALPS04.pdf`

[105] Xu, Y., Ralphs, T.K., Ladányi, L., Saltzman, M.J.: Computational experience with a software framework for parallel integer programming. The INFORMS Journal on Computing **21**, 383–397 (2009). DOI 10.1287/ijoc.1090.0347. URL `http://coral.ie.lehigh.edu/~ted/files/papers/CHiPPS-Rev.pdf`

[106] Zheng, G., Bhatelé, A., Meneses, E., Kalé, L.V.: Periodic hierarchical load balancing for large supercomputers. Int. J. High Perform. Comput. Appl. **25**(4), 371–385 (2011). DOI 10.1177/1094342010394383

Chapter 9
Parallel Constraint Programming

Jean-Charles Régin and Arnaud Malapert

Abstract Constraint programming (CP) is an efficient technique for solving combinatorial optimization problems. In CP a problem is defined over variables that take values in domains and constraints which restrict the allowed combination of values. CP uses for each constraint an algorithm that removes values of variables that are inconsistent with the constraint. These algorithms are called while a domain is modified. Then, a search algorithm such as a backtracking or branch-and-bound algorithm is called to find solutions. Several methods have been proposed to combine CP with parallelism. In this chapter, we present some of them: parallelization of the propagator, parallel propagation, search splitting, also called work-stealing, problem decomposition, also called embarrassingly parallel search (EPS), and portfolio approaches. We detail the two giving the best performances in practice: the work-stealing approach and embarrassingly parallel search. We give some experiments supporting this claim on a single multi-core machine, on a data center and on the cloud.

9.1 Introduction

Constraint Programming (CP) is an efficient technique for solving combinatorial optimization problems. It is widely used for solving real-world applications such as rostering, scheduling, car sequencing, routing, etc. CP-based solvers are general and generic tools for modeling and solving problems [12, 13, 81, 95, 83, 96, 93]. The development of such solvers is an active topic of the CP community. In this chapter, we propose to consider different approaches for parallelizing a CP-based solver. Our

Jean-Charles Régin
Université Côte d'Azur, CNRS, I3S, France, e-mail: jcregin@gmail.com

Arnaud Malapert
Université Côte d'Azur, CNRS, I3S, France e-mail: arnaud.malapert@unice.fr

Y. Hamadi und L. Sais (eds.), *Handbook of Parallel Constraint Reasoning*,
https://doi.org/10.1007/978-3-319-63516-3_9

goal is to present methods that have been used to automatically parallelize CP-based solvers. This means that no particular action of the user is required.

CP is mainly based on the exploitation of the structure of the constraints and accepts constraints whose structure is different, unlike SAT or MIP which impose certain rules on allowable models of the problem: having only boolean variables and three clauses for SAT, or having only linear constraints for MIP.

This specificity of CP allows the use of any kind of algorithm for solving a problem. We could even say that we want to exploit as much as possible the capability to use different algorithms. Currently, when a problem is modeled in CP it is possible that a large variety of algorithms are used at the same time and communicate with each other. For instance, unlike with other techniques, it is really conceivable to have at the same time flow algorithms and dynamic programming.

In CP, a problem is defined using variables and constraints. Each variable is associated with a domain containing its possible values. A constraint expresses properties that have to be satisfied by a set of variables.

In CP, a problem can also be viewed as a conjunction of sub-problems for which we have efficient resolution methods. These sub-problems can be very easy like $x < y$ or complex like the search for a feasible flow. These sub-problems correspond to constraints. Then, CP uses for each sub-problem the associated resolution method, often called a *propagator*. A propagator removes from the domains the values that cannot belong to any solution of the sub-problem. This mechanism is called *filtering*. By repeating this process for each sub-problem, so for each constraint, the domains of the variables are reduced.

After each modification of a variable domain, it is useful to reconsider all the constraints involving that variable, because that modification can lead to new deductions. In other words, the domain reduction of one variable may lead to deduce that some other values of some other variables cannot belong to a solution. So, CP calls all the propagators associated with a constraint involving a modified variable until no more modification occurs. This mechanism is called *propagation*.

Then, and in order to reach a solution, the search space will be traversed by assigning successively a value to each variable. The filtering and propagation mechanisms are, of course, triggered when a modification occurs. Sometimes, an assignment may lead to the removal of all the values of a domain: we say that a failure occurs, and the latest choice is reconsidered: there is a backtrack and a new assignment is tried. This mechanism is called *search*.

So, CP is based on three principles: filtering, propagation and search. We could represent it by reformulating Kowalski's famous definition of Algorithm (Algorithm = Logic + Control) [51] as:

$$CP = filtering + propagation + search \tag{9.1}$$

where filtering and propagation correspond to Logic and search to Control.

An objective can also be added in order to deal with optimization problems. In this case, a specific variable representing the objective is defined. When a better solution is found then this variable is updated, and this modification is permanent.

The relation between the objective variable and the other variables is usually via a constraint representing the objective function, which is often a sum constraint.

9.1.1 Filtering + Propagation

Since constraint programming is based on filtering algorithms [86], it is quite important to design efficient and powerful algorithms. Therefore, this topic caught the attention of many researchers, who discovered a large number of algorithms.

As we mentioned, a filtering algorithm directly depends on the constraint it is associated with. The advantage of using the structure of a constraint can be shown on the constraint $x \leq y$. Let $min(D)$ and $max(D)$ be respectively the minimum and the maximum value of a domain. It is straightforward to establish that all the values of x and y in the range $[min(D(x)), max(D(y)]$ are consistent with the constraint. This means that arc consistency can be efficiently and easily established by removing the values that are not in the above ranges. Moreover, the use of the structure is often the only way to avoid memory consumption problems when dealing with non-binary constraints. In fact, this approach prevents us from explicitly having to represent all the combinations of values allowed by the constraint.

One of the most famous examples is the ALLDIFFB constraint, which states that values taken by variables must be different, especially because the filtering algorithm associated with this constraint is able to establish arc consistency in a very efficient way by using matching techniques [85].

The propagation mechanism pushes the propagators associated with a variable when this variable is modified. There are usually two levels: a first level for the immediate propagation of the modification of a variable and a delayed level that aims at considering once and for all the modification of the variables involved in each propagator. The delayed level is called only when there are no more propagator to call in the first level. The delayed level is interrupted by the first level when the latter is no longer empty.

Of course, each propagator can be parallelized. However a synchronization between them is needed, so it is really difficult to obtain consistent speed up with such an approach. The propagation mechanism can also be parallelized, with the same issues.

Note that the mechanism that is used when solving a Sudoku puzzle corresponds to the application of rules, that is to the call of filtering algorithms (i.e. propagators) until we cannot make any deduction. Thus, this is a propagation mechanism.

9.1.2 Search

Solutions can be found by searching systematically through the possible assignments of values to variables. A *backtracking scheme* incrementally extends a partial assignment that specifies consistent values for some of the variables toward a complete solution, by repeatedly choosing a value for another variable. The variables are assigned sequentially.

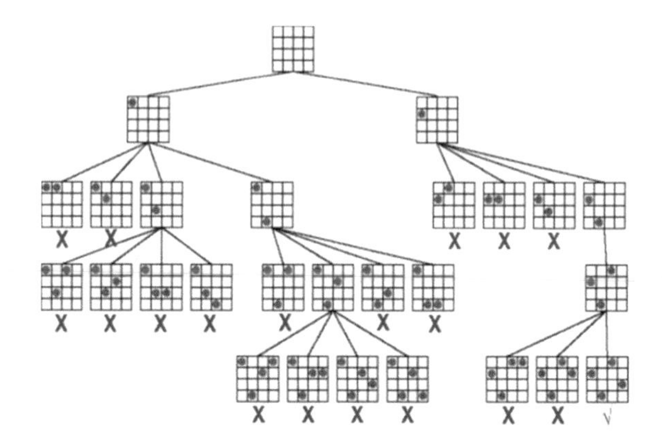

Fig. 9.1a: Search tree for the four queens problems without propagation

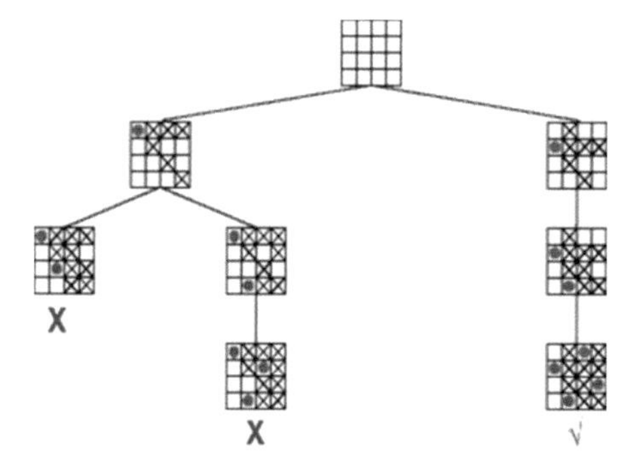

Fig. 9.1b: Search tree for the four queens problems with weak propagation

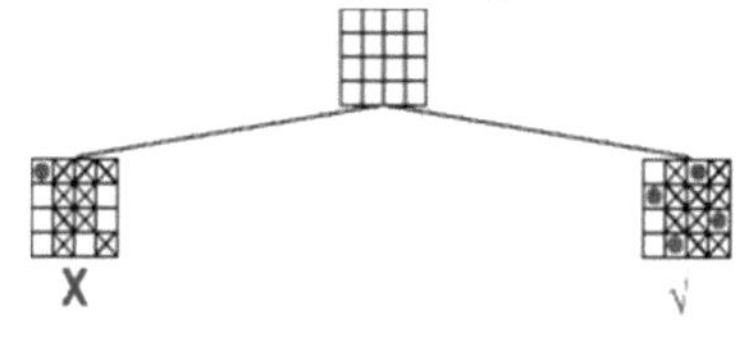

Fig. 9.1c: Search tree for the four queens problems with strong propagation

At each node of the search tree, an uninstantiated variable is selected and the node is extended so that the resulting new branches out of the node represent alternative choices that may have to be examined in order to find a solution. The branching strategy, also called the variable-value strategy, determines the next variable to be instantiated, and the order in which the values from its domain are selected. Each time a variable is assigned a value the propagation mechanism is triggered. If a partial assignment violates any of the constraints, that is if a domain becomes empty, then a backtrack is performed to the most recently assigned variable that still has alternative values available in its domain. Clearly, whenever a partial assignment violates a constraint, backtracking is able to eliminate a subspace from the Cartesian product of non-empty variable domains.

When a backtrack occurs, the refutation of the previous choice is usually added to the solver and the propagation mechanism is called. More precisely, if the assignment $x = a$ fails, then a backtrack is performed and the constraint $x \neq a$ is added.

The propagation mechanism allows the reduction of the variable domains and the pruning of the search tree whereas the branching strategy can improve the detection of solutions (or failures for unsatisfiable problems).

In the absence of specific knowledge, defining an efficient variable-value strategy for guiding the search for solutions of a given problem is not an easy task. Thus some generic variable-value strategies have been defined. They either try to apply generic principles such as the first fail principle (i.e., we should try to fail as quickly as possible) [37] or try to detect relations between variables and constraints. In the first case, we have strategies such as min-domain, which selects the variable having the minimum domain size, max-constrained, which prefers variables involved in a lot of constraints, or min-regret which selects the variable that may lead to the largest increase in the cost if it is not selected. The latter case is mainly formed by the impact-based strategy [84], the weighted degree strategy [7] and the activity-based strategy [68]. However, selecting a priori the best variable-value strategy is not an easy task, because no strategy is better than any other in general and because it is quite difficult to identify the types of problems for which a strategy is going to perform well. In addition, there is no robustness among the strategies. Any variable-value strategy can give good results for one problem and really bad results for some others. It is not rare to see the ratio of performance for a pair of strategies going to 1 to 20 (and even more sometimes) according to the problems which are solved.

The difference on the explored search spaces can be seen on the 4-queens problem: search without propagation (see Figure 9.1a), with weak propagation (see Figure 9.1b), and with strong propagation (see Figure 9.1c).

9.1.2.1 Search Methods in Solvers

In generic solvers based on constraint programming, the search mechanism is an important part. It is a generic method for controlling the solver that is used to take decisions and to introduce refutation (i.e., the negation of decisions) for solving a whole problem. Decisions can correspond to the assignment of values to variables, but they can also be more complicated. Real-life applications are usually complex and the search mechanism is used for decomposing the problem, for adding some constraints and for solving some subparts. In other words it is used for performing different tasks. For instance, in scheduling applications it is common to deal with the relation between activities, that is which one starts before the other, instead of deciding at what precise moment an activity starts. In problems in which variables are continuous it is also common that a decision splits a variable into two equal parts.

The search is usually totally controlled by the user, who specifies functions that will be called to take a decision and to refute that decision. It is important to note that, in general, there is no other way to reach a state given by a sequence of decisions and refutations than replaying from the beginning that sequence.

9.1.3 Parallelism and Constraint Programming

In this chapter, we only discuss parallel constraint solving. Some surveys have been written about parallel logic programming [21, 33], and about parallel integer programming [20, 4, 26].

The main approaches to parallel constraint solving can roughly be divided into the following main categories: parallel propagators and propagation; search-space-splitting; portfolio algorithms; distributed CSPs; problem decomposition. Most approaches require communication and synchronization, but the most important issue is load balancing, which refers to the practice of distributing approximately equal amounts of work among tasks so that all processors are kept busy all the time.

After an introduction of the main principles of each approach we will detail the two most important ones: the search space splitting method (i.e., the work-stealing approach) and problem decomposition (i.e., embarrassingly parallel search).

9.1.3.1 Parallel Propagators and Propagation

The propagators of the constraints, that is the filtering algorithms associated with constraints, can be parallelized. However, this operation is not simple, and some synchronization issues of the domains of the variables arise. In addition, one of the most important drawback is that propagation requires a specific parallelization of each constraint. So, it is not popular because of the same synchronization problems, only a few studies can be found on the parallelization of the propagation mechanism [74, 35, 91]. Thus, parallelizing propagation is challenging [44] and the scalability is limited by Amdahl's law.

9.1.3.2 Search Space Splitting

Strategies exploring the parallelism provided by the search space are common approaches [82]: when a branching is done, different branches can be explored in parallel.

The work-stealing method dynamically splits the search space during the resolution. It was originally proposed by Burton and Sleep [10] and first implemented in Lisp parallel machines [34]. When a worker has finished exploring a sub-problem, it asks other workers for another sub-problem. If another worker agrees to the demand, then it dynamically splits its current sub-problem into two disjoint sub-problems and sends one sub-problem to the starving worker. The starving worker "steals" some work from the busy one. Note that some form of locking is necessary to avoid the case that several starving workers steal the same sub-problem. The starving worker asks other workers in turn until it receives a new sub-problem. Termination of the work-stealing method must be carefully designed to reduce the overhead when almost all workers are starving, but almost no work remains. Search space splitting is an active research area [41, 67, 15]. This is also one of the methods giving the best results in practice.

Some frameworks sharing the same search tree in memory have been proposed [79, 97]. In this case, a shared list of open nodes in the search tree is maintained (nodes that have at least one child that is still unvisited) and starved processors just pick up the most promising node in the list and expand it. Although this kind of mechanism intrinsically provides excellent load balancing, it is known not to scale beyond a certain number of processors; beyond that point, performance starts to decrease. Indeed, on a shared-memory system, threads must contend with each other to communicate with the memory and the problem is exacerbated by cache consistency transactions. Thus, other approaches that do not use shared memory are preferred.

However, even if the memory is not shared it is not easy to scale up to thousands of processors, because work-stealing consumes communication, synchronization and computation time. To address these issues, Xie and Davenport allocated specific processors to coordination tasks, allowing an increase in the number of processors (linear scaling up to 256 processors) that can be used on a parallel supercomputer before performance starts to decline.

Machado et al. proposed a hierarchical works-stealing scheme adapted to a cluster physical infrastructure, in order to reduce the communication overhead [59]. A worker first tries to steal from its local node, before considering remote nodes (starting with the closest remote node). This approach achieved good scalability up to 512 cores for the n-queens and quadratic assignment problems. For constraint optimization problems, maintaining the best solution for each worker would require a large communication and synchronization overhead. However, they observed that the scalability was lowered because of the lazy dissemination of the so-far best solution, i.e., because some workers use an obsolete best solution.

General-purpose programming languages designed for multi-threaded parallel computing such as Charm++ [43] and Cilk++ [54, 9] can ease the implementation of work-stealing approaches. Otherwise, a work-stealing framework such as Bobpp [25, 52] provides an interface between solvers and parallel computers. In Bobpp, the work is shared via a global priority queue and the search tree is decomposed and allocated to the different cores on demand during the search algorithm execution. Periodically, a worker tests whether starving workers exist. In this case, the worker stops the search and the path from the root node to the right highest open node is saved and inserted into the global priority queue. Then, the worker continues the search with the left open node. Otherwise, if no starving worker exists, the worker continues the search locally using the solver. Starving workers are notified of the insertions in the global priority queue, and each one picks up a node and starts the search. Using or-tools as an underlying solver, Menouer and Le Cun observed good speedups for the Golomb Ruler problem with 13 marks (41.3 with 48 workers) and the 16-queens problem (8.63 with 12 workers) [64, 65] . Other experiments investigate the exploration overhead caused by their approach.

Bordeaux et al. proposed another promising approach based on a search-space-splitting mechanism not based on a work-stealing approach [5]. They use a hashing function implicitly allocating the leaves to the processors. Each processor applies the same search strategy in its allocated search space. Well-designed hashing constraints can address the load-balancing issue. This approach gives a linear speedup from 30 processors for the n-queens problem, but then the speedups stagnate from 30 to 64 processors. However, it only got moderate results on 100 industrial SAT instances.

Sometimes, for complex applications where very good domain-specific strategies are known, the parallel algorithm should exploit the domain-specific strategy. Moisan et al. proposed a parallel implementation of the classic backtracking algorithm, Limited Discrepancy Search (LDS) [38], that is known to be efficient in a centralized context when a good variable-value selection heuristic is provided [71, 72]. Xie and Davenport proposed that each processor locally uses LDS to search in the trees allocated to them (by a tree-splitting, work-stealing algorithm) but the global system does not replicate the LDS strategy.

9.1.3.3 Portfolio Algorithms

Portfolio algorithms explore the parallelism provided by different viewpoints on the same problem, for instance by using different algorithms or parameter tuning. This idea has also been exploited in a non-parallel context [30].

No communication is required and an excellent level of load balancing is achieved (all workers visit the same search space). Even if this approach causes a high level of redundancy between processors, it shows really good performance. It was greatly improved by using randomized restarts [58] where each worker executes its own restart strategy. More recently, Cire et al. executed the Luby restart strategy, as a whole, in parallel [17]. They proved that it achieves asymptotic linear speedups and, in practice, often obtained linear speedups. Besides, some authors proposed to allow processors to share information learned during the search [36].

One challenge is to find a scalable source of diverse viewpoints that provide orthogonal performance and are therefore of complementary interest. We can distinguish between two aspects of parallel portfolios: if assumptions can be made on the number of available processors then it is possible to handpick a set of solvers and settings that complement each other optimally. If it is not possible to make such assumptions, then we need automated methods to generate a portfolio of any size on demand [5]. So, portfolio designers became interested in feature selection [28, 29, 31, 45]. Features characterize problem instances by number of variables, domain sizes, number of constraints, constraints arities. Many portfolios select the best candidate solvers from a pool based on static features or by learning the dynamic behavior of solvers. The SAT portfolio `iSAC` [2] and the CP portfolio CP-Hydra [77] use feature selection to choose the solvers that yield the best performance. Additionally, CP-Hydra exploits the knowledge coming from the resolution of a training set of instances by each candidate solver. Then, given an instance, CP-Hydra determines the k most similar instances of the training set and determines a time limit for each candidate solver based on the constraint program maximizing the number of solved instances within a global time limit of 30 minutes. Briefly, CP-Hydra determines a switching policy between solvers (Choco, Abscon, Mistral).

In general, the main advantage of the portfolio algorithms approach is that many strategies will be automatically tried at the same time. This is very useful because defining good search strategies is a difficult task.

The best strategy can also be detected in parallel by using estimation techniques [78].

9.1.3.4 Distributed CSPs

Distributed CSPs is another idea that relates to parallelism, where the problem itself is split into pieces to be solved by different processors. The problem typically becomes more difficult to solve than in the centralized case because no processor has a complete view of the problem. So, reconciling the partial solutions of each sub-problem becomes challenging. Problem splitting typically relates to distributed CSPs, a framework introduced by Yokoo et al. in which the problem is naturally split among agents, for example for privacy reasons [102]. Other distributed CSP frameworks have been proposed [39, 14, 23, 53, 98].

9.1.3.5 Problem Decomposition

The Embarrassingly Parallel Search (EPS) method based on search space splitting with loose communications was first proposed by Régin et al. [87, 88, 90, 61].

When we have k workers, instead of trying to split the problem into k equivalent subparts, EPS proposes to split the problem into a huge number of sub-problems, for instance $30k$ sub-problems, and to give these sub-problems successively and dynamically to the workers when they need work. Instead of expecting to have equivalent sub-problems, EPS expects that *for each worker the sum of the resolution time of its sub-problems will be equivalent*. Thus, the idea is not to decompose a priori the initial problem into a set of equivalent sub-problems, but to decompose the initial problem into a set of sub-problems whose resolution time can be shared in an equivalent way by a set of workers. Note that the sub-problems that will be solved by a worker is not known in advance, because this is dynamically determined. All the sub-problems are put in a queue and a worker takes one when it needs some work.

The decomposition into sub-problems must be carefully done. Sub-problems that would have been eliminated by the propagation mechanism of the solver in a sequential search must be avoided. Thus, *only problems that are consistent with the propagation are considered.*

Fischetti et al. proposed another paradigm called SelfSplit in which each worker is able to autonomously determine, without any communication between workers, the job parts it has to process [24]. SelfSplit can be decomposed into three phases: the same enumeration tree is initially built by all workers (sampling); when enough open nodes have been generated, the sampling phase ends and each worker applies a deterministic rule to identify and solve the nodes that belong to it (solving); a single worker gathers the results from others (merging). SelfSplit exhibited linear speedups up to 16 processors and good speedups up to 64 processors on five benchmark instances. SelfSplit assumes that sampling is not a bottleneck in the overall computation whereas that can happen in practice [88].

This chapter is organized as follows. First we recall some preliminaries about parallelism and constraint programming. Then we detail the work-stealing method and the embarassingly parallel search. Next we give some results comparing the methods and showing their efficiency on different types of parallel machines. Finally, we conclude.

9.2 Background

9.2.1 *Parallelism*

For the sake of clarity, we will use the notion of *worker* instead of process, processor, core or thread. A worker is an entity which is able to perform some computations. It usually corresponds to a thread/core in a current computer.

9.2.1.1 Parallelization Measures and Amdahl's Law

Two important parallelization measures are speedup and efficiency. Let $t(c)$ be the wall-clock time of the parallel algorithm where c is the number of cores and let $t(1)$ be the wall-clock time of the sequential algorithm. The speedup $su(c) = t(1)/t(c)$ is a measure indicating how the parallel algorithm performs much faster due to parallelization. The efficiency $\mathit{eff}(c) = su(c)/c$ is a normalized version of speedup, which is the speedup value divided by the number of cores. The maximum possible speedup of a single program as a result of parallelization is known as Amdahl's law [3]. It states that a small portion of the program which cannot be parallelized will limit the overall speedup available from parallelization. Let $B \in [0,1]$ be the fraction of the algorithm that is strictly sequential. The time $t(c)$ that an algorithm takes to finish when being executed on c cores corresponds to $t(c) = t(1)\left(B + \frac{1}{c}(1-B)\right)$. Therefore, the theoretical speedup $su(c)$ is

$$su(c) = \frac{1}{B + \frac{1}{c}(1 - B)}$$

According to Amdahl's law, the speedup can never exceed the number of cores, i.e., a linear speedup. This, in terms of efficiency measure, means that efficiency will always be less than 1.

Note that the sequential and parallel branch-and-bound (B&B) algorithms do not always explore the same search space. Therefore, super-linear speedups in parallel B&B algorithms are not in contradiction with Amdahl's law because processors can access high-quality solutions in early iterations, which in turn bring a reduction in the search tree and problem size.

For the oldest approaches, scalability issues are still to be investigated because of the small number of processors, typically around 16 and up to 64 processors. One major issue is that all approaches may (and a few must) resort to communication. Communication between parallel agents is costly in general: in shared-memory models such as multi-core, this typically means an access to a shared data structure for which one cannot avoid some form of locking; the cost of message-passing cross-CPU is even significantly higher. Communication additionally makes it difficult to get insights on the solving process since the executions are highly inter dependent and understanding parallel executions is notoriously complex.

Most parallel B&B algorithms explore leaves of the search tree in a different order than they would on a single-processor system. This could be a pity in situations where we know a really good search strategy, which is not entirely exploited by the parallel algorithm.

For many approaches, experiments with parallel programming involve a great deal of non-determinism: running the same algorithm twice on the same instance, with identical number of threads and parameters, may result in different solutions, and sometimes in different runtimes.

9.2.2 Embarrassingly Parallel Computation

A computation that can be divided into completely independent parts, each of which can be executed on a separate worker, is called *embarrassingly parallel* [99].

An embarrassingly parallel computation requires none or very little communication. This means that workers can execute their task, without any interaction with other workers.

Some well-known applications are based on embarrassingly parallel computations, such as the Folding@home project, Low-level image processing, the Mandelbrot set (a.k.a. fractals) or Monte Carlo calculations [99].

Two steps must be defined: the definition of the tasks and the task assignment to the workers The first step depends on the application, whereas the second step is more general. We can either use a static task assignment or a dynamic one.

With a static task assignment, each worker does a fixed part of the problem which is known a priori.

With a dynamic task assignment, a work-pool is maintained and workers consult it to get more work. The work-pool holds a collection of tasks to be performed. Workers ask for new tasks as soon as they finish the previously assigned task. In more complex work-pool problems, workers may even generate new tasks to be added to the work-pool.

9.2.3 Internal and External Parallelization

Techniques that aim at sharing the search tree, such us work stealing can be implemented in two different ways in generic CP solvers. Either the solver integrates the capability of traversing the search by several workers at the same time, that is the parallelization is ad hoc to the solver, or the solver provides some mechanisms like monitors to control the search from outside of the solver. In the former case, we say that the parallelization is made intra solver, whereas for the latter case we say that it is an extra solver parallelization.

Usually the former case is more powerful, but requires some modifications of the source code and so is less flexible and can only be done by the author of the solver [76]. The allocation of the part of the search tree is often specific and it is difficult to control, modify or change it.

Extra solver parallelization adds an algorithm that aims at supervising and controlling the search for solutions. This algorithm also manages the work done by each worker. It interacts with the sequential solver, which provides it with some parts of the search tree. The advantage of this approach is that the sequential search is not really modified. The parallel algorithm is defined on the top of the sequential mechanism and uses the sequential search in parallel for each worker. Some functions are usually given in order to be able to define different kinds of task allocations and to have a better control of the parallelization. Unfortunately, this also has some costs: there are more communications, the protocol must be general and some functions must be provided by the solver such us the capability to give a part of the search tree and to restart the search from a given node of the search tree. Thus, this method is often dedicated to some methods of parallelization like the work stealing. For instance, the or-tools solver gives monitors that directly interact with the internal search.

Some generic frameworks have been developed in order to deal with external parallelization. Such frameworks provide the user with some features for controlling the search that are independent of the solver. The role of the framework is to implement the interface with the solver.

Bobpp [25] is a parallel framework oriented towards solving Combinatorial Optimization Problems. It provides an interface between solvers of combinatorial problems and parallel computers. It is developed in C++ and can be used as the runtime support. Bobpp provides several search algorithms that can be parallelized using different parallel programming methods. The goal is to propose a single framework for most classes of Combinatorial Optimization Problems, so that they may be solved in as many different parallel architectures as possible. Figure 9.2 shows how Bobpp interfaces with high-level applications (QAP, TSP, etc.), CP solvers, and different parallel architectures using several parallel programming environments such as Pthreads as well as MPI or more specialized libraries such as Athapascan/Kaapi.

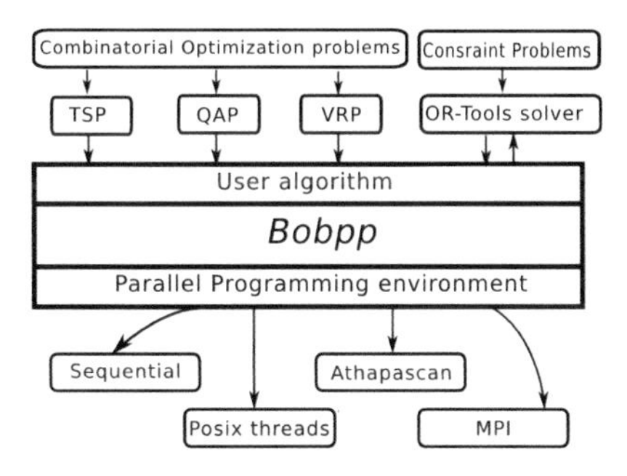

Fig. 9.2: Bobpp framework

9.2.4 Constraint Programming

A constraint network $\mathscr{CN} = (\mathscr{X}, \mathscr{D}, \mathscr{C})$ is defined by

- a set of n *variables* $\mathscr{X} = \{x_1, x_2, \ldots, x_n\}$,
- a set of n finite *domains* $\mathscr{D} = \{D(x_1), D(x_2), \ldots, D(x_n)\}$ with $D(x_i)$ the set of possible *values* for the variable x_i,
- a set of *constraints* between the variables $\mathscr{C} = \{C_1, C_2, \ldots, C_e\}$. A constraint C_i is defined on a subset of variables $X_{C_i} = \{x_{i_1}, x_{i_2}, \ldots, x_{i_j}\}$ of $\mathscr{X}$ with a subset of the Cartesian product $D(x_{i1}) \times D(x_{i2}) \times \ldots \times D(x_{ij})$ that states which combinations of values of variables $\{x_{i_1}, x_{i_2}, \ldots, x_{i_j}\}$ are compatible.

Each constraint C_i is associated with a filtering algorithm, often called a propagator, which removes values from the domains of its variables that are not consistent with it. The propagation mechanism applies the filtering algorithms of $\mathscr{C}$ to reduce the domains of variables in turn until no reduction can be done. One of the most interesting properties of a filtering algorithm is arc consistency. We say that a filtering algorithm associated with a constraint establishes arc consistency if it removes all the values of the variables involved in the constraint that are not consistent with the constraint. For instance, consider the constraint $x+3=y$ with the domain of x equal to $D(x) = \{1,3,4,5\}$ and the domain of y equal to $D(y) = \{4,5,8\}$. Then establishing arc consistency will lead to $D(x) = \{1,5\}$ and $D(y) = \{4,8\}$.

For convenience, we will use the word "problem" to designate a constraint network when it is used to represent the constraint network and not the search for a solution. We say that a problem P is consistent with the propagation if and only if running the propagation mechanism on P does not trigger a failure.

Now, we can detail the general methods giving the best results in practice.

9.3 Parallel Search Tree

One of the most popular techniques for combining constraint programming and parallelism is to define a parallel search tree. In other words, we try to traverse the search space in parallel. Usually, this result is achieved by splitting the search tree. This can be done either before the beginning of the search or dynamically during the search. The former case is named static partitioning while the latter is named dynamic partitioning

9.3.1 Static Partitioning

When we want to use k workers for solving a problem, we can split the initial search tree into k disjoint parts and give one sub-problem to each worker. Then, we gather the different intermediate results in order to produce the results corresponding to the whole problem. The advantage of this method is its simplicity. Unfortunately, it suffers from several drawbacks that arise frequently in practice: the times spent to solve each part are rarely well balanced and the communication of the objective value is not good when solving an optimization problem (the workers are independent). The main issue is that the balancing of the workload of the workers is equivalent to the balancing of the parts. Some works has been done on decomposing search trees based on their size in such a way as to equilibrate the parts to be solved [49, 18, 46]. However, the tree size is only approximated and is not strictly correlated with the solving time. In addition, we do not know how to have equivalent subtrees because the propagation mechanism will modify the tree during the search for solutions. Thus, as mentioned by Bordeaux et al. [6], it is quite difficult to ensure that each worker will receive the same amount of work. Hence, this method suffers from some issues of scalability, since the resolution time is the maximum of the resolution times of all workers. In order to remedy these issues, dynamic partitioning of the search tree is preferred.

9.3.2 Dynamic Partitioning

This strategy, called the work-stealing method, aims to partition the search tree into a set of subtrees, and schedule them during the execution of the search algorithm in order to have good load balancing between the different workers. Thus, workers each solve part of the problems and when a worker is waiting, it "steals" some work from another worker. This general mechanism can be described as follow: when a worker W no longer has any work, it asks another worker V whether it has some work to give it. If the answer is positive, then the worker V splits the search tree it is currently solving into two subtrees and gives one of them to the waiting worker W. If the answer is negative then W asks another worker U, until it gets some work to do or all the workers have been considered. The work-stealing approach partly resolves the balancing issue of the simple static decomposition method, mainly because the decomposition is dynamic. Therefore, it does not need to be able to split a search tree into well-balanced parts at the beginning.

This method has been implemented in a lot of solvers (Comet [67] or ILOG Solver [80] for instance), and in several ways [42, 103, 16] depending on whether the work to be done is centralized or not, on the way the search tree is split (into one or several parts) or on the communication method between workers.

For example, the study presented by Xie and Davenport [100] proposes the masters/workers approach. Each master has its workers. The search space is divided between the different masters, then each master puts its attributed subtrees in a work-pool to dispatch to the workers. When a node of the subtree is detected that is a root of a large subtree, the workers generate a large number of its subtrees and put them in a work-pool in order to have better load balancing. Fischetti et al. [24], propose a work-pool without communication between workers. First, the workers decompose the initial problem during a limited sampling phase, during which each worker visits nodes randomly. Thus, they can visit redundant nodes. After the sampling phase, each worker is attributed its nodes by a deterministic function. During the resolution, if a node is detected to be difficult by an estimation function, it is put into a global queue. When a worker finishes the resolution of its node, it receives a hard node from the global queue and solves it. When the queue is empty and there is no work to do, the resolution is done. Jaffar et al. [42] propose the use of a master that centralizes all pieces of information (bounds, solutions and requests). The master evaluates which worker has the largest amount of work in order to give some work to a waiting worker.

In the Bobpp framework, the work is shared thanks to a Global Priority Queue (GPQ). The search tree is decomposed and allocated to the different workers on demand and during the execution of the search algorithm. A unit of work corresponds to the solving of a subtree of the search tree. This subtree is the subtree of the search tree rooted at a given node, called the local root. This subtree is called the local search tree.

Periodically, a working worker tests whether waiting worker(s) exist(s). If this is the case, the working worker stops the search in the current node and gives a part of its local search tree, that is the subtree rooted at a node. In other words, it puts a node of the local search tree in the GPQ and continues to solve the remaining part of its local search tree. The waiting workers are notified by the insertion of a new node in the GPQ, and a waiting worker picks up the node and starts the solving of the subtree rooted at this node. This partitioning strategy has been presented in detail by Menouer and Le Cun [63].

There are two main questions that have to be answered to efficiently implement this mechanism: How do we start the search for a solution in a given subtree? And which subtree is given? In the next sections we will see that there is no perfect answer to these questions.

9.3.2.1 Local Subtree Solving

Conceptually there is no difficulty to start a search from a given node of the search tree. However, in practice, this is quite different. The main question is to be able to set the solver in the correct state corresponding to the root node. A state of a solver is defined by the domains of the variables and the internal data structure required by the propagators and some other data that may have been defined by the user. This means that some actions have been performed to reach a state. Thus, the question is how can we restore a state or how can we move from one state to another state? This is the continuation problem in computer science [89]. The restoration of a state of the search tree depends on the solver. Some solvers, such as or-tools [81] or Choco [12] or Oscar [93], use a trail mechanism. This means that they save some data when the search is going down in order to be able to restore them when the search is going up (i.e., backtracking). Some others, such as Gecode, use different mechanisms to avoid restoring the memory. The internal mechanism may lead to different strategies to move from one state to another one. Some solvers directly implement continuations [96].

There are usually three possibles methods: the state is explicitly saved, the state is recomputed from the current state or the state is recomputed from scratch, that is from the root of the search tree.

The possible implementations of these methods depend on whether the solver uses a generic search procedure or an internal one.

Internal Search Procedure.

Such a procedure means that the solver has total control over what can be done during the search for a solution. Some interactions with the search are possible but these are limited and the user cannot define its own data structures in a way which is not controlled by the solver. Usually internal search corresponds to a search method in which the only decisions are assignments or refutations. In this case, this means that nodes can be seen as sub-problems. Thus, computing a state is equivalent to restarting the search from a given sub-problem of the initial problem. This can be easily done by simply imposing the specific definition of the sub-problem and running the propagation of the solver once. Therefore the cost is not really expensive. In addition saving a sub-problem is not costly in memory so it is a good alternative to the explicit saving of the state.

The two other methods can also be used, that is we can easily backtrack to the lowest common ancestor (lca) of the current and the target node and then replay the search from the lca to the target node. We can measure whether this is more efficient than direct instantiation with the sub-problem of the target node or not. The replay from scratch is usually less interesting than restarting the problem with the constraints defining the sub-problem of the target node.

Generic Search Procedure.

As we mentioned in the Introduction, if a generic search is used then there is no way to deduce the state of a node of the search tree, mainly because we cannot know what are the data structures that are used. We have no information about the data that are defined by the user. This means that the state needs to be recomputed from the path going from the root to the target node. In this case, we say that the search is replayed. Unfortunately this has a cost.

Since we cannot define precisely the structure of a state, the memorization of a state can only be done by copying the whole memory, which is possible only when there are only a few variables and constraints. So in general the first method is not possible. The second method is only possible for the current worker. Therefore when some work is given to another worker, the third method is usually used. Replaying the search from the root node has a cost that depends on the length of the path. Therefore, it is common to study the consequences of some choices. This is the purpose of the next section.

9.3.2.2 Subtree Definition

When a worker needs some work it asks the other workers to give it a part of their current work. We discuss here how a worker can answer this request. All jobs, that is all given subtrees are not equivalent for several reasons: it can be expensive to replay the corresponding state and the solving times of the subtrees may strongly differ.

The first problem can be solved by the worker which gives some work by defining a strategy for selecting the given node of its local search tree. The simplest strategy consists of giving the current node and triggering a local backtrack in order to continue solving the local search tree (see Figure 9.3). It is also possible to give the next available open node.

However this method does not take into account the time to replay the state of a node. Thus, some other methods [62] have been developed. Notably, the node that is the closest to the root can be transferred. Some experiments have shown that this decreases the number of decision replays by a factor of 2. Figure 9.4 illustrates this approach.

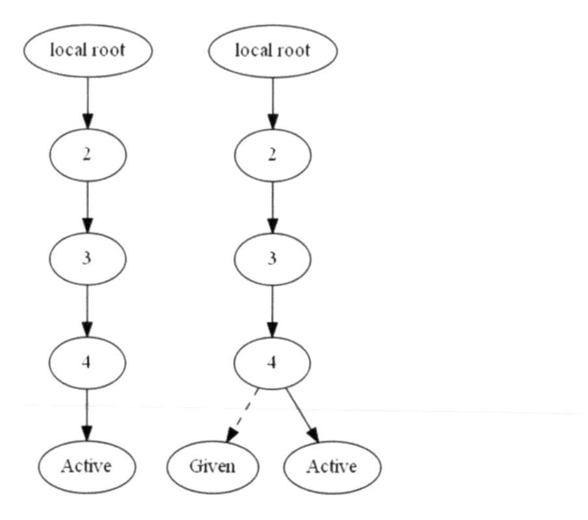

Fig. 9.3: Simple work separation

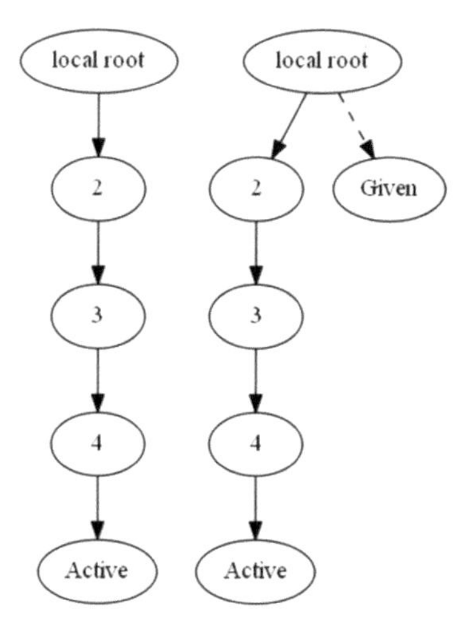

Fig. 9.4: Transmission of the subtree rooted at the node closest to the root

The second question about the amount of work that is given is more important for the global solving time. In fact, dynamic partitioning has a termination issue. When the whole search for solutions is almost done, there are more and more workers without work and so there are more and more workers asking for some work. At the same time, there are fewer and fewer workers that can give a part of their work. In addition the quantity of work that can be given is less and less important. Thus, we have more requests, fewer possible responders and less work to give. This is why we often observe a decrease in performance when the search is almost ended. Thus, we generally observe that the method scales well for a small number of workers whereas it is difficult to maintain a linear gain when the number of workers becomes larger, even though some methods have been developed to try to remedy this issue [101, 67]. Note that it is possible to have an immediate failure, that is the propagation of the new node may fail.

In order to speed up the termination of the algorithm, we should avoid giving a search tree that will be too small to be solved. Unfortunately, it is difficult to estimate the time that will be required to traverse a search tree, otherwise we would be able to have nice decompositions. One possible solution is to consider the depth of a node and to relate it to the solving time of the subtree rooted at that node. Thus, if a node is at a depth that is greater than a given threshold then the node cannot be given to another worker. This means that some workers will not be able to give any node. This idea usually improves the global behavior of the parallelization. However, it can be further improved by using a dynamic threshold that mainly depends on the depth. The choice of the value of the threshold is a difficult problem. Choosing a very small threshold makes the algorithm similar to static partitioning, with a limited number of subtrees explored by the different workers. Conversely, choosing a high threshold makes the algorithm similar to dynamic partitioning without a threshold, which makes load balancing easier between the workers but increases the exploration of redundant nodes. For instance, Menouer [62] uses a threshold equal to $2\log(\#workers)$, where $\#workers$ is the number of available workers. In addition, the threshold is increased each time a worker no longer has work. The maximum value is defined by $7\log(\#workers)$.

Even if the depth is a poor estimation of the quantity of work needed to solve a subtree, a threshold based on depth improves the work-stealing approach in practice. Figure 9.5 shows the variation in computation time according to the value of the threshold to solve the Naval Battle problem (Sb_sb_13_13_5_1) [70] using 12 cores on an Intel machine (12 cores and 48 GB of RAM). As a result, the computation time decreases with increasing threshold value until an optimal threshold (value of 25) is reached. After this optimal value the computation time increases again.

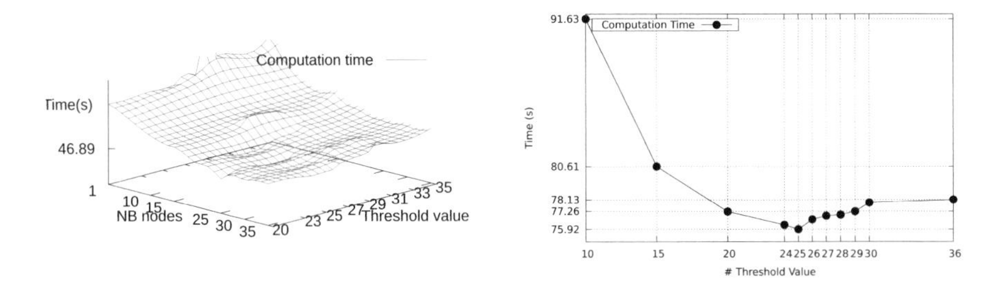

Fig. 9.5: Variation of the computation time for solving the Naval Battle problem on 12 cores according to the threshold value [66]

9.4 Problem Decomposition

The idea of Embarrassingly Parallel Search (EPS) is to statically decompose the initial problem into a huge number of sub-problems that are consistent with propagation (i.e., running the propagation mechanism on them does not detect any inconsistency). These sub-problems are added to a queue, which is managed by a master. Then, each waiting worker takes a sub-problem from the queue and solves it. The process is repeated until all the sub-problems have been solved. The assignment of the sub-problems to workers is dynamic and there is no communication between the workers. EPS is based on the idea that if there is a large number of sub-problems to solve then the resolution times of the workers will be balanced even if the resolution times of the sub-problems are not. In other words, load balancing is automatically obtained in a statistical sense.

We will detail this method in this section.

9.4.1 Principles

This approach relies on the assumption that the resolution time of disjoint sub-problems is equivalent to the resolution time of the union of these sub-problems. If this condition is not met, then the parallelization of the search of a solver (not necessarily a CP Solver) based on any decomposition method, such as simple static decomposition, work stealing or embarrassingly parallel method may be unfavorably impacted.

This assumption does not seem too strong because experiments do not show such a poor behavior with a CP Solver. However, it has been observed in some cases with a MIP Solver.

We have seen that decomposing the initial problem into the same number of sub-problems as workers may cause unbalanced resolution times for different workers. Thus, the idea of EPS is to strongly increase the number of considered sub-problems, in order to define an embarrassingly parallel computation leading to good performance.

Before going into further details of the implementation, a property can be established. While solving a problem, we will use the following terminology:

- *active time of a worker:* the sum of the resolution times of a worker (the decomposition time is excluded).
- *inactive time of a worker:* the difference between the elapsed time for solving all the sub-problems (the decomposition time is excluded) and the active time of the worker.

The EPS approach is mainly based on the following remark.

Remark 1. The active time of all the workers may be well balanced even if the resolution time of each sub-problem is not well balanced.

Since a worker may solve several sub-problems, their resolution times can be different while their sum remains equal to a given value.

The main challenge of a decomposition is not to define equivalent problems, it is to avoid having some workers without work whereas some others are running. We do not need to know in advance the resolution time of each sub-problem. We just expect that the workers will have equivalent activity time. In order to reach that goal, EPS decomposes the initial problem into a lot of sub-problems. This increases our chance to obtain well-balanced activity times for the workers, because we increase our chance to be able to obtain a combination of resolution times leading to the same activity time for each worker.

For instance, when the search space tends to be not equilibrated, there are sub-problems that will take a longer time to be solved. By having a lot of sub-problems we increase our chance to split these sub-problems into several parts having comparable resolution time and so to obtain a well-balanced load for the workers at the end. It also reduces the relative importance of each sub-problem with respect to the resolution of the whole problem.

Here is an example of the advantage of using a lot of sub-problems. Consider a problem which requires 140s to be solved sequentially and for which we have four workers. If we split the problem into four sub-problems then we have the following resolution times: $20, 80, 20, 20$. We will need 80s to solve these sub-problems in parallel. Thus, we gain a factor of $140/80 = 1.75$. Now if we split again each sub-problem into four sub-problems we might obtain the following sub-problems represented by their resolution time: $((5,5,5,5),(20,10,10,40),\ (2,5,10,3),(2,2,8,8))$. In this case, we might use the following assignment: worker1 : $5+20+2+8=35$; worker2 : $5+10+2+10=27$; worker3 : $5+10+5+3+2+8=33$ and worker4 : $5+40=45$. The elapsed time is now 45s and we gain a factor of $140/45 = 3.1$. By splitting the sub-problems again, we will reduce the average resolution time of the sub-problems and expect to break the 40s sub-problem. Note that decomposing a sub-problem further does not run away the risk of increasing the elapsed time.

Property 1. Let P be an optimization problem, or a satisfaction problem in which we search for all solutions. If P is split into sub-problems whose maximum resolution time is $tmax$, then

(i) the minimum resolution time of the whole problem is $tmax$;

(ii) the maximum inactivity time of a worker is less than or equal to $tmax$.

Proof. Suppose that a worker W has an inactivity time which is greater than $tmax$. Consider the moment where W started to wait after its activity time. At this time, there are no more available sub-problems to solve, otherwise W would be active. All active workers are then finishing their last task, whose resolution is bounded by $tmax$. Thus, the remaining resolution time of each of these other workers is less than or equal to $tmax$. Hence a contradiction.

The next section shows that the decomposition should be carefully done.

9.4.1.1 Sub-problems Generation: a Top-Down Method

We assume that we want to decompose a problem into q sub-problems.

Unlike the work-stealing approach, EPS does not aim to decompose the search tree into subtrees instead, it aims to decompose the whole problem into a set of sub-problems. These two decompositions are really different even if at first glance they look similar. Notably the relation to the sequential approach is different. There exists one rule when we try to parallelize a sequential process that should not be forgotten: *We should avoid doing something in parallel that we would not have done sequentially.*

The simplest method that can be considered does not satisfy this remark. It is a simple decomposition that is done as follows:

1. We consider any ordering of the variables $x_1,...,x_n$.
2. We define A_k to be the Cartesian product $D(x_1) \times ... \times D(x_k)$.
3. We compute the value k such that $|A_{k-1}| < q \leq |A_k|$.

Each assignment of A_k defines a sub-problem and so A_k is the sought decomposition.

This method works well for some problems such as the nqueens or the Golomb ruler, but it is really bad for some other problems, because a lot of assignments of A may be trivially not consistent. Consider for instance that x_1, x_2 and x_3 have the three values $\{a,b,c\}$ in their domains and that there is an alldiff constraint involving these three variables. The Cartesian product of the domains of these variables contains 27 tuples. Among them only six ((a,b,c), (a,c,b), (b,a,c),(b,c,a),(c,a,b), (c,b,a)) are not inconsistent with the alldiff constraint. That is, only $6/27 = 2/9$ of the generated sub-problems are not trivially inconsistent. It is important to note that most of these inconsistent problems would never be considered by a sequential search, and so we violate the previous rule. For some problems we have observed more than 99% of the generated problems were detected inconsistent by running the propagation (Figure 9.6). Thus, another method is needed to avoid this issue.

EPS solves this issue by *generating only sub-problems that are consistent with the propagation*, that is such that if we run the propagation mechanism on them then there is no failure. This means that they are not known to be inconsistent. Such sub-problems will also be considered by a sequential process, so they no longer violate the parallel-sequential rule we mentioned.

The generation of q such sub-problems becomes more complex because the number of sub-problems consistent with the propagation may not be related to the Cartesian product of some domains. A simple algorithm could be to perform a Breadth-First Search (BFS) in the search tree, until the desired number of sub-problems consistent with the propagation is reached. Unfortunately, it is not easy to perform a BFS efficiently mainly because BFS is not an incremental algorithm like Depth-First Search (DFS). Therefore, we can use a process similar to an iterative deepening depth-first search [50]: we repeat a Depth-Bounded Depth First Search (DBDFS), in other words a DFS that never visits nodes located at a depth greater than a given value, increasing the bound until we have generated the right number of sub-problems. However, even if the depth of a search tree can be precisely defined, it

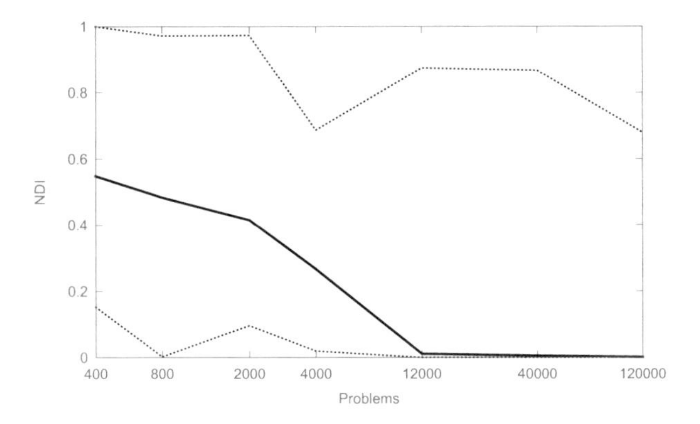

Fig. 9.6: Percentage of sub-problems consistent with the propagation (NDI) generated by the simple decomposition method for all problems. The geometric mean is in bold, dashed lines represent minimum and maximum values

is not easy to relate this notion to the number of variables already assigned. In fact, some variables may be assigned by propagation, and this is the case for the latest values of the domain of the variables. Thus, it is better to replace the depth limit by another simple limit.
We consider a set $Y \subseteq X$ of variables: we only assign the variables of Y and we stop the search when they are all assigned. In other words, we never try to assign a variable that is not in Y. This process is repeated until all assignments of Y consistent with the propagation have been found. Each branch of a search tree computed by this search defines an assignment. We will denote by A_Y the set of assignments computed with $Y \subseteq X$. To generate q sub-problems, we repeat the DBDFS by adding variables to Y if necessary until we have $|A_Y| \geq q$.

For convenience and simplicity, a static ordering of the variables is used.

This method can be improved in two ways:

1. We try to estimate some good set of variables Y in order to avoid repeating too many DBDFSs: For instance, if for a given Y we produce only $q/1000$ sub-problems and if the size of the domains of the next three non-assigned variables is 10, then we can deduce that we need to add at least three variables to Y.
2. In order to avoid repeating the same DFS for the first variables while repeating DBDFS, we store in a table constraint the previously computed assignments. More precisely, if we have computed A_Y then we use a table constraint containing all these assignments when we look for $A_{Y'}$ with $Y \subseteq Y'$.

Large Domains

This method can be adapted to large domains. A new step must be introduced in the algorithm in the latest iteration. If the domain of the latest considered variable, denoted by lx, is large then each of its values cannot be considered individually. In this case, its domain is split into a fixed number of parts and we use each part as a value. Then, either the desired number of sub-problems is generated or we have not been able to reach that number. In the latter case, the domain of lx is split again, for instance by splitting each part into two new parts (this multiplies by at most 2 the number of generated sub-problems) and we check whether the generated number of sub-problems is fine or not. This process is repeated until the right number of sub-problems is generated or the domain of lx is totally decomposed, that is each part corresponds to a value. In the latter case, we continue the algorithm by selecting a new variable.

Parallelization of the Decomposition

When there are a lot of workers, for instance 500, the decomposition into sub-problems may represent an important part of the resolution if it is done sequentially. Two reasons can explain this behavior: the ratio between a sequential method and a parallel one is large because we have 500 workers and not 6, 12 or 40. Since there are a lot of workers, there is also much more work to do because the initial problem needs to be decomposed into a larger number of sub-problems. Thus, between a sequential solution with w workers and another one with $W > w$ workers, the potential loss in term of computation power is W/w whereas we have at least W/w more work to do. So, it is can be necessary to parallelize the decomposition into sub-problems.

Experiments give some information:

1. The difference in the total work (i.e., activity time) done by the workers decreases when the number of sub-problems increases. This is not a linear relation. There is a huge difference between the activity times of the workers when there are fewer than five sub-problems per worker. These differences decrease when there are more than five sub-problems per worker.
2. A simple decomposition into sub-problems that may be inconsistent quickly causes some issues because inconsistencies are detected very quickly.
3. Splitting an initial problem into a small set of sub-problems is fast compared to the overall decomposition time and compared to the overall resolution time.

These observations show that a compromise has to be found and an iterative process decomposing the initial problem in three phases has to be defined. In the first phase, the whole problem is decomposed into only a few sub-problems because the relative cost is small even with an unbalanced workload. However, we should be careful with the first phase (i.e., starting with probably inconsistent sub-problems) because it can have an impact on the performance. Finally, the most important thing seems to be to generate five sub-problems because we could restart from these sub-problems to decompose further and such a decomposition should be reasonably well balanced.

Thus, a method in three main phases has been designed:

- An initial phase where one sub-problem per worker is generated as quickly as possible. This phase does not consume time and may remain sequential.
- A main phase which aims to generate five sub-problems per worker. Each sub-problem is consistent with the propagation. This phase can be divided into several steps to reach that goal while balancing the work among the workers.
- A final phase which consists of generating $K \geq 30$ sub-problems per worker from the set of sub-problems computed by the main phase.

9.4.1.2 Sub-problems Generation: a Bottom-Up Method

Another method for finding the requested sub-problems has been proposed by Malapert et al [61]. It is a bottom-up decomposition that tries to find in a depth-first manner the depth d at which we can generate the q sub-problems (Figure 9.7).

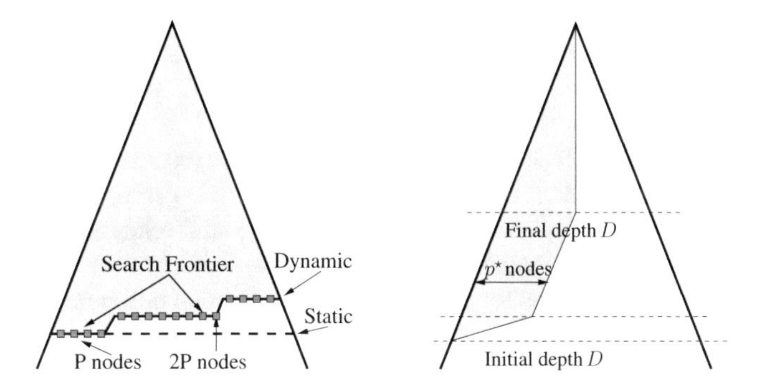

Fig. 9.7: Bottom-up decomposition and estimation

The algorithm aims to identify the topmost search frontier with approximately $p^* = q$ open nodes by sampling and estimation of the sought depth. The procedure can be divided into three phases:

1. a partial tree is built by sampling the top of the real search tree;
2. we estimate the level widths of the real tree;
3. we determine the decomposition depth d with a greedy heuristic.

Since we need to explore the top of the search tree, an upper bound D on the decomposition depth is fixed. The maximum decomposition depth D must be chosen according to the number of workers and the expected number of sub-problems per worker. If D is too small, the decomposition could generate too few sub-problems. If D is too large, the sampling time increases while the decomposition quality could decrease.

The sampling phase builds a partial tree with at most p^* assignments on a level using a depth-first search. The number of assignments (i.e., open nodes in the search tree) at each level is counted by a callback. The maximum depth D is reduced each time there are p^* assignments at a given level. If the sampling ends within its limits, then the top of the tree has been entirely visited and no estimation is needed. Otherwise, we need to estimate the widths of the topmost levels of the tree depending on the partial tree. This estimation is a straightforward adaptation of the one proposed by Cornuéjols et al. [19] to deal with n-ary search trees. In practice, the main issue is that the higher the arity is, the lower the precision of the estimation. Therefore, the heuristics that is used minimizes the absolute deviation between the estimated number of nodes and the expected number p^*. If several levels have an identical absolute deviation, then the lowest level with an estimated number of sub-problems greater than or equal to p^* is selected.

9.4.1.3 Implementation

EPS involves three tasks: the definition of the sub-problem (TaskDefinition), the task assignment of sub-problems to the workers (TaskAssignment) and a task that aims at gathering solutions and/or objective values: TaskResultGathering. In this step, the answers to all the sub-problems are collected and combined in some way to form the output (i.e., the answer to the initial problem).

For convenience, we create a master (i.e., a coordinator process) which is in charge of these operations. So, it creates the sub-problems (TaskDefinition), holds the work-pool and assigns tasks to workers (TaskAssignment) and fetches the computations made by the workers (TaskResultGathering).

We detail these operations for the satisfaction and optimization problems.

Satisfaction Problems

- The TaskDefinition operation consists of computing a partition of the initial problem P into a set S of sub-problems.
- The TaskAssignment operation is implemented by using a FIFO data structure (i.e., a queue). Each time a sub-problem is defined it is added to the back of the queue. When a worker needs some work it takes a sub-problem from the queue.
- The TaskResultGathering operation is quite simple: when searching for a solution it stops the search when one is found; when searching for all solutions, it just gathers the solutions returned by the workers.

Optimization Problems

In case of optimization problems we have to manage the best value of the objective function computed so far. Thus, the operations are slightly modified.

- The TaskDefinition operation consists of computing a partition of the initial problem P into a set S of sub-problems.
- The TaskAssignment operation is implemented by using a queue. Each time a sub-problem is defined it is added to the back of the queue. The queue is also associated with the best objective value computed so far. When a worker needs some work, the master gives it a sub-problem from the queue. It also gives it the best objective value computed so far.
- The TaskResultGathering operation manages the optimal value found by the worker and the associated solution.

Note that there is no other communication, that is when a worker finds a better solution, the other workers that are running cannot use it for improving their current resolution. So, if the absence of communication may increase our performance, this aspect may also lead to a decrease in performance. Fortunately, we do not observe this bad behavior in practice. We can see here another argument for having a lot of sub-problems in case of optimization problems: the resolution of a sub-problem should be short in order to improve the transmission of a better objective value and to avoid performing work that could have been ignored with a better objective value.

9.4.1.4 Size of the Partition

One important question is: how many sub-problems should be generated?

This is mainly an experimental question. However, in order to have good scalability, this number should be defined in relation to the number of workers that are involved. More precisely, it is more consistent to have q sub-problems per worker than a total of q sub-problems.

It appears that this number does not depend on the type of problem that is considered. Some experiments show that a minimum of 30 sub-problems per worker is required.

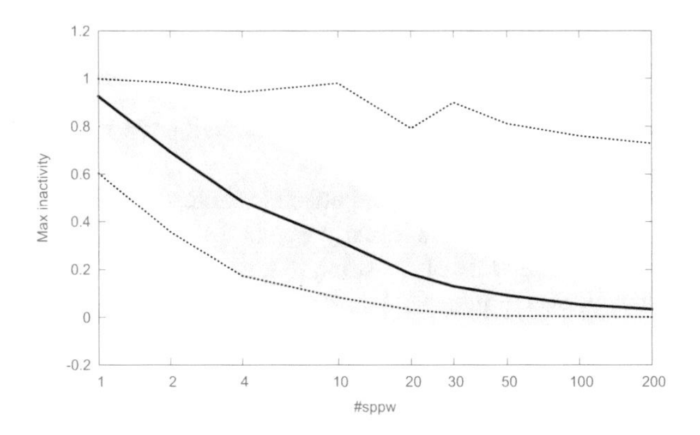

Fig. 9.8: Percentage of maximum inactivity time of the workers (geometric mean)

Figure 9.8 shows that the percentage of the maximum inactivity time of the workers decreases when the number of sub-problems per worker is increased. From 20 sub-problems per worker, we observe that on average the maximum inactivity time represents less than 20% of the resolution time.

9.4.2 Determinism

EPS can be modified to return the same solution as the sequential algorithm, which can be useful in several scenarios such as debugging or performance evaluation.

We assume that the whole problem is decomposed into the sub-problems $P_1, \ldots, P_p$ in that order and that they are selected by respecting that order.

The first solution found by the sequential algorithm belongs to the satisfiable sub-problem P_i with the smallest index, that is the leftmost solution. Consider that the parallel algorithm finds the first solution for the sub-problem P_j such that $j > i$. Then, it is not necessary to solve problems P_k such that $k > j$ and one must only wait for each problem P_k such that $k < j$ and then determine the leftmost solution, the satisfiable sub-problem with the smallest index.

This can easily be extended for optimization problems by slightly modifying the cutting constraints. Usually, when a new solution is found a cutting constraint is stated that only allows strictly better solutions. On the contrary to other constraints, the cutting constraint is always propagated while backtracking. Here, if a solution is found when solving the sub-problem P_j, then the cutting constraint only allows strictly improving solution for sub-problems $k \geq j$, but also allows equivalent solutions for sub-problems $k < j$.

So, the parallel algorithm returns the same solution as the sequential one if the sub-problems are visited in the same order. Moreover, the solution returned by the parallel algorithm does not depend on the number of workers, but only on the decomposition.

9.5 Comparison Between the Work-Stealing Approach and EPS

The EPS method has several advantages compared to the work-stealing approach. We can cite the most important ones:

- there is almost no communication between workers and the communication between the master and the workers is really weak.
- the method is independent of the solver. There is no need to know the solver in detail or to have access to internal data structures. It can be used with a generic search without any problem.
- there is no termination issue.
- very easy problems can be considered by a worker without causing any issue. There is no need for any threshold.

• there is no issue of replaying a part of the search with a generic search, because the problem is decomposed and not split.
• the method is quite simple.
• the method can be easily adapted to use distributed machines.
• by saving the order in which the sub-problems have been executed, we can simply replay a resolution in parallel. This costs almost nothing and helps a lot with the debugging of applications. Determinism is easy to achieve.

The work-stealing approach has several advantages:
• we have fine and dynamic control over the way the search is explored and split.
• the method manages the repartition of the work and if only one worker is working then its work will be shared, whereas this may not be the case with EPS.
• there is no setup time because there is no a priori decomposition of the problem

9.6 Experiments

These experiments come from Malapert et al. [61]. More information and more details can be found in [61].

9.6.1 Benchmark Instances

We consider instances of satisfaction and optimization problems. We ignore the problem of finding a first feasible solution because the parallel speedup can be completely uncorrelated to the number of workers, making the results hard to analyze. We consider optimization problems for which the same variability can be observed, but to a lesser extent because an optimality proof is required.

We perform a huge number of tests and we select the most representative ones.

The first set, called `fzn`, is a selection of 18 instances selected from more than 5000 instances either from the repository maintained by [47] or directly from the Minizinc 1.6 distribution written in the FlatZinc language [75]. Each instance is solved in more than 500 seconds and less than 1 hour with Gecode. The selection is composed of one unsatisfiable, six enumerations, and 11 optimization instances.

The set `xcsp` is composed of instances from the categories ACAD and REAL of XCSP 2.1 [92]. It consists of difficult instances that can be solved within 24 hours by `Choco2` [60]. A first subset, called `xcsp1`, is composed of five unsatisfiable and five enumeration instances whereas the second subset, called `xcsp2`, is composed of 11 unsatisfiable and three enumeration instances. The set `xcsp1` is composed of instances easier to solve than those of `xcsp2`.

9.6.1.1 Implementation Details

Three CP solvers are used: `Choco2` 2.1.5 written in Java, `Gecode` 4.2.1 and `OR-tools` rev. 3163 written in C++. Threads [73, 48] and MPI [55, 32] technologies are used. The typical difference between them is that threads (of the same process) run in a shared memory space, while MPI is a standardized and portable message-passing system to exchange information between processes running in separate memory spaces. Therefore, Thread technology does not handle multiple nodes of a cluster whereas MPI does.

In C++, Threads are implemented by using `pthreads`, a POSIX library [73, 48] used by Unix systems. In Java, the standard Java Thread technology [40] is used.

`OR-tools` uses a sequential top-down decomposition and C++ Threads. `Gecode` uses a parallel top-down decomposition and C++ Threads or MPI technologies. In fact, `Gecode` will use C++ `pthread` on the multi-core computer, OpenMPI on the data center, and MS-MPI on the cloud platform. `Gecode` and `OR-tools` both use the `lex` variable selection heuristic because the top-down decomposition requires a fixed variable ordering. `Choco2` uses a bottom-up decomposition and Java Threads. In every case, the jobs are scheduled in FIFO to mimic as much as possible the sequential algorithm so that speedups are relevant. We always take the value selection heuristic that selects the smallest value, whatever heuristic that may be.

9.6.1.2 Execution Environments

We use three execution environments that are representative of computing platforms available nowadays: multi-core, data center and cloud computing.

Multi-core is a Dell computer with 256 GB of RAM and four Intel E7-4870 2.40 GHz processors running on Scientific Linux 6.0 (each processor has 10 cores).

Data Center is the “Centre de Calcul Interactif” hosted by the Université Nice Sophia Antipolis, which provides a cluster composed of 72 nodes (1152 cores) running on CentOS 6.3, each node with 64 GB of RAM and two Intel E5-2670 2.60 GHz processors (eight cores). The cluster is managed by OAR [11], i.e., a versatile resource and task manager. As Thread technology is limited to a single node of a cluster, `Choco2` can use up to 16 physical cores whereas `Gecode` can use any number of nodes thanks to MPI.

Cloud Computing is a cloud platform managed by the Microsoft company (Microsoft Azure) that enables applications to be deployed on Windows Server technology [56]. Each node has 56 GB of RAM and Intel Xeon E5-2690E 2.6 GHz processors (eight physical cores) We were allowed to simultaneously use three nodes (24 cores) managed by the Microsoft HPC Cluster 2012 [69].

Some computing infrastructures provide hyper-threading technologies, which improves parallelization of computations (doing multiple tasks at once). For each core that is physically present, the operating system addresses two logical cores, and shares the workload among them when possible. The multi-core computer provides hyper-threading, whereas it is deactivated on the cluster, and not available on the cloud.

The time limit for solving each instance is set to 12 hours whatever be the solver. Usually, we use two workers per physical core ($w = 2c$) because hyper-threading is efficient in our experiments. The target number p of sub-problems depends linearly on the number w of workers ($p = 30 \times w$), which allows statistical balance of the workload without increasing too much the total overhead [87].

Let t be the solving time (in seconds) of an algorithm and let su be the speedup of a parallel algorithm. In the tables, a row gives the results obtained by different algorithms for a given instance. For each row, the best solving times and speedups are indicated in bold. Dashes indicate that the instance is not solved by the algorithm. Question marks indicate that the speedup cannot be computed because the sequential solver does not solve the instance within the time limit. Arithmetic means, abbreviated AM, are computed for solving times, whereas geometric means, abbreviated GM, are computed for speedups and efficiency. Missing values, i.e., dashes and question marks, are ignored when computing statistics.

9.6.2 Multi-core

In this section, we use parallel solvers based on Thread technologies to solve the instances of `xcsp1` or the nqueens problem using a multi-core computer. Let us recall that there are two workers per physical core because hyper-threading is activated ($w = 2c = 80$). We show that EPS frequently gives linear speedups, and outperforms the work-stealing approach proposed by [94] and [76].

Table 9.1 gives the solving times and speedups of the parallel solvers using 80 workers for the `xcsp1` instances. `Choco2`, `Gecode` and `OR-tools` use `lex`. They are also compared to a work-stealing approach denoted `Gecode`-WS [94, 76]. First, implementations of EPS are faster and more efficient than the work-stealing. EPS often reaches linear speedups in the number of cores whereas it never happens for the work stealing. Even worse, three instances are not solved within the 12-hour time limit using work-stealing whereas they are using the sequential solver.

Decomposition is the key to the bad performance on the instances `knights-80-5` and `lemma-100-9-mod`. The decomposition of `knights-80-5` takes more than 1,100 seconds and generates too many sub-problems, which precludes any speedup. The issue is lessened using the sequential decomposition of `OR-tools` and is resolved by the parallel top-down decomposition of `Gecode`. Note also that the sequential solving times of `OR-tools` and `Gecode` respectively are 20 and 40 times higher. Similarly, the long decomposition time of `Choco2` for `lemma-100-9-mod` leads to a low speedup. However, the moderate efficiency of `Choco2` and `Gecode` for `squares-9-9` is not caused by the decomposition.

`Gecode` and `OR-tools` are often more efficient and faster than `Choco2`. The solvers show different behaviors even when using the same variable selection heuristic because their propagation mechanisms and decompositions differ. Furthermore,

Instances	Choco2		Gecode		OR-tools		Gecode-WS	
	t	su	t	su	t	su	t	su
costasArray-14	240.0	**38.8**	62.3	19.1	**50.9**	33.4	594.0	2.0
knights-80-5	1133.1	1.5	**548.7**	**37.6**	2173.9	18.5	–	–
latinSquare-dg-8_all	328.1	39.2	251.7	**42.0**	**166.6**	35.2	4488.5	2.4
lemma-100-9-mod	123.4	4.1	6.7	10.1	**1.8**	**22.9**	3.0	22.3
ortholatin-5	249.9	36.0	421.7	13.5	**167.7**	**38.1**	2044.6	2.8
pigeons-14	899.1	15.5	**211.8**	**39.1**	730.3	18.5	–	–
quasigroup5-10	123.5	**32.5**	18.6	26.4	**17.0**	**36.9**	22.8	21.5
queenAttacking-6	**622.5**	**28.5**	15899.1	?	–	–	–	–
series-14	39.3	32.9	**11.3**	**34.2**	16.2	28.7	552.3	0.7
squares-9-9	1213.0	16.1	**17.9**	18.4	81.4	**35.0**	427.8	0.8
AM (t) or GM (su)	497.2	17.4	1745.0	24.0	378.4	28.7	1161.9	3.3

Table 9.1: Solving times and speedups (40-cores machine). Gecode and OR-tools use the lex heuristic

the parallel top-down decomposition of Gecode does not preserve the ordering of the sub-problems with regard to the sequential algorithm.

9.6.3 Data Center

In this section, we study the influence of the search strategy on the solving times and speedups, the scalability up to 512 workers, and compare EPS to a work-stealing approach.

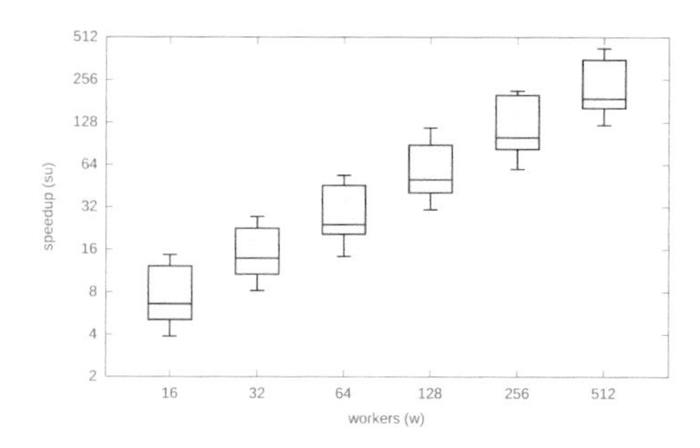

Fig. 9.9: Scalability up to 512 workers (Gecode, lex, data center)

Table 9.2 compares the Gecode implementations of EPS and work-stealing (WS) for solving xcsp instances using 16 or 512 workers. EPS is faster and more efficient than work-stealing. With 512 workers, EPS is on average almost 10 times faster than work-stealing. It is also more efficient because they both parallelize the

Instances	$w=16$				$w=512$			
	EPS		WS		EPS		WS	
	t	su	t	su	t	su	t	su
`cc-15-15-2`	–	–	–	–	–	–	–	–
`costasArray-14`	**64.4**	**13.6**	69.3	12.7	**3.6**	**243.8**	17.7	49.8
`crossword-m1c`[1]	**240.6**	**13.1**	482.1	6.6	**18.7**	**168.6**	83.1	38.0
`crossword-m1`[2]	**171.7**	**14.5**	178.5	13.9	**13.3**	**187.3**	57.8	43.0
`knights-20-9`	**5190.7**	?	38347.4	?	**153.4**	?	3312.4	?
`knights-25-9`	**7462.3**	?	–	–	**214.9**	?	–	–
`knights-80-5`	**1413.7**	**11.5**	8329.2	2.0	**49.3**	**329.8**	282.6	57.5
`langford-3-17`	24351.5	?	**21252.3**	?	**713.5**	?	7443.5	?
`langford-4-18`	**3203.2**	?	25721.2	?	**94.6**	?	5643.1	?
`langford-4-19`	**26871.2**	?	–	–	**782.5**	?	–	–
`latinSquare-dg-8_all`	**613.5**	**13.1**	621.2	13.0	**23.6**	**341.7**	124.4	64.7
`lemma-100-9-mod`	**3.4**	**14.7**	5.8	8.6	**1.0**	**51.4**	2.5	19.7
`ortholatin-5`	**309.5**	**14.1**	335.8	13.0	**10.4**	**422.0**	71.7	61.0
`pigeons-14`	**383.3**	**14.5**	6128.9	0.9	**15.3**	**363.1**	2320.2	2.4
`quasigroup5-10`	**27.1**	**13.5**	33.7	10.8	**1.7**	**211.7**	9.8	37.3
`queenAttacking-6`	42514.8	?	**37446.1**	?	**1283.9**	?	9151.5	?
`ruler-70-12-a3`	**96.6**	**15.1**	105.5	13.8	**3.7**	**389.3**	67.7	21.5
`ruler-70-12-a4`	**178.9**	**14.4**	185.2	13.9	**6.0**	**429.5**	34.1	75.5
`series-14`	**22.5**	**13.4**	56.9	5.3	**1.1**	**264.0**	8.2	36.9
`squares-9-9`	**22.8**	**11.1**	44.3	5.7	**1.3**	**191.7**	7.6	33.7
AM (t) or GM (su)	5954.8	13.5	8196.7	7.4	178.53	**246.2**	1684.6	33.5

[1]`crossword-m1-words-05-06` [2]`crossword-m1c-words-vg7-7_ext`

Table 9.2: Speedups and solving times for `xcsp` (`Gecode`, `lex`, data center, $w=16$ or 512)

same sequential solver. On the multi-core machine, `Gecode` is faster than `Choco2` on most instances of `xcsp1`. Five instances that are not solved within the time limit by `Gecode` are not reported in Table 9.2. Six instances are not solved with 16 workers whereas twelve instances are not solved with the sequential solver. By way of comparison, only five instances are not solved by `Choco2` using the `lex` heuristics whereas all instances are solved in sequential or parallel when using `dom/wdeg` or `dom/bwdeg`. Once again, this highlights the importance of the search strategy.

Figure 9.9 is a boxplot of the speedups with different numbers of workers for solving `fzn` instances. The median of speedups are around $\frac{w}{2}$ on average and their dispersion remains low.

9.6.4 Cloud Computing

The systems are deployed on the Microsoft Azure cloud platform. The available computing infrastructure is organized as follows: cluster nodes compute the appli-

Instance	EPS		WS	
	t	su	t	su
market_split_s5-02	**467.1**	**24.3**	658.6	17.3
market_split_s5-06	**452.7**	**24.4**	650.7	17.0
market_split_u5-09	**468.1**	**24.4**	609.2	18.7
pop_stress_0600	**874.8**	**10.8**	2195.7	4.3
nmseq_400	**342.4**	**8.5**	943.2	3.1
pop_stress_0500	**433.2**	**10.1**	811.0	5.4
fillomino_18	**160.2**	**13.9**	184.6	12.1
steiner-triples_09	**108.8**	**17.2**	242.4	7.7
nmseq_300	**114.5**	**6.6**	313.1	2.4
golombruler_13	**154.0**	**20.6**	210.4	15.1
cc_base_mzn_rnd_test.11	**1143.6**	**7.3**	2261.3	3.7
ghoulomb_3-7-20	**618.2**	**6.8**	3366.0	1.2
still_life_free_8x8	**931.2**	**9.6**	1199.4	7.5
bacp-6	**400.8**	**16.4**	831.0	7.9
depot_placement_st70_6	**433.9**	**18.3**	1172.5	6.8
open_stacks_01_wbp_20_20_1	**302.7**	**17.6**	374.1	14.3
bacp-27	**260.2**	**16.4**	548.4	7.8
still_life_still_life_9	**189.0**	**16.9**	196.8	16.2
talent_scheduling_alt_film117	**22.7**	**74.0**	110.5	15.2
AM (t) or GM (su)	**414.7**	**15.1**	888.4	7.7

Table 9.3: Solving times and speedups for fzn (Gecode, lex, cloud, $w = 24$)

cation; one head node manages the cluster nodes; and proxy nodes load-balance communication between cluster nodes. Unlike a data center, cluster nodes may be far from each other and communication time may take longer. Proxy nodes requires two cores and are managed by the service provider. Here, three nodes of eight cores with 56 GB of RAM memory provide 24 workers (cluster nodes) managed by MPI.

Table 9.3 compares the Gecode implementations of EPS and work stealing for solving the fzn instances with 24 workers. Briefly, EPS is always faster than work-stealing, and therefore more efficient because they both parallelize the same sequential solver. Work–stealing suffers from a higher communication overhead in the cloud than in a data center. Furthermore, the architecture of the computing infrastructure and the location of cluster nodes are mostly unknown, which precludes improvements to work-stealing such as those proposed by Machado et al. [59] or Xie and Davenport.

9.6.5 Comparison with Portfolios

Portfolio approaches exploit the variability of performance that is observed between several solvers, or several parameter settings for the same solver. We use four portfolios. The portfolio CPHydra [77] uses feature selection on top of the solvers

Mistral, Gecode and Choco2. CPHydra uses case-based reasoning to determine how to solve an unseen problem instance by exploiting a case base of problem solving experience. It aims to find a feasible solution within 30 minutes. It does not handle optimization or all-solution problems and the time limit is hard coded. The other static and fixed-size portfolios (Choco2, CAG, OR-tools) use different variable selection heuristics as well as randomization and restarts. Details about Choco2 and CAG can be found in [60]. The CAG portfolio extends the Choco2 portfolio by also using the solvers AbsCon and Gecode. So, CAG always produces better results than Choco2. The OR-tools portfolio was the gold medalist of the Minizinc challenge 2013 and 2014. It can seem unfair to compare parallel solvers and portfolios using different numbers of workers, but designing scalable portfolios (up to 512 workers) is a difficult task and almost no implementation is publicly available.

Table 9.4 gives the solving times of EPS and portfolios for the xcsp instances on the data center. First, CPHydra with 16 workers only solves two among 16 unsatisfiable instances (cc-15-15-2 and pigeons-14), but in less than 2 seconds whereas these are difficult for all other approaches. OR-tools is the second-least efficient approach because it solves fewer problems and often takes longer as confirmed by its low Borda score. The parallel Choco2 using dom/wdeg is better on average than the Choco2 portfolio even if the portfolio solves a few instances much faster such as scen11-f5 or queensKnights-20-5-mul. In this case, the diversification provided by the portfolio outperforms the speedups offered by the parallel B&B algorithm. This is emphasized for the CAG portfolio, which solves all instances and obtains several of the best solving times. The parallel Gecode with 16 workers is often slower and less robust than the portfolios Choco2 and CAG. However, increasing the number of workers to 512 clearly makes it the fastest solver, but still less robust because five instances are not solved within the time limit.

To conclude, the Choco2 and CAG portfolios are more robust thanks to their inherent diversification, but their solving times vary more from one instance to another. With 16 workers, implementations of EPS outperform the CPHydra and OR-tools portfolio, are competitive with the Choco2 portfolio, and are slightly dominated by the CAG portfolio. In fact, the good scaling of EPS is a key to beat the portfolios.

9.7 Conclusion

We have presented different methods for combining constraint programming techniques and parallelism, such as parallelization of the propagator or parallel propagation. We have detailed the most popular methods: work-stealing methods based on search tree splitting, and EPS, the embarrassingly parallel search method, which is based on problem decomposition. These methods give good results in practice. They have been tested on a single multi-core machine, on a data center and on the cloud. However, it seems that the scaling performance of EPS is better. In addition EPS is simple and easy to implement. The future will tell us whether it can replace the

Instances	EPS			Portfolio		
	Choco2	Gecode		Choco2	CAG	OR-tool
	$w = 16$	$w = 16$	$w = 512$	$w = 14$	$w = 23$	$w = 16$
cc-15-15-2	2192.1	–	–	1102.6	**3.5**	1070
costasArray-14	649.9	64.4	**3.6**	6180.8	879.4	1368
crossword-m1-words-05-06	204.6	240.6	**18.7**	512.3	512.3	22678
crossword-m1c-words-vg7-7_ext	1611.9	171.7	**13.3**	721.2	721.2	13157
fapp07-0600-7	2295.7	–	–	37.9	**3.2**	
knights-20-9	491.3	5190.7	153.4	3553.9	**0.8**	
knights-25-9	1645.2	7462.3	214.9	9324.8	**1.1**	
knights-80-5	1395.6	1413.7	**49.3**	1451.5	301.6	32602
langford-3-17	3062.2	24351.5	**713.5**	8884.7	8884.7	
langford-4-18	538.3	3203.2	**94.6**	2126.0	2126.0	
langford-4-19	2735.3	26871.2	**782.5**	12640.2	12640.2	
latinSquare-dg-8_all	294.8	613.5	**23.6**	65.1	36.4	4599
lemma-100-9-mod	145.3	3.4	**1.0**	435.3	50.1	38.
ortholatin-5	362.4	309.5	**10.4**	4881.2	4371.0	4438
pigeons-14	2993.3	383.3	**15.3**	12336.9	5564.5	12279
quasigroup5-10	451.5	27.1	**1.7**	3545.8	364.3	546
queenAttacking-6	**706.4**	42514.8	1283.9	2644.5	2644.5	
queensKnights-20-5-mul	5209.5	–	–	235.3	**1.0**	
ruler-70-12-a3	42.8	96.6	**3.7**	123.5	123.5	8763.
ruler-70-12-a4	1331.3	178.9	**6.0**	1250.2	1250.2	
scen11-f5	–	–	–	45.3	**8.5**	
series-14	338.9	22.5	**1.1**	1108.3	302.1	416.
squares-9-9	115.9	22.8	**1.3**	1223.7	254.3	138.
squaresUnsat-19-19	**3039.8**	–	–	4621.1	4621.1	
Arithmetic mean	1385.0	5954.8	**178.5**	3293.8	1902.7	7853.

Table 9.4: Solving times of EPS and portfolio (data center)

work-stealing approach in CP solvers. In any case, the obtained results show that we can efficiently combine parallelism and CP and often expect results that are almost linear.

References

[1] 16th IEEE International Conference on Tools with Artificial Intelligence (ICTAI 2004), 15-17 November 2004, Boca Raton, FL, USA. IEEE Computer Society (2004). URL http://ieeexplore.ieee.org/xpl/mostRecentIssue.jsp?punumber=9460

[2] Amadini, R., Gabbrielli, M., Mauro, J.: An Empirical Evaluation of Portfolios Approaches for Solving CSPs. In: C. Gomes, M. Sellmann (eds.) Integration of AI and OR Techniques in Constraint Programming for Combinatorial Optimization Problems, *Lecture Notes in Computer Science*, vol. 7874, pp. 316–

324. Springer Berlin Heidelberg (2013). DOI 10.1007/978-3-642-38171-3_21. URL http://dx.doi.org/10.1007/978-3-642-38171-3_21
[3] Amdahl, G.: Validity of the Single Processor Approach to Achieving Large Scale Computing Capabilities. In: Proceedings of the April 18-20, 1967, Spring Joint Computer Conference, AFIPS '67, pp. 483–485. ACM, New York, NY, USA (1967)
[4] Bader, D., Hart, W., Phillips, C.: Parallel Algorithm Design for Branch and Bound. In: H. G (ed.) Tutorials on Emerging Methodologies and Applications in Operations Research, *International Series in Operations Research & Management Science*, vol. 76, pp. 5–1–5–44. Springer, New York (2005). DOI 10.1007/0-387-22827-6_5
[5] Bordeaux, L., Hamadi, Y., Samulowitz, H.: Experiments with Massively Parallel Constraint Solving. In: Boutilier [8], pp. 443–448
[6] Bordeaux, L., Hamadi, Y., Samulowitz, H.: Experiments with massively parallel constraint solving. In: Boutilier [8], pp. 443–448
[7] Boussemart, F., Hemery, F., Lecoutre, C., Sais, L.: Boosting systematic search by weighting constraints. In: ECAI, vol. 16, p. 146 (2004)
[8] Boutilier, C. (ed.): IJCAI 2009, Proceedings of the 21st International Joint Conference on Artificial Intelligence, Pasadena, California, USA, July 11-17, 2009 (2009)
[9] Budiu, M., Delling, D., Werneck, R.: DryadOpt: Branch-and-bound on distributed data-parallel execution engines. In: Parallel and Distributed Processing Symposium (IPDPS), 2011 IEEE International, pp. 1278–1289 (2011)
[10] Burton, F.W., Sleep, M.R.: Executing Functional Programs on a Virtual Tree of Processors. In: Proceedings of the 1981 Conference on Functional Programming Languages and Computer Architecture, FPCA '81, pp. 187–194. ACM, New York, NY, USA (1981)
[11] Capit, N., Da Costa, G., Georgiou, Y., Huard, G., Martin, C., Mounie, G., Neyron, P., Richard, O.: A Batch Scheduler with High Level Components. In: Proceedings of the Fifth IEEE International Symposium on Cluster Computing and the Grid (CCGrid'05) - Volume 2 - Volume 02, CCGRID '05, pp. 776–783. IEEE Computer Society, Washington, DC, USA (2005). URL http://dl.acm.org/citation.cfm?id=1169223.1169583
[12] Choco, T.: Choco: an open source java constraint programming library. Ecole des Mines de Nantes, Research report **1**, 10–02 (2010)
[13] Choco solver
http://www.emn.fr/z-info/choco-solver/ (2013).
[14] Chong, Y.L., Hamadi, Y.: Distributed Log-Based Reconciliation. In: Proceedings of the 2006 Conference on ECAI 2006: 17th European Conference on Artificial Intelligence August 29 – September 1, 2006, Riva Del Garda, Italy, pp. 108–112. IOS Press, Amsterdam, The Netherlands (2006). URL http://dl.acm.org/citation.cfm?id=1567016.1567045
[15] Chu, G., Schulte, C., Stuckey, P.J.: Confidence-Based Work Stealing in Parallel Constraint Programming. In: Gent [27], pp. 226–241

[16] Chu, G., Schulte, C., Stuckey, P.J.: Confidence-based work stealing in parallel constraint programming. In: Gent [27], pp. 226–241
[17] Cire, A.A., Kadioglu, S., Sellmann, M.: Parallel Restarted Search. In: Proceedings of the Twenty-Eighth AAAI Conference on Artificial Intelligence, AAAI'14, pp. 842–848. AAAI Press (2014). URL http://dl.acm.org/citation.cfm?id=2893873.2894004
[18] Cornuéjols, G., Karamanov, M., Li, Y.: Early estimates of the size of branch-and-bound trees. INFORMS Journal on Computing **18**(1), 86–96 (2006)
[19] Cornuéjols, G., Karamanov, M., Li, Y.: Early Estimates of the Size of Branch-and-Bound Trees. INFORMS Journal on Computing **18**, 86–96 (2006)
[20] Crainic, T.G., Le Cun, B., Roucairol, C.: Parallel branch-and-bound algorithms. Parallel Combinatorial Optimization **1**, 1–28 (2006)
[21] De Kergommeaux, J.C., Codognet, P.: Parallel logic programming systems. ACM Computing Surveys (CSUR) **26**(3), 295–336 (1994)
[22] De Nicola, R., Ferrari, G.L., Meredith, G. (eds.): Coordination Models and Languages, 6th International Conference, COORDINATION 2004, Pisa, Italy, February 24-27, 2004, Proceedings, *Lecture Notes in Computer Science*, vol. 2949. Springer (2004)
[23] Ezzahir, R., Bessière, C., Belaissaoui, M., Bouyakhf, E.H.: DisChoco: A platform for distributed constraint programming. In: DCR'07: Eighth International Workshop on Distributed Constraint Reasoning - In conjunction with IJCAI'07, pp. 16–21. Hyderabad, India (2007). URL https://hal-lirmm.ccsd.cnrs.fr/lirmm-00189778
[24] Fischetti, M., Monaci, M., Salvagnin, D.: Self-splitting of workload in parallel computation. In: H. Simonis (ed.) Integration of AI and OR Techniques in Constraint Programming: 11th International Conference, CPAIOR 2014, Cork, Ireland, May 19-23, 2014. Proceedings, pp. 394–404. Springer International Publishing, Cham (2014). DOI 10.1007/978-3-319-07046-9_28. URL http://dx.doi.org/10.1007/978-3-319-07046-9_28
[25] Galea F., Le Cun, B.: Bob++ : a Framework for Exact Combinatorial Optimization Methods on Parallel Machines. In: International Conference High Performance Computing & Simulation 2007 (HPCS'07) and in conjunction with The 21st European Conference on Modeling and Simulation (ECMS 2007), pp. 779–785 (2007)
[26] Gendron, B., Crainic, T.G.: Parallel branch-and-bound algorithms: Survey and synthesis. Operations research **42**(6), 1042–1066 (1994)
[27] Gent, I.P. (ed.): Principles and Practice of Constraint Programming - CP 2009, 15th International Conference, CP 2009, Lisbon, Portugal, September 20-24, 2009, Proceedings, *Lecture Notes in Computer Science*, vol. 5732 (2009)
[28] Gomes, C., Selman, B.: Algorithm Portfolio Design: Theory vs. Practice. In: Proceedings of the Thirteenth Conference on Uncertainty in Artificial Intelligence, pp. 190–197 (1997)
[29] Gomes, C., Selman, B.: Search strategies for hybrid search spaces. In: Tools with Artificial Intelligence, 1999. Proceedings. 11th IEEE International Conference, pp. 359–364. IEEE (1999)

[30] Gomes, C., Selman, B.: Hybrid Search Strategies For Heterogeneous Search Spaces. International Journal on Artificial Intelligence Tools **09**, 45–57 (2000)
[31] Gomes, C., Selman, B.: Algorithm Portfolios. Artificial Intelligence **126**, 43–62 (2001)
[32] Gropp, W., Lusk, E.: The MPI communication library: its design and a portable implementation. In: Scalable Parallel Libraries Conference, Proceedings of the, pp. 160–165. IEEE (1993)
[33] Gupta, G., Pontelli, E., Ali, K.A., Carlsson, M., Hermenegildo, M.V.: Parallel execution of Prolog Programs: a Survey. ACM Transactions on Programming Languages and Systems (TOPLAS) **23**(4), 472–602 (2001)
[34] Halstead, R.: Implementation of MultiLisp: Lisp on a Multiprocessor. In: Proceedings of the 1984 ACM Symposium on LISP and Functional Programming, LFP '84, pp. 9–17. ACM, New York, NY, USA (1984)
[35] Hamadi, Y.: Optimal Distributed Arc-Consistency. Constraints **7**, 367–385 (2002)
[36] Hamadi, Y., Jabbour, S., Sais, L.: ManySAT: a Parallel SAT Solver. Journal on Satisfiability, Boolean Modeling and Computation **6**(4), 245–262 (2008)
[37] Haralick, R., Elliot, G.: Increasing tree search efficiency for constraint satisfaction problems. Artificial Intelligence **14**, 263–313 (1980)
[38] Harvey, W.D., Ginsberg, M.L.: Limited Discrepancy Search. In: Proceedings of the Fourteenth International Joint Conference on Artificial Intelligence, IJCAI 95, Montréal, Québec, Canada, August 20-25 1995, 2 Volumes, pp. 607–615 (1995)
[39] Hirayama, K., Yokoo, M.: Distributed Partial Constraint Satisfaction Problem. In: Principles and Practice of Constraint Programming-CP97, pp. 222–236. Springer (1997)
[40] Hyde, P.: Java thread programming, vol. 1. Sams (1999)
[41] Jaffar, J., Santosa, A.E., Yap, R.H.C., Zhu, K.Q.: Scalable Distributed Depth-First Search with Greedy Work Stealing. In: 16th IEEE International Conference on Tools with Artificial Intelligence [1], pp. 98–103. URL `http://ieeexplore.ieee.org/xpl/mostRecentIssue.jsp?punumber=9460`
[42] Jaffar, J., Santosa, A.E., Yap, R.H.C., Zhu, K.Q.: Scalable distributed depth-first search with greedy work stealing. In: ICTAI [1], pp. 98–103. URL `http://ieeexplore.ieee.org/xpl/mostRecentIssue.jsp?punumber=9460`
[43] Kale, L., Krishnan, S.: CHARM++: a portable concurrent object oriented system based on C++, vol. 28. ACM (1993)
[44] Kasif, S.: On the Parallel Complexity of Discrete Relaxation in Constraint Satisfaction networks. Artificial Intelligence **45**, 275–286 (1990)
[45] Kautz, H., Horvitz, E., Ruan, Y., Gomes, C., Selman, B.: Dynamic Restart Policies. 18th National Conference on Artificial Intelligence AAAI/IAAI **97**, 674–681 (2002)
[46] Kilby, P., Slaney, J.K., Thiébaux, S., Walsh, T.: Estimating search tree size. In: AAAI, pp. 1014–1019 (2006)

[47] Kjellerstrand, H.: Håkan Kjellerstrand's Blog. http://www.hakank.org/ (2014)
[48] Kleiman, S., Shah, D., Smaalders, B.: Programming with threads. Sun Soft Press (1996)
[49] Knuth, D.E.: Estimating the efficiency of backtrack programs. Mathematics of Computation **29**, 121–136 (1975)
[50] Korf, R.: Depth-first iterative-deepening: An optimal admissible tree search. Artificial Intelligence **27**, 97–109 (1985)
[51] Kowalski, R.: Algorithm = logic + control. Commun. ACM **22**(7), 424–436 (1979)
[52] Le Cun, B., Menouer, T., Vander-Swalmen, P.: Bobpp. http://forge.prism.uvsq.fr/projects/bobpp (2007)
[53] Léauté, T., Ottens, B., Szymanek, R.: FRODO 2.0: An open-source framework for distributed constraint optimization. In: Boutilier [8], pp. 160–164
[54] Leiserson, C.E.: The Cilk++ concurrency platform. The Journal of Supercomputing **51**(3), 244–257 (2010)
[55] Lester, B.: The art of parallel programming. Prentice Hall, Englewood Cliffs, NJ (1993)
[56] Li, H.: Introducing Windows Azure. Apress, Berkeley, CA, USA (2009)
[57] Lodi, A., Milano, M., Toth, P. (eds.): Integration of AI and OR Techniques in Constraint Programming for Combinatorial Optimization Problems, 7th International Conference, CPAIOR 2010, Bologna, Italy, June 14-18, 2010. Proceedings, *Lecture Notes in Computer Science*, vol. 6140. Springer (2010)
[58] Luby, M., Sinclair, A., Zuckerman, D.: Optimal Speedup of Las Vegas Algorithms. Inf. Process. Lett. **47**, 173–180 (1993)
[59] Machado, R., Pedro, V., Abreu, S.: On the Scalability of Constraint Programming on Hierarchical Multiprocessor Systems. In: ICPP, pp. 530–535. IEEE (2013)
[60] Malapert, A., Lecoutre, C.: À propos de la bibliothèque de modèles XCSP. In: 10èmes Journées Francophones de Programmation par Contraintes (JFPC'15). Angers, France (2014)
[61] Malapert, A., Régin, J., Rezgui, M.: Embarrassingly parallel search in constraint programming. J. Artif. Intell. Res. (JAIR) **57**, 421–464 (2016). DOI 10.1613/jair.5247. URL http://dx.doi.org/10.1613/jair.5247
[62] Menouer, T.: Parallélisations de Méthodes de Programmation Par Contraintes. Ph.D. thesis, Université de Versailles Saint-Quentin-en-Yvelines (2015)
[63] Menouer, T., Cun, B.L.: Anticipated dynamic load balancing strategy to parallelize constraint programming search. In: 2013 IEEE 27th International Symposium on Parallel and Distributed Processing Workshops and PhD Forum, pp. 1771–1777 (2013). DOI 10.1109/IPDPSW.2013.210. URL http://doi.ieeecomputersociety.org/10.1109/IPDPSW.2013.210
[64] Menouer, T., Le Cun, B.: Anticipated Dynamic Load Balancing Strategy to Parallelize Constraint Programming Search. In: 2013 IEEE 27th International Symposium on Parallel and Distributed Processing Workshops and PhD Forum, pp. 1771–1777 (2013)

[65] Menouer, T., Le Cun, B.: Adaptive N To P Portfolio for Solving Constraint Programming Problems on Top of the Parallel Bobpp Framework. In: 2014 IEEE 28th International Symposium on Parallel and Distributed Processing Workshops and PhD Forum (2014)
[66] Menouer, T., Rezgui, M., Cun, B.L., Régin, J.: Mixing static and dynamic partitioning to parallelize a constraint programming solver. International Journal of Parallel Programming **44**(3), 486–505 (2016). DOI 10.1007/s10766-015-0356-7. URL http://dx.doi.org/10.1007/s10766-015-0356-7
[67] Michel, L., See, A., Hentenryck, P.V.: Transparent parallelization of constraint programming. INFORMS Journal on Computing **21**(3), 363–382 (2009)
[68] Michel, L., Van Hentenryck, P.: Activity-based search for black-box constraint programming solvers. In: Integration of AI and OR Techniques in Contraint Programming for Combinatorial Optimzation Problems, pp. 228–243. Springer (2012)
[69] Microsoft Corporation: Microsoft HPC Pack 2012 R2 and HPC Pack 2012. http://technet.microsoft.com/en-us/library/jj899572.aspx (2015)
[70] Minizinc challenge http://www.minizinc.org/challenge2012/challenge.html (2012). Accessed: 14-04-2014
[71] Moisan, T., Gaudreault, J., Quimper, C.G.: Parallel Discrepancy-Based Search. In: Principles and Practice of Constraint Programming, *Lecture Notes in Computer Science*, vol. 8124, pp. 30–46. Springer (2013)
[72] Moisan, T., Quimper, C.G., Gaudreault, J.: Parallel Depth-bounded Discrepancy Search. In: H. Simonis (ed.) Integration of AI and OR Techniques in Constraint Programming: 11th International Conference, CPAIOR 2014, Cork, Ireland, May 19-23, 2014. Proceedings, pp. 377–393. Springer International Publishing, Cham (2014). DOI 10.1007/978-3-319-07046-9_27. URL http://dx.doi.org/10.1007/978-3-319-07046-9_27
[73] Mueller, F., et al.: A Library Implementation of POSIX Threads under UNIX. In: USENIX Winter, pp. 29–42 (1993)
[74] Nguyen, T., Deville, Y.: A Distributed Arc-Consistency Algorithm. Science of Computer Programming **30**(1–2), 227 – 250 (1998). DOI http://dx.doi.org/10.1016/S0167-6423(97)00012-9. URL http://www.sciencedirect.com/science/article/pii/S0167642397000129. Concurrent Constraint Programming
[75] NICTA Optimisation Research Group: MiniZinc and FlatZinc. http://www.g12.csse.unimelb.edu.au/minizinc/ (2012)
[76] Nielsen, M.: Parallel Search in Gecode. Master's thesis, KTH Royal Institute of Technology (2006)
[77] O'Mahony, E., Hebrard, E., Holland, A., Nugent, C., O'Sullivan, B.: Using case-based reasoning in an algorithm portfolio for constraint solving. In: Irish Conference on Artificial Intelligence and Cognitive Science, pp. 210–216 (2008)

[78] Palmieri, A., Régin, J., Schaus, P.: Parallel strategies selection. In: M. Rueher (ed.) Principles and Practice of Constraint Programming - 22nd International Conference, CP 2016, Toulouse, France, September 5-9, 2016, Proceedings, *Lecture Notes in Computer Science*, vol. 9892, pp. 388–404. Springer (2016). DOI 10.1007/978-3-319-44953-1_25. URL `http://dx.doi.org/10.1007/978-3-319-44953-1_25`
[79] Perron, L.: Search Procedures and Parallelism in Constraint Programming. In: Principles and Practice of Constraint Programming – CP'99: 5th International Conference, CP'99, Alexandria, VA, USA, October 11-14, 1999. Proceedings, pp. 346–360. Springer Berlin Heidelberg, Berlin, Heidelberg (1999). DOI 10.1007/978-3-540-48085-3_25. URL `http://dx.doi.org/10.1007/978-3-540-48085-3_25`
[80] Perron, L.: Search procedures and parallelism in constraint programming. In: CP, *Lecture Notes in Computer Science*, vol. 1713, pp. 346–360 (1999)
[81] Perron, L., Nikolaj, V.O., Vincent, F.: Or-Tools. Tech. rep., Google (2012)
[82] Pruul, E., Nemhauser, G., Rushmeier, R.: Branch-and-bound and Parallel Computation: A historical note. Operations Research Letters **7**, 65–69 (1988)
[83] cois Puget, J.F.: ILOG CPLEX CP Optimizer : A C++ implementation of CLP. `http://www.ilog.com/` (1994)
[84] Refalo, P.: Impact-based search strategies for constraint programming. In: M. Wallace (ed.) CP, *Lecture Notes in Computer Science*, vol. 3258, pp. 557–571. Springer (2004)
[85] Régin, J.C.: A filtering algorithm for constraints of difference in CSPs. In: Proceedings AAAI-94, pp. 362–367. Seattle, Washington (1994)
[86] Régin, J.C.: Global Constraints: a Survey. In Milano, M., Van-Hentenryck, P. eds., Hybrid Optimization. Springer (2011)
[87] Régin, J.C., Rezgui, M., Malapert, A.: Embarrassingly Parallel Search. In: Principles and Practice of Constraint Programming: 19th International Conference, CP 2013, Uppsala, Sweden, September 16-20, 2013. Proceedings, pp. 596–610. Springer Berlin Heidelberg, Berlin, Heidelberg (2013). DOI 10.1007/978-3-642-40627-0_45. URL `http://dx.doi.org/10.1007/978-3-642-40627-0_45`
[88] Régin, J.C., Rezgui, M., Malapert, A.: Improvement of the Embarrassingly Parallel Search for Data Centers. In: B. O'Sullivan (ed.) Principles and Practice of Constraint Programming: 20th International Conference, CP 2014, Lyon, France, September 8-12, 2014. Proceedings, *Lecture Notes in Computer Science*, vol. 8656, pp. 622–635. Springer International Publishing, Cham (2014). DOI 10.1007/978-3-319-10428-7_45. URL `http://dx.doi.org/10.1007/978-3-319-10428-7_45`
[89] Reynolds, J.C.: The discoveries of continuations. Lisp and Symbolic Computation. **6**(3/4), 233–248. (1993)
[90] Rezgui, M., Régin, J.C., Malapert, A.: Using Cloud Computing for Solving Constraint Programming Problems. In: First Workshop on Cloud Computing and Optimization, a conference workshop of CP 2014. Lyon, France (2014)

[91] Rolf, C.C., Kuchcinski, K.: Parallel Consistency in Constraint Programming. PDPTA '09: The 2009 International Conference on Parallel and Distributed Processing Techniques and Applications **2**, 638–644 (2009)
[92] Roussel, O., Lecoutre, C.: Xml representation of constraint networks format. http://www.cril.univ-artois.fr/CPAI08/XCSP2_1Competition.pdf (2008)
[93] Schaus, P.: Oscar, Operational Research in Scala. URL https://bitbucket.org/oscarlib/oscar/wiki/Home
[94] Schulte, C.: Parallel Search Made Simple. In Proceedings of TRICS: Techniques foR Implementing Constraint programming Systems, a post-conference workshop of CP 2000, pp. 41–57. Singapore (2000)
[95] Schulte, C.: Gecode: Generic Constraint Development Environment. http://www.gecode.org/ (2006)
[96] Van Hentenryck, P., Michel, L.: The objective-CP optimization system. In: C. Schulte (ed.) Principles and Practice of Constraint Programming - 19th International Conference, CP 2013, Uppsala, Sweden, September 16-20, 2013. Proceedings, *Lecture Notes in Computer Science*, vol. 8124, pp. 8–29. Springer (2013). DOI 10.1007/978-3-642-40627-0_5. URL http://dx.doi.org/10.1007/978-3-642-40627-0_5
[97] Vidal, V., Bordeaux, L., Hamadi, Y.: Adaptive K-Parallel Best-First Search: A Simple but Efficient Algorithm for Multi-Core Domain-Independent Planning. In: Proceedings of the Third International Symposium on Combinatorial Search. AAAI Press (2010)
[98] Wahbi, M., Ezzahir, R., Bessiere, C., Bouyakhf, E.H.: DisChoco 2: A Platform for Distributed Constraint Reasoning. In: Proceedings of the IJCAI'11 workshop on Distributed Constraint Reasoning, DCR'11, pp. 112–121. Barcelona, Catalonia, Spain (2011)
[99] Wilkinson, B., Allen, M.: Parallel Programming: Techniques and Application Using Networked Workstations and Parallel Computers, 2nd edition, Prentice-Hall Inc. (2005)
[100] Xie, F., Davenport, A.: Solving scheduling problems using parallel message-passing based constraint programming. In: Proceedings of the Workshop on Constraint Satisfaction Techniques for Planning and Scheduling Problems COPLAS, pp. 53–58 (2009)
[101] Xie, F., Davenport, A.J.: Massively parallel constraint programming for supercomputers: Challenges and initial results. In: Lodi et al. [57], pp. 334–338
[102] Yokoo, M., Ishida, T., Kuwabara, K.: Distributed Constraint Satisfaction for DAI Problems. In: Proceedings of the 1990 Distributed AI Workshop. Bandara, TX (1990)
[103] Zoeteweij, P., Arbab, F.: A component-based parallel constraint solver. In: De Nicola et al. [22], pp. 307–322

Chapter 10
Parallel Local Search

Philippe Codognet, Danny Munera, Daniel Diaz, and Salvador Abreu

Abstract Local search metaheuristics are a recognized means of solving hard combinatorial problems. Over the last couple of decades, significant advances have been made in terms of the formalization, applicability and performance of these methods. Key to the performance aspect is the increased availability of parallel hardware, which turns out to be largely exploitable by this class of procedures. As real-life cases of combinatorial optimization easily degrade into intractable territory for exact or approximation algorithms, local search metaheuristics hold undeniable interest. This situation is further compounded by the good adequacy exhibited by this class of search procedures for large-scale parallel operation. In this chapter we explore and discuss ways which lead to parallelization in local search.

10.1 Introduction

Stemming from the pioneering work on the Traveling Salesman Problem (TSP) by Flood [47] and Croes [39] in the 1950s and then Lin [75] in the 1960s, the interest in Local Search for solving large combinatorial problems has been growing since the last decade of the twentieth century and has attracted much attention from both the Operations Research and the Artificial Intelligence communities. Local search is used for finding optimal or near-optimal solutions to real-life problems when the

Philippe Codognet
University Pierre & Marie Curie/LIP6, France, e-mail: philippe.codognet@upmc.fr

Danny Munera
University of Antioquia, Medellin, Colombia, e-mail: danny.munera@udea.edu.co

Daniel Diaz
University Paris 1/CRI, France, e-mail: daniel.diaz@univ-paris1.fr

Salvador Abreu
University of Évora/LISP/CRI, Portugal, e-mail: spa@di.uevora.pt

Y. Hamadi und L. Sais (eds.), *Handbook of Parallel Constraint Reasoning*,
https://doi.org/10.1007/978-3-319-63516-3_10

search space is too large to be explored by complete search algorithms, such as Mixed Integer Programming or Constraint Solving [1, 68, 59]. Efficient general-purpose systems for local search now exist, for instance the Comet system [117], which has been parallelized for small clusters of PCs [86], or the Localsolver system [52].

Local search algorithms start from a random configuration and try to improve this configuration, little by little, by small changes in the values of the problem variables. Hence the term "local search" as, at each time step, only new configurations that are "neighbors" of the current configuration are explored. The definition of what constitutes a neighborhood will of course be problem-dependent, but basically it consists in changing the value of a few variables only (usually one or two). The advantage of local search methods is that they will usually quickly converge towards a solution (if the optimality criterion and the notion of neighborhood are defined correctly) and not exhaustively explore the entire search space. These methods naturally lead to concurrent execution, by considering the development of several configurations at the same time. This can be done sequentially by maintaining a pool of candidate configurations or in parallel if adequate hardware is available. Due to their simple algorithmic structure, local search methods therefore naturally exhibit various forms of parallelism, either with or without communication, and can be implemented on various types of parallel architectures such as multicore machines, grids or clusters, GPUs, or massively parallel machines. Indeed parallel implementation of local search methods has been studied since the early 1990s, when parallel machines started to become widely available; see [119, 118] for a general survey and concepts, or [99] for basic parallel versions of tabu search, simulated annealing, GRASP and genetic algorithms. With the increasing availability of PC clusters in the early 2000s this domain became active again [6, 38], and can further take advantage of the major advances in hardware in the last decade such as GPUs and massively parallel machines with thousands or tens of thousands of cores. However, although many methods have been developed and implemented in the last two decades, most of these experiments have been done for small-scale multiprocessors, thus giving performance evaluation for a few tens of cores at best. Only very few implementations of efficient local search solvers on larger machines have ever been reported, leaving open the question of the scalability of parallel local search in the age of exascale machines [101].

In the rest of this chapter we will present a general panorama of parallel local search methods. After a presentation of the basic mechanisms of local search methods in Section 10.2 and their sources of parallelism in Section 10.3, we will detail Single-walk approaches in Section 10.4, then Independent multi-walk methods in Section 10.5 and finally Cooperative multi-walk approaches in Section 10.6. Section 10.7 shows the effectiveness of parallel local search on two hard problems: the Stable Matching Problem and the Quadratic Assignment Problem. A short conclusion and future work end the chapter.

10.2 Local Search Metaheuristics

Metaheuristic methods aim at finding the optimal solutions (among all possible solutions) of a Combinatorial Optimization Problem. They have been proven to be very efficient on a wide variety of these problems. A metaheuristic is defined as a set of strategies for exploring the *search space* of a problem by using different methods [22]. Metaheuristics are high-level procedures using choices (i.e., heuristics) to limit the part of the search space that actually gets visited, in order to make problems tractable.

Metaheuristics generally implement two main search strategies: *intensification* and *diversification*, also called exploitation and exploration [22]. Intensification guides the solver to deeply explore a promising part of the search space. In contrast, diversification aims at extending the search into different parts of the search space [66]. In order to obtain the best performance, a metaheuristic should provide a useful balance between intensification and diversification. However, by design, some heuristics are better at intensifying the search while others are better at diversifying it. More generally, each metaheuristic has it own strengths and weaknesses. The current trend is therefore to design *hybrid* metaheuristics, by combining different metaheuristics in order to benefit from the individual advantages of each method.

In this chapter we are especially interested in local search metaheuristics; the interested reader can consult several surveys on metaheuristics [111, 100, 22, 27, 107, 108].

Local search methods (also known as trajectory methods) explore the search space by iteratively making small changes to a single solution (the current solution). These methods generally start from a randomly generated solution candidate but other strategies exist to start from a more promising initial solution constructed heuristically. At each iteration a local search method performs a single *move* (i.e., a small change to the current solution). The set of all possible moves is called the *neighborhood* (see Figure 10.1).

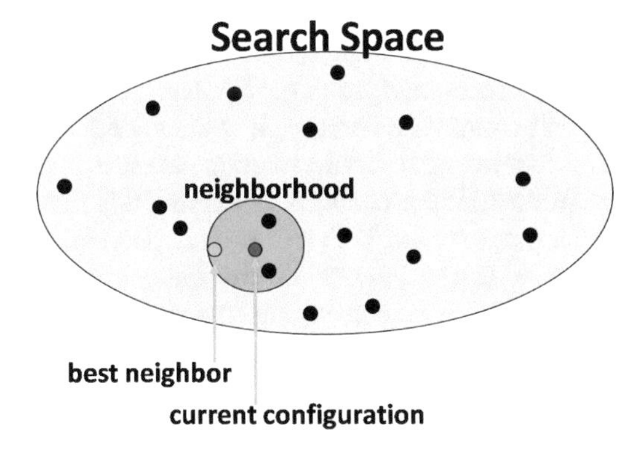

Fig. 10.1: Local search

At each iteration, a solution from the neighborhood of the current solution is selected to become the new current solution. Different strategies may be used to select the next move, for instance selecting the *best* move from the neighborhood (*hill climbing*), or the *first* move that improves the current solution (which is thus dependent on the order in which the moves are considered), or selecting a *random* improving solution.

When the neighborhood does not contain any improving solution, the metaheuristic has reached a *local optimum*[1]. Metaheuristics must provide strategies to avoid becoming trapped in local optima. They detect this situation in order to move to some other region of the search space. The simplest strategies to escape from a local optimum are to restart the search from a new (usually random) point or to perform a large perturbation of the current solution. These strategies are called multi-start local search (MLS) or iterative local search (ILS) [76]. There are other approaches and it is also possible to combine them.

By design, local search methods are very efficient at intensifying the search. However, they generally include some (simple) strategies to diversify the search (which are often executed when a local optimum is reached). Here we present the most important local search methods.

Tabu search methods [57, 55, 56] use a memory structure to avoid getting trapped in a local optimum. The main idea is to improve the basic hill-climbing algorithm by maintaining a *tabu list* of recently visited solutions (in practice, some approximations are necessary to avoid memory explosion). These solutions become prohibited (hence the term "tabu") to discourage the search from returning to previously visited places. Generally, an *aspiration criterion* is used to authorize an otherwise tabu move to be performed, in special circumstances (e.g., if it improves on the best solution found so far). The time an element remains tabu is called the *tabu tenure*. This parameter has a great influence on the efficiency of tabu search procedures and must be well tuned.

Simulated Annealing (SA) [73] is based on the annealing process of a crystalline solid used in metallurgy to improve the quality of a solid. For this, the cycles of slow cooling and heating (annealing) are alternated in order to reach a minimal energy state, which corresponds to a stable structure of the metal. Starting from a high temperature (at which the material is liquid), the cooling phase solidifies the material by a gradual decrease of the temperature. The SA method is based on this process to allow moves that result in solutions of worse quality than the current solution, in order to escape from local optima. At each iteration, it randomly selects a neighbor among its neighborhood. If it improves the current solution the move is adopted. Otherwise (a local optimum is reached) the probability of making this move is controlled by a parameter called the *temperature*. This temperature decreases during the search process; thus at the beginning of the search the probability of accepting worse moves is high but it gradually decreases, converging to a simple iterative improvement algorithm. Usually a Boltzmann distribution is used to compute the probability to

[1] The term is opposed to *global optimum* which is the best possible solution to the optimization problem. The reached local optimum may actually coincide with the global optimum, but the method is generally unable to detect this occurrence.

accept a worse quality solution (taking into account both the current temperature and how much the solution is degraded).

Variable neighborhood search (VNS) methods [88] escape from a local optimum by changing the neighborhood structure using different move types. The basic idea in VNS is that a local optimum relative to a given move type can be improved using a different move type (since the optimum is w.r.t. the neighborhood of the current solution). The search concludes when the current solution cannot be improved with all possible move types. It is thus important to correctly define the number and types of neighborhoods to be considered and the order in which they are tried. When these parameters are well tuned the VNS metaheuristic provides high-quality solutions.

Adaptive Search (AS) [34] is a generic, domain-independent, constraint-based local search method. AS takes advantage of the structure of the problem, in terms of constraints and variables, in order to guide the search more precisely than a single global cost function. Indeed, a cost is also associated with each constraint that models the problem, measuring the degree of violation of the constraint in the current solution candidate. This cost is then spread over all variables involved in the constraint (e.g., using a weight linked to the coefficient of the variable in a linear constraint). The worst variable is selected for update (i.e., to *move*), with the neighborhood being the set of all possible values for this "culprit" variable. Finally, AS maintains a tabu list of recently modified variables which led to local optima, but also implements a reset mechanism as used in ILS methods.

Extremal Optimization (EO) [24, 25, 23] is a metaheuristic inspired by self-organizing processes often found in nature. It is based on the concept of *Self-Organized Criticality* (SOC) initially proposed by Bak [18, 16], and in particular on the Bak-Sneppen model of SOC [17]. In this model of biological evolution, *species* have a *fitness* $\in [0,1]$ (0 representing the worst degree of adaptation). At each iteration, the species with the worst fitness value is updated, i.e., its fitness is replaced by a new random value. This change also affects all other species connected to this "culprit" element and their fitness value also gets updated. This results in an *extremal* process that progressively eliminates the least fit species (or forces them to mutate). Repeating this process eventually leads to a state where all species have a good fitness value, i.e., a SOC. The EO metaheuristic follows this line: it inspects the current solution, selects the worst variable (the one with the lowest fitness) and replaces its value by a random value (this corresponds to a move). However, always selecting the worst variable can lead to a deterministic behavior and the algorithm may stay blocked in a local minimum. To avoid this, the authors propose an extended algorithm; which first ranks the variables in increasing order of fitness (the worst variable has thus a rank $k = 1$) and then resorts to a probability function over the ranks k in order to introduce uncertainty in the search process: $P(\tau;k) = k^{-\tau}$. This power-law probability distribution depends on a single parameter τ, which is problem-dependent. Depending on the value of τ, EO provides a wide variety of search strategies from pure random walk ($\tau = 0$) to deterministic (greedy) search ($\tau \rightarrow \infty$). With an adequate value of τ, EO cannot be trapped in local minima since any variable is likely to mutate (even if the worst ones are privileged). This parameter can be tuned by the user.

While simple, local search procedures have been successfully used to find high-quality solutions for many Combinatorial Optimization Problems. They are also often a part of a hybrid metaheuristic to intensify the search around a promising solution found by another metaheuristic. However, there are some hard (real-life) problems for which the limit to consider the execution time as "reasonable" is rapidly reached, even using metaheuristics. It is unquestionable that the more computational resources are available, the more complex problems may be solved. It is therefore natural to consider exploiting the various forms of augmented computational power that are currently available, as conveniently as feasible.

10.3 Sources of Parallelism

Apart from domain-decomposition methods and population-based methods (such as genetic algorithms), [119] distinguishes between single-walk and multiple-walk methods for local search. Single-walk methods consist in using parallelism inside a single search process, e.g., for parallelizing the exploration of the neighborhood (see for instance [77] for such a method making use of GPUs for the parallel phase). Multiple-walk methods (parallel execution of multi-start methods) consist in developing concurrent explorations of the search space, either independently or cooperatively with some communication between concurrent processes. Sophisticated cooperative strategies for multiple-walk methods can be devised by using solution pools [37], but require shared memory or emulation of central memory in distributed clusters, thus impacting on performance.

10.3.1 Single-Walk and Multiple-Walk Methods

Figures 10.2 and 10.3 below show in a graphical way the different parallel trajectories of single-walk and multiple-walk methods.

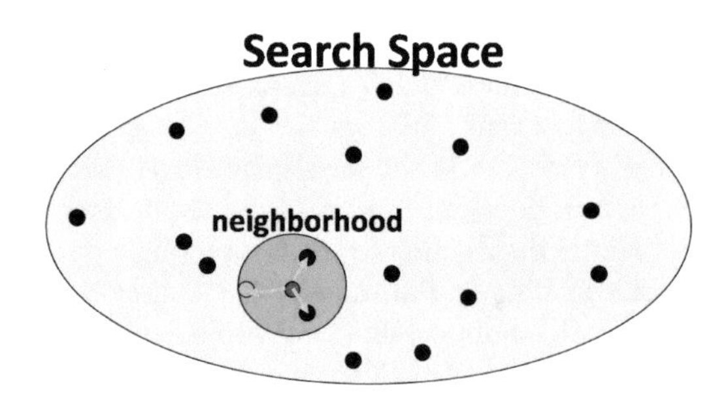

Fig. 10.2: Single-walk parallelism

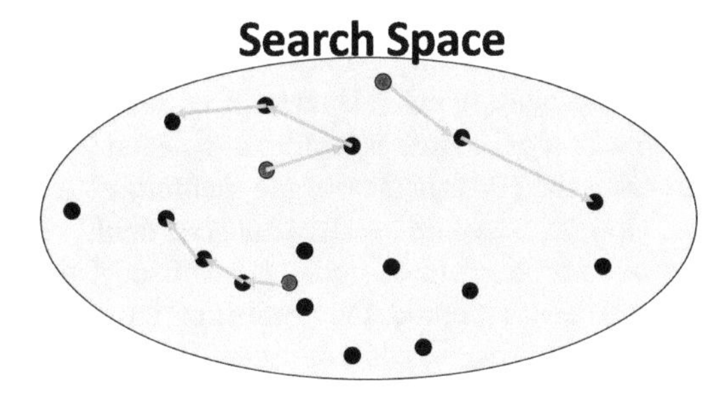

Fig. 10.3: Multiple-walk parallelism

Single-walk parallelism is limited to the neighborhood of the current solution and parallel processes need to be synchronized in order to choose the most promising neighbor and commit to the next solution. Multiple-walk parallelism explores a wider portion of the search space, limited only by the number of available concurrent processes. A key point is that independent multiple-walk methods are the easiest to implement on parallel computers, as they require no communication between processes; hence they are equivalent to parallel multi-start methods. On the other hand, one has to take care to ensure a good diversification of the search processes, which can only be achieved through communication between concurrent processes. Therefore, communication of information between concurrent processes could, if implemented without much overhead, improve the overall search. This type of parallelism is called cooperative multiple-walk parallelism. We will detail in the following sections the different methods that have been proposed in the literature for the single-walk approach, the independent multiple-walk approach and the cooperative multiple-walk approach.

10.3.2 Parallel Speedups and Runtime Distributions

Since [119, 118], it has been believed that combinatorial problems can enjoy a linear speedup when implemented in parallel by independent multiple-walks if solutions are uniformly distributed in the search space and if the method is able to diversify correctly. Thus, *in theory*, if such a method is implemented on a machine with n processors, the initial problem instance will be solved with a speedup factor of n. We will see that this is in fact not so easy to achieve in practice, especially when considering implementation on massively parallel multiprocessors, e.g., with thousands of processors. Moreover, when considering the latest cooperative methods and hybridization between different types of solvers, better performance can be achieved amounting to super-linear speedups.

But let us first see how to better analyze the execution times of local search algorithms, both sequentially and in parallel, in order to better understand the behavior and potential parallelization of such algorithms on different problem instances. Indeed the parallel speedup depends not only on the algorithm at work, but also on the structure of the problem instance which it is attempting to solve. Most papers on the performance of stochastic local search algorithms focus on the average execution time in order to measure the performance of the method, both for sequential and parallel executions. However, a more detailed analysis could be done by looking at the whole series of execution times. Indeed, because of the many stochastic choices within any local search method, the runtime on the same problem instance might vary significantly from one execution to another. Thus by considering the execution time of a local search method on a given problem instance as a random variable and by observing the execution time over many runs, the runtime behavior can be characterized by its statistical distribution. This study of so-called runtime distributions has been initially proposed in [65] for stochastic local search algorithms for the SAT problem. In this context, the property of having a linear parallel speedup in solving a given problem instance by a stochastic algorithm has been proven only under the assumption that the probability of finding a solution in a given time t follows an exponential law, that is, if the runtime behavior follows a pure exponential distribution (non-shifted). This behavior has been conjectured for local search solvers on the SAT problem in [64, 65], and shown experimentally for the GRASP metaheuristics on some combinatorial problems [4], but it is not always the case for other types of problems. Although it is very difficult to formally prove that the execution of some stochastic algorithm on a given problem instance follows an exponential distribution, it is easy to verify this experimentally. Indeed, as introduced in [5, 105], this can be done by constructing so-called *time-to-target plots*, in which the probability of having found a solution as a function of the elapsed time is measured.

However, when considering not only exponential distributions, one has to look directly at the runtime distributions and analyze them with statistical tools. Such an analysis of the scalability of independent multiple-walk local search methods has been proposed in [116] and developed in [115], where a general framework is presented in order to estimate the parallel performance of any Las Vegas algorithm [15] by analyzing the runtime behavior of the sequential version of the algorithm. Indeed, by approximating the runtime distribution of the sequential process with statistical methods, the runtime behavior of a multiple-walk parallel process can be predicted by a model based on order statistics [41]. Experiments show that the estimation is quite accurate and predicts performance close to the empirical data, with a deviation limited to about 20%. It also shows that, depending on the problem, runtime distributions can be approximated by two types of distributions, exponential (shifted and non-shifted) and lognormal, being much more complex than a pure (non-shifted) exponential distribution, which would give rise to a linear parallel speedup. In the cases of a shifted exponential distribution (the most common one) or a lognormal distribution, the speedup is no longer linear, but admits a finite limit when the number of processors goes toward infinity, and is thus bounded.

10.4 Single-Walk Approaches

Single-walk methods use parallelism *within* a single search process, e.g., by parallelizing the most computationally expensive functions of the algorithm. Runtime profiling of local search procedures reveals that one of the most resource-consuming parts is the evaluation of the neighborhood. This situation makes this function an attractive target to be parallelized with single-walk search procedures. The basic idea is to divide the neighborhood into different parts, which are then independently evaluated, in parallel. This strategy is called *neighborhood decomposition*.

In [109], Taillard presents one of the first implementations of the single-walk strategy for local search methods. He proposes a neighborhood decomposition strategy applied to the tabu search method for solving large instances of the Quadratic Assignment Problem. The implemented prototype ran on a network of Transputers.

In 1994, Garcia et al. [50] presented a new parallel version of the tabu search metaheuristic, applied to solving the vehicle-routing problem with time windows constraints. They propose a master-slave architecture where the master creates a partition of the neighborhood and assigns the portions to the available processors (slaves). Each processor then explores its own neighborhood, identifies its best move, and sends this move back to the master processor.

Note that parallel activities involved in the neighborhood decomposition task need to be performed at *each iteration* of the algorithm. These activities have to be spawned and joined several times during the main algorithm execution, thereby inducing a significant overhead due to the management of fine-grained tasks. Dealing with this overhead is considered a major challenge in single-walk parallelization. For instance, in the aforementioned work by Taillard, the authors report a maximal parallel efficiency[2] of 85% using only 10 processors.

Recent years have seen a proliferation of GPUs; which, even though they are designed to perform mostly intensive graphical operations, have significant general compute ability and relatively low cost, so as to attract research on several different applications. Such is the case for single-walk parallelization in local search methods, where GPUs have emerged as a suitable architecture to implement the neighborhood decomposition. When the operations happen to be within their reach, GPUs can effectively operate on data much faster than traditional CPU architectures: doing neighborhood decomposition in parallel on GPUs has the potential to noticeably reduce the overhead of single-walk approaches. Luong et al. in [77, 78] present a parallel local search method that uses the neighborhood decomposition strategy performed by a GPU unit. They propose guidelines to efficiently implement the parallel evaluation of the neighborhood considering the idiosyncrasies of a GPU architecture (e.g., memory management and access, thread control, mapping of neighborhood solutions to GPU threads, etc.). This approach proved to be effective in solving different optimization problems, as witness the authors' report on parallel speedups, which range from 50 when using an entry-level GPU, up to 240 with a

[2] Parallel efficiency: the division of the theoretical CPU time with an ideal speedup by the CPU time effectively observed.

higher-performance GPU board. This approach was tailored for embedding within the ParadisEO framework, as reported in [85].

Arbelaez and Codognet [11] present a parallel version of the adaptive search (AS) algorithm using both multiple-walk and single-walk parallelization. The solver takes advantage of the GPU architecture by executing multiple instances of the AS solver, but also and at the same time performing the evaluation of large neighborhoods in parallel, as previously described. The authors report a maximum speedup of 17 in solving two classical constraint satisfaction problems, and a speedup of 3 in solving the Costas Array problem.

Single-walk parallelization in GPU architectures presents rather good performance, however the implementation of local search methods on GPUs is far from trivial and the scalability of these approaches is limited, if nothing else, by Amdahl's law [8]. Amdahl's law states that the maximum speedup that may be expected from the parallelization of an algorithm is $1/s$ where s is the fraction of non-parallelizable parts of the algorithm. For instance, if a sequential algorithm is 90% parallelizable, then the theoretical maximum speedup one can ever expect by parallelizing this algorithm is 10, regardless of the number of processors in use.

10.5 Independent Multiple-Walk Approaches

Multiple-walk methods develop concurrent explorations of the search space, either *independently* or *cooperatively*. The independent multiple-walk scheme derives from the observation that local search processes, being mostly stochastic in nature, will exhibit different behavior from one run to the next. This will directly impact on the time it takes to complete an individual search, which will vary accordingly. The base insight is thus to have several instances execute concurrently, so as to collect the earliest or the best result.

Because they are concerned with processes whose execution is unrelated, independent multiple-walk methods tend to be relatively straightforward to implement on parallel computers and can lead – at least in theory – to linear speedups [119]. It should be noted, however, that this holds under the assumption that the time it takes to reach a solution obeys an exponential distribution. We will see that a more complex model may be required in order to explain the performance actually observed in larger-scale parallel executions.

10.5.1 Early Independent Multiple-Walk Methods

Early work, in 1996, by Rego and Roucairol [102] introduced a parallel variant of the tabu search metaheuristic, which they apply to the Vehicle-Routing Problem. This system uses the PVM parallel platform to perform independent parallel searches, starting from a common point but following different paths. Each search reports back to a central hub, which in turn collects solutions, looking for a local optimum, which, in turn, is used to relaunch a new batch of searches. This algorithm mixes functional with data parallelism, and it uses slightly different instances of the tabu search procedure, in the hopes that the ensuing diversity will promote better collective performance. The authors report that the parallel system begets *higher-quality* solutions, although at the expense of a sometimes significantly *slower* computation. The reason for the performance impact is not very clear, but may be related to the parallel library overheads.

In 1999, Eikelder et al. [46] proposed a Sequential and Parallel Local Search Algorithm, applied to the Job Shop Scheduling problem. In this work, the authors recognize the impact of non-determinism in performing multiple instances of a local search procedure, and establish a process whereby the parallel speedup of a simple independent multiple-walk local search algorithm may be modeled. The proposed approach takes into account the success or failure of the search procedures, as well as the quality of the solutions found, for the definition of parallel speedup. The predicted times are a good match to the observed times in the authors' experiments, scaling to about 40 large-granularity processors. The predicted and observed speedups both appear to have a largely linear section, up to about 10 processors. Beyond that, performance gains suffer a visible drop, yet there remains an undeniable benefit from running independent multiple-walk searches in parallel.

A system by Mori and Ogita [89] was proposed in 2000, which also does tabu search in parallel, applying it to the reconfiguration of power distribution systems problem. One of the driving ideas is that carrying out multiple search processes in parallel, each with just a distinct value for the tabu tenure parameter, will lead to a faster convergence on an optimal solution, because of the subsequent diversity. The authors combine this with a parallel decomposition of the neighborhood, i.e., a form of functional parallelism. The results indicate that tabu search produces the best quality solutions among several metaheuristics (which include genetic algorithms and simulated annealing), in both the sequential and parallel versions. Likewise, the parallel tabu search procedure exhibits the highest performance of the set, notably so in the case where a moderate amount of parallelism is dedicated to the parallel neighborhood decomposition (two to four sub-neighborhoods).

Finding different approaches to structure the neighborhood of a candidate solution was essential to the work of Garcia-Lopez et al. [51], published in 2002. The authors propose a parallel method to do Variable Neighborhood Search, and apply it to the p-Median problem, taking large instances from TSPLIB [103]. This proposal follows three different takes on parallelism: either the local search, the variable neighborhood search or both become subject to parallel execution. In all cases, the parallel procedures execute independently, and the runtimes reflect a near-linear speedup with up to eight processors. The prototype implementation runs on a multicore system, resorting to a shared-memory configuration using OpenMP [40], and is therefore tied to that multiprocessor organization.

Another system was described in 2003, by Bortfeldt et al. [26], which carries out multiple independent tabu search procedures, running on top of a distributed system in the form of a network of workstations. The network of parallel processes keeps tabs on the solutions found by each worker, storing them in a storage object for possible reuse by others. Even though the architecture is essentially that of independent multiple-walk parallelism, it may include various forms of information exchange among workers, as a consequence of the solution storage access pattern, by each participant. Solutions found by workers are made available to the entire network or just part of it, e.g., workers may be arranged in a ring topology. Workers may be selective as to which external solutions to look at and, should they perform better, adopt. The authors apply their prototype implementation to the Container-Loading Problem, with measurable solution quality improvements over competing approaches, namely the sequential tabu search and genetic algorithms-based solvers. Performance-wise, the parallel system actually requires more time to achieve its results and communication among workers only seems to yield minute improvements.

The 2010 work by Yazdani et al. [121] supplies another case of a parallel local search procedure: in this instance, Variable Neighborhood Search benefits from the diversification of neighborhood structures via the parallel independent exploration thereof. The parallel architecture adopted is that of shared-memory multicore processors. The authors apply their system to Flexible Job Shop Scheduling, a harder variant of the base problem, and provide experimental validation in a parallel setup with up to five processors. The results indicate that the Parallel Variable Neighborhood Search procedure computes good-quality solutions, when compared to competing approaches.

10.5.2 Recent Experiments and Performance Results

In the domain of SAT (satisfaction problem for Boolean formulas), parallel methods based on independent multi-walks have been developed under the name of the *portfolio* approach, and most of the current solvers for SAT, based either on complete or local search methods, now use portfolios for small-scale multi-core architectures. Arbelaez and Codognet experimented in [10] with multi-walks versions of several sequential local search solvers such as Sparrows, AdaptiveNovelty+ , PAWS and VW on parallel hardware up to 512 cores. Experiments were done using benchmarks from the SAT'11 competition belonging to four types of instance families: random, crafted, verification and quasigroup. The parallel speedup of each solver varies depending on the instance family but stay more or less consistent within each family. In general, nearly linear speedups are achieved on crafted and verification instances while sub-linear speedups are obtained on random and quasigroup instances. It is also worth noticing that the best sequential solver may not exhibit the best parallel speedup and therefore may not necessarily be the best one in a massively parallel context.

Work by Caniou et al. [31, 33, 32] presents a simple parallel scheme based on independent multiple-walks with no communication between processes during search, the sequential engine being based on the adaptive search metaheuristic. It was built using the MPI [48] parallel programming interface and was tested on different hardware platforms, of varying scale: up to a few hundred cores on the

GRID'5000 platform in France and the Hitachi HA8000 and Fujitsu FX10 machines at the University of Tokyo and up to 8,000 cores on the JUGENE supercomputer at Jülich Supercomputing Centre. Performance evaluation on large instances of some classical Constraint Satisfaction Problems from CSPLIB [54], such as the Magic Square, Perfect Square and All-Interval problems, shows that speedups are very good for a few tens of cores (e.g., speedup of a factor of 20-25 on 32 cores), and correct up to a few hundreds of cores (e.g., speedup of a factor of 50-60 on 256 cores), but speedup then degrades, showing that not much parallelism could be further extracted even with a larger number of cores. Figure 10.4 shows the performance results of the parallel adaptive search method on these problems in the form of runtime speedups for a given number of cores.

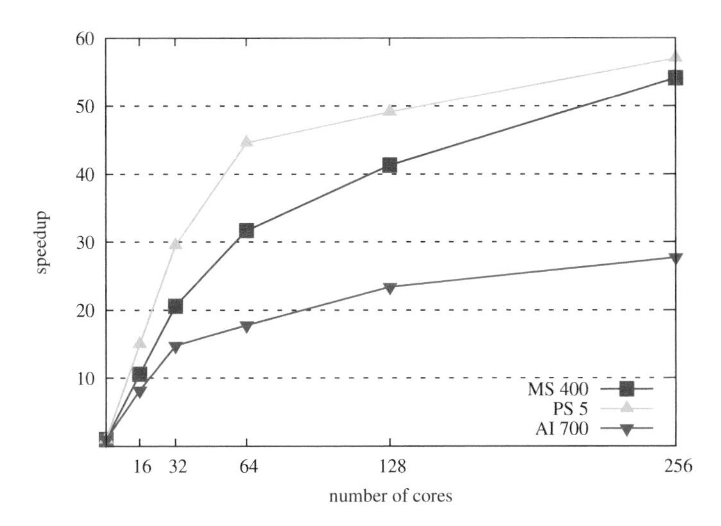

Fig. 10.4: Speedups for benchmark CSP programs on the HA8000 parallel machine, from [31]

However, another hard combinatorial benchmark, the Costas Arrays Problem (CAP), was also tested with instances of CAP up to 23 (large instances) and the experimental evaluation shows better parallel scalability. Indeed, parallel speedup scales very well (linearly) up to about 8,000 cores, on the JUGENE supercomputer. Figure 10.5 shows the performance results of the parallel adaptive search method on instances 21, 22 and 23 of CAP, in the form of runtime speedup for a given number of cores.

This can be explained by the fact that the runtime distribution of the adaptive search metaheuristic on the CAP problem exhibits a nearly pure (non-shifted) exponential distribution; see [43] for details of experimental results. The authors also experimented with a limited form of cooperation among search processes (exchanging only solution costs between processes and performing restarts), but the results were not markedly different from the independent multiple-walk strategy.

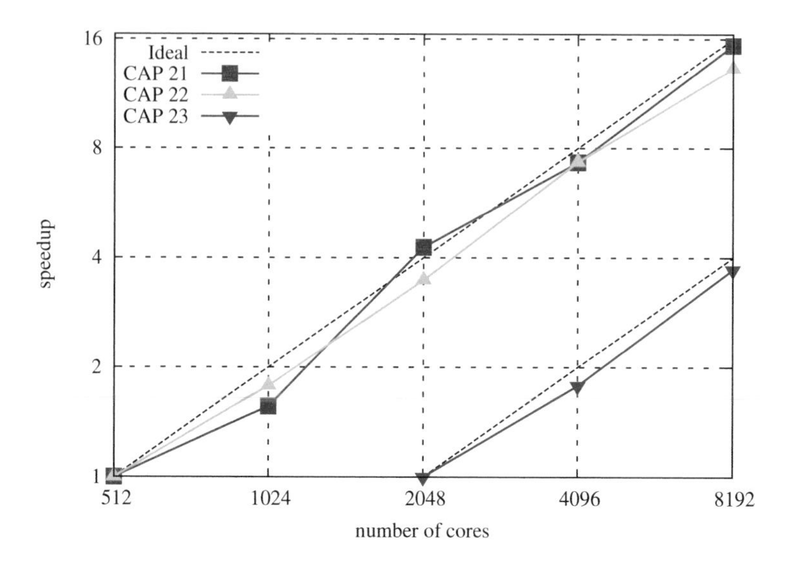

Fig. 10.5: Speedups for the Costas Array Problem on the JUGENE supercomputer, from [32]

It turns out that most independent multiple-walk procedures start off with good speedups attributable to parallel execution. However, this characteristic appears to hit a problem-dependent hard limit, which may be attributed to the lack of entropy (diversity) across the different runs which are being performed in parallel, and thereby bounds the usefulness of such a strategy, at larger scale.

Although there are stand-out exceptions, this diminishing returns situation becomes especially obvious when attempting to scale beyond a few dozen cores. This has prompted research into exploring more sophisticated parallel methods that can compensate for this performance drop, namely those that rely on some form of cooperation among worker threads, as discussed in section 10.6.

10.6 Cooperative Multiple-Walk Approaches

To overcome the limitations of the independent multiple-walk strategy, it is natural to consider a paradigm based on *cooperation*. This is the case of *Cooperative Multiple-walk* methods, which add a communication mechanism to the independent search strategy, in order to share or exchange information among solver instances during the search process. However, designing an efficient cooperative method is a very complex task, and many issues must be solved: What information is exchanged? Between which processes is it exchanged? When is the information exchanged? How is it exchanged? How is the imported data used? [114]. The work presented in [79] studies these questions, and concludes that no one cooperative configuration may efficiently tackle all problems. Indeed, most cooperative choices are problem-dependent (and even instance-dependent).

According to the literature [119, 120, 22, 111], an *efficient* cooperative method should consider four essential functionalities: flexibility, adaptability, performance and scalability. Flexibility refers to the capability of a given method to tackle different problems, using different methods and providing hybrid behavior. Adaptability is related to the ability of a given method to adjust its cooperative behavior. In addition, a method has a good performance if it can obtain a high-quality solution in a short execution time. Finally, scalability refers to the ability of a given method to efficiently use a significant number of processing units (cores).

In this chapter, we analyze several approaches using the cooperative multiple-walk strategy. We identify three different kinds of algorithms: metaheuristic parallelization, agents-based and general frameworks.

10.6.1 Metaheuristic Parallelization Approaches

We first analyze cooperative methods based on metaheuristic parallelization. One of the oldest cooperative approaches was proposed in 1993 by Hogg and Williams [63]. The basic idea is to create multiple solver entities (metaheuristics) that share partial configurations (hints) through a centralized memory (blackboard). Each entity reports to the blackboard a hint at each step (based on its current state) with a given probability p. When the entity is at an appropriate decision point, it reads a hint from the blackboard with probability p. If p is set to zero, the algorithm behaves like independent search. The implementation of the method is dedicated to solving the graph coloring problem using two different heuristics: the Berlaz algorithm and heuristic repair. The experimental evaluation is performed using 10 agents, solving graphs with 100 nodes and comparing the performance of the independent and the cooperative approaches. The cooperative version presents better performance than the independent version in terms of the execution time, however the parallel scalability is not evaluated in this work.

In 1998, Aiex et al. proposed a cooperative parallel tabu search for solving the circuit partitioning problem [3]. This method implements a master-slave model composed of search processes (slaves) which implement different combinations of the initial solution algorithm and move attribute for a specialized tabu search metaheuristic. Periodically, the search processes exchange information (elite solutions) with the master node, which maintains a centralized shared memory for elite configurations. The parallel procedure is implemented using two different parallel programming languages, PVM (based on message passing) and Linda (based on the shared-memory model). The authors test both implementations on a set of problem instances from the ISCAS benchmark on an IBM SP-2 machine with 16 processors. Only 10 processors are used in the experimental evaluation, using one master and nine search processes. The implementation improves the solution quality for all problem instances with respect to the sequential version of the algorithm, the PVM version being 20% faster than the Linda implementation. This work does not present any evaluation of the performance in terms of the execution time and the parallel scalability.

Gendreau et al. [53], in 1999, proposed another master-slave scheme to parallelize the tabu search algorithm for solving the dynamic vehicle-routing and dispatching problem. The master entity manages an adaptive memory which is fed by a set of tabu search instances (slaves). The adaptive memory is used to create new initial solutions for slave processes. The authors present a prototype implementation, which runs on a network of 17 SUN UltraSparc workstations. The proposed method is compared with other heuristic approaches and obtains a better solution quality than its competitors. An evaluation of the parallel scalability is performed using up to 16 processors, showing that the solution quality is improved by increasing the number of processors involved.

Two similar methods are proposed to implement cooperation on the GRASP (Greedy Randomized Adaptive Search Procedure) and the Path Relinking metaheuristic [2, 104]. A distributed cooperation mechanism was proposed by Aiex et al. in 2003 [2], which creates several search processes. Each process sends the best overall configuration to the other processes when the cost is improved. Each process maintains a local elite pool which possibly contains configurations from all processes. This pool is used as input for the Path Relinking phase. An experimental evaluation is carried out solving standard job shop scheduling test problems from Beasley's OR-Library. The experiments are done on an SGI Challenge computer composed of 28 R10000 MIPS processors, using 1, 2, 4, 8 and 16 processors. The prototype is coded in Fortran using the MPI library. The cooperative strategy obtains almost linear speedups, improving on the independent strategy; which, as expected, shows only a sub-linear speedup.

Ribeiro and Rosseti, in 2007, proposed another parallel cooperative approach also using GRASP and Path Relinking [104]. This method takes advantage of the multi-start behavior of the GRASP metaheuristic to implement a multiple-walks parallelization. In addition, a master-slave cooperative strategy is implemented. Slave processes send the best configurations to a master process, which maintains a centralized pool of elite solutions. Then the master can send back a new configuration to slave nodes upon request. A prototype implementation of this approach is developed using C and the MPI specification. The experiments are carried out on a cluster of 32 Pentium II 400 MHz processors solving the randomly generated instances of the 2-path network design problem. The cooperative strategy presents smaller execution times and scales better than the independent implementation, obtaining almost linear speedups and reporting a maximum speedup of 17.6 using 32 cores. Although the two previous methods present fair parallel performances, the functionality is attached to GRASP behavior and to the problem nature, thus limiting its flexibility.

A cooperative parallel approach that uses the rollout algorithm for solving the Sequential Ordering Problem was proposed in 2003 by Guerriero and Mancini [60]. This approach presents a master-slave topology in which slaves are executed in parallel, running an instance of the rollout algorithm. Slave processes periodically send the best configurations found to the master, which maintains a centralized pool of configurations. The master restarts slaves with adjusted parameters using a new initial point from the pool. The cooperative mechanism can adapt its behavior by selecting the best parameters for the base algorithm. However, this cooperative approach is strongly linked to the rollout algorithm, limiting the possibility to use this technique with other metaheuristics. The parallel version of the algorithm was implemented in C++ using the MPI library. The experiments run on a cluster of nine nodes with two Pentium 1 GHz processors, solving 14 instances of the Sequential Ordering Problem (taken from the TSPLIB). The cooperative approach obtains a good solution quality for the given set of problems. The scalability of the algorithm is evaluated using 1, 2, 4 and 8 slaves (cores). The algorithm improves either the solution quality or the execution time used to find the best solution when increasing the number of slaves. However, the authors report that the rollout-like approach obtains a higher computational time to find good solutions compared with other state-of-the-art approaches.

The 2004 work by Crainic et al. [37] presents a master-slave cooperative method to solve the p-median problem based on the Variable Neighborhood Search (VNS) metaheuristic. The master process implements a central memory to maintain the best overall solution. The master also sends the initial configuration to slaves. The slave processes (VNS processes) perform the search and notify the master when improving the overall solution. A slave process asks the master for a new search point if it cannot improve its current configuration. An MPI implementation of this approach run on a 64-processor SUN enterprise machine with 400 MHz clock. The experiments use 1, 5, 10 and 15 processors, solving a set of problems from the TSPLIB benchmark. This strategy obtains significant gains in terms of execution time, maintaining a good solution quality. However the cooperation mechanism is strongly linked to the behavior of the VNS metaheuristic and to the problem model. The principles of this approach include avoiding using parameters, which is convenient for the user but not for the adaptability of the system.

In 2012, Cordeau and Maischberger [36] proposed a parallel iterated tabu search algorithm to solve vehicle-routing problems. The basic idea is to execute in parallel several iterated tabu search solver instances using different sets of parameters. The algorithm implements a communication mechanism to share the most promising configurations found in the search process. Each process can apply a crossover operator to the received configurations (with a given probability), in order to combine information of two different received configurations. The algorithm is implemented in C++ using the MPI libraries for the parallel version. The experiments run on a cluster composed of 128 nodes, each with a 3 GHz dual Intel Xeon CPU E5472 (i.e., four cores per node). This strategy is tested solving different variants of the vehicle-routing problem, using up to 80 cores, and obtaining good performances in terms of the solution quality (allowing the identification of new best known solutions for a large set of problems).

A cooperative approach based on the execution of multiple instances of the adaptive search solver was presented in 2013 by Machado et al. [80]. A single master solver instance sends every k iterations its current configuration to the other solver instances. Since this information is stored in a shared-memory structure, all the solver instances (threads) running on the node benefit from this communication. Each solver instance decides whether it adopts the received configuration or continues its current search process. This cooperative scheme was implemented using the GPI (Global Address Space Programming Interface) API for parallel applications running on clusters. The experiments are conducted on a cluster system with 155 nodes; each node includes a dual Intel Xeon 5148LV (i.e., four cores per node). This strategy is evaluated solving two constraint satisfaction problems from the CSPLib: *all-interval* and *magic-square*; and one hard real-life problem: the Costas Array Problem (CAP). The cooperative strategy presents no gain compared to the independent strategy, when solving the CAP. For the CSPLib problems the cooperative approach presents a better speedup than the independent strategy. The parallel scalability is evaluated using up to 512 cores; however the obtained speedups are sub-linear for both cooperative and independent approaches. More recently, in 2015, Caniou et al. presented a similar approach in [32], which uses the same base algorithm (adaptive search) and the same set of CSPLib problems. The authors propose a new cooperative approach in which only single integer values are exchanged between entities (as opposed to complex data types such as vectors of variables, i.e., a configuration). The receiver entity uses this information to decide whether it is convenient to develop a restart procedure. This strategy is evaluated on the Helios cluster of the GRID'5000 platform, using up to 128 cores. The results of the experimentation cannot show an improvement in the performance using this cooperative approach.

In the domain of SAT, parallel cooperative methods based on local search have also been developped. Arbelaez and Hamadi proposed in [12] several strategies for sharing knowledge between processes, involving a pool of elite configurations containing the best configuration found so far by each process. When a restart is performed, new restart configurations are thus created on demand by agreggating those elite solutions, variable by variable. The best aggregation strategy, named *Prob-NormalizedW*, consists in weighting the influence of each process by using a probability reflecting the cost of the configuration (number of unsatisfied clauses). Small-scale experiments on 4 and 8 cores machines show that good performances could be achieved and this solver won a silver medal in the SAT'11 competition (random category, parallel track). Thoses ideas were later extended by Arbelaez and Codognet in [9] for larger-scale parallel systems (up to 256 cores), but performance then becomes very sensitive to the cost of communication and possible excessive diversification. Indeed, the best performance is achieved when defining small groups of cooperative solvers (up to 16 processes) and having no communication between different groups of solvers.

10.6.2 Agent-Based Approaches

Agent-based modeling is a powerful strategy that facilitates the implementation of cooperative approaches. In early work, in 1998, Talukdar et al. propose a multi-agent-based cooperative methodology to combine solving strategies [113]. The

A-Team (asynchronous teams) framework allows agents to cooperate through a shared memory containing a population of configurations. Agents can create, modify or delete configurations from the shared memory. Furthermore, they can obtain elite configurations from the shared memory, which have probably been created by another agent, in order to cooperate and make the initial set of configurations evolve. The A-Team framework provides a good level of flexibility, because agents can implement different algorithms, and this method can be applied to different problems. The referenced paper does not report any experimental evaluation, however this approach has been used as the basis for many agent-based cooperative solvers.

In 2004, Milano and Roli presented a multi-agent metaheuristic architecture (called MAGMA) that can describe cooperative search or hybrid metaheuristics [87]. This architecture is based on a multi-level organization in which components (agents) are classified according to their capabilities. Low-level agents describe the basic functionality of metaheuristics. A top layer manages integration and cooperation of different solvers. Agents in the top layer can store partial or complete configurations and promote changes in lower layers in response to the gathered information. This approach provides a theoretical description that can be easily adapted to tackle different problems and to use different metaheuristics, thus providing fair flexibility and adaptability. Similarly to the A-Team strategy, MAGMA is considered as a generic framework; the referenced paper only provides an experimental evaluation in the appendix, where a guided-restart iterated local search algorithm is conceived as a combination of existing components in the MAGMA framework.

A multi-agent architecture was proposed in 2006 by Bachelet and Talbi for solving large-scale instances of the Quadratic Assignment Problem [112]. This method, called COSEARCH, is composed of a set of agents that perform specific tasks: search agent, intensifying agent and diversifying agent. COSEARCH implements as the main search agent a tabu search heuristic; for the diversifying agent, it uses a genetic algorithm; and for the intensifying agent, a kick operator is used. These agents share information through an adaptive memory that stores information about the already visited areas of the search space and about the intrinsic nature of the elite solutions already found (initial and elite configurations). This strategy is evaluated solving a set of problem instances from the QAPLib benchmark. The experiments run on a heterogeneous parallel platform composed of around 150 workstations, using a significant number of cores. The results show COSEARCH presents better performance than a basic parallel multi-start strategy, in terms of execution time and solution quality.

The 2007 work by Aydin [14] proposed a study of different cooperative topologies for agent-based metaheuristics. This work tested three different schemes: A-Team, a multiple-island model and variable neighborhood search. The job shop scheduling problem is used to develop the experimental evaluation, which only considers the solution quality. All the schemes are developed using DREAM software [13] which is a Java-based framework that implements the distributed sub-population model for evolutionary algorithms by using multi-agent technology. The main objective in this experimentation is to reveal more details about each strategy.

In 2009, Cadenas et al. presented a cooperative parallel hybrid strategy that uses machine learning techniques [29]. The system is composed of two different types of agents: metaheuristic and coordinator. Multiple instances of different metaheuristics are run in parallel by metaheuristic agents, which, simultaneously, share information through a blackboard data structure. One coordinator agent is used to analyze the information in the blackboard and to adapt the metaheuristic agents' behavior. The coordinator agent incorporates knowledge from an offline machine learning process. This knowledge helps the coordinator to guide the search and to adapt the behavior of the system to different situations. The authors also proposed a Java implementation of this strategy using tabu search, simulated annealing and genetic algorithms for the metaheuristic agent. This implementation is used to solve different instances of the knapsack problem. The experiments run on an Intel core2 Quad 1.66 GHz. The parallel version of the algorithm, which consists in a parallel execution of each metaheuristic, presents better performance than the non-cooperative approach. However, no comparison with state-of-the-art methods was carried out, and the evaluation does not include a parallel scalability analysis.

A Coalition-Based Metaheuristic (CBM) was presented in 2010 by Meignan et al. [84]. This approach is based on the agent metaheuristic framework and the hyper-heuristic approach. The system architecture is composed of agents that implement a complete set of capabilities that make them suitable to perform different roles during the execution (strategist, guide, intensifier and diversifier). Agents exchange information in a decentralized and asynchronous manner. Agents use reinforcement learning and mimetism to adjust their behaviors. The authors present an implementation of the CBM in Java, running on a 3 GHz Pentium 4 processor. This implementation is tested solving the capacitated vehicle-routing problem and it shows competitive results in terms of both solution quality and execution time, using up to 20 parallel agents.

More recently, in 2014, Barbucha proposed another agent-based cooperative approach for population learning algorithms (called CPLA) [19]. This approach is based on the A-Team framework and on the population learning algorithm. The basic idea is to make a population of individuals (configurations) evolve using a process that is divided into stages. At each stage, the population is improved using dedicated algorithms and different topologies. After each stage some elite individuals are promoted to the next stage. Agents have communication capabilities and, according to the stage, can share information with other agents (through a shared elite pool). Furthermore, multiple A-Teams can be run in parallel and exchange information through a migration manager agent. An implementation of the CPLA was developed using JADE (Java Agent Development Framework) [20]. The experiments run on the HOLK cluster built of 256 Intel Itanium 2 Dual Core processors solving the vehicle-routing problem with time windows. The results show CPLA has good performance in terms of solution quality and execution time, being competitive with state-of-the-art methods. No parallel scalability is analyzed in the referenced paper.

In 2016, Martin et al. proposed another agent-based cooperative approach [82]. In this method agents implement different metaheuristics to perform the search process. Agents asynchronously exchange partial configurations; which are analyzed by machine learning techniques in order to identify patterns and to adapt the agent behavior. The experimental evaluation runs on a Linux cluster composed of eight nodes, solving three different combinatorial optimization problems: the permutation flow-shop scheduling, the capacitated vehicle-routing and the nurse-rostering problems. The results show good performance in terms of solution quality, using up to 16 cores. The referenced paper does not present information about execution times or parallel scalability.

10.6.3 Framework Approaches

In this last group we analyze cooperative methods that propose a general framework. These methods generally offer high flexibility because they can tackle different problems using different metaheuristic solvers.

Cahon, Melab and Talbi in 2004 proposed an open-source framework for parallel and distributed design of hybrid metaheuristics, ParadisEO [30]. This framework provides different hybridization mechanisms for metaheuristics including population-based and single-solution methods. ParadisEO separates the modeling of the metaheuristic formulation from the problem to be solved, using a modular architecture that allows code and design reuse. For instance, ParadisEO-MO [67] is the module dedicated to the design, analysis and implementation of local search algorithms and the ParadisEO-PEO module provides a set of classes to design and implement parallel and distributed metaheuristics. ParadisEO-PEO supports different levels of parallel metaheuristics, from neighborhood decomposition (single-walk) to independent and cooperative multiple-walk. Cooperation is implemented following the island model (from population-based methods), in which the solver instances can share information based on a migration model. ParadisEO has been successfully experimented with in a wide range of problems; for instance in [110], the ParadisEO framework is used to solve the multi-objective constrained combinatorial optimization model for a problem in radio network design.

A cooperative parallel hyper-heuristic framework was proposed in 2010 by Ouelhadj and Petrovic [96]. This framework is composed of multiple heuristic agents and one cooperative hyper-heuristic agent. Heuristic agents implement low-level heuristics performing a local search procedure. The best configuration found by the heuristic agents is sent to the cooperative hyper-heuristic agent which maintains a pool of elite configurations. This pool also stores information about low-level heuristics and the objective function. Additionally, the cooperative hyper-heuristic agent decides which low-level heuristic the heuristic agents will run and also provides them with elite configurations from the pool to diversify the search. This method clearly provides high flexibility, because it can be adapted to different problems or metaheuristics. Additionally some parameters were defined to adapt the cooperative mechanism. A prototype implementation to solve the flow shop scheduling problem is presented using C# and multi-thread libraries. The experiments run on an Intel Pentium M 1500 MHz processor obtaining good performance in terms of solution quality, however this cooperative approach does not outperform the state-of-the-art methods for the flow shop scheduling problem.

In 2014, Munera et al. [93] presented a Cooperative Parallel Local Search Framework (CPLS). This framework is both problem- and metaheuristic-independent and allows the programmer to tune the search process through an extensive set of parameters. The basic component of CPLS is an *explorer*, which executes an LS solver instance and runs on a physical core (see Figure 10.6). Several explorers are grouped into *teams*. Inside a team the explorers intensify the search, sharing the most promising solutions via an *elite pool*. The teams also communicate with one another to promote search diversification; for this a measure of the distance between teams is used to detect when two teams are exploring the same region (in which case a corrective action is taken to force one team to explore another region). Thus intra-team communication is used for intensification while inter-team communication ensures diversification.

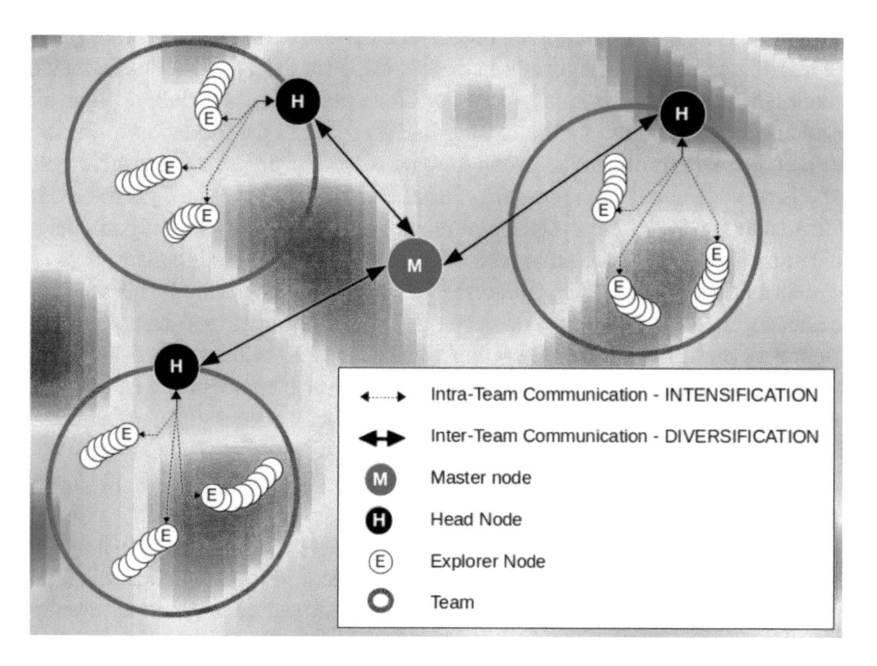

Fig. 10.6: CPLS framework

The concepts and entities involved are all subject to parametric control (e.g., tradeoff between intensification and diversification, elite pool size, communication interval, distance, corrective action, etc.). An implementation of CPLS (available as an open source library) in the X10 parallel programming language [106] has been used to solve different hard Combinatorial Optimization Problems [95], providing (super-) linear speedups up to 128 cores.

10.7 Parallelism at Work

In this section we discuss the efficacy of parallel local search methods on two hard problems, both of which have several real-world application instances: the Stable Matching (SM) and Quadratic Assignment (QAP) Problems.

10.7.1 Stable Matching Problem

The Stable Matching problem was introduced by Gale and Shapley in their seminal 1962 paper [49]. The SM problem can be stated as follows: given a set of n men and a set of n women, each of whom have ranked all members of the other set in a strict order of preference, find a *matching* (a one-to-one correspondence between the men and the women) such that there is no man-woman pair where both prefer each other than their assigned partner. This criterion is called *stability* and is a desirable property since it ensures, according to stated preferences, that there is no man-woman pair for which both have incentive to elope – such a pair is called a *blocking pair*. Gale and Shapley proved that such a stable matching always exists and proposed an $O(n^2)$ algorithm (called GS in what follows) to find one.

However, requiring *each* member to rank *all* members of the opposite sex in a *strict* order is unfeasible for many real-life, large-scale applications. A natural variant of SM is the *Stable Matching with Ties and Incomplete Lists* (SMTI) problem [70, 81]. In SMTI, the preference lists may include ties (to express indifference among several partners) and may be incomplete (to express that some partners are unacceptable). A stable matching always exists for SMTI and can be easily obtained by arbitrarily breaking the ties and applying the GS algorithm. However, with the introduction of ties and incompleteness in the preference lists, the stable matching for an instance of SMTI may have different sizes. It is thus desirable to find the stable matching of *maximal* size (that is, with the smallest number of singles). This optimization problem has been shown to be NP-hard, even for very restricted cases [70, 81]. This problem has attracted a lot of research in recent years since it is at the heart of a wide variety of important real-life applications. Indeed, matching problems can be found in several settings, such as car sharing or bipartite market sharing, job markets and social networks. Many of these applications involve very large sets, thereby ruling out the use of complete methods. SMTI has been shown to be an APX-hard problem [61] and most recent research focuses on designing efficient *approximation algorithms*, i.e., algorithms running in polynomial time yet able to guarantee solutions within a constant factor of the optimum [69, 71]. SMTI cannot be approximated within a factor of 21/19 and probably not within a factor of 4/3 either [62]. Currently, the best known algorithms are 3/2-approximations [83, 72, 98] or heuristic-based specific solutions. These algorithms produce a single solution for a given problem instance, even though it is often useful to provide multiple optimal or quasi-optimal solutions.

In [95], the authors proposed AS-SMTI, a local search procedure for SMTI based on adaptive search in the CPLS framework briefly described in Section 10.6.3.

The sequential version displays significant improvement in performance or solution quality w.r.t. the state-of-the-art exact and approximate sequential algorithms, and the independent multi-walk parallel version exhibits a significant speedup with an increasing number of cores. Moreover, the cooperative parallel version achieves super-linear speedup on average, consistently behaving very well on hard instances.

The parallel experiments were carried out on a cluster of 16 machines, each with four 16-core AMD Opteron 6376 CPUs running at 2.3 GHz and 128 GB of RAM. The nodes are interconnected with InfiniBand FDR 4× (i.e., 56 Gbps) and the experiment involved up to 128 cores (four nodes and 32 cores per node). Figure 10.7 presents log-log graphs of the speedup using independent walks (IW in red) and cooperative walks (CW in green) on 10 very hard and large instances (size $n = 1\,000$). The independent version reaches a quasi-linear speedup (91.5 for 128 cores) while the cooperative version gets super-linear speedups (492 with 128 cores).

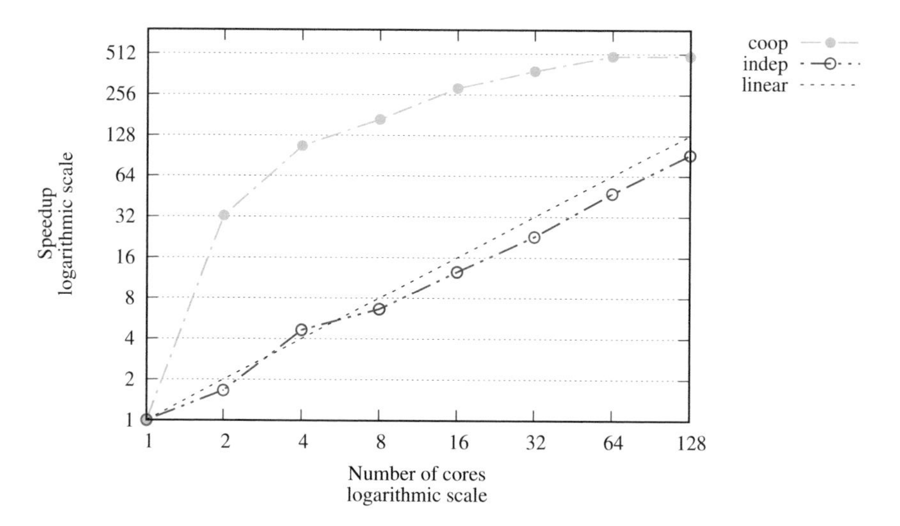

Fig. 10.7: Speedups obtained with AS-SMTI on hard instances of SMTI problems (size = 1 000)

In [94] the same authors propose an extension of their algorithm to tackle one important and hard variant of the Stable Matching problem: the Hospital/Resident problem, which is NP-hard. This problem consists of a set of n_1 residents who apply for k positions distributed among n_2 hospitals. The preference list of a resident consists of the ordered list of *acceptable* hospitals. The preference list of a hospital contains the ordered list of residents who apply to it. In the most general case, preference lists are allowed to contain ties (to express indifference) and can be incomplete (residents only apply to a subset of the hospitals and hospitals rank their corresponding candidates). In addition, each hospital has a *capacity*, which indicates the maximum number of positions it offers. The problem consists in finding a (maximum size) stable matching between residents and hospitals (thus satisfying the preference lists) that complies with the capacities (each resident being assigned to at most one hospital and the number of residents assigned to any hospital not exceeding its capacity). The HRT problem is important in the medical domain and there are national programs in various countries, the best-known ones being the National Resident Matching Program (NRMP) in the USA, the Canadian Resident Matching Service (CARMS), the Scottish Foundation Allocation Scheme (SFAS) and the Japan Residency Matching Program (JRMP). As might be expected, such programs involve very large data sets. The HRT problem also has several other application domains, e.g., assignment of applicants to positions in job markets.

The resulting cooperative parallel solver, while much simpler and more general, displays performance which is comparable to the best known specific solvers for HRT, including those which assume domain restrictions (e.g., having ties on one side only).

10.7.2 The Quadratic Assignment Problem

The Quadratic Assignment Problem (QAP) was introduced in 1957 by Koopmans and Beckmann [74] as a model for a facilities location problem. This problem consists in assigning a set of n facilities to a set of n specific locations so as to minimize the cost associated with the *flows* of items among facilities and the *distance* between them. This combinatorial optimization problem has many other real-life applications: scheduling, electronic chipset layout and wiring, process communications, turbine runner balancing and data center network topology, to cite but a few [35, 21]. This problem is known to be NP-hard and finding effective algorithms to solve it has attracted a lot of attention for many years.

Since the mid-1980s several metaheuristics have been successfully applied to the QAP: tabu search, simulated annealing, genetic algorithms, GRASP and ant-colonies [21]. For solving the hardest instances, the current trend is to resort to hybrid procedures, in order to benefit from the strengths of different classes of heuristics. Such is the case of hybrid genetic algorithms for the Quadratic Assignment Problem (a.k.a. memetic algorithms) [45]. The price to pay for this improvement is a significant increase in the complexity of the resulting solver code.

An alternative approach for constructing hybrid search methods has been presented in [91, 90], based on cooperative parallelism. The authors show additional benefits of the intra/inter-team cooperation mechanisms in order to provide hybridization behaviors. To this end, CPLS was configured with explorers running instances of *different* metaheuristics inside a team. Hybridization is obtained thanks to the collaboration between explorers through the *elite pool*. It turns out that the intra-team communication mechanism, implemented to intensify the search within a team, now also becomes a mechanism to exchange information between explorers running different metaheuristics. The whole system behaves like a hybrid solver, benefiting from cross-fertilization, which stems from the inherent diversity of the search strategies. The basic idea of running in parallel different metaheuristics that exchange elite solutions has been mentioned [7, 113] but from a general and strictly theoretical point of view. This technique may also be viewed as a *portfolio* approach [58] augmented with cooperation.

Following this line, the authors propose a parallel hybrid solver (called ParEOTS) to tackle the Quadratic Assignment Problem (QAP), combining two different metaheuristics: Taillard's Robust Tabu Search [109] and an original Extremal Optimization method [92]. This parallel hybrid solver performs very well on QAPLIB, the standard benchmark library used to assess QAP solvers [28]. For instance, linear speedups up to 128 cores can be achieved, see Figure 10.8.

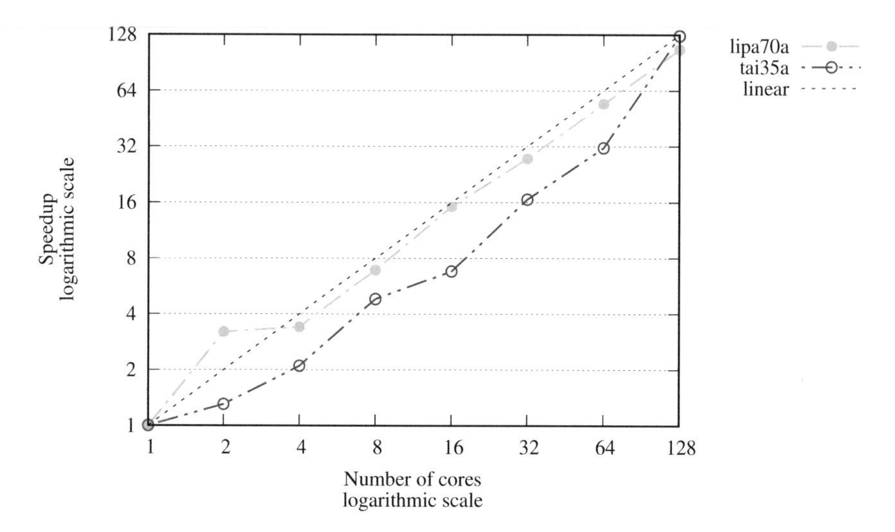

Fig. 10.8: Speedups obtained with ParEOTS on two QAPLIB instances

ParEOTS has been tested on the 33 hardest problems of QAPLIB, using 128 cores. The solver was set to stop when reaching the *Best Known Solutions* (BKS, i.e., best known optimum) as recorded in the QAPLIB archive. A (comparatively short) timeout of 5 minutes was used to limit the execution in case the BKS is not reached. Each instance was solved 10 times (results are averaged). Table 10.1 shows the performance of ParEOTS. For each problem, the table includes the current BKS (which was sometimes the optimum), the number of times the BKS was reached by the solver (#BKS), the Average Percentage Deviation (APD), which is the average of the 10 relative deviation *percentages* computed as follows: $100 \times \frac{F(sol)-BKS}{BKS}$, and the average execution time (either shown as a decimal number representing seconds or in a human-readable form as mm:ss). Even with a very short timeout, ParEOTS provided solutions of high quality. It reached the best known solution (BKS) for all but four QAPLIB instances. When the BKS was not reached, the obtained solution was nevertheless very close (less than 0.22% off, on average).

	BKS	#BKS	APD	time
els19	17212548	**10**	0.000	0.0
kra30a	88900	**10**	0.000	0.0
sko56	34458	**10**	0.000	1.5
sko64	48498	**10**	0.000	1.7
sko72	66256	**10**	0.000	8.7
sko81	90998	**10**	0.000	0:24
sko90	115534	**10**	0.000	1:32
sko100a	152002	**10**	0.000	1:09
sko100b	153890	**10**	0.000	0:45
sko100c	147862	**10**	0.000	0:56
sko100d	149576	**10**	0.000	1:03
sko100e	149150	**10**	0.000	0:47
sko100f	149036	**10**	0.000	0:57
tai40a	3139370	**10**	0.000	1:26
tai50a	4938796	**3**	0.077	4:24
tai60a	7205962	**3**	0.146	4:15
tai80a	13499184	0	0.364	5:00
tai100a	21052466	0	0.298	5:00
tai20b	122455319	**10**	0.000	0.0
tai25b	344355646	**10**	0.000	0.0
tai30b	637117113	**10**	0.000	0.1
tai35b	283315445	**10**	0.000	0.3
tai40b	637250948	**10**	0.000	0.1
tai50b	458821517	**10**	0.000	2.6
tai60b	608215054	**10**	0.000	4.6
tai80b	818415043	**10**	0.000	0:53
tai100b	1185996137	**10**	0.000	1:11
tai150b	498896643	0	0.061	5:00
tai64c	1855928	**10**	0.000	0.0
tai256c	44759294	0	0.178	5:00
tho40	240516	**10**	0.000	0.5
tho150	8133398	**1**	0.007	4:51
wil100	273038	**10**	0.000	1:37

Table 10.1: ParEOTS on the hardest instances of QAPLIB (128 cores)

This solver was also tested on even harder QAP instances from Palubeckis [97] and Drezner [44], which were designed with a known optimum but were specifically ill-conditioned in order to be difficult for many metaheuristic-based methods. Recently Carvalho & Rahmann proposed new instances, with unknown optimum, that turn out to be extremely difficult to solve [42]. For the former two classes of problems (called `palu`*XX* and `dre`*XX*) the solver was configured to reach the optimum (within a timeout of 5 minutes). For the latter (called `cr-bl`*XX* and `cr-ci`*XX*) it was configured to stop as soon as the BKS was improved (with a timeout of 6 hours). ParEOTS was able to improve the quality of several solutions. Table 10.2 summarizes the new solutions discovered by ParEOTS for these hard problems, using 128 cores.

	OPT	previous	ParEOTS		
		BKS	#OPT	new BKS	time
palu30	271092	272080	**10**	**271092**	0.1
palu40	837900	840308	**10**	**837900**	4.0
palu50	1840356	1846876	**10**	**1840356**	0:17
palu60	2967464	2978216	**10**	**2967464**	1:07
palu70	5815290	5831954	**10**	**5815290**	2:07
palu80	6597966	6618290	**10**	**6597966**	1:56
palu100	15008994	15047406	**1**	**15008994**	5:00
palu150	58352664	58468204	0	**58414888**	5:00
palu200	75405684	75543960	0	**75498892**	5:00
dre90	1838	1959	**9**	**1838**	2:47
dre110	2264	2479	**6**	**2264**	3:43
dre132	2744	3023	**1**	**2744**	4:54
cr-bl81	-	7536	-	**7532**	48:41
cr-bl100	-	9272	-	**9264**	41:33
cr-bl121	-	11412	-	**11400**	1:05:10
cr-bl144	-	13472	-	**13452**	5:32:03
cr-ci144	-	795009899	-	**794811636**	2:29:27

Table 10.2: new solutions found by ParEOTS on other hard problems (128 cores)

It becomes clear from these examples that cooperative parallel hybridization for different metaheuristics can attain very competitive results and, in some cases, sometimes achieves a clear improvement.

10.8 Conclusion

In this chapter we have tried to present a survey of parallel local search methods over the last 20 years. Although local search methods have been pioneered since the late 1950s, parallelism has only been investigated in the context of local search methods since the 1990s, when multiprocessors started to become more widely available, and this endeavor continued until the present with experiments on massively parallel supercomputers. Local search exhibits some natural opportunities for parallelism, which may be easily derived from the basic features of the search methods such as the selection of a new candidate solution within a neighborhood or the choice of an initial (random) starting solution. This observation prompted the adoption of some basic parallel schemes, such as single-walk and independent multi-walk methods, which can be effective on small-scale multiprocessor machines (e.g., with a few tens of cores). However, in order to achieve better performance on massively parallel machines, more complex schemes have to be devised, for instance cooperative multi-walks in which concurrent processes exchange information about their current search and communicate so as to guide the search towards promising areas of the search space. If information exchange and cooperation can be implemented efficiently and become effective enough to actually lead processes to parts of the search space where optimal or quasi-optimal solutions are, one may assert that cooperative strategies are instrumental in tapping the performance potential held in massively parallel computer architectures.

Encouraging results have already been achieved, e.g., super-linear speedups have been demonstrated on a few hard optimization problems, but more work is needed to develop general and efficient frameworks. Key issues to be investigated, especially in the context of the massively parallel machines with tens or hundreds of thousands of cores that are now available, are the flexibility and dynamicity of the system architecture, the scale and frequency of the communication between processes, and the nature of the information that should be exchanged.

Most, if not all, solvers that are mentioned in this text require non-trivial parameter tuning in order to attain their optimum performance. This task has been clearly identified and is the object of a significant and continued research effort, often resorting to different problem-solving techniques, such as machine learning.

Lastly, the hybrid nature of modern parallel multiprocessors poses several challenges concerning their effective use, as a significant portion of the available compute power stems from nonstandard architectures, such as GPUs or other accelerators. Making use of these multiple forms of parallelism is a high-stakes challenge, but one for which local search techniques could be a very good fit.

References

[1] Emile Aarts and Jan K Lenstra. *Local Search in Combinatorial Optimization*. John Wiley & Sons, Inc., New York, NY, USA, 1st edition, 1997.

[2] Renata Aiex, S. Binato, and Mauricio Resende. Parallel GRASP with Path-Relinking for Job Shop Scheduling. *Parallel Computing*, 29:393–430, 2003.

[3] Renata Aiex, Simone Martins, Celso Ribeiro, and Noemi De R. Rodriguez. Cooperative Multi-thread Parallel Tabu Search with an Application to Circuit Partitioning. *Lecture Notes in Computer Science Volume 1457*, 1457:310–331, 1998.

[4] Renata Aiex, Mauricio Resende, and Celso Ribeiro. Probability distribution of solution time in GRASP: An experimental investigation. *Journal of Heuristics*, 8(3):343–373, 2002.

[5] Renata Aiex, Mauricio Resende, and Celso Ribeiro. TTT plots: a Perl program to create time-to-target plots. *Optimization Letters*, 1:355–366, 2007.

[6] Enrique Alba. Special Issue on New Advances on Parallel Meta-Heuristics for Complex Problems. *Journal of Heuristics*, 10(3):239–380, 2004.

[7] Enrique Alba. *Parallel Metaheuristics: a New Class of Algorithms*. Wiley-Interscience, 2005.

[8] Gene M. Amdahl. Validity of the Single Processor Approach to Achieving Large Scale Computing Capabilities. *AFIPS Spring Joint Computer Conference, 1967. AFIPS '67 (Spring). Proceedings of the*, 30:483–485, 1967.
[9] Alejandro Arbelaez and Philippe Codognet. Massively Parallel Local Search for SAT. In *24th IEEE International Conference on Tools with Artificial Intelligence (ICTAI)*, pages 57–64, Athens, Nov 2012. IEEE Press.
[10] Alejandro Arbelaez and Philippe Codognet. From Sequential to Parallel Local Search for SAT. In *13th European Conference on Evolutionary Computation in Combinatorial Optimization (EvoCOP)*, LNCS, pages 157–168. Springer, 2013.
[11] Alejandro Arbelaez and Philippe Codognet. A GPU Implementation of Parallel Constraint-Based Local Search. In *22nd Euromicro International Conference on Parallel, Distributed and Network-Based Processing (PDP)*, volume 1, pages 648–655, Turin, Italy, 2014.
[12] Alejandro Arbelaez and Youssef Hamadi. Improving Parallel Local Search for SAT. In Carlos A. Coello Coello, editor, *5th International Conference on Learning and Intelligent Optimization (LION5)*, volume 6683 of *LNCS*, pages 46–60. Springer, 2011.
[13] M. G. Arenas, Pierre Collet, A. E. Eiben, Márk Jelasity, J. J. Merelo, Ben Paechter, Mike Preuß, and Marc Schoenauer. A Framework for Distributed Evolutionary Algorithms. In *Parallel Problem Solving from Nature - PPSN VII*, pages 665–675. Springer 2002.
[14] Mehmet E. Aydin. Metaheuristic Agent Teams for Job Shop Scheduling Problems. *Holonic and Multi-Agent Systems for Manufacturing*, 4659:185–194, 2007.
[15] László Babai. Monte-Carlo algorithms in graph isomorphism testing. Research Report D.M.S. No. 79-10, Université de Montréal, 1979.
[16] Per Bak. *How Nature Works: The Science of Self-organized Criticality*. Copernicus (Springer), 1st edition, 1996.
[17] Per Bak and Kim Sneppen. Punctuated equilibrium and criticality in a simple model of evolution. *Physical Review Letters*, 71(24):4083–4086, 1993.
[18] Per Bak, Chao Tang, and Kurt Wiesenfeld. Self-organized criticality: An explanation of the 1/f noise. *Physical Review Letters*, 59(4):381–384, 1987.
[19] Dariusz Barbucha. A Cooperative Population Learning Algorithm for Vehicle Routing Problem with Time Windows. *Neurocomputing*, 146:210–229, 2014.
[20] Fabio Luigi Bellifemine, Giovanni Caire, and Dominic Greenwood. *Developing Multi-Agent Systems with JADE*. Wiley, 2007.
[21] Ravi Kumar Bhati and Akhtar Rasool. Quadratic Assignment Problem and its Relevance to the Real World: A Survey. *International Journal of Computer Applications*, 96(9):42–47, 2014.
[22] Christian Blum and Andrea Roli. Metaheuristics in Combinatorial Optimization: Overview and Conceptual Comparison. *ACM Computing Surveys*, 35(3):268–308, 2003.

[23] Stefan Boettcher. Extremal Optimization. In Alexander K. Hartmann and Heiko Rieger, editors, *New Optimization Algorithms to Physics*, chapter 11, pages 227–251. Wiley-VCH Verlag, Berlin, 2004.
[24] Stefan Boettcher and Allon Percus. Nature's way of optimizing. *Artificial Intelligence*, 119(1–2):275–286, 2000.
[25] Stefan Boettcher and Allon Percus. Extremal Optimization: an Evolutionary Local-Search Algorithm. In *Computational Modeling and Problem Solving in the Networked World*, volume 21. Springer 2003.
[26] A. Bortfeldt, H. Gehring, and D. Mack. A Parallel Tabu Search Algorithm for Solving the Container Loading Problem. *Parallel Computing*, 29(5 SPEC.):641–662, 2003.
[27] Ilhem Boussaïd, Julien Lepagnot, and Patrick Siarry. A Survey on Optimization Metaheuristics. *Information Sciences*, 237(February):82–117, 2013.
[28] Rainer E. Burkard, S. Karisch, and F. Rendl. QAPLIB - a Quadratic Assignment Problem Library. *European Journal of Operational Research*, 55(1):115–119, 1991.
[29] J. M. Cadenas, M. C. Garrido, and E. Muñoz. Using Machine Learning in a Cooperative Hybrid Parallel Strategy of Metaheuristics. *Information Sciences*, 179(19):3255–3267, 2009.
[30] S. Cahon, N. Melab, and E. G. Talbi. ParadisEO: A Framework for the Reusable Design of Parallel and Distributed Metaheuristics. *Journal of Heuristics*, 10(3):357–380, 2004.
[31] Yves Caniou, Philippe Codognet, Daniel Diaz, and Salvador Abreu. Experiments in parallel constraint-based local search. In *EvoCOP'11, 11th European Conference on Evolutionary Computation in Combinatorial Optimisation*, volume 6622 of *Lecture Notes in Computer Science*, Torino, Italy, 2011. Springer Verlag.
[32] Yves Caniou, Philippe Codognet, Florian Richoux, Daniel Diaz, and Salvador Abreu. Large-scale Parallelism for Constraint-Based Local Search: the Costas Array Case Study. *Constraints*, 20(1):30–56, 2015.
[33] Yves Caniou, Daniel Diaz, Florian Richoux, Philippe Codognet, and Salvador Abreu. Performance Analysis of Parallel Constraint-Based Local Search. In *Symposium on Principles and Practice of Parallel Programming (PPoPP)*, PPoPP '12, New York, NY, USA, 2012. ACM. poster paper.
[34] Philippe Codognet and Daniel Diaz. Yet Another Local Search Method for Constraint Solving. In Kathleen Steinhöfel, editor, *Stochastic Algorithms: Foundations and Applications*, pages 342–344. Springer, 2001.
[35] Clayton Warren Commander. A survey of the quadratic assignment problem, with applications. *Morehead Electronic Journal of Applicable Mathematics*, 4:MATH–2005–01, 2005.
[36] Jean-Francois Cordeau and Mirko Maischberger. A Parallel Iterated Tabu Search Heuristic for Vehicle Routing Problems. *Computers and Operations Research*, 39(9):2033–2050, 2012.

[37] Teodor Crainic, Michel Gendreau, Pierre Hansen, and Nenad Mladenovic. Cooperative Parallel Variable Neighborhood Search for the p-Median. *Journal of Heuristics*, 10(3):293–314, 2004.

[38] Teodor Crainic and Michel Toulouse. Special Issue on Parallel Meta-Heuristics. *Journal of Heuristics*, 8(3):247–388, 2002.

[39] G. A. Croes. A method for solving traveling-salesman problems. *Operations Research*, 6(6):791–812, 1958.

[40] Leonardo Dagum and Ramesh Menon. OpenMP: an industry standard API for shared-memory programming. *IEEE computational science and engineering*, 5(1):46–55, 1998.

[41] H.A. David and H.N. Nagaraja. *Order Statistics*. Wiley series in probability and mathematical statistics. John Wiley, 2003.

[42] Sérgio A de Carvalho Jr. and Sven Rahmann. Microarray layout as a quadratic assignment problem. In *German Conference on Bioinformatics (GCB)*, volume 83, pages 11–20, Tübingen, Germany, 2006.

[43] Daniel Diaz, Florian Richoux, Philippe Codognet, Yves Caniou, and Salvador Abreu. Constraint-based Local Search for the Costas Array Problem. In *LION 6, Learning and Intelligent OptimizatioN Conference*, Paris, France, 2012. Springer LNCS.

[44] Zvi Drezner. The Extended Concentric Tabu for the Quadratic Assignment Problem. *European Journal of Operational Research*, 160(2):416–422, 2005.

[45] Zvi Drezner. Extensive experiments with hybrid genetic algorithms for the solution of the quadratic assignment problem. *Computers & Operations Research*, 35(3):717–736, 2008.

[46] Huub M. M. Eikelder, Bas J. M. Aarts, Marco G. A. Verhoeven, and Emile H. L. Aarts. Sequential and Parallel Local Search Algorithms for Job Shop Scheduling. In *Meta-Heuristics: Advances and Trends in Local Search Paradigms for Optimization*, pages 359–371. Springer, Boston, MA, 1999.

[47] Merrill M. Flood. The traveling-salesman problem. *Operations Research*, 4(1):61–75, 1956.

[48] Edgar Gabriel and al. Open MPI: Goals, concept, and design of a next generation MPI implementation. In *Proceedings, 11th European PVM/MPI Users' Group Meeting*, pages 97–104, Budapest, Hungary, 2004.

[49] D. Gale and L. Shapley. College Admissions and the Stability of Marriage. *American Mathematical Monthly*, 69(1):9–15, 1962.

[50] Bruno-Laurent Garcia, Jean-Yves Potvin, and Jean-Marc Rousseau. A Parallel Implementation of the Tabu Search Heuristic for Vehicle Routing Problems with Time Window Constraints. *Computers & Operations Research*, 21(9):1025–1033, 1994.

[51] F. García-López, B. Melián-Batista, J. A. Moreno-Pérez, and J. M. Moreno-Vega. The Parallel Variable Neighborhood Search for the p -Median Problem. *Journal of Heuristics*, 8(3):375–388, 2002.

[52] Frédéric Gardi and Karim Nouioua. Local search for mixed-integer nonlinear optimization: A methodology and an application. In *Evolutionary Computa-*

tion in Combinatorial Optimization - 11th European Conference, EvoCOP 2011, Torino, Italy, April 27-29, 2011. Proceedings, pages 167–178, 2011.
[53] M. Gendreau, F. Guertin, J.-Y. Potvin, and E. Taillard. Parallel Tabu Search for Real-Time Vehicle Routing and Dispatching. *Transportation Science*, 33(4):381–390, 1999.
[54] Ian Gent and Toby Walsh. CSPLib: A Benchmark Library for Constraints. CP 1999 LNCS 1713 Springer, 1999.
[55] Fred Glover. Tabu Search–Part I. *ORSA Journal on Computing*, 1(3):190–206, 1989.
[56] Fred Glover. Tabu Search–Part II. *ORSA Journal on Computing*, 2(1):4–32, 1990.
[57] Fred Glover and Manuel Laguna. *Tabu Search*. Kluwer Academic Publishers, Jul 1997.
[58] Carla Gomes and Bart Selman. Algorithm portfolios. *Artificial Intelligence*, 126(1-2):43–62, 2001.
[59] Teofilo Gonzalez, editor. *Handbook of Approximation Algorithms and Metaheuristics*. Chapman and Hall / CRC, 2007.
[60] F. Guerriero and M. Mancini. A Cooperative Parallel Rollout Algorithm for the Sequential Ordering Problem. *Parallel Computing*, 29:663–677, 2003.
[61] Magnus Halldorsson, Robert Irving, Kazuo Iwama, David Manlove, Shuichi Miyazaki, Yasufumi Morita, and Sandy Scott. Approximability Results for Stable Marriage Problems with Ties. *Theoretical Computer Science*, 306(1-5):431–447, 2003.
[62] Magnus Halldorsson, Kazuo Iwama, Shuichi Miyazaki, and Hiroki Yanagisawa. Improved Approximation of the Stable Marriage Problem. *ACM Transactions on Algorithms*, 3(3):266–277, 2007.
[63] Tad Hogg and Colin P. Williams. Solving the Really Hard Problems with Cooperative Search. In *AAAI Conference on Artificial Intelligence (AAAI-93)*, pages 231–236, 1993.
[64] Holger Hoos and Thomas Stützle. Evaluating Las Vegas algorithms: Pitfalls and remedies. In *Proceedings of the Fourteenth Conference on Uncertainty in Artificial Intelligence, UAI'98*, pages 238–245. Morgan Kaufmann, 1998.
[65] Holger Hoos and Thomas Stützle. Towards a characterisation of the behaviour of stochastic local search algorithms for SAT. *Artificial Intelligence*, 112(1-2):213–232, 1999.
[66] Holger Hoos and Thomas Stützle. *Stochastic Local Search: Foundations and Applications*. Morgan Kaufmann / Elsevier, 2004.
[67] J. Humeau, A. Liefooghe, E. G. Talbi, and S. Verel. ParadisEO-MO: From Fitness Landscape Analysis to Efficient Local Search Algorithms. Technical report, INRIA, 2013.
[68] T. Ibaraki, K. Nonobe, and M. Yagiura, editors. *Metaheuristics: Progress as Real Problem Solvers*. Springer Verlag, 2005.
[69] Robert Irving and David Manlove. Approximation Algorithms for Hard Variants of the Stable Marriage and Hospitals/Residents Problems. *Journal of Combinatorial Optimization*, 16(3):279–292, 2008.

[70] Kazuo Iwama, David Manlove, Shuichi Miyazaki, and Yasufumi Morita. Stable Marriage with Incomplete Lists and Ties. In *Proceedings of ICALP '99: the 26th International Colloquium on Automata, Languages and Programming*, number ii, pages 443–452. Springer-Verlag, 1999.

[71] Zoltán Király. Approximation of Maximum Stable Marriage. Technical report, Egervary Research Group, Budapest, Hungary, 2011.

[72] Zoltán Király. Linear Time Local Approximation Algorithm for Maximum Stable Marriage. *Algorithms*, 6(3):471—-484, aug 2013.

[73] S. Kirkpatrick, C.D. Gelatt Jr, and M.P. Vecchi. Optimization by Simulated Annealing. *Science*, 220(4598):671–680, 1983.

[74] Tjalling C. Koopmans and Martin Beckmann. Assignment Problems and the Location of Economic Activities. *Econometrica*, 25(1):53–76, 1957.

[75] S Lin. Computer solutions of the traveling salesman problem. *Bell System Technical Journal*, 44(10):2245–2269, 1965.

[76] Helena R Lourenço, Olivier C Martin, and Thomas Stützle. Iterated Local Search. In *Handbook of Metaheuristics*, pages 320–353. Kluwer Academic Publishers, Boston, 2003.

[77] Thé Van Luong, Nouredine Melab, and El-Ghazali Talbi. Local Search Algorithms on Graphics Processing Unit. A Case Study: The Permutation Perceptron Problem. In *Evolutionary Computation in Combinatorial Optimization*, pages 264–275. LNCS 6022, Springer Verlag, 2010.

[78] Thé Van Luong, Nouredine Melab, and El-Ghazali Talbi. GPU Computing for Parallel Local Search Metaheuristics. *IEEE Transactions on Computers*, 62(1):173–185, 2013.

[79] Rui Machado, Salvador Abreu, and Daniel Diaz. Parallel Local Search: Experiments with a PGAS-based programming model. In *12th International Colloquium on Implementation of Constraint and Logic Programming Systems*, pages 1–17, Budapest, Hungary, 2012.

[80] Rui Machado, Salvador Abreu, and Daniel Diaz. Parallel Performance of Declarative Programming Using a PGAS Model. In Kostis Sagonas and Gopal Gupta, editors, *Practical Aspects of Declarative Languages, PADL'2013*, Lecture Notes in Computer Science. Springer Berlin / Heidelberg, 2013.

[81] David Manlove, Robert Irving, Kazuo Iwama, Shuichi Miyazaki, and Yasufumi Morita. Hard Variants of Stable Marriage. *Theoretical Computer Science*, 276(1-2):261–279, Apr 2002.

[82] Simon Martin, Djamila Ouelhadj, Patrick Beullens, Ender Ozcan, Angel A. Juan, and Edmund K. Burke. A Multi-Agent Based Cooperative Approach to Scheduling and Routing. *European Journal of Operational Research*, 254(1):169–178, 2016.

[83] Eric McDermid. A 3/2-Approximation Algorithm for General Stable Marriage. In *International Colloquium on Automata, Languages and Programming, ICALP'2009*, pages 689–700, Rhodes, Greece, 2009.

[84] David Meignan, Abderrafiaa Koukam, and Jean Charles Créput. Coalition-based metaheuristic: A self-adaptive metaheuristic using reinforcement learning and mimetism. *Journal of Heuristics*, 16(6):859–879, 2010.

[85] Nouredine Melab, Thé Van Luong, Karima Boufaras, and El-Ghazali Talbi. ParadisEO-MO-GPU: A Framework for Parallel GPU-Based Local Search Metaheuristics. In *15th annual conference on Genetic and evolutionary computation conference GECCO '13*, pages 1189–1196, Amsterdam, The Netherlands, 2013.
[86] Laurent Michel, Andrew See, and Pascal Van Hentenryck. Distributed constraint-based local search. In Frédéric Benhamou, editor, *CP'06, 12th Int. Conf. on Principles and Practice of Constraint Programming*, Lecture Notes in Computer Science, pages 344–358. Springer Verlag, 2006.
[87] Michela Milano and Andrea Roli. MAGMA: A Multiagent Architecture for Metaheuristics. *IEEE Transactions on Systems, Man, and Cybernetics, Part B: Cybernetics*, 34(2):925–941, 2004.
[88] Nenad Mladenovic and Pierre Hansen. Variable Neighborhood Search. *Computers & Operations Research*, 24(11):1097–1100, 1997.
[89] Hiroyuki Mori and Yoshihiro Ogita. A Parallel Tabu Search Based Method for Reconfigurations of Distribution Systems. In *2000 Power Engineering Society Summer Meeting (Cat. No.00CH37134)*, volume 1, pages 73–78. IEEE, 2000.
[90] Danny Munera. *Solving Hard Combinatorial Optimization Problems using Cooperative Parallel Metaheuristics*. PhD Thesis, University Paris 1 Pantheon-Sorbonne, 2016.
[91] Danny Munera, Daniel Diaz, and Salvador Abreu. Hybridization as Cooperative Parallelism for the Quadratic Assignment Problem. In *10th International Workshop, HM 2016*, volume 9668 of *Lecture Notes in Computer Science*, pages 47–61, Plymouth, UK, 2016. Springer International Publishing.
[92] Danny Munera, Daniel Diaz, and Salvador Abreu. Solving the Quadratic Assignment Problem with Cooperative Parallel Extremal Optimization. In *The 16th European Conference on Evolutionary Computation in Combinatorial Optimisation*, Porto, 2016.
[93] Danny Munera, Daniel Diaz, Salvador Abreu, and Philippe Codognet. A Parametric Framework for Cooperative Parallel Local Search. In Christian Blum and Gabriela Ochoa, editors, *European Conference on Evolutionary Computation in Combinatorial Optimisation (EvoCOP)*, volume 8600 of *Lecture Notes in Computer Science*, pages 13–24, Granada, Spain, 2014. Springer.
[94] Danny Munera, Daniel Diaz, Salvador Abreu, Francesca Rossi, Vijay Saraswat, and Philippe Codognet. A Local Search Algorithm for SMTI and its extension to HRT Problems. In *3rd International Workshop on Matching Under Preferences*, Glasgow, UK, 2015.
[95] Danny Munera, Daniel Diaz, Salvador Abreu, Francesca Rossi, Vijay Saraswat, and Philippe Codognet. Solving Hard Stable Matching Problems via Local Search and Cooperative Parallelization. In *AAAI*, Austin, TX, USA, 2015.
[96] Djamila Ouelhadj and Sanja Petrovic. A Cooperative Hyper-heuristic Search Framework. *Journal of Heuristics*, 16(6):835–857, 2010.

[97] Gintaras Palubeckis. An Algorithm for Construction of Test Cases for the Quadratic Assignment Problem. *Informatica, Lith. Acad. Sci.*, 11(3):281–296, 2000.

[98] Katarzyna Paluch. Faster and Simpler Approximation of Stable Matchings. *Algorithms*, 7(2):176–187, Nov 2014.

[99] Panos M. Pardalos, Leonidas S. Pitsoulis, Thelma D. Mavridou, and Mauricio G. C. Resende. Parallel search for combinatorial optimization: Genetic algorithms, simulated annealing, tabu search and GRASP. In *Parallel Algorithms for Irregularly Structured Problems (IRREGULAR)*, pages 317–331, 1995.

[100] J. Antonio Parejo, Antonio Ruiz-Cortés, Sebastián Lozano, and Pablo Fernandez. Metaheuristic Optimization Frameworks: a Survey and Benchmarking. *Soft Computing*, 16(3):527–561, 2012.

[101] International Exascale Software Project. Exascale roadmap 1.0. Technical report, 2009. http://www.exascale.org/iesp/IESP:Documents.

[102] César Rego and Catherine Roucairol. A Parallel Tabu Search Algorithm Using Ejection Chains for the Vehicle Routing Problem. In *Meta-Heuristics*, pages 661–675. Springer US, Boston, MA, 1996.

[103] Gerhard Reinelt. TSPLIB–A traveling salesman problem library. *ORSA Journal on Computing*, 3(4):376–384, 1991.

[104] Celso Ribeiro and Isabel Rosseti. Efficient Parallel Cooperative Implementations of GRASP Heuristics. *Parallel Computing*, 33(1):21–35, 2007.

[105] Celso Ribeiro, Isabel Rosseti, and Reinaldo Vallejos. Exploiting run time distributions to compare sequential and parallel stochastic local search algorithms. *Journal of Global Optimization*, 54:405–429, 2012.

[106] Vijay Saraswat, Bard Bloom, Igor Peshansky, Olivier Tardieu, and David Grove. X10 Language Specification - Version 2.3. Technical report, IBM Research, 2012.

[107] Kenneth Sörensen and Fred Glover. Metaheuristics. In *Encyclopedia of Operations Research and Management Science*, pages 960–970. Springer, Boston, MA, 2013.

[108] Kenneth Sörensen, Marc Sevaux, and Fred Glover. A History of Metaheuristics. In Rafael Marti, Panos Pardalos, and Mauricio Resende, editors, *Handbook of Heuristics*. Springer, Boston, MA, 2016.

[109] Éric Taillard. Robust Taboo Search for the Quadratic Assignment Problem. *Parallel Computing*, 17(4-5):443–455, 1991.

[110] E. G. Talbi, S. Cahon, and N. Melab. Designing Cellular Networks Using a Parallel Hybrid Metaheuristic on the Computational Grid. *Computer Communications*, 30(4):698–713, 2007.

[111] El-Ghazali Talbi. *Metaheuristics: From Design to Implementation*. Wiley, 2009.

[112] El-Ghazali Talbi and Vincent Bachelet. COSEARCH: A parallel cooperative metaheuristic. *Journal of Mathematical Modelling and Algorithms*, 5(1):5–22, 2006.

[113] Sarosh Talukdar, Lars Baerentzen, Andrew Gove, and Pedro De Souza. Asynchronous Teams: Cooperation Schemes for Autonomous Agents. *Journal of Heuristics*, 4:295–321, 1998.

[114] Michel Toulouse, Teodor Crainic, and Michel Gendreau. Communication Issues in Designing Cooperative Multi-Thread Parallel Searches. In I.H. Osman and J.P. Kelly, editors, *Meta-Heuristics: Theory & Applications*, pages 501–522. Kluwer Academic Publishers, Norwell, MA., 1995.

[115] Charlotte Truchet, Alejandro Arbelaez, Florian Richoux, and Philippe Codognet. Estimating parallel runtimes for randomized algorithms in constraint solving. *J. Heuristics*, 22(4):613–648, 2016.

[116] Charlotte Truchet, Florian Richoux, and Philippe Codognet. Prediction of Parallel Speed-ups for Las Vegas Algorithms. In Jack Dongarra and Yves Robert, editors, *Proceedings of ICPP-2013, 42nd International Conference on Parallel Processing*. IEEE Press, October 2013.

[117] Pascal Van Hentenryck and Laurent Michel. *Constraint-Based Local Search*. The MIT Press, Aug 2005.

[118] Marcus Verhoeven. *Parallel Local Search*. PhD thesis, University of Eindhoven, Eindhoven, Netherlands, 1996.

[119] Marcus Verhoeven and Emile Aarts. Parallel Local Search. *Journal of Heuristics*, 1(1):43–65, 1995.

[120] Stefan Voß. Meta-heuristics: The State of the Art. In Alexander Nareyek, editor, *Local Search for Planning and Scheduling*, pages 1–23. Springer Berlin Heidelberg, 2001.

[121] M. Yazdani, M. Amiri, and M. Zandieh. Flexible Job-Shop Scheduling with Parallel Variable Neighborhood Search Algorithm. *Expert Systems with Applications*, 37(1):678–687, 2010.

Chapter 11
Parallel A* for State-Space Search

Alex Fukunaga, Adi Botea, Yuu Jinnai, and Akihiro Kishimoto

Abstract A* is a best-first search algorithm for finding optimal-cost paths in graphs. A* benefits significantly from parallelism because in many applications, A* is limited by memory usage, so distributed memory implementations of A* that use all of the aggregate memory on the cluster enable us to solve problems that can not be solved by serial, single-machine implementations. We survey approaches to parallel A*, focusing on decentralized approaches to A* which partition the state space among processors. We also survey approaches to parallel, limited-memory variants of A* such as parallel IDA*.

11.1 Introduction

This chapter surveys parallel A* for state-space search. State-space search is a very general approach to solving a broad class of problems, such as robot planning problems, domain-independent AI planning, solving puzzles, and multiple sequence alignment problems in computational biology.

Solving a problem with state-space search involves defining the *state space* as a graph where nodes represent states and edges represent actions (transitions) between states. The task is to find a sequence of actions which transforms a given initial state into a state that satisfies some goal conditions. In other words, finding a solution

Alex Fukunaga
The University of Tokyo, Tokyo, Japan, e-mail: fukunaga@idea.c.u-tokyo.ac.jp

Adi Botea
IBM Research, Dublin, Ireland, e-mail: ADIBOTEA@ie.ibm.com

Yuu Jinnai
The University of Tokyo, Tokyo, Japan, e-mail: ddyuudd@gmail.com

Akihiro Kishimoto
IBM Research, Dublin, Ireland, e-mail: AKIHIROK@ie.ibm.com

Y. Hamadi und L. Sais (eds.), *Handbook of Parallel Constraint Reasoning*,
https://doi.org/10.1007/978-3-319-63516-3_11

boils down to finding a path from the initial state to a goal state. The quality of a solution is typically measured in terms of the total cost of the path. The smaller the cost, the better the solution. Optimal search aims at finding a minimal-cost solution. We formally define concepts such as state spaces, state-space search problems, and (optimal) solutions after the next example.

15	2	8	7
1	6	9	11
13	12		4
10	5	3	14

1	2	3	4
5	6	7	8
9	10	11	12
13	14	15	

Fig. 11.1: An instance of the sliding-tile puzzle with 15 tiles. Left: initial state. Right: goal state

Consider the simple sliding-tile problem, where an $n \times n$ board is occupied by $n^2 - 1$ tiles and a "blank" space. As there are $n^2 - 1$ tiles, the problem is also called the $n^2 - 1$ puzzle. Figure 11.1 illustrates a 15 puzzle instance. Given an initial state where the tiles are out of order, the task is to find a sequence of actions which results in the goal state where the tiles are in order. The actual, physical problem is played by moving one of the tiles currently adjacent to the blank space into the blank space. As this is equivalent to moving the blank space, it is customary to treat the problem as having four actions (*up, down, left*, and *right*), which move the blank space. The most common variant of this problem in the AI literature requires finding the solution with the minimal number of moves. The problem of finding optimal solution for the sliding-tile puzzle has been used as a standard benchmark problem in the AI search algorithm literature because of the simplicity of the problem description and the difficulty of finding an optimal solution. Although puzzles of small sizes, such as 3×3 and 4×4, can be solved fairly easily, a 5×5 version of this puzzle has $(5 \times 5)!/2 \approx 7.76 \times 10^{24}$ possible configurations, providing a challenging benchmark for search algorithms.

This survey is structured as follows. First, in Section 11.2, we give a formal definition of state-space search, and review the A* algorithm [23], which is the standard, baseline approach for optimally solving state-space search problems. Next, in Section 11.3, we give an overview of the parallel-search related overheads which pose the fundamental challenges in parallelizing A*, and review the two basic approaches to parallelizing A*: the centralized and decentralized approaches. Then, in Section 11.4, we describe hash-based work distribution, the class of algorithms which is the current, state-of-the-art approach for parallelizing A* both on single, shared-memory multi-core machines as well as on large-scale clusters. Hash-based work distribution handles both load balancing and efficient detection of duplicate states. Section 11.5 reviews structure-based search space partitioning, which is another approach to decentralized search based on the concept of a duplicate detection scope which allows minimization of communications among processors. Section 11.6 surveys the various approaches to implementing hashing strategies for hash-based work distribution, including recent approaches which integrate key ideas from structure-based search space partitioning. Parallel portfolios (meta-solvers that combine multiple problem solvers) which include A*-based solvers as a component are reviewed in Section 11.7. A fundamental limitation of A* is that it can exhaust memory on hard problems, resulting in failure to solve the problem. Limited-memory variants of A* such as IDA* [41] overcome this limitation (at the cost of some search efficiency). We survey parallel, limited-memory A* variants in Section 11.8. With the emergence of cloud environments offering virtually unlimited available resources (at a cost), another approach to addressing the A* memory usage problem is to simply use more machines. Section 11.9 describes resource allocation strategies for parallel A* that are efficient with respect to cost and runtime. Recently, graphics processing units (GPUs) with thousands of cores have become widely used. Since GPUs have afundamentally different architecture compared to traditional CPUs, this provides new challenges for parallelizing A*. Section 11.10 describes recent work on parallel A* variants for GPUs. Finally, Section 11.11 reviews other approaches to parallel state-space search.

11.2 Preliminaries: Review of A*

This section provides preliminary and background material for the rest of this chapter. We first formally define state-space search, and then present the A* search algorithm.

The formal definitions presented below are adapted from Edelkamp and Schroedl's textbook on heuristic search [13].

Definition 1 (State Space Problem). A *S*tate Space Problem $P = (S, A, s_0, T)$ is defined by a set of states S, an initial state $s_0 \in S$, a set of goal states $T \subset S$, and a finite set of actions $A = a_1, \ldots, a_m$ where each $a_i : S \to S$ transforms a state into another state.

For the sliding-tile puzzle, the state space problem formulation consists of the states S, where each state corresponds to a unique configuration of the tiles, s_0 is the given initial configuration, T is a singleton set whose sole member is the configuration with the tiles in the correct order, and A corresponds to the transitions between tile configurations.

Definition 2 (State Space Problem Graph). A problem graph $G = (V, E, s_0, T)$ for the state space problem $P = (S, A, s_0, T)$ is defined by $V = S$ as the set of nodes, $s_0 \in S$ as the initial node, T as the set of goal nodes, and $E \subset V \times V$ as the set of edges that connect nodes to nodes with $(u, v) \in E$ if and only if there exists an $a \in A$ with $a(u) = v$.

Definition 3 (Solution). A solution $\pi = (a_1,...,a_k)$ is an ordered sequence of actions $a_i \in A, i \in 1,...,k$ that transforms the initial state s_0 into one of the goal states $t \in T$; that is, there exists a sequence of states $u_i \in S, i \in 0,...,k$, with $u_0 = s_0, u_k = t$, and u_i is the outcome of applying a_i to u_{i-1}, $i \in 1,...,k$.

In some problems, such as the sliding-tile puzzle, all actions have the same cost. In other problems, however, different actions can have different costs. Take pathfinding on a gridmap for example. Pathfinding refers to computing a path for a mobile agent, such as a robot or a character in a game, from an initial location to a target location. Gridmaps are a popular approach to discretizing the environment of the agent (e.g., a game map) into a search graph. A gridmap is a two-dimensional array where a cell is either traversable or blocked by an obstacle. The mobile agent occupies exactly one traversable cell at a time. In creating the problem graph, all traversable cells become states in the state space (equivalently, nodes in the problem graph). Two adjacent traversable states are connected with an edge. On a so-called 8-connected gridmap, or octile gridmap, adjacency relations are defined in 8 directions, four straight and four diagonal. Straight edges have a cost of 1, and diagonal edges have a cost of $\sqrt{2}$. State spaces where actions have different costs are called weighted state spaces.

Definition 4. A weighted state space problem $P = (S, A, s_0, T, w)$, where w is a cost function $w : A \to \mathbb{R}$. The cost of a path consisting of actions $a_1,...,a_n$ is defined as $\sum_{i=1}^{n} w(a_i)$. For a weighted state space problem, there is a corresponding *weighted problem graph* $G = (V, E, s_0, T, w)$, where w is extended to $E \to \mathbb{R}$ in the straightforward way. The graph is uniformly weighted if $w(u, v)$ is constant for all $(u, v) \in E$. The weight or cost of a path $\pi = (v_0,...,v_k)$ is defined as $w(\pi) = \sum_{i=1}^{k} w(v_{i-1}, v_i)$.

Definition 5. A solution from s_0 to a given goal state v is *optimal* if its weight is minimal among all paths between s_0 and v.

A state space is *undirected* if for every action from a state u to a state v, there exists an action from state v to state u. Otherwise, the state space is *directed*. For example, the sliding tile puzzle has an undirected state space, as every action can be reversed. On the other hand, planning an itinerary on a road map with one-way roads is a directed state space problem.

In some domains, the problem graph is sufficiently small to fit into the memory of the computer. In such cases, the search graph can be defined *explicitly*, enumerating all nodes and edges. Pathfinding on gridmaps is a typical example of a problem where the search graph can be defined explicitly. In many other problems, the search graph is very large, much larger than can fit into the memory of a modern computer. Examples include puzzles such as the Rubik's cube and the sliding tile puzzle, as well as many benchmark domains in AI domain-independent planning. In such cases, the search graph is defined *implicitly*. Defining a search graph implicitly requires three key ingredients: a specification of the initial state, a method for recognizing goal nodes, and a method for expanding any node $v \in V$. Expanding a node v refers to generating all nodes u such that (v, u) is an edge in the problem graph.

Definition 6. In an *implicit state space graph*, we have an initial node $s_0 \in V$, a set of goal nodes determined by a predicate $Goal : V \rightarrow \mathbb{B} = \{\text{false}, \text{true}\}$, and a node expansion procedure $Expand : V \rightarrow 2^V$.

Defining a graph implicitly allows us to generate portions of the search graph on demand, as a given search algorithm needs to explore new parts of the search graph.

11.2.1 The A Algorithm*

Algorithm 11.1: A*

```
1  Initialize OPEN to {s0}
2  while OPEN ≠ ∅ do
3      Get and remove from OPEN a node n with a smallest f(n)
4      Add n to CLOSED
5      if n is a goal node then
6          Return solution path from s0 to n
7      for every successor n' of n do
8          g1 = g(n) + c(n,n')
9          if n' ∈ CLOSED then
10             if g1 < g(n') then
11                 Remove n' from CLOSED and add it to OPEN
12             else
13                 Continue
14         else
15             if n' ∉ OPEN then
16                 Add n' to OPEN
17             else if g1 ≥ g(n') then
18                 Continue
19         Set g(n') = g1
20         Set f(n') = g(n') + h(n')
21         Set parent(n') = n
22 Return failure (no path exists)
```

Most of the parallel state-space search algorithms presented in this chapter are based on the serial algorithm A* [23]. A* is a best-first search algorithm whose pseudocode is illustrated in Algorithm 11.1. A* keeps two sets of nodes, called the OPEN list and the CLOSED list. The CLOSED list is the set of expanded nodes. Recall that expanding a node refers to generating its successors. The OPEN list contains the nodes that have been generated and are waiting to be expanded. At each iteration of the main while loop shown in the pseudocode, A* selects for expansion a node from the OPEN list, with the smallest f-value. The f-value of a node n is defined as $f(n) = g(n) + h(n)$. The $g(n)$ value is the cost of the best known path from the root node s_0 to the current node n. The $h(n)$ value, called the heuristic evaluation of n, is an estimation of the cost from n to a closest goal node. As such, $f(n)$ estimates the cost of a shortest solution passing through n.

A heuristic function h is *admissible* if $h(n) \leq C^*(n)$, where $C^*(n)$ is the cost of the minimal path from n to some goal, i.e., h is a lower bound on C^*. A heuristic function is *consistent* (or *monotonic*) if $h(n) \leq c(n,n') + h(n')$ for all nodes n and n' such that n' is a successor of n; and $h(t) = 0$ for all goal nodes t. A consistent heuristic is also admissible. With a consistent heuristic is used, nodes are never reopened (i.e., moved from CLOSED to OPEN as shown at line 11 of Algorithm 11.1). In other words, every node is expanded at most once. This allows us to simplify the algorithm when consistent heuristics are used, replacing lines 10–13 with a "Continue" statement.

An algorithm is *complete* if it terminates and returns a solution whenever a solution exists. An algorithm is *admissible* if it always returns an optimal solution whenever a solution exists.

Theorem 1. *A* is complete on both finite and infinite graphs [53].*

Theorem 2. *If h is an admissible function, then A* using h is admissible [23].*

Besides producing optimal solutions, another powerful feature of A* is that it is an *efficient* algorithm in terms of the number of node expansions performed. For simplicity, assume that a consistent heuristic is used, to ensure that there are no re-expansions. A* expands all nodes n with $f(n) < C^*$, where C^* is the optimal solution cost. It also expands some of the nodes n with $f(n) = C^*$, and no node with $f(n) > C^*$. A* is efficient because any other admissible algorithm using the same knowledge (e.g., the same heuristic h) must expand all nodes n with $f(n) < C^*$. The reason is that, according to the knowledge available, a node n with $f(n) < C^*$ might belong to a solution with a smaller cost than C^*. Unless extra information is available (e.g., pruning based on symmetries in the state space) the node n has to be expanded to explore whether a better solution can be found.

Besides being a powerful property of serial A*, the efficiency of A* in terms of node expansions has a special significance to parallel best-first search. It allows us to evaluate the efficiency of a parallel search algorithm, such as HDA* [37], even in very difficult instances where serial A* fails and therefore a direct comparison of the node expansions between HDA* and A* is not possible. The idea is to measure the number of expanded nodes n with $f(n) < C^*$ as a fraction of all expanded nodes. If the fraction is close to 1, then the instance at hand is solved quite efficiently [36].

11.3 Parallel Best-First Search Algorithms

Parallelization of A* heuristic search is important due to two reasons. First, effective parallelization is necessary in order to obtain good speedup on multi-core processors. However, in the case of parallelization on a cluster consisting of many machines, parallelization offers another benefit which is at least as important as speedup, which is increased aggregate memory. A* memory usage continuously increases during the run, as it must keep all expanded nodes in memory in order to guarantee the soundness (optimality of solution) and completeness of the algorithm. Running parallel A* on a cluster of machines makes the entire aggregate memory of the cluster available to A*. This allows parallel A* to solve problem instances that would not be solvable at all on a single machine (using the same heuristic function). This offers a fundamental benefit to parallelization of A*, and perhaps makes parallelization of A* an even more pressing concern than for other search algorithms.

In this section, we first describe the major technical challenges that must be addressed in parallel A*, and then describe the two basic approaches to parallelization of A*: centralized and decentralized parallelization.

11.3.1 Parallel Overheads

Efficient implementation of parallel search algorithms is challenging due to several types of overhead. *Search overhead (SO)* occurs when a parallel implementation of a search algorithm expands (or generates) more states than a serial implementation. The main cause of search overhead is partitioning of the search space among processors, which has the side effect that access to non-local information is restricted. For example, sequential A* can terminate immediately after a solution is found, because it is guaranteed to be optimal. In contrast, when a parallel A* algorithm finds a (first) solution at some processor, it is not necessarily a globally optimal solution. A better solution which uses nodes being processed in some other processor might exist.

Synchronization overhead is the idle time wasted when some processors have to wait for the others to reach synchronization points. For example, in a shared-memory environment, the idle time can be caused by mutual exclusion locks on shared data. Finally, *communication overhead (CO)* refers to the cost of inter-process information exchange. In a distributed-memory environment, this includes the cost of sending a message from one processor to another over a network. Even in a shared-memory environment, there are overheads associated with moving work from one work queue to another.

The key to achieving good speedup in parallel search is minimizing such overheads. This is often a difficult task, in part because the overheads are interdependent. For example, reducing search overhead usually increases synchronization and communication overhead.

Figure 11.2 presents a visual classification of these approaches, which summarizes the survey of approaches in the next several sections.

11.3.2 Centralized Parallel A*

Algorithms such as breadth-first or best-first search (including A*) use an open list which stores the set of states that have been generated but not yet expanded. In

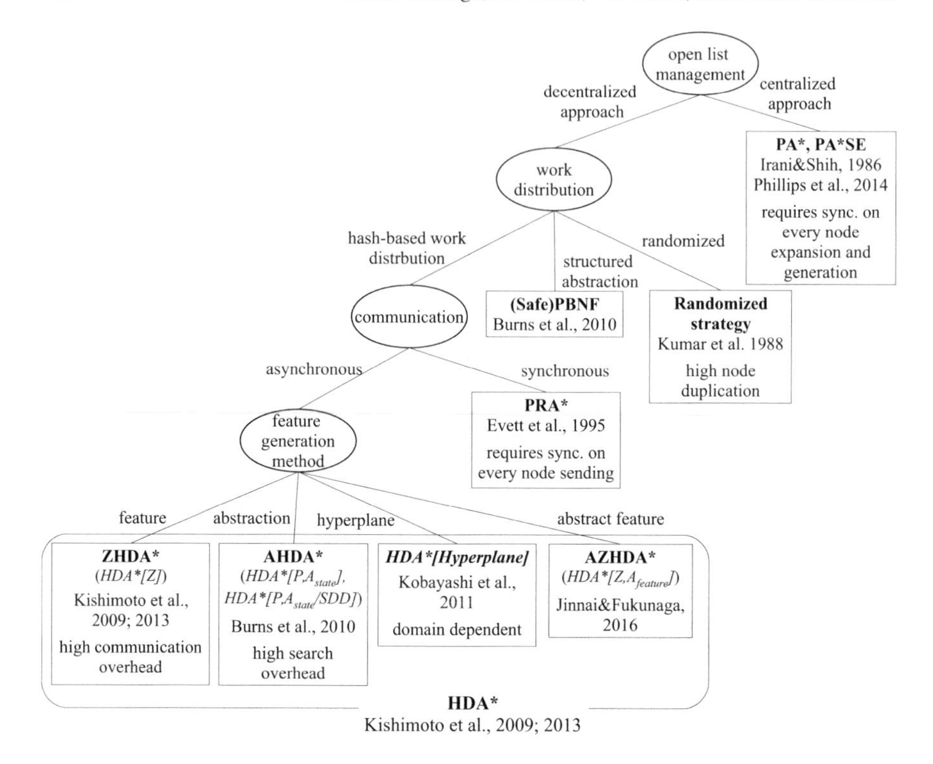

Fig. 11.2: Classification of parallel best-first search algorithms

an early study, Kumar, Ramesh, and Rao [46] identified two broad approaches to parallelizing best-first search, based on how the usage and maintenance of the open list was parallelized. We first survey centralized approaches to parallel A*.

The most straightforward way to parallelize A* on a shared-memory, multi-core machine is Simple Parallel A* (SPA*) [30], shown in Algorithm 11.2. In SPA*, a single open list is shared among all processors. Each processor expands one of the current best nodes from the globally shared open list, and generates and evaluates its children. This centralized approach introduces very little or no search overhead, and no load balancing among processors is necessary. Node re-expansions are possible in SPA* because (as with most other parallel A* variants) SPA* does not guarantee that a state has an optimal g-value when expanded. SPA* is especially simple to implement in a shared-memory architecture by using a shared data structure for the open list and closed list. However, concurrent access to the shared open list becomes a bottleneck, even if lock-free data structures are used [4] – in fact, for problems with fast node generation rates, SPA* exhibits runtimes that are *slower than single-threaded A** [4]. Thus, the scalability of the centralized approach is limited unless the time required to expand each node is extremely expensive (if the node expansion rate is slow enough, then concurrent access to the open list will not be a bottleneck).

Algorithm 11.2: Simple Parallel A* (SPA*)

```
1  Initialize OPEN_shared to {s0}
2  Initialize Lock l_o, l_i
3  Initialize incumbent.cost = ∞
4  In parallel, on each thread, execute 5-32
5  while TerminateDetection() do
6      if OPEN_shared = ∅ or Smallest f(n) value of n ∈ OPEN_shared ≥ incumbent.cost then
7          Continue
8      AcquireLock(l_o)
9      Get and remove from OPEN_shared a node n with a smallest f(n)
10     ReleaseLock(l_o)
11     Add n to CLOSED_shared
12     if n is a goal node then
13         AcquireLock(l_i)
14         if path cost from s0 to n < incumbent.cost then
15             incumbent = path from s0 to n
16             incumbent.cost = path cost from s0 to n
17         ReleaseLock(l_i)
18     for every successor n' of n do
19         g1 = g(n) + c(n, n')
20         if n' ∈ CLOSED_shared then
21             if g1 < g(n') then
22                 Remove n' from CLOSED_shared and add it to OPEN_shared
23             else
24                 Continue
25         else
26             if n' ∉ OPEN_shared then
27                 Add n' to OPEN_shared
28             else if g1 ≥ g(n') then
29                 Continue
30         Set g(n') = g1
31         Set f(n') = g(n') + h(n')
32         Set parent(n') = n
33 if incumbent.cost = ∞ then
34     Return failure (no path exists)
35 else
36     Return solution path from s0 to n
```

Vidal et al. [69] propose Parallel K-Best First Search, a multi-core version of the K-BFS algorithm [16], a satisficing (non-admissible) best-first search variant. Parallel KBFS is a centralized best-first search strategy, enhanced by the use of more threads than the number of physical cores, which improves performance on hard problems by exploiting search diversification effects. This is further improved using a restart strategy. They show that good scaling behavior can be obtained on a 4-core machine.

Phillips et al. have proposed PA*SE, a mechanism for reducing node re-expansions in SPA* [54] that only expands nodes when their g-values are optimal, ensuring that nodes are not re-expanded.

*11.3.3 Decentralized Parallel A**

As described above, SPA* suffers from severe synchronization overhead due to the need to constantly access shared open/closed lists.

Algorithm 11.3: Decentralized A* with Local OPEN/CLOSED lists

```
1  Initialize OPEN_p for each thread p
2  Initialize incumbent.cost = ∞
3  Add s_0 to OPEN_{ComputeRecipient(s_0)}
4  In parallel, on each thread p, execute 5-31
5  while TerminateDetection() do
6      while BUFFER_p ≠ ∅ do
7          Get and remove from BUFFER_p a triplet (n', g_1, n)
8          if n' ∈ CLOSED_p then
9              if g_1 < g(n') then
10                 Remove n' from CLOSED_p and add it to OPEN_p
11             else
12                 Continue
13         else
14             if n' ∉ OPEN_p then
15                 Add n' to OPEN_p
16             else if g_1 ≥ g(n') then
17                 Continue
18         Set g(n') = g_1
19         Set f(n') = g(n') + h(n')
20         Set parent(n') = n
21     if OPEN_p = ∅ or Smallest f(n) value of n ∈ OPEN_p ≥ incumbent.cost then
22         Continue
23     Get and remove from OPEN_p a node n with a smallest f(n)
24     Add n to CLOSED_p
25     if n is a goal node then
26         if path cost from s_0 to n < incumbent.cost then
27             incumbent = path from s_0 to n
28             incumbent.cost = path cost from s_0 to n
29     for every successor n' of n do
30         Set g_1 = g(n) + c(n, n')
31         Add (n', g_1, n) to BUFFER_{ComputeRecipient(n)}
32 if incumbent.cost = ∞ then
33     Return failure (no path exists)
34 else
35     Return solution path from s_0 to n
```

In contrast, in a *decentralized* approach to parallel best-first search, shown in Algorithm 11.3, each processor has its own open list. Initially, the root processor generates and distributes some search nodes among the available processors. Then, each processor starts to locally run best-first search using its local open list (as well as a closed list, in case of algorithms such as A*). Decentralizing the open list eliminates the concurrency overhead associated with a shared, centralized open list, but load balancing becomes necessary.

Kumar, Ramesh and Rao [46], as well as Karp and Zhang [34, 35] proposed a random work allocation strategy, where newly generated states are sent to random processors, i.e., in the decentralized algorithm schema in Algorithm 11.3, line 31, $ComputeRecipient(n')$ simply returns a random processor ID. In parallel architectures with non-uniform communication costs, a straightforward variant of this randomized strategy is to send states to a random neighboring processor (with low communication cost) to avoid the cost of sending to an arbitrary processor (cf., [12]).

The problem with these randomized strategies is that duplicate nodes are not detected unless they are fortuitously sent to the same processor, which can result in a tremendous amount of search overhead due to nodes that are redundantly expanded by multiple processors. In many search applications, including domain-independent planning, the search space is a graph rather than a tree, and there are multiple paths to the same state. In sequential search, duplicates can be detected and pruned by using a closed list (e.g., hash table) or other duplicate detection techniques (e.g., [44, 71]). Efficient duplicate detection is critical for performance, both in serial and parallel search algorithms, and can potentially eliminate vast amounts of redundant work.

In parallel search, duplicate state detection incurs several overheads, depending on the algorithm and the machine environment. For instance, in a shared-memory environment, many approaches, including work stealing, need to carefully manage locks on the shared open and closed lists.

11.3.3.1 Termination Detection in Decentralized Parallel Search

In a decentralized parallel A*, when a solution is discovered, there is no guarantee at that time that the solution is optimal [46]. When a processor discovers a locally optimal solution, the processor broadcasts its cost. The search cannot terminate until all processors have proved that there is no solution with a better cost. In order to correctly terminate a decentralized parallel A*, it is not sufficient to check the local open list at every processor. We must also ensure that there is no message en route to some processor that could lead to a better solution. Various algorithms to handle termination exist. A commonly used method is by Mattern [50].

Mattern's method is based on counting sent messages and received messages. If all processors were able to count simultaneously, it would be trivial to detect whether a message is still en route. However, in reality, different processors P_i will report their sent and received counters, $S(t_i)$ and $R(t_i)$, at different times t_i. To handle this, Mattern introduces a basic method where the counters are reported in two different waves. Let $R^* = \sum_i R(t_i)$ be the accumulated received counter at the end of the first wave, and $S'^* = \sum_i S(t'_i)$ be the accumulated sent counter at the end of the second wave. Mattern proved that if $S'^* = R^*$, then the termination condition holds (i.e., there are no messages en route that can lead to a better solution).

Mattern's time algorithm is a variation of this basic method that allows checking the termination condition in only one wave. Each work message (i.e., containing search states to be processed) has a time stamp, which can be implemented as a clock counter maintained locally by each processor. Every time a new termination check is started, the initiating processor increments its clock counter and sends a *control* message to another processor, starting a chain of control messages that will visit all processors and return to the first one. When receiving a control message, a processor updates its clock counter C to $\max(C,T)$, where T is the maximum clock value among processors visited so far. If a processor contains a received message m with a time stamp $t_m \geq T$, then the termination check fails. Obviously, if, at the end of the chain of messages, the accumulated sent and received counters differ, then the termination check fails as well.

11.4 Hash-Based Decentralized A*

An approach to decentralized A* which cleanly addresses both load balancing and duplicate detection assigns a unique owner processor to each search node according to a hash function. That is, in Algorithm 11.3, line 31, $ComputeRecipient(n')$ is implemented by hashing, i.e., $ComputeRecipient(n') = hash(n') \bmod numprocessors$. This maps each state to exactly one processor which "owns" the state. If the hash keys are distributed uniformly among the processors, and the time to process each state is the same, then load balancing is achieved. Furthermore, duplicate detection is performed by the "owner" state – states that are already in the local OPEN/CLOSED lists are duplicates, and by definition, nodes can never be expanded by a non-owner processor.

The idea of hash-based work distribution for parallel best-first search was first used in PRA* by Evett et al. [14], a limited-memory best-first search algorithm for a massively parallel SIMD machine (see Section 11.8.4). It was then used in a parallelization of SEQ_A*, a variant of A* that performs partial expansion of states, on a hypercube by Mahapatra and Dutt [49], who called the technique Global Hashing (GOHA). However, the hash-based work distribution mechanism itself was not studied deeply by either Evett et al. or Mahapatra and Dutt, as their work encompassed significantly more than this work distribution mechanism[1] Transposition-Table-Driven Work Scheduling (TDS) [61] is a distributed-memory, parallel IDA* with hash-based work distribution (see Section 11.8.1). Kishimoto, Fukunaga, and Botea reopened investigation into hash-based work distribution for A* by implementing

[1] PRA* has a sophisticated node retraction mechanism which allows more nodes to be searched in a limited amount of memory than A*, and GOHA was treated as a baseline for LOHA&QE, a more complex mechanism which decouples duplicate checking and load balancing and also applies a more localized hash function.

HDA*, a straightforward application of hash-based work distribution to A*, showing that it scaled quite well on both multi-core machines and large-scale clusters [37, 36]. The key to achieving good parallel speedups in hash-based work distribution is the hash function. While PRA* left the hash function undefined in the paper and GOHA used a multiplicative hash function (see Section 11.6.1), HDA* used the Zobrist hash function [77]. Unfortunately, the early work on HDA* did not quantitatively evaluate the effect of the choice of hash function, resulting in some misleading results in later work using implementations of HDA* that did not use a hash function which was as effective as the Zobrist function. Recently, Jinnai and Fukunaga compared hash distribution functions that have been used in the literature, showing that the Zobrist hash function as well as Abstract Zobrist hashing, an improved version of the Zobrist function, significantly outperforms other hash functions which have been used in the literature [31]. Further details on hash functions as well as an experimental comparison are in Section 11.6.

11.4.1 Hash Distributed A*

We now describe details of Hash Distributed A* (HDA*), a simple, decentralized parallelization of A* using hash-based work distribution. In HDA* the closed and open lists are implemented as a distributed data structure, where each processor "owns" a partition of the entire search space. The local open and closed lists for processor P are denoted $Open_P$ and $Closed_P$. The partitioning is done by hashing the state, as described below.

HDA* starts by expanding the initial state at the root processor. Then, each processor P executes the following loop until an optimal solution is found:

1. First, P checks whether one or more new states have been received in its message queue. If so, P checks for each new state s in $Closed_P$, in order to determine whether s is a duplicate, or whether it should be inserted in $Open_P$.[2]
2. If the message queue is empty, then P selects a highest priority state from $Open_P$ and expands it, resulting in newly generated states. For each newly generated state s, a hash key $K(s)$ is computed based on the state representation, and the re$K(s)$ and s is sent to the processor that owns $K(s)$. This send is asynchronous and non-blocking. P continues its computation without waiting for a reply from the destination.

In a straightforward implementation of hash-based work distribution on a shared-memory machine, each thread owns a local open/closed list implemented in shared memory, and when a state s is assigned to some thread, the writer thread obtains a lock on the target shared memory, writes s, then releases the lock. Note that whenever a thread P "sends" a state s to a destination $dest(s)$, then P must wait until the

[2] Even if the heuristic function [25] is consistent, parallel A* search may sometimes have to reopen a state saved in the closed list. For example, P may receive many identical states with various priorities from different processors and these states may reach P in any order.

lock for the shared open list (or message queue) for $dest(s)$ is available and not locked by any other thread. This results in significant synchronization overhead – for example, it was observed in [5] that a straightforward implementation of PRA* exhibited extremely poor performance on the Grid search problem, and multi-core performance for up to 8 cores was consistently *slower* than sequential A*. While it is possible to speed up locking operations by using, for example, highly optimized implementations of lock operations in inline assembly language, the performance degradation due to synchronization remains a considerable problem.

In contrast, the open/closed lists in HDA* are not explicitly shared among the processors. Thus, even in a multi-core environment where it is possible to share memory, all communications are done between separate MPI processes using non-blocking send/receive operations (e.g. `MPI_Bsend` and `MPI_Iprobe`). and relies on highly optimized message buffers implemented in MPI.

Every state must be sent from the processor where it is generated to its "owner" processor. In their work with transposition-table-driven scheduling for parallel IDA*, Romein et al. [62] showed that this communication overhead could be overcome by packing multiple states with the same destination into a single message. HDA* uses this state-packing strategy to reduce the number of messages. The relationship between performance and message sizes depends on several factors such as network configurations, the number of CPU cores, and CPU speed. In [37, 36], 100 states are packed into each message on a commodity cluster using more than 16 CPU cores and a HPC cluster, while 10 states are packed on the commodity cluster using fewer than 16 cores.

11.5 Decentralized Search Using Structure-Based Search Space Partitioning)

An alternate approach for load balancing is based on *structured abstraction*. Given a state space graph and a projection function, an abstract state graph is (implicitly) generated by projecting states from the original state space graph into abstract nodes. In many domains, a projection function can be derived by ignoring some features in the original state space. For example, an abstract space for the sliding-tile puzzle domain can be created by projecting all nodes with the blank tile at position b to the same abstract state. While the use of abstractions as the basis for heuristic functions has a long history [53], the use of abstractions as a mechanism for partitioning search states originated in Structured Duplicate Detection (SDD), an external memory search which stores explored states on disk [72]. In SDD, an n-block is defined as the set of all nodes which map to the same abstract node. SDD uses n-blocks to enable duplicate detection. For any node n that belongs to n-block B, the *duplicate detection scope* of n is defined as the set of n-blocks that can possibly contain duplicates of n, and duplicate checks can be restricted to the duplication detection scope, thereby avoiding the need to look for a duplicate of n outside this scope. SDD exploits this property for external memory search by expanding nodes within a single n-block

B at a time and keeping the duplicate detection scope of the nodes in B in RAM, avoiding costly I/O. Parallel Structured Duplicate Detection (PSDD) is a parallel search algorithm that exploits n-blocks to address both synchronization overhead and communication overhead [75]. Each processor is exclusively assigned to an n-block and its neighboring n-blocks (which are the duplication detection scopes). By exclusively assigning n-blocks with disjoint duplicate detection scopes to each processor, synchronization during duplicate detection is eliminated. While PSDD uses disjoint duplicate detection scopes to parallelize breadth-first heuristic search [73], Parallel Best-NBlock-First (PBNF) [4] extends PSDD to best-first search on multi-core machines by ensuring that n-blocks with the best current f-values are assigned to processors.

Since livelock is possible in PBNF on domains with infinite state spaces, Burns et al. proposed SafePBNF, a livelock-free version of PBNF [4]. Burns et al. [4] also proposed AHDA*, a variant of HDA* using an abstraction-based node distribution function. AHDA* is described below in Section 11.6.4.

11.6 Hash Functions for Hash-Based Decentralized Work Distribution

The performance of hash-based decentralized A* algorithms in Section 11.4 depends entirely on the characteristics of the hash function. However, early work on hash-based decentralized A* did not present empirical evaluation of candidate hash functions, and the importance of the choice of hash function was not fully understood or appreciated. Recent work has investigated the performance characteristics and tradeoffs among various hashing strategies, resulting in a significantly better understanding of previous hashing strategies, as well as new hashing strategies that combine previous methods in order to obtain superior performance [31, 33].

In this section, we first classify and review various hash functions which have been proposed for hash-based distributed A* (Sections 11.6.1-11.6.6). We then present an evaluation of some of the functions on the sliding-tile puzzle benchmark domain (Section 11.6.7). Next, we review fully automated, domain-independent methods for deriving hash functions (Section 11.6.8). Finally, we briefly review work on hash-based work distribution in the related field of model checking (Section 11.6.9).

11.6.1 Multiplicative Hashing

The multiplication method $H(\kappa)$ is a widely used hashing method that has been observed to hash a random key to P slots with almost equal likelihood [11]. Multiplicative hashing $M(s)$ uses this function to achieve good load balancing of nodes among processors [49]:

$$M(s) = H(\kappa(s)), \quad (11.1)$$

$$H(\kappa) = \lfloor p(\kappa \cdot A - \lfloor \kappa \cdot A \rfloor) \rfloor, \quad (11.2)$$

where $\kappa(s)$ is a key derived from the state s, p is the number of processors, and A is a parameter in the range $[0,1)$. Typically $A = (\sqrt{5}-1)/2$ (the golden ratio) is used since the hash function is known to work well with this value of A [39]. As $H(\kappa)$ achieves almost perfect load balance for *random* κ keys, designing $\kappa(s)$ so that it appears to be random to state s is important to its performance. However, designing such a $\kappa(s)$ for a given domain is a non-trivial problem.

11.6.2 Zobrist Hashing

Since the work distribution in HDA* is completely determined by a global hash function, the choice of the hash function is crucial to its performance. Kishimoto et al. [37, 36] noted that it is desirable to use a hash function that uniformly distribute nodes among processors, and used the Zobrist hash function [77], described below. The Zobrist hash value of a state s, $Z(s)$, is calculated as follows. For simplicity, assume that s is represented as an array of n propositions, $s = (x_0, x_1, \ldots, x_n)$. Let R be a table containing preinitialized random bit strings:

$$Z(s) := R[x_0] \; xor \; R[x_1] \; xor \; \cdots \; xor \; R[x_n] \quad (11.3)$$

In the rest of the paper, we refer to the original version of HDA* by Kishimoto et al. [37, 36], which used Zobrist hashing, as ZHDA* or $HDA^*[Z]$.

It is possible for two different states to have the same Zobrist hash key, although the probability of such a collision is extremely low with 64-bit keys. Thus, when using Zobrist hashing, checking whether a state s is a duplicate requires first checking whether the bucket for $hash(s)$ is nonempty, and if so, the state itself needs to be compared. Although this is slightly slower than comparing only the hash key, duplicate checks are guaranteed to be correct.

11.6.3 Operator-Based Zobrist Hashing

Zobrist hashing seeks to distribute nodes uniformly among all processors, without any consideration of the neighborhood structure of the search space graph. As a consequence, communication overhead is high. Assume an ideal implementation that assigns nodes uniformly among threads. Every generated node is sent to another thread with probability $1 - \frac{1}{\#threads}$. Therefore, with 16 threads, $> 90\%$ of the nodes are sent to other threads, so communication costs are incurred for the vast majority of node generations.

Operator-based Zobrist hashing (OZHDA*) [32] partially addresses this problem by manipulating the random bit strings in R, the table used to compute Zobrist hash values, such that for some selected states S, there are some operators $A(s)$ for $s \in S$ such that the successors of s that are generated when $a \in A(s)$ is applied to s are guaranteed to have the same Zobrist hash value as s, which ensures that they are assigned to the same processor as s. Jinnai and Fukunaga [32] showed that OZHDA* significantly reduces communication overhead compared to Zobrist hashing [32]. However, this may result in increased search overhead compared to *HDA**[Z], and it is not clear whether the extent of the increased search overhead in OZHDA* could be predicted *a priori*.

11.6.4 Abstraction

In order to minimize communication overhead in HDA*, Burns et al. [4] proposed AHDA*, which uses *abstraction* based node assignment. AHDA* applies the state-space partitioning technique used in PBNF [4] and PSDD [75]. Abstraction uses the abstraction strategy to project nodes in the state space to *abstract states*. A hash-based work distribution function can then be applied to the projected state. The AHDA* implementation by Burns et al. [4] assigns abstract states to processors using a perfect hashing and a modulus operator.

Thus, nodes that are projected to the same abstract state are assigned to the same thread. If the abstraction function is defined so that children of node n are usually in the same abstract state as n, then communication overhead is minimized. The drawback of this method is that it focuses solely on minimizing communication overhead, and there is no mechanism for equalizing load balance, which can lead to high search overhead.

HDA* with abstraction can be characterized by two parameters to decide its behavior – a hashing strategy and an abstraction strategy. Burns et al. [4] implemented the hashing strategy using a perfect hashing and a modulus operator, and an abstraction strategy following the construction for SDD [74] (for domain-independent planning), or a hand-crafted abstraction (for the sliding-tile puzzle and grid path-finding domains).

Jinnai and Fukunaga showed that AHDA* with a static N_{max} threshold performed poorly for a benchmark set with varying difficulty because a fixed size abstract graph results in very poor load balance, and proposed Dynamic AHDA* (DAHDA*), which dynamically sets the size of the abstract graph according to the number of features (the state space size is exponential in the number of features) [32].

11.6.5 Abstract Zobrist Hashing

Both search and communication overheads have a significant impact on the performance of HDA*, and methods that only address one of these overheads are insufficient. ZHDA*, which uses Zobrist hashing, assigns nodes uniformly to processors, achieving near-perfect load balance, but at the cost of incurring communication costs on almost all state generations. On the other hand, abstraction-based methods such as PBNF and AHDA* significantly reduce communication overhead by trying to keep generated states at the same processor as where they were generated, but this results in significant search overhead because all of the productive search may be performed at one node, while all other nodes are searching unproductive nodes that would not be expanded by A*. Thus, we need a more balanced approach that simultaneously addresses both search and communication overheads.

Abstract Zobrist hashing (AZH) is a hybrid hashing strategy which augments the Zobrist hashing framework with the idea of projection from abstraction, incorporating the strengths of both methods. The AZH value of a state, $AZ(s)$ is:

$$AZ(s) := R[A(x_0)] \; xor \; R[A(x_1)] \; xor \; \cdots \; xor \; R[A(x_n)] \tag{11.4}$$

where A is a *feature projection function*, a many-to-one mapping from each raw feature to an *abstract feature*, and R is a precomputed table for each abstract feature.

Thus, AZH is a 2-level, hierarchical hash, where raw features are first projected to abstract features, and Zobrist hashing is applied to the abstract features. In other words, we project state s to an abstract state $s' = (A(x_0), A(x_1), ..., A(x_n))$, and $AZ(s) = Z(s')$. Figure 11.3 illustrates the computation of the AZH value for an 8-puzzle state.

AZH seeks to combine the advantages of both abstraction and Zobrist hashing. Communication overhead is minimized by building abstract features that share the same hash value (abstract features are analogous to how abstraction projects state to abstract states), and load balance is achieved by applying Zobrist hashing to the abstract features of each state.

Compared to Zobrist hashing, AZH incurs less CO due to abstract feature-based hashing. While Zobrist hashing assigns a hash value to each node independently, AZH assigns the same hash value to all nodes that share the same abstract features for all features, reducing the number of node transfers. Also, in contrast to abstraction-based node assignment, which minimizes communications but does not optimize load balance and search overhead, AZH seeks good load balance, because the node assignment considers all features in the state, rather than just a subset.

AZH is simple to implement, requiring only an additional projection per feature compared to Zobrist hashing, and we can precompute this projection at initialization. Thus, there is no additional runtime overhead per node during the search. The projection function $A(x)$ can be either hand-crafted or automatatically generated.

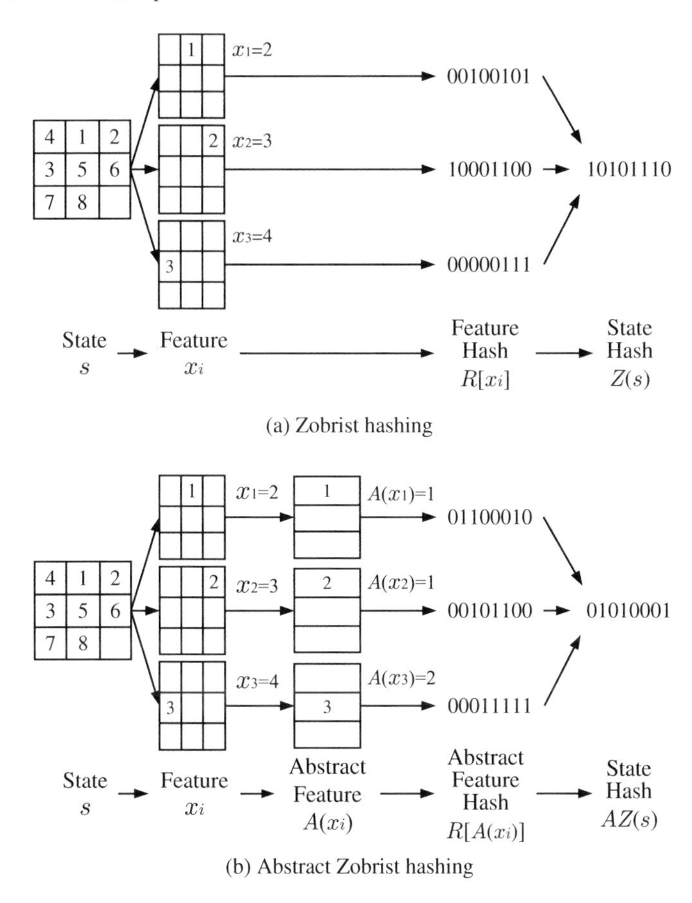

Fig. 11.3: Calculation of abstract Zobrist hash (AZH) value $AZ(s)$ for the 8-puzzle: State $s = (x_1, x_2, ..., x_8)$, where $x_i = 1, 2, ..., 9$ ($x_i = j$ means tile i is placed at position j). The Zobrist hash value of s is the result of xor'ing a preinitialized random bit vector $R[x_i]$ for each feature (tile) x_i. AZH incorporates an additional step which projects features to abstract features (for each feature x_i, look up $R[A(x_i)]$ instead of $R[x_i]$)

11.6.6 Hyperplane Work Distribution

HDA* suffers significantly from increased search overhead in the multiple sequence alignment (MSA) domain whose search space is a directed acyclic graph with non-uniform edge costs [40]. The increased search overhead is caused by reopening the nodes in the closed list to ensure solution optimality. Even with a consistent heuristic, HDA* may need to reopen a node, because HDA* selects the best node in its local open list, which is not necessarily the globally best node. On the other hand, A* with the consistent heuristic never reopens the nodes in the closed list.

Figure 11.4 illustrates an example of HDA*'s drawback. Assume that P_1 owns states a, c, and d, and P_2 owns state b. P_1 is likely to expand d via path $a \to c \to d$, since P_1 does not send a, c, and d to P_2, while b needs to be sent to P_2. Assume that P_1 saves d in the closed list with $g(d) = 1+3 = 4$ and expands d, then receives d from P_2 via $a \to b \to d$ with $g(d) = 1+1 = 2$, and saves d in the open list. Then, when choosing d for expansion, P_1 needs to regenerate the successors of d.

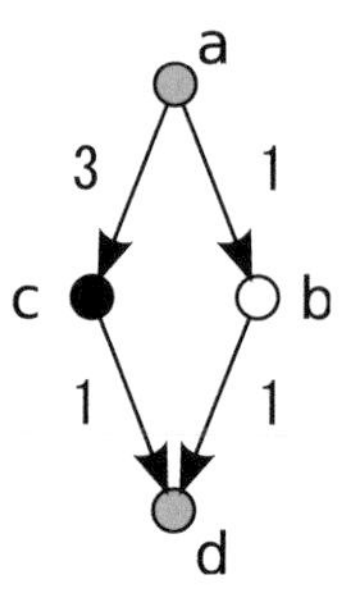

Fig. 11.4: An example from [40], showing how HDA* may expand nodes in a non-optimal order, resulting in duplicate search effort

In MSA with n sequences, a state can be represented by a location $\mathbf{x} = (x_1, x_2, \cdots, x_n)$ in the n-dimensional grid, where x_i is an integer ($0 \le x_i \le l_i$) and l_i is the length of the i-th sequence. Based on the hyperplane defined in the structural regularity in the MSA search space, the hyperplane work distribution (HWD) strategy attempts to limit the owners of successors to some processors. In HWD, the owner of state $\mathbf{x}$ is defined as:

$$\mathrm{Plane}(\mathbf{x}, d) := \begin{cases} \lfloor \frac{1}{d} \Sigma x_i \rfloor & (d \in \{1, 2, 3, ...\}) \\ \frac{1}{d} \Sigma x_i + (\mathrm{Z}(\mathbf{x}) \mod \frac{1}{d}) & (d \in \{\frac{1}{2}, \frac{1}{3}, ..., \frac{1}{p}\}) \end{cases}$$

where p is the number of processors, d is an empirically determined parameter indicating the *thickness* of the hyperplane, and Z is the Zobrist function. Then, processor P_i $(0 \le i < p)$ owns $\mathbf{x}$ where $i = P(\mathbf{x})$ and $P(\mathbf{x}) := \mathrm{Plane}(\mathbf{x}, d) \mod p$.

HWD's work localization scheme increases the chance of allocating generated successors to the same processor. The local open list of HWD orders these successors more reasonably, thus contributing to reducing the frequency of reopening the states. For example, if states b and c are allocated to the same processor and $h(b) = h(c)$ holds, the processor expands b before c. Thus, d via $a \to b \to d$ is generated first, and d via $a \to c \to d$ is successfully removed.

Assume processor P_i owns $\mathbf{x}$ and let $\mathrm{Succ}(x)$ be a set of successors of $\mathbf{x}$. Then, the following theorem indicates that HWD bounds the number of processors to which P_i sends the successors of $\mathbf{x}$.

Theorem 3.

$$\#\left(\bigcup_{x \,:\, P(x)=i} \{P(\mathbf{x}') \mid \mathbf{x}' \in \mathrm{Succ}(\mathbf{x})\} \right) \le \left\lfloor \frac{n}{d} + \max(1, \frac{1}{d}) \right\rfloor$$

There is a trade-off between load balancing and localization of the work. Choosing a good value for d is important for achieving satisfactory parallel performance (see [40] for details).

LOHA [49] distributes work with a hash function taking into account locality for the Traveling Salesperson Problem where the search space is represented as a levelized graph. LOHA is similar to HWD in the sense that both approaches limit the number of destination processors to which each processor sends work. However, there are notable differences between LOHA and HWD in the design of the hash functions. LOHA does not employ the Zobrist function, which plays an important role for uniformly distributing work. In addition, LOHA was designed for the Hypercube machine whose communication delays between subcubes are much larger than between processors inside the same subcube. As a result, LOHA first allocates coarse-grained work to a subcube, then splits such allocated work finely among the processors inside the subcube. On the other hand, HWD directly partitions fine-grained work to a restricted subset of processors, aiming to reduce search overhead incurred by reopening the states.

Both HWD and LOHA require the search space to be levelized. Their extension to non-levelized graphs such as cost-optimal planning remains an open question.

11.6.7 Empirical Comparison of Hash Functions

To illustrate the scaling behavior of the various hash functions reviewed in this section, We evaluated the performance of the following parallel A* algorithms on the 15-puzzle. See [33] for a more detailed comparison

- AZHDA*: HDA* using Abstract Zobrist hashing [31]
- ZHDA*: HDA* using Zobrist hashing [36]
- AHDA*: HDA* using abstraction based work distribution [4]
- SafePBNF: [4]
- HDA*+GOHA: HDA* using multiplicative hashing, a hash function proposed in [49]
- Randomized strategy: nodes are sent to random cores (duplicate nodes are not guaranteed to be sent to the same core) [46, 35]
- Simple Parallel A* (centralized, single OPEN list) [30]

This experiment was run on an Intel Xeon E5-2650 v2 2.60 GHz CPU with 128 GB RAM, using up to 16 cores. The code for the experiment (based on the code by [4]) is available[3].

We solved 100 randomly generated instances using Manhattan distance heuristic. Following [4], we implemented open list using a binary heap. The average runtime of sequential A* solving these instances was 52.3 seconds.

[3] `https://github.com/jinnaiyuu/Parallel-Best-First-Searches`

The features used by Zobrist hashing in ZHDA* are the positions of each tile i. The projections we used for Abstract Zobrist hashing in AZHDA* are shown in Figure 11.5. The abstraction use by AHDA* and SafePBNF ignores the positions of all tiles except tiles 1, 2, and 3. For HDA*+GOHA, we used a bit vector of the positions of the tiles for κ.

Figure 11.6 shows the speedup of each method.

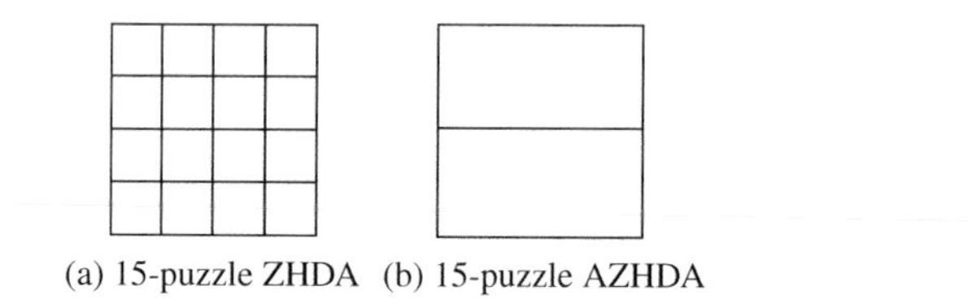

(a) 15-puzzle ZHDA (b) 15-puzzle AZHDA

Fig. 11.5: The hand-crafted abstract features used by abstract Zobrist hashing for the 15-puzzle

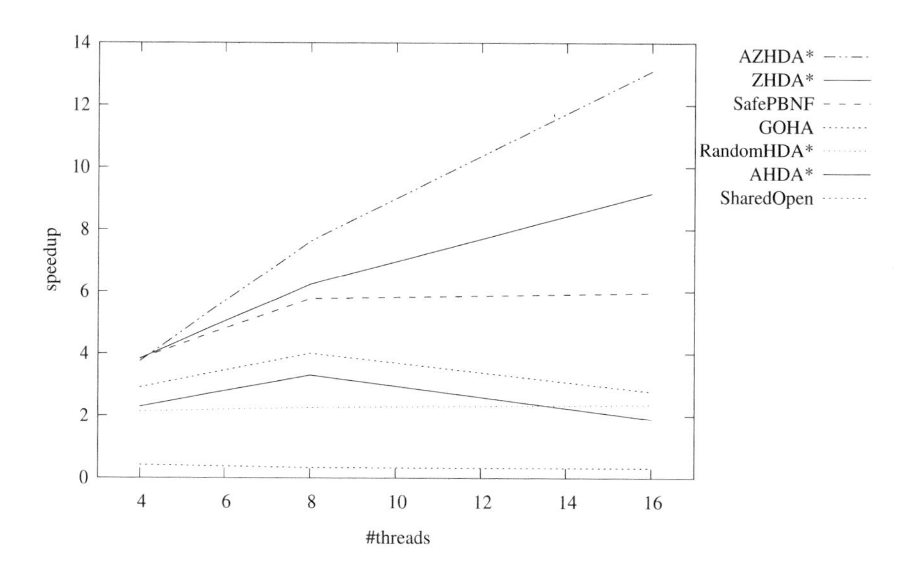

Fig. 11.6: Comparison of speedups obtained by HDA* variants using various hashing methods

11.6.8 Domain-Independent, Automatic Generation of Hash Functions

The hashing methods described above are domain-independent methods that can be applied to a wide range of problems. Although concrete implementations of hash functions for a specific problem can be hand-crafted, as in the case of the sliding-tile puzzle example above, it is possible to fully automate this process when a formal model of a domain (such as PDDL/SAS+ for classical planning) is available. For example, for ZHDA*, domain-independent feature generation for classical planning problems represented in the SAS+ representation [2] is straightforward [36]. For each possible assignment of value k to variable v_i in a SAS+ representation, e.g., $v_i = k$, there is a binary proposition $x_{i,k}$ (i.e., the corresponding STRIPS propositional representation). Each such proposition $x_{i,k}$ is a feature to which a randomly generated bit string is assigned, and the Zobrist hash value of a state can be computed by xor'ing the propositions that describe the state, as in Equation 11.3.

For AHDA*, the abstract representation of the state space can be generated by ignoring some of the features (SAS+ variables) and using the rest of the features to represent the abstraction. Burns et al. [4] used the greedy abstraction algorithms by Zhou and Hansen [74] to select the subset of features [4]. The greedy abstraction algorithm adds one atom group to the abstract graph at a time, choosing the atom group which minimizes the maximum out-degree of the abstract graph, until the graph size (number of abstract nodes) reaches the threshold given by a parameter.

For AZHDA*, the feature projection function, which generates abstract features from raw features, plays a critical role in determining the performance of AZHDA*, because AZHDA* relies on the feature projection in order to reduce communications overhead. Methods based on the domain-transition graph are proposed in [32, 33].

11.6.9 Hash-Based Work Distribution in Model Checking

While this paper focuses on parallel best-first search (more specifically, parallel A*), which is applied to standard AI search domains including domain-independent planning and the sliding-tile puzzle, distributed search, including hash-based work distribution, has also been studied extensively by the parallel model checking community. Parallel Murφ [64, 65] addresses verification tasks that involve exhaustively enumerating all reachable states in a state space, and implements a hash-based work distribution schema where each state is assigned to a unique owner processor. Kumar and Mercer [45] present a load balancing technique as an alternative to the hash-based work distribution implemented in Murφ. The Eddy Murphi model checker [51] specializes processors' tasks, defining two threads for each processing node. The worker thread performs state processing (e.g., state expansion), whereas the other thread handles communication (e.g., sending and receiving states).

Lerda and Sisto parallelized the SPIN model checker to increase the availability of memory resources [47]. Similarly to hash-based distribution, states are assigned to an owner processing node, and get expanded at their owner node. However, instead of using a hash function to determine the owner processor, only one state variable is taken into account. This is done to increase the likelihood that the processor where a state is generated is identical to the owner processor. Holzmann and Bošnački [27] introduce an extension of SPIN to multi-core, shared memory machines. Garavel et al. [20] use hash-based work distribution to convert an implicitly defined model-checking state space into an explicit file representation. Symbolic parallel model checking has been addressed in [26].

Thus, hash-based work distribution and related techniques for distributed search have been widely studied for parallel model checking. There are several important differences between previous work in model checking and this chapter. First, this chapter focuses on parallel A*. In model checking, there is usually no heuristic evaluation function, so depth-first search and breadth-first search is used instead of best-first strategies such as A*.

Second, reachability analysis in model checking (e.g., [64, 65, 47, 20]), which involves visiting all reachable states, does not necessarily require optimality. Search overhead is not an issue because both serial and parallel solvers will expand all reachable states exactly once. In contrast, A* specifically addresses the problem of finding an optimal path, a significant constraint which introduces the issue of search efficiency because distributed A* (including HDA*) searches many nodes with f-cost greater than or equal to the optimal cost, as detailed in Section 11.3.1; furthermore, node re-expansions in parallel A* can introduce search overhead.

11.7 Parallel Portfolios Using A*

An *algorithm portfolio* [29] is often employed and parallelized in other domains, such as the ManySAT solver [22] for SAT solving and ArvandHerd [67] for satisficing planning. This approach runs a set of different search algorithms in parallel. Processors execute the search algorithms mostly independently, but may periodically exchange important information with others.

A long-tailed distribution is often observed in the runtime distribution of the search algorithms [21]. The algorithm portfolio attempts to exploit such search behaviors by using a variety of algorithms that examine potentially overlapping, but different portions of the search space.

Dovetailing [38], which is a simple version of the algorithm portfolio, performs search simultaneously with different parameter settings. Valenzano et al. apply parallel dovetailing [68] to the weighted versions of IDA* [41], RBFS [42], A*, PBNF [4][4], as well as BULB [19], a suboptimal heuristic search. Their parallel dovetailing

[4] The suboptimality of these weighted algorithms is is bounded by the values of the weights.

runs search with many different weight values without exchanging information, and terminates when one of the algorithms returns a solution.

In their experiments on puzzle solving, admissible heuristics were used to evaluate the performance of parallel dovetailing. On sequential planning, weighted A* was executed with many different weights including the weight value of ∞ (i.e., identical to Greedy Best-First Search (GBFS)), one admissible heuristics and two inadmissible heuristics. In addition, the original Fast Downward planner using multiple heuristics and GBFS is included as one of the algorithms.

In both puzzle solving and sequential planning, the experimental results shown by Valenzano et al. [68] indicate that parallel dovetailing often yields good speedups and solves additional problem instances. However, unlike other approaches described in this article, parallel dovetailing does not always return optimal solutions.

11.8 Parallel, Limited-Memory A* (Parallel IDA*, TDS, PRA*)

In problem domains where the rate of node generation by A* is high, the amount of memory available becomes a significant limitation, because A* can exhaust memory and terminate before finding a solution. A*-based planners for domain-independent, classical planning such as Fast Downward [24] generate between $10^4 - 10^5$ nodes per second on standard International Planning Competition benchmark domains. If a single state requires 100 bytes to represent, this means that A*-based planning can consume $10^6 - 10^7$ bytes per second. Highly optimized solvers for specific domains such as the sliding-tile puzzle can generate over $10^6 - 10^7$ nodes per second [6], consuming memory even faster. This problem is particularly pressing for parallel A* on a single machine. Although the amount of RAM on a single machine has been steadily increasing, the number of cores on a single machine has also been rising, and the amount of memory *per core* has remained fairly constant over the past decade (around 2GB/core). If RAM is consumed at a rate of 10^7 bytes per second, then A* will exhaust 1GB in approximately 100 seconds. Thus, in domains with fast node generation rates, parallel A* can exhaust memory in a matter of minutes.

To overcome this limitation of A*, limited-memory, best-first search algorithms for finding optimal paths in implicit graphs have been extensively studied. The best-known algorithm is Iterative Deepening A* (IDA*) [41]. IDA* performs a series of depth-first searches, where each iteration is limited to an f-cost bound, which is increased on each iteration. Since each iteration of IDA* only performs a depth-first search, this requires memory which is only linear in the depth of the solution. Although each iteration revisits all of the nodes visited on all of the previous iterations, many search spaces have the property that the runtime of iterative deepening is dominated by the search performed in the last few iterations, so the overhead of repeating the work done in past iterations is relatively small as a fraction of the total search effort [41].

However, if the search space is a graph, there may be many paths to each state, which results in significant amount of wasted search effort revisiting nodes through different paths. To alleviate this problem with standard IDA*, a *transposition table*, which is a cache of lower bounds on the solution cost achievable for previously visited states, can be added so that search is pruned if a search reaches a previously visited state and it can be proven that the pruning does not result in loss of optimality [58, 1]. Other limited-memory A* variants include MA* [9], SMA* [63], and recursive best-first search [42].

Below, we review parallel, limited-memory A* variants.

11.8.1 Transposition Table-Driven Scheduling (TDS)

Transposition-Table-Driven work scheduling (TDS) [62, 60] is a distributed-memory, parallel IDA* algorithm that uses a distributed transposition table. Similarly to HDA*, TDS partitions the transposition table (TT) over processors and asynchronously distributes work using a Zobrist-based state hash function. In this way, TDS effectively allocates the large amount of distributed memory to the TT and uses the TT for efficiently detecting and pruning duplicate states that arrive at the processor. The distributed TT implementation uses the Zobrist hash function for mapping states to processors. TDS initiates parallelism within each iteration and synchronizes between iterations. In a straightforward implementation of TDS, processors need to exchange messages that convey back-propagated lower bounds, but the efficient implementation of Romein et al. eliminates such a back-propagation procedure, thus reducing communication overhead in exchange for giving up the use of more informed lower bounds (see [62, 60] for details). Due to this modification, the Mattern's algorithm (Section 11.3.3.1) is used for the termination detection of each IDA* iteration.

Romein et al. [62, 60] showed that TDS exhibits a very low (sometimes negative) search overhead and yields significant (sometimes super-linear) speedups in solving puzzles on a distributed-memory machine, compared to a sequential IDA* that runs on a single computational node with limited RAM capacity. On the other hand, for domain-independent planning (International Planning Contest benchmark instances) , Kishimoto et al. showed that HDA* was consistently faster than TDS but sometimes terminates its execution due to memory exhaustion [36]. Therefore, Kishimoto et al. proposed a simple, hybrid strategy combining HDA* and TDS. Their hybrid strategy first executes HDA* until either one of the HDA* processors exhausts its memory or the problem instance is solved. If HDA* fails to solve the problem instance, then TDS, which skips some wasteful iterations detected by HDA* search, is executed. Thus, their hybrid strategy inherits advantages of both HDA* and TDS.

11.8.2 Work Stealing for IDA*

Work stealing is a standard approach for partitioning the search space, and is used particularly for parallelizing depth-first search in shared-memory environments. In work stealing, each processor maintains a local work queue. When generating a new state, the processor places that state in its local queue. When the processor has no work in its queue, it "steals" work from the queue of a busy processor. Strategies for selecting a processor to steal the work from and determining the amount of work to steal are extensively studied (e.g., [57, 15, 17]).

Nevertheless, work stealing suffers from performance degradation in domains where detecting duplicate states plays an important role. Romein et al. implemented work stealing for IDA* with transposition tables and compared it with TDS on a distributed-memory environment. They showed that TDS was 1.6 to 12.9 times faster than work stealing in puzzle solving domains [60]. On the other hand, TDS requires the use of a reasonably low-latency, high-bandwidth network for achieving efficient parallel performance. Therefore, Romein and Bal combined TDS with work stealing [59] in a grid environment where the communication latency is high between PC clusters and is low within each cluster. They use TDS to parallelize IDA* within each cluster for carrying out efficient duplication detection. When a cluster runs out of work, it steals work from another cluster, enabling much smaller communication overhead in the presence of the high-latency network.

Variants of work stealing-based IDA* for Single Instruction, Multiple Data (SIMD) architecture machines have also been studied [55, 48]. Since all processors in a SIMD machine must execute the same instruction, these approaches used a two-phase strategy which alternates between (1) a work (search) phase where all processors perform local, IDA* search, and (2) a load balancing phase, during which all processors exchange nodes.

To our knowledge, there is no published, empirical evaluation of work stealing for A* (as opposed to IDA*) in distributed-memory environments. This is a curious gap in the literature, given that work stealing is a standard approach for other parallel search models (e.g., branch-and-bound and backtracking for integer programming and constraint programming). This may be because work stealing strategies, particularly work stealing across machines in a cluster, tend to be more complex to implement than successful hash-based decentralized approaches such as HDA*. An investigation of work stealing approaches to A* therefore remains an avenue for future work.

11.8.3 Parallel Window Search

Another approach to parallelizing IDA* is parallel-window search [56], where each processor searches from the same root node, but is assigned a different bound – that is, each processor is assigned a different, independent iteration of IDA*. When a processor finishes an iteration, it is assigned the next highest bound which has not yet been assigned to a processor. The first solution found by parallel window IDA* is not necessarily optimal. However, if, after finding a solution in the processor assigned bound b, we wait until all processors with bound less than b finish, then the optimality of the best solution found is assured.

11.8.4 Parallel Retracting A (PRA*)*

Parallel Retracting A* (PRA*) [14] simultaneously addresses the problems of work distribution and duplicate state detection. In PRA*, each processor maintains its own open and closed lists. A hash function maps each state to exactly one processor which "owns" the state (as mentioned in Section 11.4, the hash function used in PRA* was not specified in [14]). When generating a state, PRA* distributes it to the corresponding owner. If the hash keys are distributed uniformly across the processors, load balancing is achieved. After receiving states, PRA* has the advantage that duplicate detection can be performed efficiently and locally at the destination processor.

While PRA* incorporated the idea of hash-based work distribution, PRA* differs significantly from a parallel A* in that it is a parallel version of Retracting A* (RA)* [14], a limited-memory search algorithm closely related to MA* [9] and SMA* [63]. When a processor's memory becomes full, Parallel Retracting A* *retracts* states from the search frontier, and their f-values are stored in their parents, which frees up memory. Thus, unlike parallel A*, PRA* does not store all expanded nodes in memory, and will not terminate due to running out of memory in some process. On the other hand, the implementation of this retraction mechanism in [14] incurs a significant synchronization overhead: when a processor P generates a new state s and sends it to the destination processor Q, P blocks and waits for Q to confirm that s has been successfully received and stored (or whether the send operation failed due to memory exhaustion at the destination processor).

11.9 Parallel A* in Cloud Environments with Practically Unlimited Available Resources

Cloud computing resources such as Amazon EC2, which offer computational resources on demand, have become widely available in recent years. In addition to cloud computing platforms, there is an increasing availability of massive-scale, distributed grid computing resources such as TeraGrid/XSEDE, as well as massively parallel, high-performance computing (HPC) clusters. These large-scale *utility computing resources* share two characteristics that have significant implications for parallel search algorithms. First, vast (practically unlimited) aggregate memory and CPU resources are available on demand. Secondly, resource usage incurs a direct monetary cost.

Previous work on parallel search algorithms has focused on *makespan*: minimizing the runtime (wall-clock time) to find a solution, given fixed hardware resources; and *scalability*: as resource usage is increased, how are makespan and related metrics affected? However, the availability of virtually unlimited resources at some cost introduces a new context for parallel search algorithm research where an explicit consideration of cost-performance tradeoffs is necessary. For scalable algorithms, it is possible to reduce the makespan by allocating more resources (up to some point). In practice, this incurs a high cost with diminishing marginal returns. For parallel A* variants, under-allocating resources results in memory exhaustion. On the other hand, over-allocation is costly and undesirable. With the vast amounts of aggregate memory available in utility computing, the *cost* (monetary funds) can be the new limiting factor, since one can exhaust funds long before allocating all of the memory resources available from a large cloud service provider.

In utility computing services, there is some notion of an atomic unit of resource usage. A *hardware allocation unit* (HAU), is the minimal, discrete resource unit that can be requested from a utility computing service. Various HAU types can be available, each with different performance characteristics and cost. Commercial clouds such as EC2 tend to have an immediate HAU allocation model with *discrete* charges. Usage of a HAU for any fraction of an hour is rounded up. Grids and shared clusters tend to be batch-job based with a *continuous* cost model. Jobs are submitted to a centralized scheduler, with no guarantees about when a job will be run. The cost is a linear function of the amount of resources used.

11.9.1 Iterative Allocation Strategy

A scalable, *ravenous algorithm* is an algorithm that can run on an arbitrary number of processors, and whose memory consumption increases as it keeps running. HDA* is an example of a scalable, ravenous algorithm. The *iterative allocation* (IA) strategy [18] repeatedly runs a ravenous algorithm a until the problem is solved. The key detail is deciding the number of HAUs to allocate in the next iteration, if the previous iteration failed. We seek a policy that tries to minimize the total cost.

Two realistic assumptions which facilitate formal analysis are the following: Firstly, all HAUs used by IA are identical hardware configurations. Secondly, if a problem is solved on i HAUs, then it will be solved on $j > i$ HAUs (monotonicity). Monotonicity is usually (implicitly) assumed in the previous work on parallel search. Let T_v be the makespan (wall-clock) time needed to solve a problem on v HAUs. In a continuous cost model, the cost on v HAUs is $T_v \times v$. In a discrete cost model, the cost is $\lceil T_v \rceil \times v$. The *minimal width* W^+ is the minimum number of HAUs that can solve a problem with a given ravenous algorithm. Given a cost model (i.e., continuous or discrete), C^+ is the associated *min width cost*. C^* is the optimal cost to solve the problem, and the *minimal cost width* W^* is the number of HAUs that results in a minimal cost. Since W^* is usually not known a priori, the best we can hope for is to develop strategies that approximate the optimal values.

The *max iteration time* E is the maximum actual (not rounded up) time that an iteration can run before at least 1 HAU exhausts memory.

The *min-width cost ratio* R^+ is defined as $I(S)/C^+$, where $I(S)$ is the total cost of IA (using a particular allocation strategy S). The *min-cost ratio* R^* is defined as $I(S)/C^*$. The total cost $I(S)$ of IA is accumulated over all iterations. In a discrete cost model, times spent by individual HAUs are rounded up. The effect of the rounding up is alleviated by the fact that HAUs will use any spare time left at the end of one iteration to start the next iteration.

A particularly simple but useful strategy is the Geometric (b^i) Strategy, which was analyzed and evaluated by [18]. The geometric strategy allocates $\lceil b^i \rceil$ HAUs at iteration i, for some $b > 1$. For example, the 2^i (doubling) strategy doubles the number of HAUs allocated on each iteration.

Cloud platforms such as Amazon EC2 and Windows Azure typically have discrete cost models, where the discrete billing unit is 1 hour. This relatively long unit of time, combined with the fast rate at which search algorithms consume RAM, leads to the observation that many (but not all) search applications will exhaust the RAM/core in a HAU within a single billing time unit in modern cloud environments. In other words, a single iteration of IA will complete (by either solving the problem or running out of memory) within 1 billing time unit (i.e., $E \leq 1$). This observation was experimentlly validated in [18] for domain-independent planning benchmarks and sequence alignment benchmarks. In addition, HDA* has been observed to exhaust memory within 20 minutes on every planning and 24-puzzle problem studied in [36]. With a sufficiently small E, *all iterations could be executed within a single billing time unit*, entirely eliminating the repeated allocation cost overhead.

In a discrete cost model with $E \leq 1$, the cost to solve a problem on v HAUs is proportional to v. As a direct consequence, $W^+ = W^*$ and thus $R^+ = R^*$. It can be shown that in the *best case*, $R^* = R^+ = 1$, in the *worst case*, $R^*_{wo} = R^+_{wo} \leq \frac{b^2}{b-1}$, and in the average case, $R^*_{avg} = R^+_{avg} \leq \frac{2b^2}{b^2-1}$. The worst case bound $b^2/(b-1)$ is minimized by the doubling strategy ($b = 2$). As b increases above 2, the upper bound for R^*_{avg} improves, but the worst case gets worse. Therefore, *the doubling strategy is the natural allocation policy to use in practice.* For the 2^i strategy, the average case ratio is bounded by $8/3 \approx 2.67$, and the worst case cost ratio does not exceed 4. With the 2^i strategy in a discrete cost model when $E \leq 1$, we *never pay more than 4 times the optimal, but a priori unknown cost.*

11.10 Parallel A* and IDA* on Graphics Processing Units

General-purpose computing using the thousands of cores available on Graphics Processing Units (GPUs) is currently a very active area of research. Zhou and Zeng propose a GPU-based A* algorithm using many (thousands) of parallel priority queues (OPEN lists) [76]. A fundamental tradeoff successfully exploited by this approach is that by increasing the number of threads (parallel queues), they increase the effective parallelism. This results in duplicate node generations, but the duplicates are efficiently detected and eliminated using hash-based duplicate detection.

The current bottleneck with executing A* entirely in the GPU is memory capacity – the current, state-of-the-art GPU with the largest amount of RAM (Nvidia P100) has 16GB of global memory, which is an order of magnitude smaller than the amount of RAM on a current workstation. Since this GPU RAM is shared among thousands of cores, the amount of memory per core is several orders of magnitude smaller than the amount of RAM per core for the CPU, which limits the size of the search spaces that can be optimally searched.

As discussed in Section 11.8, one approach to limit memory usage is iterative deepening. Horie and Fukunaga developed Block-Parallel IDA* (BPIDA*) [28], a parallel version of IDA* [41] for the GPU. Although the single instruction, multi-thread architecture used in NVIDIA GPUs is somewhat similar to earlier SIMD architectures, Horie and Fukunaga found that simply porting earlier SIMD IDA* approaches [55, 48] to the GPU results in extremely poor performance due to warp divergence and load balancing overheads. Instead of assigning a subtree of the search to a single thread as SIMD IDA* does, BPIDA* assigns a subtree to a GPU block (a group of threads which execute on the same streaming multiprocessor and share memory), and each block has a shared, parallel open list. This was shown to significantly improve parallel efficiency on the 15-puzzle. Their implementation of BPIDA* only uses the shared memory, and completely avoids using the GPU global memory (RAM on the GPU which is shared by all streaming multiprocessors). This was possible because 15-puzzle states can be represented compatly enough that the search stacks fit entirely in shared memory; in addition, they used Mahnattan distance as the heuristic function, which requires no memory. Thus, BPIDA* achieves good parallel efficiency but the search is not efficient compared to a state-of-the-art IDA* implementation which uses a more powerful but memory-intensive heuristic function (e.g., pattern databases [43]). Using such memory-intensive heuristics (as well as other memory-intensive methods such as a transposition tables [58]) on the GPU will require using the global memory and is a direction for future work.

Heterogeneous approaches which use both the GPU as well as CPU is an open area for future work. One instance of such a hybrid GPU/CPU based approach is for best-fist search with a blind heuristic by Sulweski et al [66]. Their algorithm uses a GPU to accelerate precondition checks and successor generation, but uses the CPU for duplicate detection.

Finally, a different application of many-core GPU architectures is for multi-agent search, where each core executes an independent A* search for each agent in the simulation environment [3].

11.11 Other Approaches

One alternative to partitioning the search space among processors is to parallelize the computation done during the processing of a single search node (cf., [7, 8]). The Operator Distribution Method for parallel Planning (ODMP) [70] parallelizes the computation at each node. In ODMP, there is a single *controlling thread*, and several *planning threads*. The controlling thread is responsible for initializing and maintaining the current search state. At each step of the controlling-thread main loop, it generates the applicable operators, inserts them in an *operator pool*, and activates the planning threads. Each planning thread independently takes an operator from this shared operator pool, computes the grounded actions, generates the resulting states, evaluates the states with the heuristic function, and stores the new state and its heuristic value in a *global agenda* data structure. After the operator pool is empty, the controlling thread extracts the best new state from the global agenda, assigning it to the new, current state.

The best parallelization strategy for a search algorithm depends on the properties of the search space, as well as the parallel architecture on which the search algorithm is executed. The EUREKA system [10] used machine learning to automatically configure parallel IDA* for various problems (including nonlinear planning) and machine architectures.

Niewiadomski et al. [52] propose PFA*-DDD, a parallel version of Frontier A* with Delayed Duplicate Detection. PFA*-DDD partitions the open sets into groups (interval lists) and assigns them to processors. PFA*-DDD returns the cost of a path from start to target, not an actual path. While divide-and-conquer (DC) can be used to reconstruct a path (as in sequential frontier search), parallel DC poses non-trivial design issues that need to be addressed in future work.

Acknowledgements This work was supported in part by JSPS KAKENHI grants 25330253 and 17K00296.

References

[1] Akagi, Y., Kishimoto, A., Fukunaga, A.: On transposition tables for single-agent search and planning: Summary of results. In: Proceedings of the 3rd Symposium on Combinatorial Search (SOCS), pp. 1–8 (2010)
[2] Bäckström, C., Nebel, B.: Complexity results for SAS+ planning. Computational Intelligence **11**(4), 625–655 (1995)
[3] Bleiweiss, A.: GPU accelerated pathfinding. In: Proceedings of the EUROGRAPHICS/ACM SIGGRAPH Conference on Graphics Hardware 2008, Sarajevo, Bosnia and Herzegovina, 2008, pp. 65–74 (2008). DOI 10.2312/EGGH/EGGH08/065-074. URL `http://dx.doi.org/10.2312/EGGH/EGGH08/065-074`
[4] Burns, E., Lemons, S., Ruml, W., Zhou, R.: Best-first heuristic search for multicore machines. Journal of Artificial Intelligence Research (JAIR) **39**, 689–743 (2010)
[5] Burns, E., Lemons, S., Zhou, R., Ruml, W.: Best-first heuristic search for multi-core machines. In: Proceedings of the Twenty-First International Joint Conference on Artificial Intelligence IJCAI-09 (2009)
[6] Burns, E.A., Hatem, M., Leighton, M.J., Ruml, W.: Implementing fast heuristic search code. pp. 25–32 (2012)
[7] Campbell, M., Hoane, J., Hsu, F.: Deep Blue. Artificial Intelligence **134**(1-2), 57–83 (2002)

[8] Cazenave, T., Jouandeau, N.: On the parallelization of UCT. In: H. van den Herik et al. (ed.) Proceedings of Computers and Games CG-08, *LNCS*, vol. 5131, pp. 72–80. Springer (2008)

[9] Chakrabarti, P., Ghose, S., Acharya, A., de Sarkar, S.: Heuristic search in restricted memory. Artificial Intelligence **41**(2), 197–221 (1989)

[10] Cook, D., Varnell, R.: Adaptive parallel iterative deepening search. Journal of Artificial Intelligence Research **9**, 139–166 (1998)

[11] Cormen, T.H., Leiserson, C.E., Rivest, R.L., Stein, C.: Introduction to Algorithms, Second Edition. The MIT Press (2001). URL `http://www.amazon.ca/exec/obidos/redirect?tag=citeulike09-20{&}path=ASIN/0262531968`

[12] Dutt, S., Mahapatra, N.: Scalable load balancing strategies for parallel A* algorithms. Journal of parallel and distributed computing **22**, 488–505 (1994)

[13] Edelkamp, S., Schroedl, S.: Heuristic Search: Theory and Applications. Morgan Kaufmann Publishers Inc., San Francisco, CA, USA (2010)

[14] Evett, M., Hendler, J., Mahanti, A., Nau, D.: PRA*: Massively parallel heuristic search. Journal of Parallel and Distributed Computing **25**(2), 133–143 (1995)

[15] Feldmann, R.: Spielbaumsuche mit massiv parallelen Systemen. Ph.D. thesis, University of Paderborn (1993). English translation titled *Game tree search on massively parallel systems* is available.

[16] Felner, A., Kraus, S., Korf, R.E.: Kbfs: K-best-first search. Annals of Mathematics and Artificial Intelligence **39**, 19–39 (2003)

[17] Frigo, M., Leiserson, C.E., Randall, K.H.: The implementation of the Cilk-5 multithreaded language. In: ACM SIGPLAN Conferences on Programming Language Design and Implementation (PLDI'98), pp. 212–223 (1998)

[18] Fukunaga, A., Kishimoto, A., Botea, A.: Iterative resource allocation for memory intensive parallel search algorithms on clouds, grids, and shared clusters. In: Proceedings of the National Conference on Artificial Intelligence (AAAI) (2012). URL `http://www.aaai.org/ocs/index.php/AAAI/AAAI12/paper/view/5054`

[19] Furcy, D., Koenig, S.: Limited discrepancy beam search. In: Proceedings of the International Joint Conference on Artificial Intelligence, pp. 125–131 (2005)

[20] Garavel, H., Mateescu, R., Smarandache, I.M.: Parallel state space construction for model-checking. In: Proceedings of the 8th International SPIN Workshop, pp. 217–234 (2001)

[21] Gomes, C., Selman, B., Crato, N., Kautz, H.: Heavy-tailed phenomena in satisfiability and constraint satisfaction problems. Journal of Automated Reasoning **24**(1-2), 67–100 (2000)

[22] Hamadi, Y., Jabbour, S., Sais, L.: ManySAT: a parallel SAT solver. Journal on Satisfiability, Boolean Modeling and Computation **6**, 245–262 (2009)

[23] Hart, P., Nilsson, N., Raphael, B.: A formal basis for the heuristic determination of minimum cost paths. IEEE Transactions on System Sciences and Cybernetics **SSC-4**(2), 100–107 (1968)

[24] Helmert, M.: The Fast Downward planning system. Journal of Artificial Intelligence Research **26**, 191–246 (2006). DOI 10.1613/jair.1705

[25] Helmert, M., Haslum, P., Hoffmann, J.: Flexible abstraction heuristics for optimal sequential planning. In: Proceedings of the Seventeenth International Conference on Automated Planning and Scheduling ICAPS-07, pp. 176–183 (2007)
[26] Heyman, T., Geist, D., Grumberg, O., Schuster, A.: Achieving scalability in parallel reachability analysis of very large circuits. In: Proceedings 12th International Conference on Computer Aided Verification, pp. 20–35 (2000)
[27] Holzmann, G.J., Bošnački, D.: The design of a multicore extension of the SPIN model checker. IEEE Transactions on Software Engineering **33**(10), 659–674 (2007)
[28] Horie, S., Fukunaga, A.S.: Block-parallel IDA* for GPUs. In: Proceedings of the Tenth International Symposium on Combinatorial Search, Edited by Alex Fukunaga and Akihiro Kishimoto, 16-17 June 2017, Pittsburgh, Pennsylvania, USA., pp. 134–138 (2017). URL `https://aaai.org/ocs/index.php/SOCS/SOCS17/paper/view/15801`
[29] Huberman, B., Lukose, R., Hogg, T.: An economics approach to hard computational problems. Science **275**(5296), 51–54 (1997)
[30] Irani, K., Shih, Y.: Parallel A* and AO* algorithms: An optimality criterion and performance evaluation. In: International Conference on Parallel Processing, pp. 274–277 (1986)
[31] Jinnai, Y., Fukunaga, A.: Abstract Zobrist hashing: An efficient work distribution method for parallel best-first search. In: Proceedings of the National Conference on Artificial Intelligence (AAAI), pp. 717–723 (2016)
[32] Jinnai, Y., Fukunaga, A.: Automated creation of efficient work distribution functions for parallel best-first search. In: Proc. ICAPS (2016)
[33] Jinnai, Y., Fukunaga, A.: On work distribution functions for parallel best-first search. Journal of Artificial Intelligence Research (2017). (to appear)
[34] Karp, R., Zhang, Y.: A randomized parallel branch-and-bound procedure. In: Proceedings of the 20th ACM Symposium on Theory of Computing (STOC), pp. 290–300 (1988)
[35] Karp, R., Zhang, Y.: Randomized parallel algorithms for backtrack search and branch-and-bound computation. Journal of the Association for Computing Machinery **40**(3), 765–789 (1993)
[36] Kishimoto, A., Fukunaga, A., Botea, A.: Evaluation of a simple, scalable, parallel best-first search strategy. Artificial Intelligence **195**, 222–248 (2013). DOI 10.1016/j.artint.2012.10.007. URL `http://linkinghub.elsevier.com/retrieve/pii/S0004370212001294`
[37] Kishimoto, A., Fukunaga, A.S., Botea, A.: Scalable, parallel best-first search for optimal sequential planning. In: Proc. ICAPS, pp. 201–208 (2009). URL `http://aaai.org/ocs/index.php/ICAPS/ICAPS09/paper/view/705`
[38] Knight, K.: Are many reactive agents better than a few deliberative ones? In: Proceedings of the 13th International Joint Conference on Artificial Intelligence, pp. 432–437 (1993)

[39] Knuth, D.E.: "Sorting and Searching", The Art of Computer Programming, vol. 3. Addison-Wesley (1973)
[40] Kobayashi, Y., Kishimoto, A., Watanabe, O.: Evaluations of Hash Distributed A* in optimal sequence alignment. In: Proceedings of the 22nd International Joint Conference on Artificial Intelligence, pp. 584–590 (2011)
[41] Korf, R.: Depth-first iterative deepening: An optimal admissible tree search. Artificial Intelligence **97**, 97–109 (1985)
[42] Korf, R.: Linear-Space Best-First Search. Artificial Intelligence **62**(1), 41–78 (1993)
[43] Korf, R.E., Felner, A.: Disjoint pattern database heuristics. Artificial Intelligence **134**(1-2), 9–22 (2002)
[44] Korf, R.E., Zhang, W.: Divide-and-conquer frontier search applied to optimal sequence alignment. In: Proceedings of the 17th National Conference on Artificial Intelligence AAAI-00, pp. 910–916 (2000)
[45] Kumar, R., Mercer, E.G.: Load balancing parallel explicit state model checking. Electronic Notes in Theoretical Computer Science **128** (2005)
[46] Kumar, V., Ramesh, K., Rao, V.N.: Parallel best-first search of state-space graphs: A summary of results. In: Proceedings of the 7th National Conference on Artificial Intelligence AAAI-88, pp. 122–127 (1988)
[47] Lerda, F., Sisto, R.: Distributed-memory model checking with SPIN. In: Theoretical and Practical Aspects of SPIN Model Checking, 5th and 6th International SPIN Workshops, *Lecture Notes in Computer Science*, vol. 1680, pp. 22–39 (1999)
[48] Mahanti, A., Daniels, C.: A SIMD approach to parallel heuristic search. Artificial Intelligence **60**, 243–282 (1993)
[49] Mahapatra, N., Dutt, S.: Scalable global and local hashing strategies for duplicate pruning in parallel A* graph search. IEEE Transactions on Parallel and Distributed Systems **8**(7), 738–756 (1997)
[50] Mattern, F.: Algorithms for distributed termination detection. Distributed Computing **2**(3), 161–175 (1987)
[51] Melatti, I., Palmer, R., Sawaya, G., Yang, Y., Kirby, R.M., Gopalakrishnan, G.: Parallel and distributed model checking in Eddy. International Journal on Software Tools for Technology Transfer **11**(1), 13–25 (2009)
[52] Niewiadomski, R., Amaral, J.N., Holte, R.C.: Sequential and parallel algorithms for frontier A* with delayed duplicate detection. In: Proceedings of the 21st National Conference on Artificial Intelligence (AAAI), pp. 1039–1044 (2006)
[53] Pearl, J.: Heuristics - Intelligent Search Strategies for Computer Problem Solving. Addison–Wesley (1984)
[54] Phillips, M., Likhachev, M., Koenig, S.: PA*SE: Parallel A* for slow expansions. In: Proc. ICAPS (2014). URL `http://www.aaai.org/ocs/index.php/ICAPS/ICAPS14/paper/view/7952`
[55] Powley, C., Ferguson, C., Korf, R.: Depth-first heuristic search on a SIMD machine. Artificial Intelligence **60**, 199–242 (1993)
[56] Powley, C., Korf, R.: Single-agent parallel window search. IEEE Transactions on Pattern Analysis and Machine Intelligence **13**(5), 466–477 (1991)

[57] Rao, V.N., Kumar, V.: Parallel depth-first search on multiprocessors part I: Implementation. International Journal of Parallel Programming **16**(6), 479–499 (1987)

[58] Reinefeld, A., Marsland, T.: Enhanced iterative-deepening search. IEEE Transactions on Pattern Analysis and Machine Intelligence **16**(7), 701–710 (1994)

[59] Romein, J.W., Bal, H.E.: Wide-area transposition-driven scheduling. In: Proceedings of the 10th IEEE International Symposium on High Performance Distributed Computing, pp. 347–355 (2001)

[60] Romein, J.W., Bal, H.E., Schaeffer, J., Plaat, A.: A performance analysis of transposition-table-driven work scheduling in distributed search. IEEE Transactions on Parallel and Distributed Systems **13**(5), 447–459 (2002)

[61] Romein, J.W., Plaat, A., Bal, H.E., Schaeffer, J.: Transposition table driven work scheduling in distributed search. In: Proceedings of the National Conference on Artificial Intelligence (AAAI), pp. 725–731 (1999)

[62] Romein, J.W., Plaat, A., Bal, H.E., Schaeffer, J.: Transposition table driven work scheduling in distributed search. In: Proceedings of the National Conference on Artificial Intelligence AAAI-99, pp. 725–731 (1999)

[63] Russell, S.: Efficient memory-bounded search methods. In: Proc. ECAI (1992)

[64] Stern, U., Dill, D.L.: Parallelizing the Murphi verifier. In: Proceedings of the 9th International Conference on Computed Aided Verification, pp. 256–278 (1997)

[65] Stern, U., Dill, D.L.: Parallelizing the Murphi verifier. Formal Methods in System Design **18**(2), 117–129 (2001)

[66] Sulewski, D., Edelkamp, S., Kissmann, P.: Exploiting the computational power of the graphics card: Optimal state space planning on the GPU. In: Proceedings of the 21st International Conference on Automated Planning and Scheduling, ICAPS 2011, Freiburg, Germany June 11-16, 2011 (2011). URL `http://aaai.org/ocs/index.php/ICAPS/ICAPS11/paper/view/2699`

[67] Valenzano, R., Nakhost, H., Müller, M., Schaeffer, J., Sturtevant, N.: ArvandHerd: Parallel planning with a portfolio. In: Proceedings of the 20th European Conference on Artificial Intelligence, pp. 786–791 (2012)

[68] Valenzano, R., Sturtevant, N., Schaeffer, J., Buro, K., Kishimoto, A.: Simultaneously searching with multiple settings: An alternative to parameter tuning for suboptimal single-agent search algorithms. In: Proceedings of the 20th International Conference on Automated Planning and Scheduling, pp. 177–184 (2010)

[69] Vidal, V., Bordeaux, L., Hamadi, Y.: Adaptive k-parallel best-first search: A simple but efficient algorithm for multi-core domain-independent planning. In: Proceedings of the 3rd Symposium on Combinatorial Search (SOCS'10) (2010)

[70] Vrakas, D., Refanidis, I., Vlahavas, I.: Parallel planning via the distribution of operators. Journal of Experimental and Theoretical Artificial Intelligence **13**(3), 211–226 (2001)

[71] Zhou, R., Hansen, E.: Domain-independent structured duplicate detection. In: Proceedings of the 21st National Conference on Artificial Intelligence AAAI-06, pp. 683–688 (2006)
[72] Zhou, R., Hansen, E.A.: Structured duplicate detection in external-memory graph search. In: Proceedings of the National Conference on Artificial Intelligence (AAAI), pp. 683–689 (2004)
[73] Zhou, R., Hansen, E.A.: Breadth-first heuristic search. Artificial Intelligence **170**(4), 385–408 (2006)
[74] Zhou, R., Hansen, E.A.: Domain-independent structured duplicate detection. In: Proceedings of the National Conference on Artificial Intelligence (AAAI), pp. 1082–1087 (2006)
[75] Zhou, R., Hansen, E.A.: Parallel structured duplicate detection. In: Proceedings of the National Conference on Artificial Intelligence (AAAI), pp. 1217–1223 (2007)
[76] Zhou, Y., Zeng, J.: Massively parallel A* search on a GPU. In: Proceedings of the National Conference on Artificial Intelligence (AAAI), pp. 1248–1255 (2015). URL `http://www.aaai.org/ocs/index.php/AAAI/AAAI15/paper/view/9620`
[77] Zobrist, A.L.: A new hashing method with application for game playing. reprinted in International Computer Chess Association Journal (ICCA) **13**(2), 69–73 (1970)

Chapter 12
Parallel Model Checking Algorithms for Linear-Time Temporal Logic

Jiri Barnat, Vincent Bloemen, Alexandre Duret-Lutz, Alfons Laarman, Laure Petrucci, Jaco van de Pol, and Etienne Renault

Abstract Model checking is a fully automated, formal method for demonstrating absence of bugs in reactive systems. Here, bugs are violations of properties in Linear-time Temporal Logic (LTL). A fundamental challenge to its application is the exponential explosion in the number of system states. The current chapter discusses the use of parallelism in order to overcome this challenge. We reiterate the textbook automata-theoretic approach, which reduces the model checking problem to the graph problem of finding cycles. We discuss several parallel algorithms that attack this problem in various ways, each with different characteristics: Depth-first search (DFS) based algorithms rely on heuristics for good parallelization, but exhibit a low complexity and good on-the-fly behavior. Breadth-first search (BFS) based approaches, on the other hand, offer good parallel scalability and support distributed parallelism. In addition, we present various simpler model checking tasks, which still solve a large and important subset of the LTL model checking problem, and show how these can be exploited to yield more efficient algorithms. In particular,

Jiri Barnat
Masaryk University, Brno, Czech Republic, e-mail: xbarnat@fi.muni.cz

Vincent Bloemen
University of Twente, Enschede, The Netherlands, e-mail: v.bloemen@utwente.nl

Alexandre Duret-Lutz
LRDE, Epita, Paris, France, e-mail: adl@lrde.epita.fr

Alfons Laarman
Leiden University, Leiden, The Netherlands, e-mail: a.w.laarman@liacs.leidenuniv.nl

Laure Petrucci
LIPN, CNRS, Paris, France, e-mail: Laure.Petrucci@lipn.univ-paris13.fr

Jaco van de Pol
University of Twente, Enschede, The Netherlands, e-mail: j.c.vandepol@utwente.nl

Etienne Renault
LRDE, Epita, Paris, France, e-mail: renault@lrde.epita.fr

Y. Hamadi und L. Sais (eds.), *Handbook of Parallel Constraint Reasoning*,
https://doi.org/10.1007/978-3-319-63516-3_12

we provide simplified DFS-based search algorithms and show that the BFS-based algorithms exhibit optimal runtimes in certain cases.

12.1 Introduction

This chapter discusses parallel algorithms for model checking properties of Linear-time Temporal Logic (LTL). Model checking [30, 8] is a verification technique to establish the correctness of hardware and software systems. In contrast to theorem proving, model checking is a fully automated procedure, invented by the Turing Award winners Clarke, Emerson, and Sifakis (2007). In contrast to testing, it is a complete and exhaustive method. Nowadays, along with testing and static analysis, model checking is an indispensable industrial tool for eliminating bugs and increasing confidence in hardware designs (e.g., at Intel [45] and IBM [15]) and software products (e.g., at Microsoft [9]). For an example case study, refer to Chapter 16, An Application of Parallel Satisfiability Solving to the Verification of Complex Embedded Systems.

Formally, model checking solves the problem: "Does model M satisfy property P?" ($M \models P$). Here the model M is a finite abstraction of a hardware or software system, provided in the form of a transition system. The paths in the graph of model M consist of infinite sequences of states connected by state transitions. Paths correspond to possible runs of the system. The property P is specified in some temporal logic. In this chapter, we restrict the discussion to Linear-time Temporal Logic (LTL). An LTL property denotes a set of paths, so P can be viewed as a specification of the correct runs of the system. Section 12.2 will formalize the syntax and semantics of LTL and identify some important fragments. For this introduction, it is sufficient to view model checking as a graph search problem, where the goal is to find a bad state or, more generally, a cycle representing an infinite path violating the property.

The main obstacle to model checking is the size of the transition system, often referred to as "the state space explosion" [96]. This graph grows exponentially in the number of components and variables in the specification, mainly due to parallel interleaving in concurrent systems, and the Cartesian product of data domains. Many sequential algorithms exist to address the state space explosion, reducing the state space by exploiting symmetries [29, 41, 20, 61], restricting the interleavings to be checked [95, 63, 54, 1], or abstracting the data domains [27, 24]. Another direction is to represent state spaces symbolically, applying powerful techniques such as Binary Decision Diagrams (BDD) [23, 77] or satisfiability (SAT) [28, 16, 78]. Parallel satisfiability is discussed in Chapter 1, Parallel Satisfiability, and parallel decision diagrams in Chapter 13, Multi-core Decision Diagrams. Although these methods greatly reduce the memory and time usage of model checking, the ever-growing complexity of hardware and software designs has meant that, so far, the practical application of model checking is still hindered by memory and time resources.

Parallel Model Checking Algorithms — Pragmatics

This chapter focuses on recent advances in utilizing more hardware resources to solve the model checking problem. In distributed model checking, the memory problem is alleviated by distributing the state space over the memory of many computers in some network (cluster, cloud). Recently, several new approaches to parallel model checking emerged using multiple processors in a shared-memory machine to speed up model checking computations. Both approaches are highly non-trivial, since graph (search) algorithms must be redesigned to be fit for parallel computation. Next, we consider parallel graph algorithms from pragmatic and theoretical points of view.

From a pragmatic point of view, obtaining good parallel speedups for graph problems is notoriously hard [75, 73]. This is mainly caused by the irregularity of graphs. The efficiency of parallel programs often depends on exploiting locality, which can be predicted for regular data structures like matrices. However, state spaces are irregular sparse graphs, whose shape highly depends on the model at hand. For distributed algorithms, the consequence is that traversing a transition from a source state in the graph often requires communication with the machine where the target is stored, leading to a dramatic communication overhead. For multi-core computing, the threads are continually looking up the location of target states in main memory. Since main memory (and the memory bus) are a shared resource, memory-intensive algorithms are hard to speed up on multi-core machines. As a consequence, practical implementations pay a lot of attention to low-level details, such as local caching, evading the need for locks using atomic instructions such as compare-and-swap, and latency hiding by asynchronous communication. This chapter does not focus on these implementation details, although they are essential to demonstrate that the treated algorithms achieve speedup in practice.

Instead, we focus on the algorithmic aspects. We review the basic sequential algorithms for LTL model checking in Section 12.3. These subproblems can be solved by linear-time algorithms. However, today the only known linear time algorithms heavily depend on the Depth-First Search (DFS) strategy, which (as we will explain below) is hard to parallelize. This holds for LTL algorithms based on Nested Depth-First Search as well as for those based on the analysis of the Strongly Connected Components.

Another reason for our general preference for DFS lies in the nature of search. If we use the algorithm to search for bugs (bug hunting), we can terminate as soon as the first bug has been found. It would be a waste of resources if we were to first compute the whole state space and then search only a small part to find the bug. The DFS-based algorithms are generally well-suited for on-the-fly model checking, where computing the state space and checking the properties are intertwined. This carries over to parallel search. It is well known that parallel random search can achieve superlinear speedups when the goal states are uniformly distributed [81, 72]. In case of full verification of programs, this consideration is less important. We will present parallel DFS-based algorithms in Section 12.4.

Parallel Model Checking Algorithms — Theoretical Considerations

Finally, what does theory actually say? A well-known result [82] is that DFS is inherently sequential. To understand this, we recall the class NC (Nick's Class) of problems that admit scalable parallel algorithms [56]: A problem is in NC if it can be solved in poly-logarithmic time $(\log n)^{o(1)}$ using a polynomial amount of hardware, i.e., $n^{o(1)}$ processors. Let P be the class of problems that admit a polynomial-time algorithm. A problem is P-complete , if all problems in P can be reduced to it by an NC algorithm. The canonical P-complete problem is CVP, the circuit valuation problem (given a circuit, and its Boolean input values, determine the value of its output). Although formally open, it is widely believed that NC does not contain P-complete problems, so problems in P are "inherently sequential." Note that if NC contained a single P-complete problem, then all polynomial problems would be parallelizable. Reif [82] actually showed that lexicographic DFS is P-complete by a direct reduction to the CVP. Hence, given a graph and a fixed ordering of transitions from each state, there is probably no parallel algorithm to even check whether node x will be visited before node y in the DFS post-order, observing the fixed transition ordering.

The following intellectual positions are possible in relation to this fact from theory: First, one can decide to ignore this theoretical restriction. This is the position in Section 12.4. We introduce various parallel random DFS algorithms for which we have shown practical speedup, even though they are not poly-logarithmic. The main motivation is that, in practice, the number of processors is much smaller than the size of the graph. A practical speedup for graphs from 10^3 to 10^8 nodes does not contradict the impossibility result in the limit case of $(10^8)^k$ processors.

The second position is to take the theoretical result seriously, and avoid DFS algorithms. Parallel BFS (breadth-first search), and hence SCC decomposition, is in NC [51], which can be shown by computing transitive closure with matrix multiplication. Several BFS-based model checking algorithms and SCC decomposition methods have been designed. Although their worst-case time complexity is strictly more than linear, they behave well on practical instances, and are even linear for many model checking fragments. Moreover, since BFS-based algorithms can be parallelized, with sufficiently many processors this approach should scale (even though the increased work-complexity doesn't admit a provably efficient parallel solution). Algorithms OWCTY and MAP in Section 12.5 are an illustration of BFS-based algorithms in this category.

The third possibility is to circumvent the theoretical results. Note that it is technically still possible that non-lexicographic DFS (without fixing the ordering of the transitions in advance) is in NC. Actually, it has been proved that free DFS is in NC indeed for planar graphs [57], and for general graphs the problem is known to be in Random NC [3]. We do not claim complexity-theoretic results in this chapter. Our random parallel free DFS algorithms will not provide a single global post-order and, in the worst case, they don't run in parallel logarithmic time. However, we have proved that they provide sufficient ordering to solve the model checking problem, and we demonstrated have good speedups for practical problems. Eventually, this approach might shed some light on this intriguing 30-year-old open problem.

12.2 Preliminaries: LTL Model Checking and Automata

The current section explains the theoretical foundation of LTL model checking. The formal approach taken here is to interpret both the system and its specification as an automaton. We will show that this automaton is exponential in the size of both the system and the specification and develop the constructs required by the LTL model checking algorithms in the subsequent section to efficiently handle such large automata.

12.2.1 Automata-Theoretic Model Checking

Model checkers are tools that take two inputs: some model M of a system, and some specification φ that should be satisfied by all possible behaviors of M. For instance if M is a model of a road intersection with traffic lights and sensors, the property φ could specify that whenever a car is sensed the light of its lane should eventually become green. Note that such a property is not necessarily about the *state* of the system: in this example it is about its possible behaviors, i.e., the evolution of its state. Furthermore, the behaviors of this system are infinite.

Model checking [97] decides whether some model M satisfies some specification φ (which we denote $M \models \varphi$). In the automata-theoretic approach, the model M is first converted into an automaton K_M whose language $\mathscr{L}(K_M)$ represents the set of all (infinite) behaviors of M. The negation of the formula φ is converted into an automaton $A_{\neg\varphi}$ whose language $\mathscr{L}(A_{\neg\varphi})$ captures the forbidden behaviors. With these objects, testing whether M satisfies φ amounts to checking the emptiness of the product of the two automata: if $\mathscr{L}(K_M \otimes A_{\neg\varphi}) = \emptyset$, then $M \models \varphi$. If $\mathscr{L}(K_M \otimes A_{\neg\varphi})$ is found not to be empty, it means there exists a counterexample: a behavior of M that invalidates φ.

12.2.2 Sequences and ω-Words

We shall use $\mathbb{B} = \{\bot, \top\}$ to denote the set of Boolean values, $\omega = \{0, 1, 2, \ldots\}$ for the set of non-negative integers, and $[n] = \{0, 1, 2, \ldots, n-1\}$ the first n of those. By convention $[0] = \emptyset$.

Let AP be a finite set of (atomic) propositions. An *assignment* is a function $x : \mathrm{AP} \to \mathbb{B}$ that evaluates each proposition. We use $\mathbb{B}^{\mathrm{AP}}$ to denote the set of all assignments of AP.

An *infinite sequence* over some set Σ is a function $\sigma : \omega \to \Sigma$ that assigns an element of Σ to each possible index. We use Σ^ω to denote the set of infinite sequences over Σ.

A *finite sequence* of length n over Σ is a function $\sigma : [n] \to \Sigma$. We use Σ^* for the set of all finite sequences of any length $n \geq 0$, and Σ^+ for the set of finite sequences of length $n > 0$.

To define a particular sequence, we denote it by the concatenation of its elements $x_i \in \Sigma$ as $\sigma = x_0;x_1;x_2;\ldots$, meaning that $\sigma(i) = x_i$.

For some infinite sequence $\sigma \in \Sigma^\omega$, we use σ^i to denote the sequence obtained from σ by removing its first $i \geq 0$ elements; i.e., $\sigma^i(j) = \sigma(i+j)$ for all j. We denote by $\mathrm{Inf}(\sigma) \subseteq \Sigma$ the set of elements that appear infinitely often in σ, i.e., $\mathrm{Inf}(\sigma) = \{s \in \Sigma \mid \forall i \in \omega, \exists j > i, \sigma(j) = s\}$.

In this chapter AP is assumed to be fixed, and infinite sequences of assignments, i.e., elements of $(\mathbb{B}^{\mathrm{AP}})^\omega$, are called *$\omega$-words*. Finally, a *language* is a (possibly infinite) set of ω-words.

12.2.3 Linear-Time Temporal Logic

In model checking, ω-words are used to represent the different behaviors of the system to check.

Linear-time Temporal Logic (LTL) formulas are typically used to specify the property to verify on the system by specifying which ω-words should be accepted or rejected. LTL formulas are constructed according to the following grammar, where $a \in \mathrm{AP}$:

$$\varphi ::= \top \mid \bot \mid a \mid \neg\varphi \mid \varphi \vee \varphi \mid \varphi \wedge \varphi \mid \varphi \,\mathsf{U}\, \varphi \mid \varphi \,\mathsf{R}\, \varphi \mid \mathsf{F}\varphi \mid \mathsf{G}\varphi \mid \mathsf{X}\varphi$$

Given an ω-word $\sigma \in (\mathbb{B}^{\mathrm{AP}})^\omega$ and an LTL formula φ, we say that σ satisfies φ (denoted $\sigma \models \varphi$) according to the following semantics. For any $a \in \mathrm{AP}$ and any LTL formulas φ_1 and φ_2,

$$\begin{array}{ll}
\sigma \models \top & \\
\sigma \not\models \bot & \\
\sigma \models a & \text{iff } \sigma(0)(a) = \top \\
\sigma \models \neg\varphi_1 & \text{iff } \sigma \not\models \varphi_1 \\
\sigma \models \varphi_1 \vee \varphi_2 & \text{iff } (\sigma \models \varphi_1) \vee (\sigma \models \varphi_2) \\
\sigma \models \varphi_1 \wedge \varphi_2 & \text{iff } (\sigma \models \varphi_1) \wedge (\sigma \models \varphi_2) \\
\sigma \models \varphi_1 \,\mathsf{U}\, \varphi_2 & \text{iff } \exists i \geq 0, (\sigma^i \models \varphi_2) \wedge (\forall j < i, \sigma^j \models \varphi_1) \\
\sigma \models \varphi_1 \,\mathsf{R}\, \varphi_2 & \text{iff } \forall i \geq 0, (\sigma^i \models \varphi_2) \vee (\exists j < i, \sigma^j \models \varphi_1) \\
\sigma \models \mathsf{F}\varphi_1 & \text{iff } \exists i \geq 0, \sigma^i \models \varphi_1 \\
\sigma \models \mathsf{G}\varphi_1 & \text{iff } \forall i \geq 0, \sigma^i \models \varphi_1 \\
\sigma \models \mathsf{X}\varphi_1 & \text{iff } \sigma^1 \models \varphi_1
\end{array}$$

The language of a formula φ is the set of words that satisfy it: $\mathscr{L}(\varphi) = \{\sigma \in (\mathbb{B}^{\mathsf{AP}})^\omega \mid \sigma \models \varphi\}$. Two LTL formulas are *equivalent* iff they have the same language: $\varphi_1 \equiv \varphi_2 \iff \mathscr{L}(\varphi_1) = \mathscr{L}(\varphi_2)$. For example one can see that $\neg\mathsf{FG}a \equiv \mathsf{GF}\neg a$.

The size of an LTL formula φ, denoted $|\varphi|$, is the number of symbols in φ. For example $|\neg\mathsf{FG}a| = 4$.

12.2.4 Kripke Structures

A Kripke structure is an automaton with states labeled by assignments.

Definition 1 (Kripke Structure). A Kripke structure is a tuple $K = (Q, \iota, \delta, \ell)$ where

- Q is a finite set of states,
- $\iota \in Q$ is the initial state,
- $\delta \subseteq Q \times Q$ is a set of transitions,
- $\ell : Q \to \mathbb{B}^{\mathsf{AP}}$ is a function labeling each state with an assignment.

The runs of K, denoted $\mathrm{Runs}(K)$, are the infinite sequences of states $\rho \in Q^\omega$ that start with ι and follow transitions in δ:

$$\mathrm{Runs}(K) = \{\rho \in Q^\omega \mid \rho(0) = \iota \text{ and } \forall i \geq 0, (\rho(i), \rho(i+1)) \in \delta\}$$

If we naturally extend the labeling function ℓ to runs, then each run ρ is associated with an ω-word $\ell(\rho)$ defined by $\ell(\rho)(i) = \ell(\rho(i))$. The language $\mathscr{L}(K)$ of the Kripke structure is the set of words associated with all its runs: $\mathscr{L}(K) = \{\ell(\rho) \mid \rho \in \mathrm{Runs}(K)\}$.

Definition 2 (Deadlock-Free Kripke Structure). A Kripke structure is said to be *deadlock-free* if all its states have at least one successor. In other words $K = (Q, \iota, \delta, \ell)$ is *deadlock-free* if $\forall s \in Q, \exists d \in Q, (s, d) \in \delta$.

12.2.5 Büchi Automata

Büchi automata can represent ω-regular languages. We shall define different flavors of Büchi automata that correspond to combinations of the following two options:

- transition-based or state-based acceptance
- classical Büchi acceptance, or generalized Büchi acceptance.

While all the resulting automata have the same expressive power, they can have different degrees of conciseness, and may require different emptiness-check procedures. Hence from the model checking point of view, these choices can affect memory consumption and emptiness-check complexity.

Definition 3 (TGBA). A *Transition-based Generalized Büchi Automaton* is a tuple $A = (Q, \iota, \delta, n, M)$ where

- Q is a finite set of states,
- $\iota \in Q$ is the initial state,
- $\delta \subseteq Q \times \mathbb{B}^{\mathrm{AP}} \times Q$ is a set of transitions,
- n is an integer specifying a number of accepting marks,
- $M : \delta \to 2^{[n]}$ is a *marking* function that specifies a subset of marks associated with each transition.

For a transition $t \in \delta$ we write t^s for its source, t^ℓ for its label, and t^d for its destination: $t = (t^s, t^\ell, t^d)$.

The runs of A are infinite sequences of consecutive transitions:

$$\mathrm{Runs}(A) = \{\rho \in \delta^\omega \mid \rho(0)^s = \iota \text{ and } \forall i \geq 0,\, \rho(i)^d = \rho(i+1)^s\}$$

The *accepting runs* of A are those that have, for each acceptance mark, infinitely many transitions with that mark:

$$\mathrm{Acc}(A) = \left\{\rho \in \mathrm{Runs}(A) \;\middle|\; [n] = \bigcup_{t \in \mathrm{Inf}(\rho)} M(t)\right\}$$

Let us also define the word $\ell(\rho)$ associated with a run ρ by $\ell(\rho)(i) = \rho(i)^\ell$. Now the language $\mathscr{L}(A)$ of the automaton A is the set of words associated with its accepting runs:

$$\mathscr{L}(A) = \{\ell(\rho) \mid \rho \in \mathrm{Acc}(A)\}$$

For convenience, we will also overload the δ notation and write $\delta(q)$ for the set of outgoing transitions of any state $q \in Q$: $\delta(q) = \{(s,x,d) \in \delta \mid s = q\}$.

Definition 4 (SGBA). A *State-based Generalized Büchi Automaton* is also a tuple $A = (Q, \iota, \delta, n, M)$, with identical definitions for Q, ι, δ, and n, but this time the marking function M associates marks with states: $M : Q \to 2^{[n]}$. The runs are defined similarly. The accepting runs are those that have infinitely many states marked with each acceptance mark:

$$\mathrm{Acc}(A) = \left\{\rho \in \mathrm{Runs}(A) \;\middle|\; [n] = \bigcup_{t \in \mathrm{Inf}(\rho)} M(t^s)\right\}$$

and then the automaton's language is still defined as $\mathscr{L}(A) = \{\ell(\rho) \mid \rho \in \mathrm{Acc}(A)\}$.

Definition 5 (SBA and TBA). *State-based* and *Transition-based Büchi Automata* are particular cases of the above definitions where $n = 1$.

Figure 12.1 shows four automata with different acceptance conditions, all recognizing the language of the LTL formula $\mathsf{GF}a \wedge \mathsf{GF}b$: a and b should each hold infinitely often, but not necessary at the same time. As usual, multiple transitions of the form (s,x,d) and (s,y,d) are pictured as a single edge $(s) \xrightarrow{x,y} (d)$.

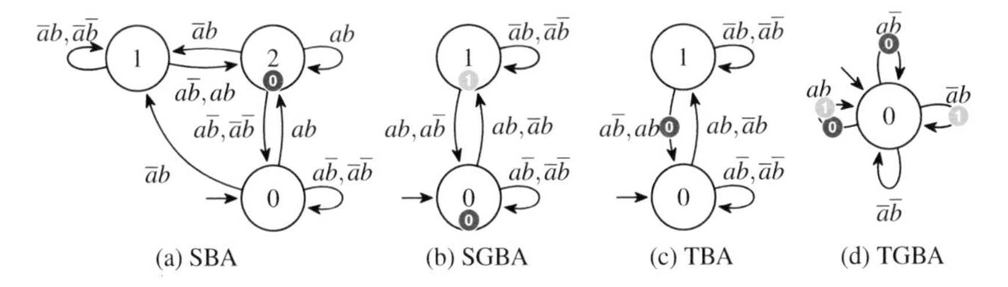

Fig. 12.1: Minimal deterministic automata recognizing $\mathscr{L}(\mathsf{GF}a \land \mathsf{GF}b)$. The SGBA and TGBA use $n = 2$ accepting marks, while the SBA and TBA have $n = 1$ by definition

Marked states and transitions are denoted using colored bullets such as ⓿ or ①. So the fact that $M\big((s,x,d)\big) = \{0,1\}$ is pictured as (s)-⓿$\overset{x}{-}$①→(d). Looking at the automaton of Figure 12.1(b), the run $\rho_1 = (0,\bar{a}b,1);(1,\bar{a}b,1);(1,a\bar{b},0);(0,\bar{a}b,1);(1,\bar{a}b,1);(1,a\bar{b},0);\ldots$ is an accepting run for the word $\bar{a}b;\bar{a}b;a\bar{b};\bar{a}b;\bar{a}b;a\bar{b};\ldots$ as it visits ⓿ and ① infinitely often. The run $\rho_2 = (0,\bar{a}b,1);(1,\bar{a}b,1);(1,\bar{a}b,1);(1,\bar{a}b,1);\ldots$ is not accepting because it only visits ① infinitely often. By comparing the two definitions of Acc, it is clear that an SGBA $A = (Q,\iota,\delta,n,M)$ can be converted into a language-equivalent TGBA $B = (Q,\iota,\delta,n,M')$ by defining $M'(t) = M(t^s)$. This amounts to pushing the acceptance marks onto the outgoing transitions, as in Figure 12.2.

The automata of Figure 12.1 are minimal in the sense that there does not exist language-equivalent automata with the same acceptance condition and fewer states. This figure is therefore an example showing how TGBAs *can be* more concise than the other types of automata presented, but in Section 12.2.8 we will also discuss some classes of properties for which using SBAs is sufficient, i.e., no reduction can be obtained by using generalized or transition-based acceptance.

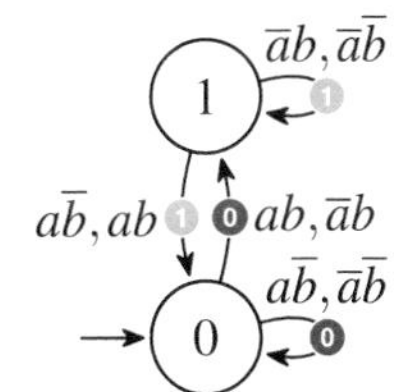

Fig. 12.2: How to interpret the SGBA of Fig. 12.1(b) as a TGBA

Property 1. Any TGBA (Q,ι,δ,n,M) can be "degeneralized" into a language-equivalent SBA with at most $(n+1)\,|Q|$ states, or into a language-equivalent TBA with at most $n \cdot |Q|$ states.

There exist several variants of degeneralization constructions, discussed for instance by Gastin and Oddoux [49], or Giannakopoulou and Lerda [53], and improved by Babiak et al. [7]. The automata of Figures 12.1(a) and (c) are typically what one could obtain by degeneralizing the TGBA of Figure 12.1(d).

Property 2. For any LTL formula φ, there exists a language-equivalent TGBA with $\mathrm{O}(2^{|\varphi|})$ states and $n = \mathrm{O}(|\varphi|)$ acceptance marks.

Numerous *translations* from LTL to TGBAs exist, and are implemented in tools such as `ltl2ba` [49], `ltl3ba` [6], or Spot's `ltl2tgba` [37]. Now, combining Properties 1 and 2, we get

Property 3. For any LTL formula φ, there exists a language-equivalent SBA with $O(|\varphi| \cdot 2^{|\varphi|})$ states.

These upper bounds are rarely reached in practice. For instance Dwyer et al. [39] define 55 LTL formulas[1] that represent 11 intents (Absence, Response, Precedence, etc) combined with five different scopes (Before, Between, After, etc). These 55 formulas have an average size of 16.75 (maximum 40), but the SBAs produced by `ltl2tgba` (from Spot 2.1) have on average only 3.945 states (maximum 13). Using TGBAs instead of SBAs is only marginally better: `ltl2tgba` produces TGBAs with an average of 3.782 states (maximum 10); we will discuss this point in Section 12.2.8.

These small automata, representing the negation of a property we want to check, will be combined with a (potentially very large) Kripke structure representing the state space of the model to verify.

Property 4 (Synchronized product). Let $K = (Q_1, \iota_1, \delta_1, \ell)$ be a Kripke structure, and $A = (Q_2, \iota_2, \delta_2, n, M)$ be a TGBA. Then the TGBA $K \otimes A = (Q', \iota', \delta', n, M')$ where

- $Q' = Q_1 \times Q_2$,
- $\iota' = (\iota_1, \iota_2)$,
- $((s_1, s_2), x, (d_1, d_2)) \in \delta' \iff (s_1, d_1) \in \delta_1 \wedge \ell(s_1) = x \wedge (s_2, x, d_2) \in \delta_2$,
- $M'\big(((s_1, s_2), x, (d_1, d_2))\big) = M\big((s_2, x, d_2)\big)$,

is such that $\mathscr{L}(K \otimes A) = \mathscr{L}(K) \cap \mathscr{L}(A)$.

The product between a Kripke structure and a SGBA can be defined similarly, with $M'\big((s_1, s_2)\big) = M(s_2)$ as the only change.

Clearly $|Q'| = |Q_1| \cdot |Q_2|$. However the states reachable from ι' can be a subset of that, and only that subset needs to be explored to decide whether $\mathscr{L}(K \otimes A)$ is empty.

12.2.6 The Emptiness-Check Problem

The emptiness-check problem can be presented as follows:

Given an automaton $B = (Q, \iota, \delta, n, M)$, decide whether $\mathscr{L}(B) = \emptyset$.

The automaton B could be any type of automaton presented previously. We will focus on TGBA, the more compact ones, as well as SBA, more frequently used because of their simple structure.

[1] `http://patterns.projects.cs.ksu.edu/documentation/patterns/ltl.shtml`

Property 5. If $\mathscr{L}(B) \neq \emptyset$, then there exists a *lasso-shaped* accepting run, i.e., a run $\rho \in \mathrm{Acc}(B)$ for which there exist $i \geq 0$ and $j \geq i$ such that $\rho(i) = \rho(j)$. (Figure 12.3.)

To show the existence of such a run, consider an automaton B (a TGBA or SGBA) and assume that $\mathscr{L}(B) \neq \emptyset$. Then by definition of $\mathscr{L}(B)$, there exists an accepting run $\pi \in \mathrm{Acc}(B)$, but that run is not necessarily lasso-shaped. The set $\mathrm{Inf}(\pi)$ contains transitions of B that (1) are visited infinitely often by π, (2) cover all acceptance marks (since π is accepting), (3) are all reachable from one another, and (4) are reachable from the initial state. Then a lasso-shaped run ρ can be constructed by building a prefix connecting the initial state of B to any transition $t \in \mathrm{Inf}(\pi)$, and then building a cycle around t that visits all transitions of $\mathrm{Inf}(\pi)$. Note that for the lasso-shaped run ρ, the set $\mathrm{Inf}(\rho)$ corresponds exactly to the transitions that appear on the cycle. We therefore have $\mathrm{Inf}(\rho) \supseteq \mathrm{Inf}(\pi)$, which entails that ρ is also accepting.

Definition 6 (Accepting cycle). Given a TGBA (Q, ι, δ, n, M), and a finite sequence of transitions $c \in \delta^+$ of length k. We say that c is a *cycle* if its transitions actually form a cycle: $\forall i < k,\ c(i)^d = c(i+1 \mod k)^s$.

We say that a cycle c is an *elementary cycle* if additionally $|\{c(i)^s \mid i < k\}| = k$, i.e., if c goes through k different states.

We say that a cycle c is an *accepting cycle* if its transitions visit each acceptance mark at least once: $\forall i \in [n], \exists j < k, i \in M(c(j))$. Accepting cycles for SGBA are defined likewise, replacing $M(c(i))$ by $M(c(i)^s)$.

Note that the cycle part of any lasso-shaped accepting run is an accepting cycle. Combining this with Property 5 allows us to reduce the emptiness-check problem to the search for an accepting cycle.

Property 6. For an automaton B, we have $\mathscr{L}(B) \neq \emptyset$ if and only if B contains an accepting cycle reachable from the initial state.

However the number of cycles can be infinite, so it is useful to consider the simpler case where only *elementary cycles* need to be checked for acceptance:

Property 7. For an automaton B with $n \leq 1$ acceptance marks, we have $\mathscr{L}(B) \neq \emptyset$ if and only if B contains an accepting *elementary* cycle reachable from the initial state.

The case with $n = 0$ is obvious, since any cycle would be accepting, and if a cycle exists, an elementary cycle also exists. For $n = 1$, any accepting cycle c contains some

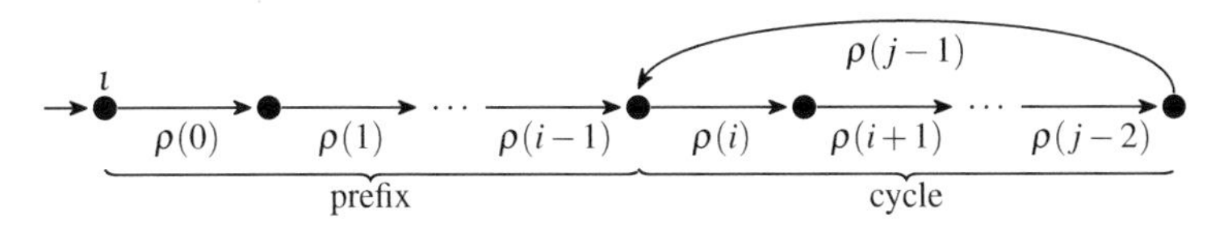

Fig. 12.3: A lasso-shaped run can be built from two finite sequences of transitions: a (possibly empty) prefix and a (non-empty) cycle

transition $c(i)$ such that $M(c(i)) = 1$, and there necessarily exists some elementary accepting cycle around this transition. Note that this does not hold for $n \geq 2$, as in the example of Figure 12.4 where the only two elementary cycles are rejecting, but they can be combined to form an infinite number of accepting cycles.

The goal of all emptiness-check algorithms presented in the sequel is to establish the existence or absence of such accepting cycles. Finding an accepting lasso-shaped run is one direct way to prove the existence of a reachable accepting cycle, but it is not the only one. Another one, which is especially useful with generalized acceptance ($n \geq 2$), is to prove that the automaton has a (reachable) strongly connected component that covers all acceptance marks. This is formalized by Definition 7 and Property 8.

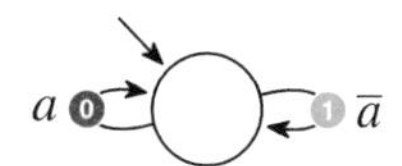

Fig. 12.4: This TGBA has an infinite number of accepting cycles; none are elementary

Definition 7 (SCC). In an automaton (Q, ι, δ, n, M), a *partial strongly connected component* (partial SCC) is a nonempty set of states $C \subseteq Q$ such that any ordered pair of states of C can be connected by a sequence of consecutive transitions. If additionally C is maximal with respect to set inclusion, we call it a *maximal strongly connected component* (maximal SCC). Let us use $C_\delta = \{(s,x,d) \in \delta \mid s \in C, d \in C\}$ to denote the set of transitions induced by C.

We call an SCC C *trivial* if $C_\delta = \emptyset$. In a TGBA we say that a non-trivial SCC C is accepting if C_δ covers all acceptance marks, i.e., $\forall i \in [n], \exists t \in C_\delta, i \in M(t)$. In an SGBA a non-trivial SCC C is accepting if C covers all acceptance marks, i.e., $\forall i \in [n], \exists s \in C, i \in M(s)$.

A *rejecting* SCC is either a trivial SCC, or a non-trivial SCC that does not cover all acceptance marks.

Property 8. For an automaton B, we have $\mathscr{L}(B) \neq \emptyset$ if and only if the initial state can reach an accepting SCC.

Note that it does not matter whether the accepting SCC is partial. SCC-based emptiness checks usually maintain a set of partial SCCs, to which they add new states when cycles are discovered. For each (reachable) partial SCC C they maintain the set of acceptance marks seen in C (that is $S_C = \bigcup_{t \in C_\delta} M(t)$ in the case of TGBAs, or $S_C = \bigcup_{s \in C} M(s)$ for SGBAs), and they can report the non-emptiness of the automaton as soon as one of these sets equals $[n]$.

In the context of model checking, the automaton B to be checked for emptiness is actually the product of a Kripke structure (representing the state space of the model under verification) with an automaton capturing the behaviors invalidating an LTL formula φ (the specification to check).

Theorem 1. *Let φ be an LTL formula, $A_{\neg\varphi}$ an automaton with n acceptance marks such that $\mathscr{L}(\neg\varphi) = A_{\neg\varphi}$, and K a Kripke structure. The following statements are equivalent:*

1. $\mathscr{L}(K) \subseteq \mathscr{L}(\varphi)$,
2. $\mathscr{L}(K) \cap \mathscr{L}(A_{\neg\varphi}) = \emptyset$,

3. $\mathscr{L}(K \otimes A_{\neg\varphi}) = \emptyset$,
4. $K \otimes A_{\neg\varphi}$ *has no reachable, accepting cycle;*
or in case $n \leq 1$ *no reachable accepting elementary cycle,*
5. $K \otimes A_{\neg\varphi}$ *has no reachable, accepting SCC.*

The emptiness checks we will present either look for accepting elementary cycles (when $n \leq 1$) or accepting SCCs. However an important point is that they search for those in the product $K \otimes A_{\neg\varphi}$. Because the Kripke structure K can be pretty large, a classical optimization is to generate both the Kripke structure K and the product $K \otimes A_{\neg\varphi}$ on the fly, as required by the needs of the emptiness-check procedure. Doing so avoids generating any part of K that would never be reached in the product, and it may also save a lot of time in case an accepting cycle is discovered early: the emptiness check can then exit immediately without exploring the rest of the product. For this on-the-fly construction to work, the emptiness check should only move *forward*, i.e., from a given state (s_1, s_2) of the product, one may only compute its successors, but not its predecessors. Originally, only the initial state (ι_1, ι_2) is known, and the emptiness check may explore the successors of this state, as well as the successors of any new state discovered this way. In such a setup, any cycle or SCC we discover is necessarily *reachable*.

12.2.7 Implicit Models and Automata

We have seen in Property 2 that the size of the Büchi automaton can be exponential in the size of the LTL formula, i.e., the number of symbols it contains. Not much has been said about the size of the model M. To expand on this, we first need to make some assumptions about its representation.

Definition 8. A *model* is a tuple $M = (D, \theta, \text{state-labels}, \text{next-state})$ where

- $D = V_1 \times \cdots \times V_k$ is the data of the model composed of k Boolean variables,
- $\theta \in D$ is the initial state,
- $\text{state-labels}: D \rightarrow \mathbf{2}^{\text{AP}}$ is a state label function, and
- $\text{next-state}: D \rightarrow \mathbf{2}^{D}$ is a next-state function.

The data D of the model can be thought of as the values of all variables and program (thread) counters in some imperative language. The set D represents all *potential* states of the model. The next-state function provides an implicit encoding of all transitions in the system from a given state. It is typically an implementation of the system semantics of the individual program statements; for an example see [62].

The actual Kripke structure can be computed as an explicit representation of the data that the model represents implicitly.

Definition 9. The Kripke structure $K_M = (Q, \iota, \delta, \ell)$ of a model $M = (D, \theta, \text{state-labels}, \text{next-state})$ is defined as follows:

- $\iota = \theta$,

- Q is the smallest fixpoint of next-state that includes θ,
- $\delta = \{(s,d) \in D^2 \mid d \in \text{next-state}(s)\}$, and
- $\ell = \text{state-labels}$.

The introduction mentioned that the graph of the system (the Kripke structure of the model) is exponential in the number of components and variables. We can now be more exact. Let n be an upper bound on the data domains, i.e., $|V_i| \leq n$ $(0 \leq i \leq k)$.

Property 9. The number of states in the Kripke structure $K = (Q, \iota, \delta, \ell)$ is exponential in the number of variables in the model (k): $|Q| \in \mathrm{O}(n^k)$.

The implicit definition of the Kripke structure can be extended to the product automaton as well.

Definition 10 (Implicit Product Automaton). The *implicit product automaton* of a model $M = (D, \theta, \text{state-labels}, \text{next-state})$ and a TGBA $A = (Q, \iota, \delta, n, M)$ is the implicit TGBA $C = (Q', \iota', \text{next-product}, n, M')$ where

- $\iota' = (\theta, \iota)$,
- $Q' = D \times Q$,
- $(x, (d_1, d_2)) \in \text{next-product}((s_1, s_2)) \iff (d_1) \in \text{next-state}(s_1) \wedge \text{state-labels}(s_1) = x \wedge (s_2, x, d_2) \in \delta$, and
- $M'(((s_1, s_2), x, (d_1, d_2))) = M((s_2, x, d_2))$.

Definition 11. The TGBA $(Q'', \iota', \delta', n, M')$ *generated from* the implicit product automaton $(Q', \iota', \text{next-product}, n, M')$ is defined by taking:

- Q'' is the smallest fixpoint of next-product that includes ι',
- $\delta' = \{(s, x, d) \in D^2 \mid (x, d) \in \text{next-product}(s)\}$.

Property 10. By definition, the product TGBA of M and A in Definition 11 is the same as $K_M \otimes A$ from Property 4.

Property 11. The number of states in the product structure $K_M \otimes A_{\neg\varphi} = (Q, \iota, \delta, n, M)$ of a model $M = (D, \theta, \text{state-labels}, \text{next-state})$ and a TGBA $A_{\neg\varphi}$ can be exponential in the number of variables in the model ($|D| = l$, with data domains bounded by n) and in the formula φ: $|Q| \in \mathrm{O}(n^l \times \mathbf{2}^{|\varphi|})$.

The implicit definition helps us to avoid storing all transitions of the Kripke structure and its product, by recomputing them from the states. Moreover, entire parts of the Kripke structure might never have to be generated as they are suppressed by the synchronization of the product. The algorithms in the subsequent section will therefore use the implicit definition. While this definition prevents algorithms from doing backwards traversals (the inverse of next-state is not always computable), we will see that this is not required.

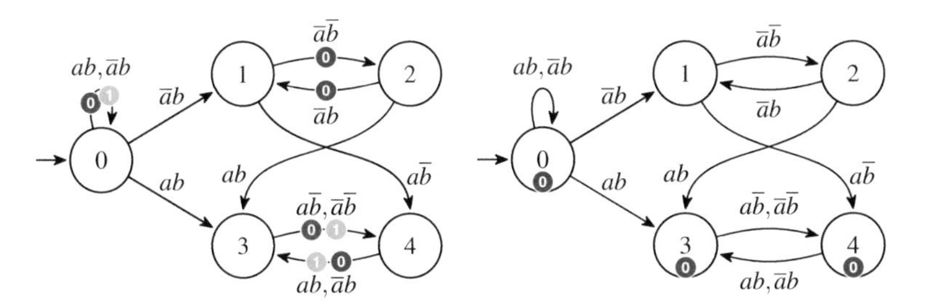

Fig. 12.5: A weak TGBA (left) and an equivalent weak SBA (right). Both have two accepting SCCs and one rejecting SCC. Inside each SCC, all transitions or states bear the same marks. Their language is that of the formula $(\mathsf{F}a \wedge \mathsf{G}((b \wedge \mathsf{X}\neg b) \vee (\neg b \wedge \mathsf{X}b)))\,\mathsf{R}\,b$, which is an LTL persistence

12.2.8 Simpler Subclasses

In 1990, Manna and Pnueli [76] presented a classification of temporal properties (i.e., languages expressed either as LTL or automata), into a hierarchy. Two subclasses are of particular interest in the context of model checking [25]: *guarantee* and *persistence* properties. The reason is that they can be represented by automata with additional constraints that simplify their emptiness checks.

Let us call an LTL guarantee (φ_G) and an LTL persistence (φ_P) any property that can be defined as an LTL formula using the following grammar, where $a \in AP$ is any atomic proposition. (φ_S and φ_R correspond to the dual classes of safety and recurrence.)

$$\varphi_G ::= \bot \mid \top \mid a \mid \varphi_G \vee \varphi_G \mid \varphi_G \wedge \varphi_G \mid \mathsf{X}\varphi_G \mid \mathsf{F}\varphi_G \mid \varphi_G\,\mathsf{U}\,\varphi_G \mid \neg\varphi_S$$
$$\varphi_S ::= \bot \mid \top \mid a \mid \varphi_S \vee \varphi_S \mid \varphi_S \wedge \varphi_S \mid \mathsf{X}\varphi_S \mid \mathsf{G}\varphi_S \mid \varphi_S\,\mathsf{R}\,\varphi_S \mid \neg\varphi_G$$
$$\varphi_P ::= \varphi_S \mid \varphi_G \mid \varphi_P \vee \varphi_P \mid \varphi_P \wedge \varphi_P \mid \mathsf{X}\varphi_P \mid \mathsf{F}\varphi_P \mid \varphi_P\,\mathsf{U}\,\varphi_P \mid \varphi_P\,\mathsf{R}\,\varphi_S \mid \neg\varphi_R$$
$$\varphi_R ::= \varphi_S \mid \varphi_G \mid \varphi_R \vee \varphi_R \mid \varphi_R \wedge \varphi_R \mid \mathsf{X}\varphi_R \mid \mathsf{G}\varphi_R \mid \varphi_R\,\mathsf{R}\,\varphi_R \mid \varphi_R\,\mathsf{U}\,\varphi_G \mid \neg\varphi_P$$

For instance, $\mathsf{GF}a$ is a recurrence formula (φ_R), $\mathsf{FG}b$ is a persistence formula (φ_P), but the conjunction of these two formulas $\mathsf{GF}a \wedge \mathsf{FG}b$ does not belong to any of the above classes.

LTL guarantee and LTL persistence formulas can be represented respectively by *terminal* and *weak* automata.

Definition 12 (Weak Automaton). A TGBA (or SGBA) is *weak* if in any of its SCCs all transitions (or states) have the same marks.

This definition implies that in each SCC of a weak automaton, either all cycles are accepting, or all cycles are rejecting. Because of that, any weak TGBA (Q,ι,δ,n,M) can be trivially converted into an equivalent SBA $(Q,\iota,\delta,1,M')$, with the same transition structure, but defining M' by $M'(s) = [1]$ if there exists a transition $t \in \delta(s)$

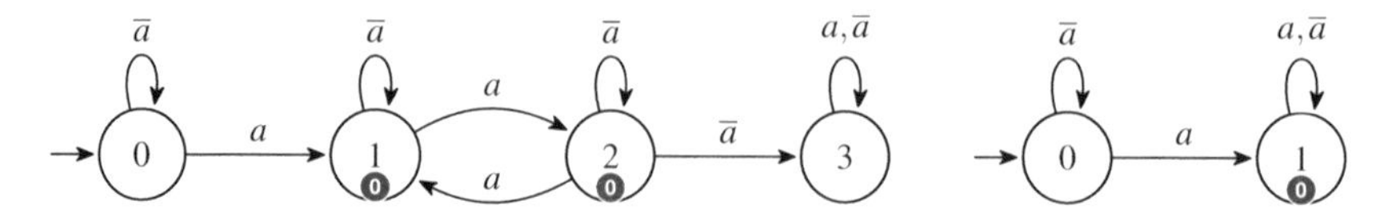

Fig. 12.6: Two terminal SBAs recognizing $\mathscr{L}(\mathsf{F}a)$. The left one was made artificially more complex to illustrate how any terminal automaton can be simplified by compacting all accepting SCCs into a single (and unique) state, and removing any SCCs that are only reachable via an accepting SCC

such that $M(t) = [n]$ and t^s and t^d belong to the same SCC; or $M'(s) = \emptyset$ otherwise. Figure 12.5 illustrates this.

Weak automata can still express a large subclass of LTL properties. Many properties encountered in practice turn out to be weak or even simpler [11, 69].

An even simpler subclass of weak automata is terminal automata.

Definition 13 (Terminal Automaton). A TGBA (SGBA) (Q, ι, δ, n, M) is *terminal* if it is weak, and if any of its accepting SCCs is complete: that is, for any accepting SCC $C \subseteq Q$, any pair of states $s, d \in C$ within that SCC, and any assignment $x \in \mathbb{B}^{\mathrm{AP}}$, there exists $(s, x, d) \in \delta$.

The states that belong to accepting SCCs are called *terminal states*.

Note that because the accepting SCCs of terminal automata are complete, they will accept all suffixes. Therefore any terminal automaton can be simplified into an equivalent terminal automaton with a single terminal state looping over all possible assignments. Figure 12.6 illustrates this.

Property 12. From any LTL guarantee (φ_G on page 471) one can build an equivalent terminal automaton. Similarly, one can build a weak automaton equivalent to any LTL persistence (φ_P).

The subclass of LTL guarantees is simple enough that typical LTL translation algorithms [49, 6, 37] produce terminal automata naturally. A construction of weak automata from LTL persistence properties is given by Černá and Pelánek [25], and is implemented for instance in `ltl2tgba`.

The usefulness of terminal automata for model checking comes from the fact that to prove the existence of an accepting run, we only need to reach a terminal state. This fact also applies to the product with a Kripke structure, provided that the Kripke structure is known to be *deadlock-free* (Definition 2).

Property 13. Let $K = (Q_1, \iota_1, \delta_1, \ell)$ be a *deadlock-free* Kripke structure, and $A = (Q_2, \iota_2, \delta_2, n, M)$ a terminal automaton. $\mathscr{L}(A \otimes K) \neq \emptyset$ if and only if there exists a reachable state $(s_1, s_2) \in Q_1 \times Q_2$ where s_2 is a terminal state.

Indeed, the fact that K is dead lock-free implies that any prefix from ι_1 to s_1 can be continued into a lasso-shaped accepting run on K, and the fact that s_2 belongs to an accepting and complete SCC means that any suffix can be accepted from there. Therefore, upon reaching (s_1, s_2) it is clear that an accepting run can be found in $A \otimes K$.

In the subsequent section, we show that these simpler classes of automata also allow for simpler algorithms to solve the emptiness-check problem.

12.3 Basic Sequential LTL Model Checking Algorithms

The current section presents sequential algorithms for checking emptiness of Büchi automata. As discussed in the previous section, this problem can be solved in the case of $n \leq 1$ by showing that none of the elementary cycles are accepting. In the generalized case with $n \geq 2$, however, all cycles need to be considered according to Theorem 1. Therefore, we present a specialized algorithm called Nested Depth-First Search for the case where $n \leq 1$ and an SCC-based algorithm for the general case. We will show that the generality of the second algorithm comes at the cost of a slightly higher resource consumption.

We also saw that the automaton to check is the *product* between the property automaton $A_{\neg\varphi}$ and the Kripke structure K_M. Since this product can be large, a classical technique these algorithms employ is to compute this product *on the fly*. Before presenting the algorithms, we first discuss the on-the-fly technique and its advantages.

12.3.1 On-the-Fly Algorithms

While the automaton $A_{\neg\varphi}$ representing the specification is usually quite small (often fewer than 10 states), the automaton K_M can have billions of states, and the product of these two automata is a Cartesian product of their states in the worst case (i.e., $|K_M \otimes A_{\neg\varphi}| \leq |K_M| \otimes |A_{\neg\varphi}|$).

For efficiency reasons model checkers will therefore compute K_M and $K_M \otimes A_{\neg\varphi}$ *on the fly*, using the implicit definitions from Section 12.2.7. So instead of using the static definition of product transitions δ, we use its implicit counterpart next-product. This approach has various advantages:

- any part of K_M that does not synchronize with $A_{\neg\varphi}$ is not computed,
- we do not need to store the transitions of K_M and $K_M \otimes A_{\neg\varphi}$ since these can be recomputed when needed, and
- states can be deleted and recomputed, at the expense of re-explorations of the automaton, thus allowing for trading of computation time for memory use.[2]

[2] Various state space caching techniques have been invented that also ensure termination of the model checking algorithm [55, 89].

The advantages are especially important when we recall that the number of states in the product automaton is exponential in both the property and the system (see Property 11). As memory is often a bottleneck for model checking, it would be disastrous to store those as well since there might be up to quadratically more transitions than states.

An important consequence is that these emptiness-check algorithms are only allowed to move forward in the automaton: from a state of A, one can compute the successors, but not the predecessors. This restriction comes from the fact that the actions of the original model might not be reversible (it might be intractable to compute the inverse of next-product). While respecting this constraint, the emptiness check needs to explore the product automaton to find information about cycles or SCCs.

12.3.2 Depth-First Search

This exploration can be done using one of the two classical graph traversal algorithms: *breadth-first search* (BFS) or *depth-first search* (DFS). These algorithms iterate over vertices of a graph (or states of an automaton). The evolution of both DFS and BFS may be described as a process by which every state in the automaton is colored. At the beginning a state has no color and, at some point, it becomes "activated" and receives its color. In the general description of DFS below, we use $\perp$ for "no color" and $\top$ for "a color". These algorithms only differ by the order in which states are colored. In depth-first search, when choosing which state to explore next, children are favored over siblings. In contrast, in a breadth-first search siblings are favored over children. Even if both DFS and BFS have running time that is linear in the size of the product automaton (i.e., the number of states plus the number of transitions), most sequential emptiness checks are based on a DFS exploration since it can be used to detect cycles easily.

Algorithm 12.1: Depth-First Search Algorithm

```
1 function SETUP (A = (Q, ι, next-product, n, M))
2 └ DFS(A, ι)
3 function DFS (A = (Q, ι, next-product, n, M), s)
4 │ s.color := ⊤
5 │ forall t ∈ next-product(s) do
6 │ │ if t^d.color = ⊥ then
7 └ └ └ DFS(A, t^d)
```

Algorithm 12.1 presents a DFS exploration for an implicit automaton $A = (Q, \iota, \text{next-product}, n, M)$. Lines 1–2 only set up the exploration and launch the DFS exploration with the initial state ι of the automaton A. The main procedure (lines 3–

7) maintains for each state a Boolean *color*, initially set to $\bot$, that keeps track of "activated" states. Every time a state is visited, its field *color* is set to $\top$ (line 4). At line 5 all successors of the currently visited state are processed: only new ones, i.e., with $color = \bot$, are recursively visited (line 7). The stack of recursive calls is also called the *DFS stack*. A state that is colored $\top$ and is on the stack is called *scheduled* or *stacked*. Once all its successors have been considered, it is popped off the stack, or *backtracked*.

A closer look at this algorithm shows that DFS exploration by its nature supports on-the-fly processing: only the initial state is used at the beginning (line 2) and the predecessors of a state are never computed (line 5).

The emptiness of a terminal automaton $A = (Q, \iota, \text{next-product}, n, M)$ (see Section 12.2.8) can easily be verified using the above DFS. All we have to do is to check whether $M(t) = [n]$ (for transition-based acceptance) or $M(t^s) = [n]$ (for the state-based case) in the *for* loop. The check is so simple that it can be done by a BFS algorithm as well.

To detect elementary cycles of the automaton, the DFS algorithm has to be extended to keep track of the states on the stack. Algorithm 12.2 does this. It first marks the state s that is about to be explored *gray* at line 5. When backtracking over a state (removing it from the (program) stack), its color is set to *black* (line 11). When exploring the successor t^d of s at line 6, if t^d is in the DFS stack, a cycle has been found. Indeed, the states in the DFS stack between t^d and s form a path and t^d is a successor of s. Otherwise, if t^d is not on the DFS stack, no information about cycles can be inferred.

The algorithm exploits this to check the accepting condition in the weak case (Definition 12). Since in this case either all states on the cycle are accepting or none are, the following solution is correct. At line 2, the automaton is first converted into an equivalent state-based version. Then at line 7, the check for elementary cycles is performed by checking whether $t^d.color = gray$. If additionally the state t^d is accepting ($M(t^d) = [1]$), non-emptiness of the automaton is reported at line 8. We only need to check the accepting mark on t^d (or s), and not the marks of other states on the cycle, as all states in one SCC have the same mark by Definition 12 and consequently all states on the same cycle also carry the same mark.

Edelkamp et al. [40] show how such simple algorithms can be used even in the case when only part of the automaton is weak or terminal. In Section 12.4.1, we discuss similar parallel variants.

Since the Büchi emptiness-check problem requires an inspection of all cycles to exclude accepting cycles, most algorithms rely on a DFS exploration (with some more elaborate cycle checks for general, non-weak TGBAs/SBAs as we will show in the subsequent section on Nested-DFS). These algorithms either use DFS directly to conclude emptiness by inspecting elementary cycles, exploiting Property 7, or decompose the automaton into SCCs, exploiting Property 8. *Nested-DFS* falls in the former category, while the SCC algorithm falls in the latter.

In contrast, a BFS exploration cannot easily detect cycles. Consequently, using BFS as exploration strategy requires a redesign of the LTL model checking algorithms, as we will illustrate in Section 12.5.

Algorithm 12.2: Sequential Emptiness Check for Weak TGBAs Based on DFS

```
1  function SETUP (A = (Q, ι, next-product, n, M))
2      Convert A to an equivalent SBA A′ (e.g. Figure 12.5)
3      DFS(A′, ι)
4  function DFS (A = (Q, ι, next-product, 1, M), s)
5      s.color := gray
6      forall t ∈ next-product(s) do
7          if t^d.color = gray ∧ M(t^d) = [1] then
8              report non-empty
9          if t^d.color = ⊥ then
10             DFS(A, t^d)
11     s.color := black
```

12.3.3 Nested-DFS

The Nested-DFS algorithm (NDFS) was originally proposed by Courcoubetis et al. [31] and relies on the detection of accepting elementary cycles reachable from the initial state. This algorithm focuses on SBA with $n \leq 1$ and runs in time linear with respect to the size of the graph. The algorithm accomplishes this by using DFS. Its use of DFS is however not as simple as we have seen in the previous section, because we cannot simply check the acceptance criterion on any state in the cycle as is sufficient in the case of weak automata.

NDFS uses a first DFS to detect *accepting states*, i.e., states of the automaton holding the unique acceptance mark. Traditionally this DFS is called *blue*-DFS since it colors in blue all the states encountered during the exploration. When an accepting state is about to be backtracked during this search, a second DFS is then invoked with the accepting state as a *seed*. This DFS colors all states in red and thus it is often called *red*-DFS. The goal of this second exploration is again to reach the seed state. If this state, which is accepting, can be reached itself, an accepting run is reported proving that the automaton has a non-empty language. Because the version in Algorithm 12.3 contains several improvements, we first discuss its details.

The BLUEDFS function (lines 4–15) is similar to the DFS presented in Algorithm 12.1. Nonetheless some improvements have been added to transform it into an emptiness check. First of all, this algorithm uses two bits per state to keep track of the associated colors. Four colors are used:

- *white*: the initial color of a state. We assume that states are white when they are generated for the first time.
- *cyan*: the state is still in the DFS stack of the blue search.
- *blue*: all the direct successors of the state have been visited by the blue-DFS but not yet by a red one.
- *red*: states that have been considered in both the blue- and the red-DFS.

Algorithm 12.3: Nested Depth-First Search Algorithm

```
1  function NDFS (A = (Q, ι, next-product, n, M))
2  |   assert(n = 1)
3  |   DFSBLUE(A, ι)
4  function DFSBLUE (A = (Q, ι, next-product, 1, M), s)
5  |   s.color := cyan
6  |   forall t ∈ next-product(s) do
7  |   |   if t^d.color = cyan ∧ (M(t^s) = [1] ∨ M(t^d) = [1]) then
8  |   |   |   report non-empty
9  |   |   else if t^d.color = white then
10 |   |   |   DFSBLUE(A, t^d)
11 |   if M(s) = [1] then
12 |   |   DFSRED(A, s)
13 |   |   s.color := red
14 |   else
15 |   |   s.color := blue
16 function DFSRED (A = (Q, ι, next-product, 1, M), s)
17 |   forall t ∈ next-product(s) do
18 |   |   if t^d.color = cyan then
19 |   |   |   report non-empty
20 |   |   else if t^d.color = blue then
21 |   |   |   s.color := red
22 |   |   |   DFSRED(A, t^d)
```

The BLUEDFS function starts by coloring any new state in cyan (line 5). This color helps to detect accepting cycles directly inside the BLUEDFS (lines 7 and 8): during this search, if the successor t^d of an accepting state s is cyan an accepting run exists since there is a path from d to s and vice versa. Similarly, if t^d is accepting and cyan, an accepting run exists. Otherwise, if t^d has not yet been visited (line 9) a recursive call is performed (line 10).

Two cases are of interest when all the successors of a state have been visited, i.e., just before backtracking it from the blue search. If the state is not accepting (line 15), its color becomes blue and the state is backtracked. Otherwise, the state is accepting (line 11) and the algorithm launches a nested exploration using the REDDFS function. This function uses the accepting state as a *seed*, which is treated specially: it remains cyan during the red search and becomes red afterwards (line 13). This is required to limit the algorithm to four colors (which can be stored in two bits). The REDDFS function only looks for a state with the cyan color, i.e., a state that belongs to the DFS stack of the blue exploration. Because the stack of the blue search terminates in the seed, this condition is sufficient to demonstrate the reachability of a cycle over an accepting state. Therefore, if a cyan state is detected in the red search (line 18) then an accepting run exists and the automaton is reported to have a non-empty language (line 19).

Because the red search therefore never crosses the stack of the blue search, it will only explore blue states.

One can also note that all states visited by the REDDFS are marked red (line 21) and thus will be ignored by other (blue or red) explorations. This makes NDFS linear in the size of the input automaton (in terms of states and transitions). But why does the red search not have to reset its visited states like the inner search of the previous algorithm? It turns out that the DFS order of the blue search plays a crucial role here. Consider the case where the red search is started from a seed s and it encounters a red state. It can be shown that this state can never lead back to the cyan stack, because that would contradict the depth-first order of the blue search. An intuition for this property can be found in [48] and a detailed proof in [64].

Note that if the automaton has no accepting state the NDFS is optimal since states and transitions are visited only by the blue-DFS.

Many improvements of this algorithm have been proposed [59, 50, 40] to faster detect non-emptiness, reduce the size of accepting runs if they exist, or to reduce memory footprint. Algorithm 12.3, derived from the work of Schwoon and Esparza [87], presents a combination of all these optimizations.

12.3.4 Algorithms Based on SCC Decomposition

The algorithm presented in the previous section works only if the automaton to check is a non-generalized Büchi automaton. If the input automaton is a generalized one, the emptiness check of Tauriainen [94] can be used. This algorithm derives from the NDFS and repeats the inner DFS several times (at worst n times, with n the number of acceptance marks). The main drawback of this algorithm is that its complexity depends of the number of acceptance marks: this reduces all the benefits of using a generalized Büchi automaton.

Another idea to check for the emptiness of a generalized Büchi automaton is to degeneralize this automaton (as described by Property 1) before checking its emptiness. In this approach, the degeneralized automaton may have $n \cdot |Q|$ states, with $|Q|$ the number of states of the input automaton and n the number of acceptance marks. Once again, this approach is not optimal since it depends of the number of acceptance marks.

Another emptiness-check approach is to compute the accepting strongly connected components of the generalized Büchi automaton. SCC-based emptiness checks [32, 52, 33, 4, 48] are still based on a DFS exploration of the automaton; they do not require another nested DFS, have a linear time complexity and directly support TGBA. These emptiness checks are based on the classical SCC decomposition algorithm for directed graphs by Tarjan [90], which partitions the set of states according to the SCC equivalence classes. Each partition is then associated with the set of acceptance marks that appears inside the corresponding SCC to facilitate the emptiness check.

Intuitively, Tarjan's algorithm maintains a separate stack (apart from the search stack) of partial SCCs. Partial SCCs are enlarged when the DFS finds a cycle by adding its states to the secondary SCC stack. Each partial SCC is associated with a *potential root*, i.e., the state of the partial SCC that is the lowest on the stack. Thus, every time the partial SCC is enlarged, a new potential root may be selected. When the root is backtracked, the DFS order guarantees that the entire SCC was visited and is on the secondary stack. This is the moment when it is popped off the stack and the SCC can be reported even before the algorithm finishes traversing the entire graph (i.e., on the fly). To identify current roots the algorithm uses indices. Therefore, it uses slightly more memory per state than the NDFS algorithm, which requires only two bits per state.

We focus on a version of Tarjan's algorithm that maintains partial SCCs in a database, as it forms the basis of communicating partial SCCs in our parallel algorithm (see Section 12.4.3). It was developed by Purdom [80] even before Tarjan's algorithm, and later optimized by Munro [79]. Like Tarjan's algorithm it uses DFS, but this is not explicitly mentioned (Tarjan was the first to do so). In this algorithm, the secondary stack only stores roots as the partial SCC is kept in the database. We also add the ability to collapse cycles into partial SCCs immediately (as in Dijkstra [35, 47]).

The database with partial SCCs is implemented using a union-find data structure. As its name suggests, a union-find is a data structure that represents sets and provides efficient union and membership-check procedures. The union-find structure partitions a set E of elements and associates a unique representative (an element of E) with each partition. This structure offers the following methods on elements $x, y \in E$:

- MAKESET(x): creates a new partition containing the element x if x is not already in the union-find.
- FIND(x): returns null if x is not in the union-find, otherwise returns the actual representative of the partition containing x.
- SAMESET(x, y): returns a Boolean indicating whether x and y are in the same partition.
- UNITE(x, y): merges the partitions containing x and y.

With this structure, the set E of elements is partitioned into disjoint subsets $\{S_1, \ldots, S_m\}$ where m corresponds to the number of disjoint subsets. The underlying data structure of each subset S_i is typically a reverse arborescence (an in-tree), represented by a *parent* function $p(x) \in S_i$ for each $x \in S_i$. A unique representative y is appointed as the root of this in-tree. It is often designated with a self-pointer $p(y) = y$.

The parent function is usually implemented using an array of size $|E|$ that stores, for each element in $|E|$, the index of its parent in the tree. The array elements are initialized to $\perp$ representing the empty subset. The operation MAKESET(x) then creates a singleton set consisting of its root $p(x) := x$. If two sets are merged with UNITE(x, y), first the representativity of $r_x = \text{FIND}(x)$ and $r_y = \text{FIND}(y)$ is identified. Then one of them, e.g., r_y, is designated the new root by setting $p(r_x) := r_y$.

By compacting the paths in the in-tree, i.e., making leaves point directly to the root, the operations on the structure can all be solved in quasi-constant, amortized time [92]. Many variants on compaction schemes and unite strategies have been studied by Tarjan and van Leeuwen [93].

Algorithm 12.4 presents the emptiness check [83] for TGBA. Two global variables are used:

Algorithm 12.4: SCC-Based Emptiness Check

```
1  Union-find of ⟨Q ⊎ {Dead}⟩ : uf
2  Stack of ⟨q ∈ Q, a ∈ 2^[n], ingoing ∈ 2^[n]⟩ : roots
3
4  function SETUP (A = (Q, ι, next-product, n, M))
5      uf.MAKESET(Dead)
6      SCCBASED(A, ι, ∅)
7  function SCCBASED (A = (Q, ι, next-product, n, M), s, acc)
8      uf.MAKESET(s)
9      roots.PUSH(⟨s, ∅, acc⟩)
10     forall t ∈ next-product(s) do
11         if uf.SAMESET(t^d, Dead) then
12             continue
13         else if uf.FIND(t^d) = null then
14             SCCBASED(A, t^d, M(t))
15         else
16             roots.TOP().a ← roots.TOP().a ∪ M(t)
17             while ¬uf.SAMESET(t^d, s) do
18                 ⟨r, a, i⟩ ← roots.POP()
19                 roots.TOP().a ← roots.TOP().a ∪ i ∪ a
20                 uf.UNITE(r, roots.TOP().q)
21             if roots.TOP().a = [n] then
22                 report non-empty
23     if roots.TOP().q = s then
24         roots.POP()
25         uf.UNITE(s, Dead)
```

1. The union-find *uf* (line 1), which stores the various partitions corresponding to the SCCs discovered so far by the exploration. This structure maintains a special partition *Dead*, which holds all states of already completed SCCs (without accepting run), i.e., all states that cannot be part of an accepting run.
2. The roots stack *roots* (line 2), which contains tuples composed of: q the potential root, a the set of acceptance marks (visited so far) associated with the SCC containing q, and a special field *ingoing*. This special field keeps track of the acceptance marks held by the ingoing transition. This information must be kept since it is not directly available on TGBAs.

Lines 4 to 6 only set up the union-find with the special partition *Dead*, and then call the recursive exploration through the SCCBASED function. This function takes three parameters: the automaton to check, the state to explore, and the acceptance mark held by the ingoing transition.

Lines 8 and 9 respectively insert the state into the union-find and the roots stack. Lines 10 to 22 process all the successors of the current state s. If the destination t^d of a transition is already *Dead* (lines 11–12) then the transition is just skipped since it cannot lead to an accepting run. If the destination has not yet been visited (lines 13–14) the function is called recursively. Finally, the destination can be a part of an

SCC (trivial or not) that is not yet marked *Dead*. In this case, a cycle has been found and partial SCCs stored in the roots stack (lines 16–20) must be merged. During this merge the acceptance marks in the SCC are also merged (line 19). When all partial SCCs have been merged, an accepting run exists iff the field a of the top of the roots stack contains all acceptance marks. Note that this test could also be done during the merge.

Finally, when the root of an SCC is about to be backtracked, all states belonging to this SCC must be marked *Dead*. Line 25 performs this operation in quasi-constant time, by virtue of the union-find data structure.

12.4 Multi-core, DFS-Based Solutions

12.4.1 Terminal and Weak Acceptance

In Section 12.3, we saw that the simplest classes of Büchi automata often allow for simpler and more efficient algorithms. Here we show that checking emptiness of weak and terminal automata can be done using a parallel version of DFS that preserves enough of the depth-first order to still be able to find all elementary cycles. First, we show how a simple parallel search can detect emptiness of terminal automata, as it illustrates nicely what low-level ingredients are required for shared-memory parallel algorithms.

Terminal Acceptance

Algorithm 12.5 shows a parallel search algorithm with a shared state set. To simplify the acceptance condition, the algorithm first converts the terminal automaton, which is by extension also a weak automaton, into an equivalent SBA A' at line 4. Then it schedules the initial state in the stack or the queue of the first worker *Queues*[0]. The first worker will start exploring from this state and generate new states, as we will see later, while a load balancer will take care that work arrives in the queues of the other workers. When the initializations are completed, the algorithm launches the actual search procedure in parallel at line 7. At the first encounter of an accepting state the algorithm terminates at line 15, just like the sequential algorithm for terminal acceptance discussed in Section 12.3.2.

Each worker perpetually calls the load balancer at line 10. When its queue is non-empty ($Q[p] \neq \emptyset$), the *load-balance* function will merely return true. When a worker has run out of work ($Q[p] = \emptyset$), however, the function takes some work from the queue of another thread and adds it to the local queue $Q[p]$. Only when the load balancer detects termination, using a specialized termination detection algorithm [85], will the load balancer return false, allowing the worker thread to exit the SEARCH function.

Algorithm 12.5: A Parallel Search Algorithm for Checking the Emptiness of Terminal Automata

```
1  global Queues[P]
2  global StateSet
3  function PAR-TERMINAL-CHECK (A = (Q, ι, next-product, n, M), P)
4      Convert A into an equivalent SBA A′ (e.g. Figure 12.5).
5      Queues[0] := {ι}
6      StateSet := ∅
7      SEARCH_1(A′) || ... || SEARCH_P(A′)
8      report no-cycle
9  function SEARCH_p (A = (Q, ι, next-product, 1, M))
10     while load-balance(Queues[p]) do
11         s := Queues[p].dequeue()
12         if StateSet.find-or-put(s) then
13             forall t ∈ next-product(s) do
14                 if M(t^d) = [1] then
15                     report cycle and terminate
16                 Queues[p].queue(t^d)
```

The use of a load balancer has the advantage that no communication occurs while workers still have work locally available (their queue is non empty). Only in the extreme cases when a worker is without work, e.g., right after initialization and when most of the state space has been processed, will the algorithm experience overhead from additional synchronization. Specialized concurrent "deque" data structures allow the load balancer to be particularly efficient [19].

For the rest, the parallel search function operates as expected: A state is taken from the local queue at line 11, its successors are considered at line 13, and when a new state is encountered it is added to the local queue at line 16. The worker thus traverses the state space more or less independently, with one exception: visited states are entered into a shared set *StateSet*. To atomically add states, this set implementation has a *find-or-put* operation, which at the same time checks whether a state s is already contained in the set, and when this is not the case, adds it to the set. It can be used to "grab" new states and thus exclusively assign them to the worker that encounters a state first.

The state set can be implemented efficiently as a concurrent hash table or tree table data structure [71, 68]. Because the set of visited states accounts for almost all memory use of the algorithm (recall from the previous section that transitions do not need to be stored), and because workers diverge into different parts of the (huge) state space, most lookups in the table do not collide, i.e., they access different parts of the table. This is another efficient aspect of the algorithm; it exploits the random memory characteristic of model checking algorithms (as also discussed in the introduction) to increase parallelism.

In the sequential case, the algorithm yields a strict DFS order when implementing *Queues* as a stack, and a strict BFS order when implementing *Queues* as a fifo-queue. This parallel algorithm variant however violates a strict order as soon as workers start encountering the same states. Because only one of them will win the race in the *find-or-put* call, the others are forced to violate the order. For this reason, the algorithm might just as well immediately try to "grab" each generated state t^d inside the for loop by moving line 12 right before line 16 (the state set should be initialized to $\{\iota\}$). While this causes a more abnormal search order, it limits all duplication of states on local stacks.

Various researchers have found ways to approach BFS more precisely in parallel algorithms, while also limiting communication by introducing separate queues [2, 58]. A more precise order can have practical benefits, e.g., it allows the model checker to find the shortest counterexample, but also mitigates the on-the-fly behavior of the procedure. It is unknown yet whether (non-lexicographic) DFS can be preserved efficiently as well (recall from the introduction that lexicographic DFS, with fixed transition ordering, likely is not parallelizable according to theory). Nonetheless, we now show that with a simple parallel algorithm, we can preserve enough of the DFS order to find all elementary cycles, which is sufficient to tackle the LTL model checking problem as the following sections show.

Weak Acceptance

Emptiness of weak automata is a little harder to compute than for terminal automata because the algorithm still needs to inspect all elementary cycles. In Section 12.3.2, we showed how DFS can solve it sequentially. Algorithm 12.6 does the same in parallel. Again, to simplify the acceptance condition, the algorithm first converts the terminal automaton to an equivalent SBA A' at line 2. Then, the algorithm launches the actual search procedure in parallel at line 3. All workers start searching from the same initial state.

Algorithm 12.6: A parallel DFS algorithm for checking emptiness of weak automata

```
1  function PAR-WEAK-CHECK (A = (Q, ι, next-product, n, M), P)
2      Convert A to an equivalent SBA A' (e.g. Figure 12.5)
3      PAR-DFS_1(A', ι) || ... || PAR-DFS_P(A', ι)
4      report no-cycle
5  function PAR-DFS_p (A = (Q, ι, next-product, 1, M), s)
6      s.gray[p] := true
7      forall t ∈ RANDOMIZE(next-product(s)) do
8          if t^d.gray[p] ∧ M(t^d) = [1] then
9              report cycle and terminate
10         if ¬t^d.gray[p] ∧ ¬t^d.black then
11             PAR-DFS_p(t^d)
12     s.black := true
13     s.gray[p] := false
```

The search procedure resembles the sequential DFS procedure of Algorithm 12.2, with the exception that the stack states are now colored *gray* locally. This means that workers' stacks might overlap while searching through the state space. When backtracked, however, the states are colored globally *black*, pruning the search space for other workers. This is where the speedup of the parallel algorithm comes from. To obtain the best performance, the search order of each parallel worker should be randomized, so that workers are guided into different parts of the state space [65]. Although redundant due to the set inclusion, we nonetheless emphasize this with the RANDOMIZE function.

To detect cycles, the algorithm uses the same stack-based check as its sequential counterpart. It will not miss any cycles because of the parallel search for the following reasons:

- It is possible to show that all black states always have black or gray states as successors (gray for some worker).
- When a worker p ignores a state t^d for being black, and that state actually has a path to its gray stack, then by induction on the cycle, it can be shown that there is some other worker in a similar situation or able to find a path back to its stack.
- Because there are a finite number of workers, one will eventually find the cycle.

A full proof of correctness can be found in Laarman and Faragó [69].

Because of the use of DFS, the weak emptiness check algorithm looks simpler than Algorithm 12.5. Indeed, it does not require a load-balancer, because work distribution is achieved by letting stacks (partly) overlap. While it may be the case that workers exclude each other from parts of the state space, there are easy ways to remedy that [69]. Because of the lack of a load balancer, the stack can be completely local (here it is maintained as part of the program stack). However, it is not the case that the algorithm does without a global state set. The set is hidden behind the color variables and implicitly accessed when these are referenced in the algorithm. Therefore, an efficient concurrent hash table or tree data structure is again crucial for its performance.

To detect non-progress properties, another subset of LTL, Laarman and Faragó [69] introduce DFS-FIFO, an algorithm that utilizes a similar parallel DFS. It can be used for checking emptiness of weak automata as well and improves the parallel scalability by combining the search with a highly scalable BFS. A similar approach was taken for parallel checking of weak LTL properties on timed automata in [34]. The parallel DFS approach has the additional benefit that it combines well with state space reduction techniques, as these can implemented with the same on-the-fly algorithm [70].

12.4.2 CNDFS

Two algorithms were presented simultaneously (LNDFS by Laarman et al. [66] and ENDFS by Evangelista et al. [42]) that adapted the Nested-DFS (NDFS) algorithm to multi-core architectures. Both share the principle of launching multiple instances of NDFS that synchronize themselves to avoid useless state revisits, just like the algorithm for checking emptiness of weak automata discussed in the previous section. Although they are heuristic algorithms in the sense that, in the worst case, they reduce to spawning multiple unsynchronized instances of NDFS, the experiments reported by Laarman et al. [66, 65] show good practical speedups.

They were then combined and improved in the CNDFS algorithm by Evangelista et al. [43]. This algorithm is both much simpler and uses less memory, making it more compatible with exact compression techniques such as tree compression [68] that can compress large states down to two integers.

CNDFS is presented in Alg. 12.7 for P threads. It is based on the principle of SWARM worker threads (indicated by subscript p here), sharing information via colors stored in the visited states: here *blue* and *red*. After randomly visiting all successors (lines 13–15), a state is marked blue at line 16 (meaning "globally visited"), causing the (other) blue-DFS workers to lose the strict postorder property.

If the state s is accepting, as in the sequential NDFS algorithm, a red-DFS is launched at line 19 to find a cycle. At this point, state s is called "the seed." All states visited by DFSRED$_p$ are collected in $\mathscr{R}_p$. If no cycle is found in the red-DFS, none exists for the seed. Still, because the red-DFS was not necessarily called in postorder, other (non-seed, non-red) accepting states may be encountered about which we know nothing, except the fact that they are out of order and reachable from the seed. These are handled after completion of the red-DFS at line 20 by simply waiting for them to become red.

In this scenario there is always another worker that can color such a state red. The intuition behind this is that there has to be another worker to cause the out-of-order red search in the first place (by coloring blue) and, in the second place, this worker can continue its execution because cyclic waiting configurations can only happen for accepting cycles. These accepting cycles would however be encountered first, causing termination and a cycle report (line 8). After completion of the waiting procedure, CNDFS marks all states in $\mathscr{R}_p$ globally red, pruning other red-DFSs.

An efficient parallelization of the blue-DFS is absolutely essential for scalability, since the number of blue states (all reachable states) typically exceeds the number of red states (visited by the red-DFS). Since it was impossible to color both blue and red while backtracking from the respective DFS procedures, CNDFS uses an intermediate solution, using a wait statement as a compromise, leaving enough parallelism to maintain scalability.

CNDFS only uses $P+2$ bits per state plus the sizes of $\mathscr{R}$. In the theoretical worst case (an accepting initial state), each worker $p \in [P]$ could collect all states in $\mathscr{R}_p$. According to extensive experiments, the set rarely contains more than one state and never more than thousands, which is still negligible compared to $|Q|$.

Algorithm 12.7: CNDFS, a Multi-Core Algorithm for LTL Model Checking

```
1  function CNDFS (ι,P)
2  |  DFSBLUE_1(ι) || ... || DFSBLUE_P(ι)
3  |_ return no-cycle
4  function DFSRED_p(A = (Q, ι, next-product, n, M), s)
5  |  ℛ_p := ℛ_p ∪ {s}
6  |  forall t ∈ RANDOMIZE(next-product(s)) do
7  |  |  if t^d.cyan[p] then
8  |  |  |_ return cycle and terminate
9  |  |  if t^d ∉ ℛ_p ∧ ¬t^d.red then
10 |_ |_ |_ DFSRED_p(A, t^d)
11 function DFSBLUE_p(A = (Q, ι, next-product, n, M), s)
12 |  s.cyan[p] := true
13 |  forall t ∈ RANDOMIZE(next-product(s)) do
14 |  |  if ¬t^d.cyan[p] ∧ ¬t^d.blue then
15 |  |_ |_ DFSBLUE_p(A, t^d)
16 |  s.blue := true
17 |  if M(s) ≠ ∅ then
18 |  |  ℛ_p := ∅
19 |  |  DFSRED_p(A, s)
20 |  |  await ∀s' ∈ ℛ_p s.t. M(s') ≠ ∅ : s ≠ s' ⇒ s'.red
21 |  |  forall s' ∈ ℛ_p do
22 |  |_ |_ s'.red := true
23 |_ s.cyan[p] := false
```

12.4.3 Multi-core/DFS-Based SCC Decomposition

To handle emptiness checking of TGBAs, a parallel SCC-based algorithm is required as Theorem 1 indicates. Traditional parallel SCC algorithms [86, 46, 13, 98, 60, 88] are BFS-based implementations of divide-and-conquer approaches, which are not on the fly [18]. Also, these algorithms often exhibit an $n \times \log(n)$ or quadratic-time worst-case complexity. We therefore rely on DFS to detect SCCs in parallel since DFS-based SCC detection can be both on the fly and linear time. The main difficulty here, like in the previous section, is that a sufficient amount of the DFS order must be preserved for correctly detecting cycles.

We first briefly discuss a fully synchronized approach and show how bottlenecks impose limitations on the algorithm's performance. Then we present a random search/swarmed approach that performs linearly and show how this technique scales for multiple workers.

Fully Synchronized Parallel SCC Algorithm

The general idea of the fully synchronized algorithm [74] is to have multiple non-overlapping search instances. Every reachable state is visited by exactly one worker, who globally takes ownership of the state. Searches are spawned from unvisited successor states. Upon encountering a state taken by a different worker, the search suspends until the state is marked as being completely explored. Otherwise, the search proceeds similarly to Tarjan's algorithm [91].

A cycle of suspended searches can occur as a consequence. In case no further actions are taken, the algorithm may never finish. A map of suspended searches is used to detect such cycles. If a worker suspends a search and detects a cycle of suspended searches, it transfers all relevant states from the suspended searches to one search and proceeds normally. For example, suppose that a worker visits edge $v \to w$ and detects that w is part of a different search. Before suspending, it checks whether the path $w \to^* v$ can be found by traversing states from the suspended searches. If so, a cycle is detected, which should be resolved by the current worker.

Maintaining the suspended map and resolving cycles of suspended searches is a costly process. The sequential linear-time performance of Tarjan's algorithm reduces to a quadratic worst-case performance in the synchronized variant. For the practical performance of the algorithm, two important cases can be distinguished: graphs containing relatively *large* SCC sizes ($|C| \sim |Q|$), often consisting of many interconnections; and *small* SCCs, consisting of only a few states ($|C| \sim 1$). The synchronized algorithm exhibits good scalability for graphs containing only small SCCs, since the different searches do not tend to interfere with each other. For graphs with large SCCs, a fully synchronizing algorithm can pay a large performance penalty if the worst-case time complexity is attained due to the wait-cycle checks. On the other hand, this algorithm totally avoids any redundant explorations as searches never overlap. Hence, while in appearance similar to the multi-core NDFS approaches discussed in the previous subsection, the fully synchronous algorithm has characteristics similar to the BFS-based algorithms that will be discussed in Section 12.5.

Swarmed Parallel SCC Algorithm

A different approach is to detect SCCs in a *swarmed* fashion, similarly to CNDFS (Section 12.4.2). The general idea of the algorithm is to spawn multiple instances of a sequential DFS algorithm and communicate the fully explored SCCs in a shared data structure [84]. An SCC is considered to be fully explored when all its successors (direct or indirect) have been explored. As a consequence, an instance of the algorithm can ignore all states belonging to a fully explored SCC. Thus, communicating fully explored SCCs allows us to prune other DFSs since an instance will never traverse a state that belongs to a fully explored SCC.

In this approach, two instances can still visit the same SCC (in a swarmed fashion) until one of the instances detects that it has been fully explored. If the SCC contains an accepting run, we want to be able to speed up its discovery. The multiple instances can then share the acceptance marks discovered so far for each (partial) SCC. This information helps us to find whether an accepting run exists. Suppose that we have two instances of a classical SCC-based algorithm running on the example of

Figure 12.7 without sharing acceptance marks. Neither of these instances can detect an accepting cycle before $\delta_0, \delta_1, \delta_2$, and δ_3 have all been visited. Let us now suppose that they share acceptance marks and that the first instance i_0 has visited δ_1 and δ_2 while the other instance i_1 has visited δ_0. When instance i_1 discovers the transition δ_3 it also discovers that s_0 and s_1 are in the same SCC. In this case, since i_0 and i_1 share information about acceptance marks, they can detect the existence of an accepting cycle.

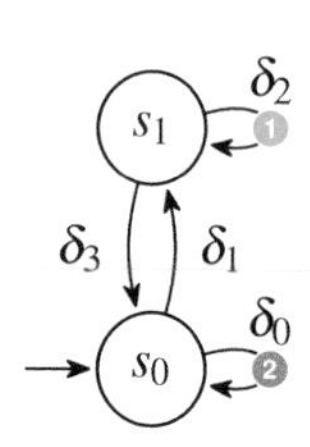

Fig. 12.7: Sharing acceptance marks

In the sequential SCC-based emptiness check (Algorithm 12.4) the information about fully explored SCCs is already stored inside a union-find data structure with a dedicated partition *Dead*. *Lock-free* versions of the union-find structure exist [5]. A simple implementation of this structure is presented in Algorithm 12.8. As mentioned in Section 12.3.4 each element stored by the union-find maintains a field *parent*, which represents a forest of reverse arborescences. In a parallel setting this field must not be updated concurrently by two threads. This can be done using a *compare-and-swap* (CAS) operation (line 13 and 15). This operation is an atomic instruction used in multithreading to achieve synchronization: CAS(m, v_1, v_2) compares the contents of a memory location m to a given value v_1 and, only if they are the same, modifies the contents of that memory location to a given new value v_2. The CAS operation returns true if the modification was successful and false otherwise. A closer look to Algorithm 12.8 shows that this structure is only *lock-free* and not *wait-free* because of the spin-wait loops of lines 8 and 19. The rest of this union-find remains similar to the sequential version apart from the use of atomic operations.

This union-find can then be shared among the multiple instances to communicate fully explored SCCs. This structure can also be extended to store, for each partition, a set of acceptance marks. This modification slightly impacts the interface of union-find:

- When MAKESET(e) effectively creates a partition for e (because it did not exist before), the associated acceptance set is $\emptyset$.
- The UNITE function takes an extra argument representing the set of acceptance marks that occur in the (partial) SCC. During this operation, the union-find must propagate the acceptance set to the representative of the partition. This is costless since this representative is already computed by the FIND function. Also note that for implementation details, a union with the partition containing *Dead* always returns $\emptyset$.

This swarmed emptiness check is presented in Algorithm 12.9 and mostly relies on the sequential SCC-based emptiness check presented in Algorithm 12.4. It performs a DFS, maintains a roots stack, and uses a union-find to store partitions representing partial SCC and *Dead* states. Nonetheless, some minor changes have been made:

- Only the union-find is shared among the threads. The roots stack is local to each instance.

Algorithm 12.8: Concurrent Union-Find Data Structure

```
1  function FIND (a)
2      if a.parent ≠ a then
3          a.parent := FIND(a.parent)
4      return a.parent
5  function UNITE (a,b)
6      x := a
7      y := b
8      while true do
9          x := FIND(x)
10         y := FIND(y)
11         if x = y then return
12         else if x < y then
13             if CAS(x.parent, x, y) then return
14         else
15             if CAS(y.parent, y, x) then return
16 function SAMESET (a,b)
17     x := a
18     y := b
19     while TRUE do
20         x := FIND(x)
21         y := FIND(y)
22         if x = y then return TRUE
23         else if x.parent = x then return FALSE
```

- Each instance p now maintains a local integer $counter_p$ (line 3). This integer is only incremented (line 10) so it can be used to (locally) order states that have been visited.
- For an instance p, each state s is associated with a *live number*, i.e., an integer accessible via $s.livenum_p$. This live number is given according to $counter_p$ the first time the state is visited by the thread p (line 11).
- Line 21 has been changed since SAMESET cannot be used to pop the roots stack until the new root is discovered. Indeed, since the union-find is shared among all threads, no assumptions about its internal state can be made.

It is worth noting that the union-find structure collects the acceptance marks that are discovered by all threads. Thus, at line 23 the algorithm uses the global *uf* structure to detect which acceptance marks have been found, by any worker, in the partial SCC. This helps speed up reporting the existence of an accepting run. Nonetheless, if an SCC is not accepting, its states cannot be marked *Dead* before a thread has visited all the states and all the transitions of this SCC. This is a serious drawback of this algorithm when the automaton to check is composed of a single large SCC: in this case, the expected speedup is null. The next algorithm solves this problem.

Algorithm 12.9: Swarmed SCC-Based Algorithm

```
1  Shared Union-find of ⟨Q ∪ {Dead}, a ∈ 2^[n]⟩ : uf
2  Local Stack of ⟨q ∈ Q, ingoing ∈ 2^[n]⟩ : roots_p
3  Local Integer counter_p
4
5  function SETUP (A = (Q, ι, next-product, n, M))
6      uf.MAKESET(Dead)
7      counter_1 ← 0; ...; counter_P ← 0
8      SWARMEDSCCBASED_1(ι, ∅) || ... || SWARMEDSCCBASED_P(ι, ∅)
9  function SWARMEDSCCBASED_p (A = (Q, ι, next-product, n, M), s, acc)
10     counter_p := counter_p + 1
11     s.livenum_p := counter_p
12     uf.MAKESET(s)
13     roots_p.PUSH(⟨s, acc⟩)
14     forall t ∈ RANDOMIZE(next-product(s)) do
15         if uf.SAMESET(t^d, Dead) then
16             continue
17         else if uf.FIND(t^d) = null then
18             SWARMEDSCCBASED_p(A, t^d, M(t))
19         else
20             uf.FIND(s).a := uf.FIND(s).a ∪ M(t)
21             while t^d.livenum_p < roots_p.TOP().q.livenum_p do
22                 ⟨r, i⟩ := roots_p.POP()
23                 uf.UNITE(r, roots_p.TOP().q, i)
24             if uf.FIND(s).a = [n] then
25                 report non-empty
26     if roots_p.TOP().q = s then
27         roots_p.POP()
28         uf.UNITE(s, Dead, ∅)
```

Improved Parallel Swarmed SCC Algorithm

The key aspects of the improved algorithm are to communicate partially found SCCs and globally track the remaining work left for each SCC. The SCC algorithm is presented in [18] and is applied to LTL model checking in [17]. It is presented in Algorithm 12.10 and differs slightly from Algorithm 12.9.

The local *counter* and *livenum* have been replaced by globally tracking worker IDs in the union-find structure. This *worker set*, $w \in 2^P$, is a bitset that tracks which worker threads are *active* in the current SCC. The MAKESET(p,s) is extended to set the bit for worker p in the partial SCC of s, which is tracked in the representative of the set. This worker set is used in line 14 to detect a cycle. Note that if worker p has visited some state s in a partial SCC, every state of this partial SCC is considered to have been visited before. This is valid since there is a path from every other state in the SCC to s. Also note that multiple workers aid each other by concurrently adding more states to the set, thus increasing the number of states that have been "visited before."

Algorithm 12.10: UFSCC Algorithm: Improved Swarmed SCC Algorithm

1 **Shared Union-find** of $\langle Q \cup \{Dead\}, a \in 2^{[n]}, w \in 2^{P}, list \in 2^{Q} \rangle : uf$
2 **Local Stack** of $\langle q \in Q, ingoing \in 2^{[n]} \rangle : roots_p$
3
4 **function** SETUP $(A = (Q, \iota, \text{next-product}, n, M))$
5   uf.MAKESET$(Dead)$
6   IMPROVEDSCC$_1(\iota, \emptyset)$ || ... || IMPROVEDSCC$_P(\iota, \emptyset)$
7 **function** IMPROVEDSCC$_p$ $(A = (Q, \iota, \text{next-product}, n, M), s, acc)$
8   uf.MAKESET(p, s)
9     $roots_p$.PUSH$(\langle s, acc \rangle)$
10   **while** $s' \in uf$.PICKFROMLIST(s) **do**
11   **forall** $t \in$ RANDOMIZE(next-product(s')) **do**
12   **if** uf.SAMESET$(t^d, Dead)$ **then**
13   **continue**
14   **else if** $p \notin uf$.FIND$(t^d).w$ **then**
15   IMPROVEDSCC$_p(A, t^d, M(t))$
16   **else**
17   uf.FIND$(s).a := uf$.FIND$(s).a \cup M(t)$
18   **while** $\neg$SAMESET(s, t^d) **do**
19                     $\langle r, i \rangle := roots_p$.POP()
20   UNITE$(r, roots_p$.TOP$().q, i)$
21   **if** uf.FIND$(s).a = [n]$ **then**
22   **report non-empty**
23   uf.REMOVEFROMLIST(s')
24   uf.UNITE$(s, Dead, \emptyset)$
25   **if** $roots_p$.TOP$() = s$ **then**
26         $roots_p$.POP()

In order to collaborate in detecting when an SCC has been fully explored, the union-find structure has been further extended to track a list of *Busy* states in each partial SCC. The idea is to initially keep a global list consisting of every state in the SCC. Then, after concluding that no new knowledge can be obtained from a state, it gets removed from the list and another state is chosen. In the algorithm this is shown in lines 10 and 23. When all successors of state s' have been handled (lines 11–22) we can conclude that for every successor d of s' we either have: (1) d is part of a *Dead* SCC, or (2) d is part of the same SCC as s'. In the latter case, d has been added to the list of *Busy* states and therefore s' can be removed from the list. Multiple workers pick states from the list, explore them, and correspondingly remove them from the list to cooperatively reduce the number of states in the list. Once the list is empty (exit condition for line 10), every state of the SCC has been fully explored and the SCC can be marked *Dead*.

In the implementation, the union-find structure is extended such that every state contains a worker set of size $|P|$, which is maintained (similarly to the acceptance set) in the UNITE procedure. Every state in the structure also contains a list-next pointer such that a cyclic list is formed of all states in the partial SCC. See Figure 12.8 for an illustration. Combining two lists in the UNITE procedure is then realized by swapping

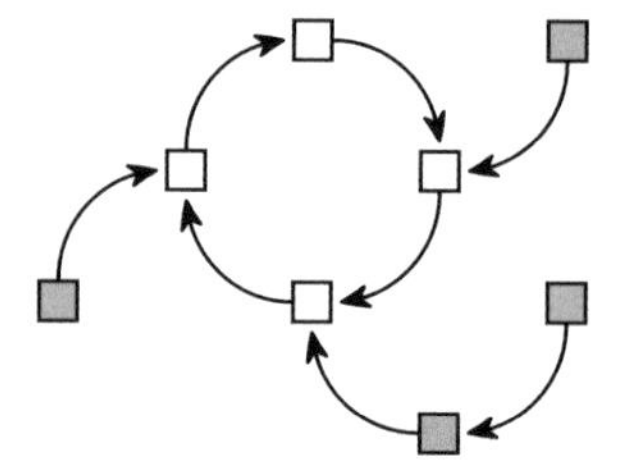

Fig. 12.8: Cyclic list of *Busy* states. White nodes are *Busy* and gray nodes have been removed from the list

two list pointers with a fine-grained lock to form a single list containing all states. A Boolean flag is used to mark a state to be removed from the list. Workers then traverse the list to find *Busy* states and update the next pointers such that the *removed* states are detached from the list.

12.5 Distributed, BFS-Based Solutions

In shared-memory, parallel algorithms can exploit relatively fast accesses to concurrent data structures to dynamically distribute the search procedure over the processor cores, as achieved in the previous section through the use of a shared state set. In the distributed setting, such synchronous communication would be too costly. To solve this problem, distributed algorithms statically partition the states over the workers, using so called *hash-based partitioning*. Under this scheme, every state of the graph to be stored is assigned to a single workstation that is responsible for its storage. The function to assign an owner of a state is referred to as the *partition* or *owner* function.

This section discusses two algorithms suitable for distributed computation. Because the static partitioning works best in combination with the highly scalable breadth-first search, the emptiness-check problem is first rephrased so that it can be solved by an iterative approach. In the worst case, each iteration represents one pass over the state space, but can be implemented with BFS. Nonetheless, for many inputs the time complexity of this approach is still optimal and we demonstrate that the emptiness check can even be made partially on the fly. At the end of this section, we show how this approach compares to the DFS approaches in the previous section.

12.5.1 *One-Way-Catch-Them-Young*

The emptiness-check algorithm discussed in this section is built on top of a procedure for topological sorting. It relies on the fact that vertices of a directed graph may be topologically sorted if and only if the graph is acyclic. The topological sort procedure may be effectively adapted for parallel processing without any increase in the theoretical time complexity. While topological sort can directly detect the presence of a cycle in a directed graph, it cannot distinguish between accepting and non-accepting cycles. Therefore, it must be accompanied by another technique in order to be used as a Büchi automata emptiness check. One of the options to achieve accepting-cycle detection is to combine the topological sort procedure with a forward reachability analysis that eliminates states not reachable from accepting states. The algorithm relying on this combination is referred to as the OWCTY algorithm (One-Way-Catch-Them-Young) [44, 26].

The idea of the OWCTY algorithm is to remove leading rejecting SCCs (SCCs without accepting states) from the graph of the product Büchi automaton, then use the topological sort procedure to remove leading accepting states that do not lie on a cycle. This process is iterated until a fixpoint is reached. When the remaining graph is empty, it contains no accepting cycle. When the remaining graph is non-empty, the presence of an accepting cycle in the graph is guaranteed.

The OWCTY algorithm therefore uses two removal procedures, ELIM-NO-ACCEPTING and ELIM-NO-PREDECESSORS, which alternate. See Algorithm 12.11 for details. ELIM-NO-ACCEPTING is a procedure that computes all states that are reachable from an accepting state in the graph and removes the rest. Efficiently, this procedure removes all leading SCCs that contain no accepting states at all. Obviously, these SCCs must be rejecting. After that, all leading SCCs in the rest of the graph contain an accepting state; however, they might all be trivial SCCs (contain no edges). To detect whether there is a non-trivial leading SCC with an accepting state in the graph, the trivial leading SCCs must be removed. For that, the ELIM-NO-PREDECESSORS procedure is used. Note that after the removal of a leading trivial SCC another trivial SCC may become leading. To deal with that the ELIM-NO-PREDECESSORS procedure proceeds iteratively, and removes all trivial SCCs from the top of the graph (mimicking the topological sort procedure). After ELIM-NO-PREDECESSORS finishes, all leading SCCs in the remaining part of the graph are non-trivial, hence a new round of the elimination is executed, starting again with the ELIM-NO-ACCEPTING procedure.

To learn whether a state is a trivial leading component, the algorithm needs to detect not-yet-removed predecessors of the state. To do so, the algorithm maintains an integer value associated with every vertex to keep the number of not-yet-removed direct predecessors, the so called *indegree*. The unique feature of the OWCTY algorithm is that the indegrees are updated without the need to enumerate the predecessors of a state. In fact, the algorithm only performs forward traversal procedures to maintain the indegrees. This is exactly what the *One-Way* in the name of the algorithm stands for. While this does not immediately make the algorithms on the fly (we do so in the subsequent section), it does already avoid the costly need to store all edges of the graph for reverse traversals as discussed in Section 12.3.1. To emphasize this fact, we again use the implicit definition of the Büchi automaton, i.e., with next-product instead of δ, as defined in Section 12.2.7.

The pseudo code of the OWCTY algorithm depends on the following notational conventions. Distributed data structures $R, Open$, and $OldR$ are referred to either in a global way, in which case no subscript is used, or in a local (partitioned) way, in which case the subscript denotes which part of a distributed data structure is accessed. For example, a set of states R is a union over distributed data parts of R denoted by $R_1, \ldots, R_n$. R_p is used in procedure ELIM-NO-PREDECESSORS$_p$ to denote that the algorithm accesses the local part of the data structure. The indegrees are denoted as fields of the states, but in reality should be stored in the state set R_p, which can be implemented as a hash map. At line 20, the indegree is set to 0 for (newly encountered) accepting states, as these are the roots of the search tree, but to 1 for other states, indicating that these are reachable from one accepting state. At line 17, the indegree is incremented when other incoming edges of the state s are found. Termination detection is implemented by TERMINATION.

12.5.2 MAP

Yet another approach to accepting-cycle detection in distributed memory is taken by the algorithm MAP [21]. The main idea behind the algorithm is based on the fact that each accepting state lying on an accepting cycle is its own predecessor. When the algorithm computes the set of all accepting predecessors for every accepting state, it is sufficient to check, whether any of the accepting states is present in its own predecessor set. However, to compute and store all this information would be rather expensive. The algorithm instead stores only a single unique representative of the set of all accepting predecessors per state. Let us assume a linear ordering $\prec$ of vertices (given; e.g., by their representation in memory), then the unique representative could just be the *maximal accepting predecessor* (MAP). Let $\perp$ be a unique value that is the lowest in the order. We will present here a sequential version of the MAP algorithm and explain in a subsequent section how it can be integrated into the OWCTY algorithm to achieve a parallel version with on-the-fly properties. See Algorithm 12.12 for the pseudocode of the sequential version of the MAP algorithm.

For a state u, we denote its maximal accepting predecessor in the graph G by $map_G(u)$. Clearly, if an accepting state is its own maximal accepting predecessor ($map_G(u) = u$), then it lies on an accepting cycle. Unfortunately, the converse does not hold in general. Assume that u is the largest accepting state on some accepting cycle. It can happen that the maximal accepting predecessor of u lies outside the cycle, i.e., $map_G(u) = v$ for some accepting state v. However, for this accepting state v either $map_G(v) = v$, in which case the presence of an accepting cycle can be detected on v, or $map_G(v) \prec v$, in which case v is not part of any cycle in the graph. In the latter case, v can safely be removed from the set of accepting states (or marked as non-accepting). However, removing v from the set of accepting states invalidates the value of $map_G(u)$, which has to be recomputed.

The basic workflow of the algorithm is thus to compute maximal accepting predecessors for accepting states in the graph, and when no accepting cycle can be proved, to *shrink* the set of accepting states. These two steps are alternated until either a cycle is found, or there are no more accepting states in the graph to be removed.

Algorithm 12.11: OWCTY Algorithm

```
1  global Open, R, OldR
2  function OWCTY(A = (Q, ι, next-product, n, M))
3      R := Open := ∅
4      o := owner(ι)
5      Open_o := {ι}
6      REACH(A, R_p)
7      OldR := ∅
8      while (R ≠ OldR) ∧ (R ≠ ∅) do
9          OldR := R
10         ELIM-NO-ACCEPTING_1(A, R_1) || ... || ELIM-NO-ACCEPTING_n(A, R_n)
11         ELIM-NO-PREDECESSORS_1(A, R_1) || ... || ELIM-NO-PREDECESSORS_n(A, R_n)
12     if R ≠ ∅ then report "accepting cycle" else report "no accepting cycle"
13 function REACH_p (A = (Q, ι, next-product, n, M), R_p)
14     while Open_p ≠ ∅ ∧ ¬TERMINATION(Open) do
15         s := Open_p.dequeue()
16         if s ∈ R_p then
17             s.indegree := s.indegree + 1
18         else
19             R_p.add(s)
20             s.indegree := if M[s] = [1] then 0 else 1
21             forall t ∈ next-product(s) do
22                 o := owner(t^d)
23                 Open_o.queue(t^d)
24 function ELIM-NO-ACCEPTING_p (A = (Q, ι, next-product, n, M), R_p)
25     forall s ∈ R_p do
26         if M[s] = [1] then
27             Open_p.queue(s)
28     R'_p := ∅
29     BARRIER()                    // Wait until all workers reinitialized R'_p
30     REACH(A, R'_p)
31 function ELIM-NO-PREDECESSORS_p (A = (Q, ι, next-product, n, M), R_p)
32     forall s ∈ R_p do
33         if s.indegree = 0 then
34             Open_p.queue(s)
35     while Open_p ≠ ∅ ∧ ¬TERMINATION(Open) do
36         s := Open_p.dequeue()
37         s.indegree := s.indegree − 1
38         if s.indegree ≤ 0 then
39             R_p.remove(s)
40             forall t ∈ next-product(s) do
41                 o := owner(t^d)
42                 Open_o.queue(t^d)
```

To compute the value of the map_G function, Algorithm 12.12 proceeds by the principle of value propagation. Note that whenever some value is propagated to a

Algorithm 12.12: MAP Algorithm

```
1  function MAP(A = (Q, ι, next-product, n, M))
2      Waiting.add(ι)
3      oldmap(Q) := ⊥
4      ShrinkM := ∅
5      while Waiting ≠ ∅ do
6          while Waiting ≠ ∅ do
7              seed := Waiting.dequeue()
8              PROPAGATE-MAP(A, seed, ShrinkM)
9          Waiting := ShrinkM
10         ShrinkM := ∅
11     report "no accepting cycle"
12 function PROPAGATE-MAP(A = (Q, ι, next-product, n, M), seed, ShrinkM)
13     oldmap(seed) := seed
14     map(seed) := ⊥
15     Seeds.queue(seed)
16     while Seeds ≠ ∅ do
17         u := Seeds.dequeue()
18         if M[u] = [1] ∧ (u ≠ oldmap(u)) then
19             if map(u) ≤ u then
20                 propagate := u
21                 ShrinkM.add(u)
22             else
23                 propagate := map(u)
24                 ShrinkM.remove(u)
25         else
26             propagate := map(u)
27         forall t ∈ next-product(u) do
28             if propagate = t^d then
29                 report "accepting cycle"
30             if map(t^d) = oldmap(t^s) then
31                 oldmap(t^d) := oldmap(t^s)
32                 map(t^d) := propagate
33                 Seeds.queue(t^d)
34             else if (propagate > map(t^d)) ∧ (oldmap(t^d) = oldmap(t^s)) then
35                 map(t^d) := propagate
36                 Seeds.queue(t^d)
```

state from which a low value had been propagated before, the new higher value must be repropagated. Due to these duplicate propagations, this procedure requires quadratic time with respect to the size of the graph.

An interesting property of the map_G function is that once computed, the values of map_G partition the graph into subgraphs. More precisely, states that share the same value of map_G may lie on a cycle, however, states that do not share the same value of map_G cannot lie on the same cycle (they cannot be part of the same SCC). The algorithm takes advantage of this observation and in the propagation phase it restricts the propagation to only the subgraphs given by the same value of the map_G function from the previous iteration. In particular, when exploring a transition t within a given subgraph for the first time, it is the case that $map_G(t^d) = oldmap_G(t^s)$ (line 30); later on, $oldmap_G(t^d) = oldmap_G(t^s)$ is used to localise the exploration to a single subgraph (line 34).

To do so, the algorithm maintains *oldmap* values for all states. Also note that the subgraphs where the next iteration of *map* propagation is about to be computed are rooted in the accepting states that were just shrunk. Note that some accepting states may be temporarily recorded as roots of a subgraph, but later on they may become dominated by some other accepting state, in which case they are no longer considered to be roots (see line 24).

An interesting question is how to define the ordering with respect to which the maximal accepting state is determined. It has been shown [22] that for every graph an optimal ordering exists, however, to find it is as difficult as to define a DFS postorder, which is hard to parallelize, and would bring us back to the algorithms in Section 12.4.

12.5.3 Combining OWCTY and MAP

Algorithm MAP works on the fly, i.e., it is capable of reporting the presence of accepting cycles without the need to explore the whole underlying graph. This is not the case with algorithm OWCTY, as to properly compute the indegrees, the whole graph has to be traversed. On the other hand, the time complexity of the OWCTY algorithm is quadratic, while the time complexity of MAP is cubic. In [11] a combination of the two algorithms has been presented to obtain the best of both worlds. In particular, while performing the ELIM-NO-ACCEPTING procedure in the OWCTY algorithm, it is possible to perform limited propagation of *map* values at the

	Complexity	Scalability	Optimal	On-the-fly	TGBA
CNDFS	$\mathcal{O}(V+E)$	+	Yes	Yes	No
UFSCC	$\mathcal{O}(V+E)$	+	Yes	Yes	Yes
OWCTY					
general Büchi	$\mathcal{O}(V.(V+E))$	++	No	No	?
weak Büchi	$\mathcal{O}(V+E)$	++	Yes	No	?
MAP	$\mathcal{O}(V^2 \cdot (V+E))$	++	No	Partially	?
OWCTY + MAP					
general Büchi	$\mathcal{O}(V \cdot (V+E))$	++	No	Partially	?
weak Büchi	$\mathcal{O}(V+E)$	++	Yes	Partially	?

Table 12.1: Overview of distributed-memory algorithms for accepting-cycle detection. Complexity is expressed in the number of vertices V, the number of edges E, and the number of processes P

same time. The propagation is limited to a single visit of a state (no repropagation is allowed). Still, if the algorithm finds an accepting state that is its own predecessor, the accepting cycle may be reported and the algorithm may terminate without the need for the whole exploration of the graph.

Table 12.1 provides an overview of all emptiness algorithms discussed in this chapter. We use a subjective scale for the scalability of these algorithms, as a theoretical treatise on the matter is out of the scope of this handbook. The DFS algorithms feature optimal runtimes, but not necessarily good scalability. The BFS algorithms on the other hand sacrifice the optimality property to attain better scalability. However, in practice, both approaches have been shown to scale well on multi-core machines [43, 65, 12]. Moreover, in many important cases, i.e., for weak automata, the OWCTY algorithm and its combination with MAP also achieve optimal runtime in theory.

The integration of MAP into the OWCTY algorithm further yields some on-the-fly behavior. While not completely on the fly, OWCTY tends to deliver shorter counterexamples because of its use of BFS. Short counterexamples are important for repairing errors in the model as they simplify error diagnosis. In practice, CNDFS has been shown to also be able to yield similarly short counterexamples with increasing parallelism [43], but it provides no guarantees about counterexample length. Thus far, only the SCC algorithms are suitable for direct use on TGBAs. CNDFS likely cannot be adapted to support TGBAs without increasing the complexity, but we consider the combination of the BFS algorithms with TGBAs to be an open problem.

12.6 Conclusion

This chapter has revisited the automata-theoretic approach to LTL model checking in Section 12.2. The starting point is a translation of an LTL formula into a (Transition-based Generalized) Büchi Automaton. Fragments of LTL lead to weak or even terminal automata. The LTL model checking algorithm is reduced to emptiness checking of automata, which boils down to detecting accepting cycles.

To speed up cycle detection, we have introduced parallel algorithms for shared-memory multi-core machines in Section 12.4. These algorithms are based on Depth-First Search and come in two flavors: those based on Nested-DFS, and those based on SCC detection. We showed instances of both.

Based on the observation that DFS is hard to parallelize, an alternative is to design BFS-based algorithms to detect accepting cycles. We have done so in Section 12.5. This type of algorithm is used in shared-memory machines, but was originally designed for distributed clusters of machines connected by a fast communication network.

Although this chapter has focused on the algorithmic ideas behind the various parallel LTL model checking algorithms, we would like to stress that the algorithms that we have explained are also available to the community in open-source tools. The translations from LTL to automata are available in the Spot toolset[3] [36, 38]. Various DFS-based multi-core algorithms are available in the LTSmin toolset[4] [67, 62]. Finally, the distributed and multi-core implementation of the BFS-based algorithms are available through the DiVinE toolset[5] [10, 14].

The scientific papers connected to the algorithms implemented in these tools report on extensive experiments to investigate the practical efficiency and parallel speedup on various benchmark suites of realistic examples, and on their performance in international model checking competitions.

References

[1] P. Abdulla, S. Aronis, B. Jonsson, and K. Sagonas. Optimal dynamic partial order reduction. In *Proceedings of the 41st ACM SIGPLAN-SIGACT Symposium on Principles of Programming Languages*, POPL '14, pages 373–384, New York, NY, USA, 2014. ACM. ISBN 978-1-4503-2544-8. doi: 10.1145/2535838.2535845. URL `http://doi.acm.org/10.1145/2535838.2535845`.

[2] V. Agarwal, F. Petrini, D. Pasetto, and D. A. Bader. Scalable Graph Exploration on Multicore Processors. In *Proceedings of the 2010 ACM/IEEE International Conference for High Performance Computing, Networking, Storage and Analysis*, SC '10, pages 1–11, Washington, DC, USA, 2010. IEEE Computer Society. ISBN 978-1-4244-7559-9. doi: 10.1109/SC.2010.46.

[3] A. Aggarwal, R. J. Anderson, and M. Kao. Parallel depth-first search in general directed graphs. *SIAM J. Comput.*, 19(2):397–409, 1990. doi: 10.1137/0219025. URL `http://dx.doi.org/10.1137/0219025`.

[4] R. Alur, S. Chaudhuri, K. Etessami, and P. Madhusudan. On-the-fly reachability and cycle detection for recursive state machines. In N. Halbwachs and L. Zuck, editors, *Proceedings of the 11th International Conference on Tools and Algorithms for the Construction and Analysis of Systems (TACAS'05)*, volume 3440 of *Lecture Notes in Computer Science*, pages 61–76. Springer Berlin Heidelberg, April 2005.

[5] R. Anderson and H. Woll. Wait-free Parallel Algorithms for the Union-find Problem. In *Proceedings of the Twenty-third Annual ACM Symposium on Theory of Computing*, STOC '91, pages 370–380, New York, NY, USA, 1991. ACM. ISBN 0-89791-397-3. doi: 10.1145/103418.103458. URL `http://doi.acm.org/10.1145/103418.103458`.

[6] T. Babiak, M. Křetínský, V. Řehák, and J. Strejček. LTL to Büchi automata translation: Fast and more deterministic. In *Proc. of the 18th Int. Conf. on*

[3] `https://spot.lrde.epita.fr`

[4] `http://fmt.cs.utwente.nl/tools/ltsmin`

[5] `https://divine.fi.muni.cz`

Tools and Algorithms for the Construction and Analysis of Systems (TACAS'12), volume 7214 of *LNCS*, pages 95–109. Springer, 2012.

[7] T. Babiak, T. Badie, A. Duret-Lutz, M. Křetínský, and J. Strejček. Compositional approach to suspension and other improvements to LTL translation. In *Proceedings of the 20th International SPIN Symposium on Model Checking of Software (SPIN'13)*, volume 7976 of *Lecture Notes in Computer Science*, pages 81–98. Springer, July 2013. doi: 10.1007/978-3-642-39176-7_6.

[8] C. Baier and J.-P. Katoen. *Principles of Model Checking*. The MIT Press, 2008.

[9] T. Ball, V. Levin, and S. K. Rajamani. A decade of software model checking with SLAM. *Commun. ACM*, 54(7):68–76, 2011. doi: 10.1145/1965724.1965743. URL `http://doi.acm.org/10.1145/1965724.1965743`.

[10] J. Barnat, L. Brim, and P. Rockai. DiVinE 2.0: High-performance model checking. In *Proceedings of the International Workshop on High Performance Computational Systems Biology (HiBi'09)*, pages 31–32. IEEE Computer Society Press, 2009.

[11] J. Barnat, L. Brim, and P. Ročkai. A time-optimal on-the-fly parallel algorithm for model checking of weak LTL properties. In *Proceedings of the 11th International Conference on Formal Engineering Methods (ICFEM'09)*, volume 5885 of *LNCS*, pages 407–425, Berlin, Heidelberg, 2009. Springer-Verlag.

[12] J. Barnat, L. Brim, and P. Ročkai. Scalable shared memory LTL model checking. *International Journal on Software Tools for Technology Transfer*, 12(2):139–153, 2010.

[13] J. Barnat, P. Bauch, L. Brim, and M. Cežka. Computing strongly connected components in parallel on cuda. In *2011 IEEE International Parallel Distributed Processing Symposium*, pages 544–555, May 2011. doi: 10.1109/IPDPS.2011.59.

[14] J. Barnat, L. Brim, V. Havel, J. Havlícek, J. Kriho, M. Lenco, P. Rockai, V. Still, and J. Weiser. Divine 3.0 - an explicit-state model checker for multithreaded C & C++ programs. In N. Sharygina and H. Veith, editors, *Computer Aided Verification - 25th International Conference, CAV 2013, Saint Petersburg, Russia, July 13-19, 2013. Proceedings*, volume 8044 of *Lecture Notes in Computer Science*, pages 863–868. Springer, 2013. ISBN 978-3-642-39798-1. doi: 10.1007/978-3-642-39799-8_60. URL `http://dx.doi.org/10.1007/978-3-642-39799-8_60`.

[15] S. Ben-David, C. Eisner, D. Geist, and Y. Wolfsthal. Model checking at IBM. *Formal Methods in System Design*, 22(2):101–108, 2003. doi: 10.1023/A:1022905120346. URL `http://dx.doi.org/10.1023/A:1022905120346`.

[16] A. Biere, A. Cimatti, E. M. Clarke, M. Fujita, and Y. Zhu. Symbolic model checking using sat procedures instead of bdds. In *Proceedings of the 36th Annual ACM/IEEE Design Automation Conference*, DAC '99, pages 317–320, New York, NY, USA, 1999. ACM. ISBN 1-58113-109-7. doi: 10.1145/309847.309942. URL `http://doi.acm.org/10.1145/309847.309942`.

[17] V. Bloemen and J. van de Pol. Multi-core SCC-Based LTL Model Checking. In R. Bloem and E. Arbel, editors, *Hardware and Software: Verification and Testing: 12th International Haifa Verification Conference, HVC 2016, Haifa, Israel, November 14-17, 2016, Proceedings*, pages 18–33, Cham, 2016. Springer International Publishing. ISBN 978-3-319-49052-6. doi: 10.1007/978-3-319-49052-6_2. URL `http://dx.doi.org/10.1007/978-3-319-49052-6_2`.

[18] V. Bloemen, A. Laarman, and J. van de Pol. Multi-core On-the-fly SCC Decomposition. In *Proceedings of the 21st ACM SIGPLAN Symposium on Principles and Practice of Parallel Programming*, PPoPP '16, pages 8:1–8:12, New York, NY, USA, 2016. ACM. ISBN 978-1-4503-4092-2. doi: 10.1145/2851141.2851161. URL `http://doi.acm.org/10.1145/2851141.2851161`.

[19] R. D. Blumofe and C. E. Leiserson. Scheduling multithreaded computations by work stealing. *Journal of the ACM (JACM)*, 46(5):720–748, 1999.

[20] D. Bošnački. A nested depth first search algorithm for model checking with symmetry reduction. In D. A. Peled and M. Y. Vardi, editors, *Formal Techniques for Networked and Distributed Sytems*, volume 2529 of *LNCS*, pages 65–80. Springer Berlin Heidelberg, 2002. ISBN 978-3-540-00141-6. doi: 10.1007/3-540-36135-9_5.

[21] L. Brim, I. Černá, P. Moravec, and J. Šimša. Accepting predecessors are better than back edges in distributed LTL model-checking. In A. J. Hu and A. K. Martin, editors, *Proceedings of the 5th International Conference on Formal Methods in Computer-Aided Design (FMCAD'04)*, volume 3312 of *Lecture Notes in Computer Science*, pages 352–366. Springer, November 2004.

[22] L. Brim, I. Černá, P. Moravec, and J. Šimša. How to Order Vertices for Distributed LTL Model-Checking Based on Accepting Predecessors. In *Proceedings of the 4th International Workshop on Parallel and Distributed Methods in verifiCation (PDMC 2005)*, pages 1–12, Lisboa, Portugal, 2005. TU Munchen.

[23] R. E. Bryant. Graph-based algorithms for boolean function manipulation. *IEEE Transactions on Computers*, 35(8):677–691, Aug. 1986.

[24] J. R. Burch, E. M. Clarke, K. L. McMillan, D. L. Dill, and L. Hwang. Symbolic model checking: 10^{20} states and beyond. In *Proceedings of the Fifth Annual IEEE Symposium on Logic in Computer Science*, pages 1–33, Washington, D.C., 1990. IEEE Computer Society Press.

[25] I. Černá and R. Pelánek. Relating hierarchy of temporal properties to model checking. In B. Rovan and P. Vojtáă, editors, *Proceedings of the 28th International Symposium on Mathematical Foundations of Computer Science (MFCS'03)*, volume 2747 of *Lecture Notes in Computer Science*, pages 318–327, Bratislava, Slovak Republic, Aug. 2003. Springer-Verlag.

[26] I. Černá and R. Pelánek. Distributed explicit fair cycle detection (set based approach). In T. Ball and S. Rajamani, editors, *Proceedings of the 10th International SPIN Workshop on Model Checking of Software (SPIN'03)*, volume 2648 of *Lecture Notes in Computer Science*, pages 49–73. Springer Berlin Heidelberg, May 2003.

[27] E. Clarke, O. Grumberg, S. Jha, Y. Lu, and H. Veith. *Counterexample-Guided Abstraction Refinement*, pages 154–169. Springer Berlin Heidelberg, Berlin, Heidelberg, 2000. ISBN 978-3-540-45047-4. doi: 10.1007/10722167_15. URL http://dx.doi.org/10.1007/10722167_15.

[28] E. Clarke, A. Biere, R. Raimi, and Y. Zhu. Bounded model checking using satisfiability solving. *Formal Methods in System Design*, 19(1):7–34, 2001.

[29] E. M. Clarke, E. A. Emerson, S. Jha, and A. P. Sistla. *Symmetry reductions in model checking*, pages 147–158. Springer, 1998. ISBN 978-3-540-69339-0. doi: 10.1007/BFb0028741. URL http://dx.doi.org/10.1007/BFb0028741.

[30] E. M. Clarke, O. Grumberg, and D. A. Peled. *Model Checking*. The MIT Press, 2000.

[31] C. Courcoubetis, M. Y. Vardi, P. Wolper, and M. Yannakakis. Memory-efficient algorithm for the verification of temporal properties. *Formal Methods in System Design*, 1:275–288, 1992.

[32] J.-M. Couvreur. On-the-fly verification of temporal logic. In J. M. Wing, J. Woodcock, and J. Davies, editors, *Proceedings of the World Congress on Formal Methods in the Development of Computing Systems (FM'99)*, volume 1708 of *Lecture Notes in Computer Science*, pages 253–271, Toulouse, France, Sept. 1999. Springer-Verlag. ISBN 3-540-66587-0.

[33] J.-M. Couvreur, A. Duret-Lutz, and D. Poitrenaud. On-the-fly emptiness checks for generalized Büchi automata. In P. Godefroid, editor, *Proceedings of the 12th International SPIN Workshop on Model Checking of Software (SPIN'05)*, volume 3639 of *Lecture Notes in Computer Science*, pages 143–158. Springer, Aug. 2005.

[34] A. Deshpande, F. Herbreteau, B. Srivathsan, T. Tran, and I. Walukiewicz. Fast detection of cycles in timed automata. *CoRR*, abs/1410.4509, 2014. URL http://arxiv.org/abs/1410.4509.

[35] E. W. Dijkstra. EWD 376: Finding the maximum strong components in a directed graph. http://www.cs.utexas.edu/users/EWD/ewd03xx/EWD376.PDF, May 1973.

[36] A. Duret-Lutz. Manipulating LTL formulas using Spot 1.0. In *Proceedings of the 11th International Symposium on Automated Technology for Verification and Analysis (ATVA'13)*, volume 8172 of *Lecture Notes in Computer Science*, pages 442–445, Hanoi, Vietnam, Oct. 2013. Springer. doi: 10.1007/978-3-319-02444-8_31.

[37] A. Duret-Lutz. LTL translation improvements in Spot 1.0. *International Journal on Critical Computer-Based Systems*, 5(1/2):31–54, Mar. 2014. doi: 10.1504/IJCCBS.2014.059594.

[38] A. Duret-Lutz, A. Lewkowicz, A. Fauchille, T. Michaud, E. Renault, and L. Xu. Spot 2.0 — a framework for LTL and ω-automata manipulation. In *Proceedings of the 14th International Symposium on Automated Technology for Verification and Analysis (ATVA'16)*, volume 9938 of *Lecture Notes in Computer Science*, pages 122–129. Springer, 2016.

[39] M. B. Dwyer, G. S. Avrunin, and J. C. Corbett. Property specification patterns for finite-state verification. In M. Ardis, editor, *Proceedings of the 2nd Workshop on Formal Methods in Software Practice (FMSP'98)*, pages 7–15, New York, Mar. 1998. ACM Press.

[40] S. Edelkamp, A. L. Lafuente, and S. Leue. Directed explicit model checking with HSF-SPIN. In *Proceedings of the 8th international Spin workshop on model checking of software (SPIN'01)*, volume 2057 of *Lecture Notes in Computer Science*, pages 57–79. Springer-Verlag, 2001.

[41] E. A. Emerson and T. Wahl. *Dynamic Symmetry Reduction*, pages 382–396. Springer, 2005. ISBN 978-3-540-31980-1. doi: 10.1007/978-3-540-31980-1_25. URL `http://dx.doi.org/10.1007/978-3-540-31980-1_25`.

[42] S. Evangelista, L. Petrucci, and S. Youcef. Parallel nested depth-first searches for LTL model checking. In *Proceedings of the 9th international conference on Automated technology for verification and analysis (ATVA'11)*, volume 6996 of *Lecture Notes in Computer Science*, pages 381–396. Springer-Verlag, 2011.

[43] S. Evangelista, A. Laarman, L. Petrucci, and J. van de Pol. Improved multi-core nested depth-first search. In *Proceedings of the 10th international conference on Automated technology for verification and analysis (ATVA'12)*, volume 7561 of *Lecture Notes in Computer Science*, pages 269–283. Springer-Verlag, 2012.

[44] K. Fisler, R. Fraer, G. Kamhi, M. Y. Vardi, and Z. Yang. Is there a best symbolic cycle-detection algorithm? In *Proceedings of the fourth International Conference on Tools and Algorithms for the Construction and Analysis of Systems (TACAS'01)*, volume 2031 of *LNCS*, pages 420–434. Springer-Verlag, 2001.

[45] L. Fix. Fifteen years of formal property verification in Intel. In O. Grumberg and H. Veith, editors, *25 Years of Model Checking - History, Achievements, Perspectives*, volume 5000 of *Lecture Notes in Computer Science*, pages 139–144. Springer, 2008. ISBN 978-3-540-69849-4. doi: 10.1007/978-3-540-69850-0_8. URL `http://dx.doi.org/10.1007/978-3-540-69850-0_8`.

[46] L. K. Fleischer, B. Hendrickson, and A. Pınar. *On Identifying Strongly Connected Components in Parallel*, pages 505–511. Springer Berlin Heidelberg, Berlin, Heidelberg, 2000. ISBN 978-3-540-45591-2. doi: 10.1007/3-540-45591-4_68. URL `http://dx.doi.org/10.1007/3-540-45591-4_68`.

[47] H. N. Gabow. Path-based depth-first search for strong and biconnected components. *Information Processing Letters*, 74(3-4):107–114, February 2000.

[48] A. Gaiser and S. Schwoon. Comparison of algorithms for checking emptiness on Büchi automata. In P. Hlinený, V. Matyás, and T. Vojnar, editors, *Procedings of Annual Doctoral Workshop on Mathematical and Engineering Methods in Computer Science (MEMICS'09)*, volume 13 of *OASICS*. Schloss Dagstuhl, Leibniz-Zentrum fuer Informatik, Germany, Nov. 2009.

[49] P. Gastin and D. Oddoux. Fast LTL to Büchi automata translation. In G. Berry, H. Comon, and A. Finkel, editors, *Proceedings of the 13th International Con-*

ference on Computer Aided Verification (CAV'01), volume 2102 of *Lecture Notes in Computer Science*, pages 53–65, Paris, France, 2001. Springer-Verlag.

[50] P. Gastin, P. Moro, and M. Zeitoun. Minimization of counterexamples in SPIN. In S. Graf and L. Mounier, editors, *Proceedings of the 11th International SPIN Workshop on Model Checking of Software (SPIN'04)*, volume 2989 of *Lecture Notes in Computer Science*, pages 92–108, Apr. 2004.

[51] H. Gazit and G. L. Miller. An improved parallel algorithm that computes the BFS numbering of a directed graph. *Inf. Process. Lett.*, 28(2):61–65, 1988. doi: 10.1016/0020-0190(88)90164-0. URL `http://dx.doi.org/10.1016/0020-0190(88)90164-0`.

[52] J. Geldenhuys and A. Valmari. More efficient on-the-fly LTL verification with Tarjan's algorithm. *Theoretical Computer Science*, 345(1):60–82, Nov. 2005. Conference paper selected for journal publication.

[53] D. Giannakopoulou and F. Lerda. From states to transitions: Improving translation of LTL formulæ to Büchi automata. In D. Peled and M. Vardi, editors, *Proceedings of the 22nd IFIP WG 6.1 International Conference on Formal Techniques for Networked and Distributed Systems (FORTE'02)*, volume 2529 of *Lecture Notes in Computer Science*, pages 308–326, Houston, Texas, Nov. 2002. Springer-Verlag.

[54] P. Godefroid. Using partial orders to improve automatic verification methods. In *Computer-Aided Verification*, pages 176–185. Springer, 1991.

[55] P. Godefroid, G. Holzmann, and D. Pirottin. State-space caching revisited. *Formal Methods in System Design*, 7(3):227–241, Nov. 1995.

[56] R. Greenlaw, H. J. Hoover, and W. L. Ruzzo. *Limits to Parallel Computation: P-Completeness Theory*. Oxford University Press, 1995.

[57] X. He and Y. Yesha. A nearly optimal parallel algorithm for constructing depth first spanning trees in planar graphs. *SIAM J. Comput.*, 17(3):486–491, 1988. doi: 10.1137/0217028. URL `http://dx.doi.org/10.1137/0217028`.

[58] G. Holzmann. Parallelizing the spin model checker. In A. Donaldson and D. Parker, editors, *SPIN'12*, volume 7385 of *LNCS*, pages 155–171. Springer, 2012. ISBN 978-3-642-31758-3. URL `http://dx.doi.org/10.1007/978-3-642-31759-0_12`.

[59] G. J. Holzmann, D. A. Peled, and M. Yannakakis. On nested depth first search. In J.-C. Grégoire, G. J. Holzmann, and D. A. Peled, editors, *Proceedings of the 2nd Spin Workshop*, volume 32 of *DIMACS: Series in Discrete Mathematics and Theoretical Computer Science*. American Mathematical Society, May 1996.

[60] S. Hong, N. C. Rodia, and K. Olukotun. On fast parallel detection of strongly connected components (scc) in small-world graphs. In *2013 SC - International Conference for High Performance Computing, Networking, Storage and Analysis (SC)*, pages 1–11, Nov 2013. doi: 10.1145/2503210.2503246.

[61] T. Junttila. *On the Symmetry Reduction Method for Petri Nets and Similar Formalisms*. PhD thesis, Helsinki University of Technology, Laboratory for Theoretical Computer Science, Espoo, Finland, 2003.

[62] G. Kant, A. Laarman, J. Meijer, J. Pol, S. Blom, and T. Dijk. LTSmin: High-performance language-independent model checking. In C. T. Christel Baier, editor, *Proceedings of the 21st International Conference on Tools and Algorithms for the Construction and Analysis of Systems (TACAS'15)*, volume 9035 of *Lecture Notes in Computer Science*, pages 692–707. Springer-Berlin, 2015.
[63] S. Katz and D. Peled. *An efficient verification method for parallel and distributed programs*, pages 489–507. Springer, 1989. ISBN 978-3-540-46147-0. doi: 10.1007/BFb0013032. URL http://dx.doi.org/10.1007/BFb0013032.
[64] A. Laarman. *Scalable multi-core model checking*. PhD thesis, University of Twente, 2014.
[65] A. Laarman and J. van de Pol. Variations on multi-core nested depth-first search. In *PDMC'11*, pages 13–28, 2011.
[66] A. Laarman, R. Langerak, J. van de Pol, M. Weber, and A. Wijs. Multi-core nested depth-first search. In T. Bultan and P.-A. Hsiung, editors, *Proceedings of the Automated Technology for Verification and Analysis, 9th International Symposium (ATVA'11)*, volume 6996 of *Lecture Notes in Computer Science*, pages 321–335, Taipei, Taiwan, October 2011. Springer.
[67] A. Laarman, J. van de Pol, and M. Weber. Multi-core LTSmin: Marrying modularity and scalability. In M. Bobaru, K. Havelund, G. Holzmann, and R. Joshi, editors, *NFM 2011, Pasadena, CA, USA*, volume 6617 of *LNCS*, pages 506–511, Berlin, July 2011. Springer. doi: 10.1007/978-3-642-20398-5_40.
[68] A. Laarman, J. van de Pol, and M. Weber. Parallel Recursive State Compression for Free. In A. Groce and M. Musuvathi, editors, *SPIN 2011*, LNCS, pages 38–56, London, July 2011. Springer. URL http://doc.utwente.nl/77024/.
[69] A. W. Laarman and D. Faragó. Improved on-the-fly livelock detection. In G. Brat, N. Rungta, and A. Venet, editors, *NFM 2013*, volume 7871 of *LNCS*, pages 32–47. Springer, 2013. ISBN 978-3-642-38087-7. doi: 10.1007/978-3-642-38088-4_3.
[70] A. W. Laarman and A. J. Wijs. Partial-Order Reduction for Multi-core LTL Model Checking. In E. Yahav, editor, *HVC 2014*, volume 8855 of *LNCS*, pages 267–283. Springer, 2014. ISBN 978-3-319-13337-9. doi: 10.1007/978-3-319-13338-6_20. URL http://dx.doi.org/10.1007/978-3-319-13338-6_20.
[71] A. W. Laarman, J. C. van de Pol, and M. Weber. Boosting Multi-Core Reachability Performance with Shared Hash Tables. In N. Sharygina and R. Bloem, editors, *FMCAD 2010*. IEEE Computer Society, 2010. URL http://dl.acm.org/citation.cfm?id=1998496.1998541.
[72] T. Lai and S. Sahni. Anomalies in parallel branch-and-bound algorithms. *Commun. ACM*, 27(6):594–602, 1984. doi: 10.1145/358080.358103. URL http://doi.acm.org/10.1145/358080.358103.
[73] A. Lenharth, D. Nguyen, and K. Pingali. Parallel graph analytics. *Commun. ACM*, 59(5):78–87, Apr. 2016. ISSN 0001-0782. doi: 10.1145/2901919. URL http://doi.acm.org/10.1145/2901919.

[74] G. Lowe. Concurrent depth-first search algorithms based on Tarjan's Algorithm. *International Journal on Software Tools for Technology Transfer*, pages 1–19, 2015. ISSN 1433-2779. doi: 10.1007/s10009-015-0382-1. URL http://dx.doi.org/10.1007/s10009-015-0382-1.

[75] A. Lumsdaine, D. Gregor, B. Hendrickson, and J. Berry. Challenges in parallel graph processing. *Parallel Processing Letters*, 17(1):5–20, 2007. doi: 10.1142/S0129626407002843. URL http://dx.doi.org/10.1142/S0129626407002843.

[76] Z. Manna and A. Pnueli. A hierarchy of temporal properties. In *Proceedings of the sixth annual ACM Symposium on Principles of distributed computing (PODC'90)*, pages 377–410, New York, NY, USA, 1990. ACM.

[77] K. L. McMillan. *Symbolic Model Checking*, pages 25–60. Springer US, Boston, MA, 1993. ISBN 978-1-4615-3190-6. doi: 10.1007/978-1-4615-3190-6_3. URL http://dx.doi.org/10.1007/978-1-4615-3190-6_3.

[78] K. L. McMillan. *Interpolation and SAT-Based Model Checking*, pages 1–13. Springer, Berlin, Heidelberg, 2003. ISBN 978-3-540-45069-6. doi: 10.1007/978-3-540-45069-6_1. URL http://dx.doi.org/10.1007/978-3-540-45069-6_1.

[79] I. Munro. Efficient determination of the transitive closure of a directed graph. *Information Processing Letters*, 1(2):56–58, 1971.

[80] P. Purdom. A transitive closure algorithm. *BIT Numerical Mathematics*, 10(1):76–94, 1970.

[81] N. V. Rao and V. Kumar. Superlinear speedup in parallel state-space search. *Foundations of Software Technology and Theoretical Computer Science*, pages 161–174, 1988. URL http://dx.doi.org/10.1007/3-540-50517-2_79.

[82] J. H. Reif. Depth-first search is inherently sequential. *Information Processing Letters*, 20:229–234, 1985.

[83] E. Renault, A. Duret-Lutz, F. Kordon, and D. Poitrenaud. Three SCC-based emptiness checks for generalized Büchi automata. In K. McMillan, A. Middeldorp, and A. Voronkov, editors, *Proceedings of the 19th International Conference on Logic for Programming, Artificial Intelligence, and Reasoning (LPAR'13)*, volume 8312 of *Lecture Notes in Computer Science*, pages 668–682. Springer, Dec. 2013. doi: 10.1007/978-3-642-45221-5_44.

[84] E. Renault, A. Duret-Lutz, F. Kordon, and D. Poitrenaud. Variations on parallel explicit model checking for generalized Büchi automata. *International Journal on Software Tools for Technology Transfer (STTT)*, 19(6): 653-673, Apr. 2016.

[85] P. Sanders. Lastverteilungsalgorithmen für parallele Tiefensuche. number 463. In *Fortschrittsberichte, Reihe 10. VDI*. Verlag, 1997.

[86] W. Schudy. Finding strongly connected components in parallel using o(log2n) reachability queries. In *Proceedings of the Twentieth Annual Symposium on Parallelism in Algorithms and Architectures*, SPAA '08, pages 146–151, New York, NY, USA, 2008. ACM. ISBN 978-1-59593-973-9. doi: 10.1145/1378533.1378560. URL http://doi.acm.org/10.1145/1378533.1378560.

[87] S. Schwoon and J. Esparza. A note on on-the-fly verification algorithms. In N. Halbwachs and L. Zuck, editors, *Proceedings of the 11th International Conference on Tools and Algorithms for the Construction and Analysis of Systems (TACAS'05)*, volume 3440 of *Lecture Notes in Computer Science*, Edinburgh, Scotland, Apr. 2005. Springer.

[88] G. M. Slota, S. Rajamanickam, and K. Madduri. Bfs and coloring-based parallel algorithms for strongly connected components and related problems. In *2014 IEEE 28th International Parallel and Distributed Processing Symposium*, pages 550–559, May 2014. doi: 10.1109/IPDPS.2014.64.

[89] U. Stern and D. L. Dill. Combining state space caching and hash compaction. In *Methoden des Entwurfs und der Verifikation digitaler Systeme, 4. GI/ITG/GME Workshop*, pages 81–90. Shaker Verlag, 1996.

[90] R. Tarjan. Depth-first search and linear graph algorithms. In *Conference records of the 12th Annual IEEE Symposium on Switching and Automata Theory*, pages 114–121. IEEE, Oct. 1971. Later republished [91].

[91] R. Tarjan. Depth-first search and linear graph algorithms. *SIAM Journal on Computing*, 1(2):146–160, 1972.

[92] R. E. Tarjan. Efficiency of a good but not linear set union algorithm. *Journal of the ACM (JACM)*, 22(2):215–225, Apr. 1975.

[93] R. E. Tarjan and J. van Leeuwen. Worst-case analysis of set union algorithms. *Journal of the ACM*, 31(2):245–281, Mar. 1984.

[94] H. Tauriainen. Nested emptiness search for generalized Büchi automata. In *Proceedings of the 4th International Conference on Application of Concurrency to System Design (ACSD'04)*, pages 165–174. IEEE Computer Society, June 2004.

[95] A. Valmari. Stubborn sets for reduced state space generation. In *Proceedings of the 10th International Conference on Applications and Theory of Petri Nets (ICATPN'91)*, volume 618 of *Lecture Notes in Computer Science*, pages 491–515, London, UK, 1991. Springer-Verlag.

[96] A. Valmari. The state explosion problem. In W. Reisig and G. Rozenberg, editors, *Lectures on Petri Nets 1: Basic Models*, volume 1491 of *Lecture Notes in Computer Science*, pages 429–528. Springer-Verlag, 1998.

[97] M. Y. Vardi. Automata-theoretic model checking revisited. In *Proceedings of the 8th International Conference on Verification, Model Checking and Abstract Interpretation (VMCAI'07)*, volume 4349 of *Lecture Notes in Computer Science*, Nice, France, Jan. 2007. Springer. Invited paper.

[98] A. Wijs, J.-P. Katoen, and D. Bošnački. *GPU-Based Graph Decomposition into Strongly Connected and Maximal End Components*, pages 310–326. Springer International Publishing, Cham, 2014. ISBN 978-3-319-08867-9. doi: 10.1007/978-3-319-08867-9_20. URL http://dx.doi.org/10.1007/978-3-319-08867-9_20.

Chapter 13
Multi-core Decision Diagrams

Tom van Dijk and Jaco van de Pol

Abstract Decision diagrams are fundamental data structures that revolutionized fields such as model checking, automated reasoning and decision processes. As performance gains in the current era mostly come from parallel processing, an ongoing challenge is to develop data structures and algorithms for modern multi-core architectures. This chapter describes the parallelization of decision diagram operations as implemented in the parallel decision diagram package Sylvan, which allows sequential algorithms that use decision diagrams to exploit the power of multi-core machines.

13.1 Introduction

Decision diagrams are fundamental data structures in computer science and find applications in many areas. They are extensively used in symbolic model checking [15, 16], logic synthesis [40, 41, 55], Boolean satisfiability, fault tree analysis [52, 12], test generation [6, 1] and even to represent access control lists [26]. A recent survey paper by Minato [44] provides an accessible history of research into decision diagrams, listing applications to data mining [38], Bayesian networks and probabilistic inference models [45, 32], and game theory [53].

In the past, the processing power of computers increased mostly by improvements in the clock speed and the efficiency of processors, which often do not require adaptations to algorithms. However, as physical constraints seem to limit such improvements, further increases in processing power of modern machines inevitably

Tom van Dijk
Institute for Formal Methods and Verification, Johannes Kepler University, Linz, Austria
e-mail: tom.vandijk@jku.at

Jaco van de Pol
Formal Methods and Tools, University of Twente, Enschede, The Netherlands
e-mail: j.c.vandepol@utwente.nl

Y. Hamadi und L. Sais (eds.), *Handbook of Parallel Constraint Reasoning*,
https://doi.org/10.1007/978-3-319-63516-3_13

come from using multiple cores. To make optimal use of the processing power of multi-core machines, algorithms must be adapted.

This chapter discusses the techniques that we used to parallelize decision diagram algorithms in the parallel decision diagram library Sylvan [61, 64, 59]. These techniques are based on two main ingredients. The first ingredient is work-stealing to perform task-based algorithms such as decision diagram operations in parallel. The second ingredient consists of two concurrent data structures: a single shared hash table that stores all nodes of the decision diagrams, and a single concurrent operation cache that stores the intermediate results of operations for reuse.

This chapter is largely based on the research related to the parallel decision diagram library Sylvan, which is described in [66] and in the PhD thesis of Van Dijk [59]. Sylvan implements parallelized operations on binary decision diagrams (BDDs), list decision diagrams (LDDs), which are used in the model checking toolset LTSMIN [33], and multi-terminal binary decision diagrams (MTBDDs) [5, 22]. Sylvan can replace existing non-parallel implementations to bring the processing power of multi-core machines to non-parallel applications.

The remainder of this chapter is organized in the following way:

Section 13.2 gives a high-level overview of decision diagrams and decision diagram operations.

Section 13.3 discusses how decision diagram operations can be parallelized using work-stealing.

Section 13.4 discusses the main concurrent data structures: the hash table that contains the nodes of the decision diagrams, and the operation cache that stores the intermediate results of the operations.

Section 13.5 presents parallel garbage collection.

Section 13.6 briefly reviews the performance of parallel decision diagram operations for a number of applications. We discuss previously reported case studies on using decision diagrams in model checking, bisimulation reduction and probabilistic model checking.

Section 13.7 finally concludes the chapter.

13.2 Preliminaries

This section gives a high-level overview of decision diagrams and decision diagram operations. We discuss Boolean logic and the most well-known form of decision diagrams, binary decision diagrams, in Sections 13.2.1 and 13.2.2, as well as one popular extension of binary decision diagrams with non-binary leaves in Section 13.2.3. Section 13.2.4 describes how typical decision diagram operations are implemented. Section 13.2.5 discusses lock-free programming. Finally, Section 13.2.6 aims to provide the reader with an overview of parallelized decision diagram operations in earlier literature.

13.2.1 Boolean Logic and Notation

Boolean logic is fundamental in computer science, especially as all digital data can be expressed in binary form. Boolean variables are either `true` or `false`. Boolean formulas are defined on Boolean variables and have operators such as conjunction ($x \wedge y$), disjunction ($x \vee y$), negation ($\neg x$) and quantification ($\exists$ and $\forall$). Boolean functions are functions $\mathbb{B}^N \to \mathbb{B}$ (on N inputs), with a Boolean formula representing the relation between the inputs and the output of the Boolean function.

In this chapter, we also use 0 to denote `false` and 1 to denote `true`. We use the notation $f_{x=v}$ for a Boolean function f where the variable x is given value v. For example, given a function f defined on N variables:

$$f(x_1,\ldots,x_i,\ldots,x_N)_{x_i=0} \equiv f(x_1,\ldots,0,\ldots,x_N)$$
$$f(x_1,\ldots,x_i,\ldots,x_N)_{x_i=1} \equiv f(x_1,\ldots,1,\ldots,x_N)$$

This notation is especially relevant for decision diagrams, as they are recursively defined on the value of a Boolean variable.

13.2.2 Binary Decision Diagrams

Binary decision diagrams (BDDs) are a concise and canonical representation of Boolean functions $\mathbb{B}^N \to \mathbb{B}$ [3, 14] and are a basic structure in discrete mathematics and computer science.

A (reduced, ordered) BDD is a rooted directed acyclic graph with leaves 0 and 1. Each internal node has a variable label x_i and two outgoing edges labeled 0 and 1, called the "low" and the "high" edge. Variables are encountered along each directed path according to a fixed variable ordering. Equivalent nodes (two nodes with the same label and outgoing edges) and nodes with two identical outgoing edges (redundant nodes) are forbidden. It is well known that, given a fixed ordering, every Boolean function is represented by a unique BDD [14].

The following figure shows the BDDs for several Boolean functions. Internal nodes are drawn as circles with variables, and leaves as boxes. High edges are drawn solid, and low edges are drawn dashed. Given a valuation of the variables, BDDs are evaluated by following the high edge when the variable x is `true`, or the low edge when it is `false`.

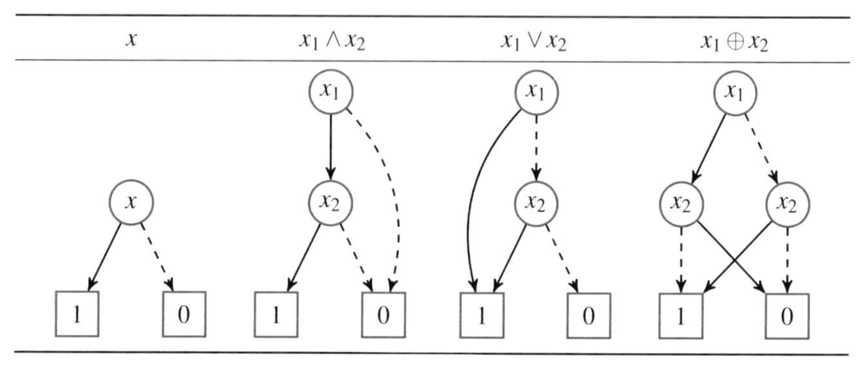

There are various equivalent ways to interpret a binary decision diagram, leading to the same Boolean function:

1. Consider every distinct path from the root of the BDD to the terminal 1. Every such path assigns `true` or `false` to the variables encountered along that path, by following either the high edge or the low edge. In this way, every path corresponds to a conjunction of literals, sometimes called a cube. For example, the cube $x_0\overline{x_1}x_3x_4\overline{x_5}$ corresponds to a path that follows the high edges of nodes labeled x_0, x_3 and x_4, and the low edges of nodes labeled x_1 and x_5. If the cubes $c_1,\ldots,c_k$ correspond to the k distinct paths in a BDD, then this BDD encodes the function $c_1 \vee \cdots \vee c_k$.
2. Alternatively, after computing $f_{x=1}$ and $f_{x=0}$ by interpreting the BDDs obtained by following the high and the low edges, a BDD node with variable label x represents the Boolean function $xf_{x=1} \vee \overline{x}f_{x=0}$.

In addition, we use complemented edges [13] as a property of an edge to denote the negation of a BDD, i.e., the leaf 1 in the BDD will be interpreted as 0 and vice versa, or in general, each terminal node will be interpreted as its negation. This is a well-known technique. We write $\neg$ to denote toggling this property on an edge. The following figure shows the BDDs for the same simple examples as above, but with complemented edges:

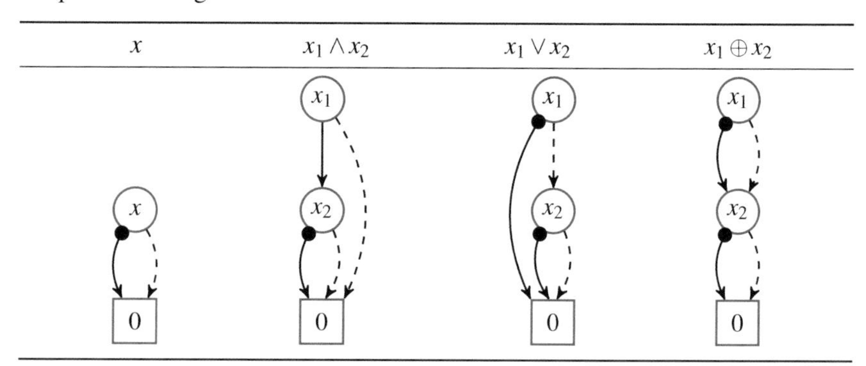

As this example demonstrates, always strictly fewer nodes are required, and there is only one ("false") terminal node. The terminal "true" is simply a complemented edge to "false". We only allow complement marks on the high edges to maintain the property that BDDs uniquely represent Boolean functions (see also below).

The interpretation of a BDD with complemented edges is as follows:

1. Count the complemented edges on each path to the terminal 0. Since negation is an involution ($\neg\neg x = x$), each path with an odd number of complemented edges is a path to "true", and with cubes $c_1, \ldots, c_k$ corresponding to all such paths, the BDD encodes the Boolean function $c_1 \vee \cdots \vee c_k$.
2. If the high edge has a complement mark, then the BDD node represents the Boolean function $x \neg f_{x=1} \vee \overline{x} f_{x=0}$, otherwise $x f_{x=1} \vee \overline{x} f_{x=0}$.

With complemented edges, the following BDDs are identical:

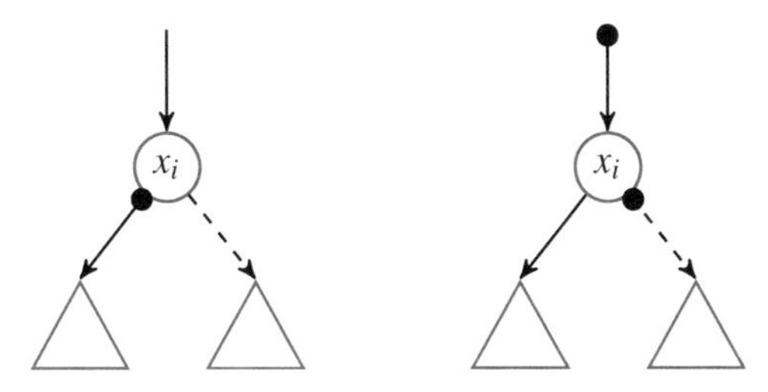

Complemented edges thus introduce a second representation of a Boolean function: if we toggle the complement mark on the two outgoing edges and on all incoming edges, we find that it encodes the same Boolean function. By forbidding a complement on one of the outgoing edges, for example the low edge, BDDs remain canonical representations of Boolean functions, since then the representation without a complement mark on the low edge is always used [13].

13.2.3 Multi-terminal Binary Decision Diagrams

In addition to BDDs with leaves 0 and 1, multi-terminal binary decision diagrams (MTBDDs) have been proposed [5, 22] with arbitrary leaves, representing functions from the Boolean space $\mathbb{B}^N$ into any set. For example, MTBDDs can have leaves representing integers (encoding $\mathbb{B}^N \to \mathbb{N}$), floating-point numbers (encoding $\mathbb{B}^N \to \mathbb{R}$) or rational numbers (encoding $\mathbb{B}^N \to \mathbb{Q}$). In our implementation of MTBDDs, we also allow for partially defined functions, using a leaf $\bot$. See Figure 13.1 for a simple example of such an MTBDD.

Similar to the interpretation of BDDs, MTBDDs are interpreted as follows:

1. An MTBDD encodes functions from a Boolean domain $D \subseteq \mathbb{B}^N$ onto some codomain C, such that for each path to a leaf $V \in C$, all inputs matching the corresponding cube c map to V. Also, given all such cubes $c_1, \ldots, c_k$, the domain D equals $c_1 \vee \cdots \vee c_k$. All paths corresponding to cubes not in D, i.e., for which the function is not defined, lead to the leaf $\bot$.

2. If an MTBDD is a leaf with the label V, then it represents the function $f(x_1,\ldots,x_N) \equiv V$. Otherwise, it is an internal node with label x. After recursively computing $f_{x=1}$ and $f_{x=0}$ by interpreting the MTBDDs obtained by following the high and the low edges, the node represents a function $f(x_1,\ldots,x_N) \equiv \text{if } x \text{ then } f_{x=1} \text{ else } f_{x=0}$.

Like BDDs, MTBDDs can have complement edges. This works only for leaf types for which negation is properly defined, i.e., each leaf x has a unique negated counterpart $\neg x$, such that $\neg\neg x = x$ and $\neg x \neq x$. In general, this does not work for numbers as $0 = -0$ in ordinary arithemetic. In addition, this also does not work for partially defined functions, as the negation of $\bot$ is not properly defined. In practice this means that complement edges are not typically used with MTBDDs.

13.2.4 Algorithms on Decision Diagrams

Many BDD packages implement the basic BDD operations `and`, `not` and `xor`, the if-then-else (`ite`) operation, and `exists` (Table 13.1). Negation $\neg$ is performed using complemented edges (Section 13.2.2) and is basically free. See Algorithm 13.1 for a typical implementation of `and`.

This algorithm showcases all features of a typical decision diagram operation. Most decision diagram operations first check whether the operation can be applied immediately to x and y (lines 2–4). This is typically the case when x and y are leaves. Often there are also other trivial cases that can be checked first. In Algorithm 13.1, this is the case when $x = y$ or when $x = \neg y$.

Often, the parameters of an operation can be normalized in some way to increase the cache efficiency. For example, $a \wedge b$ and $b \wedge a$ are the same operation. Normalization rules can then rewrite the parameters to some standard form in order to increase cache utilization, as at line 5. A well-known example is the if-then-else algorithm, which rewrites using rewrite rules called "standard triples" as described in [13].

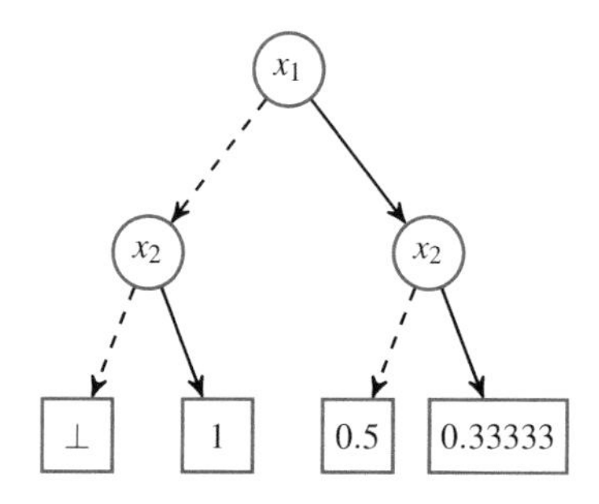

Fig. 13.1: A simple MTBDD for a function which maps $\overline{x_1}x_2$ to 1, $x_1\overline{x_2}$ to 0.5 and x_1x_2 to 0.33333. The function is undefined for the input $\overline{x_1}\overline{x_2}$

Operation	Implementation
$x \wedge y$	and(x,y)
$x \vee y$	not(and(not(x),not(y)))
$\neg(x \wedge y)$	not(and(x,y))
$\neg(x \vee y)$	and(not(x),not(y))
$x \oplus y$	xor(x,y)
$x \leftrightarrow y$	not(xor(x,y))
$x \rightarrow y$	not(and(x,not(y)))
$x \leftarrow y$	not(and(not(x),y))
if x then y else z	ite(x,y,z)
$\exists v\colon x$	exists(x,v)
$\forall v\colon x$	not(exists(not(x),v))

Table 13.1: Basic BDD operations on the input BDDs x, y, z

Algorithm 13.1: The BDD Algorithm and, with the BDDs x and y as Parameters

```
1  def and (x, y):
2      if x = 1 : return y
3      if y = 1 ∨ x = y : return x
4      if x = 0 ∨ y = 0 ∨ x = ¬y : return 0
5      if x > y : swap x and y
6      if result ← cache[(x, y)] : return result
7      v ← topvar (x,y)
8      low ← and (x_{v=0}, y_{v=0})
9      high ← and (x_{v=1}, y_{v=1})
10     result ← lookupBDDnode (v, low, high)
11     cache[(x, y)] ← result
12     return result
```

We consult the operation cache (line 6) to see whether this (sub)operation has been computed earlier. The operation cache is required to reduce the time complexity of BDD operations from exponential to polynomial in the size of the BDDs.

If x and y are not leaves and the operation is not trivial or in the cache, we use a function topvar (line 7) to determine the first variable of the root nodes of x and y. If x and y have different variables in their root nodes, topvar returns the first one in the variable ordering of x and y. We then compute the recursive application to the cofactors of x and y with respect to variable v at lines 8–9.

We write $x_{v=i}$ to denote the cofactor of x where variable v takes value i. Since x and y are ordered according to the same fixed variable ordering, we can easily obtain $x_{v=i}$. If the root node of x has the variable v, then $x_{v=i}$ is obtained by following the low ($i = 0$) or high ($i = 1$) edge of x. Otherwise, $x_{v=i}$ equals x.

After computing the suboperations, we compute the result by either reusing an existing or creating a new BDD node (line 10). This is done by a function lookupBDDnode, which, given a variable v and the BDDs of $\text{result}_{v=0}$ and $\text{result}_{v=1}$, returns the BDD for result by consulting the unique table.

When the result has been computed, we store it in the operation cache (line 11) and return the result (line 12).

13.2.5 Parallelism

A major goal in computing is to perform ever larger calculations and to improve their performance and efficiency. This can be accomplished using various techniques that are often orthogonal to each other, such as better algorithms, faster processors and parallel computing using multiple processors. Faster hardware increases the performance of most computations, often regardless of the algorithm, although some algorithms benefit more from processor speed while others benefit more from faster memory access. For suitable algorithms, parallel processing can considerably improve the performance, on top of what is possible just by increased processor speeds.

For some algorithms, efficient parallelism is almost trivial. It is no coincidence that graphics cards contain thousands of small processors, resulting in massive speedups for very particular applications. Other algorithms are more difficult to parallelize. For example, some algorithms are inherently sequential, with few opportunities for the parallel execution of independent calculation paths. Other algorithms have enough independent paths for parallelization in theory, but are difficult to parallelize in practice, for example because they are irregular and continually require load balancing, moving work between processors. Some algorithms are memory-intensive, i.e., they spend most of their time manipulating data in memory, which can result in bottlenecks due to the limited bandwidth between the processors and the memory, as well as time spent waiting in locks.

This chapter discusses the parallelization of algorithms for decision diagrams, which are large directed acyclic graphs. They are typically irregular and mainly consist of unpredictable memory accesses with high demands on memory bandwidth. Decision diagrams are often used as the underlying operations of other algorithms. If the underlying decision diagram operations are parallelized, then sequential algorithms that use them may also benefit from the parallelization.

Lock-Free Programming

In parallel programs, memory accesses can result in race conditions or data corruption, for example when multiple threads write to the same location in memory. Typically data structures are protected against race conditions using locking techniques. While locks are relatively easy to implement and reason about, they often severely cripple parallel performance, especially as the number of threads increases. Threads have to wait until the lock is released, and locks can be a bottleneck when many threads try to acquire the same lock. Also, locks can sometimes cause spurious delays that smarter data structures could avoid, for example by recognizing that some operations do not interfere even though they access the same resource.

A standard technique that avoids locks uses the atomic `compare-and-swap` (`cas`) operation, which is supported by many modern processors.

```
1 def compare-and-swap(location, expected, newvalue):
2     value ← *location
3     if value ≠ expected : return False
4     *location ← newvalue
5     return True
```

This operation atomically compares the contents of a given location in shared memory to some given expected value and, if the contents match, changes the contents to a given new value. If multiple processors try to change the same bytes in memory using `cas` at the same time, then only one succeeds.

Data structures that avoid locks are called non-blocking or lock-free. Such data structures often use the atomic `cas` operation to make progress in an algorithm, rather than protecting a part that makes progress. For example, when modifying a shared variable, an approach using locks would first acquire the lock, then modify the variable, and finally release the lock. A lock-free approach would use atomic `cas` to modify the variable directly. This requires only one memory write rather than three, but lock-free approaches are typically more complicated to reason about, and prone to bugs that are more difficult to reproduce and debug.

13.2.6 Historical Perspective

This section describes various approaches have been tried in the past for parallel processing of decision diagrams, as discussed in [59].

Massively Parallel Computing (early 1990s)

In the early 1990s, researchers tried to speed up BDD manipulation by parallel processing. The first paper [34] views BDDs as automata, and combines them by computing a product automaton followed by minimization. Parallelism arises by handling independent subformulas in parallel: the expansion and reduction algorithms themselves are not parallelized. They use locks to protect the global hash table, but this still results in a speedup that is almost linear with the number of processors. Most other work in this era implemented BFS algorithms for vector machines [46] or massively parallel SIMD machines [17, 28] with up to 64K processors. Experiments were run on supercomputers, such as the Connection Machine. Given the large number of processors, the speedup (around 10 to 20) was disappointing.

Parallel Operations and Constructions

An interesting contribution in this period is the paper by Kimura et al. [35]. Although they focus on the construction of BDDs, their approach relies on the observation that suboperations of a logic operation can be executed in parallel and the results can be merged to obtain the result of the original operation. Our solution to parallelizing BDD operations follows the same line of thought, although the work-stealing method for efficient load balancing that we use was first published two years later [10]. Similarly to [35], Parasuram et al. implement parallel BDD operations for distributed systems, using a "distributed stack" for load balancing, with speedups from 20–32 on a CM-5 machine [50]. Chen and Banerjee implement the parallel construction of BDDs for logic circuits using lock-based distributed hash tables, parallelizing on the structure of the circuits [18]. Yang and O'Hallaron [71] parallelize breadth-first BDD construction on multi-processor systems, resulting in reasonable speedups of up to $4\times$ with eight processors, although there is a significant synchronization cost due to their lock-protected unique table.

Distributed Memory Solutions (late 1990s)

Attention shifted towards Networks of Workstations, based on message passing libraries. The motivation was to combine the collective memory of computers connected via a fast network. Both depth-first [4, 58, 7] and breadth-first [54] traversal have been proposed. In the latter, BDDs are distributed according to variable levels. A worker can only proceed when its level has a turn, so these algorithms are inherently sequential. The advantage of distributed memory is not that multiple machines can perform operations faster than a single machine, but that their memory can be combined in order to handle larger BDDs. For example, even though [58] reports a nice parallel speedup, the performance with 32 machines is still $2\times$ slower than the non-parallel version. BDDNOW [43] is the first BDD package that reports some speedup compared to the non-parallel version, but it is still very limited.

Parallel Symbolic Reachability (after 2000)

After 2000, research attention shifted from parallel implementations of BDD operations towards the use of BDDs for symbolic reachability in distributed [29, 19] or shared memory [23, 21]. Here, BDD partitioning strategies such as horizontal slicing [19] and vertical slicing [31] were used to distribute the BDDs over the different computers. Also the saturation algorithm [20], an optimal iteration strategy in symbolic reachability, was parallelized using horizontal slicing [19] and using the work-stealer Cilk [23], although it is still difficult to obtain good parallel speedup [21].

Multi-core BDD Algorithms

There is some recent research on multi-core BDD algorithms. There are several implementations that are thread-safe, i.e., they allow multiple threads to use BDD operations in parallel, but they do not offer parallelized operations. In a thesis on the BDD library JINC [49], Chapter 6 describes a multi-threaded extension. JINC's parallelism relies on concurrent tables and delayed evaluation. It does not parallelize the basic BDD operations, although this is mentioned as possible future research. Also, a recent BDD implementation in Java called BeeDeeDee [39] allows execution of BDD operations from multiple threads, but does not parallelize single BDD operations. Similarly, the well-known sequential BDD implementation CUDD [57] supports multi-threaded applications, but only if each thread uses a different "manager," i.e., unique table to store the nodes in. Except for our contributions [62, 61, 64] related to Sylvan, there is no recent published research on modern multi-core shared-memory architectures that parallelizes the actual operations on BDDs. Recently, Oortwijn et al. [47, 48] continued our work by parallelizing BDD operations on shared-memory abstractions of distributed systems using remote direct memory access. Work by Velev et al. [68] implements BDD operations on GPUs for a small case study with promising results.

13.3 Parallel Decision Diagrams

The requirements for the efficient parallel implementation of decision diagrams are not the same as for a non-parallel implementation. We refer to Somenzi [56] for a general discussion on the implementation of non-parallel decision diagrams. Somenzi already established several aspects of a BDD package. The two central data structures of a BDD package are the *unique table* (or *nodes table*) and the *computed table* (or *operation cache*). Furthermore, garbage collection is essential for a BDD package, as most BDD operations continuously create and discard BDD nodes. The two central data structures are discussed in Section 13.4 and garbage collection in Section 13.5. The current section presents the parallelization of decision diagram operations by work-stealing.

13.3.1 Work-Stealing

Operations on decision diagrams are typically recursively defined on the structure of the inputs. To parallelize decision diagram operations, we consider each subproblem as a separate task and execute independent tasks in parallel. This type of parallelism is called *task-based parallelism*.

For task parallelism that fits a "strict" fork-join model, i.e., each task creates the subtasks that it depends on, work-stealing is well known to be an effective load-

Algorithm 13.2: The Algorithm (left) is Implemented (right) Using SPAWN, SYNC and CALL

```
1 do in parallel:                 1 SPAWN(F1, x, y, z)
2 |  K ← F1(x, y, z)              2 SPAWN(F2, a, b, c)
3 |  L ← F2(a, b, c)              3 M ← CALL(F3, g, h)
4 |_ M ← F3(g, h)                 4 L ← SYNC
                                  5 K ← SYNC
```

balancing method [10], with implementations such as Cilk [11, 27] and Wool [24, 25] that allow parallel programs to be written in a style similar to sequential programs [2]. Work-stealing has been proven to be optimal for a large class of problems and has tight memory and communication bounds [10].

In work-stealing, tasks are executed by a fixed number of workers, typically equal to the number of processor cores. Each worker owns a task pool into which it inserts new subtasks created by the task it currently executes. Idle workers steal tasks from the task pools of other workers. Worker are idle either because they do not have any tasks to perform (e.g., at the start of a computation), or because all their subtasks have been stolen and they have to wait for the result of the stolen subtasks to continue the current task. Typically, one worker starts executing a root task and the other workers perform work-stealing to acquire subtasks.

We use **do in parallel** to denote that tasks are executed in parallel. Programs in the Cilk/Wool style are then implemented like in Algorithm 13.2. The SPAWN keyword creates a new task. The SYNC keyword matches with the last unmatched SPAWN, i.e., operating as if spawned tasks are stored on a stack. It waits until that task is completed and retrieves the result. Every SPAWN during the execution of the program must have a matching SYNC. The CALL keyword skips the task stack and immediately executes a task.

One important aspect of the work-stealing algorithm is victim selection. For example in systems with hierarchy, e.g., a network of workstations, it might be useful to steal from local workers first before trying to steal from a remote worker. Another strategy would be to remember how much work other workers have after a steal attempt, and use this to intelligently select targets. In our implementation, workers with an empty task pool steal from random victims.

When synchronizing with a stolen task, a possible strategy for the victim is to steal from the thief until the stolen task is completed. By stealing back from the thief, a worker executes subtasks of the stolen task. This technique is called leapfrogging [69]. When stealing from random workers instead, the size of the task pool of each worker could grow beyond the size needed for complete sequential execution [25], since stealing will build a new stack on top of the blocked join. Using leapfrogging rather than stealing from random workers thus limits the space requirement of the task pools to that of sequential execution, although in practice it is expensive to guarantee that the tasks that are stolen from the thief are really subtasks of the original task. It might be possible that the thief finished the original task and stole a different branch of the

Work-stealing operations	Task pool operations
`spawn`(task)	`push`(task)
`sync`	`peek`, `pop`
`steal-and-run`(victim)	`steal`

Table 13.2: Operations of the work-stealing algorithm and matching operations of the task pool of each worker

task tree after the victim checked the status of the stolen task. Our implementation also uses the leapfrogging strategy.

Another concern is which task(s) to steal. A simple algorithm is to steal the first unstolen task from the bottom of the stack. A variation could be to steal multiple tasks, or to steal a random task from anywhere in the stack. In our implementation, thieves steal the first unstolen task from the bottom of the stack.

See Table 13.2 for an overview of the work-stealing operations and how they match with operations on the task pool. The methods `spawn` and `sync` implement the keywords `SPAWN` and `SYNC`. The method `steal-and-run` tries to steal a task from the given victim and, if successful, executes the task and communicates the result back to the owner of the task. The methods `push`, `peek`, `pop` and `steal` are implemented by the task pool:

- The `push`, `peek` and `pop` operations are only used by the owner of the stack, and the `steal` operation only by thieves.
- The `push` operation puts a task on the stack.
- The `peek` operation fixes the status of the task at the top of the stack: either stolen or available as work. After `peek`, the top task, if not stolen, cannot be stolen until the next `push` (or if `peek` is called again).
- The `pop` operation removes the topmost task from the stack. Furthermore we assume that the task data remains in the task pool until overwritten by a `push` operation.
- The `steal` operation steals a task from the bottom of the stack, changing its status from available work to stolen work. Stolen tasks are kept on the stack so the results of tasks can be communicated back to the original owner of the task.

Different implementations of the work-stealing stack can be used, as long as they implement the described functionality. Experiments show that the difference in performance between the private deque by Acar et al. [2], the shared deque in Wool [24, 25] and the shared deque we implemented in Lace [63] are relatively small; they all have sufficient scalability, although Lace also implements a stop-the-world feature required for garbage collection (Section 13.5).

Algorithm 13.3: The Implementation of Work-Stealing Using Leapfrogging when Waiting for a Stolen Task to Finish, i.e., steal from the thief

```
1  def spawn(task):
2      push(task)
3  def sync():
4      res ← peek()
       // res is Work(task) or Stolen(task)
5      if res = Work(task):
6          pop()
7          return task.execute()
8      else:
9          while task.thief = None: (loop)
10         while ¬ task.done: steal-and-run(task.thief)
11         pop-stolen()
12         return task.result
13 def steal-and-run(victim):
14     if victim.steal() = Task(stolentask):
15         stolentask.thief ← me
16         result ← stolentask.execute()
17         stolentask.result ← result
18         stolentask.done ← True
19 thread worker(id, roottask):
20     done ← False
21     if id = 0:
22         roottask.execute()
23         done ← True
24     else: while done is False: steal-and-run(random victim)
```

13.3.2 Parallel Operations with Work-Stealing

Decision diagram operations such as and (Algorithm 13.1) are parallelized by executing the subtasks (lines 9–10) in parallel:

```
8  do in parallel:
9      low ← and(x_{v=0}, y_{v=0})
10     high ← and(x_{v=1}, y_{v=1})
```

This is equivalent to the following:

```
8  SPAWN(and, x_{v=0}, y_{v=0})
9  high ← CALL(and, x_{v=1}, y_{v=1})
10 low ← SYNC
```

A more involved example is the parallelized algorithm exists (Algorithm 13.4), which computes existential quantification. This algorithm receives the input parameters x and V, where x is the BDD representing the function to which quantification is applied, and V is the BDD representing the conjunction of the variables that are

Algorithm 13.4: Parallelized BDD Algorithm `exists`, with the BDD x and V the Cube of Variables that are Abstracted via Existential Quantification

```
1  def exists(x, V):
2      if x = 0 ∨ x = 1 ∨ V = ∅ : return x
3      v = var(x)
4      while V ≠ ∅ ∧ var(V) < v: V ← next(V)
5      if V = ∅ : return x
6      if result ← cache[(x,V)] : return result
7      if v = var(V) :
8          if x_{v=0} = 1 ∨ x_{v=1} = 1 ∨ x_{v=0} = ¬x_{v=1} : result ← 1
9          else:
10             low ← exists(x_{v=0}, next(V))
11             if low = 1 : result ← 1
12             else:
13                 high ← exists(x_{v=1}, next(V))
14                 result ← or(low, high)
15     else:
16         do in parallel:
17             low ← exists(x_{v=0}, V)
18             high ← exists(x_{v=1}, V)
19         result ← lookupBDDnode(v, low, high)
20     cache[(x,V)] ← result
21     return result
```

abstracted away from x. After the trivial cases (line 2), we check whether V actually contains variables that are in the BDD (lines 3–5), exploiting the fact that V is also an ordered BDD. This is also a normalization step for the cache, which is checked at line 6. Now, there are two cases: either the current root variable v is in V (lines 7–14) or it is not in V (lines 15–19). In the second case, we simply perform the two suboperations in parallel and compute the result. In the first case, after checking some trivial cases, we can either 1) perform the two suboperations in parallel; 2) perform the "low" suboperation first; or 3) perform the "high" suboperation first. If either of these suboperations returns 1, then the other does not need to be computed. The advantage of option 1 is that there is more opportunity for parallelization, at the cost of possible extra work. However, this extra independent work might not be necessary, since there is already a lot of independent work from the parallelization at lines 17–18 and inside the `or` operation. In Algorithm 13.4, we compute the "low" suboperation first.

In model checking using decision diagrams, relational products play a central role. Relational products compute the successors or the predecessors of (sets of) states. Typically, states are encoded using Boolean variables $\mathbf{x} = x_1, x_2, \ldots, x_N$. Transitions between these states are represented using Boolean variables $\mathbf{x}$ for the source states and variables $\mathbf{x}' = x_1', x_2', \ldots, x_N'$ for the target states. Given a set of states S_i encoded as a BDD on variables $\mathbf{x}$, and a transition relation R encoded as a BDD on variables $\mathbf{x} \cup \mathbf{x}'$, the set of states S_{i+1}' encoded on variables $\mathbf{x}'$ is obtained by computing

Algorithm 13.5: The Parallel Algorithm `relnext`, which Given the BDDs S (representing a set of states), R (representing a transition relation) and V (the cube of interleaved variables $\mathbf{x} \cup \mathbf{x}'$) Computes the Set of Successor States Defined on $\mathbf{x}$, i.e., $\big(\exists \mathbf{x}\colon (S \wedge R)\big)[\mathbf{x}' := \mathbf{x}]$. We Assume that all Variables in R are also in V

```
1  def relnext (S, R, V):
2      if S = 0 ∨ R = 0 : return 0
3      if S = 1 ∧ R = 1 : return 1
4      v = topvar (S,R)
5      while var (V) < v: V ← next (V)
       // if V = ∅, we assume R is irrelevant
6      if V = ∅ : return S
7      if result ← cache[(S,R,V)] : return result
8      if v = var (V) :
9          x, x' ← unprimed v, primed v
10         V' ← V without x and x'
11         do in parallel:
12             a ← relnext (S_{x=0}, R_{x=0,x'=0}, V')
13             b ← relnext (S_{x=1}, R_{x=1,x'=0}, V')
14             c ← relnext (S_{x=0}, R_{x=0,x'=1}, V')
15             d ← relnext (S_{x=1}, R_{x=1,x'=1}, V')
16         do in parallel:
17             low ← or (a, b)
18             high ← or (c, d)
19         result ← lookupBDDnode (x, low, high)
20     else:
           // v is not in R, by assumption
21         do in parallel:
22             low ← relnext (S_{v=0}, R, V)
23             high ← relnext (S_{v=1}, R, V)
24         result ← lookupBDDnode (v, low, high)
25     cache[(S,R,V)] ← result
26     return result
```

$S'_{i+1} = \exists \mathbf{x}\colon (S_i \wedge R)$. BDD packages typically implement an operation `and_exists` that combines $\exists$ and $\wedge$ to compute S'_{i+1}.

Typically we want the BDD of the successor states defined on the unprimed variables $\mathbf{x}$ instead of the primed variables $\mathbf{x}'$, so the `and_exists` call is then followed by a variable substitution that replaces all occurrences of variables from $\mathbf{x}'$ with the corresponding variables from $\mathbf{x}$. Furthermore, the variables are typically interleaved in the variable ordering, like $x_1, x'_1, x_2, x'_2, \ldots, x_N, x'_N$, as this often results in smaller BDDs. This combination of `and_exists` and variable renaming can be done with a specialized operation `relnext`, which computes the successors of sets of states, where the transition relation is encoded with the interleaved variable ordering.

See Algorithm 13.5 for the parallel implementation of `relnext`. This function takes as input a set S, a transition relation R and the set of variables V, which is the union of the interleaved sets $\mathbf{x}$ and $\mathbf{x}'$ (the variables on which the transition relation is defined). We first check for terminal cases (lines 2–3). These are the same cases as for the $\wedge$ operation. Then we process the set of variables V to skip variables that are not in S and R (lines 5–6). After consulting the cache (line 7), either the current variable is in the transition relation, or it is not. If it is not, we perform the usual recursive calls and compute the result (lines 21–24). If the current variable is in the transition relation, then we let x and x' be the two relevant variables (either of these equals v) and compute four subresults, namely for the transitions (a) from 0 to 0, (b) from 1 to 0, (c) from 0 to 1, and (d) from 1 to 1 in parallel (lines 11–15). We then abstract from x' by computing the existential quantifications in parallel (lines 16–18), and finally compute the result (line 19). This result is stored in the cache (line 25) and returned (line 26).

13.3.3 Conclusion

This section discussed using work-stealing to perform operations on decision diagrams in parallel. We looked at three operations in particular: `and`, which is a prototype for many simple decision diagram operations; `exists`, which adds the complexity that the subtasks are not completely independent (if "low" returns 1, "high" does not need to be computed); and `relnext`, which adds the complexity of having two phases with independent subtasks.

13.4 Concurrent Data Structures

To efficiently parallelize decision diagram operations, we must perform memory operations in a scalable manner, i.e., using optimized scalable data structures. This section describes the organization of decision diagram nodes in memory, as well as the design of the unique table and the operation cache.

13.4.1 Representation of Nodes

The representation of BDD and MTBDD nodes in memory is important for both the sequential and the parallel performance of decision diagram implementations. We use 16 bytes for all types of nodes, so we can use the same unique table for all nodes and have a fixed node size. With 16 bytes per node, exactly four nodes fit in a cacheline of 64 bytes (the size of the cacheline for many current computer architectures, in particular the x86 family that we use). If the unique table is properly aligned in memory, then only one cacheline needs to be accessed when accessing a node.

We use 40 bits to store the index of a node in the unique table. This is sufficient to store up to 2^{40} nodes, i.e., 16 terabytes of nodes, excluding overhead costs.

Sylvan defines the type MTBDD as a 64-bit integer, representing an edge to an MTBDD node. The lowest 40 bits represent the location of the node in the nodes table, and the most significant bit stores the complement mark [13], mainly used by BDDs. The BDD 0 is reserved for the leaf `false`, with the complemented edge to 0 (i.e., `0x8000000000000000`) meaning `true`.

Internal BDD and MTBDD nodes store the variable label (24 bits), the low edge (40 bits), the high edge (40 bits), the complement bit of the high edge (1 bit, the first bit below) and the fact they are not a leaf (1 bit, the second bit below, set to 0):

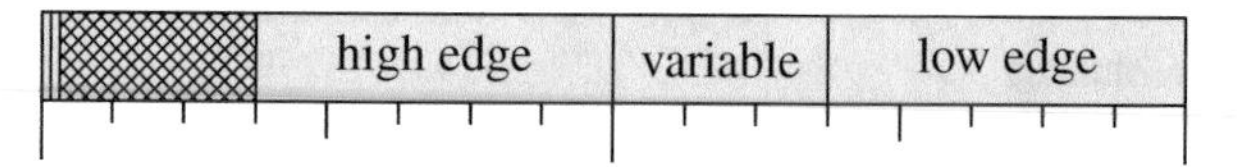

MTBDD leaves store the leaf type (32 bits), the leaf value (64 bits) and the fact that they are a leaf (1 bit, set to 1):

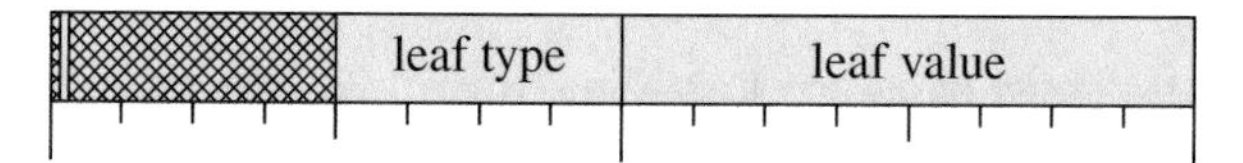

The unused space bits are set to 0. They can also be used by the decision diagram library for other node types or for temporary marking of nodes in algorithms, which is beyond the scope of this chapter.

13.4.2 Unique Table

The unique table stores all decision diagram nodes and is essential to avoid duplicate nodes. This table is typically implemented as a hash table, in particular because the `find-or-insert` operation is performed in time O(1) on average (amortized) by a hash table.

The unique table can either be one shared table, or be split into multiple parts somehow. For example, Somenzi [56] argues for a subtable for each variable level, as this makes the implementation of variable reordering easier. The disadvantage of subtables is that their sizes must be adjusted dynamically, thus requiring the different parallel processes to cooperate on performing garbage collection and resizing when subtables are full. In addition, there is some overhead to compute the correct size for each table, which can be avoided by using a single table. Finally, subtables require the additional complexity of decreasing subtable sizes and compressing decision diagrams, which we avoid by using a single table that only increases in size when this is needed.

In the past, there have been various proposals to split the unique table into several parts for parallel applications, for example to assign parts of the decision diagrams to certain processors or workstations. This is a consideration that can be orthogonal to parallelism. As we use work-stealing to perform the load balancing of the decision diagram operations, we have no control over which processor performs specific operations. Therefore, we use a single continuous block of memory, and we let the operating system take care of allocating memory blocks on all available memories in the system.

The unique table essentially requires the following operations, which must be highly scalable:

- a `find-or-insert` method, which, given a 16-byte node, either finds the existing node in the table, or creates a new node.
- a method to delete nodes for garbage collection. Our implementation has a separate "data array" containing the nodes and a "hash array" containing the metadata. We require three operations:
  - `clear` removes all entries from the hash array;
  - `mark` marks a given node for reinsertion in the hash array; and
  - `rehash` reinserts a given node in the hash array.

Our design strictly separates lookup and insertion of nodes from a stop-the-world garbage collection phase, during which the table may be resized. From the perspective of the nodes table algorithms (and correctness), all threads of the program are in one of two phases:

1. During *normal operation*, threads only call the `find-or-insert` operation, which takes as input the 16-byte data and either returns a unique identifier for the data, or raises the TableFull signal if the algorithm fails to insert the data.
2. During *garbage collection*, the `find-or-insert` operation is never called. Instead, methods `clear`, `mark` and `rehash` (described in Section 13.5) are called to perform garbage collection.

This simplifies the requirements for the hash tables. The `find-or-insert` operation must have the following property: if the operation returns a value for some given data, then other `find-or-insert` operations may not return the same value for a different input, or return a different value for the same input. This property must hold between garbage collections; garbage collection obviously breaks the property for nodes that are not kept during garbage collection, as nodes are removed from the table to make room for new data.

The unique table we use in Sylvan is based on the hash table in [36], which is designed to store visited states in model checking. This hash table incorporates two ideas that we also use in our design:

- Using a probe sequence called "walking-the-line" that is efficient with respect to transferred cachelines.
- Separating the stored data in a "data array" and the hash of the data in the "hash array" to avoid directly comparing the data.

Furthermore, to manage the "data array" we use bit arrays as a convenient parallel allocator, although other scalable parallel allocation mechanisms for fixed-size (16 bytes) memory blocks could be used to manage the data array.

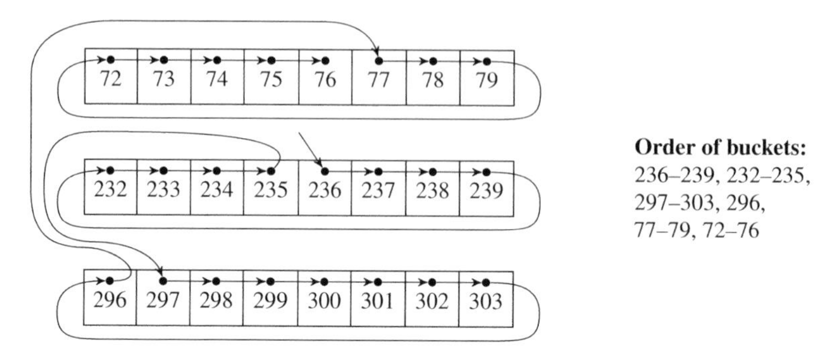

Fig. 13.2: Example of the walking-the-line probe sequence, with the starting buckets 236, 297 and 77 based on the first three hash values of the data

The Walking-the-Line Probe Sequence

Every hash table needs to implement a strategy to deal with hash table collisions, i.e., when different data hashes to the same location in the table. To find a location for the data in the hash table, some hash tables use open addressing: they visit buckets in the hash table in a deterministic order called the probe sequence, to either detect that the data is already in the hash table, or to find an empty bucket, which indicates that the data can be inserted into that bucket. One of the simplest probe sequences is linear probing, where the data is hashed once to obtain the first bucket (e.g., bucket 61), and the probe sequence consists of all buckets from that first bucket (e.g., 61, 62, 63, ...).

An alternative to linear probing is walking-the-line, proposed in [36]. Since data in a computer is transferred in blocks called cachelines, it is more efficient to use the entire cacheline instead of only a part of the cacheline. For example, if there are eight buckets per cacheline and we assume that the buckets are properly aligned so that the first cacheline starts with bucket 0, then linear probing starting at bucket 61 would only check buckets 61–63 of the first accessed cacheline. In walking-the-line, the other buckets in that cacheline are also checked, so after buckets 61–63, also buckets 56–60 would be checked. Then, a new hash value is obtained for the data using a hash function to obtain the next starting bucket. In theory, this procedure could be repeated forever; in practice, after a certain number of cachelines the procedure terminates with the result that the table is full. See also Figure 13.2 for an example of walking-the-line.

Separated Arrays

The hash table stores the hash of the data in each bucket in a separate array. The idea is that the `find-or-insert` algorithm does not need to access the stored data if the stored hash does not match with the hash of the data given to `find-or-insert`. This reduces the number of accessed cachelines during `find-or-insert`.

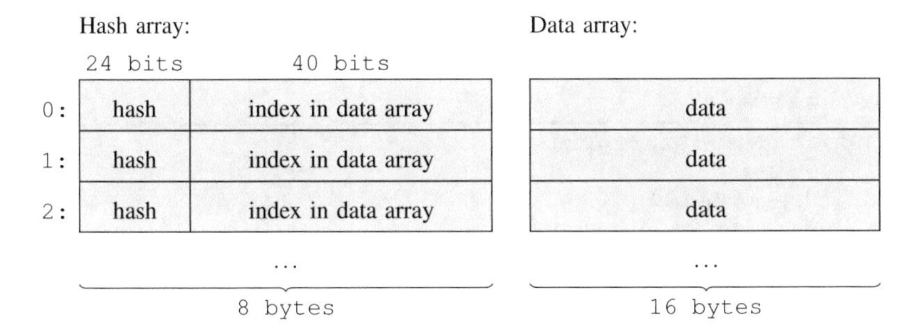

Fig. 13.3: Layout of the hash array and data array

Bit Arrays for Data Management

We use a separate bit array `databits` to implement a parallel allocator for the data array. Furthermore, to avoid having to use `cas` for every change to `databits`, we divide this bit array into regions, such that every region matches exactly with one cacheline of the `databits` array, i.e., 512 buckets per region if there are 64 bytes in a cacheline, which is the case for most current architectures. Every worker has exclusive access to one region, which is managed with a second bit array `regionbits`. Only changes to `regionbits` (to claim a new region) require an atomic `cas`. We therefore only use normal writes for insertion and uninsertion into the data array, and only occasionally an atomic `cas` during speculative insertion to obtain exclusive access to the next region of 512 buckets.

A claimed region is not given back until garbage collection, which resets claimed regions. On startup and after garbage collection, the `regionbits` array is cleared and all threads claim an initial region using the `claim-next-region` method in Algorithm 13.6. All threads start at a different position (distributed over the entire table) for their first claimed region, to minimize the interactions between threads. The `databits` array is empty at startup and during garbage collection threads use atomic `cas` to set the bits in `databits` of decision diagram nodes that must be kept in the table. In addition, the bit of the first bucket is always set to 1 to avoid using the index 0 since this is a reserved value in Sylvan.

The layout of the hash array and the data array is given in Figure 13.3. We use a hash function that never hashes to 0 and we forbid nodes with the index 0 because 0 is a reserved value in Sylvan. The fields hash and index are therefore never 0, unless the hash bucket is `empty`, so the field H to indicate that hash and index have valid values is not necessary. Manipulating the hash array bucket is also simpler, since we no longer need to take into account changes to the field D.

Inserting data into the hash table consists of three steps. First the algorithm tries to find whether the data is already in the table. If this is not the case, then a new bucket in the data array is reserved in the current region of the thread with the `reserve-data-bucket` function. If the current region is full, then the thread

Algorithm 13.6: Algorithm for Parallel `find-or-insert` of the Hash Table, with 512 Buckets per Region. The Variable `myregion` is a Thread-Specific Variable

```
1  def find-or-insert (data):
2      index ← 0
3      h ← hash(data)
4      for s ∈ probe-sequence(data) :
5          V ← harray[s]
6          if V = 0 :
7              if index = 0 :
8                  index ← reserve-data-bucket ()
9                  darray[index] ← data
10             if cas(harray[s], 0, {h,index}) : return index
11             else: V ← harray[s]
12         if V.hash = h ∧ darray[V.index] = data :
13             if index ≠ 0 : free-data-bucket (index)
14             return V.index
15     raise TableFull
16 def reserve-data-bucket ():
17     loop:
18         if myregion has a bit set to 0 :
19             i ← first bit in myregion that is 0
20             set-bit(databits, 512 × myregion + i, 1)
21             return 512 × myregion + i
22         else: myregion ← claim-next-region (myregion)
23 def free-data-bucket (d):
24     set-bit(databits, d, 0)
25 def claim-next-region (oldregion):
26     newregion ← (oldregion + 1) mod (tablesize/512)
27     while newregion ≠ oldregion :
28         loop:
29             if the bit for newregion is 1 : break
30             if set-bit-cas(regionbits, newregion, 0, 1) : return newregion
31         newregion ← (newregion + 1) mod (tablesize/512)
32     raise TableFull
```

claims a new region with the `claim-next-region` function. Note that it may be possible that the next region contains used buckets, if there has been a garbage collection earlier. Afterwards the new bucket is inserted into the hash array. Sometimes, the data has been inserted concurrently (by another thread) and then the bucket in the data array is freed again with the `free-data-bucket` function, so it is available the next time the thread wants to insert data.

The main method of the hash table is `find-or-insert`. See Algorithm 13.6. The algorithm uses the local variable "index" to keep track of whether the data is inserted into the data array. This variable is initialized to 0 (line 2), which signifies that data is not yet inserted into the data array. For every bucket in the probe sequence, we first check whether the bucket is empty (line 6). In that case, the data is not yet in the table. If we did not yet write the data in the data array, then we reserve the next bucket and write the data (lines 7–9). We use atomic `cas` to insert the hash and index into the hash array (line 10). If this is succesful, then the algorithm is done and returns the location of the data in the data array. If the `cas` operation fails, some other thread inserted data here and we refresh our knowledge of the bucket (line 11) and continue at line 12. If the bucket is not empty, then we compare the stored hash with the hash of our data, and if this matches, we compare the data in the data array with the given input (line 12). If this matches, then we may need to free the reserved bucket (line 13) and we return the index of the data in the data array (line 14). If we finish the probe sequence without inserting the data, we raise the TableFull signal (line 15).

The `find-or-insert` method relies on `reserve-data-bucket` and on `free-data-bucket`, which are also given in Algorithm 13.6. They are fairly straightforward.

The `claim-next-region` method searches in the `regionbits` array for the first 0-bit. The value `tablesize` here represents the size of the entire table. We use a simple linear search and a `cas`-loop to actually claim the region. Note that we may be competing with threads that are trying to set the bit of a different region, since the smallest range for the atomic `cas` operation is 1 byte or 8 bits.

13.4.3 Computed Table

The operation cache is a hash table that stores intermediate results of BDD operations. It is well known that an operation cache is required to reduce the worst-case time complexity of BDD operations from exponential time to polynomial time [56]. As with the unique table, we use only one shared operation cache for all operations, because we want to minimize interaction between workers, such as synchronization when shared parts of memory are resized.

In [56], Somenzi writes that a lossless computed table guarantees polynomial cost for the basic synthesis operations, but that lossless tables (which do not throw away results) are not feasible when manipulating many large BDDs and in practice lossy computed tables (which may throw away results) are implemented. If the cost of recomputing subresults is sufficiently small, it can pay to regularly delete results or even prefer to sometimes skip the cache to avoid data races. We design the operation cache to abort operations as early as possible when there may be a data race or the data may already be in the cache.

We use an operation cache that consists of two arrays: the hash array and the data array. See Figure 13.4 for the layout.

Since we implement a lossy cache, the design of the operation cache is extremely simple. We do not implement a special strategy to deal with hash collisions, but simply overwrite the old results. There is a trade-off between the cost of recomputing operations and the cost of synchronizing with the cache. For example, the caching

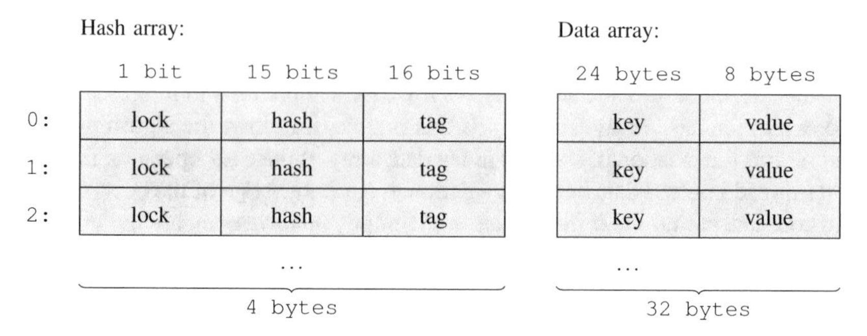

Fig. 13.4: Layout of the operation cache

Algorithm 13.7: The `cache-put` Algorithm

```
1 def cache-put (key, value) :
2     h, location ← hash(key)
3     s ← harray[location]
4     if s.lock : return
5     if s.hash = h : return
6     if not cas(harray[location], s, {1,h,s.tag+1}) : return
7     darray[location] ← {key,value}
8     harrray[location] ← {0,h,s.tag+1}
```

granularity (see Section 13.4.3) increases the number of recomputed operations but improves the performance in practice.

The most important concern for correctness is that every result obtained via `cache-get` was inserted earlier with `cache-put`, and the most important concern for performance is that the number of memory accesses is as low as possible. To ensure this, we use a 16-bit "tag" counter that increments (modulo 4096) with every update to the bucket, and check this value before reading the cache and after reading the cache to check that the obtained result is valid. The chance that this tag counter is the same for a different result is astronomically small, as this requires exactly 4096 `cache-put` operations on the same bucket by other workers between the first and the second time the tag is read in `cache-get`, and the last of these 4096 other operations must have the same hash value but different data.

We reserve 24 bytes of the bucket for the operation and its parameters. We use the first 64-bit value to store a BDD parameter and the operation identifier. The remaining 128 bits store other parameters, such as up to two 64-bit values, or up to three BDDs (123 bits, with 41 bits per BDD with a complement edge). The same holds for MTBDDs and LDDs. The result of the operation can be any 64-bit value or a BDD. Note that with 32 bytes per bucket and a properly aligned array, accessing a bucket requires only one cacheline transfer.

See Algorithms 13.7 and 13.8 for the `cache-put` and `cache-get` algorithms.

The algorithms are quite straightforward. We use a 64-bit hash function that returns sufficient bits for the 15-bit `h` value and the `location` value. The `h` value is

Algorithm 13.8: The `cache-get` Algorithm

```
1 def cache-get (key):
2     h, location ← hash(key)
3     s ← harray[location]
4     if s.lock : return ⊥
5     if s.hash ≠ h : return ⊥
6     storedkey, value ← darray[location]
7     if storedkey ≠ key : return ⊥
8     if s ≠ harray[location] : return ⊥
9     return value
```

used for the hash in the hash array, and the `location` for the location of the bucket in the table. The `cache-put` operation aborts as soon as some problem arises, i.e., if the bucket is locked (line 4), or if the hash of the stored key matches the hash of the given key (line 5), or if the `cas` operation fails (line 6). If the `cas` operation succeeds, then the bucket is locked. The key-value pair is written to the cache array (line 7) and the bucket is unlocked (line 8, by setting the locked bit to 0).

In the `cache-get` operation, when the bucket is locked (line 4), we abort instead of waiting for the result. We also abort if the hashes are different (line 5). We read the result (line 6) and compare the key to the requested key (line 7). If the keys are identical, then we verify that the cache bucket has not been manipulated by a concurrent operation by comparing the "tag" counter (line 8).

It is theoretically possible that between lines 6–8 of the `cache-get` operation, exactly 4096 `cache-put` operations are performed on the same bucket by other workers, with at least one of these such that the comparison at line 7 succeeds. The chances of this occurring are astronomically small. The reason we choose this design is that this implementation of `cache-get` only reads from memory and never writes. Memory writes cause additional communication between processors and with the memory when writing to the cacheline, and also force other processor caches to invalidate their copy of the bucket. We also want to avoid locking buckets for reading, because locking often causes bottlenecks. Since there are no loops in either algorithm, both algorithms are wait-free.

13.5 Garbage Collection

Operations on decision diagrams typically create many new nodes and discard old nodes. Nodes that are no longer referenced are typically called "dead nodes." Garbage collection, which removes dead nodes from the unique table, is essential for the implementation of decision diagrams. Since dead nodes are often reused in later operations, garbage collection should be delayed as long as possible [56].

There are various approaches to garbage collection. For example, a *reference count* could be added to each node, which records how often the node is referenced.

Nodes with a reference count of zero are either immediately removed when the count decreases to zero, or during a separate garbage collection phase. Another approach is *mark-and-sweep*, which marks all nodes that should be kept and removes all unmarked nodes. We refer to [56] for a more in-depth discussion of garbage collection.

For a parallel implementation, reference counts can incur a significant cost, as accessing nodes implies continuously updating the reference count, increasing the amount of communication between processors, as writing to a location in memory requires all other processors to refresh their view on that location. This is not a severe issue when there is only one processor, but with many processors this results in excessive communication, especially for nodes that are commonly used.

When parallelizing decision diagram operations, we can choose to perform garbage collection "on the fly", allowing other workers to continue inserting nodes, or we can "stop-the-world" and have all workers cooperate on garbage collection. We use a separate garbage collection phase, during which no new nodes are inserted. This greatly simplifies the design of the hash table, and we see no major advantage to allowing some workers to continue inserting nodes during garbage collection.

Some decision diagram implementations maintain a counter that counts how many buckets in the nodes table are in use and triggers garbage collection when a certain percentage of the table is in use. We want to avoid global counters like this and instead use a bounded "probe sequence" (see Section 13.4) for the nodes table: when the algorithm cannot find an empty bucket in the first K buckets, garbage collection is triggered. In simulations and experiments, we find that this occurs when the hash table is between 80% and 95% full.

As described in Section 13.4, decision diagram nodes are stored in a "data array," separated from the metadata of the unique table, which is stored in the "hash array." Nodes can be removed from the hash table without deleting them from the data array, simply by clearing the hash array. The nodes can then be reinserted during garbage collection, without changing their location in the data array, thus preserving the identity of the nodes.

We use a mark-and-sweep approach, where we keep track of all nodes that must be kept during garbage collection. Our approach of parallel garbage collection consists of the following steps:

1. Initiate the operation using the work-stealing framework (e.g., as supported by Lace) to arrange the "stop-the-world" interruption of all ongoing tasks. This feature is described below.
2. Clear the hash array of the unique table, and clear the operation cache. The operation cache is cleared instead of checking each entry individually after garbage collection, although that would also be possible.
3. Mark all nodes that we want to keep, allowing various mechanisms that keep track of the decision diagram nodes that we want to keep (see below).
4. Count the number of kept nodes and optionally increase the size of the unique table. Also optionally change the size of the operation cache.
5. Rehash all marked nodes in the hash array of the unique table.

The garbage collection process itself is also executed in parallel using task parallelism. Removing all nodes from the hash table and clearing the operation cache is an instant operation that is amortized over time by the operating system by reallocating the memory (see below). Marking nodes that must be kept occurs in parallel, mainly by implementing the marking operation as a recursive task. Counting the number of used nodes and rehashing all nodes (steps 4–5) is also parallelized using a standard binary divide-and-conquer approach , which distributes the memory pages over all workers.

Various mechanisms can be used to store the set of nodes to be kept in step 3. Operations must often temporarily store subresults that may not be removed; we use thread-local stacks to store these subresults, which minimizes worker interactions. External references (outside of operations) are less sensitive to these interactions; one can use any kind of set implementation (we use a simple hash table) to implement this; an important optimization is to not store references to nodes directly, but pointers to the variables; this way, updating a variable does not incur calls to remove and add references.

One helpful feature for garbage collection in Sylvan that we implemented in the work-stealing framework Lace is a feature that suspends all current tasks and starts a new task tree. Lace implements a macro `NEWFRAME(...)` that starts a new task tree, where one worker executes the given task and all other workers perform work-stealing to help execute this task in parallel. The exact implementation depends on the queue and involves several steps, where workers regularly check a flag in shared memory and use barriers to coordinate starting a new task tree. Further details are beyond our scope here, as they strongly depend on the used queue implementation. Interested readers are referred to [59].

13.6 Empirical Results

This section showcases the performance of parallel decision diagram operations in a number of applications, as reported in the literature. We briefly introduce model checking using decision diagrams in Section 13.6.1. We show the performance for symbolic on-the-fly reachability in the LTSMIN toolset as discussed in [62, 61, 64, 33, 59] in Section 13.6.2. For symbolic bisimulation minimization, which is related to symbolic model checking, we obtained good performance results in [65], which we report in Section 13.6.3. Finally, in Section 13.6.4 we discuss a performance comparison with other decision diagram implementations [60], showing that decision diagrams can be parallelized effectively without much overhead.

13.6.1 Symbolic Model Checking

As modern society increasingly depends on automated and complex systems, the safety demands on such systems increase as well. We depend on automated systems for basic infrastructure, to clean our water, to supply energy, to control our cars and trains, to monitor and process our financial transactions and for the internet. We use systems for entertainment when watching TV or using the phone, or for cooking with modern stoves, microwaves and fridges. Failure or unexpected behavior in these ubiquitous systems can have many consequences, from mild annoyances to fatal accidents. This motivates research into the formal verification of such systems, as well as computing properties such as failure rates and time to recovery.

In model checking, systems are modeled as sets of possible states of the system and transitions between these states. System states are typically represented by Boolean vectors. Fixed-point algorithms, which are procedures that repeatedly apply some operation until a fixed point is reached, play a central role in many model checking algorithms. An example of a fixed-point algorithm is state space exploration ("reachability"), which computes all states reachable from the initial state of the system. Many model checking algorithms depend on state space exploration to determine the number of states, to check whether an invariant is always true, to find cycles and deadlocks, and so forth.

A major challenge in model checking is that the space and time requirements of these algorithms increase exponentially with the size of the models. One technique to alleviate this problem is symbolic model checking [15, 16]. Symbolic model checking operates on sets of states and transitions, rather than individual states and transitions. These sets are then represented by their characteristic (Boolean) functions, which can be stored using BDDs. One advantage of using BDDs for fixed point computations is that equivalence testing is a trivial check, since BDDs uniquely represent Boolean functions. As small Boolean formulas can describe very large state spaces, symbolic model checking has been very successful at pushing the limits of model checking in the past [15]. Symbolic representations are also quite natural for the composition of multiple transition systems, e.g., when composing systems from subsystems.

13.6.2 Symbolic On-the-Fly Reachability

LTSMIN is a model checking toolset that provides a language-independent Partitioned Next-State Interface (PINS), which connects various input languages to model checking algorithms [9, 37, 62, 33, 42]. In PINS, the states of a system are represented by vectors of N integer values. Furthermore, transitions are distinguished in K disjunctive "transition groups," i.e., each transition in the system belongs to one of these transition groups. The transition relation of each transition group usually only depends on a subset of the entire state vector called the "short vector," further distinguished by the variables that are "read" and the variables that are "written" [42]. This enables the efficient encoding of transitions that only affect some integers of the state

Experiment	T_1	T_{48}	T_1/T_{48}
`firewire_link.1`	4.24	0.48	8.8
`anderson.1`	8.93	6.21	1.4
`firewire_tree.1`	4.23	0.30	14.1
`blocks.4`	635.86	17.27	36.8
`collision.5`	341.57	10.99	31.1
`lifts.8`	416.04	13.05	31.9
`exit.4`	494.85	13.95	35.5
`telephony.8`	915.61	28.18	32.5
Sum of all 269 models	16231	896	18.1

Table 13.3: Benchmark results (runtimes in seconds) for symbolic on-the-fly reachability with the LTSMIN toolset. Each data point is the average of at least five measurements

vector. Exploiting this information lets the PINS interface work in a quasi-symbolic way, as a single pair of short vectors can represent many transition relations on the full state vector. Initially, LTSMIN does not have knowledge of the transitions in each transition group, and only the initial state is known. The transition system is explored by learning new transitions via the PINS interface, which are then added to the transition relation.

We evaluated the application of parallelization to LTSMIN [64, 59]. The experimental evaluation was based on the BEEM model database [51]. We performed the benchmarks on 269 benchmark models on a 48-core machine, consisting of four AMD Opteron™ 6168 processors with 12 cores each and 128 GB of internal memory. A summary of results is given in Table 13.3.

As is clear from these results, obtained speedups (T_1/T_{48}) strongly depend on the models; for some models, we obtain speedups above 30×, up to 36.8× for the `blocks.4` model.

See Figure 13.5 for a speedup graph of a selection of these models. This speedup graph was obtained using list decision diagrams, which are discussed in [59] and are beyond the scope of this chapter. The speedup graph suggests that most likely further speedups would be obtained after 48 cores for the selected models.

13.6.3 Symbolic Bisimulation Minimisation

One of the main challenges for model checking is that the space and time requirements of model checking algorithms increase exponentially with the size of the models. One technique that helps combat this challenge is called bisimulation minimization. Given an input model, bisimulation minimization computes the smallest equivalent model, also called the maximal bisimulation, under some notion of equiva-

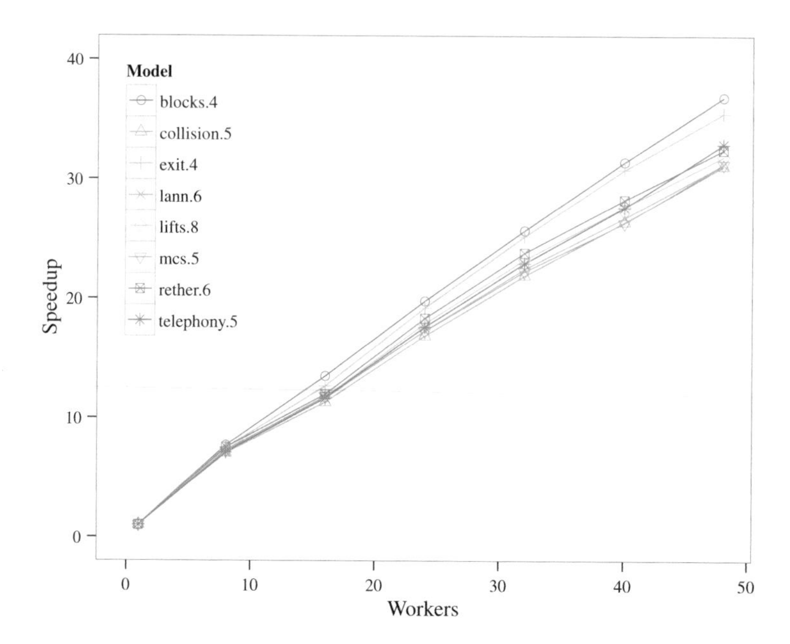

Fig. 13.5: Speedup graphs of several well-performing models. Each data point is an average of at least five measurements

lence. This can significantly reduce the number of states. This technique is also used to abstract models from internal behavior, when only observable behavior is relevant.

The maximal bisimulation of a model is typically computed using partition refinement. Starting with an initially coarse partition (e.g., all states are equivalent), the partition is refined until states in each equivalence class can no longer be distinguished. The result is the maximal bisimulation with respect to the initial partition. Blom et al. [8] introduced a signature-based method, which assigns states to equivalence classes according to a characterizing signature. This method easily extends to various types of bisimulation.

In [65, 67], we studied bisimulation minimization for labeled transition systems (LTSs), continuous-time Markov chains (CTMCs) and interactive Markov chains (IMCs), which combine the features of LTSs and CTMCs. These allow the analysis of quantitative properties, e.g., performance and dependability. We implemented strong bisimulation and branching bisimulation in the SIGREFMC tool. Strong bisimulation preserves both internal behavior (τ-transitions) and observable behavior, while branching bisimulation abstracts from internal behavior. The SIGREFMC tool also connects to the LTSMIN tool described in Section 13.6.2, enabling the minimization of models described with various input languages.

Bisimulation minimization is performed in two steps. The first step is computing the maximal bisimulation, which is a partition computed using signature refinement.

SIGREF LTS models			Signature refinement			Quotient computation		
Model	States	Blocks	T_1	T_{48}	Sp.	T_1	T_{48}	Sp.
kanban03	1024240	85356	10.09	0.88	11.52×	6.72	0.35	19.08×
kanban04	16020316	778485	148.15	11.37	13.03×	106.22	5.38	19.73×
kanban05	16772032	5033631	1284.86	73.57	17.47×	740.53	33.80	21.91×
SIGREF CTMC models			Signature refinement			Quotient computation		
Model	States	Blocks	T_1	T_{48}	Sp.	T_1	T_{48}	Sp.
cycling-4	431101	282943	26.72	2.60	10.29×	59.51	3.32	17.90×
cycling-5	2326666	1424914	170.28	19.42	8.77×	294.15	13.48	21.83×
fgf	80616	38639	8.86	0.88	10.04×	7.42	0.73	10.20×
p2p-5-6	2^{30}	336	26.96	2.99	9.03×	10.25	1.41	7.29×
p2p-6-5	2^{30}	266	9.49	1.21	7.82×	3.67	0.55	6.71×
p2p-7-5	2^{35}	336	24.01	2.97	8.08×	9.26	1.19	7.79×
polling-16	1572864	98304	118.50	10.18	11.64×	66.25	4.49	14.75×
polling-17	3342336	196608	303.65	22.58	13.45×	161.74	10.02	16.14×
polling-18	7077888	393216	705.22	49.81	14.16×	359.49	21.68	16.58×
LTSMIN LTS models			Signature refinement			Quotient computation		
Model	States	Blocks	T_1	T_{48}	Sp.	T_1	T_{48}	Sp.
brp-3-4-4	40,592	10,326	13.50	0.92	14.75×	2.45	0.14	17.53×
brp-4-4-4	109,422	27,106	38.91	2.23	17.43×	9.84	0.52	18.93×
franklin-3-3	41,401	883	24	1.24	19.40×	3.13	0.19	16.46×
franklin-4-2	272,241	10,706	330.56	14.67	22.53×	28.04	1.43	19.63×
hesselink-4	142,081,536	6,036	51.41	3.56	14.44×	7.01	1.21	5.78×
hesselink-5	883,738,000	11,005	179.85	12.61	14.26×	22.32	3.64	6.14×
swp-2-4	2,589,056	69,555	267.46	11.33	23.60×	30.78	1.39	22.21×
swp-3-3	1,652,724	65,025	142.60	6.13	23.26×	24.89	1.11	22.39×
swp-4-3	7,429,632	264,708	630.73	25.92	24.34×	111.69	4.55	24.56×

Table 13.4: Computation time in seconds for partition refinement and quotient computation on various benchmarks provided with the original SIGREF tool and generated by LTSMIN

The second step is computing the quotient of the original model and the partition, resulting in the minimized system.

See Table 13.4 for a selection of the benchmark results from [67]. We used benchmark models provided with the original SIGREF tool by Wimmer et al. [70] and from process algebra (in the MCRL2 language) prepared using LTSMIN. See further [67] for a description of the models. The benchmarks were performed on a 48-core machine, consisting of four AMD Opteron™ 6168 processors with 12 cores each and 128 GB of internal memory. Table 13.4 shows that the parallel speedup varies with the model used, similarly to the results we obtained with symbolic reachability in Table 13.3. We obtained speedups of up to 24× for both the signature refinement step and the quotient computation step.

13.6.4 Probabilistic Model Checking

Sylvan has also been used as a symbolic back-end in the model checker ISCASMC, a probabilistic model checker [30] written in Java. A recent study [60] compared the performance of the BDD libraries CUDD, BuDDy, CacBDD, JDD, Sylvan and BeeDeeDee when used as the symbolic back-end of ISCASMC and performing symbolic reachability.

They summarize the overall runtimes by the following table [60]:

back-end	time (s)	back-end	time (s)
sylvan-7	608	buddy	2156
cacbdd	1433	jdd	2439
cudd-bdd	1522	beedeedee	2598
sylvan-1	1838	cudd-mtbdd	2837

This result was produced with variant 2 of the nodes table in Sylvan. As the results show, Sylvan is competitive with other BDD implementations when used sequentially (with one worker) and benefits from parallelism (with seven workers).

13.7 Conclusions

This chapter has discussed the two basic ingredients to achieve scalable binary decision diagrams in a multi-core shared-memory environment. The first ingredient is a fine-grained work-stealing framework that provides parallel execution and load balancing of the decision diagram operations. The second ingredient consists of the concurrent, lock-free hash tables for the unique table and the operation cache.

We discussed Sylvan, a parallel implementation of decision diagrams. Sylvan offers an easy-to-use, sequential interface like a traditional BDD package, but with a parallel implementation of its operations. Thus, existing sequential algorithms that depend on decision diagram operations benefit from the multi-core parallelization offered by Sylvan. In addition, sequential algorithms can further profit from the parallel work-stealing framework embedded in Sylvan by implementing parallel tasks that call decision diagram operations in parallel. For example, a transition system can be partitioned and the partitioned transition relations can be applied in parallel via tree-like reductions.

The approach presented in this chapter is versatile. As shown with the different types of decision diagrams implemented by Sylvan and used in the specific applications, the principles of parallel decision diagram operations can be applied to BDDs, MTBDDs, list decision diagrams, multi-way decision diagrams, zero-suppressed decision diagrams, etc. As decision diagrams are heavily used in many application domains, we foresee that parallel decision diagram operations can be a practical tool to bring parallelization to these domains. Future directions also include tackling the challenges that other applications bring, such as efficient dynamic variable reordering and tentative execution of decision diagram operations. Furthermore, the development of parallel decision diagram operations for heterogeneous systems such as clusters of multi-core computers [47, 48] and systems with many cores and highly specialized hierarchies such as GPUs [68] offers additional challenges for BDD operations that need to be addressed in the future.

References

[1] Magdy S. Abadir and Hassan K. Reghbati. Functional Test Generation for Digital Circuits Described Using Binary Decision Diagrams. *IEEE Trans. Computers*, 35(4):375–379, 1986.
[2] Umut A. Acar, Arthur Charguéraud, and Mike Rainey. Scheduling parallel programs by work stealing with private deques. In *PPOPP*, pages 219–228. ACM, 2013.
[3] S.B. Akers. Binary Decision Diagrams. *IEEE Trans. Computers*, C-27(6):509–516, 6 1978.
[4] Prakash Arunachalam, Craig M. Chase, and Dinos Moundanos. Distributed binary decision diagrams for verification of large circuit. In *ICCD*, pages 365–370, 1996.
[5] R. Iris Bahar, Erica A. Frohm, Charles M. Gaona, Gary D. Hachtel, Enrico Macii, Abelardo Pardo, and Fabio Somenzi. Algebraic decision diagrams and their applications. In *ICCAD 1993*, pages 188–191, 1993.
[6] Debashis Bhattacharya, Prathima Agrawal, and Vishwani D. Agrawal. Test Generation for Path Delay Faults Using Binary Decision Diagrams. *IEEE Trans. Computers*, 44(3):434–447, 1995.
[7] F. Bianchi, Fulvio Corno, Maurizio Rebaudengo, Matteo Sonza Reorda, and Roberto Ansaloni. Boolean function manipulation on a parallel system using BDDs. In *HPCN Europe*, pages 916–928, 1997.

[8] Stefan Blom and Simona Orzan. Distributed Branching Bisimulation Reduction of State Spaces. *ENTCS*, 89(1):99–113, 2003.
[9] Stefan Blom, Jaco van de Pol, and Michael Weber. LTSmin: Distributed and Symbolic Reachability. In *CAV*, volume 6174 of *LNCS*, pages 354–359. Springer, 2010.
[10] Robert D. Blumofe. Scheduling multithreaded computations by work stealing. In *FOCS*, pages 356–368. IEEE Computer Society, 1994.
[11] Robert D. Blumofe, Christopher F. Joerg, Bradley C. Kuszmaul, Charles E. Leiserson, Keith H. Randall, and Yuli Zhou. Cilk: An Efficient Multithreaded Runtime System. *J. Parallel Distrib. Comput.*, 37(1):55–69, 1996.
[12] Marco Bozzano, Alessandro Cimatti, and Francesco Tapparo. Symbolic fault tree analysis for reactive systems. In *ATVA 2007*, volume 4762 of *LNCS*, pages 162–176. Springer, 2007.
[13] Karl S. Brace, Richard L. Rudell, and Randal E. Bryant. Efficient implementation of a BDD package. In *DAC*, pages 40–45, 1990.
[14] Randal E. Bryant. Graph-Based Algorithms for Boolean Function Manipulation. *IEEE Trans. Computers*, C-35(8):677–691, 8 1986.
[15] Jerry R. Burch, Edmund M. Clarke, Kenneth L. McMillan, David L. Dill, and L. J. Hwang. Symbolic model checking: 10^20 states and beyond. *Inf. Comput.*, 98(2):142–170, 1992.
[16] J.R. Burch, E.M. Clarke, D.E. Long, K.L. McMillan, and D.L. Dill. Symbolic model checking for sequential circuit verification. *IEEE Transactions on Computer-Aided Design of Integrated Circuits and Systems*, 13(4):401–424, 4 1994.
[17] G.P. Cabodi, S. Gai, and M. Sonza Reorda. Boolean function manipulation on massively parallel computers. In *Proc. of 4th Symp. on Frontiers of Massively Parallel Computation*, pages 508–509. IEEE, 10 1992.
[18] Jer-Sheng Chen and P. Banerjee. Parallel construction algorithms for BDDs. In *ISCAS 1999*, pages 318–322. IEEE, 1999.
[19] Ming-Ying Chung and Gianfranco Ciardo. Saturation NOW. In *QEST*, pages 272–281. IEEE Computer Society, 2004.
[20] Gianfranco Ciardo, Gerald Lüttgen, and Radu Siminiceanu. Saturation: An Efficient Iteration Strategy for Symbolic State-Space Generation. In *TACAS*, volume 2031 of *LNCS*, pages 328–342, 2001.
[21] Gianfranco Ciardo, Yang Zhao, and Xiaoqing Jin. Parallel symbolic state-space exploration is difficult, but what is the alternative? In *PDMC*, pages 1–17, 2009.
[22] Edmund M. Clarke, Kenneth L. McMillan, Xudong Zhao, Masahiro Fujita, and J. Yang. Spectral Transforms for Large Boolean Functions with Applications to Technology Mapping. In *DAC*, pages 54–60, 1993.
[23] Jonathan Ezekiel, Gerald Lüttgen, and Gianfranco Ciardo. Parallelising symbolic state-space generators. In *CAV*, volume 4590 of *LNCS*, pages 268–280, 2007.
[24] Karl-Filip Faxén. Wool–A work stealing library. *SIGARCH Computer Architecture News*, 36(5):93–100, 2008.

[25] Karl-Filip Faxén. Efficient work stealing for fine grained parallelism. In *ICPP 2010*, pages 313–322. IEEE Computer Society, 2010.
[26] Kathi Fisler, Shriram Krishnamurthi, Leo A. Meyerovich, and Michael Carl Tschantz. Verification and change-impact analysis of access-control policies. In *ICSE 2005*, pages 196–205. ACM, 2005.
[27] Matteo Frigo, Charles E. Leiserson, and Keith H. Randall. The Implementation of the Cilk-5 Multithreaded Language. In *PLDI*, pages 212–223. ACM, 1998.
[28] S. Gai, M. Rebaudengo, and M. Sonza Reorda. An improved data parallel algorithm for Boolean function manipulation using BDDs. In *Proc. Euromicro Workshop on Par. and Distrib. Processing*, pages 33–39. IEEE, 1 1995.
[29] Orna Grumberg, Tamir Heyman, and Assaf Schuster. A work-efficient distributed algorithm for reachability analysis. *Formal Methods in System Design*, 29(2):157–175, 2006.
[30] Ernst Moritz Hahn, Yi Li, Sven Schewe, Andrea Turrini, and Lijun Zhang. iscasmc: A web-based probabilistic model checker. In *FM*, volume 8442 of *LNCS*, pages 312–317. Springer, 2014.
[31] Tamir Heyman, Danny Geist, Orna Grumberg, and Assaf Schuster. Achieving Scalability in Parallel Reachability Analysis of Very Large Circuits. In *Computer Aided Verification*, volume 1855 of *Lecture Notes in Computer Science*, pages 20–35. Springer Berlin / Heidelberg, 2000.
[32] Masakazu Ishihata, Taisuke Sato, and Shin-ichi Minato. Compiling Bayesian networks for parameter learning based on shared BDDs. In *AI 2011*, volume 7106 of *Lecture Notes in Computer Science*, pages 203–212. Springer, 2011.
[33] Gijs Kant, Alfons Laarman, Jeroen Meijer, Jaco van de Pol, Stefan Blom, and Tom van Dijk. LTSmin: High-Performance Language-Independent Model Checking. In *TACAS 2015*, volume 9035 of *LNCS*, pages 692–707. Springer, 2015.
[34] S. Kimura and E.M. Clarke. A parallel algorithm for constructing binary decision diagrams. In *Proc. of IC on Computer Design: VLSI in Computers and Processors ICCD*, pages 220–223, 9 1990.
[35] S. Kimura, T. Igaki, and H. Haneda. Parallel Binary Decision Diagram Manipulation. *IEICE Transactions on Fundamentals of Electronics, Communications and Computer Science*, E75-A(10):1255–62, 10 1992.
[36] Alfons Laarman, Jaco van de Pol, and Michael Weber. Boosting multi-core reachability performance with shared hash tables. In *FMCAD 2010*, pages 247–255. IEEE, 2010.
[37] Alfons W. Laarman, Jaco van de Pol, and Michael Weber. Multi-Core LTSmin: Marrying Modularity and Scalability. In *NASA Formal Methods - Third International Symposium, NFM 2011, Pasadena, CA, USA, April 18-20, 2011. Proceedings*, volume 6617 of *LNCS*, pages 506–511. Springer, 2011.
[38] Elsa Loekito and James Bailey. Fast mining of high dimensional expressive contrast patterns using zero-suppressed binary decision diagrams. In *SIGKDD 2006*, pages 307–316. ACM, 2006.

[39] Alberto Lovato, Damiano Macedonio, and Fausto Spoto. A Thread-Safe Library for Binary Decision Diagrams. In *SEFM*, volume 8702 of *LNCS*, pages 35–49. Springer, 2014.

[40] Sharad Malik, Albert R. Wang, Robert K. Brayton, and Alberto L. Sangiovanni-Vincentelli. Logic verification using binary decision diagrams in a logic synthesis environment. In *ICCAD 1998*, pages 6–9, 1988.

[41] Yusuke Matsunaga and Masahiro Fujita. Multi-level logic optimization using binary decision diagrams. In *ICCAD 1989*, pages 556–559. IEEE, 1989.

[42] Jeroen Meijer, Gijs Kant, Stefan Blom, and Jaco van de Pol. Read, Write and Copy Dependencies for Symbolic Model Checking. In Eran Yahav, editor, *HVC*, volume 8855 of *Lecture Notes in Computer Science*, pages 204–219. Springer, 2014.

[43] Kim Milvang-Jensen and Alan J. Hu. BDDNOW: A parallel BDD package. In *FMCAD*, pages 501–507, 1998.

[44] Shin-ichi Minato. Techniques of BDD/ZDD: Brief History and Recent Activity. *IEICE Transactions*, 96-D(7):1419–1429, 2013.

[45] Shin-ichi Minato, Ken Satoh, and Taisuke Sato. Compiling Bayesian networks by symbolic probability calculation based on zero-suppressed BDDs. In *IJCAI 2007*, pages 2550–2555, 2007.

[46] Hiroyuki Ochi, Nagisa Ishiura, and Shuzo Yajima. Breadth-first manipulation of SBDD of Boolean functions for vector processing. In *DAC*, pages 413–416, 1991.

[47] Wytse Oortwijn. Distributed Symbolic Reachability Analysis. Master's thesis, University of Twente, Dept. of C.S., 2015.

[48] Wytse Oortwijn, Tom van Dijk, and Jaco van de Pol. Distributed Binary Decision Diagrams for Symbolic Reachability. In *SPIN*, pages 21–30. ACM, 2017.

[49] Jörn Ossowski. *JINC – A Multi-Threaded Library for Higher-Order Weighted Decision Diagram Manipulation.* PhD thesis, Rheinische Friedrich-Wilhelms-Universität Bonn, 10 2010.

[50] Yegnashankar Parasuram, Edward P. Stabler, and Shiu-Kai Chin. Parallel implementation of BDD algorithms using a distributed shared memory. In *HICSS (1)*, pages 16–25, 1994.

[51] Radek Pelánek. BEEM: benchmarks for explicit model checkers. In *SPIN*, pages 263–267, Berlin, Heidelberg, 2007. Springer-Verlag.

[52] Karen A. Reay and John D. Andrews. A fault tree analysis strategy using binary decision diagrams. *Rel. Eng. & Sys. Safety*, 78(1):45–56, 2002.

[53] Yuko Sakurai, Suguru Ueda, Atsushi Iwasaki, Shin-ichi Minato, and Makoto Yokoo. A compact representation scheme of coalitional games based on multi-terminal zero-suppressed binary decision diagrams. In *PRIMA 2011*, volume 7047 of *Lecture Notes in Computer Science*, pages 4–18. Springer, 2011.

[54] Jagesh V. Sanghavi, Rajeev K. Ranjan, Robert K. Brayton, and Alberto L. Sangiovanni-Vincentelli. High performance BDD package by exploiting memory hiercharchy. In *DAC*, pages 635–640, 1996.

[55] Mathias Soeken, Laura Tague, Gerhard W. Dueck, and Rolf Drechsler. Ancilla-free synthesis of large reversible functions using binary decision diagrams. *J. Symb. Comput.*, 73:1–26, 2016.

[56] Fabio Somenzi. Efficient manipulation of decision diagrams. *STTT*, 3(2):171–181, 2001.

[57] Fabio Somenzi. CUDD: CU decision diagram package release 3.0.0. `http://vlsi.colorado.edu/~fabio/CUDD/`, 2015.

[58] Tony Stornetta and Forrest Brewer. Implementation of an efficient parallel BDD package. In *DAC*, pages 641–644, 1996.

[59] Tom van Dijk. *Sylvan: Multi-core Decision Diagrams*. PhD thesis, University of Twente, 7 2016.

[60] Tom van Dijk, Ernst Moritz Hahn, David N. Jansen, Yong Li, Thomas Neele, Mariëlle Stoelinga, Andrea Turrini, and Lijun Zhang. A Comparative Study of BDD Packages for Probabilistic Symbolic Model Checking. In *SETTA*, volume 9409 of *LNCS*, pages 35–51. Springer, 2015.

[61] Tom van Dijk, Alfons Laarman, and Jaco van de Pol. Multi-core BDD operations for symbolic reachability. *ENTCS*, 296:127–143, 2013.

[62] Tom van Dijk, Alfons W. Laarman, and Jaco van de Pol. Multi-core and/or Symbolic Model Checking. *ECEASST*, 53, 2012.

[63] Tom van Dijk and Jaco van de Pol. Lace: Non-blocking Split Deque for Work-Stealing. In *MuCoCoS*, volume 8806 of *LNCS*, pages 206–217. Springer, 2014.

[64] Tom van Dijk and Jaco van de Pol. Sylvan: Multi-Core Decision Diagrams. In *TACAS*, volume 9035 of *LNCS*, pages 677–691. Springer, 2015.

[65] Tom van Dijk and Jaco van de Pol. Multi-Core Symbolic Bisimulation Minimisation. In *TACAS*, volume 9636 of *LNCS*, pages 332–348. Springer, 2016.

[66] Tom van Dijk and Jaco van de Pol. Sylvan: multi-core framework for decision diagrams. *International Journal on Software Tools for Technology Transfer*, 19(6) pp 675–696, 2017.

[67] Tom van Dijk and Jaco van de Pol. Multi-core symbolic bisimulation minimisation. *International Journal on Software Tools for Technology Transfer*, 2017. Published online, August 2017.

[68] Miroslav N. Velev and Ping Gao. Efficient parallel GPU algorithms for BDD manipulation. In *ASP-DAC*, pages 750–755. IEEE, 2014.

[69] David B. Wagner and Brad Calder. Leapfrogging: A portable technique for implementing efficient futures. In *PPOPP*, pages 208–217. ACM, 1993.

[70] Ralf Wimmer, Marc Herbstritt, Holger Hermanns, Kelley Strampp, and Bernd Becker. Sigref – A Symbolic Bisimulation Tool Box. In *ATVA*, volume 4218 of *LNCS*, pages 477–492. Springer, 2006.

[71] Bwolen Yang and David R. O'Hallaron. Parallel breadth-first BDD construction. In *PPOPP*, pages 145–156, 1997.

Chapter 14
Parallel Model-Based Diagnosis

Kostyantyn Shchekotykhin, Dietmar Jannach, and Thomas Schmitz

Abstract Model-Based Diagnosis (MBD) is a general-purpose computational approach to determine why a system under observation, e.g., an electronic circuit or a software program, does not behave as expected. MBD approaches utilize knowledge about the system's expected behavior if all of its components work correctly. In case of an unexpected behavior they systematically explore the possible reasons, i.e., diagnoses, for the misbehavior. Such diagnoses are determined through systematic or heuristic search procedures which often use MBD-specific rules to prune the search space. In this chapter we review approaches that rely on parallel or distributed computations to speed up the diagnostic reasoning process. Specifically, we focus on recent parallelization strategies that exploit the capabilities of modern multi-core computer architectures and report results from experimental evaluations to shed light on the speedups that can be achieved by parallelization for various MBD applications.

14.1 Introduction

14.1.1 Background

Model-Based Diagnosis (MBD) is a subfield of Artificial Intelligence that focuses on automated reasoning methods that are capable of generating hypotheses and

Kostyantyn Shchekotykhin
Institute for Applied Informatics, Alpen-Adria-Universität Klagenfurt, Austria,
e-mail: konstantin.schekotihin@aau.at

Dietmar Jannach
Department of Computer Science, TU Dortmund, Germany,
e-mail: dietmar.jannach@tu-dortmund.de

Thomas Schmitz
Department of Computer Science, TU Dortmund, Germany,
e-mail: thomas.schmitz@tu-dortmund.de

Y. Hamadi und L. Sais (eds.), *Handbook of Parallel Constraint Reasoning*,
https://doi.org/10.1007/978-3-319-63516-3_14

explanations why a system under observation is not behaving as expected. The term "model-based" means that the diagnostic inference process is based on knowledge (i.e., a model) of how the system and its components work, which makes it possible to simulate the system's behavior in order to test alternative hypotheses.

MBD techniques were pioneered in the 1980s and many of the proposals at that time were centered on the application domain of digital circuits [1, 2, 3]. The model in such cases consists of knowledge about (a) the normal and expected behavior of the components of the circuit (e.g., how an AND-gate works when it functions correctly), and (b) how the components are interconnected. This knowledge can then be used to simulate the behavior of the circuit and to compute the *expected outputs* for a given set of inputs. The resulting simulated behavior is then contrasted with the real and *observed behavior* of the system. Whenever there is a discrepancy between the expected and the observed outputs, the task of an MBD system is to determine which parts of the analyzed system *can be* responsible for the observed outputs.

The predominant algorithmic approach to find one or more sets of components that can be responsible for the observed faulty outputs is to systematically test different hypotheses about the (binary) health state of each of the components. One main and computationally complex part of the diagnostic reasoning process is therefore a search process in which the search space in principle consists of all possible subsets of the components of the analyzed system. In much research work the search process itself is guided by so-called "conflicts," which are typically comparatively small subsets of the system's components, which – according to the simulation – cannot all be working correctly, i.e., at least one of them must be faulty. These conflicts can help to significantly reduce the search space. Their computation can however also be computationally demanding. But even if all conflicts were known in advance, finding all possible diagnoses using the known conflicts corresponds to finding a solution to a set cover (hitting set) problem, leading to an NP-hard search problem.

In this chapter we review existing work that approaches the problem of the computational complexity of model-based reasoning processes by parallelizing parts of the reasoning and search processes.

14.1.2 Outline of the Chapter

The chapter is organized as follows. Next, in Section 14.2, we will review the formal and logic-based characterization of the Model-Based Diagnosis problem and a conflict-directed, sound and complete tree-based search method as introduced by Reiter in [3]. Reiter's domain-independent problem formalization is the basis for most of the works discussed in the chapter. Its generality is also one of the reasons for the success of MBD techniques and why they are still relevant today. MBD techniques are in fact not limited to electronic circuits, but have been applied over the last three decades in particular to a variety of software artifacts including logic programs, ontologies, process specifications, special purpose and general-purpose languages such as VHDL or Java, and recently also to spreadsheet programs [4, 5, 6, 7, 8, 9, 10, 11, 12, 13, 14].

Section 14.3 then discusses alternative ways of finding diagnoses more quickly than with Reiter's approach. Typically, these methods make some sort of assumptions to make the problem easier to solve, e.g., by requiring that the diagnosed system has a certain structure or by focusing only on certain subsets of all existing diagnoses, thereby sacrificing the completeness of the algorithm. In this section we will also discuss so called "direct" methods that encode the diagnosis problem in a form that can be directly processed by specific inference engines such as a SAT solver without explicitly creating a diagnosis search tree.

Since diagnostic reasoning is predominantly a tree search problem, we will review general strategies for search parallelization (e.g., tree decomposition or window-based processing) and analyze the applicability of parallelized versions of general search strategies such as A^* to the Model-Based Diagnosis problem in Section 14.4.

Section 14.5 then presents recent techniques for parallelizing the MBD reasoning process on multi-core machines. The main focus will be on sound and complete diagnosis approaches and on methods that parallelize Reiter's tree search method in different ways. Selected results of empirical evaluations that were made using a number of benchmark problems are then presented in Section 14.6. These results help us to quantify the possible gains that can be obtained by running the search process on parallel threads on one machine.

Finally, in Section 14.7, we will discuss a number of alternative parallelization approaches that are not based on Reiter's HS-tree algorithm.

14.2 Reiter's Diagnosis Framework

In this section, we will summarize the formal characterization of the MBD problem as proposed by Reiter in [3]. The goal of Reiter's work was to provide a generic formalization that allows one to diagnose any system whose behavior can be modeled by a set of first-order sentences. Given a formal model describing the "normal" behavior of the system under observation, the general task of an MBD algorithm is to analyze the possible fault reasons, whenever the system does not behave as expected. The starting points for this analysis are therefore discrepancies between the system's outputs as predicted by the model and the observed outputs.

14.2.1 Example: A Diagnosis Problem Instance

Let us consider the binary half adder shown in Figure 14.1 as an example. This simple digital circuit whose behavior we know when it works normally, has two inputs A and B, two outputs S and C, an AND-gate A_1, and an XOR-gate X_1.

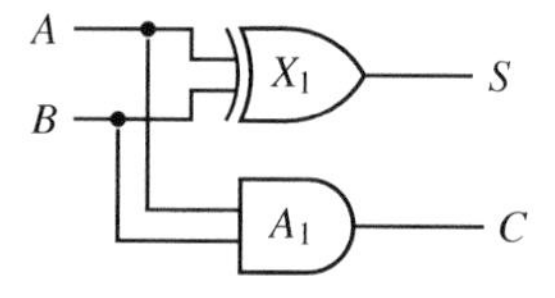

Fig. 14.1: Half-adder with inputs A and B, and outputs S and C

The half-adder can be described by first-order sentences as follows. First, we describe the expected behavior of the two types of logic gates, where *in(x,1)* denotes the first input of component x, *in(x,2)* its second input, and *out(x)* refers to the output of the component. The *and* and *xor* functions implement boolean conjunction and exclusive disjunction respectively.

$$\forall x : \text{AND}(x) \rightarrow [out(x) = and(in(x,1), in(x,2))]$$
$$\forall x : \text{XOR}(x) \rightarrow [out(x) = xor(in(x,1), in(x,2))]$$

The components A_1 and X_1 and their wiring are modeled next.

$$\text{AND}(A_1), \quad \text{XOR}(X_1)$$
$$in(A_1,1) = in(X_1,1), \quad in(A_1,2) = in(X_1,2)$$

Finally, we state that the inputs of X_1 can only be 0 or 1 (which then also applies to A_1 according to the wiring of our circuit).

$$in(X_1,1) = 0 \vee in(X_1,1) = 1, \quad in(X_1,2) = 0 \vee in(X_1,2) = 1$$

Now let us assume that the following inputs to the system were provided, which are again described as a set of first-order sentences: $in(X_1,1) = 1,\ in(X_1,2) = 0$. The observed outputs were however $out(X_1) = 0,\ out(A_1) = 0$. This is obviously unexpected, since the output of the XOR-gate X_1 should be 1.

Having completed our description of the system and the inputs and outputs, we can feed all the first-order sentences modeled so far into a *theorem prover* (TP), which will find that the model is inconsistent. Consequently, an abnormal behavior has been detected.

However, the presented way of modeling does not easily allow us to find the true cause of the problem, i.e., that gate X_1 is broken. In order to find the possible causes of a problem, typical MBD systems test different assumptions about the faultiness of individual components. To be able to systematically explore the possible causes, Reiter proposes a modeling approach that uses a unary predicate $\text{AB}(\cdot)$ to denote that a component is "abnormal". Following this approach, we reformulate the first set of sentences of our model as follows.

$$\forall x : \neg\text{AB}(x) \rightarrow (\text{AND}(x) \rightarrow [out(x) = and(in(x,1), in(x,2))])$$
$$\forall x : \neg\text{AB}(x) \rightarrow (\text{XOR}(x) \rightarrow [out(x) = xor(in(x,1), in(x,2))])$$

The resulting model – in combination with the observations (i.e., the inputs and outputs) – allows us to test different assumptions about the correctness of the components. For instance, if we assume that all components are working properly $\{\neg\text{AB}(X_1), \neg\text{AB}(A_1)\}$, the theorem prover will find that the model is inconsistent given the observations. However, under the assumption $\{\text{AB}(X_1), \neg\text{AB}(A_1)\}$, i.e., X_1 is faulty, we will find that the model is consistent with the observations. Assuming that X_1 is not working correctly therefore explains the observed outputs and the set $\{X_1\}$, which is a subset of the system's components, would thus be what is called a diagnosis in Reiter's framework.

Finally, the main advantage of the described modeling approach using "abnormal" predicates is that we can determine such diagnoses by systematically or heuristically varying our assumptions about the faultiness of the components.

14.2.2 Diagnoses and Conflicts

Formally, the principled approach to Model-Based Diagnosis by Reiter can be summarized as follows.

Definition 1. (Diagnosis problem instance) Let $(\text{SD}, \text{COMPS}, \text{OBS})$ be a triple, where SD and OBS are finite sets of first-order sentences which encode a system description and observations, respectively, and COMPS is a finite set of constants that represent the system's components.

$(\text{SD}, \text{COMPS}, \text{OBS})$ is a *diagnosis problem instance* (DPI) iff (1) SD is consistent, (2) OBS is consistent, and (3) $\text{SD} \cup \text{OBS} \cup \{\neg\text{AB}(c) \mid c \in \text{COMPS}\}$ is inconsistent.

Definition 2. (Diagnosis problem) Given a DPI $(\text{SD}, \text{COMPS}, \text{OBS})$, the diagnosis problem is to find a subset-minimal set $\Delta \subseteq \text{COMPS}$, called a *diagnosis*, such that $\text{SD} \cup \text{OBS} \cup \{\text{AB}(c) \mid c \in \Delta\} \cup \{\neg\text{AB}(c) \mid c \in \text{COMPS} - \Delta\}$ is consistent.

According to Definition 2, we are only interested in *minimal* diagnoses, i.e., diagnoses which contain no superfluous elements and which are thus not supersets of other diagnoses. Whenever we use the term *diagnosis* in the remainder of the chapter, we mean *minimal diagnosis*. Whenever we want to refer to non-minimal diagnoses, we will explicitly mention this fact.

Finding a diagnosis can in theory be done by simply trying out all possible subsets of COMPS and checking their consistency with the observations. A simple tree search algorithm that enumerates the possible assumptions in a breadth-first manner can be used for that purpose. Consider Algorithm 14.1, which takes a problem instance DPI as an input and returns one single diagnosis. The algorithm starts with the simple assumption that all components work properly. In every iteration it uses Definition 2 to test whether the current assumption Δ is a diagnosis. If this is not the case, the algorithm extends the search frontier by all supersets of Δ by adding one component from the set $\text{COMPS} - \Delta$. The breadth-first order guarantees the subset-minimality of the returned set of faulty components Δ.

Algorithm 14.1: Tree Search Algorithm

```
    Input: DPI (SD, COMPS, OBS)
    Output: A diagnosis Δ
 1  closed ← ∅                                   /* Maintain list of work already done */
 2  frontier ← {∅}                                /* Initialize the search frontier */
 3  while frontier ≠ ∅ :
        /* Get and remove the first element of the frontier */
 4      Δ ← Pop(frontier)
 5      if Δ ∉ closed :                                             /* Skip the node */
 6          closed ← closed ∪ {Δ}
 7          if SD ∪ OBS ∪ {AB(c) | c ∈ Δ} ∪ {¬AB(c) | c ∈ COMPS − Δ} is consistent :
 8              return Δ                                       /* A diagnosis is found */
            /* Generate successors of Δ and add them to the frontier */
 9          frontier ← frontier ∪ {Δ ∪ {c} | c ∈ (COMPS − Δ)}
10  return failure                                   /* Provided triple is not a DPI */
```

Algorithm 14.1 is obviously very inefficient because it exhaustively enumerates all possible assumptions. In many real-world scenarios such an algorithm would make many assumptions about the faultiness of components that are in fact irrelevant. This can for instance be the case if the unexpected behavior is observed only for a certain part of a physical system. Therefore, Reiter [3] proposes a more efficient procedure based on the concept of *conflicts*. The main idea is to focus only on those components that are actually involved in an inconsistency and to ignore all others.

Definition 3. (Conflict) A conflict for $(\textsc{sd}, \textsc{comps}, \textsc{obs})$ is a set $CS \subseteq \textsc{comps}$ such that $\textsc{sd} \cup \textsc{obs} \cup \{\neg\textsc{ab}(c) \mid c \in CS\}$ is inconsistent.

A conflict corresponds to a subset of components for which it would not be consistent to assume that they all work correctly given the observations. A conflict CS is considered to be *minimal* if no proper subset of CS exists that is also a conflict.

In the original approach by Reiter the conflicts are computed through calls to a theorem prover TP. The TP component is considered to be a "black box" and no assumptions are made about how the conflicts are determined or whether they are minimal or not. In practice, however, researchers often use specific algorithms such as QUICKXPLAIN (QXP) [15], Progression [16] or MERGEXPLAIN (MXP) [17] to efficiently find the conflicts (see later sections for more details). These conflict detection algorithms, in contrast to the original assumptions by Reiter, furthermore have the advantage that they can guarantee that the returned conflict sets are minimal.

Reiter then describes the relationship between conflicts and diagnoses and shows that the set of diagnoses for a collection of minimal conflicts **CS** is equivalent to the set **H** of minimal hitting sets[1] of **CS**.

---

[1] Let S be a finite set and $\mathbf{C}$ be a family of subsets of S, then a subset-minimal set $H \subseteq S$ is a hitting set for $\mathbf{C}$ iff for any $C \in \mathbf{C}$ it holds that $H \cap C \neq \emptyset$. This corresponds to the set cover problem.

14.2.3 The Hitting Set Tree Algorithm

To determine the minimal hitting sets and therefore the diagnoses, Reiter proposes a breadth-first search procedure for the computation of a hitting set tree (HS-tree), whose construction is guided by conflicts (Algorithm 14.2). Furthermore, the algorithm implements different techniques to prune the search space. Algorithm 14.2 is sound and complete when it is guaranteed that `getConflicts` used in Algorithm 14.3 only returns minimal conflicts. Soundness and completeness in this context means that all returned solutions are guaranteed to be minimal diagnoses and no diagnosis for the given set of conflicts will be missed.[2]

Algorithm 14.2: HS-TREE ALGORITHM

Input: DPI $(\textsc{sd}, \textsc{comps}, \textsc{obs})$
Output: All minimal diagnoses Δ

```
1 D ← ∅                                  /* Initialize set of known diagnoses */
2 CS ← ∅                                 /* Initialize set of known conflicts */
3 frontier ← {(∅,∅)}                       /* Initialize the search frontier */
4 while frontier ≠ ∅ :
      /* Get and remove the first element of the frontier                   */
5     (CS,Δ) ← Pop(frontier)
6     frontier ← frontier ∪ processNode(D, CS, (CS,Δ))
7 return D                                 /* Return the set of all diagnoses */
```

The main principle of the HS-tree algorithm is to create a search tree where each node corresponds to a pair (CS, Δ). The first element CS represents a conflict with which a node is labeled. The second element Δ represents the set of components which are supposed to be faulty at the current node.

When the next node is retrieved from the frontier and forwarded to Algorithm 14.3, the set CS of the retrieved node is empty and a new label must be computed. Algorithm 14.3 can either reuse one of the known conflicts stored in **CS** (line 3) that is not hit by Δ or use `getConflicts` to determine one or more new conflicts (line 5). When no conflict can be reused and CS remains empty, the set Δ hits all conflicts of the given DPI and, therefore, is a diagnosis.

To guarantee the subset-minimality of the computed hitting sets, Algorithm 14.3 includes a pruning rule in line 2. This rule forces the algorithm to ignore all nodes where the set Δ is a superset of one of the already known minimal hitting sets.

[2] An algorithm variant called HS-DAG (Directed Acyclic Graph) is proposed in [18] for cases when the returned conflicts are not minimal.

Algorithm 14.3: PROCESSNODE

Input: Sets of diagnoses **D** and conflicts **CS** as well as the node (CS, Δ)
Output: A set of new nodes *frontier*

```
1  frontier ← ∅
2  if ∀Δ' ∈ D : Δ' ⊈ Δ :                          /* If not superset of known diagnosis */
3      if ∃CS' ∈ CS : CS' ∩ Δ = ∅ : CS ← CS'                        /* Reuse a conflict */
4      else:                              /* Compute a set of new conflicts not hit by Δ */
5          CS ← CS ∪ getConflicts(COMPS, SD ∪ OBS ∪ {AB(c) | c ∈ Δ})
6          CS ← Pop({CS' | CS' ∈ CS, CS' ∩ Δ = ∅})                  /* Retrieve a conflict */
7      if CS = ∅ : D ← D ∪ {Δ}                                 /* No new conflict is found */
8      else:
           /* Generate successors of (CS,Δ) and add them to the frontier               */
9          frontier ← frontier ∪ {(∅, Δ ∪ {c}) | c ∈ CS}
10 return frontier
```

14.2.4 Example: Hitting Set Tree Construction

In the following example we show how Reiter's approach can be applied to locate a fault in a specification of a Constraint Satisfaction Problem (CSP). The example, adapted from [19], also illustrates the generality of the proposed consistency-based MBD framework.

A CSP instance I is defined as a tuple (V, D, C), where $V = \{v_1, \ldots, v_n\}$ is a set of variables, $D = \{D_1, \ldots, D_n\}$ is a set of domains for the variables in V, and $C = \{C_1, \ldots, C_k\}$ is a set of constraints. An assignment to any subset $X \subseteq V$ is a set of pairs $A = \{\langle v_1, d_1 \rangle, \ldots, \langle v_m, d_m \rangle\}$ where $v_i \in X$ is a variable and $d_i \in D_i$ is a value from the domain of this variable. An assignment comprises exactly one variable-value pair for each variable in X. Each *constraint* $C_i \in C$ is defined over a list of variables S, called its scope, and forbids or allows certain simultaneous assignments to the variables in its scope. An assignment A to S *satisfies* a constraint C_i if A comprises an assignment allowed by C_i. An assignment A is a *solution* to I if it satisfies all constraints C.

Consider a CSP instance I with variables $V = \{a, b, c\}$ where each variable has the domain $\{1, 2, 3\}$ and where the following constraints are defined:

$$C_1 : a > b, \quad C_2 : b > c, \quad C_3 : c = a, \quad C_4 : b < c$$

Obviously, no solution for I exists and our diagnosis problem consists in finding subsets of the constraints whose definition is faulty. The engineer who has modeled the CSP could, for example, have made a mistake when writing down C_2, which should have been $b < c$. Eventually, C_4 was added later on to correct the problem, but the engineer forgot to remove C_2.

The problem can be represented as a DPI as follows by defining SD as

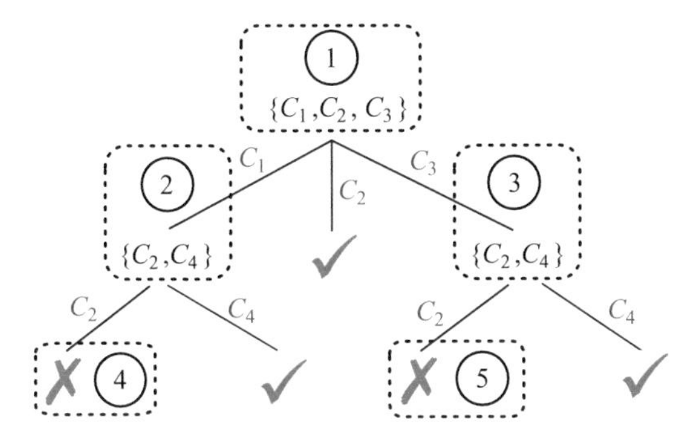

Fig. 14.2: Example of HS-tree construction, adapted from [19]

$$\neg\text{AB}(C_1) \rightarrow (a > b), \qquad \neg\text{AB}(C_2) \rightarrow (b > c)$$
$$\neg\text{AB}(C_3) \rightarrow (c = a), \qquad \neg\text{AB}(C_4) \rightarrow (b < c)$$

the set of components COMPS $= \{C_1, C_2, C_3, C_4\}$, and OBS $= \emptyset$.

Given the faulty definition of I, two minimal conflicts exist. Namely, the sets $\{\{C_1,C_2,C_3\}, \{C_2,C_4\}\}$ can be found by `getConflicts` using, for instance, QUICKXPLAIN as a conflict detection algorithm. Given these two conflicts, the HS-tree algorithm determines three minimal hitting sets $\{\{C_2\},\{C_1,C_4\}, \{C_3,C_4\}\}$, which are diagnoses for the problem instance. The set of diagnoses contains the true cause of the error, the definition of C_2.

Let us now review in more detail how the HS-tree is constructed when using QUICKXPLAIN (QXP) as a conflict detection technique for the example problem. The tree construction process is illustrated in Figure 14.2.

Since no conflict can be reused in the first iteration, a call to QXP is made, which returns the conflict $CS = \{C_1,C_2,C_3\}$. This conflict is used to label the root node ① of the tree and is added to the set of conflicts **CS**. For each element of the conflict, a child node is created and the respective conflict element is used as a path label (line 9). Hence, in the next iteration of the main loop the frontier comprises the three nodes $(\emptyset, \{C_1\})$, $(\emptyset, \{C_2\})$, and $(\emptyset, \{C_3\})$.

Next, node $(\emptyset, \{C_1\})$ is retrieved from the frontier, shown as ② in Figure 14.2. Since the known conflict cannot be reused, a new conflict is computed by QXP under the assumption that C_1 is abnormal. This call returns the conflict $\{C_2,C_4\}$. The set containing only C_1 is therefore not a diagnosis and the new conflict is used as a label for node ②. The algorithm then proceeds in breadth-first style and retrieves the node $(\emptyset, \{C_2\})$. For this node none of the known conflicts can be reused and no new conflict can be found. Therefore, $\Delta = \{C_2\}$ is a diagnosis and is added to the set **D**. The corresponding node is marked with ✓ in the figure and not further expanded. At node ③, which does not correspond to a diagnosis, the already known conflict $\{C_2,C_4\}$ can be reused as it has no overlap with the node's path label. Consequently, no call to TP (QXP) is required. At the last tree level, the nodes ④ and ⑤ are not further expanded ("closed" and marked with ✗) because $\{C_2\}$ has already been identified as a diagnosis at the previous level and the resulting diagnoses would be supersets of $\{C_2\}$. Finally, the sets $\{C_1, C_4\}$ and $\{C_3, C_4\}$ are identified as additional diagnoses.

14.2.5 Complexity Considerations

Finding the hitting sets for a given collection of sets – in our case a given set of conflicts – is known to be an NP-hard problem [20]. Furthermore, in many of the mentioned applications of MBD techniques to practical problems, we cannot assume that the set of minimal conflicts is given in advance. Therefore, in these applications (as well as in Algorithm 14.2), the conflicts are computed "on demand," i.e., during tree construction. Depending on the problem setting, finding the conflicts can be the computationally most demanding part of the entire diagnosis process.

Deciding whether an *additional diagnosis* exists when conflicts are computed on demand is also NP-complete, even for propositional Horn theories [21]. Therefore, a number of heuristics-based, approximate and thus incomplete, as well as problem-specific diagnosis algorithms have been proposed over the years. We will discuss such approaches next in Section 14.3.

Later on, in Section 14.5, however, we will focus on application scenarios where the goal is to find *all minimal diagnoses* for a given problem, i.e., we focus on *complete* algorithms. Application scenarios in which finding just one or a few diagnoses is insufficient include, for example, the MBD-based approach for finding errors in spreadsheet programs described in [6].

14.3 Alternative Approaches to Compute Diagnoses

Since determining diagnoses in Reiter's framework is a computationally demanding problem, a number of alternatives to the logic-based framework and to the sound and complete HS-tree diagnosis method have been proposed over the years. Three main categories of approaches can be identified: (I) approaches that limit the search to certain diagnoses, (II) approaches that use other formalisms and problem encodings than the logic-based one, and (III) approaches that trade soundness and/or completeness for efficiency.

(I) Approaches that limit the search to certain diagnoses

Approaches of this type aim to find only certain subsets of all existing diagnoses of a given DPI. The most prominent examples include approaches that (a) focus on the computation of one "minimum-cardinality" diagnosis or (b) find the k best diagnoses.

(a) Finding minimum-cardinality diagnoses conceptually means extending Reiter's definition of a diagnosis (Definition 2) with an additional requirement: Δ is a minimum-cardinality diagnosis iff there is no diagnosis Δ' such that $|\Delta| > |\Delta'|$. A minimum-cardinality diagnosis therefore comprises the smallest possible number of system components that, if assumed to be faulty, explain the observed misbehavior [22, 23, 24]. These types of diagnoses are important when analyzing very large systems for which the computation of diagnoses can be very time consuming.

(b) Approaches of this type compute k diagnoses that are optimal with respect to some predefined measure. For instance, finding the *k most probable* diagnoses is one of the most prominent examples found in the literature, see, e.g., [25, 26] or [27]. One can easily modify the HS-tree algorithm from Section 14.2.3 to compute the k most probable diagnoses. We only have to rewrite the condition in line 4 to $(\mathit{frontier} \neq \emptyset \wedge |\mathbf{D}| \leq k)$ and extend the function `Pop` to return the most probable node of a tree from the frontier. The latter modification turns the breadth-first scheme of the HS-tree algorithm into a uniform-cost scheme.

Parallelization approaches: Algorithms of this group can be parallelized in the same way as any uninformed search algorithm, including breadth-first or depth-first search. How these general search strategies can be parallelized is discussed in more detail in Section 14.4.

(II) Approaches that use alternative formalisms and problem encodings

The main idea of these approaches is to use special types of problem encodings that support a more efficient computation of the diagnoses. Siddiqi and Huang [28], for example, suggest to solve the diagnosis problem by using Bayesian networks. To speed up the computations they apply a differential approach to the reasoning in these networks [29].

Pill and Quaritsch [30], on the other hand, propose to transform a given DPI into an algebraic expression. The diagnoses are then computed by calling a recursive function H, which systematically selects a part of the provided algebraic expression and applies one of five pre-defined modification operators. One of them, for example, extracts individual elements of the selected sub-expressions that are also elements of at least one diagnosis for the given DPI.

The function H is designed in such a way that the method guarantees its termination after a finite number of rewritings. The algebraic expression returned as an output finally represents all or k diagnoses of the given DPI.

However, the currently most common and very efficient alternative way of encoding the problem is to use so-called "direct" methods, see, e.g., [23, 24, 31, 32, 33, 34]. Direct approaches transform and encode diagnosis problem instances in such a way that every model output (solution) returned by an inference engine that can process the target encoding corresponds to a diagnosis. Typical approaches encode the DPIs as constraint satisfaction (CSP) or as boolean satisfiability (SAT) problems. These methods also support the generation of multiple diagnoses in one single call to the inference engine.

In [35], Nica et al. report the results of a series of experiments in which they compared conflict-directed search with direct encodings. Their findings indicate that for several problem settings, using the direct encoding was advantageous. However, direct methods can be applied only if there is a knowledge representation and reasoning system which is (i) expressive enough to encode the given problem setting (input system description), and in which (ii) the computation of minimal hitting sets can be embedded in some form.

In case of direct SAT encodings, one can for example weave the computation of the diagnoses into the system description [32]. Alternatively, one can reformulate the diagnosis problem as a MaxSAT problem [24]. However, there are also application areas of Model-Based Diagnosis such as the diagnosis of description logic (DL) ontologies [36], for which such direct encodings cannot be applied because existing DL reasoners are not able to generate models that correspond to diagnoses.

Parallelization approaches: The applicability of parallelization strategies for direct diagnosis methods largely depends on the approach that is used to reformulate the DPI. For instance, the methods suggested in [32] and [24] can be parallelized simply by using a SAT/MaxSAT solver that is able to compute multiple models in parallel. The parallelization of SAT and MaxSAT methods is covered in Chapter 1, Parallel Satisfiability and Chapter 3, Parallel Maximum Satisfiability.

(III) Approaches that trade soundness and/or completeness for efficiency

This family of methods uses "approximate" algorithms to increase the efficiency of the search process, but in exchange cannot make guarantees on soundness and/or completeness. Typical strategies include the application of stochastic search techniques such as genetic algorithms, simulated annealing, or greedy approaches.

For instance, the method presented in [37] uses a two-step greedy approach. In the first step, a random and possibly non-minimal diagnosis candidate is determined by a modified DPLL[3] algorithm. In the second step, the algorithm minimizes the candidate returned by the DPLL technique by repeatedly applying random modifications.

The approach by Li et al. [38], as another example, uses a genetic algorithm that takes a number of conflict sets as input and generates a set of bit-vectors (chromosomes), where every bit encodes a truth value of an atom over the $\textsc{ab}(\cdot)$ predicate. In each iteration the algorithm applies genetic operations such as mutation, crossover, etc., to obtain new chromosomes. Subsequently, all obtained bit-vectors are evaluated by a "hitting set" fitting function which eliminates bad candidates. The algorithm stops after a predefined number of iterations and returns the best diagnosis.

Parallelization approaches: So far, no proposals have been made to parallelize such approximate techniques for MBD problems. Parallel algorithms for search approaches such as simulated annealing exist [39], but it is still open so far whether

[3] Davis-Putnam-Logemann-Loveland.

these parallelization schemes can be applied to the approximate MBD algorithms described above. In particular, this is unclear because these approximate techniques usually relax the definition of the diagnosis problem in different ways.

Since no parallel versions of approximate or incomplete diagnosis algorithms have been proposed so far, the remaining sections of this chapter will focus on the parallelization of sound and complete tree-based diagnosis algorithms that can be applied to different variants of the diagnosis problem. In particular, various approaches to parallelize the HS-tree algorithm will be discussed in Section 14.5. It will turn out that most of these algorithms implement basic strategies from uninformed search algorithms and the difference mainly lies in the used heuristics and/or pruning rules. In the next section we will therefore first discuss general approaches for search parallelization.

14.4 Parallelization of Tree Search Algorithms

In this section, we summarize different general strategies to parallelize tree search processes and discuss their applicability to Model-Based Diagnosis problems.

14.4.1 General Parallelization Strategies

Historically, the parallelization of tree search algorithms has been approached in three different ways [40]:

(I) *Parallelization of node processing*: When applying this type of parallelization, the tree is expanded by one single process, but the computation of labels or the evaluation of heuristics is done in parallel.

(II) *Window-based processing*: In this approach, sets of nodes, called "windows," are processed by different threads in parallel. The windows are formed by the search algorithm according to some predefined criteria.

(III) *Tree decomposition approaches*: The main idea of these methods is to identify a node in a search tree such that all sub-trees originating in this node are independent and can be processed in parallel [41, 42].

In principle, all three types of parallelization can be applied in some form to the problem of generating a hitting set tree.

(I) Parallelization of node processing

Applying this strategy to the MBD problem setting means parallelizing the process of conflict computation. That is, the method `getConflicts` (Algorithm 14.3, line 5) relies on a theorem prover that either makes use of multiple threads for consistency checking or that implements a parallel conflict search algorithm.

The first variant is usually simple to implement as most of the conflict computation algorithms, as in [15, 16, 43, 44], are not dependent on a particular method used in TP for consistency checking. Depending on the implemented strategy, such algorithms select a set of components $CS \subseteq$ COMPS and ask TP whether $\textsc{sd} \cup \textsc{obs} \cup \{\neg \textsc{ab}(c) \mid c \in CS\}$ is inconsistent. Depending on the result, the algorithm returns a conflict set or selects another subset of components to be checked.

In the latter case – parallel conflict search – one can use, for instance, a slightly modified version of the recently proposed MERGEXPLAIN method [17], which will be discussed in more detail later in section 14.5.2.1. In every step, this algorithm splits the set of components into two non-empty disjoint subsets, which can then be processed in parallel. For other well-known conflict or prime implicate computation algorithms like the ones listed above, parallelizing the process is not as easy. Therefore, more research is required on parallel conflict computation when the goal is to speed up node processing through parallelization.

(II) Window-based processing

This strategy, in which multiple sets of *independent* nodes of the search tree (windows) are computed in parallel, was for example applied by Powley et al. [45]. In their work the windows are determined by different thresholds of a heuristic function of *Iterative Deepening A**.

In principle we can apply such a strategy also to the HS-tree construction problem. This would however mean that we have to categorize the nodes to be expanded according to some criterion, e.g., the probability of finding a diagnosis, and then allocate the different groups to individual threads. In the absence of such criteria, we could use a constant window size such that each open node is allocated to a separate thread. In this case the number of parallel threads (windows) should not exceed the number of physically available computing threads to obtain the best performance.

A general problem when applying a window-based strategy is that in the general case it is hard to find a window function that assigns sets of tree nodes to different threads. Ideally such a function should guarantee that the decisions made by an algorithm for nodes of different windows are independent. In case of MBD, there are two types of such decisions: labeling of nodes and pruning. For the first type, the window function has to guarantee that no two TP calls for two nodes of different windows return the same conflict. For the second type, it has to be guaranteed that the results of tree pruning for nodes of one window are irrelevant for nodes of other windows.

Unfortunately, for many MBD problem instances the computations at different levels of the tree are not independent of each other. Moreover, there is no general way to split the components of a DPI into subsets in such a way that the computation of *different* conflicts for every subset is ensured. Therefore, the parallel MBD algorithms discussed in Section 14.5 do not rely on window-based parallelization.

(III) Tree decomposition approaches

The idea of these approaches is to determine sub-trees in the search tree that can be processed independently of each other by different threads. Any decision made within one sub-tree must therefore not influence the behavior or decision of threads working on other parts of the search tree.

For example, Anglano et al. [46] suggest a tree decomposition approach for MBD problems in which they parallelize the diagnosis problem based on structural problem characteristics. In their work, they first map a given diagnosis problem to a Behavioral Petri Net (BPN). Then, the obtained BPN is manually partitioned into subnets and every subnet is submitted to a different Parallel Virtual Machine (PVM) for parallel processing. A major drawback of this approach is that a manual problem decomposition step is required and that such a decomposition has to be possible in the first place.

The main problem with structure-based parallelization algorithms is that they impose a number of requirements on the DPI. For example, in order to apply the decomposition approach in Binary HS-tree (BHS) algorithms [30], all conflict sets for the given diagnosis problem instance must be known in advance. Given this additional information, the method can split the search tree into two sub-trees for which the resulting sets of diagnoses are disjoint. This makes the computations in both sub-trees independent and, therefore, easy to execute in parallel. However, in the majority of cases the set of conflicts is unknown for a given DPI. The structure-based approach proposed in [47] on the other hand requires that the diagnosis problem has a tree-like structure. Overall, the applicability of tree decomposition approaches to parallel MBD is therefore limited to specific types of problem settings.

14.4.2 Applying Domain-Independent Parallelized Search Techniques

In principle, parallelized versions of domain-independent search algorithms such as A^* can be applied to MBD settings as well (see Chapter 11, Parallel A* For State Space Search). However, the MBD problem has different particularities that make the application of some of these algorithms difficult. For instance, the PRA* method and its variant HDA* discussed in the work of Burns et al. [40] use a mechanism to minimize the memory requirements by retracting parts of the search tree. These "forgotten" parts are later regenerated when required. In our MBD setting, the generation of nodes is however the most costly part, which is why the applicability of HDA* seems limited. Similarly, duplicate detection algorithms such as PBNF [40] require the existence of an abstraction function that partitions the original search space into blocks. In general MBD settings, however, we cannot assume that such a function is given.

In order to improve the performance of a parallel algorithm one should in general try to avoid the duplicate generation of nodes by different threads. A promising starting point for this research could be the work by Phillips et al. [48]. The authors suggest a variant of the A* algorithm that generates only independent nodes in order to reduce the costs of node generation. Two nodes are considered to be independent if the generation of one node does not lead to a change of the value returned by the heuristic function for the other node. The generation of independent nodes can then be done in parallel without the risk of the repeated generation of an already known state. The main difficulty when adopting this algorithm for MBD is the formulation of an admissible heuristic required to evaluate the independence of the nodes for arbitrary diagnosis problems. However, for specific problems that can be encoded as CSPs, Williams and Ragno [26] present a heuristic that depends on the number of unassigned variables at a particular search node.

Finally, parallelization has also been used in the literature to speed up the search in very large search trees that do not fit in memory. Korf [49], for instance, suggests an extension of a hash-based delayed duplicate detection algorithm that allows a search algorithm to continue search while other parts of the search tree are written to or read from the hard drive. Such methods can in theory be used in combination with parallel MBD algorithms in case of complex diagnosis problems.

14.5 Parallelized Hitting Set Tree Construction Schemes

In this section, we review two algorithms that implement parallelization strategies for Reiter's sound and complete HS-tree algorithm. As discussed in Section 14.2.3, in every iteration the HS-tree algorithm picks a node from the queue, labels it with a conflict set, generates a set of successor nodes, and applies the pruning rules. The two approaches to parallelize this process considered in this section follow the general idea of breadth-first search parallelization. In particular, they assign every iteration step of the main loop to a different thread. The main problem of such a scheme, however, is that the generation of node labels, i.e., the computation of conflict sets, can be time consuming. Therefore, we also show three possible extensions to these schemes which move more of the processing power to the conflict computation task.

14.5.1 Computing Multiple Hitting Set Tree Nodes in Parallel

The two algorithms presented next aim to maintain the breadth-first tree exploration scheme of the HS-tree algorithm and implement strategies that utilize multiple threads without sacrificing the soundness and completeness of the diagnosis process. Furthermore, both algorithms do not make any assumptions about specific properties of the diagnosis problem instances to be solved. Therefore, they can be applied to different variations of the diagnosis problem definition from Section 14.2. Conceptually, the presented algorithms can also be seen as special cases of the window-based parallelization scheme described in Section 14.4.1, where every window comprises exactly one node.

14.5.1.1 Level-Wise Parallelization

The strategy of the first parallelization scheme is to examine all nodes *of one tree level* in parallel and to proceed with the next level only when all elements of the current level have been processed.

In the example shown in Figure 14.2, this would mean that the computations (TP calls) required for the three first-level nodes labeled with $\{C_1\}$, $\{C_2\}$, and $\{C_3\}$ can be processed in three parallel threads. The nodes of the next level are processed when all threads of the previous level are finished.

Using this Level-Wise Parallelization (LWP) scheme, the breadth-first order is strictly maintained. The parallelization of the computations is generally feasible because the consistency checks for each node can be done independently of those done for the other nodes on the same level. Synchronization is only required to make sure that no thread starts exploring a path that is already under examination by another thread.

Algorithm 14.4: DIAGNOSELW: Level-Wise Parallelization

Input: DPI $(\text{SD}, \text{COMPS}, \text{OBS})$
Output: All minimal diagnoses Δ

```
1  D ← ∅                                        /* Initialize set of known diagnoses */
2  CS ← ∅                                       /* Initialize set of known conflicts */
3  frontier ← {(∅,∅)}                           /* Initialize the search frontier */
4  level ← ∅                                    /* Initialize level-wise processing */
5  while frontier ≠ ∅ :
       /* Get and remove the first element of the frontier                        */
6      (CS,Δ) ← Pop(frontier)
7      level ← level ∪ runParallel({processNode(D,CS,(CS,Δ))})
8  wait()
9  if level ≠ ∅ : frontier ← level; goto 4
10 return D                                     /* Return the set of all diagnoses */
```

Algorithm 14.4 shows how the sequential method implemented in Algorithm 14.2 can be adapted to support this parallelization approach.[4] The statement `runParallel` takes a function as a parameter and schedules it for execution in a pool of threads of a given size. With a thread pool of, e.g., size 2, the generation of the first two nodes is done in parallel and then the main thread waits until one of the threads finishes. Only then will the third node be processed. Using this mechanism we can ensure that the number of threads executed in parallel is less than or equal to the number of hardware threads or CPUs.

[4] Differences to the original Algorithm 14.2 are highlighted with a gray background.

In addition, during the execution of the algorithm all changes to global structures such as **D** and **CS** have to be synchronized. That is, two threads may read the structures concurrently, but writing to one of these data structures is exclusive. While one thread is writing, all other threads must wait until the operation has finished regardless of whether they want to read or write.[5] Finally, the statement `wait` is used for synchronization and blocks the execution of the subsequent code until all scheduled threads are finished.

Theorem 1 ([19]). *Level-Wise Parallelization is sound and complete.*

14.5.1.2 Full Parallelization

The LWP scheme maintains the breadth-first order of the original algorithm and, therefore, inherits all its properties, such as soundness and completeness. However, in some situations the level-wise processing procedure might get stuck at the end of a level when the computation of a conflict for one of the nodes takes a long time. In this case, all threads that have already finished processing their nodes have to wait for the one thread still working.

The Full Parallelization (FP) approach presented in Algorithm 14.5 immediately schedules the first node of the frontier for execution as soon as some thread becomes idle. In this way, the FP scheme avoids the problem observed for the LWP procedure. The main loop continues until the frontier is empty and no more nodes are processed in parallel.

Algorithm 14.5: DIAGNOSEFP: Full Parallelization

Input: DPI (SD, COMPS, OBS)
Output: All minimal diagnoses Δ

```
1 D ← ∅                                   /* Initialize set of known diagnoses */
2 CS ← ∅                                  /* Initialize set of known conflicts */
3 frontier ← {(∅,∅)}                       /* Initialize the search frontier */
4 while frontier ≠ ∅ ∨ runningThreads > 0 do
      /* Get and remove the first element of the frontier */
5     (CS,Δ) ← Pop(frontier)
6     frontier ← frontier ∪ runParallel({processNode(D,CS,(CS,Δ))})
7     checkMinimality(D)
8 return D                                 /* Return the set of all diagnoses */
```

Since the nodes are expanded as soon as possible, FP might not follow the breadth-first strategy of the HS-tree algorithm anymore. Consequently, it may find non-minimal diagnoses during the search. Therefore, in every iteration the elements stored in the set of known diagnoses **D** must be checked for minimality. The method

[5] Implementing such concurrency aspects is comparatively simple in modern programming languages such as Java, e.g., by using the `synchronized` keyword.

`checkMinimality` removes all Δ_j from the set **D** if there exists $\Delta_i \in \mathbf{D}$ such that $\Delta_i \subset \Delta_j$. Note that the check for minimality has to be synchronized on **D**. That is, all other threads are not allowed to modify **D** while its elements are removed.

From the performance perspective, the FP method ensures that all its threads are busy as long as there is at least one node in the frontier. On the other hand it needs more time to synchronize the threads and to remove redundant hitting sets. If the computation of conflicts does not take long and the last nodes of a level are finished simultaneously, then LWP can be advantageous. These aspects will be discussed in more detail in Section 14.6 where we present results of an empirical comparison of FP and LWP for different problems.

Theorem 2 ([19]). *Full Parallelization is sound and complete, if applied to find all diagnoses up to some cardinality.*

Theorem 3 ([19]). *Full Parallelization cannot guarantee completeness and soundness when applied to find the first k diagnoses, i.e., $1 \leq k < N$, where N is the total number of diagnoses of a problem.*

Note that in cases when FP is used to search for only k diagnoses, every computed hitting set must additionally be minimized by applying an algorithm such as INV-QUICKXPLAIN [50]. Similarly to QUICKXPLAIN, INV-QUICKXPLAIN applies a divide-and-conquer strategy , but in this case to find one minimal diagnosis for a given diagnosis problem instance. Applied to a given, possibly non-minimal hitting set H, this algorithm can find a minimal hitting set $H' \subseteq H$ requiring only $O(|H'| + |H'| \log(|H|/|H'|))$ calls to the theorem prover TP. The first part, $|H'|$, corresponds to the number of TP calls required to determine whether or not H' is minimal. The second part indicates the number of subproblems that must be considered by INV-QUICKXPLAIN's divide-and-conquer strategy to find the minimal hitting set H'.

14.5.2 Computing Nodes and Conflicts in Parallel

In all versions of the HS-tree algorithm presented above, the TP call (`getConflicts`) corresponds to an invocation of QXP. Whenever a new node of the HS-tree is created, QXP returns a set of elements that represents exactly one new conflict. This strategy has the advantage that the call to TP immediately returns after one conflict has been determined. This in turn means that the other parallel execution threads immediately "see" this new conflict in the shared data structures and can, in the best case, reuse it when constructing new nodes.

A disadvantage of computing only one conflict at a time with QXP is that the search for conflicts is *restarted* on each invocation. A recently proposed new conflict detection technique called MERGEXPLAIN (MXP) [17] is capable of computing multiple conflicts in one call. The general idea of MXP is to continue the search after the identification of the first conflict and look for additional conflicts in the remaining constraints (or logical sentences) in a divide-and-conquer approach.

When combined with a sequential HS-tree algorithm, the effect is that during tree construction more time is initially spent on conflict detection before the construction continues with the next node. In exchange, the chances of having a conflict available for reuse increase for the next nodes. At the same time, the identification of some of the conflicts is less time-intensive as smaller sets of constraints have to be investigated due to the divide-and-conquer approach of MXP. An experimental evaluation on various benchmark problems shows that substantial performance improvements are possible in a sequential HS-tree scenario when the goal is to find *a few leading diagnoses* [17].

14.5.2.1 Background: QUICKXPLAIN and MERGEXPLAIN

The QXP method implemented in Algorithm 14.6 is a conflict detection technique which was originally applied to find one minimal conflict set for a set of unsatisfiable constraints.[6] However, over the last decade it was often applied to find conflicts for Model-Based Diagnosis problems.

QXP implements a recursive divide-and-conquer strategy that operates on two sets of constraints B and C. The set B – called "background theory" – comprises all constraints that are considered to be correct in the current recursive call. When QXP starts, this set is initialized with SD, OBS, and $\{\neg \text{AB}(c) \mid c \in L\}$, where L is a set of components used as arc labels on the path from the root to the current node in the HS-tree (*visited nodes*). The set C comprises all constraints that are possibly faulty and in which we search for a conflict.

In case the set C is consistent with the background theory or C is empty, then no conflict can be found and the algorithm immediately returns. Otherwise, QXP calls `computeConflict`, which corresponds to Junker's QUICKXPLAIN' function in [15]. The only difference between these methods is that `computeConflict` does not require a strict partial order for the set of constraints C. We omit the requirement for a strict partial order here, as in many applications of MBD prior information about fault probabilities is not available.

Roughly, QXP applies a divide-and-conquer strategy that has two modes: "search" and "extraction." The algorithm starts in the search mode, in which every recursive call partitions the set of faulty constraints C into two sets C_1 and C_2. If C_1 is inconsistent, then QXP will continue to search for a conflict within this set and partitions C_1 in the next recursive call. The algorithm switches into the extraction mode if the partitioning process has split all conflicts of C into two parts, i.e., C_1 is consistent. In this mode QXP finds the first part of a conflict in the set C_2 and then the second part in the set C_1.

The recently proposed MXP algorithm extends the ideas of QXP by searching for conflicts not only in C_1, but also in C_2 in one call. This results in the computation

[6] Hereafter we use the term constraints as it was done in the original paper [15]. However, note that QXP uses the theorem prover only for consistency checking and is independent of the underlying reasoning technique. Therefore, the elements of the sets could be any set of logic sentences for which sound and complete reasoning methods exist.

Algorithm 14.6: QUICKXPLAIN (QXP)

Input: A diagnosis problem (SD, COMPS, OBS), a set *pathNodes* of labels on the path from the root to the current node
Output: A set containing one minimal conflict $CS \subseteq \mathsf{C}$

```
1  B = SD ∪ OBS ∪ {AB(c)|c ∈ pathNodes}
2  C = {¬AB(c)|c ∈ COMPS \ pathNodes}
3  if isConsistent(B ∪ C) : return 'no conflict'
4  elif C = ∅ : return ∅
5  return {c|¬AB(c) ∈ computeConflict(B, B, C)}

   function computeConflict(B, D, C)
6      if D ≠ ∅ ∧ ¬isConsistent(B) : return ∅
7      if |C| = 1 : return C
8      Split C into disjoint, non-empty sets C1 and C2
9      D2 ← computeConflict(B ∪ C1, C1, C2)
10     D1 ← computeConflict(B ∪ D2, D2, C1)
11     return D1 ∪ D2
```

of multiple conflicts, if they exist. Algorithm 14.7 presents the general procedure of MXP. First, the algorithm checks whether the sets of input constraints actually include at least one conflict. Next, it calls the function `findConflicts`, which returns a tuple $\langle \mathsf{C}', \mathbf{CS} \rangle$, where C' is a set of remaining consistent constraints and **CS** is a set of found conflicts. Similarly to QXP's `computeConflict` this function first recursively partitions the set C into two subsets. However, in contrast to `computeConflict`, `findConflicts` continues the search for conflicts in both subsets. This allows the algorithm to identify conflict sets in both C_1 and C_2. Every found conflict is stored in the set **CS** and is resolved by removing one of its elements from the set C_1. Finally, after all conflicts in the subsets C_1 and C_2 have been found and resolved, the function checks whether the union of these two sets is consistent. If not, it searches for a conflict in $\mathsf{C}_1' \cup \mathsf{C}_2'$ (and the background theory) in the style of QXP.[7]

14.5.2.2 Strategies for Combining Node and Conflict Computation

The main idea of the following strategies is to invest more processing power in the task of conflict computation while the nodes of the HS-tree are constructed in parallel using LWP or FP. Our expectation is that higher conflict reuse levels can be achieved during tree construction as we know more conflicts at the beginning of the process.

The desired effect can be achieved by embedding variants of MXP as a conflict detection strategy, because in MXP we invest more time in looking for *additional* conflicts in one call before we proceed with the next node. In principle, computing one conflict with MXP requires at least one consistency check more for every partition than QXP. However, MXP should still be advantageous because it can

[7] Please see [17] for more details. The paper also contains the results of an in-depth experimental analysis for different problems.

Algorithm 14.7: MERGEXPLAIN (MXP)

```
Input: A diagnosis problem (SD, COMPS, OBS), a set pathNodes of labels on the path
       from the root to the current node
Output: CS, a set of minimal conflicts
1  B = SD ∪ OBS ∪ {AB(c) | c ∈ pathNodes}
2  C = {¬AB(c) | c ∈ COMPS \ pathNodes}
3  if ¬isConsistent(B) : return 'no solution'
4  if isConsistent(B ∪ C) : return ∅
5  ⟨_, CS⟩ ← findConflicts(B, C)
6  return {c | ¬AB(c) ∈ CS}

   function findConflicts(B, C) returns tuple ⟨C', CS⟩
7      if isConsistent(B ∪ C) : return ⟨C, ∅⟩
8      if |C| = 1 : return ⟨∅, {C}⟩
9      Split C into disjoint, non-empty sets C1 and C2
10     ⟨C'1, CS1⟩ ← findConflicts(B, C1)
11     ⟨C'2, CS2⟩ ← findConflicts(B, C2)
12     CS ← CS1 ∪ CS2
13     while ¬isConsistent(C'1 ∪ C'2 ∪ B) :
14         X ← computeConflict(B ∪ C'2, C'2, C'1)
15         CS ← X ∪ computeConflict(B ∪ X, X, C'2)
16         C'1 ← C'1 \ {α} where α ∈ X
17         CS ← CS ∪ {CS}
18     return ⟨C'1 ∪ C'2, CS⟩
```

avoid the potential overheads that can happen when the same conflict is computed simultaneously by LWP and FP.

In the following we discuss different ways of incorporating MXP within the full parallelization scheme FP:

Strategy (1)

One first strategy is simply to call MXP instead of QXP during node generation. Whenever MXP finds a conflict, it is added to the global list of known conflicts and can be (re-)used by other parallel threads. The thread that executes MXP during node generation continues with the next node when MXP returns.

Strategy (2)

This strategy implements a variant of MXP that is slightly more complex. Once MXP finds the first conflict, the method immediately returns this conflict so that the calling thread can continue exploring additional nodes. At the same time, a new background thread is started which continues the search for additional conflicts, i.e., it completes the work of the MXP call. In addition, whenever MXP finds a new conflict it checks whether any other already running node generation thread could have reused the conflict if it had been available beforehand. If this is the case, the search for conflicts in this *other* thread is stopped as no new conflict is needed anymore. Strategy (2) could in theory result in better CPU utilization, as we do not have to wait for an MXP call to finish before we can continue building the HS-tree. However, the strategy also leads to higher synchronization costs between the threads as, for instance, we have to potentially notify the working threads about newly identified conflicts.

Strategy (3)

A final strategy is to parallelize the conflict detection procedure of MXP itself. Whenever the set C of constraints is split into two parts, the first recursive call of `findConflicts` is queued for execution in a thread pool and the second call is executed in the current thread. When both calls are finished, the algorithm can continue. An empirical evaluation of this approach in [19] however indicated that the additional gains that can be obtained through this strategy are limited. Nonetheless, parallelizing the conflict detection algorithm – which can be any sort of Theorem Prover – in principle represents a possible strategy to speed up the overall tree construction process.

14.6 Effectiveness of Computing Multiple Nodes in Parallel

The goal of this section is to quantify the possible gains that can be obtained through parallelization for typical Model-Based Diagnosis problems. In this section, we will analyze the effectiveness of the two approaches presented in Section 14.5.1 to parallelize Reiter's HS-tree algorithm (Level-Wise Parallelization and Full Parallelization) as examples. The presented results are taken from [19], which also contains a detailed discussion of a number of additional experiments.

14.6.1 General Considerations

In our experiments, we will use *wall times* as a basic measure for our evaluation, because the comparison of wall times is a common approach in the literature to assess the improvements that one can obtain through parallelization. Wall times represent a start-to-end measurement approach for a given task, which means that also times are included when processors have to wait for resources etc. Using wall time instead of CPU time is particularly appropriate for the given problem setting, because the synchronization between threads in particular in the FP algorithm can take a significant amount of time.

The differences in the wall times are often reported with the help of two measures, *speedup* and *efficiency*, that take the amount of available computing resources into account. Speedup S_p is computed as $S_p = T_1/T_p$, where T_1 is the wall time when using one thread (the sequential algorithm) and T_p the wall time when p parallel threads are used. The efficiency E_p is defined as S_p/p and compares the speedup with the theoretical optimum.

While speedup and efficiency are well defined, one has to be careful when interpreting or trying to generalize the results, because the speedups that can be achieved depend not only on the algorithms, but can also be influenced by the underlying hardware architecture. With Intel's Hyper-Threading Technology, for example, virtual computing cores can be used, which are, however, mapped to the same physical cores. For a parallel program, these virtual cores appear like real physical computing nodes, but the results that are obtained when using virtual cores can be different from those that one would see with more physical cores. In addition, an algorithm can perform differently when executed on a single CPU with multiple cores or on a server architecture with multiple CPUs on a single main board. Furthermore, running the same algorithm on multiple computers connected in a network might lead to yet different results. The results that are reported below were obtained with specific hardware configurations. For alternative hardware architectures, the speedups might be different and other algorithms might even be better suited.

Besides the hardware on which the experiments are executed, also the choice of the benchmark problems influences the obtained results. Obviously, problems that we only need a few milliseconds to solve in a single-threaded system are not a good subject to study the possible benefits of parallelization. In such cases, the speedups that might eventually be achieved can easily be eaten up by the overhead costs of parallelization. Starting a new execution thread in Java, for example, is often considered expensive, e.g., due to the costs of thread initialization and lifecycle management. In addition to the general complexity of the individual benchmark problems, we should furthermore also look at certain other characteristics of the individual problems that might impact the benefits of parallelization. Which of the characteristics are relevant, however, might depend on the specific parallelization algorithm. For example, the average width of the search tree can impact the performance of the level-wise approach LWP, i.e., if the tree is not very wide, the degree of parallelization that can be achieved will be limited.

Overall, the discussions show that a number of aspects, both hardware-related ones and problem-specific ones, can impact the effectiveness of different parallelization strategies. In [19], the proposed parallelization approaches were therefore tested on a variety of different benchmark problems, and we will summarize some of the results next. The experiments were limited to two different types of standard hardware architectures and did not require any special computing equipment for massive parallelization or the utilization of the processing power of Graphics Processing Units (GPUs). The obtained results will show that even with standard hardware and general-purpose programming languages significant speedups can be achieved.

14.6.2 Results for Standard Electronic Circuit Benchmark Problems

In order to evaluate the usefulness of the LWP and FP parallelization strategies in comparison with Reiter's original single-threaded method, a number of diagnosis problems from three different application domains were used in [19]. In this chapter we report the detailed results of one of these domains, the electronic circuit benchmarks from the DX Competition 2011 Synthetic Track [51], and summarize the overall findings. The detailed results for the other problem domains – faulty descriptions of Constraint Satisfaction Problems (CSPs) as well as problems from the domain of ontology debugging – can be found in [19].

For the evaluation on the DX benchmarks the first five systems of the competition dataset were used (see Table 14.1). For each system, the competition specifies 20 scenarios with injected faults that result in different faulty output values. The system descriptions and the given input and output values were used for the diagnosis process, while the additional information about the injected faults was of course ignored. The problems were converted into Constraint Satisfaction Problems, which allowed us to simulate the behavior of the circuits. In the experiments Choco [52] served as a constraint solver and QXP was used for conflict detection.

System	#C	#V	#F	#D	$\overline{\#D}$	$\overline{\lvert D \rvert}$
74182	21	28	4 - 5	30 - 300	139.0	4.66
74L85	35	44	1 - 3	1 - 215	66.4	3.13
74283*	38	45	2 - 4	180 - 4,991	1,232.7	4.42
74181*	67	79	3 - 6	10 - 3,828	877.8	4.53
c432*	162	196	2 - 5	1 - 6,944	1,069.3	3.38

Table 14.1: Characteristics of the selected DXC benchmarks

Table 14.1 shows the characteristics of the systems in terms of the number of constraints (#C) and the problem variables (#V).[8] The number of injected faults (#F) and the number of calculated diagnoses (#D) vary strongly because of the different scenarios for each system. For both columns we show the ranges of values over all scenarios. The columns $\overline{\#D}$ and $\overline{\lvert D \rvert}$ indicate the average number of diagnoses and their average cardinality. As can be seen, the search tree for the diagnosis can become extremely broad with up to 6,944 diagnoses with an average diagnosis size of only 3.38 for the system c432.

Table 14.2 shows the averaged results when searching for all minimal diagnoses in the DXC benchmarks. We first list the running times in milliseconds for the sequential version (Seq.) and then the improvements of LWP or FP in terms of speedup $\mathbf{S}_4$ and efficiency $\mathbf{E}_4$ with respect to the sequential version. The fastest algorithm for each system is highlighted in bold.

[8] For systems marked with *, the search depth was limited to the actual number of faults to ensure that the sequential algorithm terminated within a reasonable time frame.

System	Seq. [ms]	LWP		FP	
		S_4	E_4	S_4	E_4
74182	65	2.23	0.56	**2.28**	**0.57**
74L85	209	2.55	0.64	**2.77**	**0.69**
74283*	371	2.53	0.63	**2.66**	**0.67**
74181*	21,695	1.22	0.31	**3.19**	**0.80**
c432*	85,024	1.47	0.37	**3.75**	**0.94**

Table 14.2: Observed performance gains for the DXC benchmarks

In all tests, both parallelization approaches outperform the sequential algorithm. Furthermore, the difference between the sequential algorithm and one of the parallel approaches was statistically significant ($p < 0.05$) in 95 of the 100 tested scenarios. For all systems, FP was more efficient than LWP and the speedups range from 2.28 to 3.75 (i.e., up to a reduction in running time of more than 70%). In 59 of the 100 scenarios the difference between LWP and FP was statistically significant. A trend that can be observed is that the efficiency of FP was higher for the more complex problems. The reason is that for these problems the time needed for node generation is much larger in absolute numbers than the additional overhead times that are required for thread synchronization.

As mentioned above, additional experiments for other problem domains were reported in [19]. The obtained results show that parallelizing the HS-tree algorithm is also advantageous for these domains. For the CSP and ontology debugging problems, however, FP was not consistently faster than LWP. This indicates that the advantage of FP over LWP can depend on the characteristics of the problems. In addition, for some scenarios the speedups of the parallelized approaches were not as high as the speedups achieved for the DXC benchmark problems.

14.6.3 Systematic Variation of Problem Characteristics

The empirical analyses reported in the previous section for typical MBD benchmark problems show that computing multiple nodes in parallel can help to significantly speed up the diagnosis process. Both parallelization techniques were faster than the sequential algorithm in all tests. However, the full parallelization approach FP was not consistently faster than the level-wise approach LWP across all tested problem instances.

14.6.3.1 Method

To obtain a better understanding of how different problem characteristics impact the performance of the parallelization techniques, a series of additional experiments with synthetic problem instances was performed in [19]. For these experiments, a suite of hitting set computation problems was created where the following characteristics were systematically varied: number of components (#Cp), number of conflicts (#Cf), and average size of the conflicts ($\overline{|\mathrm{Cf}|}$).

To create these diagnosis problem instances, a problem generation algorithm was designed which constructs a set of minimal conflicts of the specified average size for the given number of components. To obtain realistic scenarios, not all generated conflicts were of equal size but their size was varied in a randomized process according to a Gaussian distribution with the desired size as a mean. Similarly, since not all components should be equally likely to be part of a conflict, again a randomized process was used to assign failure probabilities to the components.

An additional aspect that can impact the performance of the different parallelization techniques is the time that is required to compute one conflict to label a new node in the search tree. For the suite of synthetic benchmark problems, the conflicts are known in advance (as they were used to construct the diagnosis problems in the first place). A call to the theorem prover would therefore simply mean looking up one of the known conflicts from memory, which requires almost no computation time. To be still able to measure the impact of varying conflict computation times, artificial processing delays were introduced into the diagnosis process to simulate varying conflict detection times. Technically, this can be done by adding artificial and slightly randomized *waiting times* (Wt) upon each consistency check inside the theorem prover. Of course, the consistency-checking method is only called if no already retrieved conflict can be reused for the current node.

#Cp, #Cf, $\overline{\|\mathrm{Cf}\|}$	#D	Wt [ms]	**Seq.** [ms]	**LWP** $\mathbf{S_4}$	**LWP** $\mathbf{E_4}$	**FP** $\mathbf{S_4}$	**FP** $\mathbf{E_4}$
Varying computation times Wt							
50, 5, 4	25	**0**	23	2.26	0.56	**2.58**	**0.64**
50, 5, 4	25	**10**	483	2.98	0.75	**3.10**	**0.77**
50, 5, 4	25	**100**	3,223	2.83	0.71	**2.83**	**0.71**
Varying conflict sizes							
50, 5, **6**	99	10	1,672	3.62	0.91	**3.68**	**0.92**
50, 5, **9**	214	10	3,531	3.80	0.95	**3.83**	**0.96**
50, 5, **12**	278	10	4,605	3.83	0.96	**3.88**	**0.97**
Varying numbers of components							
50, 10, 9	201	10	3,516	**3.79**	**0.95**	3.77	0.94
75, 10, 9	105	10	2,223	**3.52**	**0.88**	3.29	0.82
100, 10, 9	97	10	2,419	3.13	0.78	**3.45**	**0.86**
#Cp, #Cf, $\overline{\|\mathrm{Cf}\|}$	#D	Wt [ms]	**Seq.** [ms]	**LWP** $\mathbf{S_8}$	**LWP** $\mathbf{E_8}$	**FP** $\mathbf{S_8}$	**FP** $\mathbf{E_8}$
Adding more threads (8 instead of 4)							
50, 5, **6**	99	10	1,672	6.40	0.80	**6.50**	**0.81**
50, 5, **9**	214	10	3,531	7.10	0.89	**7.15**	**0.89**
50, 5, **12**	278	10	4,605	7.25	0.91	**7.27**	**0.91**

Table 14.3: Simulation results

14.6.3.2 Results

The results of the systematic variation of the problem characteristics are shown in Table 14.3. The table shows the effects of varying the conflict computation times, effects of different conflict sizes, effects of different problem sizes in terms of system components, and finally the effects of using more parallel computation threads.

The following observations can be made. First, if conflicts can be found in almost no time ($\text{Wt} = 0$) parallelizing the computation of multiple nodes still helps to speed up the overall diagnosis process, but due to the overhead of thread creation and synchronization the benefits of parallelization are greater for cases in which the conflict computation actually takes at least a few milliseconds.

Second, larger conflicts ($\overline{|\text{Cf}|}$) and correspondingly broader HS-trees are better suited for parallel processing. On the other hand, increasing the number of components with an unchanged number and size of conflicts leads to larger diagnoses. Searching for diagnoses up to a pre-defined search depth in this case leads to fewer found diagnoses and a narrower search tree. As a result, the relative performance gains of the parallelized algorithms are lower than when there are fewer components. Finally, when there are larger conflicts, using more threads leads to further improvements as in these cases even higher levels of parallelization can be achieved.

Overall, the simulation experiments clearly demonstrate that parallelization is advantageous for a variety of problem configurations. For all tests, the speedups of LWP and FP are statistically significant. The results also reveal how the different problem characteristics of the underlying problem impact the possible performance gains. Finally, regarding the comparison of the LWP and the FP scheme, the additional gains of not waiting at the end of each search level (as done by the FP method) typically led to small further improvements.

14.7 Alternative Model-Based Diagnosis Parallelization Approaches

The parallelization approaches presented in the previous sections maintain the generic and problem-independent nature of the original HS-tree algorithm. With slight and straightforward adaptations they are furthermore applicable to different variations of the general definition of the diagnosis problem introduced in Section 14.2. In this section we briefly review existing alternative parallelization approaches from the recent literature that were developed for specific diagnosis problem settings.

14.7.1 Tree-Based Approaches To Find One or Few Diagnoses

The first situation we consider here is when computing all minimal diagnoses is extremely challenging in cases when, e.g., a system to be diagnosed is huge or the computation of even one minimal conflict is too complex and takes unacceptably long. Also, there could be application scenarios where the allowed response time to return the diagnoses is very limited. In such settings, one can try to focus on a specific subset of all existing diagnoses which might be easier to compute and, e.g., search for a predefined number of diagnoses or for diagnoses of a limited cardinality (some examples are given in Section 14.3).

Different heuristic, stochastic, or approximative algorithms have been suggested for such situations in the literature [22, 32, 37]. The internal designs of these approaches are quite diverse. Therefore, it is challenging to analyze them with respect to the potential benefits of parallelization in a general manner.

As a simplification and approximation of such algorithms, one can however look at the possible benefits of parallelizing a depth-first strategy to find a limited number of diagnoses. If parallelizing such a strategy proves beneficial, we can see this as an indicator that parallelizing other strategies such as those mentioned above could be worth investigating. In [19], the results of a number of experiments are reported in which two variants of such tree-based approaches were compared with the Full Parallelization approach from Section 14.5.

Parallel Random Depth-First Search (PRDFS)

The PRDFS method aims at the fast computation of *one single diagnosis*, using a depth-first strategy of expanding the search tree. Given the root node of the search tree, every parallel execution thread of the algorithm greedily searches for a diagnosis by expanding random nodes of the tree depth-first. Whenever a node has been labeled with a conflict, each PRDFS thread – in contrast to the HS-tree algorithm – randomly selects one component of the conflict and generates only one successor. This node is then expanded in the next iteration of the thread's main loop.

Whenever a diagnosis is found with this greedy strategy, it is obviously not guaranteed that the diagnosis is minimal. As in the situation when applying the FP strategy to search for a limited number of diagnoses, every returned diagnosis therefore has to be minimized, i.e., the redundant elements have to be removed.

A Hybrid Strategy

This algorithm, similar to PRDFS, focuses on the computation of one single diagnosis. The algorithm however considers that depending on the cardinality of the existing minimal diagnoses it can either be advantageous to quickly descend in the search tree or to exhaustively look for diagnoses of very small sizes first. In the proposed hybrid strategy, half of the available threads therefore follow the PRDFS scheme to descend in the search tree and the other half of them explore the tree in breadth-first manner using the FP algorithm.

The coordination between the two algorithms can be done with the help of shared data structures that contain the known conflicts and diagnoses. When enough diagnoses (e.g., one) are found, all running threads can be terminated and the results are returned.

Insights

In the experiments from [19], the same benchmark problems were used as for the evaluation of the LWP and FP techniques. The goal this time however was to find one arbitrary minimal diagnosis. The obtained results can be summarized as follows. When only one diagnosis is needed, using a depth-first strategy is in most cases faster than using the FP technique and thus faster than Reiter's single-threaded HS-tree algorithm. Using multiple threads in this depth-first search (PRDFS) is in almost all cases beneficial. Finally, the hybrid strategy represents a good compromise whose performance on average is between the breadth-first FP scheme and the PRDFS method.

14.7.2 Distributed Hitting Set Algorithms with Known Conflicts

Several research papers in the MBD literature make the assumption that the set of (non-minimal) conflicts is already given at the beginning of the diagnosis process. Consequently, the diagnosis problem is reduced to the construction of hitting sets.

For such situations, Cardoso and Abreu in [53] suggested a distributed version of their STACCATO algorithm [54], which is based on the popular MapReduce computation scheme [55]. The proposed algorithm computes the minimal hitting sets, and thus the minimal diagnoses, in a distributed manner, given the (non-minimal) set of conflicts as an input.

In every execution step, the algorithm builds a hitting set d' by adding a component to it that hits at least one of the so far *unhit* minimal conflict sets. The selection of the component is done from a queue R that comprises all components that are not in d'. In addition, the elements of R are ranked according to the Ochiai coefficient.

The mapping step implements two functions that assign elements of the queue to one of the n available processes. The mapping can be done with one of two possible functions, *stride* and *random*. The first function assigns elements in a cyclical manner, i.e., process k gets a component l if $(l \mod n) = k - 1$. The second function assigns components by randomly sampling from a uniform distribution.

The authors compared their distributed version of the algorithm with the single-threaded STACCATO method using a variety of artificially created benchmark problems. Their analyses showed that the new algorithm is faster than the previous one in both a distributed and a single-CPU setup. Furthermore, the required additional overhead for distributing the problem across computing nodes seemed to diminish in particular for the large problem instances.

In [56], Zhao and Ouyang suggest two further algorithms that can be used in distributed settings. Given a set of conflicts, the first algorithm starts with the partitioning of this set such that any two conflicts from different partitions share no components. Next, it computes minimal hitting sets for every partition and finds the set of diagnoses. To compute a diagnosis in this set, the algorithm selects one minimal hitting set for every partition and then joins them.

The second proposed algorithm is used in cases when additional conflicts arise after the diagnoses are already computed. The main idea is to update the families of conflict sets with new elements and find diagnoses in a distributed way, which is done in a similar way to the first algorithm.

The parallelization approach in this work relies on the fact that hitting sets of such partitions can be computed in parallel. This resulted in a considerable reduction of the required computation times compared to the single-threaded Boolean [57] and Boolean-HS-Tree [30] algorithms.

14.8 Summary

The computation of possible explanations of an unexpected system behavior using Model-Based Diagnosis approaches can be computationally challenging, in particular in application scenarios in which it is not sufficient to know only *some* heuristically determined diagnoses.

Even though multi-core computers are common today, and in different domains relying on the processing power of Graphics Processing Units has proven to be useful, limited research exists so far on parallel computation approaches in the context of Model-Based Diagnosis.

In this chapter, we have focused on different strategies for parallelizing Reiter's classical HS-tree algorithm on multi-core computers. While the algorithm in principle follows a breadth-first tree search strategy, using the concept of conflicts is essential to prune the search space, which is why the presented parallelization approaches try to maintain the basic character of the algorithm.

We believe that the presented algorithms therefore only represent a first step toward a better usage of the computing power that we have available today. Instead of being parallel versions of existing single-threaded algorithms, future Model-Based Diagnosis techniques should be designed with the concept of parallelization in mind from the beginning.

Acknowledgements

The authors were supported by the Carinthian Science Fund (KWF) under contract KWF-3520/26767/38701, the Austrian Science Fund (FWF) and the German Research Foundation (DFG) under contract numbers I 2144 N-15 and JA 2095/4-1 (Project "Debugging of Spreadsheet Programs").

References

[1] de Kleer, J., Mackworth, A.K., Reiter, R.: Characterizing Diagnoses and Systems. Artificial Intelligence **56**(2-3) (1992) 197–222
[2] de Kleer, J., Williams, B.C.: Diagnosing Multiple Faults. Artificial Intelligence **32**(1) (April 1987) 97–130
[3] Reiter, R.: A Theory of Diagnosis from First Principles. Artificial Intelligence **32**(1) (1987) 57–95
[4] Felfernig, A., Friedrich, G., Jannach, D., Stumptner, M.: Consistency-based Diagnosis of Configuration Knowledge Bases. Artificial Intelligence **152**(2) (2004) 213–234
[5] Mateis, C., Stumptner, M., Wieland, D., Wotawa, F.: Model-Based Debugging of Java Programs. In: AADEBUG'00. (2000)
[6] Jannach, D., Schmitz, T.: Model-based Diagnosis of Spreadsheet Programs: A Constraint-based Debugging Approach. Automated Software Engineering **23**(1) (2016) 105–144
[7] Wotawa, F.: Debugging Hardware Designs Using a Value-Based Model. Applied Intelligence **16**(1) (2001) 71–92
[8] Felfernig, A., Friedrich, G., Isak, K., Shchekotykhin, K.M., Teppan, E., Jannach, D.: Automated Debugging of Recommender User Interface Descriptions. Applied Intelligence **31**(1) (2009) 1–14
[9] Console, L., Friedrich, G., Dupré, D.T.: Model-Based Diagnosis Meets Error Diagnosis in Logic Programs. In: IJCAI'93. (1993) 1494–1501
[10] Friedrich, G., Shchekotykhin, K.M.: A General Diagnosis Method for Ontologies. In: ISWC'05. (2005) 232–246
[11] Stumptner, M., Wotawa, F.: Debugging Functional Programs. In: IJCAI'99. (1999) 1074–1079
[12] Friedrich, G., Stumptner, M., Wotawa, F.: Model-Based Diagnosis of Hardware Designs. Artificial Intelligence **111**(1-2) (1999) 3–39
[13] White, J., Benavides, D., Schmidt, D.C., Trinidad, P., Dougherty, B., Cortés, A.R.: Automated Diagnosis of Feature Model Configurations. Journal of Systems and Software **83**(7) (2010) 1094–1107
[14] Friedrich, G., Fugini, M., Mussi, E., Pernici, B., Tagni, G.: Exception Handling for Repair in Service-Based Processes. IEEE Transactions on Software Engineering **36**(2) (2010) 198–215
[15] Junker, U.: QUICKXPLAIN: Preferred Explanations and Relaxations for Over-Constrained Problems. In: AAAI'04. (2004) 167–172
[16] Marques-Silva, J., Janota, M., Belov, A.: Minimal Sets over Monotone Predicates in Boolean Formulae. In: Computer Aided Verification. (2013) 592–607
[17] Shchekotykhin, K., Jannach, D., Schmitz, T.: MergeXplain: Fast Computation of Multiple Conflicts for Diagnosis. In: IJCAI'15. (2015) 3221–3228
[18] Greiner, R., Smith, B., Wilkerson, R.: A Correction to the Algorithm in Reiter's Theory of Diagnosis. Artificial Intelligence **41**(1) (1989) 79–88

[19] Jannach, D., Schmitz, T., Shchekotykhin, K.: Parallel Model-Based Diagnosis On Multi-Core Computers. Journal of Artificial Intelligence Research (JAIR) **55** (2016) 835–887
[20] Garey, M.R., Johnson, D.S.: Computers and Intractability: A Guide to the Theory of NP-Completeness. W. H. Freeman & Co. (1979)
[21] Eiter, T., Gottlob, G.: The Complexity of Logic-Based Abduction. Journal of the ACM **42**(1) (1995) 3–42
[22] de Kleer, J.: Hitting Set Algorithms for Model-based Diagnosis. In: DX'11. (2011) 100–105
[23] Stern, R., Kalech, M., Feldman, A., Provan, G.: Exploring the Duality in Conflict-Directed Model-Based Diagnosis. In: AAAI'12. (2012) 828–834
[24] Marques-Silva, J., Janota, M., Ignatiev, A., Morgado, A.: Efficient Model Based Diagnosis with Maximum Satisfiability. In: IJCAI'15. (2015) 1966–1972
[25] de Kleer, J., Williams, B.C.: Diagnosing Multiple Faults. Artif. Intell. **32**(1) (apr 1987) 97–130
[26] Williams, B.C., Ragno, R.J.: Conflict-directed A* and its Role in Model-based Embedded Eystems. Discrete Applied Mathematics **155**(12) (2007) 1562–1595
[27] Darwiche, A.: Model-Based Diagnosis using Structured System Descriptions. Journal of Artificial Intelligence Research **8** (1998) 165–222
[28] Siddiqi, S., Huang, J.: Sequential Diagnosis by Abstraction. Journal of Artificial Intelligence Research **41** (2011) 329–365
[29] Darwiche, A.: A Differential Approach to Inference in Bayesian Networks. Journal of the ACM **50**(3) (May 2003) 280–305
[30] Pill, I., Quaritsch, T.: Optimizations for the Boolean Approach to Computing Minimal Hitting Sets. In: ECAI'12. (2012) 648–653
[31] Feldman, A., Provan, G., de Kleer, J., Robert, S., van Gemund, A.: Solving Model-Based Diagnosis Problems with Max-SAT Solvers and Vice Versa. In: DX'10. (2010) 185–192
[32] Metodi, A., Stern, R., Kalech, M., Codish, M.: A Novel SAT-Based Approach to Model Based Diagnosis. Journal of Artificial Intelligence Research **51** (2014) 377–411
[33] Mencia, C., Marques-Silva, J.: Efficient Relaxations of Over-constrained CSPs. In: ICTAI'14. (2014) 725–732
[34] Mencía, C., Previti, A., Marques-Silva, J.: Literal-Based MCS Extraction. In: IJCAI'15. (2015) 1973–1979
[35] Nica, I., Pill, I., Quaritsch, T., Wotawa, F.: The Route to Success: A Performance Comparison of Diagnosis Algorithms. In: IJCAI'13. (2013) 1039–1045
[36] Shchekotykhin, K., Friedrich, G., Fleiss, P., Rodler, P.: Interactive Ontology Debugging: Two Query Strategies for Efficient Fault Localization. Journal of Web Semantics **12–13** (2012) 88–103
[37] Feldman, A., Provan, G., van Gemund, A.: Approximate Model-Based Diagnosis Using Greedy Stochastic Search. Journal of Artifcial Intelligence Research **38** (2010) 371–413
[38] Li, L., Yunfei, J.: Computing Minimal Hitting Sets with Genetic Algorithm. In: DX'02. (2002) 1–4

[39] Ram, D.J., Sreenivas, T.H., Subramaniam, K.G.: Parallel Simulated Annealing Algorithms. Journal of Parallel and Distributed Computing **37**(2) (1996) 207 – 212
[40] Burns, E., Lemons, S., Ruml, W., Zhou, R.: Best-First Heuristic Search for Multicore Machines. Journal of Artificial Intelligence Research **39** (2010) 689–743
[41] Ferguson, C., Korf, R.E.: Distributed Tree Search and its Application to alpha-beta Pruning. In: AAAI'88. (1988) 128–132
[42] Brüngger, A., Marzetta, A., Fukuda, K., Nievergelt, J.: The Parallel Search Bench ZRAM and its Applications. Annals of Operations Research **90**(0) (1999) 45–63
[43] Kalyanpur, A., Parsia, B., Horridge, M., Sirin, E.: Finding All Justifications of OWL DL Entailments. In: ISWC 2007 + ASWC 2007. (2007) 267–280
[44] Previti, A., Ignatiev, A., Morgado, A., Marques-Silva, J.: Prime Compilation of Non-Clausal Formulae. In: IJCAI'15. (2015) 1980–1987
[45] Powley, C., Korf, R.E.: Single-agent Parallel Window Search. IEEE Transactions on Pattern Analysis and Machine Intelligence **13**(5) (1991) 466–477
[46] Anglano, C., Portinale, L.: Parallel Model-based Diagnosis using PVM. In: EuroPVM'96. (1996) 331–334
[47] Wotawa, F.: A Variant of Reiter's Hitting-set Algorithm. Information Processing Letters **79**(1) (2001) 45–51
[48] Phillips, M., Likhachev, M., Koenig, S.: PA*SE: Parallel A* for Slow Expansions. In: ICAPS'14. (2014)
[49] Korf, R.E., Schultze, P.: Large-scale Parallel Breadth-first Search. In: AAAI'05. (2005) 1380–1385
[50] Shchekotykhin, K.M., Friedrich, G., Rodler, P., Fleiss, P.: Sequential Diagnosis of High Cardinality Faults in Knowledge-Bases by Direct Diagnosis Generation. In: ECAI'14. (2014) 813–818
[51] Kurtoglu, T., Feldman, A.: Third International Diagnostic Competition (DXC 11). `https://sites.google.com/site/dxcompetition2011` (2011) Accessed: 2016-03-15.
[52] Prud'homme, C., Fages, J.G., Lorca, X.: Choco Documentation. (2015) `http://www.choco-solver.org`.
[53] Cardoso, N., Abreu, R.: A Distributed Approach to Diagnosis Candidate Generation. In: EPIA'13. (2013) 175–186
[54] Abreu, R., van Gemund, A.J.C.: A Low-Cost Approximate Minimal Hitting Set Algorithm and its Application to Model-Based Diagnosis. In: SARA'09. (2009) 2–9
[55] Dean, J., Ghemawat, S.: MapReduce: Simplified Data Processing on Large Clusters. Communications of the ACM **51**(1) (2008) 107–113
[56] Zhao, X., Ouyang, D.: Deriving All Minimal Hitting Sets Based on Join Relation. IEEE Transactions on Systems, Man, and Cybernetics: Systems **45**(7) (2015) 1063–1076
[57] Lin, L., Jiang, Y.: The computation of Hitting Sets: Review and New Algorithms. Information Processing Letters **86**(4) (2003) 177–184

Part II
Tools and Applications

Chapter 15
Selection and Configuration of Parallel Portfolios

Marius Lindauer, Holger Hoos, Frank Hutter, and Kevin Leyton-Brown

Abstract In recent years the availability of parallel computation resources has grown rapidly. Nevertheless, even for the most widely studied constraint programming problems such as SAT, solver development and applications remain largely focussed on sequential rather than parallel approaches. To ease the burden usually associated with designing, implementing and testing parallel solvers, in this chapter, we demonstrate how methods from automatic algorithm design can be used to construct effective parallel portfolio solvers from sequential components. Specifically, we discuss two prominent approaches for this problem. (I) *Parallel portfolio selection* involves selecting a parallel portfolio consisting of complementary sequential solvers for a specific instance to be solved (as characterised by cheaply computable instance features). Applied to a broad set of sequential SAT solvers from SAT competitions, we show that our generic approach achieves nearly linear speedup on application instances, and super-linear speedups on combinatorial and random instances. (II) *Automatic construction of parallel portfolios via algorithm configuration* involves a parallel portfolio of algorithm parameter configurations that is optimized for a given set of instances. Applied to gold-medal-winning parameterized SAT solvers, we show that our approach can produce significantly better-performing SAT solvers than state-of-the-art parallel solvers constructed by human experts, reducing time-outs by 17% and running time (PAR10 score) by 13% under competition conditions.

Marius Lindauer
University of Freiburg, Germany, e-mail: `lindauer@cs.uni-freiburg.de`

Holger Hoos
University of British Columbia, Canada & Leiden University, The Netherlands, e-mail: `hh@liacs.nl`

Frank Hutter
University of Freiburg, Germany, e-mail: `fh@cs.uni-freiburg.de`

Kevin Leyton-Brown
University of British Columbia, Canada, e-mail: `kevinlb@cs.ubc.ca`

Y. Hamadi und L. Sais (eds.), *Handbook of Parallel Constraint Reasoning*,
https://doi.org/10.1007/978-3-319-63516-3_15

15.1 Introduction

Given the prevalence of multi-core processors and the ready availability of large compute clusters (e.g., in the cloud), parallel computation continues to grow in importance. This is particularly true in the vibrant area of propositional satisfiability (SAT), where over the last decade, parallel solvers have received increasing attention and shown impressive performance in the influential SAT competitions. Nevertheless, development and research efforts remain largely focused on sequential rather than parallel designs; for example, 29 sequential solvers participated in the main track of the 2016 SAT Competition, compared to 14 parallel solvers.

One key reason for this focus on sequential solvers lies in the complexity of designing, implementing and testing effective parallel solvers. This involves a host of challenges, including coordination between threads or processes, efficient communication strategies for information sharing, and non-determinism due to asynchronous computation. As a result, it is typically difficult to effectively parallelise a sequential solver; in most cases, fundamental redesign is required to harness the power of parallel computation. Methods that can produce effective parallel solvers from one or more sequential solvers automatically (or with minimal human effort) are therefore very attractive, even if they cannot generally be expected to reach the performance levels of a carefully hand-crafted parallel solver design. In this chapter, we give an overview of several such automatic approaches. We illustrate these for SAT solvers, in part because SAT is one of the most widely studied NP-hard problems, but also because these approaches, although not limited to SAT solving, were first developed in this context.

One of the simplest automatic methods for constructing a parallel solver is to run multiple sequential solvers independently in parallel on the same input; this is called a *parallel algorithm portfolio*. For SAT, this approach has been applied with considerable success. A well-known example is *ppfolio* [70], which, despite the simplicity of the approach, won several categories of the 2011 SAT Competition; *ppfolio* runs several sequential SAT solvers (including *CryptoMiniSat* [74], *Lingeling* [13], *clasp* [24], *TNM* [54], and *march_hi* [33]) as well as one parallel solver (*Plingeling* [13]) in parallel, without any communication between the solvers, except that all portfolio components are terminated as soon as the first solves the given SAT instance. This works well when the component solvers have complementary strengths. For example, *CryptoMiniSat* and *Lingeling* perform well on application instances, *clasp* excels on "crafted" instances, and *TNM* and *march_hi* are particularly effective on randomly generated SAT instances. The *ppfolio* portfolio was constructed manually by experts with deep insights into the performance characteristics of SAT solvers, drawing from a large set of sequential SAT solvers and using limited computational experiments to assemble hand-picked components into an effective parallel portfolio.

In the following, we focus on generic methods that automate the construction of effective parallel solvers from given sequential components. Such methods can be seen as instances of *programming by optimization* [35] and *search-based software engineering* [30]. There are several advantages to using automatic methods for parallel solver construction: substantially reduced need for rare and costly human expertise; easier exploitation of new component solvers; and easier adaptation to different sets or distributions of problem instances. Broadly speaking, there are two automatic methods for parallel solver construction:[1]

Parallel Portfolio Selection. Parallel portfolio selection focuses on combining a set of algorithms by means of per-instance algorithm selection or algorithm schedules. In per-instance algorithm selection [69, 39, 51], we select one solver from a given set based on features of that instance, with the goal of optimizing performance on the given instance. Per-instance selection can be generalised to produce a parallel portfolio of solvers rather than a single solver [56]; this portfolio consists of sequential solvers that run concurrently on a given problem instance. Algorithm schedules exploit solver complementarity through a sequence of runs with associated time budgets. This strategy can be parallelised by concurrently running multiple sequential schedules, each on a separate processing unit [36].

Automatic Construction of Parallel Portfolios (ACPP). In automated algorithm configuration [45], the goal is to set the parameters of a given algorithm (e.g., a SAT solver) to optimise performance for a given set or distribution of problem instances. Automatic configuration can also be used to determine a set of configurations [79, 50] that jointly perform well when combined into a parallel portfolio [58].

These two approaches address orthogonal problems: the former allows us to effectively use an existing set of solvers for each *instance*, while the latter builds an effective portfolio for a given *instance set* from complementary components drawn from a large (often infinite) configuration space of solvers.

Both of these approaches are based on the assumption that different solvers or solver configurations exhibit sufficient *performance complementarity*: i.e., they differ substantially in efficacy relative to each other depending on the problem instance to be solved. Solver complementarity is known to exist for many NP-hard problems—notably SAT, where it has been studied by Xu et al. [82]—and is also reflected in the excellent performance of many portfolio-based solvers [28, 70, 25, 15, 6]. While in the following we focus on SAT, solver complementarity has also been demonstrated and exploited for a broad range of other problems, including MAXSAT [4], quantified Boolean formulas [68, 53], answer set programming [64], constraint satisfaction [67], AI planning [31, 73], and mixed integer programming [41, 81]; we thus expect that the techniques we describe could successfully be applied to these problems.

This chapter is organized as follows. We discuss parallel portfolio selection in Section 15.2 and automatic construction of parallel portfolios from parameterized solvers in Section 15.3. We conclude the chapter by discussing limitations as well as possible extensions and combinations of the two approaches (Section 15.4). The material in this chapter builds on and extends previously published work on parallel portfolio selection [56] and automatic construction of parallel portfolios [58].

[1] We note that parallel resources can also be used for parallel algorithm configuration [44]; while this is an important area of study, in this chapter, we focus on methods that produce parallel portfolios as an output.

15.2 Per-Instance Selection of Parallel Portfolios

Well-known per-instance algorithm selection systems for SAT include *SATzilla* [65, 80, 83], *3S* [49], *CSHC* [62], and *AutoFolio* [57]. The algorithm portfolios such systems construct have been very successful in past SAT competitions, regularly outperforming the best non-portfolio solvers.[2] Algorithm selection systems perform particularly well on heterogeneous instance sets, for which no single solver (or parameter configuration of a solver) performs well overall [71]. For example, the instance sets used in SAT competitions include problems from packing, argumentation, cryptography, hardware verification, planning, scheduling, and software verification [12].

In *parallel portfolio selection*, we select a set of algorithms to run together in a parallel portfolio. This offers robustness against errors in solver selection, can reduce dependence on instance features, and provides a simple yet effective way of exploiting parallel computational resources.

15.2.1 Problem Statement

Formally, the algorithm selection problem is defined as follows.

Definition 1 (Sequential Algorithm Selection). An instance of the *per-instance algorithm selection problem* is a 4-tuple $\langle I, \mathscr{D}, \mathscr{A}, m \rangle$, where

- I is a set of instances of a problem,
- $\mathscr{D}$ is a probability distribution over I,
- $\mathscr{A}$ is a set of algorithms for solving instances in I, and
- $m : \mathscr{A} \times I \rightarrow \mathbb{R}$ quantifies the performance of algorithm $A \in \mathscr{A}$ on instance $\pi \in I$.

The objective is to construct an *algorithm selector*, i.e., a mapping $\phi : I \rightarrow \mathscr{A}$, such that the expected performance measure $\mathbb{E}_{\pi \sim \mathscr{D}}[m(\phi(\pi), \pi)]$ across all instances is optimised. In this chapter, we will consider a performance measure based on minimizing running time.

The mapping ϕ is computed by extracting features $f(\pi) \in F$ from a given instance π, which are subsequently used to determine the algorithm to be selected [66, 80, 48]; a mapping from this feature space F to algorithms is typically constructed using machine learning techniques. Instance features for algorithm selection must be cheap to compute (normally costing at most a few seconds) to avoid taking too much time away from actually solving the instance.

---

[2] New SAT competition rules limit portfolio systems to two SAT solving engines. Nevertheless, algorithm selection systems have remained quite successful; e.g., *Riss BlackBox* [3] won 3 medals in 2014.

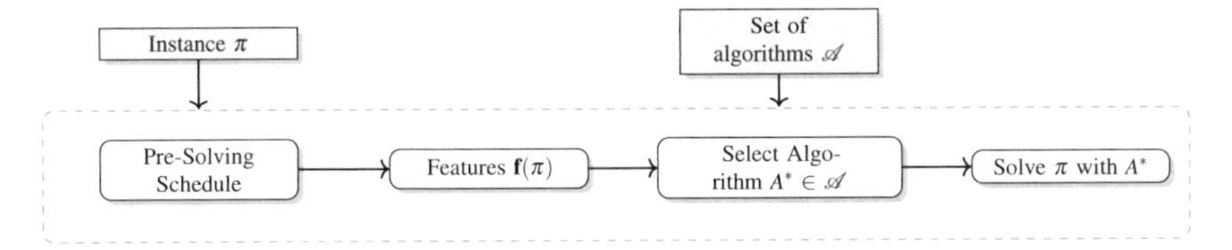

Fig. 15.1: Sequential algorithm selection

Some important examples for instance features include:

- *Size features*, such as the number of variables and clauses, or their ratio [20];
- *CNF graph features* based on the variable-clause graph, variable graph [32], or clause graph;
- *Balance features*, such as the fraction of unary, binary or ternary clauses [66, 80];
- *Proximity to Horn formula features*, such as statistics on horn clauses [66];
- *Survey propagation features*, which estimate variable bias with probabilistic inference [38];
- *Probing features*, which are computed by running, e.g., DPLL solvers, stochastic local search solvers, LP solvers or CDCL solvers for a short amount of time to obtain insights in their solving behavior [66], such as the number of unit propagations at a given search tree depth;
- *Timing features*, the time required to compute other features [48].

For a full list of currently used SAT features, we refer the interested reader to Hutter et al. [48] and to Alfonso et al. [2].

Some performance metrics based on running time penalize solvers for spending seconds to solve instances that can be solved in milliseconds. (A complex performance metric of this type has been used in some past SAT competitions.) In such cases, evaluating features for every instance can lead to unacceptable penalties. Such penalties can be mitigated via static presolving schedules [80, 49, 37]. Based on the observation that many solvers solve a given instance either quickly or not all, a presolving schedule runs a sequence of complementary solvers, each for a small fraction of the overall running time cutoff. If the given instance is solved in any of these runs, the remainder of the presolving and algorithm selection workflow is skipped. Furthermore, the presolving schedule is static, meaning that it does not vary between instances. Beyond saving the time to compute features, static presolving schedules also have another benefit: by running more than the finally selected algorithm, to some degree we hedge against suboptimal selection outcomes based on instance features.

Parallel portfolio selection takes this idea further, selecting a whole set of solvers to run in parallel. Thus, instead of learning a single mapping $\phi : I \rightarrow \mathscr{A}$ to select a solver, we learn a mapping $\phi_k : I \rightarrow \mathscr{A}^k$ to select a portfolio with k components for a given number of processing units k.

Formally, the parallel portfolio selection problem [56] is defined as follows.

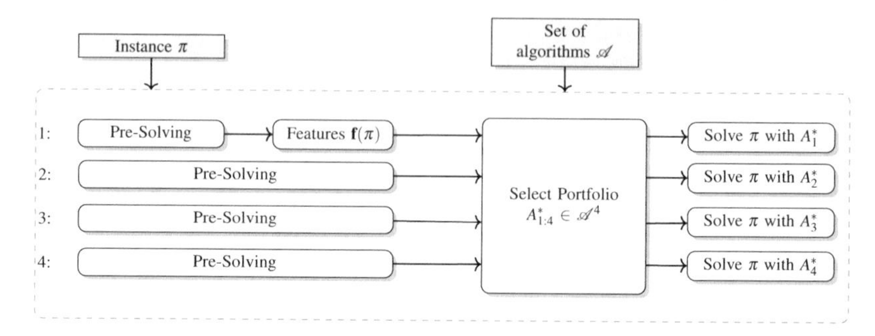

Fig. 15.2: Parallel portfolio selection with presolving on four processing units

Definition 2 (Parallel Portfolio Selection). An instance of the *per-instance parallel portfolio selection problem* is a 5-tuple $\langle I, \mathscr{D}, \mathscr{A}, m, k \rangle$, where

- I is a set of instances of a problem,
- $\mathscr{D}$ is a probability distribution over I,
- $\mathscr{A}$ is a set of algorithms for instances in I,
- k is the number of available processing units, and
- $m : \mathscr{A}^l \times I \to \mathbb{R}$ quantifies the performance of a portfolio $A_{1:l}$ on an instance $\pi \in I$ for any given portfolio size l.

The objective is to construct a *parallel portfolio selector*, i.e., a mapping $\phi_k : I \to \mathscr{A}^k$, such that the expected performance measure $\mathbb{E}_{\pi \sim \mathscr{D}}[m(\phi_k(\pi), \pi)]$ across all instances is optimised. If the concurrently running algorithms in the selected portfolio neither interact nor communicate, the objective can be written as $\mathbb{E}_{\pi \sim \mathscr{D}}[\text{MIN}_{A^* \in \phi_k(\pi)} m(A^*, \pi)]$.

As in the case of selecting a single solver, we can extend parallel portfolio selection to include a static presolving schedule. Figure 15.2 shows the workflow of a parallel portfolio selection procedure. First, we run a parallel presolving schedule on all processing units. Since feature computation is currently still a sequential process, we run a short presolving schedule on the first unit and then start feature computation if necessary. On all other units, we presolve until feature computation finishes. We then select an algorithm for each processing unit.

15.2.2 Parallelization of Sequential Algorithm Selectors

We now discuss a general strategy for parallelizing sequential algorithm selection methods. This approach is motivated by the availability of a broad range of effective sequential selection approaches that use an underlying scoring function $s : \mathscr{A} \times I \to \mathbb{R}$ to rank the candidate algorithms for a given instance to be solved, such that the putatively best algorithm receives the lowest score value, the second best the second lowest score, etc. [52]. The key idea is to use this scoring function to produce portfolios of algorithms to run in parallel by simply sorting the algorithms in $\mathscr{A}$ based on their scores (breaking ties arbitrarily) and choosing the n best-ranked algorithms. In the following, we discuss five existing algorithm selection approaches, their scoring functions and how we can efficiently extend them to parallel portfolio selection.

15.2.2.1 Performance-Based Nearest Neighbor (PNN)

The algorithm selection approach in *3S* [62] in its simplest form uses a k-nearest neighbour approach. For a new instance π with features $\mathbf{f}(\pi)$, it finds the k nearest training instances $I_k(\pi)$ in the feature space F and selects the algorithm that has the best training performance on them. Formally, given a performance metric $m : \mathscr{A} \times I \rightarrow \mathbb{R}$, we define $m_k(A,\pi) = \sum_{\pi' \in I_k(\pi)} m(A,\pi')$ and select algorithm $\arg\min_{A \in \mathscr{A}} m_k(A,\pi)$.

To extend this approach to parallel portfolios, we determine the same k nearest training instances $I_k(\pi)$ and simply select the n algorithms with the best performance on I_k. Formally, our scoring function in this case is simply

$$s_{PNN}(A,\pi) = m_k(A,\pi). \quad (15.1)$$

In terms of complexity, identifying the k nearest instances costs time $O(\#f \cdot |I| \cdot \log|I|)$, with $\#f$ denoting the number of used instance features; averaging the performance values over the k instances costs time $O(k \cdot |\mathscr{A}|)$.

15.2.2.2 Distance-Based Nearest Neighbor (DNN)

ME-ASP [64] implements an interface for different machine learning approaches used in its selection framework, but its released version uses a simple nearest neighbour approach with neighbourhood size 1, which also worked best empirically in experiments by the authors of *ME-ASP* [64]. At training time, this approach memorizes the best algorithm $A^*(\pi')$ on each training instance $\pi' \in I$. For a new instance π, it finds the nearest training instance π' in the feature space using Euclidean distance and selects the algorithm $A^*(\pi')$ associated with that instance.

To extend this approach to parallel portfolios, for a new test instance π, we score each algorithm A by the minimum of the distances between π and any training instance associated with A. Formally, letting $d(\mathbf{f}(\pi),\mathbf{f}(\pi'))$ denote the distance in feature space between instance π and π', we have the following scoring function

$$s_{DNN}(A,\pi) = \text{MIN}\{d(\mathbf{f}(\pi),\mathbf{f}(\pi')) \mid \pi' \in I \wedge A^*(\pi') = A\}. \quad (15.2)$$

Intuitively, an algorithm is preferred to be in a parallel portfolio if it performed best on a problem instance similar to the instance at hand (where similarity is judged by Euclidean distance in feature space).

If $\{\pi' \in I \mid A^*(\pi') = A\}$ is empty (because algorithm A was never the best algorithm on an instance) then $s_{DNN}(A,\pi) = \infty$ for all instances π. Since we memorize the best algorithm for each instance in the training phase, the time complexity of this method is dominated by the cost of computing the distance of each training instance to the test instance, $O(|I| \cdot \#f)$, where $\#f$ is the number of features.

15.2.2.3 Clustering

The selection part of *ISAC* [50][3] uses a technique similar to *ME-ASP*'s distance-based NN approach, with the difference that it operates on clusters of training instances instead of on single instances. Specifically, *ISAC* clusters the training instances, memorizing the cluster centers Z (in the feature space) and the best algorithms $\hat{A}(z)$ for each cluster $z \in Z$. For a new instance, similarly to *ME-ASP*, it finds the nearest cluster z in the feature space and selects the algorithm associated with z.

To extend this approach to parallel portfolios, for a new test instance π, we score each algorithm A by the minimum of the distances between π and the clusters associated with A. Formally, using $d(\mathbf{f}(\pi), z)$ to denote the distance in feature space between instance π and cluster center z, we have the following scoring function:

$$s_{Clu}(A,\pi) = \text{MIN}\{d(\mathbf{f}(\pi),z) \mid z \in Z \wedge \hat{A}(z) = A\}. \tag{15.3}$$

As for DNN, if $\{z \in Z \mid \hat{A}(z) = A\}$ is empty (because algorithm A was not the best algorithm on any cluster) then $s_{Clu}(A,\pi) = \infty$ for all instances π. The time complexity is as for DNN, replacing the number of training instances $|I|$ with the number of clusters $|Z|$.

15.2.2.4 Regression

The first version of *SATzilla* [65] used a regression approach, which, for each $A \in \mathscr{A}$, learns a regression model $r_A : F \to \mathbb{R}$ to predict performance on new instances. For a new instance π with features $\mathbf{f}(\pi)$, it selected the algorithm with the best predicted performance, i.e., $\arg\min_{A \in \mathscr{A}} r_A(\mathbf{f}(\pi))$.

This approach trivially extends to parallel portfolios; we simply use scoring function

$$s_{Reg}(A,\pi) = r_A(\mathbf{f}(\pi)) \tag{15.4}$$

to select the A algorithms predicted to perform best. The complexity of model evaluations differs across models, but it is polynomial for all models in common use; we denote this polynomial by P_{reg}. Since we need to evaluate one model per algorithm, the time complexity to select a parallel portfolio is then $O(P_{reg} \cdot |\mathscr{A}|)$.

---

[3] In its original version, *ISAC* is a combination of algorithm configuration and selection, but only the selection approach was used in later publications.

15.2.2.5 Pairwise Voting

The most recent *SATzilla* version [82] uses cost-sensitive random forest classification to learn for each pair of algorithms $A_1 \neq A_2 \in \mathscr{A}$ which of them performs better for a given instance; each such classifier $c_{A_1,A_2} : F \to \{0,1\}$ votes for A_1 or A_2 to perform better, and *SATzilla* then selects the algorithms with the most votes from all pairwise comparisons. Formally, let $v(A,\pi) = \sum_{A' \in \mathscr{A} \setminus \{A\}} c_{A,A'}(\mathbf{f}(\pi'))$ denote the sum of votes algorithm A receives for instance π; then, *SATzilla* selects $\arg\max_{A \in \mathscr{A}} v(\pi, A)$.

To extend this approach to parallel portfolios, we simply select the n algorithms with the most votes by defining our scoring function to be minimized as

$$s_{Vote}(A,\pi) = -v(A,\pi). \tag{15.5}$$

As for regression models, the time complexity for evaluating a learned classifier differs across classifier types, but it is polynomial for all commonly used types, in particular random forests; we denote this polynomial function by P_{class}. Since we need to evaluate pairwise classifiers for all algorithm pairs, the time complexity to select a parallel portfolio is then $O(P_{class} \cdot |\mathscr{A}|^2)$.

15.2.3 Parallel Presolving Schedules

As mentioned previously, our approach for parallel portfolios does not only consider parallel portfolios selected on a per-instance basis, but also uses parallel presolving schedules (see Figure 15.2). Fortunately, Hoos et al. [36] already proposed an effective system to compute a static parallel schedule for a given set of instances, called *aspeed*. This system is based on an answer set programming (ASP) encoding of the NP-hard problem of algorithm scheduling, and we only need to add one further constraint in this encoding to shorten the schedule in the first processing unit to allow for feature computation. We approximate the required time for feature computation by the allowed upper bound.

Computationally, it is not a problem that finding the optimal algorithm schedule is NP-hard, since this step is performed offline during training and not online in the solving process. Furthermore, the empirical results of Hoos et al. [36] indicated that the problem of optimizing parallel schedules gets easier with more processing units such that it also scales well with an increasing number of processing units.

15.2.4 Empirical Study on Satisfiability Benchmarks

To study the performance of our selected parallel portfolios, we show results on the SAT scenarios of the algorithm selection library (ASlib [17]). ASlib scenarios define a cross validation split scheme, i.e., the instances are split into 10 equally sized

Scenario	$\lvert I \rvert$	$\lvert U \rvert$	$\lvert \mathscr{A} \rvert$	$\#f$	$\#f_g$	$\varnothing t_f$	t_c	Ref.
SAT11-INDU	300	47	18	115	10	135.3	5000	[82, 48]
SAT11-HAND	296	77	15	115	10	41.2	5000	[82, 48]
SAT11-RAND	600	108	9	115	10	22.0	5000	[82, 48]
SAT12-INDU	1167	209	31	115	10	80.9	1200	[83, 48]
SAT12-HAND	767	229	31	115	10	39.0	1200	[83, 48]
SAT12-RAND	1362	322	31	115	10	9.0	1200	[83, 48]

Table 15.1: The *ASlib* algorithm selection scenarios for SAT solving – information on the number of instances $|I|$, number of unsolvable instances $|U|$ ($U \subset I$), number of algorithms $|\mathscr{A}|$, number of features $\#f$, number of feature groups $\#f_g$, the average feature computation cost of the used default features $\varnothing t_f$, and running time cutoff t_c

subsets, and in each iteration, one of the splits is used as a test set and the remaining ones are used as a training set.

In particular, we study the performance of parallel portfolio selection systems on two different SAT scenarios. As in the SAT competitions, each scenario is divided into application, crafted (a.k.a. handmade or hard combinatorial) and random tracks.

1. *SAT11**. The SAT11 scenarios consider the SAT solvers, instances and measured runtimes from the SAT Competition 2011. As features, we used the features from the *SATzilla* [83] feature generator.
2. *SAT12**. The SAT12 scenarios include all instances used in competitions prior to the SAT Competition 2012; the solvers are from all tracks of the previous SAT Competition 2011. The instance features are the same as in the SAT11 scenarios. The data was used to train *SATzilla* [83] for the SAT Competition 2012.

To run these experiments, we extended the flexible algorithm selection framework *claspfolio 2* [37] (see also Chapter 7, Parallel Answer Set Programming) to parallel portfolio selection.

Table 15.1 shows the details of the used scenarios. The main differences are that the SAT11 scenarios have fewer instances and fewer algorithms with a larger running time budget in comparison to the SAT12 scenarios. Comparing the different tracks, the time to compute the instance features is largest for industrial instances, followed by crafted instances and random instances. However, in our experiments we use only the 54 "base" features that do not include any probing features and are much cheaper to compute.

Table 15.2 shows the speedup of our parallel portfolio selection approaches based on PAR10 scores[4] depending on the number of processing units k. Since all approaches have different sequential performance, we use the performance of the sequential single best solver (*SB*, i.e., the solver with the best performance across all training instances) as the baseline for the speedup computation; for example,

[4] PAR10 [45] is the penalized average running time, counting each timeout as 10 times the running time cutoff.

k	1	2	4	8
SAT11-INDU (*VBS*: 21.4)				
PNN	1.1	1.5	2.6	**5.2**
DNN	1.4	**1.9**	**2.6**	**7.8**
clustering	1.3	**1.9**	**3.3**	**5.3**
pairwise-voting	**2.0**	**2.4**	**3.6**	**4.7**
regression	1.3	**2.0**	**3.6**	**7.8**
SB	1.0	1.7	**2.9**	**7.2**
SAT11-HAND (*VBS*: 37.2)				
PNN	2.3	2.8	**8.4**	10.8
DNN	**3.2**	**5.2**	**9.6**	**23.9**
clustering	1.6	2.9	4.2	7.0
pairwise-voting	**3.4**	**4.8**	**8.6**	**10.9**
regression	**2.9**	**4.5**	**8.4**	**12.5**
SB	1.0	1.2	1.9	6.2
SAT11-RAND (*VBS*: 65.7)				
PNN	**6.5**	**9.3**	10.7	60.2
DNN	3.8	**11.0**	**42.2**	60.5
clustering	**6.1**	**9.5**	32.3	42.7
pairwise-voting	4.4	8.3	11.4	60.4
regression	5.9	7.8	8.3	60.3
SB	1.0	5.9	6.8	**64.8**

k	1	2	4	8
SAT12-INDU (*VBS*: 15.4)				
PNN	1.6	2.3	**3.9**	**5.7**
DNN	2.0	2.4	**3.4**	**5.0**
clustering	1.3	2.1	2.8	4.6
pairwise-voting	**2.4**	**3.0**	**3.8**	**5.4**
regression	1.9	2.5	**3.5**	**6.3**
SB	1.0	1.5	2.5	4.8
SAT12-HAND (*VBS*: 34.7)				
PNN	2.0	2.8	4.9	7.5
DNN	**3.7**	**6.2**	**11.4**	**14.3**
clustering	1.8	2.3	3.3	4.6
pairwise-voting	**4.2**	**5.4**	**9.0**	**12.4**
regression	2.9	4.2	7.0	**9.8**
SB	1.0	1.0	1.4	1.9
SAT12-RAND (*VBS*: 12.1)				
PNN	**1.2**	**2.1**	**4.8**	**7.3**
DNN	0.8	1.5	**4.7**	**8.6**
clustering	**1.3**	1.7	2.7	4.9
pairwise-voting	1.1	1.7	2.8	6.4
regression	**1.3**	**1.8**	**5.2**	8.3
SB	1.0	1.5	**4.0**	**6.8**

Table 15.2: Speedup on PAR10 (wallclock) in comparison to *SB* with one processing unit ($k = 1$). Entries shown in bold-face are statistically indistinguishable from the best speedups obtained for the respective scenario and number of processing units (according to a permutation test with 100 000 permutations and $\alpha = 0.05$)

a speedup of 1.0 corresponds to the same performance as the *SB*. We applied a paired statistical test (i.e., a permutation test) with significance level $\alpha = 0.05$ to mark statistically indistinguishable performance from the best-performing system for each number of processing units. We note that algorithm selection ($k = 1$) already performs better than the *SB* in all settings except DNN on *SAT12-RAND*.

Since we do not consider clause sharing in our experiments, the maximal possible speedup is limited by the virtual best solver (*VBS*, i.e., running the best solver for each instance, or running all available solvers in parallel). The performance of the *VBS* depends on the complementarity of the component solvers. The set of all SAT solvers in a SAT competition tends to be quite complementary [82], but since this complementarity is not always the same across different instance sets and available algorithms, the maximal speedup that can be achieved differs between the scenarios. The extremes in our experiments were *SAT11-RAND* with a maximal speedup factor of 65.7 using 8 cores and *SAT12-RAND* with "only" a speedup factor of 12.1 using 8 cores.

Overall, the speedups were quite large (sometimes superlinear, particularly for the random and crafted instances) and there was no clear winner amongst the different approaches. On the industrial and crafted scenarios, the pairwise-voting approaches from *SATzilla* [83] and DNN performed consistently well. Surprisingly, in contrast, on the random instances pairwise-voting had amongst the worst performances, but simply selecting statically the n best-performing solvers (*SB*) from the training instances performed well.[5] We note that the performance with 8 processing units on *SAT11-RAND* nearly saturates, since we select 8 out of the 9 available solvers.

15.2.5 Other Parallel Portfolio Selection Approaches

A relevant medal-winning system in the SAT Competition 2013 was the parallel portfolio selector *CSHCpar* [63], which is based on the algorithm selection of cost-sensitive hierarchical clustering (*CSHC* [62]). It always runs, independently and in parallel, the parallel SAT solver *Plingeling* with 4 threads, the sequential SAT solver *CCASat*, and three solvers that are selected on a per-instance basis. These per-instance solvers are selected by three models that are trained on application, crafted and random SAT instances, respectively. While *CSHCpar* is particularly designed for the SAT Competition with its 8 available cores, it does not provide an obvious way of adjusting the number of processing units and does not support use cases without explicitly identified, distinct instance classes (such as industrial, crafted and random).

The extension of *3S* [49] to parallel portfolio selection, dubbed *3Spar* [61], selects a parallel portfolio using k-NN to find the k most similar instances in instance feature space. Using integer linear programming (ILP), *3Spar* constructs a static presolving schedule offline and a per-instance parallel algorithm schedule online, based on

[5] We note that the solvers in SAT*-RAND are randomized, but the scenarios in ASlib do not reflect this; thus, these performance estimates are probably optimistic [19].

training data of the k most similar instances. The ILP problem that needs to be solved for every instance is NP-hard and its time complexity grows exponentially with the number of parallel processing units and the number of available solvers. Unlike our approach, during the feature computation phase, *3Spar* runs in a purely sequential manner. Since feature computation can require a considerable amount of time (e.g., more than 100 seconds on industrial SAT instances), this can leave important performance potential untapped.

EISAC [60] clusters the training instances in the feature space and provides a method for selecting parallel portfolios for each cluster of instances by searching over all $\binom{|\mathscr{A}|}{k}$ combinations of $|\mathscr{A}|$ algorithms and k processing units. As this approach quickly becomes infeasible for growing $|\mathscr{A}|$ and k, Yuri Malitsky, author of *EISAC*, recommends to limit its use to at most 4 processing units (README file[6]).

The *aspeed* system [36] solves a similar scheduling problem to *3Spar*, but generates a static algorithm schedule during an offline training phase, thus avoiding overhead in the solving phase. Unlike *3Spar*, *aspeed* does not support including parallel solvers in the algorithm schedule, and the algorithm schedule is static and not selected on a per-instance basis. For this reason, *aspeed* is not directly applicable to per-instance selection of parallel portfolios; however, our approach uses it to effectively compute parallel presolving schedules.

RSR-WG [84] combines a case-based-reasoning approach from *CP-Hydra* [67] with greedy construction of parallel portfolio schedules via *GASS* [75] for CSPs. Since the schedules are constructed on a per-instance basis, *RSR-WG* relies on instance features. In the first step, a schedule is greedily constructed to maximize the number of instances solved within a given cutoff time, and in the second step, the components of the schedule are distributed over the available processing units. In contrast to our approach, *RSR-WG* optimizes the number of timeouts and is not directly applicable to arbitrary performance metrics. Since the schedules are optimized online on a per-instance base, *RSR-WG* has to solve an NP-hard problem for each instance, which is done heuristically. Finally, there are also different possible extensions of algorithm schedules to per-instance schedules [49, 55], which aim to select an algorithm schedule on an instance-by-instance basis.

15.3 Automatic Construction of Parallel Portfolios from Parameterized Solvers

So far, we have assumed that we are given a set of solvers for a given problem, such as SAT, and that for a problem instance to be solved, we select a subset of these solvers to be run as a parallel portfolio. Now, we focus on a different approach for constructing parallel portfolios, starting from the observation that solvers for computationally challenging problems typically expose parameters, whose settings can have a very substantial impact on performance. For SAT solvers, these parameters control key

[6] https://sites.google.com/site/yurimalitsky/downloads

aspects of the underlying search process (e.g., the variable selection mechanism, clause deletion policy and restart frequency); by choosing their values specifically for a given instance set, performance can often be increased by orders of magnitude over that obtained using default parameter settings [40, 45, 43, 23, 47]. The task of automatically determining parameter settings such that performance on a given instance set is optimised is known as *algorithm configuration* [45].

Based on the success of algorithm selection and configuration systems, we conjecture that there is neither a single best algorithm nor a single best parameter configuration for all possible instances. Therefore, by combining complementary parameter configurations into a parallel portfolio solver more robust behaviour can be achieved on a large variety of instances. In fact, many parallel SAT solvers already exploit this idea by using different parameter settings in different threads, e.g., *ManySAT* [28], *clasp* [25] or *Plingeling* [16]. However, these portfolios are hand-designed, which requires a tedious, error-prone and time-consuming manual parameter optimization process.

Combining the ideas of parallel portfolios of different parameter settings and automatic algorithm configuration leads to our approach of *automatic construction of parallel portfolios* (ACPP). In its simplest form, the only required input is a single parameterized sequential SAT solver. Using an automatic algorithm configuration procedure, we determine a set of complementary parameter configurations and run them in parallel to obtain a robust and efficient parallel portfolio solver.

15.3.1 Problem Statement

The traditional algorithm configuration task consists of determining a parameter configuration with good performance on a set of instances from the configuration space of a given algorithm. Formally, this gives rise to the following problem.

Definition 3 (Algorithm Configuration; AC). An instance of the *algorithm configuration problem* is a 6-tuple $(A,\Theta,I,\mathscr{D},\kappa,m)$, where

- A is a parameterized target algorithm,
- Θ is the parameter configuration space of A,
- I is a set of instances of a problem,
- $\mathscr{D}$ is a probability distribution over I,
- $\kappa \in \mathbb{R}^+$ is a cutoff time, after which each run of A will be terminated if still running, and
- $m : \Theta \times I \to \mathbb{R}$ quantifies the performance of configuration $\theta \in \Theta$ on instance $\pi \in I$ w.r.t. a given cutoff time κ.

The objective is to determine a configuration $\theta^* \in \Theta$ that achieves near-optimal performance across instances $\pi \in I$ drawn from $\mathscr{D}$. As in the previous section, we consider a performance measure based on running time, which we aim to minimise; therefore, we aim to determine $\theta^* \in \arg\min_{\theta\in\Theta} \mathbb{E}_{\pi\sim\mathscr{D}}[m(\theta,\pi)]$.

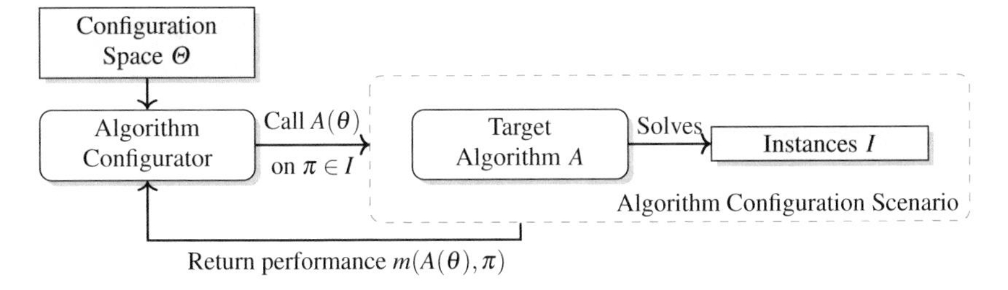

Fig. 15.3: Algorithm configuration workflow

The workflow of algorithm configuration is visualized in Figure 15.3. An AC procedure iteratively determines algorithm runs to be performed by selecting appropriate pairs of configurations $\langle \theta$ and instances $\pi \rangle$, executing the corresponding algorithm runs, and observing their performance measurements. Finally, after a given configuration budget—usually a given amount of computing time—has been exhausted, the AC procedure returns its *incumbent* parameter configuration $\hat{\theta}$ at that time, i.e., its best known configuration.

For several reasons, AC is a challenging problem. First, the only mode of interaction between the AC procedure and the target algorithm A is to run A on some instances and observe its performance. Thus, A is treated as a black box, and no specific knowledge about its inner workings can be directly exploited. As a result, automatic algorithm configuration procedures are broadly applicable, but have to work effectively with very limited information.

Second, the configuration space of many solvers is large and complex. These spaces typically involve parameters of different types, such as categorical and continuous parameters, and often exhibit structure in the form of conditional dependencies between parameters or forbidden parameter configurations. For example, the configuration space of *Lingeling* [15] in the configurable SAT solver challenge [47] had 241 parameters giving rise to 10^{974} possible parameter configurations.

Third, particularly when solving *NP*-hard problems (such as SAT), even a single evaluation of a target algorithm configuration on one problem instance can be costly in terms of running time. Therefore, AC procedures typically can only evaluate a small number of pairs $\langle \theta, \pi \rangle$ in a high-dimensional search space—often, only hundreds (and sometimes, thousands) of evaluations are possible even within typical configuration budgets of 12–48 hours of computing time.

Nevertheless, in recent years, algorithm configuration systems have been able to substantially improve the performance of SAT solvers on many types of SAT instances [47]. Well-known algorithm configuration systems include (i) *ParamILS* [46, 45], which performs iterated local search in the configuration space; (ii) *GGA* [5, 4], which is based on a genetic algorithm; (iii) *irace* [59], which uses F-race [9] for racing parameter configurations against each other; and (iv) *SMAC* [43, 42], which makes use of an extension of Bayesian optimization [18] to handle potentially heterogeneous sets of problem instances. For some more details on the mechanisms used in these configuration procedures, we refer the interested reader to the report on the Configurable SAT Solver Challenge [47].

Our extension of algorithm configuration to parallel problem solving is called parallel portfolio construction. The task consists of finding a parallel portfolio $\theta_{1:k}$ of k parameter configurations whose performance (e.g., wallclock time) is evaluated by the first component of $\theta_{1:k}$ that solves a given instance π. We formally define the problem as follows.

Definition 4 (Parallel Portfolio Construction). An instance of the *parallel portfolio construction problem* is a 7-tuple $(A, \Theta, I, \mathscr{D}, \kappa, m, k)$, where

- A is a parameterized target algorithm,
- Θ is the parameter configuration space of A,
- I is a set of problem instances,
- $\mathscr{D}$ is a probability distribution over I,
- $\kappa \in \mathbb{R}^+$ is a cutoff time, after which each run of A will be terminated if still running,
- k is the number of available processing units, and
- $m : \Theta^l \times I \rightarrow \mathbb{R}$ quantifies the performance of a portfolio $\theta_{1:l} \in \Theta^l$ on instance $\pi \in I$ w.r.t. a given cutoff time κ for any given portfolio size l.

The objective is to construct a parallel portfolio $\theta^*_{1:k} \in \Theta^k$ from the k-fold configuration space Θ^k that optimizes the expected performance across instances $\pi \in I$ drawn from $\mathscr{D}$; in the case of minimising a performance metric based on running time, as considered here, we aim to find

$$\theta^*_{1:k} \in \underset{\theta_{1:k} \in \Theta^k}{\arg\min} \, \mathbb{E}_{\pi \sim \mathscr{D}} \left[m(\theta_{1:k}, \pi) \right].$$

If the configurations in the portfolio $\theta^*_{1:k}$ are run independently, without any interaction (e.g., in the form of clause sharing), and the overhead from running configurations in parallel is negligible, this is identical to identifying

$$\theta^*_{1:k} \in \underset{\theta_{1:k} \in \Theta^k}{\arg\min} \, \mathbb{E}_{\pi \sim \mathscr{D}} \left[\mathrm{MIN}_{i \in \{1 \ldots k\}} m(\theta_i, \pi) \right].$$

Compared to algorithm configuration, parallel portfolio construction involves even larger configuration spaces. For a portfolio with k parameter configurations, an algorithm configuration procedure has to search in a space induced by k times the number of parameters of A, and therefore of total size $|\Theta|^k$.

15.3.2 Automatic Construction of Parallel Portfolios (ACPP)

In the following, we explain two methods to address automatic construction of parallel portfolios (ACPP). Since this problem is an extension of the algorithm configuration

---

Algorithm 15.1: Portfolio Configuration Procedure GLOBAL

Input : parametric algorithm with configuration space Θ; desired number k of component solvers; instance set I; performance metric m; configuration procedure AC; number n of independent configurator runs; total configuration time t

Output : parallel portfolio solver with portfolio $\hat{\theta}_{1:k}$

1 **for** $j := 1 \ldots n$ **:**

2 obtain portfolio $\theta_{1:k}^{(j)}$ by running AC for time t/n on configuration space Θ^k on I using m

3 choose $\hat{\theta}_{1:k} \in \arg\min_{\theta_{1:k}^{(j)} | j \in \{1 \ldots n\}} \sum_{\pi \in I} m(\theta_{1:k}^{(j)}, \pi)$ that achieved best performance on I according to m

4 **return** $\hat{\theta}_{1:k}$

---

problem, we consequently build upon an existing algorithm configuration procedure and extend it for ACPP.

15.3.2.1 Multiplying Configuration Space: GLOBAL

Algorithm 15.1 shows the most straightforward method for using algorithm configuration for ACPP. The GLOBAL approach consists of using the algorithm configuration procedure AC on Θ^k, the k-fold Cartesian product of the configuration space Θ.[7] The remaining parts of the procedure follow the standard approach for algorithm configuration: instead of running AC only once with configuration budget t, we perform n runs of AC with a budget of t/n each (Lines 1 and 2). Each of these AC runs ultimately produces one portfolio of size k. Performing these n runs in parallel reduces the wallclock time required for the overall configuration process by leveraging parallel computation. Of the n portfolios obtained from these independent runs, we select the one that performed best on average on the given instance set I (Lines 3 and 4).

In principle, this method can find the best parallel portfolio, but the configuration space grows exponentially with portfolio size k to a size of $|\Theta|^k$, which can become problematic even for small k.

15.3.2.2 Iterative Approach: PARHYDRA

To avoid the complexity of GLOBAL, the iterative, greedy ACPP procedure outlined in Algorithm 15.2 can be used. Inspired by *Hydra* [79, 81], PARHYDRA determines one parameter configuration in each iteration and adds it to the final portfolio. The

---

[7] The product of two configuration spaces X and Y is defined as $\{x||y \mid x \in X, y \in Y\}$, with $x||y$ denoting the concatenation (rather than nesting) of tuples.

Algorithm 15.2: Portfolio Configuration Procedure PARHYDRA

Input : parametric algorithm with configuration space Θ; desired number k of component solvers; instance set I; performance metric m; configurator AC; number n of independent configurator runs; total configuration time t

Output : parallel portfolio solver with portfolio $\hat{\theta}_{1:k}$

1 let θ_{init} be the default configuration in Θ
2 **for** $i := 1 \ldots k$ **:**
3 **for** $j := 1 \ldots n$ **:**
4 obtain portfolio $\theta_{1:i}^{(j)} := \hat{\theta}_{1:i-1} || \theta^{(j)}$ by running AC on configuration space $\{\hat{\theta}_{1:i-1}\} \times \{(\theta) \mid \theta \in \Theta\}$ and initial incumbent $\hat{\theta}_{1:i-1} || \theta_{init}$ on I using m for time $t/(k \cdot n)$
5 let $\hat{\theta}_{1:i} \in \arg\min_{\theta_{1:i}^{(j)} | j \in \{1 \ldots n\}} \sum_{\pi \in I} m(\theta_{1:i}^{(j)}, \pi)$ be the portfolio which achieved the best performance on I according to m
6 let $\theta_{init} \in \arg\min_{\theta^{(j)} | j \in \{1 \ldots n\}} \sum_{\pi \in I} m(\hat{\theta}_{1:i} || \theta^{(j)}, \pi)$ be the configuration that has the largest marginal contribution to $\hat{\theta}_{1:i}$
7 **return** $\hat{\theta}_{1:k}$

configuration to be added is determined such that it best complements the configurations that have previously been added to the portfolio.

In detail, our PARHYDRA approach runs for k iterations (Line 2) to construct a portfolio with k components. In each iteration, we fix one further parameter configuration of our final portfolio. As before, we perform n AC runs in each PARHYDRA-iteration i (Lines 3–5). The configuration space consists of the Cartesian product of the (fixed) portfolio $\hat{\theta}_{1:i-1}$ constructed in the previous $i-1$ iterations with the full configuration space Θ. Each AC run effectively determines a configuration to be added to the portfolio such that the overall portfolio performance is optimised. As configuration budget, each AC run is allocated $t/(k \cdot n)$, where t is the overall budget.

An extension in comparison to *Hydra* [79, 81] is that the initial parameter configuration θ_{init} for the search is adapted in each iteration. For the first iteration, we simply use the default parameter configuration (Line 1)—if no default parameter configuration is known, we could simply use the mean parameter value from the parameter domain ranges or randomly sample an initial configuration. At the end of each iteration, we determine which returned parameter configuration $\theta^{(j)}$ from the last n AC runs ($j \in \{1, \ldots, n\}$) would improve the current portfolio $\hat{\theta}_{1:i}$ the most (Line 6). This configuration is used to initialize the search in the next iteration. This avoids discarding all of each iteration's unselected configurations, keeping at least one to guide the search in future iterations.[8]

[8] Note that this strategy assumes multiple configuration runs per iteration (e.g., n independent runs of a sequential algorithm configuration procedure) and would not be directly applicable if we used a parallel algorithm configuration procedure [44] that only returned a single configuration. Whether one can gain more from using parallel algorithm configuration or from having a good initializiation strategy is an open question.

	Lingeling ala (application)			*clasp (hard combinatorial)*		
Solver Set	#TOs	PAR10	PAR1	#TOs	PAR10	PAR1
DEFAULT-SP	72	2317	373	137	4180	481
CONFIGURED-SP	68	2204	368	140	4253	473
DEFAULT-MP(8)-CS	64	2073	345	96	2950	358
DEFAULT-MP(8)+CS	**53***	**1730***	**299***	**90***	**2763***	**333***
GLOBAL-MP(8)	**52***	**1702***	**298***	98	3011	365
PARHYDRA-MP(8)	**55***†	**1788***†	**303***†	**96***†	**2945***†	**353***†

Table 15.3: Running time statistics on the test set from *application* and *hard combinatorial* SAT instances achieved by single-processor (SP) and 8-processor (MP(8)) versions. DEFAULT-MP(8) was *Plingeling* in case of *Lingeling* and `clasp -t 8` for *clasp* where we show results with (+CS) and without (-CS) clause sharing. The performance of a solver is shown in boldface if it was not significantly different from the best performance, and is marked with an asterisk (*) if it was not significantly worse than DEFAULT-MP(8)+CS (according to a permutation test with 100 000 permutations and significance level $\alpha = 0.05$). The best ACPP portfolio on the training set is marked with a dagger (†)

15.3.2.3 Comparing GLOBAL and PARHYDRA

On the one hand, in comparison to GLOBAL, PARHYDRA has the advantage that it only needs to search the original space Θ in each iteration (in contrast to the exponentially larger $|\Theta|^k$). On the other hand, PARHYDRA has k times less time per iteration, and the configuration tasks may get harder in each iteration because fewer configurations will be complementary for growing portfolio size. It is also not guaranteed that PARHYDRA will find the optimal portfolio because of its greedy nature; for example, if our instance set I consists of two homogeneous subsets $I_1 \cup I_2 = I$, in principle GLOBAL can directly find a well-performing configuration for each of the two subsets. In contrast, PARHYDRA will optimize the configuration on the entire instance set I in the first iteration and can only focus on one of the two subsets in the second iteration. Therefore, PARHYDRA may return suboptimal solutions.

We note, however, that this suboptimality is bounded, since PARHYDRA's portfolio performance is a submodular set function (the effect of adding a further parameter configuration to a smaller portfolio of an early iteration will be larger than adding it to a larger portfolio of a later iteration). This property can be exploited to derive bounds for the performance of *Hydra*-like approaches [73], such as PARHYDRA.

15.3.2.4 Empirical Study on SAT 2012 Challenge

We studied the effectiveness of our two proposed ACPP methods, i.e., GLOBAL and PARHYDRA, on two award-winning solvers from the 2012 SAT Challenge: *Lingeling* [14] and *clasp* [25]. To this end, we compared the default sequential solver settings (DEFAULT-SP), the configured sequential solvers (CONFIGURED-SP), the default parallel counterparts of both solvers (i.e., *Plingeling* for *Lingeling*) without (DEFAULT-MP(8)-CS) and with clause sharing (DEFAULT-MP(8)+CS) and finally, with GLOBAL and PARHYDRA. As instance sets, we used the instances from the application track and hard combinatorial track of the 2012 SAT Challenge for *Lingeling* and *clasp*, respectively. Both instance sets were split into a training set for configuration and test set to obtain an unbiased performance estimate. The parallel solvers used eight processing units ($k = 8$). We used *SMAC* [43, 42], a state-of-the-art algorithm configuration procedure, to minimise penalized average running time, and every configuration approach (i.e., CONFIGURED-SP, GLOBAL and PARHYDRA) was given the same configuration budget t.

Table 15.3 summarizes our results. First of all, we note that algorithm configuration on heterogeneous instance sets, such as the instance sets from SAT competitions and challenges, is challenging, because various instances are solved best by potentially very different configurations, which can pull the search process in different directions. Therefore, the configured sequential version (CONFIGURED-SP) of *Lingeling* performed only slightly better than the default, and the performance of *clasp* even slightly deteriorated due to overtuning [45], i.e., it showed a performance improvement on the training instances that did not generalize to the test instances. The default parallel versions of *Lingeling* and *clasp* performed consistently better than their sequential counterparts. Enabling clause sharing (CS) for both solvers improved their performance even further.

Our automatically constructed parallel portfolio solvers performed well in comparison to the manually hand-crafted parallel solvers. To verify whether the observed performance differences were statistically significant, we used a permutation test to compare the best-performing approach against all others. This analysis revealed that the portfolios manually built by human experts did not perform significantly better than those automatically constructed using PARHYDRA. We emphasize that our ACPP portfolios do not use any clause-sharing strategies, and the configuration process was initialised with the parameter setting of DEFAULT-SP. Therefore, our methods had no hint how to construct a parallel solver. Nevertheless, our automatic approach produced parallel solvers performing as well as those designed manually by experts within a few days of computing time on a small cluster.

15.3.2.5 ACPP with Multiple Solvers

Even though it is appealing to automatically generate a parallel solver from a sequential solver, our ACPP methods are not limited to a single solver as an input. Using more than one solver often increases the opportunity for leveraging performance com-

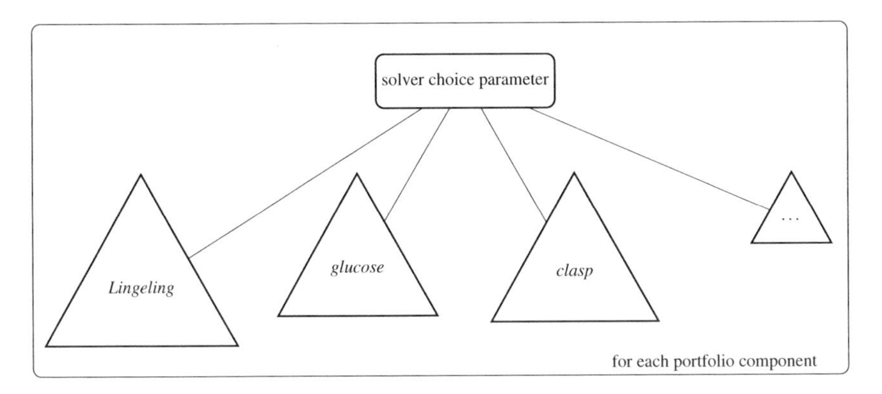

Fig. 15.4: Conditional configuration space involving multiple solvers

plementarities, since SAT solvers often implement complementary strategies [82]. To construct a parallel portfolio solver from a set of parameterized solvers as an input using our ACPP methods, we only need to adapt our configuration space Θ. Following the idea of Thornton et al. [76], we introduce a top-level parameter that indicates which solver to use. The parameters of the individual solvers are then conditionally dependent on this new selector parameter, leading to structured configuration spaces as illustrated in Figure 15.4.

15.3.3 Automatic Construction of Parallel Portfolios from Parallel Parameterized Solvers

The ACPP methods presented thus far always assumed that one or more sequential SAT solvers are given. However, over the course of the last decade, many parallel SAT solvers have been developed (e.g., [28, 72, 25, 8, 16]). On the one hand, these solvers often expose performance-critical parameters (e.g., controlling clause sharing); on the other hand, these solvers can also be used in our ACPP methods as components to include in a parallel portfolio. In the following, we discuss both approaches to further improve the performance of parallel SAT solvers by using algorithm configuration.

15.3.3.1 Configuration of Clause Sharing

Clause sharing is an important strategy to reduce redundant work in parallel SAT solving and hence, to improve the performance of parallel SAT solvers. However, clause sharing also has many open implementation options, e.g., communication topology, how often to share learned clauses, which learned clauses to share, which clauses to integrate in the clause database, etc. The best configuration of these

clasp variant	#TOs	PAR10	PAR1
No Clause Sharing	96	2945	353
Default Clause Sharing	90	2777	347
Configured Clause Sharing	88	2722	346

Table 15.4: Comparison of different clause sharing strategies on top of our constructed PARHYDRA-MP(8) portfolio with *clasp* on the test instances of the *hard combinatorial* set

parameters can have a crucial impact on performance and depends on the nature of the instances to be solved. Therefore, we can use algorithm configuration to optimise the settings of the parameters that control clause sharing.

The results shown in Table 15.4 have been obtained using the same experimental setup as already described in Section 15.3.2.4 for *clasp* on hard-combinatorial instances. As a starting point, we used the parallel *clasp* portfolio found by PARHYDRA—without using any clause sharing. Adding the default clause sharing policy on top of the PARHYDRA portfolio lead to solving 6 more instances, which is equivalent to the performance of DEFAULT-MP(8)+CS (see Table 15.3). However, *clasp* allows adjustment of the clause sharing distribution and integration policies. Using automatic algorithm configuration to optimize these policies, the *clasp* portfolio was able to solve two additional instances. We note that it is suboptimal to first configure a parallel portfolio without any communication between component solvers, and then add clause sharing to the portfolio thus obtained. In principle, configuring the portfolio and the clause sharing mechanism jointly should result in better performance; therefore, the results presented here only give a lower bound on what can be achieved.

15.3.3.2 Portfolio Construction Using Parallel Solvers

Another way of using existing parallel solvers is to allow them to be part of an automated parallel portfolio solver. Similarly to *ppfolio*, we could run a parallel solver, such as *Plingeling*, in some execution threads and some sequential solvers in others. To do this, we can use the trick of adding a top-level parameter to decide between different sequential and parallel solvers (see Section 15.3.2.5). If a parallel solver gets selected l times by top-level parameters of each portfolio component, we merge these components into one call of the parallel solver with l threads. By using this approach, we can directly apply the GLOBAL methods to determine a well-performing automatically constructed parallel portfolio including other parallel SAT solvers.

Unfortunately, PARHYDRA cannot be directly applied to this setting because its reliance on a greedy algorithm makes it suboptimal. For example, if the portfolio $\theta_{1:i}$ already includes a configuration of sequential solver A_s in iteration i, PARHYDRA will never add the parallel counterpart A_p of A_s, because in each iteration, PARHYDRA can

Algorithm 15.3: Portfolio Configuration Procedure PARHYDRA$_b$

Input : set of parametric solvers $A \in \mathscr{A}$ with configuration spaces Θ_A; desired number k of component solvers; number b of component solvers simultaneously configured per iteration; instance set I; performance metric m; configurator AC; number n of independent configurator runs; total configuration time t

Output : parallel portfolio solver with portfolio $\hat{\theta}_{1:k}$

1 $i := 1$
2 let θ_{init} be a portfolio with b times the default configuration in Θ of a default solver $A \in \mathscr{A}$.
3 **while** $i < k$ **:**
4 $\quad i' := i + b - 1$
5 $\quad$ **for** $j := 1..n$ **:**
6 $\quad\quad$ obtain portfolio $\theta^{(j)}_{1:i'} := \hat{\theta}_{1:i-1} || \theta^{(j)}_{i:i'}$ by running AC for time $t \cdot b/(k \cdot n)$ on configuration space $\{\hat{\theta}_{1:i-1}\} \times (\prod^b \bigcup_{A \in \mathscr{A}} \{(\theta) \mid \theta \in \Theta_A\})$ and initial incumbent $\hat{\theta}_{1:i-1} || \theta_{init}$ on I using m
7 $\quad$ let $\hat{\theta}_{1:i'} \in \arg\min_{\theta^{(j)}_{1:i'} | j \in \{1...n\}} \sum_{\pi \in I} m(\theta^{(j)}_{1:i'}, I)$ be the portfolio that achieved best performance on I according to m
8 $\quad$ let $\theta_{init} \in \arg\min_{\theta^{(j)}_{i:i'} | j \in \{1...n\}} \sum_{\pi \in I} m(\hat{\theta}_{1:i'} || \theta^{(j)}_{i:i'}, \pi)$ be the portfolio that has the largest marginal contribution to $\hat{\theta}_{1:i'}$
9 $\quad i := i + b$
10 **return** $\hat{\theta}_{1:k}$

only pick A_p for one thread, which is outperformed by A_s.[9] As a concrete example, let us consider the highly parameterized sequential solver *Lingeling* and its non-parameterized parallel counterpart, *Plingeling*. After a configuration of *Lingeling* was added to the portfolio, PARHYDRA never added *Plingeling* with a single thread in later iterations, because the optimized *Lingeling* outperformed *Plingeling*.

To permit a trade-off between the problems of GLOBAL (exponential increase of the search space) and PARHYDRA (suboptimality in portfolio construction), we propose an extension of PARHYDRA, called PARHYDRA$_b$, which adds not just one, but b configurations to the portfolio in each iteration. Algorithm 15.3 shows an outline of the PARHYDRA$_b$ approach. The main idea is the same as that of PARHYDRA, but we use an additional variable i' to keep track of the parameter configurations added in each iteration. For example, if we have already fixed a portfolio $\theta_{1:4}$ with 4 components and want to add two configurations ($b = 2$) per iteration, we are in iteration $i = 5$, in which we will determine the fifth and sixth configuration $\theta_{i=5:6=5+2-1=i'}$ (Lines 4 and 6) to be added to $\theta_{1:i-1=4}$. Furthermore, the starting point for the configuration process in the following iteration is now obtained by adding a portfolio of size $i' - i + 1 = b$ (Line 8) to $\theta_{1:4}$ from the previous iteration. Other than that, PARHYDRA$_b$ is the same as PARHYDRA.

[9] In principle, one could imagine grouping A_s and A_p to effectively treat them as the same solver, allowing PARHYDRA to add A_p and join this with A_s into a 2-thread version of A_p. However, this kind of grouping is not supported by PARHYDRA.

Solver	#TOs	PAR10	PAR1
Single-threaded solvers: DEFAULT-SP			
glucose 2.1	55	1778	293
Parallel solvers: DEFAULT-MP(8)			
Plingeling(ala)+CS	53	1730	299
pfolioUZK-MP8+CS	35	1168	223
ACPP solvers including a parallel solver			
PARHYDRA-MP(8)	34	1143	225
PARHYDRA$_2$-MP(8)	32	1082	218
PARHYDRA$_4$-MP(8)	**29**	**992**	**209**
GLOBAL-MP(8)	35	1172	227

Table 15.5: Comparison of parallel solvers with 8 processors on the test set of *application*. The performance of a solver is shown in boldface if its performance was at least as good as that of any other solver, up to statistically insignificant differences (according to a permutation test with 100 000 permutations at significance level $\alpha = 0.05$)

15.3.3.3 Empirical Study on 2012 SAT Challenge

Again, we demonstrate the effect of our ACPP methods using parallel SAT solvers on the industrial instance set of the 2012 SAT Challenge. The winning parallel solver in this challenge was *pfolioUZK* [78], a hand-designed portfolio consisting of sequential and parallel portfolios. In particular, *pfolioUZK* uses *satUZK*, *glucose*, *contrasat* and *Plingeling* with 4 threads, leaving one of the 8 available CPU cores unused; however, the set of solvers considered during the design of *pfolioUZK* involved additional solvers that do not appear in the final design. To fairly compare with this manually constructed portfolio, we used the same underlying set of solvers as the starting point for our ACPP methods:

- *contrasat* [26]: 15 parameters;
- *glucose* 2.0 [7]: 10 parameters for *satelite* preprocessing and 6 for *glucose*;
- *Lingeling* 587 [14]: 117 parameters;
- *Plingeling* 587 [14]: 0 parameters;
- *march_hi* 2009 [33]: 0 parameters;
- *MPhaseSAT_M* [21]: 0 parameters;
- *satUZK* [27]: 1 parameter;
- *sparrow2011* [77]: 0 parameters[10]; and
- *TNM* [54]: 0 parameters.

We note that of these, *Plingeling* is the only parallel SAT solver and the only one to make use of clause sharing.

[10] Although *sparrow2011* should be parameterized [77], the source code and binary provided with *pfolioUZK* does not expose any parameters.

Table 15.5 shows the performance of *glucose* 2.1 (which won the main application SAT+UNSAT track of the 2012 SAT Challenge), *Plingeling*(ala) with clause sharing, *pfolioUZK* (which won the parallel application SAT+UNSAT track) and our ACPP methods. Surprisingly, on 8 cores, *Plingeling* performed only slightly better than *glucose*. However, *pfolioUZK* solved 18 instances more than *Plingeling* within the cutoff time used in the competition. By applying GLOBAL (i.e., PARHYDRA$_b$ with $b = k = 8$), we obtained a parallel portfolio performing as well as *pfolioUZK*. PARHYDRA$_b$ with $b = 4$ performed statistically better than *pfolioUZK* by solving 6 instances more.

Looking at the performance achieved by PARHYDRA$_b$ for different values of b reveals that b is an important parameter of our method. One might be concerned that PARHYDRA$_4$-MP(8) performed as well as it did as a result of over-tuning on b. We note, however, that PARHYDRA$_4$-MP(8) also performed best on the training instances used for configuration, which are different from the test instance results shown in Table 15.5.

15.4 Conclusions and Future Work

In this chapter, we presented two generic approaches for automatically generating parallel portfolio solvers for computationally challenging problems from one or more sequential solvers. While our focus was on SAT, the techniques we discussed are in no way specific to this particularly well-studied constraint programming problem, and can be expected to give rise to similarly strong performance when applied to a broad range of CP problems, and, indeed, to many other NP-hard problems. We note that there are three fundamental assumptions that need to be satisfied in order for these generic parallelisation methods to scale well with the number of processing units k.

1. *Performance complementarity:* within a given set $\mathscr{A}$ of solvers that are available (as in algorithm selection) or within the parameter space of a single solver (as in algorithm configuration), there is sufficient performance complementarity. In algorithm selection with deterministic algorithms, algorithm selectors cannot perform better than the virtual best solver (VBS) of the given algorithm portfolio. Therefore, a parallel portfolio selector can scale at most to a number of processing units that equals the number of candidate solvers in $\mathscr{A}$. Unfortunately, this upper bound will usually not be attained, because in most sets $\mathscr{A}$, some solvers will have little or no contribution to the virtual best solver [82].
   In parallel portfolio configuration, the given parameter space Θ is often infinite; still, in our experiments, little or no performance improvement was obtained beyond a modest number of portfolio components (e.g., using PARHYDRA, the performance of our automatically constructed parallel portfolio based on *Lingeling* improved only for the first 4 portfolio components – for details, see [58]). This may indicate that our current approaches are too weak to find better and larger portfolios (since the complexity of the search problems increases with the

size of the portfolio), or that such portfolios simply do not exist for the instance sets we considered, and that the smaller portfolios we found basically exhaust the complementarity of the parameter space. Which of these two explanations holds is an interesting subject for future research.

2. *Heterogeneity of instances:* the given instance set I is sufficiently heterogeneous given a set of solvers or parameter configurations. If the instance set is perfectly homogeneous, a single solver or configuration is dominant on all instances, and a parallel portfolio (without communication between component solvers) cannot perform better. In contrast, if each instance in I requires a different solver or configuration to be solved most effectively, our generic parallel portfolio construction methods can in principle scale to a number of processing units equal to the size of the instance set. Therefore, in practice, the performance potential of these approaches depends on characteristics of the set or distribution of problem instances of interest in a given application context—the more diverse that set, the larger the potential for large speedups due to parallelisation. How to assess the heterogeneity of an instance set in an effective yet computationally cheap way is an open problem.
3. *Minimal interference between runs:* when sequential solvers are run concurrently, there is only minimal impact on performance due to detrimental interference. If each solver runs on a separate system, this assumption can easily be guaranteed, and because neither of our approaches requires much communication between portfolio component solvers, this scenario is quite feasible.
   However, since modern machines are equipped with multi-core CPUs, it is generally desirable to run more than one solver on a single machine with the component solvers sharing resources, such as RAM and CPU cache. Since solver performance can substantially depend on the available CPU cache [1], running several solvers on multiple CPU cores with shared cache can lead to significant slowdown due to cache contention.
   The extent to which this happens depends on the characteristics of the execution environment and on the solvers in question. For example, in the experiments reported in Section 15.3, we observed that *Lingeling* suffered more from this effect on the larger industrial instances than *clasp* did on the smaller crafted instances. Furthermore, we have observed that *Lingeling*'s performance is less affected on newer CPUs with larger amounts of cache. Therefore, we believe that in the future, with the advent of CPUs with even more cache memory, this issue might become less critical.

There are many prominent avenues for future work on generic parallelisation techniques, and we see much promise in the combination of the two approaches discussed in this chapter. For example, one could run PARHYDRA$_b$ to generate many complementary configurations of one or more parameterised solvers and then use parallel portfolio selection on those configurations to create a per-instance parallel portfolio selector for a given number of processing units. Another interesting extension is the automatic configuration of parallel portfolio selectors, analogously to AutoFolio [57]. Similarly, we see promise in the configuration of parallel algorithm schedules, similarly to Cedalion [73]. It might also be interesting to use an approach such as *aspeed* [36] to post-optimize an automatically generated parallel portfolio into a parallel algorithm schedule.

We see substantial promise in exploring instance features specifically designed for parallel portfolio selection, *e.g.*, probing features of parallel solvers possibly related to the communication between solver components. Finally, it would be interesting to improve the construction of portfolios that include randomized parallel component solvers with clause sharing by estimating the potential risks and gains of adding such component solvers based on their running time distributions.

Acknowledgement

M. Lindauer was supported by the DFG (German Research Foundation) under Emmy Noether grant HU 1900/2-1 and project SCHA 550/8-3, H. Hoos and K. Leyton-Brown by NSERC Discovery Grants, K. Leyton-Brown also by an NSERC E.W.R. Steacie Fellowship, and F. Hutter also by the DFG under Emmy Noether grant HU 1900/2-1.

References

[1] Aigner, M., Biere, A., Kirsch, C., Niemetz, A., Preiner, M.: Analysis of portfolio-style parallel SAT solving on current multi-core architectures. In: Berre, D.L. (ed.) Proceedings of the Fourth Pragmatics of SAT workshop. EPiC Series in Computing, vol. 29, pp. 28–40. EasyChair (2014)

[2] Alfonso, E., Manthey, N.: New CNF features and formula classification. In: Berre, D.L. (ed.) Proceedings of the Fifth Pragmatics of SAT workshop. EPiC Series in Computing, vol. 27, pp. 57–71. EasyChair (2014)

[3] Alfonso, E., Manthey, N.: Riss 4.27 BlackBox. In: Belov, A., Diepold, D., Heule, M., Järvisalo, M. (eds.) Proceedings of SAT Competition 2014. Department of Computer Science Series of Publications B, vol. B-2014-2, pp. 68–69. University of Helsinki, Helsinki, Finland (2014)

[4] Ansótegui, C., Malitsky, Y., Sellmann, M.: MaxSAT by improved instance-specific algorithm configuration. In: Brodley, C., Stone, P. (eds.) Proceedings of the Twenty-eighth National Conference on Artificial Intelligence (AAAI'14). pp. 2594–2600. AAAI Press (2014)

[5] Ansótegui, C., Sellmann, M., Tierney, K.: A gender-based genetic algorithm for the automatic configuration of algorithms. In: Gent, I. (ed.) Proceedings of the Fifteenth International Conference on Principles and Practice of Constraint Programming (CP'09). Lecture Notes in Computer Science, vol. 5732, pp. 142–157. Springer-Verlag (2009)

[6] Audemard, G., Hoessen, B., Jabbour, S., Lagniez, J.M., Piette, C.: Penelope, a parallel clause-freezer solver. In: Balint et al. [10], pp. 43–44

[7] Audemard, G., Simon, L.: Glucose 2.1. in the SAT challenge 2012. In: Balint et al. [10], pp. 23–23
[8] Audemard, G., Simon, L.: Lazy clause exchange policy for parallel SAT solvers. In: Sinz, C., Egly, U. (eds.) Proceedings of the Seventeenth International Conference on Theory and Applications of Satisfiability Testing (SAT'14). Lecture Notes in Computer Science, vol. 8561, pp. 197–205. Springer (2014)
[9] Balaprakash, P., Birattari, M., Stützle, T.: Improvement strategies for the F-Race algorithm: Sampling design and iterative refinement. In: Bartz-Beielstein, T., Aguilera, M., Blum, C., Naujoks, B., Roli, A., Rudolph, G., Sampels, M. (eds.) International Workshop on Hybrid Metaheuristics. Lecture Notes in Computer Science, vol. 4771, pp. 108–122. Springer (2007)
[10] Balint, A., Belov, A., Diepold, D., Gerber, S., Järvisalo, M., Sinz, C. (eds.): Proceedings of SAT Challenge 2012: Solver and Benchmark Descriptions. University of Helsinki (2012)
[11] Balint, A., Belov, A., Heule, M., Järvisalo, M. (eds.): Proceedings of SAT Competition 2013: Solver and Benchmark Descriptions, Department of Computer Science Series of Publications B, vol. B-2013-1. University of Helsinki (2013)
[12] Belov, A., Diepold, D., Heule, M., Järvisalo, M.: The application and the hard combinatorial benchmarks in SAT competition 2014. In: Belov, A., Diepold, D., Heule, M., Järvisalo, M. (eds.) Proceedings of SAT Competition 2014. Department of Computer Science Series of Publications B, vol. B-2014-2, pp. 80–83. University of Helsinki, Helsinki, Finland (2014)
[13] Biere, A.: Lingeling, Plingeling, PicoSAT and PrecoSAT at SAT race 2010. Tech. Rep. 10/1, Institute for Formal Models and Verification. Johannes Kepler University (2010)
[14] Biere, A.: Lingeling and friends at the SAT competition 2011. Technical Report FMV 11/1, Institute for Formal Models and Verification, Johannes Kepler University (2011)
[15] Biere, A.: Lingeling, Plingeling and Treengeling entering the SAT competition 2013. In: Balint et al. [11], pp. 51–52
[16] Biere, A.: Lingeling and friends entering the SAT race 2015. Tech. rep., Institute for Formal Models and Verification, Johannes Kepler University (2015)
[17] Bischl, B., Kerschke, P., Kotthoff, L., Lindauer, M., Malitsky, Y., Frechétte, A., Hoos, H., Hutter, F., Leyton-Brown, K., Tierney, K., Vanschoren, J.: ASlib: A benchmark library for algorithm selection. Artificial Intelligence 237, 41–58 (2016)
[18] Brochu, E., Cora, V., de Freitas, N.: A tutorial on Bayesian optimization of expensive cost functions, with application to active user modeling and hierarchical reinforcement learning. Computing Research Repository (CoRR) abs/1012.2599 (2010)
[19] Cameron, C., Hoos, H., Leyton-Brown, K.: Bias in algorithm portfolio performance evaluation. In: Kambhampati, S. (ed.) Proceedings of the Twenty-Fifth International Joint Conference on Artificial Intelligence (IJCAI'16). pp. 712–719. IJCAI/AAAI Press (2016)

[20] Cheeseman, P., Kanefsky, B., Taylor, W.: Where the really hard problems are. In: Mylopoulos, J., Reiter, R. (eds.) Proceedings of the 12th International Joint Conference on Artificial Intelligence. pp. 331–340. Morgan Kaufmann (1991)

[21] Chen, J.: Phase selection heuristics for satisfiability solvers. CoRR abs/1106.1372 (v1) (2011)

[22] Cimatti, A., Sebastiani, R. (eds.): Proceedings of the Fifteenth International Conference on Theory and Applications of Satisfiability Testing (SAT'12), Lecture Notes in Computer Science, vol. 7317. Springer-Verlag (2012)

[23] Falkner, S., Lindauer, M., Hutter, F.: SpySMAC: Automated configuration and performance analysis of SAT solvers. In: Heule, M., Weaver, S. (eds.) Proceedings of the Eighteenth International Conference on Theory and Applications of Satisfiability Testing (SAT'15). pp. 1–8. Lecture Notes in Computer Science, Springer-Verlag (2015)

[24] Gebser, M., Kaufmann, B., Schaub, T.: Conflict-driven answer set solving: From theory to practice. Artificial Intelligence 187-188, 52–89 (2012)

[25] Gebser, M., Kaufmann, B., Schaub, T.: Multi-threaded ASP solving with clasp. TPLP 12(4-5), 525–545 (2012)

[26] van Gelder, A.: Contrasat - a contrarian SAT solver. Journal on Satisfiability, Boolean Modeling and Computation 8(1/2), 117–122 (2012)

[27] Grinten, A., Wotzlaw, A., Speckenmeyer, E., Porschen, S.: satUZK: Solver description. In: Balint et al. [10], pp. 54–55

[28] Hamadi, Y., Jabbour, S., Sais, L.: ManySAT: a parallel SAT solver. Journal on Satisfiability, Boolean Modeling and Computation 6, 245–262 (2009)

[29] Hamadi, Y., Schoenauer, M. (eds.): Proceedings of the Sixth International Conference on Learning and Intelligent Optimization (LION'12), Lecture Notes in Computer Science, vol. 7219. Springer-Verlag (2012)

[30] Harman, M., Jones, B.: Search-based software engineering. Information and Software Technology 43(14), 833–839 (2001)

[31] Helmert, M., Röger, G., Karpas, E.: Fast Downward Stone Soup: A baseline for building planner portfolios. In: ICAPS-2011 Workshop on Planning and Learning (PAL). pp. 28–35 (2011)

[32] Herwig, P.: Using graphs to get a better insight into satisfiability problems. Master's thesis, Delft University of Technology, Department of Electrical Engineering, Mathematics and Computer Science (2006)

[33] Heule, M., Dufour, M., van Zwieten, J., van Maaren, H.: March_eq: Implementing additional reasoning into an efficient look-ahead SAT solver. In: Hoos, H., Mitchell, D. (eds.) Proceedings of the Seventh International Conference on Theory and Applications of Satisfiability Testing (SAT'04). Lecture Notes in Computer Science, vol. 3542, pp. 345–359. Springer-Verlag (2004)

[34] Holte, R., Howe, A. (eds.): Proceedings of the Twenty-second National Conference on Artificial Intelligence (AAAI'07). AAAI Press (2007)

[35] Hoos, H.: Programming by optimization. Communications of the ACM 55(2), 70–80 (2012)

[36] Hoos, H., Kaminski, R., Lindauer, M., Schaub, T.: aspeed: Solver scheduling via answer set programming. Theory and Practice of Logic Programming 15, 117–142 (2015)
[37] Hoos, H., Lindauer, M., Schaub, T.: claspfolio 2: Advances in algorithm selection for answer set programming. Theory and Practice of Logic Programming 14, 569–585 (2014)
[38] Hsu, E., Muise, C., Beck, C., McIlraith, S.: Probabilistically estimating backbones and variable bias: Experimental overview. In: Stuckey, P. (ed.) Proceedings of the Fourteenth International Conference on Principles and Practice of Constraint Programming (CP'08). Lecture Notes in Computer Science, vol. 5202, pp. 613–617. Springer (2008)
[39] Huberman, B., Lukose, R., Hogg, T.: An economic approach to hard computational problems. Science 275, 51–54 (1997)
[40] Hutter, F., Babić, D., Hoos, H., Hu, A.: Boosting verification by automatic tuning of decision procedures. In: O'Conner, L. (ed.) Formal Methods in Computer Aided Design (FMCAD'07). pp. 27–34. IEEE Computer Society Press (2007)
[41] Hutter, F., Hoos, H., Leyton-Brown, K.: Automated configuration of mixed integer programming solvers. In: Lodi, A., Milano, M., Toth, P. (eds.) Proceedings of the Seventh International Conference on Integration of AI and OR Techniques in Constraint Programming (CPAIOR'10). Lecture Notes in Computer Science, vol. 6140, pp. 186–202. Springer-Verlag (2010)
[42] Hutter, F., Hoos, H., Leyton-Brown, K.: Bayesian optimization with censored response data. In: NIPS workshop on Bayesian Optimization, Sequential Experimental Design, and Bandits (BayesOpt'11) (2011)
[43] Hutter, F., Hoos, H., Leyton-Brown, K.: Sequential model-based optimization for general algorithm configuration. In: Coello, C. (ed.) Proceedings of the Fifth International Conference on Learning and Intelligent Optimization (LION'11). Lecture Notes in Computer Science, vol. 6683, pp. 507–523. Springer-Verlag (2011)
[44] Hutter, F., Hoos, H., Leyton-Brown, K.: Parallel algorithm configuration. In: Hamadi and Schoenauer [29], pp. 55–70
[45] Hutter, F., Hoos, H., Leyton-Brown, K., Stützle, T.: ParamILS: An automatic algorithm configuration framework. Journal of Artificial Intelligence Research 36, 267–306 (2009)
[46] Hutter, F., Hoos, H., Stützle, T.: Automatic algorithm configuration based on local search. In: Holte and Howe [34], pp. 1152–1157
[47] Hutter, F., Lindauer, M., Balint, A., Bayless, S., Hoos, H., Leyton-Brown, K.: The configurable SAT solver challenge (CSSC). Artificial Intelligence Journal (AIJ) 243, 1–25 (2017)
[48] Hutter, F., Xu, L., Hoos, H., Leyton-Brown, K.: Algorithm runtime prediction: Methods and evaluation. Artificial Intelligence 206, 79–111 (2014)
[49] Kadioglu, S., Malitsky, Y., Sabharwal, A., Samulowitz, H., Sellmann, M.: Algorithm selection and scheduling. In: Lee, J. (ed.) Proceedings of the Seventeenth International Conference on Principles and Practice of Constraint Program-

ming (CP'11). Lecture Notes in Computer Science, vol. 6876, pp. 454–469. Springer-Verlag (2011)

[50] Kadioglu, S., Malitsky, Y., Sellmann, M., Tierney, K.: ISAC - instance-specific algorithm configuration. In: Coelho, H., Studer, R., Wooldridge, M. (eds.) Proceedings of the Nineteenth European Conference on Artificial Intelligence (ECAI'10). pp. 751–756. IOS Press (2010)

[51] Kotthoff, L.: Algorithm selection for combinatorial search problems: A survey. AI Magazine pp. 48–60 (2014)

[52] Kotthoff, L.: Ranking algorithms by performance. In: Pardalos, P., Resende, M. (eds.) Proceedings of the Eighth International Conference on Learning and Intelligent Optimization (LION'14). Lecture Notes in Computer Science, Springer-Verlag (2014)

[53] Kotthoff, L., Gent, I., Miguel, I.: An evaluation of machine learning in algorithm selection for search problems. AI Communications 25(3), 257–270 (2012)

[54] Li, C.M., Wei, W., Li, Y.: Exploiting historical relationships of clauses and variables in local search for satisfiability. In: Cimatti and Sebastiani [22], pp. 479–480

[55] Lindauer, M., Bergdoll, D., Hutter, F.: An empirical study of per-instance algorithm scheduling. In: Festa, P., Sellmann, M., Vanschoren, J. (eds.) Proceedings of the Tenth International Conference on Learning and Intelligent Optimization (LION'16). pp. 253–259. Lecture Notes in Computer Science, Springer-Verlag (2016)

[56] Lindauer, M., Hoos, H., Hutter, F.: From sequential algorithm selection to parallel portfolio selection. In: Dhaenens, C., Jourdan, L., Marmion, M. (eds.) Proceedings of the Ninth International Conference on Learning and Intelligent Optimization (LION'15). pp. 1–16. Lecture Notes in Computer Science, Springer-Verlag (2015)

[57] Lindauer, M., Hoos, H., Hutter, F., Schaub, T.: Autofolio: An automatically configured algorithm selector. Journal of Artificial Intelligence Research 53, 745–778 (Aug 2015)

[58] Lindauer, M., Hoos, H., Leyton-Brown, K., Schaub, T.: Automatic construction of parallel portfolios via algorithm configuration. Artificial Intelligence 244, 272–290 (2017)

[59] López-Ibáñez, M., Dubois-Lacoste, J., Caceres, L.P., Birattari, M., Stützle, T.: The irace package: Iterated racing for automatic algorithm configuration. Operations Research Perspectives 3, 43–58 (2016)

[60] Malitsky, Y., Mehta, D., O'Sullivan, B.: Evolving instance specific algorithm configuration. In: Helmert, M., Röger, G. (eds.) Proceedings of the Sixth Annual Symposium on Combinatorial Search (SOCS). AAAI Press (2013)

[61] Malitsky, Y., Sabharwal, A., Samulowitz, H., Sellmann, M.: Parallel SAT solver selection and scheduling. In: Milano, M. (ed.) Proceedings of the Eighteenth International Conference on Principles and Practice of Constraint Programming (CP'12). Lecture Notes in Computer Science, vol. 7514, pp. 512–526. Springer-Verlag (2012)

[62] Malitsky, Y., Sabharwal, A., Samulowitz, H., Sellmann, M.: Algorithm portfolios based on cost-sensitive hierarchical clustering. In: Rossi, F. (ed.) Proceedings of the 23rd International Joint Conference on Artificial Intelligence (IJCAI'13). pp. 608–614 (2013)

[63] Malitsky, Y., Sabharwal, A., Samulowitz, H., Sellmann, M.: Parallel lingeling, CCASat, and CSCH-based portfolio. In: Balint et al. [11], pp. 26–27

[64] Maratea, M., Pulina, L., Ricca, F.: A multi-engine approach to answer-set programming. Theory and Practice of Logic Programming 14, 841–868 (2014)

[65] Nudelman, E., Leyton-Brown, K., Andrew, G., Gomes, C., McFadden, J., Selman, B., Shoham, Y.: Satzilla 0.9 (2003), solver description, International SAT Competition

[66] Nudelman, E., Leyton-Brown, K., Hoos, H., Devkar, A., Shoham, Y.: Understanding random SAT: beyond the clauses-to-variables ratio. In: Wallace, M. (ed.) Proceedings of the international conference on Principles and Practice of Constraint Programming. Lecture Notes in Computer Science, vol. 3258, pp. 438–452. Springer (2004)

[67] O'Mahony, E., Hebrard, E., Holland, A., Nugent, C., O'Sullivan, B.: Using case-based reasoning in an algorithm portfolio for constraint solving. In: Bridge, D., Brown, K., O'Sullivan, B., Sorensen, H. (eds.) Proceedings of the Nineteenth Irish Conference on Artificial Intelligence and Cognitive Science (AICS'08) (2008)

[68] Pulina, L., Tacchella, A.: A self-adaptive multi-engine solver for quantified boolean formulas. Constraints 14(1), 80–116 (2009)

[69] Rice, J.: The algorithm selection problem. Advances in Computers 15, 65–118 (1976)

[70] Roussel, O.: Description of ppfolio (2011), available at `http://www.cril.univ-artois.fr/~roussel/ppfolio/solver1.pdf`

[71] Schneider, M., Hoos, H.: Quantifying homogeneity of instance sets for algorithm configuration. In: Hamadi and Schoenauer [29], pp. 190–204

[72] Schubert, T., Lewis, M., Becker, B.: Pamiraxt: Parallel SAT solving with threads and message passing. JSAT 6(4), 203–222 (2009)

[73] Seipp, J., Sievers, S., Helmert, M., Hutter, F.: Automatic configuration of sequential planning portfolios. In: Bonet, B., Koenig, S. (eds.) Proceedings of the Twenty-ninth National Conference on Artificial Intelligence (AAAI'15). AAAI Press (2015)

[74] Soos, M., Nohl, K., Castelluccia, C.: Extending SAT solvers to cryptographic problems. In: Kullmann, O. (ed.) Proceedings of the Twelfth International Conference on Theory and Applications of Satisfiability Testing (SAT'09). Lecture Notes in Computer Science, vol. 5584, pp. 244–257. Springer (2009)

[75] Streeter, M., Golovin, D., Smith, S.: Combining multiple heuristics online. In: Holte and Howe [34], pp. 1197–1203

[76] Thornton, C., Hutter, F., Hoos, H., Leyton-Brown, K.: Auto-WEKA: combined selection and hyperparameter optimization of classification algorithms. In: I. Dhillon, Koren, Y., Ghani, R., Senator, T., Bradley, P., Parekh, R., He, J., Grossman, R., Uthurusamy, R. (eds.) The 19th ACM SIGKDD International

Conference on Knowledge Discovery and Data Mining (KDD'13). pp. 847–855. ACM Press (2013)
[77] Tompkins, D., Balint, A., Hoos, H.: Captain Jack – new variable selection heuristics in local search for SAT. In: Sakallah, K., Simon, L. (eds.) Proceedings of the Fourteenth International Conference on Theory and Applications of Satisfiability Testing (SAT'11). Lecture Notes in Computer Science, vol. 6695, pp. 302–316. Springer (2011)
[78] Wotzlaw, A., van der Grinten, A., Speckenmeyer, E., Porschen, S.: pfolioUZK: Solver description. In: Balint et al. [10], p. 45
[79] Xu, L., Hoos, H., Leyton-Brown, K.: Hydra: Automatically configuring algorithms for portfolio-based selection. In: Fox, M., Poole, D. (eds.) Proceedings of the Twenty-fourth National Conference on Artificial Intelligence (AAAI'10). pp. 210–216. AAAI Press (2010)
[80] Xu, L., Hutter, F., Hoos, H., Leyton-Brown, K.: SATzilla: Portfolio-based algorithm selection for SAT. Journal of Artificial Intelligence Research 32, 565–606 (2008)
[81] Xu, L., Hutter, F., Hoos, H., Leyton-Brown, K.: Hydra-MIP: Automated algorithm configuration and selection for mixed integer programming. In: RCRA workshop on Experimental Evaluation of Algorithms for Solving Problems with Combinatorial Explosion at the International Joint Conference on Artificial Intelligence (IJCAI) (2011)
[82] Xu, L., Hutter, F., Hoos, H., Leyton-Brown, K.: Evaluating component solver contributions to portfolio-based algorithm selectors. In: Cimatti and Sebastiani [22], pp. 228–241
[83] Xu, L., Hutter, F., Shen, J., Hoos, H., Leyton-Brown, K.: SATzilla2012: improved algorithm selection based on cost-sensitive classification models. In: Balint et al. [10], pp. 57–58
[84] Yun, X., Epstein, S.: Learning algorithm portfolios for parallel execution. In: Hamadi and Schoenauer [29], pp. 323–338

Chapter 16
An Application of Parallel Satisfiability Solving to the Verification of Complex Embedded Systems

Orlando Ferrante, Alberto Ferrari, Christos Sofronis, Leonardo Mangeruca, and Luca Benvenuti

Abstract Model checking has reached a maturity level that allows its techniques to be applied to the verification of industrial systems. Several algorithms and methods have been proposed to increase its effectiveness to tackle models of increasing complexity. In this chapter we present an application of Parallel Satisfiability Solving to the verification of embedded control systems. The adopted toolchain is part of the Formal Specs Verifier framework for the formal verification of Simulink/Stateflow models. The experiments we performed show that the use of a parallel satisfiability solver allows for an average speedup of an order of magnitude or more on industrial strength models.

16.1 Introduction

Model checking has reached a high maturity level that allows its techniques to be applied to the verification of complex embedded systems. Several techniques and tools have been proposed to tackle industrial-sized models. In [1] the authors describe the verification of a Flight Control System modeled in MATLAB Simulink using the NuSMV model checker. In [2] the verification of embedded avionics software is performed using three different model checkers (namely NuSMV [3], SAL [4] and PROVER [5]). In addition, model checkers have been successfully applied to the verification of software using both static and dynamic analysis [6, 7, 8]. In recent decades several techniques have been developed in order to tackle the state explosion

Orlando Ferrante, Alberto Ferrari, Christos Sofronis, Leonardo Mangeruca
Advanced Laboratory on Embedded Systems - United Technologies Research Center
e-mail: name.surname@utrc.utc.com
.
Luca Benvenuti
"Sapienza" University of Rome
e-mail: luca.benvenuti@uniroma1.it

Y. Hamadi und L. Sais (eds.), *Handbook of Parallel Constraint Reasoning*,
https://doi.org/10.1007/978-3-319-63516-3_16

problem, which limits the application of formal methods. The use of binary decision diagrams allowed the application of model checking to industrial case studies [9]. Bounded model checking [10] introduced the use of SAT solvers in the context of symbolic model checking and provided the basis for its extension to the unbounded case. During the last decade increased research and industrial efforts have been spent on the area of Satisfiability Modulo Theories (SMT) [11] trying to improve the efficiency of formal verification tools by exploiting the integration of SAT-based reasoning methods with specific theories; promising results have been obtained in particular in the field of software model checking. However, the application of such theories to complex industrial embedded systems which usually expose nonlinear dynamics and complex numeric control algorithms, is still an open research area. Parallel algorithms have been shown to be capable of providing a significant speedup in solving hard combinatorial problems by leveraging multi-core computers and clusters. In Chapter 1, Parallel Satisfiability an overview of the evolution of such algorithms is provided. In this chapter we present some experimental results on the application of parallel solvers to the verification of embedded control systems modeled as Simulink models, we present a toolchain obtained by composing the Formal Specs Verifier toolset and a modified version of the NuSMV2 model checker that integrates the ManySAT parallel SAT solver [12]. The Formal Specs Verifier is a framework for the formal verification of Simulink/Stateflow models. The experimental results have been collected using a rich set of industrial use cases and they show the existence of a relevant speedup of the verification process when the bounded model checking technique is used for checking invariants. The designed toolchain allows for the exploitation of recent advances in SAT solver techniques with the strength of the Formal Specs Verifier environment and the Bounded Model Checking verification. In this chapter we present the application of the toolchain to a cruise control system and an additional set of logic models of industrial size. The main contributions of the chapter are (1) the description of the integration of the ManySAT parallel SAT solver (2) the description of experimental results that show the speedup for the verification of invariant properties of industrial sized embedded systems compared to the use of classical SAT solvers when using the bounded-model checking technique and (3) the integration of the tool with the Formal Specs Verifier framework for the verification of Simulink/Stateflow models and its application to a cruise control case study.

16.2 FormalSpecs Verifier Verification Framework

The experiments described in this chapter were performed using the FormalSpecs Verifier (FSV) framework for the verification of discrete systems for the MATLAB Simulink environment [13]. The framework supports several operative modes. In this work we focus on the capability of verifying properties described as invariants via temporal logic formulae. The FSV tool can be seen as a translator from a Simulink model and specification to the NuSMV tool's native language. The transformation process produces a semantically equivalent NuSMV representation of the input model taking into account the non-determinism that may be introduced during the transformation step. In Figure 16.1 the flow is described in detail. As a first step the Simulink textual file is parsed. Then the parsed Simulink model is processed, generating a semantically equivalent NuSMV model that is used to generate the concrete NuSMV artifact with a model-to-text step.

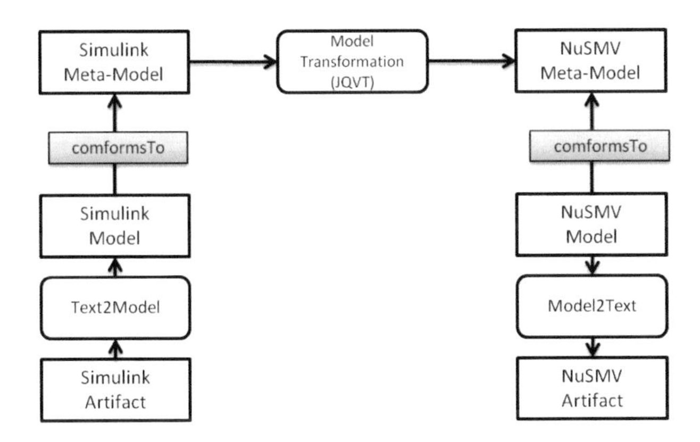

Fig. 16.1: FormalSpecs Verifier transformation flow

The technology used to perform the model transformation step is an internally developed Java embodiment of the OMG Query/View/Transformation [14] language called JQVT. The JQVT library aims at providing an industry-level operational implementation of the QVT language. It supports the definition of QVT mappings and the definition of mapping inheritance, disjunction and merging. JQVT allows the capture of the mapping relation that links a source-model element to a target-model element, and it supports the resolve and resolveIn operators to retrieve the set of mapping source model elements from a given mapped target model element. JQVT does not support the entire QVT specification. However, it has been extensively used as translation infrastructure of different tools for the translation of industrial-sized models [15]. The Formal Specs Verifier has been used in both industrial and research projects [16, 17, 18], and several additional components are available for the verification of systems [19], synthesis of failure scenarios [20], automatic test generation [21, 22] and requirements validation [16].

16.3 Integration of the ManySAT Solver

In this section we discuss how we integrated the ManySAT parallel SAT solver into the toolchain used to perform the experiments. The overall picture of the integration is provided in Figure 16.2.

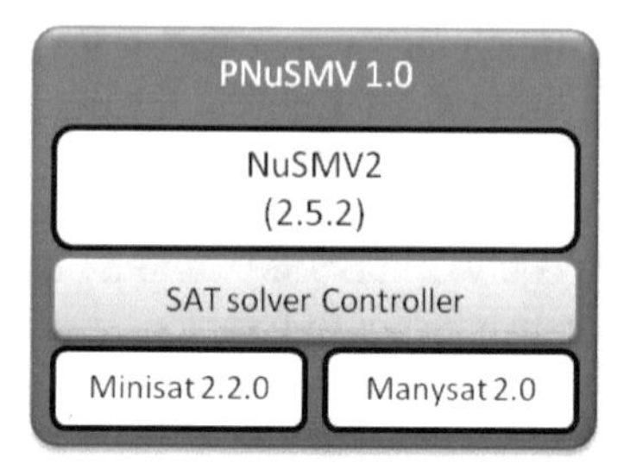

Fig. 16.2: Parallel NuSMV layered structure

The integration was performed modifying the NuSMV 2.5.2 open-source model checker, which is a state-of-the art tool for the formal verification of discrete systems. NuSMV2 provides both BDD (binary decision diagrams) and SAT-based model checking, providing a flexible API for the integration of several SAT solvers such as Minisat [23] and zChaff [24]. To integrate the ManySAT solver, a Facade component (the SAT Solver Controller) was developed to act as an intermediate layer between the model checker problem formulation component and the SAT solver. The role of the controller is to properly instantiate, initialize and coordinate several SAT solvers and to provide a glue layer between the SAT solver component public API and the NuSMV2 SAT solver interface. Currently we successfully integrated the MiniSat v2.2.0 and the ManySAT 2.0. The latter is the last iteration of the parallel SAT solver that won SAT-Race 2008 and SAT Competion 2009. The availability of multicore platforms allows the efficient exploitation of the parallel nature of the solver, easily obtaining an average speedup of an order of magnitude for several industrial-level models as described in Section 16.6.

16.4 Cruise Control Use Case

To show the performance of the tool using a concrete application we describe a cruise control system modeled using MATLAB Simulink and translated with the Formal Specs Verifier tool. Cruise control is the term used to describe a control system that regulates the speed of an automobile. The basic operation of a cruise controller is to sense the speed of the vehicle, compare this speed to a desired reference, and then accelerate or decelerate the car as required. A simple control algorithm for controlling the speed is to use a "proportional plus integral" feedback based on the error between the current and the desired speed. The model of the truck is based on a force balance for the body as depicted in Figure 16.3. For a detailed description of the example please refer to [25]. Let v be the speed of the truck, m the total mass, F_T the traction force generated by the engine on the wheels, and F_d the force generated by additional elements (such as gravity and aerodynamic drag). The mathematical model of the system is given by the equation

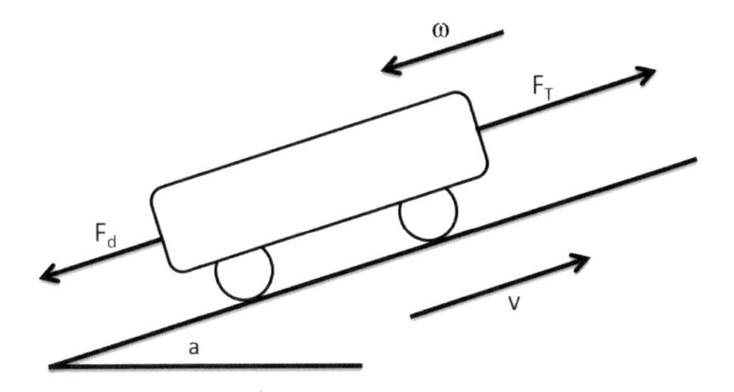

Fig. 16.3: Cruise control model

$$m\frac{dv}{dt} = F_T - F_d \tag{16.1}$$

where $m = 3450$ kg and F_T is the force of the engine. The force F_d is composed of the gravitational force (F_g), the rolling friction (F_r) and the aerodynamic drag (F_a) as follows

$$F_g = mg\sin(a) \tag{16.2}$$

with a the road slope and g the gravitational constant. The rolling friction is

$$F_r = mgC_r sign(v) \tag{16.3}$$

with C_r the friction coefficient. Finally

$$F_a = \frac{1}{2}\rho C_x A(v+\omega)^2 \tag{16.4}$$

where ρ is the air density, C_x is the aerodynamic drag coefficient, A is the area of the truck and ω models the wind gusts. In our model the values of the parameters of the equations are $\rho = 1.228\ kg/m^3$, $C_x = 0.55$ and $A = 2.4\ m^2$.

The control algorithm regulates the traction force F_T by acting on the throttle aperture on the basis of the error $e = v_{REF} - v$ between the desired speed v_{REF} and the current speed v of the car. The algorithm consists of a proportional integral control as follows:

$$\alpha(kT) = k_P e(kT) + k_I \sum_{h=0}^{k} e(hT) \tag{16.5}$$

and

$$F_T(kT) = F_{MAX}(kT) * \alpha(kT) \tag{16.6}$$

where $\alpha(kT)$ is the throttle aperture, $F_{MAX}(kT)$ is the maximum available tractive force, $T = 40$ ms is the ECU sampling rate and $k_P = 2$, $k_I = 2$ are the proportional and integral gain, respectively.

16.5 Simulink Model and Specification

The cruise control system was modeled in MATLAB Simulink using the following workflow:

- First a non-linear infinite state system model was developed. This model represents the mathematical model of the cruise control for both the plant and the control law with the intent of validating the design of the controller using a higher-fidelity model.
- After validating the correct behavior of the controlled system, a discrete-time and discrete-value version of the model was developed. This model can be translated into a finite-state machine and is amenable to analysis using model checking techniques.
- Finally, a set of invariant properties was defined to enable the verification of the discrete model using the latest advances in parallel SAT-solving to search for potential violation of the properties.
- If a counterexample is found to the discrete-time version it is also used to evaluate the behavior of the original higher-fidelity model in order to evaluate wether the counterexample is an artifact of the discretization (spurious counterexample) or an effective redesign of the controller is required.

In the following sub-section a detailed description of the model is provided.

16.5.1 Continuous-Time Non-linear Model

The original model captures the controller and the plant using a continuous time, non-linear formulation of the physical laws that governs the dynamical system. The model is decomposed into three subsystems, namely the ECU, the Engine and the Vehicle. Each component will be described in the following subsections.

16.5.1.1 ECU Subsystem

The ECU subsystem has as input the cruise reference speed, the actual speed and the initial throttle aperture. Its output is the effective throttle aperture which is controlled in order to obtain an actual vehicle speed that is equal to the reference speed. The controller is based on a Proportional-Integration control scheme (PI) that sets the

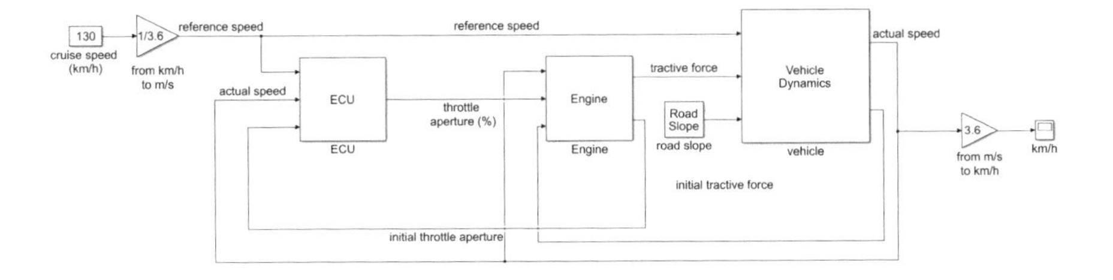

Fig. 16.4: Continuous-time non-linear model

throttle value based on an error signal computed as the difference between the reference speed and the actual speed. (Figure 16.5)

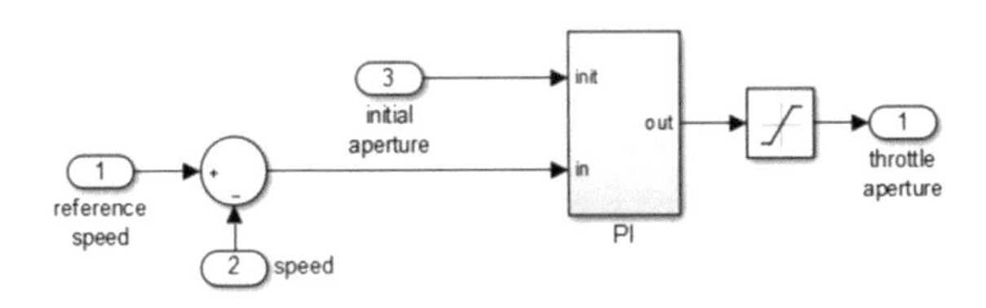

Fig. 16.5: ECU component

16.5.1.2 Engine Subsystem

The engine subsystem models the effective engine dynamic as represented in Figure 16.6

The system computes the effective tractive force taking into account the maximum tractive force and the throttle aperture. Note that the maximum tractive force depends on the speed of the car, which in turn, depends on the throttle aperture through the vehicle dynamics captured in the vehicle component model. In addition a first non-linear computation is performed in this part of the model since the maximum tractive force is obtained by multiplying the current throttle aperture and the maximum tractive force which depends on the current vehicle speed. Note also that all these variables are physical values (real values) approximated as double-precision floating-point variables in the Simulink environment.

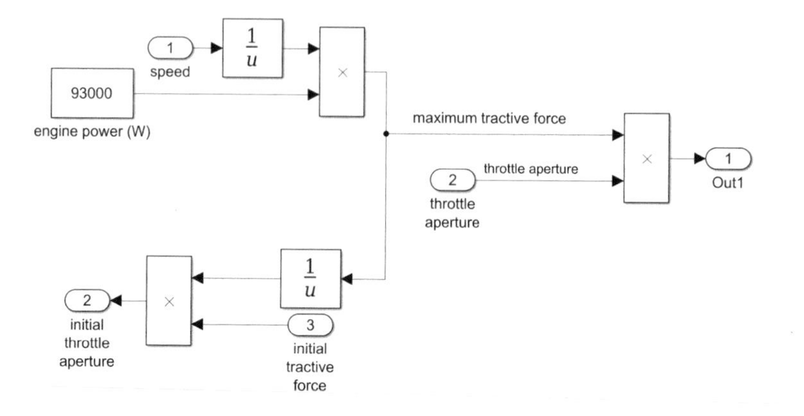

Fig. 16.6: Engine Component

16.5.1.3 Vehicle Dynamics Subsystem

The vehicle sub-system model computes how the tractive force and the road affect the vehicle speed. In particular a model of the road slope and the rolling force is used to compute how the road slope (input of the subsystem) affects the vehicle speed (output of the subsystem) given the tractive force. Note that in order to compute this value a non-linear multiplication is required by the physical model, which add another element of non-linearity.

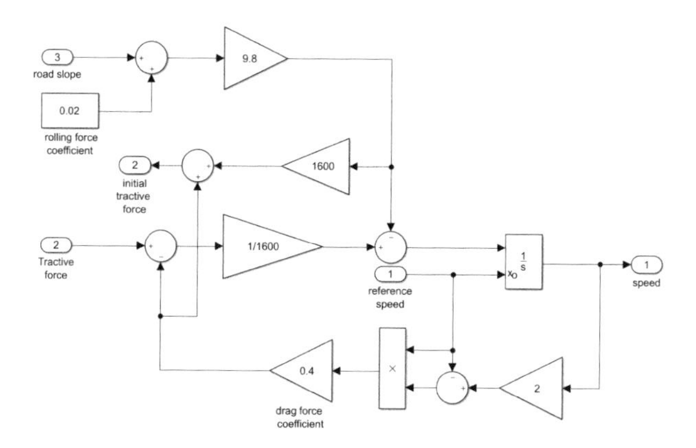

Fig. 16.7: Vehicle Component

Also for this component all the physical variables and related constants (road slope, tractive force, speed etc.) are real values approximated as a double-precision floating-point. The overall model has been instrumental in providing a correct design of the PI coefficients in order to have a controlled system that behaves as desired. As an example of the expected behavior of the system, consider the case in which the cruise reference speed is set to a constant (130 km/h) and a road slope changes from -2% to 6%. The graphical representation of this is captured in Figure 16.8.

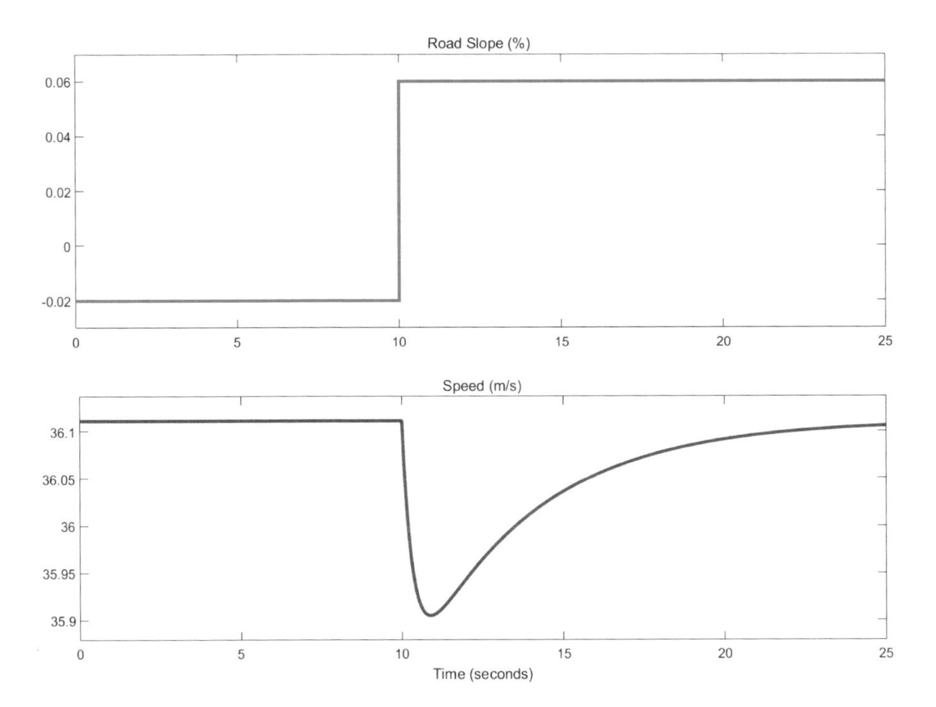

Fig. 16.8: Road and Speed profiles

16.5.2 Discrete-Time Discrete-Value Model

The original cruise control model is a continuous-time, infinite-state non-linear system. This system cannot in general be validated using finite-state model checking techniques. In order to enable the application of standard finite-state model checking techniques and leverage recent advances in SAT solving, the model was modified by applying several transformations. In this section we briefly introduce the final model obtained by applying the different discretization transformations. It is important to stress that each transformation has been performed to obtain a model that is conservative w.r.t. counterexample generation. As a first step a discrete-time version of the model was produced in which the continuous-time dynamics of the model (e.g. continuous-time integrators) were replaced by the equivalent discrete-time components. This discretization step needs to be performed taking into account the

ECU sampling rate of 40 ms. A second transformation was performed to introduce fixed point representation of all the real variables, substituting when necessary the floating-point representation. During the second discretization step the structure of the model was rearranged to simplify some blocks and propagate the constant values whenever possible. The new simplified model is represented in Figures 16.9, 16.10 and 16.11. This model has an additional input (wind gusts) that was added to take into account the effect of wind on the vehicle.

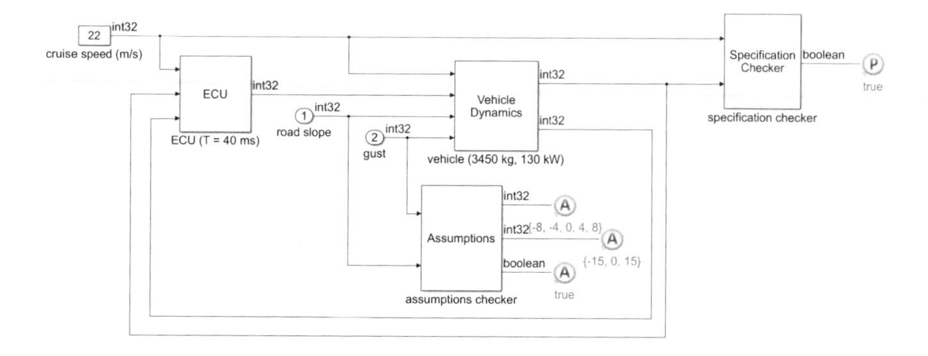

Fig. 16.9: Discrete model

The ECU block has the same input as the original block (reference speed, speed and throttle initial values) and implements a fixed-point, discrete-time PI controller.

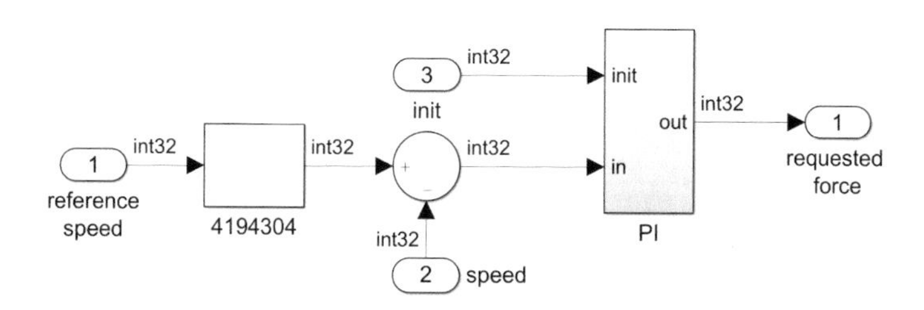

Fig. 16.10: ECU

Executing the discretization process, the plant was abstracted by collapsing the dynamics of the vehicle into one sub-system only. Also for the discretized plant model the real variables of the system were represented using fixed-point notation and adopting a bit vector representation to represent the fixed point variables. The dynamic of the plant has been modeled using discrete-time blocks. Note that the original non-linearity is present, with the difference that all the non-linear computations are represented as bit vector operations. This discrete model due to its finite-state nature can be translated into a finite state machine representation and validated using model checkers to check its correctness with respect to the high-level requirements.

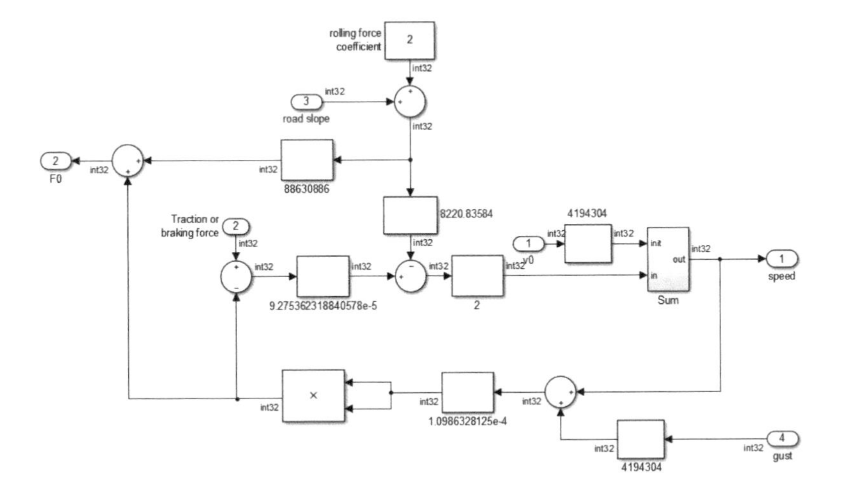

Fig. 16.11: Discrete Vehicle and Dynamics

16.5.2.1 Verification Subsystems

The specification of the properties to be checked was captured using the Formal Specs Verifier properties toolbox following the contract-based methodology that allows for the specification of requirements in terms of contracts $C = (A;G)$ where A is the assumption and G the guarantee (or promise). Intuitively a contract is a requirement of the form *A implies G* where the guarantee represents the set of possible system behaviors under the hypothesis that the environment behaves as described in the assumption (i.e., A represents the set of acceptable environments). The contract-based theory and its application to the verification of complex distributed embedded systems has been investigated by the authors in the context of several EU projects such as SPEEDS, SPRINT, MBAT and DANSE projects [26].

In this section we briefly describe one contract to be verified against the system. It can be informally stated as follows:

- *Assumptions*: the road slope assumes values in the set $\{-8,-4,0,4,8\}$ m/s, the wind gusts are in the set $\{-15,0,15\}$ m/s and the derivative of the wind gusts is bounded to be at most 15 m/s^2

- *Guarantee*: under any admissible wind gusts, and for every admissible road slope, the closed-loop system will always be capable of maintaining the effective speed within the bounds $[95\%, 105\%]$ of the reference speed value.

The verification of this property using model checking techniques consists of evaluating whether the invariant defined by the property $\phi = A \rightarrow G$ can be violated by searching for a sequence of values of the road slope and wind gusts that does not allow the cruise control to maintain the speed within the given range. An example of the encoding of the guarantee is provided in Figure 16.12.

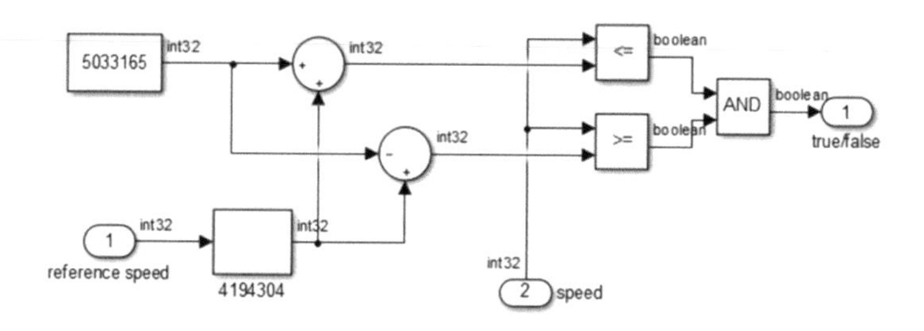

Fig. 16.12: Guarantee model

16.6 Experimental Results

In this section we describe the performance results obtained by executing a set of experiments with the designed toolchain using both the ManySAT and MiniSAT solvers. The host machines used for the execution of the experiments are an Intel iCore 7@1.87 GHz with 8 GB RAM platform hosting a Linux 64-bit Ubuntu 10.04 operating system (platform A) and an Intel(R) Xeon(R) CPU X5550 @ 2.67 GHz with 50 GB RAM platform hosting an Ubuntu Linux 10.04 64-bit operating system (platform B).

16.6.1 Cruise Control Model

For the cruise control we developed several experiments in order to exercise the verification tools in different ways. The model was designed to falsify the property previously described after 33 execution steps. The translated NuSMV model has size 88 bits, which that is small compared to the size of the other models used in another set of experiments (thousands of bits). However, the model represents a good

Table 16.1: First experiment results

Steps	ManySAT (s)	MiniSAT (s)	AV. SP.	Plat
8	6.53 ± 0.65	34.47 ± 7.01	5.28	*A*
15	40.02 ± 5.92	856.41 ± 275.07	21.19	*A*
20	121.86 ± 119.72	6220.72 ± 2127.13	51.05	*B*
25	516.89 ± 118.37	49149.79 ± 219.51	95.09	*B*
33	5446.32 ± 456.08	N/A	N/A	*B*

benchmark for automotive applications models, which usually contains 32-bit signals, complex arithmetic operators and multiple feedback control loops. In addition, the model is hard to verify and the satisfiability problem produced by the encoding of the invariant checking as bounded model checking problem has not been found SAT or UNSAT by the MiniSAT solver after several days of computation, hence it represents an interesting benchmark for the quantitative evaluation of ManySAT speedup.

16.6.1.1 Bounded Model Checking Verification with Incremental Bounds

As a first experiment we executed a bounded model checking verification of the property using different bound lengths. The verification was performed using the `check_invar_bmc_inc` command with the `forward` strategy. The average execution time of each set of runs is summarized in Table 16.1. For each length bound a set of executions were performed using both MiniSAT and ManySAT and the average value and standard deviation of execution times have been computed. The collected data give us the opportunity to propose some comments. The speedup factor (computed as MiniSat execution time/ManySAT execution time) is always greater than one and it increases with the dimension of the bound length. This is in line with the ManySAT solver's expected performance. In particular we notice that the parallel solver fully exploits the available CPUs. A noticeable drawback of the use of the ManySAT solver is the increasing consumption of memory that, however, did not explode exponentially, making the approach usable for industrial applications.

16.6.1.2 Bounded Model Checking Verification with Fixed Bound Value

As a second experiment we performed a bounded model checking verification of the property to try to show its violation. The ManySAT solvers was able to prove the satisfiability of the formula produced by the BMC (hence producing a valid counterexample) in an average time of 5446 seconds (approximately 1 hour and 30 minutes). The same model was processed using the MiniSAT software but a valid counter example has not been found before the 11000 minutes timeout. In

Figure 16.13 we represent the execution time for both ManySAT and MiniSAT and we extrapolate an approximate super-linear trend function between the number K of BMC step performed and the average execution time. We can notice how the speedup gain is approximately of the form ae^{bK} with $b \simeq 0.16$.

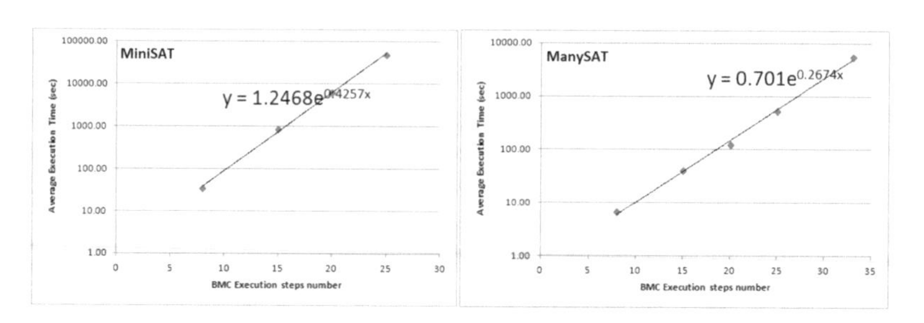

Fig. 16.13: ManySAT and MiniSAT performance graph

16.6.2 Additional Experiments

The cruise control model is an open-source model developed for the evaluation of the ManySAT's performance on automotive domain applications and we collected promising results in terms of verification speedup. In order to better evaluate the tool performance for a larger set of applications we performed additional experiments using another set of models based on synthetic logic systems. Compared to cruise control model these contains mainly logical operators and the reduced number of arithmetic blocks allows for the efficient verification of models with sizes in the thousands of bit in few hours (in contrast with the cruise control model, which is two order of magnitudes smaller in size but requires two orders of magnitude more time to verify the property). In our experiments, two models were analyzed both with size in the thousands of bits and we performed a BMC incremental verification of an invariant property falsified in fixed number of steps using both MiniSAT and ManySAT. The value of the speedup factors are summarized in Table 16.2 (all experiments were executed on platform B). For this class of models we noticed significant variability in the speedup factor. This is in line with the reported behavior of the ManySAT engine w.r.t. the reproducibility of the performance. As a final remark let us observe that the average speedup value grows by a factor of 3 (from 7 to 20) with the increase of the size of the model. This is in line with what we experienced in the first set of experiments.

Table 16.2: Logic-based tests execution results

Model	ManySAT (s)	MiniSAT (s)	AV. SP.
I	440 ± 75	3094 ± 925	7
II	639 ± 208	14074 ± 12629	20

16.7 Conclusions

In this chapter we described a toolchain that allows for the analysis of the speedup of the ManySAT 2.0 parallel SAT solver. The experimental results report a promising speedup for both control-based models, such as the cruise control model described in this paper, and logic-based models.

References

[1] Miller, S., Anderson, E., Wagner, L., Whalen, M., Heimdahl, M.: Formal verification of flight critical software. In: Proceedings of the AIAA Guidance, Navigation and Control Conference and Exhibit. (2005) 15–18
[2] Miller, S., Whalen, M., Cofer, D.: Software model checking takes off. Communications of the ACM **53**(2) (2010) 58–64
[3] Cimatti, A., Clarke, E., Giunchiglia, E., Giunchiglia, F., Pistore, M., Roveri, M., Sebastiani, R., Tacchella, A.: Nusmv 2: An opensource tool for symbolic model checking. (2002) 359–364
[4] `http://sal.csl.sri.com/`
[5] `http://www.prover.com/`
[6] Ball, T., Levin, V., Rajamani, S.K.: A decade of software model checking with SLAM, Communications of the ACM, 54(7), pp 68–76, 2011.
[7] Ball, T., Cook, B., Levin, V., Rajamani, S.K.: SLAM and Static Driver Verifier: Technology transfer of formal methods inside Microsoft. pp 1–20, IFM, 2004.
[8] Godefroid, P.: Compositional dynamic test generation (extended abstract), POPL 2007.
[9] Burch, J.R., Clarke, E.M., Mc Millan, K.L., Dill, D.L., Hwang, L.J.: Symbolic model checking: 10^20 states and beyond, LICS, 1990.
[10] Biere, A., Cimatti, A., Clarke, E., Zhu, Y.: Symbolic model checking without BDDs. Tools and Algorithms for the Construction and Analysis of Systems (1999) 193–207
[11] Barrett, C., Sebastiani, R., Seshia, S., Tinelli, C.: Satisfiability modulo theories. Handbook of Satisfiability **4** (2009)
[12] Hamadi, Y., Sais, L.: ManySAT: a parallel SAT solver. Journal on Satisfiability, Boolean Modeling and Computation (JSAT) (2009)
[13] `http://www.mathworks.com/products/simulink/`

[14] http://www.omg.org/spec/QVT/index.htm
[15] Ferrari, A., Mangeruca, L., Ferrante, O., Mignogna, A.: DesyreML: a SysML profile for heterogeneous embedded systems. In: Embedded Real Time Software and Systems (ERTS). (2012)
[16] Mangeruca, L., Ferrante, O., Ferrari, A.: Formalization and completeness of evolving requirements using contracts. In: 8th IEEE International Symposium on Industrial Embedded Systems (SIES). (2013)
[17] Carloni, M., Ferrante, O., Ferrari, A., Massaroli, G., Orazzo, A., Petrone, I., Velardi, L.: Contract-based analysis for verification of communication-based train control (CBTC) system. In: SAFECOMP. (2014)
[18] Carloni, M., Ferrante, O., Ferrari, A., Massaroli, G., Orazzo, A., Petrone, I., Velardi, L.: Contract modeling and verification with formal specs verifier toolsuite - application to Ansaldo STS rapid transit metro system use case. In: SAFECOMP. (2015)
[19] Ferrante, O., Benvenuti, L., Mangeruca, L., Sofronis, C., Ferrari, A.: Parallel NuSMV: a NuSMV extension for the verification of complex embedded systems. Lecture Notes in Computer Science: Computer Safety, Reliability, and Security **7613** (2012) 409–416
[20] Marazza, M., Ferrante, O., Ferrari, A.: Automatic generation of failure scenarios for sytems-on-chip. In: Real Time Software and Systems (ERTS). (2014)
[21] Ferrante, O., Ferrari, A., Marazza, M.: An algorithm for the incremental generation of high coverage test suites. In: 19th IEEE European Test Symposium. (2014)
[22] Ferrante, O., Ferrari, A., Marazza, M.: Formal Specs Verifier ATG: a tool for model-based generation of high coverage test suites. In: ERTS. (2016)
[23] Een, N., Sörensson, N.: An extensible SAT-solver [ver 1.2] (2003)
[24] Herbstritt, M.: zChaff: Modifications and extensions. (2001)
[25] Murray, R.M., et al.: Feedback Systems An Introduction for Scientists and Engineers. Princeton University Press (2009)
[26] http://www.danse-ip.eu/home/

Chapter 17
Parallel Constraint-Based Local Search: An Application to Designing Resilient Long-Reach Passive Optical Networks

Alejandro Arbelaez, Deepak Mehta, Barry O'Sullivan, and Luis Quesada

Abstract Many network design problems arising in areas as diverse as VLSI circuit design, QoS routing, traffic engineering, and computational sustainability require clients to be connected to a facility under path-length constraints and budget limits. These problems can be seen as instances of the Rooted Distance-Constrained Minimum Spanning-Tree problem (RDCMST), which is NP-hard. An inherent feature of these networks is that they are vulnerable to a failure. Therefore, it is often important to ensure that all clients are connected to two or more facilities via edge-disjoint paths. We call this problem the Edge-disjoint RDCMST (ERDCMST). Previous work on RDCMST has focused on dedicated algorithms and, therefore, it is difficult to use these algorithms to tackle ERDCMST. We present a constraint-based parallel local search algorithm for solving ERDCMST. Traditional ways of extending a sequential algorithm to run in parallel perform either portfolio-based search in parallel or parallel neighbourhood search. Instead, we exploit the semantics of the constraints of the problem to perform multiple moves in parallel by ensuring that they are mutually independent. The ideas presented in this chapter are general and can be adapted to other problems as well. The effectiveness of our approach is demonstrated by experimenting with a set of problem instances taken from real-world passive optical network deployments in Ireland, Italy, and the UK. Our results show that performing

Alejandro Arbelaez
Insight Centre for Data Analytics, University College Cork, Ireland
e-mail: alejandro.arbelaez@insight-centre.org

Deepak Mehta
Insight Centre for Data Analytics, University College Cork, Ireland
e-mail: deepak.mehta@insight-centre.org

Barry O'Sullivan
Insight Centre for Data Analytics, University College Cork, Ireland
e-mail: barry.osullivan@insight-centre.org

Luis Quesada
Insight Centre for Data Analytics, University College Cork, Ireland
e-mail: luis.quesada@insight-centre.org

Y. Hamadi und L. Sais (eds.), *Handbook of Parallel Constraint Reasoning*,
https://doi.org/10.1007/978-3-319-63516-3_17

moves in parallel can significantly reduce the elapsed time and improve the quality of the solutions of our local search approach.

17.1 Introduction

Many network design problems arising in areas as diverse as VLSI circuit design [20], QoS routing [32], traffic engineering [27], and computational sustainability [11] require clients to be connected to a facility under path-length constraints and budget limits. Here the length of the path can be interpreted as distance, delay, signal loss, etc. For example, in a multicast communication [12] setting where a single node is broadcasting to a set of clients, it is important to restrict the path delays from the server to each client. In Long-Reach Passive Optical Networks (LR-PON) [23], a metro node (MN) is connected to a set of exchange sites via optical fibres; the length of the fibre between an exchange site (ES) and its metro node is bounded by the signal loss. The goal is to minimise the cost resulting from the total length of fibres [21]. In VLSI circuit design, path delay is a function of the maximum interconnection path length while power consumption is a function of the total interconnection length [19]. In package shipment services, guarantee constraints are expressed as restrictions on total travel time from an origin to a destination, and the organisation wants to minimise the transportation costs [24].

Many of these network design problems are instances of the Rooted Distance-Constrained Minimum Spanning-Tree Problem (RDCMST) [19], which is NP-hard. The objective is to find a minimum cost spanning tree with the additional constraint that the length of the path from a specified root node (or facility) to any other node (client) must not exceed a given threshold. Many networks are complex systems that are vulnerable to a failure. A major fault occurrence would be a complete failure of the facility, which would affect all the clients connected to the facility. Therefore it is important to provide network resilience. We restrict our attention to networks where all clients are required to be connected to two facilities via two edge-disjoint paths so that whenever a single facility fails or a single link fails all clients are still connected to at least one facility. We define this problem as the Edge-disjoint Rooted Distance-Constrained Minimum Spanning-Tree Problem (ERDCMST). Given a set of facilities and a set of clients such that each client is associated with two facilities, the problem is to find a set of distance-constrained spanning trees rooted at each facility with minimum total cost. Additionally, each client is connected to its two facilities via two edge-disjoint paths. This would effectively mean that each pair of distance-bounded spanning trees would be mutually disjoint in terms of edges.

Previous work on RDCMST [16, 25] has focused on dedicated algorithms, which are hard to extend with side constraints, and therefore it is difficult to use these algorithms to tackle ERDCMST. We present a mixed integer programming formulation of ERDCMST and a constraint-based local search algorithm, which can easily be extended to apply widely. We present two efficient local move operators and an incremental way of maintaining the objective function, which is often a key element for the efficiency of a local search algorithm. Our local search algorithm is able to solve both RDCMST and ERDCMST. These move operators were proposed and evaluated with respect to edge-disjointness, capacity, and distance constraints in [6] and [5].

We extend our sequential algorithm and propose a parallel version of our constraint-based local search algorithm. The traditional way of extending a sequential algorithm to run in parallel is to perform either portfolio-based search in parallel or parallel neighbourhood search. Instead, we exploit the semantics of the constraints of the problem to perform multiple moves in parallel by ensuring that they do not conflict with each other. The effectiveness of our approach is demonstrated by experimenting with a set of problem instances taken from real-world passive optical network deployments in Ireland, the UK, and Italy. Our results show that performing moves in parallel can significantly reduce the time required to find a target solution and improves the anytime behaviour of our local search algorithm.

17.2 Formal Specification and Complexity

Let G be a directed graph with set of nodes $\mathscr{N}$ and set of edges $\mathscr{L}$. We assume that G is a complete graph without self-loops, so $|\mathscr{L}| = |\mathscr{N}| \times (|\mathscr{N}| - 1)$ in our case. Each edge has a cost and a distance value associated with it. Let $\mathscr{M}$ be a set of facilities and let $u_i \in \mathscr{U}$ be a set of (users or) clients. We define $\mathscr{N}$ as the disjoint union of $\mathscr{M}$ and $\mathscr{U}$. Let $U_i \subseteq \mathscr{U}$ be the set of clients that are associated with facility $m_i \in \mathscr{M}$. We use T_i to denote the tree network associated with facility i. We also use $N_i = U_i \cup \{m_i\}$ to denote the set of nodes in the tree T_i associated with the facility m_i, and a set of edges $L_i \subseteq N_i^2$.

T_i is a subgraph of G that contains a directed path from facility m_i to all of its clients and contains no cycles. The length of a path (or path-length) between two nodes is the sum of the distances of the edges connecting the two nodes, and the cost of T_i is the sum of the cost of its edges. In this chapter, without loss of generality, we assume that the cost is symmetrical, i.e., the cost of an edge $\langle i, j\rangle$ is equal to the cost of the edge $\langle j, i\rangle$. As mentioned before, we also assume that the graph is complete since non-existing edges in the original graph can be represented by edges with a very large distance with respect to the distance threshold. Additionally, we remove edges from nodes to themselves.

Definition 17.1 *Rooted Minimum Spanning Tree Problem (RMST). Given a graph* $\mathscr{G} = (N_i, L_i)$ *with a facility* $m_i \in \mathscr{M}$ *and a real value* c_{jk} *denoting the cost of each edge* $(j,k) \in L_i$*, RMST is to find a spanning tree of minimum cost in* $\mathscr{G}$*.*

Definition 17.2 *Rooted Distance-Constrained Minimum Spanning-Tree Problem (RDCMST). Given a graph* $\mathscr{G} = (N_i, L_i)$ *with a facility* $m_i \in \mathscr{M}$*, the set of clients* U_i*, two real values* c_{jk} *and* d_{jk} *denoting the cost and the distance of each edge* $(j,k) \in L_i$*, and a real value* λ*, RDCMST is to find a spanning tree* T_i *with minimum cost in* $\mathscr{G}$

such that the length of the path from the facility m_i to any client $u_j \in U_i$ is not greater than λ.

Definition 17.3 *Edge-Disjoint Rooted Distance-Constrained Minimum Spanning-Trees Problem (ERDCMST). Given a graph $\mathcal{G} = (\mathcal{N}, \mathcal{L})$ with a set of facilities $\mathcal{M}$, a set of clients $\mathcal{U}$, a set of edges $\mathcal{L}$, two real values c_{jk}, and d_{jk} denoting the cost and distance of each edge $(j,k) \in \mathcal{L}$, an association of clients with two facilities $\pi : \mathcal{U} \rightarrow \mathcal{M}^2$, and a real value λ, ERDCMST is to find a spanning tree T_i of $\mathcal{G}$ for each facility m_i such that*

1. *The length of the unique path for all m_i to any of its clients is not greater than λ.*
2. *For each client u_k, the two paths connecting u_k to m_i and m_j, where $\pi(u_k) = \langle m_i, m_j \rangle$, are edge disjoint.*
3. *The sum of the costs of the edges in all the spanning trees is minimum.*

Figure 17.1 shows an example with two facilities F_1 and F_2 and $N = \{a, b, c, d, e, f\}$, black (respectively grey) edges denote the set of edges used to reach F_1 (respectively F_2), the value of λ is 16 and the total cost of the solution is 46 for this illustrative example. Figure 17.1a shows a valid solution satisfying both distance and edge-disjointness constraints. The distance from the facilities to any node is less than or equal to $\lambda = 16$. The paths connecting the set of nodes to the facilities are edge disjoint. Figure 17.1b shows a cheaper solution replacing a grey edge $\langle F_2, b \rangle$ with $\langle a, b \rangle$, but this solution is not edge disjoint since a failure in $\langle a, b \rangle$ would disconnect node b from both facilities.

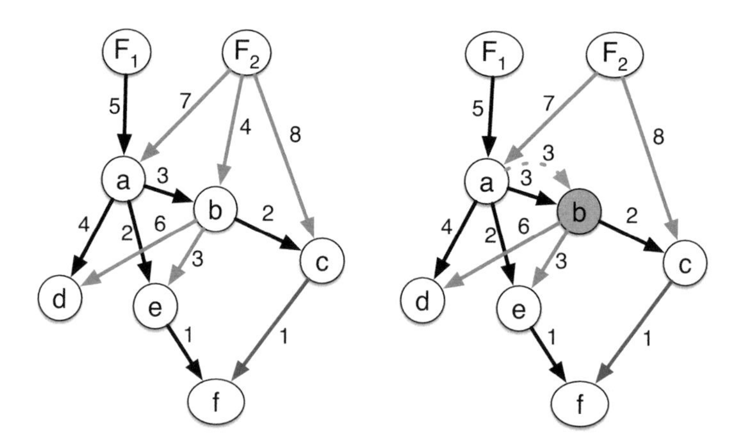

(a) Satisfying both distance and edge disjointness

(b) Satisfying distance constraint but violating edge disjointness

Fig. 17.1: Example of an instance of the ERDCMST problem. λ=16

Complexity. ERDCMST involves finding a rooted distance-bounded spanning tree for every facility whose total cost is minimum. This problem is known to be NP-hard [19].

17.3 A Mathematical Model for ERDCMST

We present a mixed integer programming formulation of ERDCMST.

Variables

- Let x^i_{jk} be a variable over $\{0,1\}$ that denotes whether an edge from node $j \in N_i$ to $k \in N_i$ of facility $i \in \mathscr{M}$ is selected (1) or not (0).
- Let f^i_j be a variable that denotes the upper bound on the length of the path from the facility i to its client j.

We note that the partial order enforced by f helps to rule out cycles in the solution.

Constraints

Each facility is directly connected to at least one of its clients:

$$\forall_{m_i \in \mathscr{M}} : \quad \sum_{u_j \in U_i} x^i_{ij} \geq 1.$$

The total number of edges in any tree T_i is equal to $|U_i|$:

$$\forall_{m_i \in \mathscr{M}} : \quad \sum_{u_j \in N_i} \sum_{u_k \in U_i, u_j \neq u_k} x^i_{jk} = |U_i|.$$

The graph cannot have an edge between a node and itself and the distance from the root node (or facility) to itself is 0:

$$\forall_{m_i \in \mathscr{M}} \forall_{u_j \in N_i} : x^i_{jj} = 0.$$

$$\forall_{m_i \in \mathscr{M}} : \quad f^i_i = 0.$$

The length of the path from any client to its facility is bounded by λ:

$$\forall_{m_i \in \mathscr{M}} \forall_{u_j \in N_i} : \quad f^i_j \leq \lambda.$$

If there is an edge from $u_j \in U_i$ to $u_k \in U_i$ then the length of the path from m_i to u_k is equal to the sum of the length of the path from m_i to u_j plus the distance between u_j and u_k. We use the big-M method to model this implication as a linear constraint.

The value of the constant $\mathscr{C}$ has to be greater than λ plus the maximum distance between any pair of edges in order to be consistent with the implication:

$$\forall_{m_i \in \mathscr{M}} \forall_{\{u_j,u_k\} \in N_i}: \quad \mathscr{C}(1 - x^i_{jk}) + f^i_k \geq f^i_j + d_{jk}.$$

If m_i and $m_{i'}$ are the facilities of the client j, and if there exists any path in the subnetwork associated with facility i that includes the edge $\langle u_j, u_k \rangle$, then facility i' cannot use the same edge. Therefore, we enforce the following constraint:

$$\forall_{\{m_i,m_{i'}\} \in \mathscr{M}} \forall_{\{u_j,u_k\} \in U_i \cap U_{i'}}: x^i_{jk} + x^{i'}_{jk} \leq 1.$$

Objective

The objective is to minimise the total cost:

$$\text{MIN} \sum_{m_i \in \mathscr{M}} \sum_{\{u_j,u_k\} \in N_i} c_{jk} \cdot x^i_{jk}.$$

17.4 Iterated Constraint-Based Local Search

For large networks containing more than 10,000 clients and 100 facilities we cannot expect to solve the previously presented problems using systematic search. In this section we present an iterated constraint-based local search approach for solving our problem.

The Iterated Constraint-Based Local Search (ICBLS) [14, 29] framework depicted in Algorithm 17.1 comprises two phases. First, in a local search phase, the algorithm improves the current solution, little by little, by performing small changes. The algorithm employs a move operator in order to move from one solution to another in the hope of improving the value of the objective function. Second, in the perturbation phase, the algorithm perturbs the incumbent solution (s^*) in order to escape from difficult regions of the search (e.g., a local minimum). The acceptance criterion decides whether to update s^* or not. The algorithm accepts s'^* with probability p, and s^* otherwise. Furthermore, we limit the space of candidate solutions to valid solutions satisfying the constraints of the problem.

Algorithm 17.1: Iterated Constraint-Based Local Search (*move-op*, *s*)

```
1 s* := ConstraintBasedLocalSearch(move-op, s)
2 repeat
3 |   s' := Perturbation(s*)
4 |   s'* := ConstraintBasedLocalSearch(move-op, s')
5 |   s* := AcceptanceCriterion(s*, s'*)
6 until a given stopping criterion is met
```

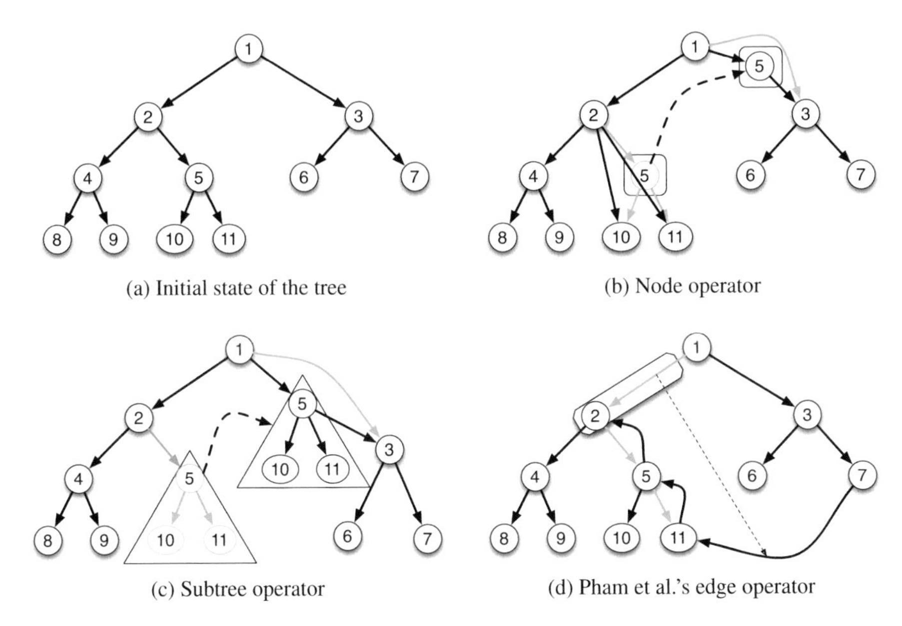

Fig. 17.2: Move Operators. Grey arrows indicate edges removed from the current solution

Our algorithm requires two parameters: s the initial solution where all clients are able to reach their facilities while satisfying all constraints (i.e., the upper bound in the length and disjointness), and the move operation (*move-op*) which is a function. We switch from the local search phase to the perturbation phase when a local minimum is observed. In the perturbation phase we perform a given number of random moves. In this chapter, unless otherwise stated, we use the trivial solution of connecting all clients directly to their respective facilities as the initial solution. The stopping criterion is either a timeout or a given number of iterations.

17.4.1 Move Operators

We describe the *node*, *subtree*, and *edge* move operators. We use T_i to denote the tree associated with facility i. An edge between two clients u_j and u_k is denoted by $\langle u_j, u_k \rangle$. We have defined *node* and *subtree* in [6], and we take the `edge` operator from the literature [10]. Informally speaking the location of a node u_j (or the subtree emanating from it) in a tree is a tuple (u_{p_j}, S_j) where u_{p_j} denotes the predecessor of u_j and S_j denotes the list of immediate successors of u_j in the tree.

Node Operator

In Figure 17.2b we move a given node u_i from the current location to another in the tree. The node (or subtree) changes location if either the predecessor or any successor of the node is different. As a result of the move, all successors of u_i will be directly connected to the predecessor node of u_i. u_i can be placed as a new successor for another node or in the middle of an existing edge in the tree.

Subtree Operator

In Figure 17.2c we move a given node u_i and the subtree emanating from u_i from the current location to another in the tree. As a result of this, the predecessor of u_i is no longer connected to u_i, and all successors of u_i are still directly connected to u_i. u_i can be placed as a new successor for another node or in the middle of an existing edge.

Edge Operator

In this chapter we limit our attention to moving a node or a complete subtree. In [10] the authors proposed to move edges in the context of the Constrained Optimum Path problem. The Pham et al. move operator (Figure 17.2d) chooses an edge in the tree and finds another location for it without breaking the flow.

17.4.2 Operations and Complexities

We first present the complexities of the node and subtree operators as they share similar features. For an efficient implementation of the move operators, it is necessary to maintain b^i_j: the length of the path from u_j to the farthest leaf associated with the tree T_i, and the previously described variable f^i_j: the distance from the facility to the client. Let u_{p_j} be the immediate predecessor of u_j and let S_j be the set of immediate successors of u_j in a given tree.

In order to complete a move the node and subtree operators require to execute the following four steps:

1. Randomly select a node (u_j) from a facility (m_i) from the current solution;
2. Delete u_j from T_i if the node operator is used, or u_j and the emanating subtree of the node in T_i if the subtree operator is used;
3. Identify the best location, i.e., a new predecessor u_{p_j} and a potential new successor u_{ns_j} for u_j in T_i satisfying all constraints;
4. Insert u_j as a new successor of u_{p_j}, and if there is a new successor, add u_{ns_j} as a new successor of u_{p_j}.

Table 17.1 summarises the complexities of the move operators. In this table n denotes the maximum number of clients associated with a single facility. The last row (move) indicates the time complexity of completing a move with a given move operator, i.e., completing the four previously mentioned steps.

Table 17.1: Complexities of different operations

	Node	Subtree	Edge
Delete	$O(n)$	$O(n)$	$O(1)$
Feasible delete	$O(n)$	$O(1)$	$O(1)$
Feasible insert	$O(1)$	$O(1)$	$O(n)$
Best location	$O(n)$	$O(n)$	$O(n^3)$
Insert	$O(n)$	$O(n)$	$O(n)$
Move	$O(n)$	$O(n)$	$O(n^3)$

Feasible Delete

Checking whether a solution is feasible after the node operator deletes a node u_j in T_i requires linear complexity with respect to the number of clients in S_j since it is necessary to check whether the new distance from the root to the furthest leaf for all nodes $u_k \in S_j$ satisfies the distance constraint:

$$f^i_{p_j} + b^i_k + d_{p_j,k} \leq \lambda.$$

Delete

Deleting a node or the subtree emanating from the node u_j in T_i requires linear time with respect to the number of clients of facility m_i. For both operators, it is necessary to update $b^i_{j'}$ for all the nodes j' in the path from the facility m_i to the client u_{p_j} in T_i. In addition, the node operator updates $f^i_{j'}$ for all the nodes j' in the subtree emanating from u_j. After deleting a node u_j or a subtree emanating from u_j, the objective function is updated as follows:

$$obj := obj - c_{j,p_j}.$$

Furthermore, the node operator needs to add to the objective function the cost of disconnecting each successor element of u_j and reconnecting them to u_{p_j}:

$$obj := obj + \sum_{k \in S_j} (c_{k,p_j} - c_{kj}).$$

Feasible Insert

Checking feasibility of a move can be performed in constant time by using f^i_j and b^i_j. If u_j is inserted between the two nodes of edge $\langle u_p, u_q \rangle$ then we check the following:

$$f^i_p + d_{pj} + d_{jq} + b^i_q \leq \lambda.$$

If u_j is inserted as a new successor of u_p we check the distance constraint as follows:

$$f^i_p + d_{pj} + b^i_j \leq \lambda.$$

Best Location

Selecting the best location involves traversing all clients associated with the facility and selecting the one with the maximum reduction in the objective function. Both operators need to traverse the tree in order to evaluate the cost of breaking all existing edges or adding a new node or subtree to the current solution. In this chapter we use a depth-first exploration of the tree.

Insert

A move can be performed in linear time. We recall that this move operator might replace an existing edge $\langle u_p, u_q \rangle$ with two new edges $\langle u_p, u_j \rangle$ and $\langle u_j, u_q \rangle$. This operation requires us to update f^i_j for all nodes in the subtree emanating from u_j, and b^i_j in all nodes in the path from the facility acting as a root node down to the new location of u_j. The objective function must be updated as follows:

$$obj := obj + c_{pj} + c_{jq} - c_{pq}.$$

Now we switch our attention to the edge operator. This operator does not benefit from using b^i_j. The reason is that moving a given edge from one location to another might require changing the direction of a certain number of edges in the tree as shown in Figure 17.2d. Deleting an edge requires constant time. This operation generates two separated subtrees and no data structures need to be updated. Checking the feasibility of adding an edge $\langle u_{p'}, u_{q'} \rangle$ to connect the two subtrees requires linear time. It is necessary to traverse the new tree to obtain the distance from $u_{q'}$ to the farthest leaf in the tree associated with facility i. Performing a move requires linear time. It involves updating f^i_j for the new emanating tree of $u_{q'}$. Finding the best location requires cubic time complexity as the number of possible locations is bounded by n^2 (total number of possible edges for connecting the two subtrees) and for each possible move it is necessary to check feasibility. Due to the high complexity ($O(n^3)$) of completing a move with the edge operator, hereafter we limit our attention to the node and subtree operators.

As pointed out earlier in this chapter, maintaining the backward distance does not provide any reduction in the complexity of the moves for the edge operator. Notice that deleting an edge does not necessarily require us to update the distance of the affected nodes in the tree. In this case, only the affected edge is removed from the solution.

Disjointness

To ensure disjointness among spanning trees we maintain a $2{\cdot}|\mathscr{U}|$ matrix, where $|\mathscr{U}|$ represents the number of clients. Checking disjointness for every client requires constant time complexity and only involves checking that the two integers indicating the predecessors in the primary and secondary facilities are different. Therefore the complexities presented in Table 17.1 remain valid when disjointness is considered.

Move

As pointed out in Table 17.1 completing a move for the node and subtree operators requires linear time complexity and it involves deleting a node u_j (or the complete subtree emanating from u_j), checking feasibility for all potential candidate locations, and then inserting the node or subtree in the most suitable location. We randomly select u_j in order to balance the greediness and the complexity of the move. Alternatively, one can select the best deletion-insertion pair. However, this option make the complexity of completing a move $O(n^2)$ as it involves deleting all nodes in the current solution and re-inserting them. Additionally, the use of the best deletion-insertion pair may lead to premature convergence to local minima.

In [25] the authors proposed two move operators for RDCMST. *Edge-Replace* is similar to the edge operator described above. The difference is that in this case the authors only take into consideration the distance constraint, and therefore, the authors use the cheapest alternative edge to reconnect the two trees. Similarly to the edge operator, *Edge-Replace* might change the direction of some affected edges and therefore the new solution might not satisfy the edge-disjointness constraint. An alternative solution might be to limit the set of candidate neighbours to those not changing the direction of the tree. This would be a particular case of the subtree operator where the subtree is reconnected to an existing leaf of the current solution. *Component-Renew* removes an edge in the tree. Nodes that are separated from the root node are sequentially added to the tree using Prim's algorithm. Nodes that violate the length constraint are added to the tree using a pre-computed route from the root node to any node in the tree. It is worth noticing that the *Component-Renew* operator cannot be applied to ERDCMST as the pre-computed route from the root node to a given node might not be available due to the disjointness constraint.

17.5 Sequential Algorithm

We describe a constraint-based local search algorithm, which is parameterised by a move operator. A (node or subtree) move is composed of four sub-operations: *delete*, *best location*, *insert*, and *feasibility check*. A move involves removing a node or a subtree from the tree and adding it in the best location. In order to find the best location a sequence of feasibility checks are carried out.

The location of a node in a tree is defined by its parent node and its set of successor nodes. If the node operator is used then a new location for a node u_j can be found by either (1) selecting an existing edge $\langle u_a, u_b \rangle$ and inserting u_j between the two nodes such that the parent node of u_j is u_a and the set of successors is the singleton set $\{u_b\}$ or (2) selecting any node u_a and adding a new edge $\langle u_a, u_j \rangle$ such that the parent node of u_j is u_a and the set of successor nodes is empty. Similarly, if the subtree operator is used then a new location for a subtree emanating from a node u_j can be found by either (1) selecting an existing edge $\langle u_a, u_b \rangle$ and inserting u_j between the two nodes such that the parent node of u_j is u_a and the set of successors of u_j is updated by adding the node u_b; or (2) selecting a node u_a and adding a new edge $\langle u_a, u_j \rangle$ such that the parent node of u_j is u_a and the set of successor nodes of u_j remains unchanged. We use $Locations(u_j, T_i, move\text{-}op)$ to denote all the possible locations of a node (or a subtree emanating from) u_j in tree T_i using *move-op*.

Let *list* be a set of potential nodes for which we might be able to find better locations. If the *list* is empty then it means that the algorithm has reached a local minimum for the chosen move operator. We also compute a graph, which we call the *facility connectivity graph*, denoted by *fcg*, where the vertices represent the facilities and an edge between a pair of facilities represent that a change in the tree of a facility might help in finding better locations for some nodes in the tree associated with the other facility. An edge between two vertices in *fcg* is added if the facilities share at least two clients. Notice that if two facilities do not share at least two clients then they are independent from the disjointness point of view. Figure 17.3 shows an example of an *fcg* of the problem with four facilities and 11 clients: clients $\{u_1, u_2, u_3, u_4\}$ are common to m_1 and m_2; clients $\{u_7, u_8, u_9\}$ are common to m_2 and m_4; clients $\{u_5, u_6\}$ are common to m_1 and m_4; and clients $\{u_{10}, u_{11}\}$ are common to m_1 and m_3.

The pseudo-code of our constraint-based local search is shown in Algorithm 17.2. It starts by initialising *list* and *fcg* (Lines 2 and 3). It repeatedly selects a client u_j of a facility m_i randomly from Tree T_i (Line 4), saves its current location and deletes it from the tree (Lines 6-7). The delete operation depends on the move operation (*move-op*) and deletes a single node or the entire subtree emanating from the node. The cost of the current location is saved in *cost* (Line 8). For a chosen node or subtree, in each iteration (Lines 9-16), the algorithm identifies the best location. Line 10 verifies that the new move does not break any constraint and *CostLoc* returns the cost of using the new location using the given move operator (Line 11). If the cost of the new location is better then the best set of candidates is reinitialised to that location (Lines 12-14). If the cost is equal to the best known cost so far then the best set of candidates is updated by adding that location (Lines 15-16). The new location for a

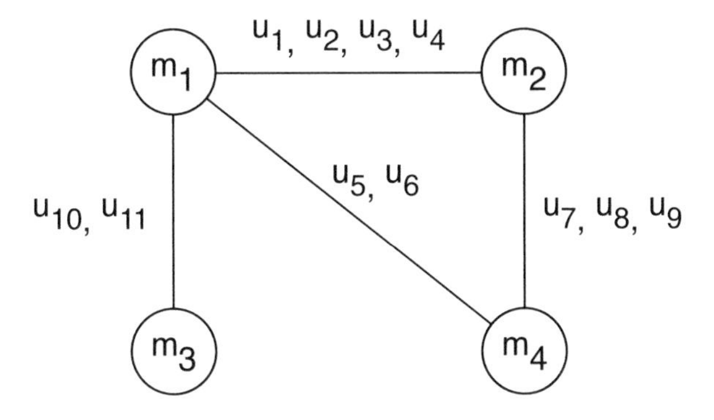

Fig. 17.3: Example of the facility connectivity graph (*fcg*) for a problem with 4 facilities (i.e., m_1, m_2, m_3, and m_4), and 11 clients

given node is randomly selected from the best candidates if there are more than one (Line 18).

Instead of verifying that a local minimum is reached by exhaustively checking all moves for all clients of all facilities, we maintain a list of pairs of clients and facilities. The list is initialised at Line 1 with all pairs of facilities and clients. In each iteration a pair consisting of a facility and a client, (m_i, u_j), is selected and removed from the list (Line 25). When *list* is empty the algorithm reaches a local minimum. If an improvement is observed then one simple way to ensure the correctness of the local minimum is to populate *list* with all pairs of facilities and clients. However, this can be very expensive in terms of time. We instead exploit the *facility connectivity graph* (*fcg*) where an edge between two facilities is added (in Line 2) if they share at least two clients. If an edge between m_i and m_k exists in *fcg* then it means that if there is a change in T_i then it might be possible to make another change in T_k such that the total cost can be improved. Consequently, we only need to consider the clients of affected facilities. This mechanism helps to significantly reduce the time taken by reducing the number of useless moves. We further strengthen the condition for populating *list* by observing the fact that the set of clients that can be affected in a facility m_k is a subset of $\{u_j\} \cup S_j$. The reason is that all the other clients of m_k are not subject to any constraint from the node u_j of T_i (Line 20). Furthermore all the clients of T_i are also added to the *list*.

The perturbation phase of the algorithm works similarly to Algorithm 17.2, by selecting a random node, deleting this node or its subtree in the current solution, and then instead of selecting the best location, the algorithm selects a random valid location. In particular, for the perturbation phase, we replace Lines 10-16 with Line 16, i.e., *BestLoc* := *BestLoc* $\cup \{(u'_{p_j}, S'_j)\}$. Therefore, the algorithm selects a new random location for u_j.

We note that Algorithm 17.2 can tackle both RDCMST and ERDCMST. In the case of RDCMST, there is only one facility, and *FeasibleInsert* only checks the path-length constraint.

Algorithm 17.2: Constraint-Based Local Search (*move-op*, $\{T_1,\ldots,T_{|\mathscr{M}|}\}$)

```
1  list := {(m_i, u_j) | m_i ∈ M ∧ u_j ∈ U_i}
2  fcg := {(m_i, m_j) | |N_i ∩ N_j| ≥ 2}
3  while list ≠ ∅ :
4      Select (m_i, u_j) randomly from list
5      if FeasibleDelete(T_i, u_j, move-op) :
6          oldLoc := (u_{p_j}, S_j)
7          Delete(T_i, u_j, move-op)
8          cost := CostLoc((u_{p_j}, S_j), u_j, move-op)
9          for (u'_{p_j}, S'_j) in Locations(u_j, T_i, move-op) − {oldLoc} :
10             if FeasibleInsert((u'_{p_j}, S'_j), u_j, move-op) :
11                 cost' := CostLoc((u'_{p_j}, S'_j), u_j, move-op)
12                 if cost' < cost :
13                     BestLoc := {(u'_{p_j}, S'_j)}
14                     cost := cost'
15                 elif cost' = cost :
16                     BestLoc := BestLoc ∪ {(u'_{p_j}, S'_j)}
17         if BestLoc ≠ ∅ :
18             Select (u'_{p_j}, S'_j) randomly from BestLoc
19             loc := (u'_{p_j}, S'_j)
20             list := list ∪ {(m_k, u_l) | (m_i, m_k) ∈ fcg ∧ u_l ∈ N_k ∩ (S_j ∪ {u_j})}
21             list := list ∪ {(m_i, u_l) | u_l ∈ N_i}
22         else:
23             loc := oldLoc
24         T_i := Insert(T_i, loc, u_j, move-op)
25     list := list − {(m_i, u_j)}
26 return {T_1, ..., T_n}
```

17.6 Parallel Algorithm

Parallelisation has been widely studied to speedup and improve the performance of local search algorithms to tackle a large variety of problems including TSP [7], Capacitated Network Design [9], Steiner Tree [31], SAT [17], and CSPs [8] just to name a few. These approaches employ the Multi-walk and/or Single-walk framework [30] to devise the parallel algorithm. In particular we focus our attention on constraint-based local search solvers.

17.6.1 Multi-Walk and Single-Walk

Multi-walk (also known as parallel portfolio) involves executing several algorithms (or different copies of the same one with different random seeds) in parallel, with or without cooperation, until a solution is found or a given timeout is reached. The implicit assumption is that different processes handle different parts of the search space. The multi-walk method has two important properties. First, no load balancing is required to parallelise the sequential algorithm. Second, the speedup of the parallel algorithm strictly depends on the performance of the sequential one. As discussed in the literature (e.g., [28] and [26]) a high variance of the sequential algorithm usually means good parallel speedup factors.

The *Single-walk* methods involves using parallelism inside a single search process. In this approach, a typical way to develop the algorithm involves parallelising the exploration of the neighbourhood, e.g., dividing the neighbourhood into several sub-neighbourhoods and searching them in parallel to find the best move.

We observe the two mentioned levels of parallelism in the context of SAT (see [3] for a recent survey). An elaboration on multi-walk approaches with and without cooperation can be found in [1, 4], where the cooperation is implemented by sharing the best solution in order to properly craft a new starting point. In SAT, the single-walk approaches are implemented by flipping multiple variables at the same time [22].

The Comet solver [18] has been proposed in the context of constraint-based local search. Comet provides abstractions for implementing multi-walk parallelism (with and without cooperation) and single-walk parallelism.

17.6.2 Parallel Moves for ERDCMST

The aim of our parallel approach is to reduce the elapsed time for finding a target solution (i.e., optimal or near-optimal solution) by taking advantage of the availability of multiple cores and processors. In order to accomplish this, we propose a novel approach to perform multiple moves in parallel, which can be applied both in single-walk and multi-walk settings.

A move for ERDCMST can be defined as selecting and removing a node from a tree and adding it back to the tree; preferably in a different location. The general idea is to partition the set of all moves in such a way that when multiple moves are performed by selecting them from different elements of the partition no constraint is violated. We use this approach to develop a parallel algorithm for the ERDCMST problem.

Informally speaking the algorithm takes into account the disjointness constraint to divide the problem space into mutually exclusive subproblems and asynchronously perform multiple moves in parallel. We use the *fcg* to decompose the problem. Let us recall that if N facilities do not share any client we can use the LS algorithm to optimise the local solution of individual facilities in parallel. For instance, in Figure 17.3 m_3 and m_2 do not share clients, so we can improve the solution by executing two parallel copies of Algorithm 17.2 limiting the input solution to m_3 for one core, and m_2 for the other one. In this section we describe two methods to decompose the problem. The first method computes a set of independent facilities. The second method randomly selects a set of facilities and resolves the conflicts between clients before executing the LS algorithm.

Algorithm 17.3: Random Independent set(*fcg*, *card*)

```
1 S := ∅
2 while fcg ≠ ∅ and |S| < card :
3     v := random vertex in fcg
4     S := S ∪ v
5     Remove v and its neighbours from fcg
6 return S
```

Let Γ_{ij} be the set of all potential predecessors of client u_j in T_i when considering either a node move or a subtree move. Ideally, we would like to find a set of nodes (or clients) whose sets of locations are pairwise mutually exclusive so that moving all those nodes simultaneously in their trees is conflict-free. The advantage is that finding a best location for all such nodes can be done in parallel without restricting access to the data structures or creating duplicate copies of the same data structure.

As the sets of locations for the selected nodes must be independent, we select at most one node from a tree. It is indeed possible to find two nodes in the same tree whose sets of potential predecessors are mutually exclusive, but finding such nodes is not straightforward. Therefore, the number of moves that can be performed simultaneously is bounded by the number of facilities.

As mentioned before, changing the location of a node within a tree T_i is not only constrained by the other nodes of T_i but also by the nodes of the other trees sharing nodes with T_i because of the disjointness constraint. In order to determine the number of subproblems, we use the previously defined facility connectivity graph. In particular, we explore the following two mechanisms:

1. *Independent set* defines partitions by computing independent sets in *fcg*. In this approach, we know beforehand that each client has at most one predecessor, so all elements in the set can be safely executed in parallel without violating the disjointness constraint. Algorithm 17.3 computes a random set of independent elements in *fcg*. These elements will then be used in the parallel section of the algorithm to solve the problem. *card* refers to the cardinality of the independent set to be computed. Ideally, *card* should match the number of cores available. However, the degree of parallelism of the algorithm is bounded by the cardinality of the maximum independent set in *fcg*. For instance, let us revisit the *fcg* in Figure 17.3. In this case, we will have at most two independent facilities (i.e., m_3 and m_4). Therefore, even if *card* is greater than two, Algorithm 17.3 will return at most two vertices. In practice we expect sparse graphs in real networks, so the cardinality of independent sets should be more than a few tens of elements for realistically sized networks.
2. *Random conflict* selects, uniformly at random, n facilities and resolves the conflicts between clients a priori. It is recalled that two facilities can be in conflict if and only if they share at least two clients. Let us say that two facilities m_i and m_i' are selected, and the clients u_j and $u_{j'}$ are connected to both facilities. Let $C = \Gamma_{ij} \cap \Gamma_{i'j'}$ be a non-empty set. To resolve the conflict we modify the sets

Algorithm 17.4: Iterated Constraint-based Parallel Local Search (*move-op*, t)

```
1  s* := Initial Solution
2  repeat
3      {s*_1, ..., s*_n} := s*
4      {s'_1, ..., s'_n} := s*
5      P := CreatePartition(s*)
6      for p_i ∈ P do in parallel
7          while local time limit t for parallelism has not been reached :
8              if s*_i is internally in a local minimum :
9                  s'_i := Perturbation(s*_i)
10             s'*_i := ConstraintBasedLocalSearch(move-op, s'_i)
11             s*_i := AcceptanceCriterion(s*_i, s'*_i)
12     s* := {s*_1, ..., s*_n}
13 until a given stopping criterion is met
```

Γ_{ij} and $\Gamma_{i'j'}$ such that they become mutually exclusive. We recall that Γ_{ij} ($\Gamma_{i'j'}$) represents the set of potential predecessors for u_j ($u_{j'}$) in the tree T_i (T_i').

- If $u_k \in C$ is already a predecessor of u_j in T_i then we remove u_k from $\Gamma_{i'j'}$, or viceversa.
- If $u_k \in C$ is a predecessor of neither u_j in T_i nor $u_{j'}$ in T_i' then we remove u_k randomly from either Γ_{ij} or $\Gamma_{i'j'}$. We can say that the algorithm decides beforehand to which set u_k should belong.

Unlike *independent set* where the degree of parallelism is limited by the size of the maximum independent set, *random conflict* allows as many processes as the number of facilities in the problem, which in practice goes up to a few hundred cores. The solution of a problem whose *fcg* is the one in Figure 17.3 will start by randomly selecting a set of facilities, e.g., m_1, m_3, and m_4, and then the *random conflict* method resolves conflicts beforehand for conflicting clients. For instance, if edge $\langle u_{10}, u_{11} \rangle$ is present neither in the current solution of m_1 nor in the current solution of m_3, the algorithm randomly decides whether to forbid the edge in m_1 or in m_3. The algorithm repeats this process for all pairs of conflicting clients.

The Iterated Constraint-based Parallel Local Search algorithm (ICPLS) works in two phases. First, the algorithm selects a set of facilities, denoted by P. If the set is independent then the locations of the clients of the set facilities are mutually exclusive. If the set is in conflict then the locations of the clients of the set facilities are modified by restricting their locations to resolve the conflicts. Second, for each facility $p_i \in P$ it performs, in parallel, the sequential local search algorithm for a given amount of time t to explore the search space. As is the case for the sequential algorithm, s^* denotes the current incumbent solution of the problem, s' denotes the solution after perturbing the incumbent solution, s'^* denotes the best solution obtained after executing the constraint-based local search framework. s_i^* (respectively s_i' and $s_i'^*$) denotes the solution (or tree) associated with partition p_i. In the parallel section of the algorithm (Lines 6-11) we differentiate between the local minimum for each partition and the local minimum of the problem. In the sequential algorithm we diversify the current incumbent solution after finding a local minimum of the problem, i.e., a state in which no neighbour solution leads to an improvement in the objective. In the parallel algorithm we diversify when the solution s_i associated with partition p_i is in a local minimum but the global state of the whole problem is still unknown.

The sequential algorithm scans the list of active clients (*list*), deleting clients from the list when they cannot improve the objective function. The perturbation starts when a local minimum in the problem is reached (i.e., *list* = { }). Following the same approach in the parallel algorithm may introduce significant processor idle time, in particular when approaching a local minimum. Therefore, we start the perturbation locally for each tree as soon as an internal local minimum for a given tree is reached to minimise idle time. Moreover, after applying a move in a tree, only nodes of the same tree are added to *list* to reduce synchronisation among processors (Lines 23-24 in Algorithm 17.2).

Algorithm 17.4 shows the ICPLS proposed in this chapter. The algorithm starts with an initial solution. CreatePartitions computes P using either *independent set* of *random conflict*. We recall that *random conflict* computes the set of potential predecessors for all nodes connected to the selected facilities. In the parallel section of the algorithm (Lines 6-11), we check whether the tree associated with p_i is internally in a local minimum and if so perturb the solution associated with the partition (Lines 8-9). Then, the Constraint-based Local Search procedure is invoked with the current solution s_i associated with p_i using the acceptance criterion of the sequential algorithm. It is worth noticing that unlike the sequential algorithm, where a local minimum is sure to be reached after invoking the local search procedure, in the parallel algorithm it might be the case that the parallel algorithm has not reached the local minimum due to the local time limit of each parallel execution.

17.7 Application: Long-Reach Passive Optical Networks

To demonstrate the effectiveness of our approach we consider a real-world problem arising in the domain of optical networks. Long-Reach Passive Optical Networks (LR-PONs) are attracting increasing interest as they provide a low-cost and economically viable solution for fibre-to-the-home network architectures [21]. An example of a Long-Reach PON is shown in Figure 17.4. In LR-PON each metro node is connected to tens of thousands of customers via tens of hundreds of exchange sites. A major fault occurrence would be a complete failure of the metro node that terminates the LR-PON, which could affect tens of thousands of customers. The *dual homing* protection mechanism for LR-PON enables customers to be connected to two metro nodes via a local exchange site, so that whenever a single node fails all customers are still connected to a backup [15]. Simply connecting two metro nodes to an exchange site is not sufficient to guarantee the connectivity because if a link is common in

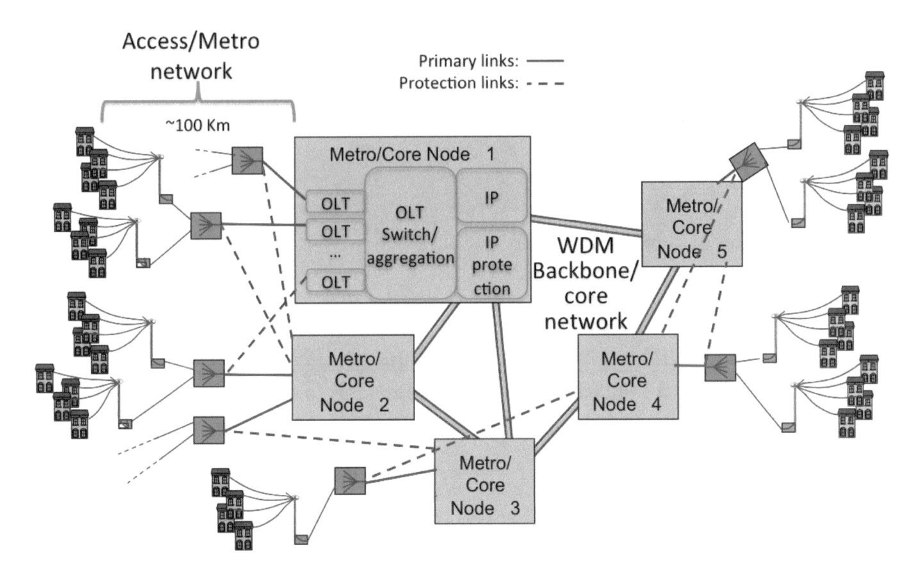

Fig. 17.4: Example of a Long-Reach Passive Optical Network

the routes of the fibre going from the exchange site to its two metro nodes then both metro nodes would be disconnected. The part of the LR-PON that one wants to protect is between the metro node and the old local exchange site. An important property of such a network is resilience to a single metro node failure. Nevertheless, connecting an exchange site to an additional metro node has a cost overhead. Here a metro node is a facility and an exchange site is a client.

In the optical network fibres are distributed from the metro nodes to the exchanges through cables that form a tree distribution network. As the association between metro nodes and exchange sites is already given, we could treat each one-to-many relation (i.e., tree) independently if disjointness were not an issue. However, the paths from the metro nodes to the exchange sites may share edges since a pair of exchange sites may be associated with the same pair of metro nodes. Therefore, we need to make sure that the routes that we choose for connecting the exchange sites to their metro nodes (i.e., main metro node and backup metro node) are disjoint. Otherwise, this would void the purpose of having double coverage. In Figure 17.5 we show two ways of connecting a given set of exchange sites to a metro node. In the first case (Figure17.5a) we simply connect each exchange site. Certainly the option of connecting each exchange site directly to the metro node leads to shorter connection paths. However, the drawback of connecting each exchange site directly is the total amount of cable used. In the second case (Figure 17.5b) we compute a minimum spanning tree rooted at the metro node. Certainly this option minimises the total length of cable but the drawback is that we might be violating the maximum cable distance allowed between the metro node and any of its exchange sites.

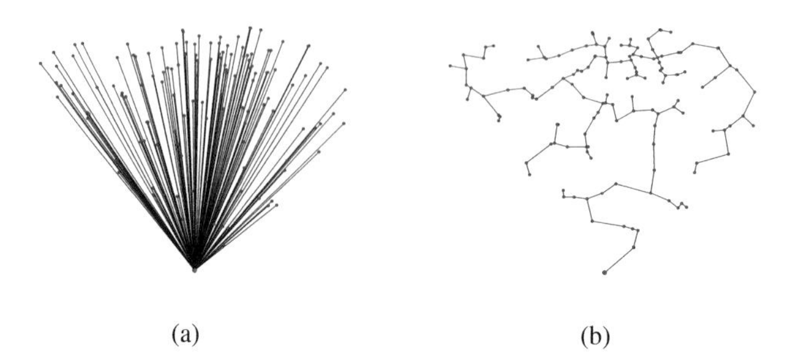

Fig. 17.5: (a) Local exchanges are directly connected to the metro node. (b) Local exchanges are connected to the metro node through a spanning tree.

We are interested in both restricting the length of the paths and the total amount of cable used. Keeping both requirements is known to be a hard problem [13]. As mentioned before this problem is computationally complex [19]. Notice that our problem is even more complicated than the bounded spanning tree problem. The objective of our problem is to determine the optimal routes of cables in the context of an already existing association of metro nodes with exchange sites such that the total cable length required for connecting each exchange site to two metro nodes is minimised subject to the maximum distance constraint and disjointness constraint. In other words, the idea is to find two edge-disjoint paths for all exchange sites to their respective metro nodes by maximising sharing (since we want to minimise the cost of digging, for example), and reducing the amount of cable.

17.8 Empirical Evaluation

In this section we present experimental results for the sequential and parallel versions of the local search algorithm proposed in this chapter to tackle ERDCMST. Moreover, we also demonstrate the effectiveness of the algorithm when compared to a dedicated algorithm for the RDCMST. We present results for random and real-world instances. The real-world instances correspond to the optical networks of three EU countries: Ireland, with 1,121 exchange sites and 18, 20, 22, and 24 Metro Nodes; the UK, with 5,393 exchange sites and 75, 80, 85, and 90 Metro Nodes; and Italy, with 10,709 exchange sites and 140, 150, 160, and 170 Metro Nodes. We present experimental results for the ERDCMST problem using both the sequential and parallel LS algorithms.

All the experiments were performed on a four-node cluster; each node features two Intel Xeon E5-2640 processors at 2.5 GHz, and 64 GB of RAM memory. Each processor has six cores for a total of 12 cores per node. The local search algorithm was implemented in C++ and used OpenMP to implement the parallel version using shared memory. In all the experiments we use the following parameters for Algorithm 17.1: $p = 5\%$, 20 random moves in the perturbation phase of the algorithm, and the algorithm stops after a given time limit is reached. Additionally, we use the trivial solution, i.e., direct connections from the MNs to the ESs, as an initial solution for the algorithm.

We now switch our attention to the ERDCMST problem. To this end, we evaluate the proposed algorithms using the following scenarios:

- *Real-life*: We consider real-life instances from our industrial partners with real networks in Ireland, the UK, and Italy. Figure 17.6 shows the histogram with the distribution of the distances from the ESs to their MNs. In the following experiments we use $\lambda = 67$ for the Ireland and $\lambda = 62.5$ for the UK and Italy. Additionally, Figure 17.7 shows the resulting facility connectivity graph for the three countries.
- *Random*: We generated 10 random instances extracted from the previous real-life network in Ireland. Each instance is generated by using 18 facilities and for each facility we randomly selected $|U| \in \{100, 200, \ldots, 1000\}$ nodes. Instances were generated by iteratively selecting a random client from the Irish dataset and balancing the load of clients per metro node.

17.8.1 ERDCMST Results: Sequential LS

Table 17.2 reports results for the *Random* experimental scenario, which depict the median value across 11 independent executions of the node and subtree operators; the best solution obtained with CPLEX; the best solution obtained with CPLEX using the solution of the first execution of LS with the subtree operator as warm start (LS+CPLEX); and the best known LBs for each instance obtained with CPLEX using a larger time limit.[1] The time limit for each local search experiment was set to 30 minutes, and 4 hours for the CPLEX-based approaches.

In these experiments we observe that the subtree operator generally outperforms the node operator. We attribute this to the fact that moving a complete subtree helps to maintain the structure of the tree in a single iteration of the algorithm. The node operator might eventually reconstruct the structure, however, more iterations would be required. For the CPLEX-based approaches we report the optimal solution for 100 and 200 clients, while the median execution of the local search approaches reported the optimal solution for 100 clients, and the subtree operator reached the optimal solution in five out of the 11 executions for 200 clients. After $|U| = 500$ LS dominates the performance and a margin ranging between 1% ($|U| = 500$) to 12% ($|U|=900$). LS+CPLEX was only able to improve the average performance of the subtree operator (reached after executing LS for one hour) by a very small factor, i.e., up to 0.4% for $|U| = 800$, after running CPLEX for four hours. We also experimented

[1] In this chapter CPLEX corresponds to solving the MIP model with IBM ILOG CPLEX Optimisation Studio version 12.5.1.

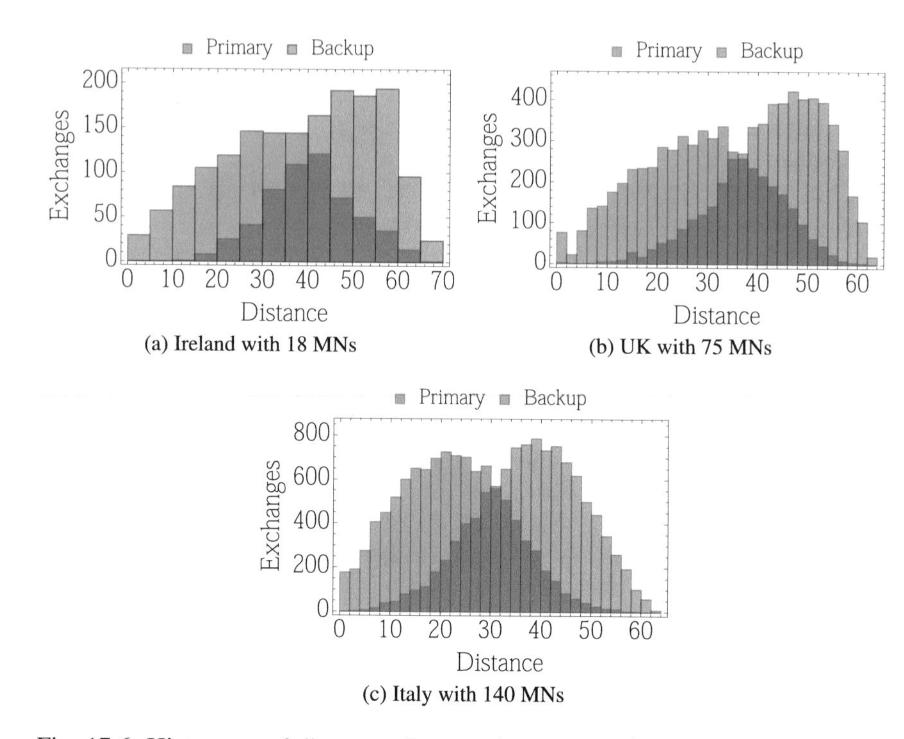

Fig. 17.6: Histogram of distances from exchanges to primary and backup metro nodes

Table 17.2: Results for the small-sized instances of ERDCMST problem where $|M| = 18$, $\lambda = 67$, 30 minutes time limit for LS approaches, 4 hours for CPLEX, and 5 hours for LS+CPLEX

$\|U\|$	LS (Subtree)	LS (Node)	CPLEX	LS+CPLEX	LB
100	**4674**	**4674**	**4674**	**4674**	4674
200	6966	6988	**6962**	**6962**	6962
300	8419	8575	**8404**	**8404**	8152
400	9728	10008	9728	**9721**	9329
500	**11203**	11672	11318	**11203**	10298
600	**11885**	12559	12276	11924	10517
700	13148	13981	13812	**13140**	11485
800	14040	15133	15118	**13977**	12402
900	**14770**	16098	16438	14839	12860
1000	**15962**	17479	18174	16009	13943

with instances with $|U| < 100$ and $|U| > 1000$. In the first case the three algorithms and the mixed approach (LS+CPLEX) reported similar results. In the second case,

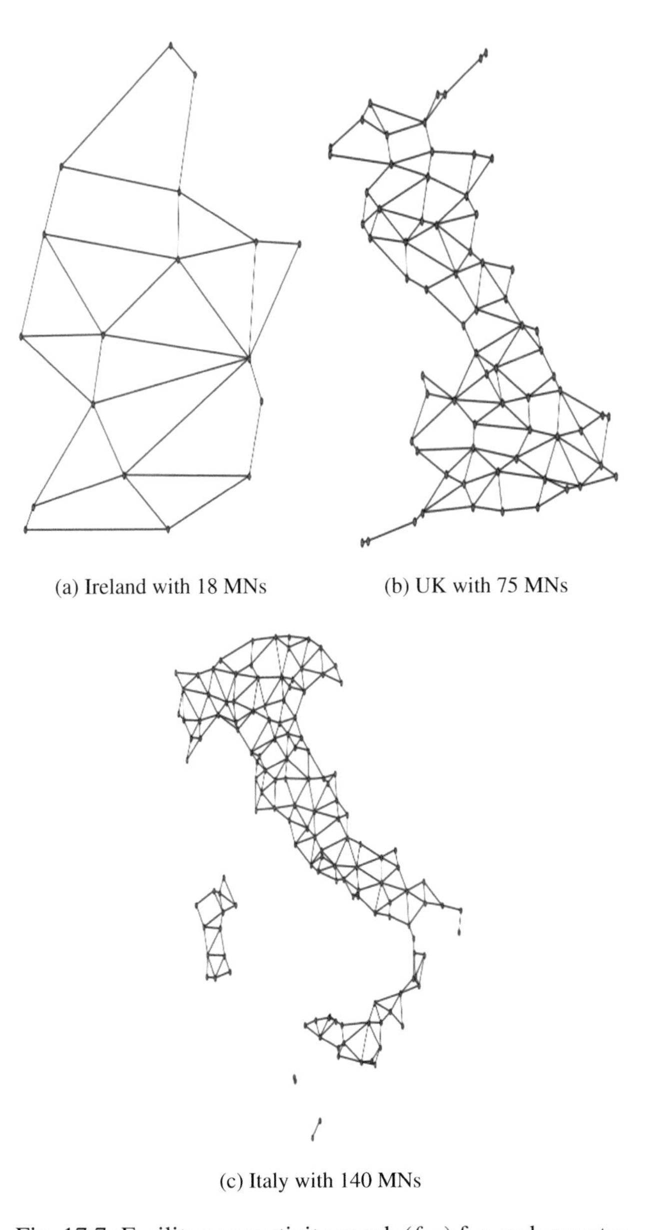

(a) Ireland with 18 MNs

(b) UK with 75 MNs

(c) Italy with 140 MNs

Fig. 17.7: Facility connectivity graph (*fcg*) for each country

Table 17.3: Results for Ireland, UK, and Italy with 30 minutes time limit (wall time) for the LS algorithm and 4 hours time limit for CPLEX (wall time)

Country	$\|M\|$	Subtree	CPLEX	LB	Gap-Subtree	Gap-CPLEX
	18	**17155**	26787	14809	13.67	44.71
Ireland	20	**16884**	83746	14845	12.07	82.27
$\|U\|$=1121	22	**16715**	79919	14990	10.32	81.24
	24	**16173**	26918	14570	9.91	45.87
	75	**66367**	285014	54720	17.54	80.80
UK	80	**65380**	301190	54975	15.91	81.74
$\|U\|$=5393	85	**64189**	281546	55035	14.26	80.45
	90	**62763**	220041	55087	12.23	74.96
	140	**90796**	–	76457	15.79	–
Italy	150	**89519**	–	76479	14.56	–
$\|U\|$=10708	160	**89497**	–	76794	14.19	–
	170	**88497**	–	77013	12.97	–

only LS with the subtree operator was able to provide good quality solutions with a gap of 10% with respect to the LB.

Our second set of experiments for the sequential version of the constraint-based local search algorithms are showed in Table 17.3 where we report results for real ERDCMST instances from Ireland, Italy, and the UK. In this case, we used a time limit of 30 minutes for LS (using the subtree move operator), and four hours for CPLEX. As can be observed, LS dominates the performance in all these experiments and, once again, the solution quality of LS does not degrade with the problem size. Indeed, the gap with respect to the LB for local search varies from 9.9% to 13.6% for Ireland, 12.2% to 17.5% for UK, and 12.9% to 15.7% for Italy. CPLEX ran out of memory when solving instances from Italy. We report '–' when no valid solution was obtained. For the UK instances CPLEX also ran out of memory before the time limit. Once again we would like to recall that algorithms such as BKRUS, PBH, and KBH cannot be used for the ERDCMST problem as they are dedicated algorithms for the RDCMST that rely on the use of shortest paths to build valid solutions. However, in the ERDCMST problem such paths might not be available due to disjointness.

17.8.2 ERDCMST Results: Parallel LS

In this section we evaluate the performance of the proposed parallel local search algorithm. To this end we use the same real-world instances used for the sequential algorithm for Ireland, the UK, and Italy. We limit our attention to the subtree operator as it greatly outperforms the node operator in the sequential setting.

We define the *gain* of the parallel algorithm as the relative percentage gain with respect to the sequential algorithm in the cost solution after a given time limit and using a given number of cores. Let $C(t, inst, c)$ be the cost of the best solution obtained after t seconds using c cores to solve $inst$. Let $T(cost, inst, c)$ be the time to reach a solution whose cost is at least as good as $cost$ for a given instance $inst$ using c cores.

$$Gain(inst, c, t) = \frac{C(t, inst, 1) - C(t, inst, c)}{C(t, inst, c)} \times 100.$$

Tables 17.4, 17.5, and 17.6 show the results of the empirical evaluation of the parallel algorithm. In these tables we present the cost of the solution for the sequential algorithm, the parallel algorithms, and the relative cost gain of the parallel algorithm after the 30-minute time limit. We use the multi-walk (MW) framework, i.e., executing multiple copies of the algorithm with different random seeds, as a baseline for comparison. We also include the proposed parallel algorithms using both *random conflict* (RC) and *independent set* (IS) for selecting multiple moves. For each instance and each approach we report the median value across 11 executions with a time-limit of 30 minutes.

Figures 17.8, 17.9, and 17.10 show the performance evolution of the algorithms (parallel and sequential) to tackle one instance for each dataset, i.e., Ireland with 18 facilities, the UK with 75 facilities, and Italy with 140 facilities. We remark that similar results have been observed for the remaining instances. The y-axis indicates the quality of the solution after certain time is reached (x-axis).

Ireland instance (Table 17.4 and Figure 17.8). The local search algorithms find a very good solution within a very short time window (gap of up to 13% with respect to the lower bound), and the variance between independent executions of the algorithm is very low. For this reason, when increasing the number of cores we observe very little difference in the performance of the algorithms.[2] We observe that

Table 17.4: Performance summary of the parallel algorithms (Multi-walk (MW), *random conflict* (RC) and *independent set* (IS)), with a 30-minute time limit (wall time)

Country	$\|M\|$	Seq	Alg	4 Cores		8 Cores		12 Cores	
				Cost	Gain	Cost	Gain	Cost	Gain
Ireland $\|U\|$=1121	18	17155	MW	17110	+0.26	17092	+0.37	17085	+0.41
			RC	17293	-0.83	17287	-0.86	17266	-0.77
			IS	17276	-0.69	17324	-0.76	17307	-0.85
	20	16884	MW	16841	+0.26	16829	+ 0.33	16828	+0.33
			RC	17001	-0.68	16998	-0.78	17015	-0.75
			IS	17007	-0.73	17014	-0.69	17006	-0.76
	22	16715	MW	16704	+0.07	16691	+0.14	16686	+0.17
			RC	16891	-1.01	16888	-1.05	16896	-1.04
			IS	16886	-1.00	16896	-1.02	16884	-0.98
	24	16173	MW	16152	+0.13	16148	+0.15	16136	+0.23
			RC	16315	-0.93	16318	-1.05	16347	-1.05
			IS	16327	-0.86	16327	-0.89	16323	-0.86

[2] Similar behaviour for other local search algorithms has been observed in [2] in the context of the satisfiability problem.

Table 17.5: Performance summary of the parallel algorithms (Multi-walk (MW), *random conflict* (RC) and *independent set* (IS)), with a 30-minute time limit (wall time)

Country	$\|M\|$	Seq	Alg	4 Cores		8 Cores		12 Cores	
				Cost	Gain	Cost	Gain	Cost	Gain
UK $\|U\|$=5393	75	66367	MW	66153	+0.32	66126	+0.36	66093	+0.41
			RC	65083	+1.99	64988	+1.73	64972	+1.82
			IS	65083	+1.96	64980	+1.81	64981	+1.89
	80	65380	MW	65328	+0.08	65282	+0.15	65265	+0.18
			RC	64290	+1.65	64227	+1.35	64217	+1.53
			IS	64328	+1.62	64207	+1.45	64195	+1.55
	85	64189	MW	64168	+0.03	64146	+0.07	64131	+0.09
			RC	63487	+1.11	63433	+0.97	63421	+1.02
			IS	63486	+1.10	63403	+1.01	63414	+1.04
	90	62763	MW	62726	+0.06	62651	+0.18	62641	+0.19
			RC	62210	+0.86	62171	+0.73	62153	+0.84
			IS	62260	+0.81	62191	+0.73	62140	+0.87

Table 17.6: Performance summary of the parallel algorithms (Multi-walk (MW), *random conflict* (RC) and *independent set* (IS)), with a 30-minute time limit (wall time)

Country	$\|M\|$	Seq	Alg	4 Cores		8 Cores		12 Cores	
				Cost	Gain	Cost	Gain	Cost	Gain
Italy $\|U\|$=10709	140	90796	MW	90669	+0.14	90633	+0.18	90621	+0.19
			RC	88573	+2.51	88382	+2.71	88332	+2.80
			IS	88529	+2.56	88358	+2.74	88330	+2.71
	150	89519	MW	89427	+0.1	89357	+0.18	89309	+0.24
			RC	87517	+2.27	87411	+2.42	87414	+2.41
			IS	87526	+2.26	87462	+2.37	87379	+2.43
	160	89537	MW	89421	+0.13	89360	+0.20	89309	+0.26
			RC	87679	+2.15	87579	+2.24	87525	+2.29
			IS	87666	+2.13	87564	+2.26	87528	+2.31
	170	88497	MW	88433	+0.07	88359	+0.16	88359	+0.16
			RC	86954	+1.76	86925	+1.83	86862	+1.89
			IS	86955	+1.77	86869	+1.86	86869	+1.88

the sequential algorithm is slightly better than the parallel ones with a percentage gain of between 0.68% to 1.05% within the 30-minute time limit. However, we note that the parallel algorithm reaches good solutions faster than the sequential one as depicted in Figure 17.8. This figure also helps to illustrate the difference between using *independent sets* and *random conflict* for computing the partitions. Notice that eight-core random conflict algorithm reports a better performance than 12-core *independent set*. That is because the cardinality of the maximum independent set for this problem is nine and the number of parallel processes is bounded by that number, thus voiding the advantage of having three more cores. *Random conflicts* allows as many parallel processes as the number of metro nodes in the problem.

UK instance (Table 17.5 and Figure 17.9). The UK instance is about four times bigger (with respect to the number of clients) than the Irish instance. Except for the multi-walk approach (where no significance improvement is seen), we observe that the parallel algorithms improve the quality of the solutions when increasing the number of cores. Summing up, the observed performance gain ranges from 0.81% to 1.99% (four cores), from 0.73% to 1.81% (eight cores), and 0.84% to 1.89% (12 cores). Moreover, as depicted in Figure 17.9, the parallel algorithm based on single walk also reaches a very high-quality solution much faster than the sequential algorithm, and the performance increases as the number of cores increases. However, in this case, we observe similar performances between *independent set* and *random conflict*. That is because the independent sets are always larger than 12 and therefore both approaches exploit parallelism as much as possible.

*Italy instance (Table 17.6 and Figure 17.10).*The largest performance improvement of the parallel algorithm is observed for Italy (Table 17.6). We attribute this to the size of the problem: the larger the problem the better the parallel algorithms perform. Here we observe a performance gain ranging from 1.76% to 2.56 (four cores), 1.83% to 2.74% (eight cores), and 1.88% to 2.80% (12 cores). Similarly to the Irish and the UK datasets, Figure 17.10 shows the performance in time of the parallel algorithm, and once again it can be observed that the parallel version reaches very good solutions faster than the sequential algorithm.

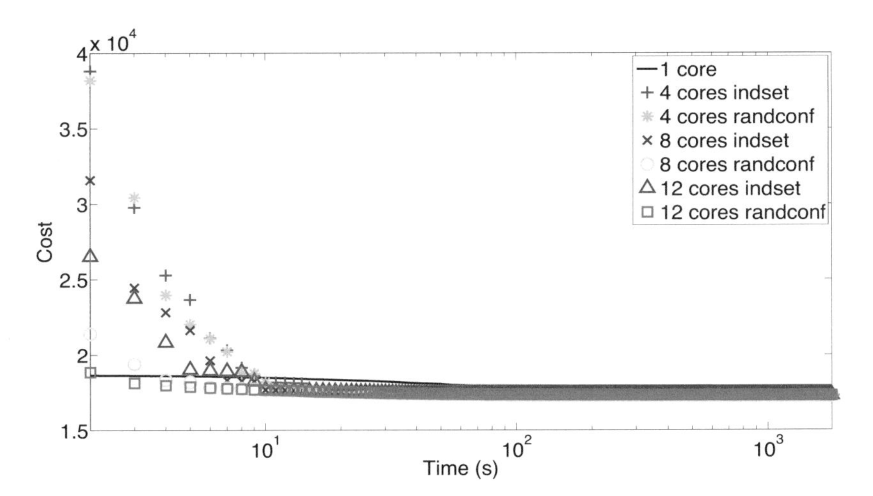

Fig. 17.8: Ireland ($|M| = 18$): Solution cost vs. wall clock time (range [2, 1800] seconds)

As pointed out before, the sequential algorithm obtains better solutions for the Irish dataset after the 30-minute time limit, however, the parallel algorithm computes near-optimal solutions faster than the sequential algorithm. For instance, as shown in Table 17.7, the average gain (out of the four scenarios for each country with respect to the sequential algorithm) after 100 seconds using four cores is 1.62% (independent

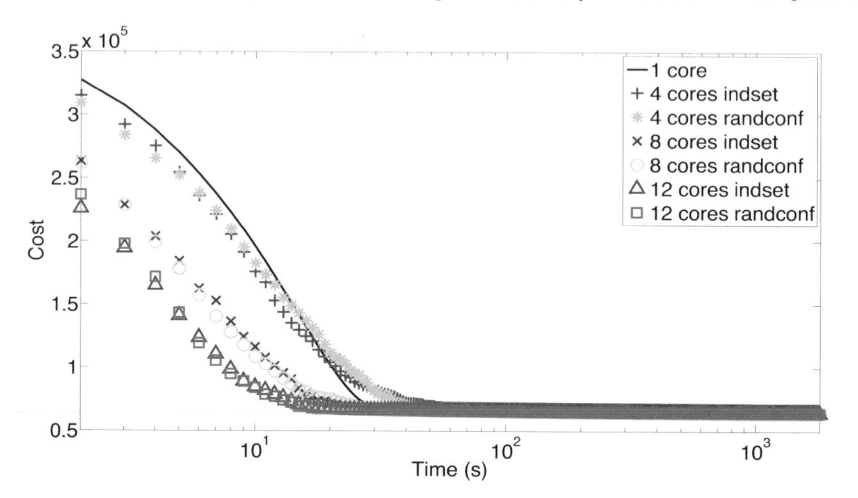

Fig. 17.9: UK ($|M| = 75$): Solution cost vs. wall clock time (range [2, 1800] seconds)

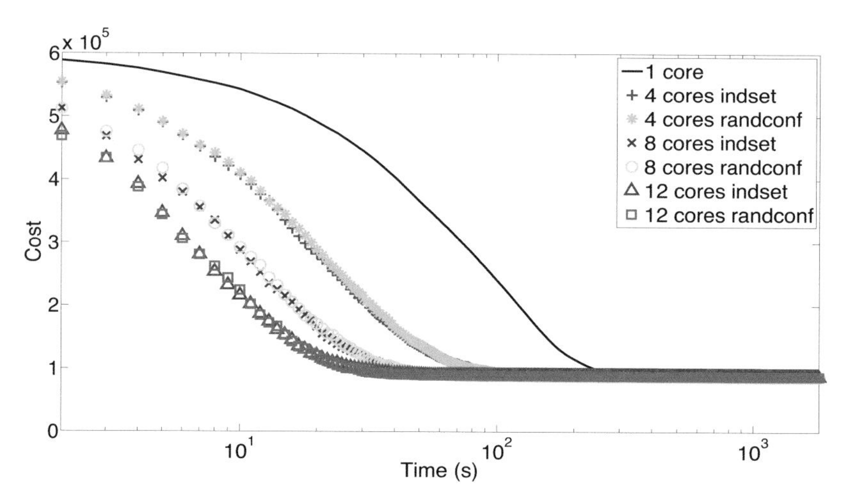

Fig. 17.10: Italy ($|M| = 140$): Solution cost vs. wall clock time (range [2, 1800] seconds)

set) and 1.69% (*random conflict*), and after 10 seconds we observe a gain of up to 4.29% for *random conflict* and 1.39 % for *independent set*. Once again we observe that when the cardinality of the maximum independent set is small with respect to the number of cores *random conflict* performs better than *independent set*. Interestingly, the relative gain of the parallel algorithm with respect to the sequential one is more than 100% on two occasions for the UK (up to 130% for RC with 12 cores and 100 secs) and on eight occasions for Italy (up to 182% for IS and RC with 12 cores and 100 secs).

Table 17.7: Average gain with different time settings for each country

Country	Time Secs	4 cores IS	4 cores RC	8 cores IS	8 cores RC	12 cores IS	12 cores RC
Ireland	10	-3.70	-6.05	1.39	4.00	0.65	4.29
	100	1.64	1.56	1.62	1.69	1.66	1.64
	1000	-0.75	-0.79	-0.82	-0.78	-0.78	-0.82
	1800	-0.82	-0.86	-0.83	-0.93	-0.86	-0.90
UK	10	6.88	5.29	66.53	77.37	125.85	130.32
	100	3.86	3.82	4.70	4.58	5.06	5.01
	1000	2.13	2.15	2.37	2.34	2.42	2.40
	1800	1.37	1.40	1.25	1.19	1.33	1.30
Italy	10	30.99	30.34	78.36	79.13	131.27	128.50
	100	167.63	168.96	181.79	180.87	182.30	182.17
	1000	2.75	2.71	2.94	2.92	3.00	2.99
	1800	2.19	2.16	2.30	2.3	2.33	2.34

Finally, Table 17.8 concludes the experiments reporting the average speedup factor (out of the four scenarios for each country) to reach the best solution after a given amount of time, i.e., 10, 100, 1000, 1800 seconds. We recall that the speedup factor is the gain in the speed of the parallel algorithm. We compute the speedup factor of the parallel algorithm with c cores after t seconds to solve a given instance *inst* as follows:

$$Speedup(inst,c,t) = \frac{T(C(t,inst,1),inst,1)}{T(C(t,inst,1),inst,c)}.$$

Table 17.8: Average speedup factor with different time settings for each country

Country	Time Secs	4 cores IS	4 cores RC	8 cores IS	8 cores RC	12 cores IS	12 cores RC
Ireland	10	0.86	0.79	1.19	1.76	1.06	2.62
	100	6.90	5.86	9.45	10.34	8.74	12.78
	1000	–	–	–	–	–	–
	1800	–	–	–	–	–	–
UK	10	1.07	1.02	2.13	2.13	3.37	3.00
	100	1.86	2.03	3.58	3.54	5.25	5.31
	1000	10.02	9.81	18.43	16.80	26.31	24.26
	1800	7.49	7.31	13.84	12.54	18.96	17.94
Italy	10	3.00	3.00	4.50	4.50	4.50	4.5
	100	4.00	4.03	7.82	7.68	11.00	10.79
	1000	7.71	7.97	15.54	15.65	21.18	22.30
	1800	12.22	11.72	22.98	20.57	28.60	30.38

In Table 17.8 we report '–' in those cases where a parallel solution was not obtained within the timeout. In the 10 seconds time-limit scenario, we observe a speedup factor close to 1 for the UK and Italy. The speedup factor improves considerably after 1000 seconds. It goes up to a factor of 26 for the UK (IS with 12 cores and 1000 secs) to a factor of 30 with eight cores for Italy (RC with 12 cores and 1800 secs).

17.9 Conclusions and Future Work

We have presented an efficient local search algorithm for solving the Edge-disjoint Rooted Distance-Constrained Minimum Spanning-Tree problem. We presented two move operators along with their complexities and an incremental evaluation of the neighbourhood and the objective function. Furthermore, we have proposed a parallelisation scheme for the local search algorithm, which significantly reduces the time required by the sequential version to reach high-quality solutions. Any problem involving tree structures could benefit from these ideas and the techniques presented are relevant for a constraint-based local search framework where this type of incrementality is needed for network design problems. The effectiveness of our approach is demonstrated by experimenting with a set of problem instances taken from real-world long-reach passive optical network deployments in Ireland, Italy, and the UK.

In the future we would like to extend ERDCMST with the notion of optional nodes, since this extension is a common requirement in several applications of ERDCMST. Effectively this means that we would compute for every facility a Minimum Steiner Tree where all clients are covered but the path to them may follow some optional nodes. We also plan to investigate alternative heuristics for selecting the most suitable node and subtree for deletion at each iteration of the local search algorithm.

Acknowledgments

This work was supported by DISCUS (FP7 Grant Agreement 318137), and Science Foundation Ireland (SF) Grant No. 10/CE/I1853. The Insight Centre for Data Analytics is also supported by SFI under Grant Number SFI/12/RC/2289.

References

[1] A. Arbelaez and P. Codognet. Massivelly parallel local search for SAT. In *24th IEEE International Conference on Tools with Artificial Intelligence, ICTAI'12*, pages 57–64, Athens, Greece, November 2012. IEEE Computer Society.

[2] A. Arbelaez and P. Codognet. From sequential to parallel local search for SAT. In *13th European Conference on Evolutionary Computation in Combinatorial Optimisation, EvoCOP'13*, volume 7832 of *Lecture Notes in Computer Science*, pages 157–168. Springer, 2013.

[3] A. Arbelaez and P. Codognet. A survey of parallel local search for SAT. In *Theory, Implementation, and Applications of SAT Technology. Workshop at JSAI'13*, Toyama, Japan, June 2013.
[4] A. Arbelaez and Y. Hamadi. Improving parallel local search for SAT. In *5th International Conference on Learning and Intelligent Optimization LION 5*, volume 6683 of *Lecture Notes in Computer Science*, pages 46–60. Springer, 2011.
[5] A. Arbelaez, D. Mehta, B. O'Sullivan, and L. Quesada. Constraint-based local search for the distance-and capacity-bounded network design problem. In *26th IEEE International Conference on Tools with Artificial Intelligence, ICTAI'14*, pages 178–185. IEEE, 2014.
[6] A. Arbelaez, D. Mehta, B. O'Sullivan, and L. Quesada. A constraint-based local search for edge disjoint rooted distance-constrained minimum spanning tree problem. In *12th International Conference on Integration of AI and OR Techniques in Constraint Programming, CPAIOR'15*, volume 9075 of *Lecture Notes in Computer Science*, pages 31–46. Springer, 2015.
[7] R. Baraglia, J. I. Hidalgo, and R. Perego. A parallel hybrid heuristic for the TSP. In *Applications of Evolutionary Computing, EvoWorkshops 2001: EvoCOP, EvoFlight, EvoIASP, EvoLearn, and EvoSTIM*, volume 2037 of *Lecture Notes in Computer Science*, pages 193–202. Springer, 2001.
[8] Y. Caniou, D. Diaz, F. Richoux, P. Codognet, and S. Abreu. Performance analysis of parallel constraint-based local search. In *17th ACM SIGPLAN Symposium on Principles and Practice of Parallel Programming, PPOPP'12*, pages 337–338. ACM, 2012.
[9] T. G. Crainic and M. Gendreau. Cooperative parallel tabu search for capacitated network design. *J. Heuristics*, 8(6):601–627, 2002.
[10] P. Q. Dung, Y. Deville, and P. Van Hentenryck. Constraint-based local search for constrained optimum paths problems. In *9th International Conference on Integration of AI and OR Techniques in Contraint Programming for Combinatorial Optimzation Problems, CPAIOR'19*, volume 7298 of *Lecture Notes in Computer Science*, pages 267–281. Springer, 2010.
[11] E. Eaton, C. P. Gomes, and B. C. Williams. Computational sustainability. *AI Magazine*, 35(2):3–7, 2014.
[12] C. Gao, Y. Shi, Y. T. Hou, H. D. Sherali, and H. Zhou. Multicast communications in multi-hop cognitive radio networks. *IEEE Journal on Selected Areas in Communications*, 29(4):784–793, 2011.
[13] J. M. Ho and D. T. Lee. Bounded diameter minimum spanning trees and related problems. In *Proceedings of the Fifth Annual Symposium on Computational Geometry*, SCG '89, pages 276–282, New York, USA, 1989. ACM.
[14] H. Hoos and T. Stützle. *Stochastic Local Search: Foundations and Applications*. Morgan Kaufmann, 2005.
[15] D. K. Hunter, Z. Lu, and T. H. Gilfedder. Protection of long-reach PON traffic through router database synchronization. *Journal of Optical Communications and Networking*, 6(5):535–549, 2007.

[16] M. Leitner, M. Ruthmair, and G. R. Raidl. Stabilized branch-and-price for the rooted delay-constrained Steiner tree problem. In J. Pahl, T. Reiners, and S. Voß, editors, *INOC*, volume 6701 of *Lecture Notes in Computer Science*, pages 124–138. Springer, 2011. ISBN 978-3-642-21526-1.

[17] R. Martins, V. M. Manquinho, and I. Lynce. An overview of parallel SAT solving. *Constraints*, 17(3):304–347, 2012.

[18] L. Michel, A. See, and P. Van Hentenryck. Parallel and distributed local search in comet. *Computers and Operations Research*, 36:2357–2375, 2009.

[19] J. Oh, I. Pyo, and M. Pedram. Constructing minimal spanning/Steiner trees with bounded path length. *Integration*, 22(1-2):137–163, 1997.

[20] S. Pant. *Design and Analysis of Power Distribution Networks in VLSI Circuits*. PhD thesis, The School of Electrical Engineering in The University of Michigan, 2008.

[21] D. B. Payne. FTTP deployment options and economic challenges. In *Proceedings of the 36th European Conference and Exhibition on Optical Communication (ECOC 2009)*, 2009.

[22] A. Roli. Criticality and parallelism in structured SAT instances. In *8th International Conference on Principles and Practice of Constraint Programming, CP'02*, volume 2470 of *Lecture Notes in Computer Science*, pages 714–719, Ithaca, NY, USA, 2002. Springer.

[23] M. Ruffini, L. Wosinska, M. Achouche, J. Chen, N. J. Doran, F. Farjady, J. Montalvo-Garcia, P. Ossieur, B. O'Sullivan, N. Parsons, T. Pfeiffer, X. Qiu, C. Raack, H. Rohde, M. Schiano, P. D. Townsend, R. Wessäly, X. Yin, and D. B. Payne. DISCUS: an end-to-end solution for ubiquitous broadband optical access. *IEEE Communications Magazine*, 52(2):24–56, 2014.

[24] M. Ruthmair and G. R. Raidl. A Kruskal-based heuristic for the rooted delay-constrained minimum spanning tree problem. In R. Moreno-Díaz, F. Pichler, and A. Quesada-Arencibia, editors, *EUROCAST*, volume 5717 of *Lecture Notes in Computer Science*, pages 713–720. Springer, 2009. ISBN 978-3-642-04771-8.

[25] M. Ruthmair and G. R. Raidl. Variable neighborhood search and ant colony optimization for the rooted delay-constrained minimum spanning tree problem. In R. Schaefer, C. Cotta, J. Kolodziej, and G. Rudolph, editors, *PPSN (2)*, volume 6239 of *Lecture Notes in Computer Science*, pages 391–400. Springer, 2010. ISBN 978-3-642-15870-4.

[26] O. V. Shylo, T. Middelkoop, and P. M. Pardalos. Restart Strategies in Optimization: Parallel and Serial Cases. *Parallel Computing*, 37(1):60–68, 2011.

[27] R. Sigua. *Fundamentals of Traffic Engineering*. University of the Philippines Press, 2008.

[28] C. Truchet, A. Arbelaez, F. Richoux, and P. Codognet. Estimating parallel runtimes for randomized algorithms in constraint solving. *J. of Heuristics*, 22 (4):613–648, 2016.

[29] P. Van Hentenryck and L. Michel. *Constraint-based local search*. The MIT Press, 2009.

[30] M. Verhoeven and E. Aarts. Parallel local search. *Journal of Heuristics*, 1(1): 43–65, 1995.
[31] M. Verhoeven and M. Severens. Parallel local search for Steiner trees in graphs. *Annals of Operations Research*, 90:185–202, 1999.
[32] X. Yuan and A. Saifee. Path selection methods for localized quality of service routing. In *10th International Conference on Computer Communications and Networks, ICCCN'01*, pages 102–107. IEEE, 2001.

List of Algorithms

1.1 A Generic Local Search Algorithm ... 6
1.2 The DPLL Algorithm ... 7
1.3 The CDCL Algorithm ... 8
2.1 The General Framework of the Procedure CreateCubes ... 39
2.2 The Procedure CreateCubes* with the Cutoff Mechanism ... 41
2.3 The Pseudo-Code of SolveCubes Using the Partition ... 43
3.1 Linear Search SAT-UNSAT Algorithm ... 67
3.2 Linear Search UNSAT-SAT Algorithm ... 68
3.3 WMSU3 Algorithm ... 69
3.4 Fu-Malik for Weighted MaxSAT Algorithm ... 70
4.1 Pseudocode of QCDCL ... 109
4.2 Splitting Algorithm for QBF Evaluation ... 112
5.1 The CS-SDSMT Algorithm ... 150
5.2 An Interpolation-based Reconciliation Algorithm ... 159
7.1 Naive Computation of the Least Model ... 241
7.2 Basic SMODELS Procedure ... 243
7.3 Parallel Grounding on Beowulf Cluster (from [6]) ... 248
7.4 Component Level Parallelism ... 249
7.5 Rule Level Parallelism (adapted from [68]) ... 250
7.6 Single-Rule Level Parallelism (adapted from [68]) ... 251
7.7 Overall Structure of a Parallel Search ASP Computation ... 256
7.8 Naive Lookahead ... 263
7.9 Parallel Lookahead ... 263
7.10 GPU-ASP-Computation ... 267
7.11 Stratified Datalog Computation ... 272
8.1 A Generic Tree Search Algorithm ... 284
8.2 A Generic Branch-and-Bound Algorithm ... 287
8.3 Basic Racing Algorithm ... 309
8.4 Static Load-Balancing Algorithm ... 310
8.5 Master (Master-Worker) ... 312
8.6 Worker (Master-Worker) ... 312

Y. Hamadi und L. Sais (eds.), *Handbook of Parallel Constraint Reasoning*,
https://doi.org/10.1007/978-3-319-63516-3

8.7 Supervisor (Supervisor-Worker) 314
8.8 Worker (Supervisor-Worker) 315
8.9 Master (Master-Hub-Worker) 316
8.10 Hub Master (Master-Hub-Worker) 317
8.11 Worker (Master-Hub-Worker) 318
8.12 Self Coordination Algorithm 318
11.1 A* 423
11.2 Simple Parallel A* (SPA*) 427
11.3 Decentralized A* with Local OPEN/CLOSED lists 428
12.1 Depth-First Search Algorithm 474
12.2 Sequential Emptiness Check for Weak TGBAs Based on DFS 476
12.3 Nested Depth-First Search Algorithm 477
12.4 SCC-Based Emptiness Check 480
12.5 A Parallel Search Algorithm for Checking the Emptiness of Terminal Automata 482
12.6 A parallel DFS algorithm for checking emptiness of weak automata 483
12.7 CNDFS, a Multi-Core Algorithm for LTL Model Checking 486
12.8 Concurrent Union-Find Data Structure 489
12.9 Swarmed SCC-Based Algorithm 490
12.10 UFSCC Algorithm: Improved Swarmed SCC Algorithm 491
12.11 OWCTY Algorithm 495
12.12 MAP Algorithm 496
13.1 The BDD Algorithm `and`, with the BDDs x and y as Parameters ... 515
13.2 The Algorithm (left) is Implemented (right) Using `SPAWN`, `SYNC` and `CALL` 520
13.3 The Implementation of Work-Stealing Using Leapfrogging when Waiting for a Stolen Task to Finish, i.e., steal from the thief 522
13.4 Parallelized BDD Algorithm `exists`, with the BDD x and V the Cube of Variables that are Abstracted via Existential Quantification . 523
13.5 The Parallel Algorithm `relnext`, which Given the BDDs S (representing a set of states), R (representing a transition relation) and V (the cube of interleaved variables $\mathbf{x} \cup \mathbf{x}'$) Computes the Set of Successor States Defined on $\mathbf{x}$, i.e., $(\exists \mathbf{x}\colon (S \wedge R))[\mathbf{x}' := \mathbf{x}]$. We Assume that all Variables in R are also in V 524
13.6 Algorithm for Parallel `find-or-insert` of the Hash Table, with 512 Buckets per Region. The Variable `myregion` is a Thread-Specific Variable 530
13.7 The `cache-put` Algorithm 532
13.8 The `cache-get` Algorithm 533
14.1 Tree Search Algorithm 552
14.2 HS-TREE ALGORITHM 553
14.3 PROCESSNODE 554
14.4 DIAGNOSELW: Level-Wise Parallelization 563
14.5 DIAGNOSEFP: Full Parallelization 564
14.6 QUICKXPLAIN (QXP) 567

14.7 MERGEXPLAIN (MXP) 568
15.1 Portfolio Configuration Procedure GLOBAL 599
15.2 Portfolio Configuration Procedure PARHYDRA 600
15.3 Portfolio Configuration Procedure PARHYDRA$_b$ 605
17.1 Iterated Constraint-Based Local Search (*move-op*, *s*) 638
17.2 Constraint-Based Local Search (*move-op*, $\{T_1, \ldots, T_{|\mathscr{M}|}\}$) 646
17.3 Random Independent set(*fcg*, *card*) 648
17.4 Iterated Constraint-based Parallel Local Search (*move-op*, *t*) 649

Index

ω-regular language, 463

A* search algorithm, 423, 562
abstraction
  algorithm, 295
  communication, 306
  implementation, 296
  interface, 295
accepting run
  definition, 464
  lasso-shaped, 467
accepting SCC, 468
ACPP: Global, 599
ACPP: ParHydra, 600
ACPP: parHydra$_b$, 605
adaptivity, 298
admissible heuristic, 424
agent-based modeling, 398
algorithm
  abstract parallel, 308
  abstraction, 295
  comparison, 326
  correctness, 290
  deterministic, 17, 85, 290, 304, 319
  effectiveness, 290, 291
  framework, 295, 308, 321
  integration, 295
  parallel, 290, 326
  phase, 292
  separation, 298
  sequential, 286, 325
  underlying sequential, 293
algorithm configuration, 596, 597
algorithm parameters, 595
answer set, 242
  computation, 246, 256, 267
  constraint, 243, 275
  grounder, 239, 247–252
antecedent clause, 39, 109
Aquarius, 180, 181, 202, 207
ASlib, 591
asserting clause, 110
assignment cache, 121
assignment tree, 103, 106, 108, 109, 113, 119
associative-commutative symbol, 208, 209
assumption-based reasoning, 121
automatic construction of parallel portfolios, 585, 596
automaton
  Büchi, 463
  degeneralization, 465
  terminal, 471
  weak, 471

backjumping, 9, 121, 212, 246
backtrack, 8, 108, 184, 255, 339, 445, 474
backward contraction, 186–188, 190, 195–197, 203, 205–208, 213
Beowulf, 248
binary decision diagram, 458, 509–541, 618, 620
bisimulation, 510, 535
bisimulation minimization, 537
blocking, 209
bloqqer, 117–119, 126, 128
bounded expansion, 104
bounding, 286, 288, 299
branching, 8, 16, 36, 39, 51, 79, 287–289, 319–322, 325, 340–342, 538
  method, 288
  pseudocost, 299
  strategy, 285, 288
  strong, 297
breadth-first search, 433, 474, 492, 553

Y. Hamadi und L. Sais (eds.), *Handbook of Parallel Constraint Reasoning*,
https://doi.org/10.1007/978-3-319-63516-3

C-reduction, 183
caching, 126, 183, 184, 473, 532
callback, 320
caqe, 115, 116, 118, 125
cardinality constraints, 66
CDCL, 7, 81, 103, 104, 108, 109, 212, 213, 215–217
CEGAR, 125, 129
CL-SDSAT, 214
clasp, 239, 247, 267, 270
claspfolio, 274
clausal simplification, 185, 186, 189, 206
clause diffusion, 180, 181, 190, 202–212, 214, 215
clause learning, 33, 81, 108, 128, 212, 255
clause sharing, 16, 17, 22–24, 44, 61, 72, 81, 82, 93, 594, 598, 602, 603, 609
clingo, 247
column generation, 319
communication protocol
 MPI, 307
 OpenMP, 307
 PVM, 307
completion procedure, 187, 192, 196, 208
computational platform, *see* platform
concurrent rewriting, 192
configuration space, 596
conflict, 81, 245, 267–270, 552
 analysis, 121, 269, 289
 graph, 289
 MERGEXPLAIN, 568
 QUICKXPLAIN, 567
 search algorithms, 566–567
conflict clause, 35, 39, 50, 126, 212–214, 216
contraction-based strategies, 181, 186, 187, 190, 201, 215
cooperative parallelism, 394, 404, 405
coordination, 303
coordination mechanism, 308
 master-hub-worker, 313
 master-worker, 310
 multiple-master-worker, 313
 parallel racing, 308
 self coordination, 315
 supervisor-worker, 311
CPTHEO, 199, 201
cube, 213, 214, 217
cube learning, 128
CUDA parallelism, 265
cut, *see* cutting plane
cutting plane, 287, 298

Datalog, 239, 244, 248, 252, 265, 271, 272
decomposition, 10, 157, 298, 345, 478, 561
degeneralization, 465
delta debugging, 131
dependency graph, 243, 244, 247, 249, 250, 253, 265
DepQBF, 115–121
depth-first search, 8, 184, 358, 474, 487, 575
determinism, 87, 303, 304, 319, 364
deterministic parallelism
 strong, 304
 weak, 305
deterministic solver, 85
diagnosis, 551
 parallel algorithms
  Boolean-HS-Tree, 577
  evaluation, 569–574
  full parallelization, 564
  hybrid strategy, 575
  join relation, 576
  leading diagnoses, 574
  level-wise parallelization, 563
  MapReduce, 576
  node and conflict search, 567
  parallel random depth-first search, 575
 parallelization strategies
  node processing, 559
  tree decomposition, 561
  window-based processing, 560, 562
distributed fairness, 206, 207
distributed global contraction, 206, 207
distributed proof reconstruction, 206, 207
distributed search, 93, 191, 198, 202, 203, 206, 208–211, 213, 215, 217
distributed-memory algorithms, 9, 257, 315, 444, 445, 492
divide and conquer, 10, 151, 179, 211, 450, 486, 535, 565, 566
DLV, 239, 247, 248
DPLL, 5, 103, 108, 124, 189, 211–213, 215, 242, 245
duality-aware reasoning, 103, 129
dynamic synchronization, 91

efficiency, *see* parallel
 parallel, 324
emptiness check
 parallel scc based, 487
 problem statement, 466
EQP, 208–210
existential quantification, 522–523
existential reduction, 107, 110, 111
expansion-based QBF solving, 104, 111, 129
expansion-based solving, 124
expansion-oriented strategies, 181, 186–188, 190, 213

factoring, 184, 185, 205, 206
fairness, 189, 201, 207
feasible region, 284
folding-up, 183, 184
forward contraction, 186, 188, 189, 194, 195, 197, 203, 204, 206, 207, 213, 214
Fu-Malik algorithm, 70
fuzz testing, 131

gap
  absolute, 287
  optimality, 287
  relative, 288
garbage collection, 533–535
geometric mean, 328, 363
  shifted, 328
given-clause algorithm, 185, 194–196, 200, 208, 209
Google File System, 270
GPU parallelism, 264–265
  ASP, 266
  Datalog, 265
GPU thread, 265, 389
granularity, *see* task
gringo, 239, 247
grounding, 239, 247–252
guarantee formula, 471
guiding path, 79, 113, 114, 121, 126, 212, 213

hard clauses, 65
hash distributed A* (HDA*), 431
hashing, abstract Zobrist, 436
hashing, abstraction-based, 435
hashing, hyperplane work distribution, 438
hashing, operator-based Zobrist, 435
hashing, Zobrist, 434
helpful master scheduling, 125
Herbrand
  function, 130
  model, 241, 272
heterogeneous systems, 198, 199, 201, 202
heuristic
  function, 424
  primal, 287, 289, 298
hiqqer, 117
hiqqerfork, 111, 116, 117
hitting set tree search, 553–556
homogeneous systems, 198, 202
HordeQBF, 111, 116, 120, 121
HordeSAT, 111, 120
Horn clause, 182, 241
HPDS, 199
hqspre, 128
HS-Tree search, *see* hitting set tree search

hyperresolution, 185, 186, 199, 206, 215, 217

idle time, *see* overhead
implication graph, 81, 117
independent parallelism, 390
inequality
  valid, 287, 299
infeasible, 287
initialization
  direct, 301
  enumerative, 300
  racing ramp-up, 301
  root, 300
  selective, 301
  spiral, 301
  two-level root, 301
inprocessing, 48, 49, 104, 116
instance
  features, 586
  heterogeneous, 586
  selection, 327
  strategies, 180, 181, 189, 190, 194, 197, 215
integration, *see* algorithm
interface
  abstraction, 295
  communication, 307
interval splitting, 75
iterative deepening A* (IDA*), 443, 560

knowledge, 299, 320
  broker, 322
  global, 299
  local, 299
  sharing, 24, 116, 120–122, 128, 297, 299, 320, 584
Kripke structure
  definition, 463
  on-the-fly computation, 469
  product with TGBA, 466

Lace, 521
language
  ω-regular, 463
  of a TGBA, 464
lasso-shaped accepting run, 467
learning, 7, 33, 81, 121, 151, 202, 255
lemmatization, 183, 212
lexicographic DFS, 459
linear optimization problem, 284, 298
linear resolution, 182, 184
linear search MaxSAT, 67
linear-time temporal logic, *see* LTL, *see* LTL
literal watching, 121
load balancing, 79, 248, 250, 267, 269, 299

asynchronous round-robin, 302
dynamic, 301
nearest neighbor, 302
pure static, 310
quality, 301
quantity, 301
random polling, 302
static, 300
work-sharing, 302
local search methods, 381
lock-free programming, 516
lookahead, 262, 263
loop formula, 245
lower bound, 68
LP, *see* linear optimization problem
LP relaxation, *see* relaxation
lparse, 239, 247, 248
LTL, 462
subclasses, 471
translation to TGBA, 465

ManySAT, 19, 53, 213, 442, 596, 618, 619, 628, 630, 631
MAP algorithm, 494, 497
Map-Reduce parallelism, 270
ASP, 272
well-founded model, 272
master control object, 126
master process, 310
master-hub-worker, 313
master-worker, 310
maximal accepting predecessor, *see* MAP algorithm
ME-ASP, 275
memory
contention, 306
lock, 306
MERGEXPLAIN, 568
metaheuristic methods, 383, 395
METEOR, 191, 194
MILP, *see* mixed integer linear optimization problem
Minisat, 19, 37, 118, 620
mixed integer linear optimization problem, 284
model checking, 458, 536–537
automata-theoretic approach, 461
model elimination, 180, 182, 183, 199, 215
model-based diagnosis, 547
complexity, 556
conflict, 552
diagnosis, 551
hitting set tree search, 553–556
modeling, 549
tree search, 552
model-based reasoning, 180, 181, 211, 213–216, 547
model-based testing, 131
model-elimination tableaux, 180, 182, 215
Moufang identities, 210
MPI, 118, 120, 121, 123, 126, 209, 307, 315, 321, 324, 349, 366, 396, 397, 432
MPIDepQBF, 114, 116, 118, 119, 121, 127
MTBDD, 513
multi-search, 191, 198–202, 208–211, 213–215, 217
multi-terminal binary decision diagram, 513
mutiple-master-worker, 313

nearest neighbor, *see* load balancing
nested depth-first search, 476
CNDFS algorithm, 485, 486
ENDFS algorithm, 484
LNDFS algorithm, 484
NDFS algorithm, 484
Nick's Class (NC), 459
node, 286, 299
child, 286
leaf, 286
parent, 286
terminal, 286
nogood, 245, 246, 267, 269
completion nogood, 245
forgetting, 270
learning, 255, 267
loop nogood, 245
propagation, 242, 246, 267, 268
non-variable overlap, 192
normalization, 185, 190, 193, 205, 206
NP-completeness, 101, 242

on-the-fly computation
Kripke structure, 469
product automaton, 470, 473
One-Way-Catch-Them-Young, *see* OWCTY
ordering-based strategies, 180, 181, 185, 187, 190, 191, 194, 197, 198, 200, 202, 206, 214, 215
OTTER, 194, 195, 201, 204, 207–209
overhead
communication, 293
idle time, 293, 294
parallel, 293, 324
redundant work, 294
OWCTY algorithm, 492, 497

P-completeness, 459, 460
pairs algorithm, 208, 209
PaMiraXT, 213

PAQuBE, 114, 116, 121, 122, 124, 127
par-pd-depqbf, 111, 116–118, 130
parallel
  performance, 291, 294
  scalability, 291, 292
  speed-up, 325
parallel algorithm, *see* algorithm
parallel linear search algorithms, 74
parallel overhead, *see* overhead
parallel portfolio construction, 598
parallel portfolio selection, 585–588
parallel presolving schedules, 591
parallel random depth-first search
  fully synchronized, 486
  swarming, 487
parallel retracting A* (PRA*), 446, 561
parallel rewriting, 190, 192, 193, 197
parallel speedup, 365, 387, 390–393, 396, 402–404, 406, 459, 499, 540, 647
parallel structured duplicate detection, 433
parallel unsatisfiability-based algorithms, 73
parallel window search, 445
parallelism
  distributed memory, 306
  node, 297
  shared memory, 306
  strong deterministic, 304
  subnode, 297
  subtree, 297
  tree, 297
  weak deterministic, 305
parallelism at the clause level, 190, 193, 194, 196, 197
parallelism at the search level, 190, 197, 198
parallelism at the term/literal level, 190, 191
paramodulation, 180, 184, 185, 205, 206, 208, 209
Parthenon, 191, 194
PARTHEO, 191, 194, 199
partial assignment, 105
partial MaxSAT, 65
PBNF, 433
pcaqe, 115–118, 125
PCNF, *see* prenex conjunctive normal form
Peers, 181, 202, 205, 208
Peers-mcd, 181, 202, 208–210
performance
  measurement, 324
  profile, 328
  variability, 308, 325
performance complementarity, 585
period synchronization, 90
persistence formula, 471
phase, 292
  primary, 293
  ramp-down, 293
  ramp-up, 292
Picosat, 118
pivot variable, 107
platform, 305
  computational, 291
  solution, 291
  solver, 308
PMSat, 214
polynomial hierarchy, 102
Portfolio parallelism
  ASP, 274
portfolio solving, 72, 111, 116, 117, 120, 181, 198, 202, 211, 406, 584
PQSAT, 115, 116, 123
PQSolve, 114–116, 122, 124, 125, 127
PQUABS, 115, 116, 118, 125
prenex conjunctive normal form, 105
prenex negation normal form, 125
preprocessing, 104, 116, 121, 126, 127, 193, 289
presolving schedule, 587
primary phase, *see* phase
product automaton
  of TGBA and Kripke structure, 466
  on-the-fly computation, 470
program
  ASP program, 239, 242
  CUDA program, 265
  Datalog program, 252, 265
  definite program, 241, 242
  normal program, 242
  program completion, 244, 245
  range-restricted program, 244
  stratified program, 252, 272
  tight ASP program, 244, 245
programming by optimization, 584
Prolog Technology Theorem Proving, 184, 199
propositional satisfiability, 65, 101–102, 189, 211
PSATO, 179, 211, 212
pseudo-Boolean constraints, 66
PSPACE, 116
PSPACE-completeness, 103
PVM, 255, 307, 320, 390, 395, 561

Q-resolution, 106–110, 114
  proof, 107, 130
Q-resolution calculus, 106
Q2CNF, 116
QBCP, 109
QBF, 102–132
  assignment, 105–106

assumption-based reasoning, 118
blocked clause elimination, 120
clause, 105
clause learning, 106, 108, 110
closed formula, 105
conflict, 103, 109
conjunctive normal form, 105
countermodel, 102, 106
cube, 105
cube learning, 106, 108, 110
decision making, 108–110
disjunctive normal form, 105
existential reduction, 107
expansion-based solving, 104, 115, 116, 129
free formula, 105
incremental solving, 116, 129
inprocessing, 116
knowledge sharing, 113, 116
learning, 108, 116
matrix, 105
model, 102, 106
negation normal form, 105
preprocessing, 104, 116, 119
pure literal, 108, 109, 126
quantifier scope, 105
restart, 120
satisfiability-equivalent, 106
search-based solving, 103, 108–111, 129
semantics, 106
existential player, 124
game, 124
universal player, 124
solution, 103, 109
strategy, 130
syntax, 105
unit clause, 111
unit literal, 109, 110
unit propagation, 108
variable assignment, 125
QBFEVAL, 104, 116, 117, 131, 132
QCDCL, 103, 104, 106, 108–111, 113, 114, 116, 118, 120, 126, 130
QCIR, 111
QMiraXT, 114, 116, 122, 124, 126, 127
QSAT, 102, 115, 116, 123
QSolve, 115, 116, 124
quabs, 115, 116, 125
quantified Boolean formulas, *see* QBF
quantifier elimination, 123
quantifier inversion, 124
Quantor, 126, 128
QuBE, 115, 116, 121
QUICKXPLAIN, 567
qxbf, 117

racing ramp-up, *see* initialization
ramp-down phase, *see* phase
ramp-up phase, *see* phase
random polling, *see* load balancing
reachability, 442, 477, 493, 535, 536, 540
redundancy, 181, 183, 188, 197, 207, 210, 214, 217
redundant work, *see* overhead
regressive merging, 183
rejecting SCC, 468
relational product, 523–525
relaxation
LP, 284
relaxation variables, 67
resolution, 7, 18, 35, 39, 50, 106, 107, 159, 180, 184, 185, 199, 205, 211, 212, 217, 238, 269, 342
resolution refutation, 7
restart, 16, 46, 75, 120, 169, 270, 344, 565
Robbins algebras, 209, 210
ROO, 181, 194–196
round-robin, *see* load balancing
runtime distribution, 387, 388

SAT local search, 5, 398
SAT solver, 109, 125, 179, 181, 190, 211, 213–215, 558, 618
SBA, 464
scalability, 92, *see* parallel
SCC, 468
computation algorithms, 478
scheduling, 255, 259–261
search, 288
best bound, 288
depth-first, 311
diving, 288
strategy, 284, 288
tree, 284
search overlap, 198, 204, 205, 210, 212, 214
search space splitting, 75, 111
search-based QBF solving, 103, 108–111, 129
selection heuristics, 269
self coordination, 315
semantic guidance, 185, 186, 190, 215, 216
semantic resolution, 185, 215
sequential algorithm, *see* algorithm
sequential algorithm selection, 586
sequential MaxSAT, 66
set of support, 185, 186, 194, 215
SGBA, 464
SGGS, 180, 216, 217
shared clauses, 17, 81, 126
simplification, 185, 187, 193, 206, 207, 209
SIMT parallelism, 265, 266

single quantification level scheduling, 127
Skolem function, 130
smodels, 239, 243, 247, 253, 262
SMP parallelism, 248
SMT, 141, 180
SMT solver, 180, 181, 215
soft clauses, 65
solution, 284, 298, 299
solution analysis, 121
solution learning, 108
solution platform, *see* platform
solution quality, 391, 403, 406, 407
speedup, 209, 210, *see* parallel
SQLS, 122, 124
SqueezBF, 121, 122, 128
stable model, 238
standard synchronization, 89
state-space partitioning, 492
state-space search, 419
strongly connected component, 117, 247, 249, 468
structure
  Kripke, *see* Kripke structure
structured duplicate detection, 432
subclasses
  LTL, 471
subgoal-reduction strategies, 180, 181, 183, 184, 190, 191, 193, 197, 198, 201, 213
subsumption, 185, 186, 189, 199, 206
superposition, 184, 185, 187, 192, 193, 205, 206, 209
supervisor-worker, 311
SWARM, 485
Sylvan, 509–541
symbolic bisimulation minimization, 537
symbolic reachability, 536
synchronization, 303
  barrier, 303
synchronization point, 85

task, 319
  granularity, 297
  sharing, 255, 258
  stealing, 193
TBA, 464
TBGA
  product with Kripke structure, 466
Team-Work, 181, 200–202, 214
TECHS, 201
temporal hierarchy, 471
terminal automaton, 471
TGBA, 464
  degeneralization, 465
  translation from LTL, 465
thread divergence, 267
transposition-driven scheduling, 444
trivial falsity, 124
trivial SCC, 468
trivial truth, 124
truth assignment, 4, 33, 149, 169, 268

UIP, 270
union-find, 478, 487
unit-resulting resolution, 199, 206
universal reduction, 107, 109, 110
unsatisfiable subformulas, 69
upper bound, 67

variable dependency, 104
variable-activity scaling, 121
VSIDS, 19, 79, 83, 126

weak automaton, 471
weighted MaxSAT, 66
WMSU3 algorithm, 69
work-sharing, *see* load balancing
work-stealing, 79, *see* load balancing, 337, 342, 343, 346, 351, 355, 358, 364, 365, 367, 368, 370, 371, 429, 444, 519–526, 534, 535, 540

young brothers wait scheduling, 125

Printed in the United States
By Bookmasters